AREAWIDE PEST MANAGEMENT

Theory and Implementation

AREAWIDE PEST MANAGEMENT

Theory and Implementation

Edited by

Opender Koul

*Insect Biopesticide Research Centre,
Jalandhar, India*

Gerrit Cuperus

*Stillwater,
Oklahoma, USA*

and

Norman Elliott

*USDA-ARS Plant Science Research Laboratory
Stillwater,
Oklahoma, USA*

www.cabi.org

CABI is a trading name of CAB International

CABI Head Office
Nosworthy Way
Wallingford
Oxfordshire OX10 8DE
UK

Tel: +44 (0)1491 832111
Fax: +44 (0)1491 833508
E-mail: cabi@cabi.org
Website: www.cabi.org

CABI North American Office
875 Massachusetts Avenue
7th Floor
Cambridge, MA 02139
USA

Tel: +1 617 395 4056
Fax: +1 617 354 6875
E-mail: cabi-nao@cabi.org

A catalogue record for this book is available from the British Library, London, UK.

Library of Congress Cataloging-in-Publication Data
Areawide pest management: theory and implementation/edited by Opender Koul, Gerrit Cuperus and Norman Elliott.
    p. cm.
  Includes bibliographical references.
  ISBN 978-1-84593-372-2 (alk. paper)
  1. Pests--Integrated control. I. Koul, Opender. II. Cuperus, Gerrit W. III. Elliott, Norman. IV. Title.

  SB950.A74 2008
  632'.9--dc22

ISBN-13: 978 1 84593 372 2   2007029243

Typeset by AMA DataSet Ltd, UK.
Printed and bound in the UK by Biddles, King's Lynn.

Contents

The colour plate section may be found following p.274.

About the Editors

Opender Koul, Fellow of the National Academy of Agricultural Sciences and the Indian Academy of Entomology, is an insect toxicologist/physiologist/chemical ecologist and currently the Director of the Insect Biopesticide Research Centre, Jalandhar, India. After obtaining his PhD in 1975 he joined the Regional Research Laboratory (CSIR), Jammu and then became Senior Group Leader of Entomology at Malti-Chem Research Centre, Vadodara, India (1980–1988). He has been a visiting scientist at the University of Kanazawa, Japan (1985–1986), University of British Columbia, Canada (1988–1992) and Institute of Plant Protection, Poznan, Poland (2001). His extensive research experience concerns insect–plant interactions, spanning toxicological, physiological and agricultural aspects.

Honoured with an Indian National Science Academy medal (INSA), the Kothari Scientific Research Institute award, KEC Science Society award and the Recognition award of the National Academy of Agricultural Sciences of India (2003–2004) for outstanding contribution in the field of insect toxicology/physiology and plant protection, he has authored over 160 research papers and articles, and is the author/editor of the books *Insecticides of Natural Origin* (1997), *Phytochemical Biopesticides* (2001), *Microbial Biopesticides* (2002), *Biopesticides and Pest Management* volumes I and II (2003), *Predators and Parasitoids* (2003), *Integrated Pest Management: Potential, Constraints and Challenges* (2004), *Neem: Today and in the New Millennium* (2004), *Transgenic Crop Protection: Concepts and Strategies* (2004), *Insect Antifeedants* (2005), *Biopesticides and Pest Management: Conventional and Biotechnological Approaches* (2007) and *Ecologically Based Integrated Pest Management* (2007), published by leading publishers globally. Dr Koul is on expert panels in many committees and leading international and national journals. He has also been an informal consultant to BOSTID, NRC of USA at ICIPE, Nairobi.

Gerrit Cuperus was a Regent's Professor and Integrated Pest Management Coordinator at Oklahoma State University for over 20 years. Dr Cuperus obtained his PhD in 1982, joined the Department of Entomology at Oklahoma State University and has since been involved in national IPM programmes in the USA aiming at an interdisciplinary focus in solving management issues. Dr Cuperus has chaired and served in different capacities in various state and national committees on food safety and pest management. He has made specific contributions in extension/research and has won distinguished service awards from the USDA. His research efforts focused on stored product pest management have helped to build the Stored Product Research and Education Center (SPREC) at Oklahoma State University. He has authored over 60 research papers and articles and is an editor of *Successful Implementation of IPM for Agriculture Crops* (1992), *Stored Product Management* (1995), *Integrated Pest Management: Potential, Constraints and Challenges* (2004) and *Ecologically Based Integrated Pest Management* (2007).

Norman Elliott is a Research Biologist at the US Department of Agriculture, Agricultural Research Service, Plant Science Research Laboratory, Stillwater, Oklahoma. Dr Elliott obtained his PhD in 1985 from Michigan State University and joined USDA-ARS upon graduation. Dr Elliott has served in various capacities on state and national committees on biological control and pest management. His research has focused mainly on the ecology and management of cereal aphids and their natural enemies, and on classical and conservation biological control. Dr Elliott has authored or co-authored over 200 scientific research papers and articles.

Contributors

Charles T. Allen, *Program Director, Texas Boll Weevil Eradication Foundation and Extension Specialist, Texas Cooperative Extension, PO Box 5089, Abilene, Texas 79608, USA*

Gerald L. Anderson, *USDA-ARS, Northern Plains Agricultural Research Center, Sidney, Montana 59270, USA*

M.D. Aubuchon, *USDA-ARS Center for Medical and Veterinary Entomology, 1600 SW 23rd Drive, Gainesville, Florida 32608, USA; e-mail: MattAubuchon-ARS@ars.usda.gov*

Bruno Bagnoli, *CRA-Istituto Sperimentale per la Zoologia Agraria, via Lanciola, 12/A, 50125 Florence, Italy*

Michael J. Brewer, *Integrated Pest Management Program, CIPS Building, Michigan State University, East Lansing, Michigan 48824, USA*

Paul A. Burgener, *Research Coordinator, Agricultural Economics, Panhandle Research and Extension Center, University of Nebraska-Lincoln, 4502 Avenue I, Scottsbluff, Nebraska 69361, USA*

David N. Byrne, *Department of Entomology, University of Arizona, Forbes 410, PO Box 2100: (36), Tucson, Arizona 85721-0036, USA; e-mail: byrne@ag.arizona.edu*

J.D. Carlson, *Oklahoma State University, Biosystems and Agricultural Engineering, 217 Agriculture Hall, Stillwater, Oklahoma 74078-6016, USA; e-mail: jd.carlson@okstate.edu*

G.R. Carner, *Regional Advisor, Environmental Services Program, Medan, North Sumatra, Indonesia*

Raymond I. Carruthers, *USDA-ARS, Western Regional Research Center, 800 Buchanan St, Albany, California 94710, USA; e-mail: ric@pw.usda.gov*

Vasile Catana, *Oklahoma State University, Department of Entomology and Plant Pathology, 127 Noble Research Center, Stillwater, Oklahoma 74078–3033, USA; e-mail: vasile.catana@okstate.edu*

Laurence D. Chandler, *USDA-ARS-NPA, 2150 Center Ave., Building D, Suite 300, Fort Collins, Colorado 80526, USA; e-mail: larry.chandler@ars.usda.gov*

James R. Coppedge, *USDA-ARS-SPA, 1001 Holleman Drive East, College Station, Texas 77840, USA*

Gerrit W. Cuperus, *1008 E. Franklin, Stillwater, Oklahoma 74074, USA*

C. Jack DeLoach, *USDA-ARS, Grassland, Soil and Water Research Laboratory, 808 E. Blackland Road, Temple, Texas 76502, USA*

Christina D. DiFonzo, *Department of Entomology, Michigan State University, 243 Natural Science Building, East Lansing, Michigan 48824, USA*

R. Dilts, *Department of Plant Protection, Bogor Agricultural University, Bogor, West Java, Indonesia*

C. Richard Edwards, *Department of Entomology, Purdue University, West Lafayette, Indiana 47907, USA*

Norman C. Elliott, *US Department of Agriculture, Agricultural Research Service, Plant Science Research Laboratory, Stillwater, Oklahoma 74075, USA; e-mail: Norman.elliott@ars. usda.gov*

Robert M. Faust, *US Department of Agriculture, Agricultural Research Service, National Program Staff, 5601 Sunnyside Avenue, Room 4-2228, Mailstop 5139, Beltsville, Maryland 20705, USA: e-mail: rml@ars.usda.gov*

Paul W. Flinn, *USDA-ARS, Grain Marketing and Production Research Center, 1515 College Avenue, Room 101, Manhattan, Kansas 66502, USA*

Kristopher Giles, *Department of Entomology and Plant Pathology, Oklahoma State University, 127 Noble Research Center, Stillwater, Oklahoma 74078-3033, USA; e-mail: kris.giles@olastate.edu*

David W. Hagstrum, *USDA-ARS, Grain Marketing and Production Research Center, 1515 College Avenue, Room 101, Manhattan, Kansas 66502, USA (retired)*

M.D. Hammig, *Department of Applied Economics and Statistics, Clemson University, Clemson, South Carolina 29634, USA; e-mail: mhammig@clemson.edu*

Gary Hein, *Department of Entomology, University of Nebraska Panhandle R&E Center, 4502 Ave I, Scottsbluff, Nebraska 69361, USA*

John C. Herr, *USDA-ARS, Western Regional Research Center, 800 Buchanan St, Albany, California 94710, USA*

Erin E. Hladilek, *Department of Entomology, University of Kentucky, S-225 Agricultural Science Center North, Lexington, Kentucky 40546, USA*

Claudio Ioriatti, *Istituto Agrario di San Michele a/A, via E. Mach, 1, 38010 San Michele a/A (Trento), Italy; e-mail: claudio.Ioriatti@iasma.it*

Eric B. Jang, *US Department of Agriculture, Agricultural Research Service, Pacific Basin Agricultural Research Center, PO Box 4459, Hilo, Hawaii 96720, USA*

Sean P. Keenan, *Postdoctoral Fellow, Department of Entomology and Plant Pathology, Oklahoma State University, 127 Noble Research Center, Stillwater, Oklahoma 74078-3033, USA; e-mail: keenansp@comcast.net*

Alan L. Knight, *Yakima Agricultural Research Laboratory, Agricultural Research Service, USDA, 5230 Konnowac Pass Rd., Wapato, Washington 98951, USA; e-mail: aknight@yarl.usda.gov*

Allen E. Knutson, *Texas AdM University and Texas Agricultural Experiment Station, Dallas, Texas, USA*

Opender Koul, *Insect Biopesticide Research Centre, 30 Parkash Nagar, Jalandhar 144003, India; e-mail: koul@jla.vsnl.net.in*

James A. Litsinger, *1365 Jacobs Place, Dixon, California 95620, USA; e-mail: jlitsinger@thegrid.net*

Andrea Lucchi, *Dip. C.D.S.L. Sez. Entomologia Agraria, Università di Pisa, via San Michele, 2, 56124 Pisa, Italy*

Ronald F.L. Mau, *University of Hawaii at Manoa, College of Tropical Agriculture and Human Resources, Department of Plant and Environmental Protection Sciences, 3050 Maile Way, Gilmore 611, Honolulu, Hawaii 96822, USA*

Hendrik J. Meyer, *USDA-CSREES, National Program Leader for Entomology, 3470 Waterfront Center, 800 9th St SW., Washington, DC 20024, USA*

Jerry Michels, *Texas A&M University, Agricultural Experiment Station, 2301 Experiment Station Rd., Bushland, Texas 79012-0010, USA*

Mustafa Mirik, *Texas A&M University, Agricultural Experiment Station, 2301 Experiment Station Rd., Bushland, Texas 79012-0010, USA*

Takuji Noma, *Integrated Pest Management Program, Department of Entomology, CIPS Building, Michigan State University, East Lansing, Michigan 48824, USA*

Robert M. Nowierski, *USDA-CSREES, National Program Leader for Bio-Based Pest Management, 3443 Waterfront Center, 800 9th St, SW, Washington, DC 20024, USA; e-mail: rnowierski@csrees.usda.gov*

David W. Onstad, *278 National Soybean Research Center, MC-634, 1101 West Peabody Drive, University of Illinois, Urbana, Illinois 61801, USA*

Frank Peairs, *Department of Bioagricultural Sciences and Pest Management, Colorado State University, Fort Collins, Colorado 80523, USA*

Thomas W. Phillips, *Department of Entomology and Plant Pathology, Oklahoma State University, Stillwater, Oklahoma 74078, USA*

David Porter, *US Department of Agriculture, Agricultural Research Service, Plant Science Research Laboratory, Stillwater, Oklahoma 74075, USA*

Edward B. Radcliffe, *Department of Entomology, University of Minnesota, 219 Hodson Hall, 1980 Folwell Ave., St Paul, Minnesota 55108, USA; e-mail: radcl001@umn.edu*

David W. Ragsdale, *Department of Entomology, University of Minnesota, 219 Hodson Hall, 1980 Folwell Ave., St Paul, Minnesota 55108, USA*

A. Rauf, *Department of Plant Protection, Bogor Agricultural University, Bogor, West Java, Indonesia*

Carl R. Reed, *Department of Grain Science and Industry, Kansas State University, Kansas 66506, USA*

B. Merle Shepard, *Department of Entomology, Soils, and Crop Science, Clemson University, Clemson, South Carolina 29634, USA*

Robert A. Suranyi, *MGK® McLaughlin Gormley King Company, 8810 Tenth Ave. N., Minneapolis, Minnesota 55427, USA*

Albert Sutherland, *Oklahoma State University, Biosystems and Agricultural Engineering, National Weather Center, 120 David L Boren Boulevard, Suite 2900, Norman, Oklahoma 73072-7305, USA*

Jon J. Tollefson, *Department of Entomology, Iowa State University, Ames, Iowa 50011, USA*

K. Vander Meer, *USDA-ARS, Center for Medical and Veterinary Entomology, 1600 SW 23rd Drive, Gainesville, Florida 32608, USA*

Roger I. Vargas, *US Department of Agriculture, Agricultural Research Service, Pacific Basin Agricultural Research Center, PO Box 4459, Hilo, Hawaii 96720, USA; e-mail: Roger.Vargas@ars.usda.gov*

Gerald R. Wilde, *Department of Entomology, Kansas State University, Manhattan, Kansas 66506, USA*

Lyle Wong, *Hawaii Department of Agriculture, Division of Plant Industry, PO Box 221659, Honolulu, Hawaii 96823, USA*

Preface

The conventional approach to pest management has been to treat a crop or commodity on an individual management unit basis before an economically damaging infestation of the pest develops. While there have been many successes at managing pests using the individual management unit approach, especially when an integrated pest management approach is used, it is recognized that management could sometimes be more effective if the pest was suppressed over a broad spatial area (larger than an individual management unit). That is the essence of the areawide pest management (AWPM) approach. AWPM contrasts with conventional pest management in that management tactics are applied over a broad spatial area, often treating the whole area simultaneously, to maintain the pest below economic levels or, in some cases, to completely eradicate it.

The number of pest management programmes that can be classified as AWPM has increased dramatically over the last decade. AWPM has potential advantages over the conventional approach: suppression across a broad area may result in reduced reinfestation by migration from unmanaged areas into previously treated areas, and the pest management tactics employed may be more effective – particularly ecologically based tactics – when applied areawide.

The purpose of this book is threefold. The first is to lay out the historical underpinnings of AWPM and to highlight current activity in the field. In 1993, the USDA-Agricultural Research Service in concert with a USDA IPM Working Group developed a partnership framework for a national AWPM initiative that would include the federal, state and private sectors as partners. The introductory chapter of this book is written by Dr Robert Faust, USDA-ARS, who has served as National Program Leader for AWPM programmes since initiation of the national initiative and who elegantly accomplishes the first objective and lays the groundwork for the rest of the book.

The second objective is to delve into concepts that have direct impact on the successful implementation of AWPM. These include: (i) biological and ecological concepts important for understanding the dynamics of populations in spatially heterogeneous environments; (ii) the critical role of inter-agency and multidisciplinary interactions in the development and implementation of AWPM programmes, which are often complex inter-agency and intergovernmental endeavours; (iii) the roles of modelling, meteorology and databases in AWPM programmes which, by their nature, are information intensive; and (iv) the importance of economic and sociological evaluation in successful AWPM implementation.

The third objective is to compile recent case examples of pest management programmes that have used the AWPM approach. We survey a wide variety of programmes developed for protecting agricultural and natural resource systems and which use a wide range of pest management tactics. We hope we have met our objectives, and that this book presents the current state of knowledge of AWPM to all those interested in using ecologically sound AWPM approaches. Furthermore, we hope the book proves useful for helping identify when AWPM is likely to be more applicable and successful than conventional pest management.

We received tremendous response and support from the authors and greatly appreciate their effort in writing very interesting and highly informative chapters. We also thank Sarah Hulbert of CABI for her patience and assistance at various stages of book preparation.

1 General Introduction to Areawide Pest Management

ROBERT M. FAUST

USDA-Agricultural Research Service, Beltsville, Maryland, USA

Welcome to the realm of areawide pest management (AWPM). This book represents one of the first comprehensive 'treatises' on the AWPM concept and approach, and should be of interest and use to many types of readers, from research scientists in government, university and industry to pest control advisors and extension personnel, growers, pest control and integrated pest management (IPM) practitioners, students, teachers, natural resource managers and others interested in environmentally sound pest control. There is a range of topics included in the subject area. The book is grouped into three parts. Chapters 2–8 discuss the foundation of areawide pest management; Chapters 9–20 describe case examples of recent areawide pest management programmes and projects; and Chapter 21 is a synthesis of the book's contents that integrates the theory and concepts presented in the various chapters into common themes that arise from the case examples. Chapter 21 also contains a discussion on the future potential of the areawide approach and how it augments and expands upon the traditional IPM strategy.

Historically, the AWPM concept in some form or another has been practised since the late 1800s. The overall premise is that a number of serious economic pests can be effectively managed using an organized and coordinated attack on their populations over large areas rather than by using a field-by-field approach (Knipling, 1978, 1979; Rabb, 1978; Knipling and Stadelbacker, 1983; Bellows, 1987; Myers *et al.*, 1998). The entomological literature contains numerous examples of large-scale, highly coordinated programmes that fit into the areawide concept. Chandler and Faust (1998) have given a number of historic examples of AWPM programmes in a previous publication, and they will be highlighted here only for the purpose of this introduction, with a few additional examples added. The reader is referred to the publication by Chandler and Faust (1998), as well as to the various chapters in this book, for more detailed historic information.

Very early programmes targeting a key pest over a wide area are mentioned in the scientific literature. One programme was against the grape phylloxera, *Daktulosphaira vitifoliae*, in Europe during the 1870s and 1880s, using resistant grapevines (Kogan, 1982).

© CAB International 2008. *Areawide Pest Management: Theory and Implementation* (eds O. Koul, G. Cuperus and N. Elliott)

The pest was fully under control by 1890. Classical biological control was used for the cottony cushion scale, *Icerya purchasi*, a pest that seriously affected the California citrus industry in the 1880s. Two biological control agents were introduced from Australia, the vedalia ladybeetle, *Rodolia cardinalis*, and the parasitic fly, *Cryptochaetum iceryae*. The vedalia ladybeetle brought about the complete suppression of this scale insect by the end of 1889 (Doutt, 1958), and this has been attributed to an AWPM strategy that used coordinated efforts and a broad distribution of the two biological control agents.

Several eradication programmes have been highly successful using areawide concepts as an integral part of the programme, with the goal of bringing the populations down to zero: those for the cattle ticks, *Boophilus annulatus* and *Boophilus annulatus* var. *microplus*, and the screwworm, *Cochliomyia hominovorax*. The two species of cattle tick had been eradicated from most of the USA by the 1950s (Cole and MacKeller, 1956) via a cooperative federal and state cattle-dipping protocol commencing in 1906 across 15 southern and south-western states. Using a sterile male technique, the screwworm was eradicated from the USA, Mexico and portions of Central America (Knipling, 1979; Bushland, 1985; Baumhover, 2002). Since 1991, the screwworm also has been eliminated from Belize (1994), Guatemala (1994), El Salvador (1995) and Honduras (1995) (USDA-APHIS, 1998).

The sterile male tactic has also been used to eradicate the melon fruit fly from Okinawa and the southern islands of Japan, as well as against the tsetse fly on the island of Unguja, Zanzibar (Vreysen *et al.*, 2000). The US Animal and Plant Health Inspection Service (APHIS) also uses the sterile male technique to eradicate recurring infestations of the Mediterranean fruit fly from the continental USA, in partnership with the affected state(s). A number of other AWPM programmes have been in progress throughout the world and will be summarized briefly later in this introduction. The description of AWPM examples, which makes up Chapters 9–20 of this book, provides more detailed information concerning several projects.

There is consensus that the recent interest in AWPM is related to the great success of the screwworm eradication programme, with Dr Edward F. Knipling, US Department of Agriculture (USDA), Agricultural Research Service (ARS), having been a strong proponent of the screwworm effort going back to at least 1955. A more definitive AWPM concept was published by Dr Knipling (Knipling, 1980), referring to it as 'regional management'; this probably helped to lay the theoretical foundation for the concept and the criteria for implementing AWPM projects, and since then the numerous discussions and planning activities around the concept that will be discussed in this introduction have built upon this foundation. Even as early as 1966, Dr Knipling (Knipling, 1966) envisioned the advantages of 'areawide management' as opposed to a 'field-to-field' approach. However, Knipling recognized that not all pests are good candidates for areawide tactics, necessitating reliance on a field-by-field control approach. Klassen (2003) has published a detailed account of Knipling's thoughts and activities in areawide and eradication applications, and the reader is referred to this excellent article for more information.

In September 1992, Knipling and G.G. Rohwer presented a proposal to the North American Plant Protection Organization (NAPPO) entitled 'Area-wide Pest Management' (E.F. Knipling, Maryland, 1993, personal communication). Their vision of the process was that AWPM programmes must be: (i) conducted on large geographical areas; (ii) should be coordinated by organizations rather than by individual producers; (iii) may involve eradication, if practical and advantageous, but should focus on reducing and maintaining a pest population at an acceptably low density; and (iv) must involve a mandatory component to ensure project success within the entire geographic area, because 'voluntary programs historically have not provided the desired level of pest management'.

Areawide pest management was defined as the systematic reduction of a target pest(s), to predetermined levels by uniformly applied mandatory pest mitigation measures over geographical areas clearly defined by biologically based criteria (e.g. pest colonization, dispersal potential). 'Pest' as used in the definition can include weeds, pathogens of animals and plants, and insects or other organisms (e.g. mites, ticks) that have an economic impact on the agricultural industry or human health. The stated advantage of managing pests on an areawide basis is that AWPM can offer a long-term solution to agricultural pest problems as opposed to quick-fix solutions on individual crops or small acreage. Properly implemented, the methodology could prevent major pest outbreaks and provide a more permanent control procedure for pests.

Areawide pest management and IPM were seen as similar, distinct and potentially complementary. The two approaches could be complementary in that when a key pest is effectively managed in an areawide programme, the potential to manage other key pests and secondary pests by alternative approaches becomes more readily achievable. Although AWPM generally targets a key pest or small group of pests, the strategy should consider other pests (e.g. secondary pests) in the system in a holistic fashion. On the other hand, IPM is often applied to individual farms or cropping systems and is generally voluntary in nature. As the reader will see throughout the various chapters, the mandatory requirement suggested in the proposal to NAPPO for AWPM programmes has not always been strongly adhered to in some programmes initiated in recent times, but these have been quite successful without such a requirement, given a vigorous outreach effort. The boll weevil eradication programme in the USA is an example of an effective 'mandatory' AWPM programme (Dickerson and Haney, 2001). A caveat here is that a federally implemented boll weevil eradication programme was not seen as the desired option, but that state regulatory authority, combined with USDA support and local grower leadership, provided the preferred option for the programme. Most of the funding support for this programme now comes from the cotton producers within each region.

Integrated pest management generally addresses the complex of pests in a production system and the pest problems associated with multi-commodity production systems intercropped or in crop rotation systems. Close to 70 definitions of IPM have been proposed (Bajwa and Kogan, 2002), with them all sharing a common theme: IPM is a sustainable, environmentally friendly approach to managing pests by combining biological, cultural, physical and chemical tools in a way that minimizes economic, health and environmental risks. This includes anticipating pest problems and preventing pests from reaching economically damaging levels. All appropriate techniques can be used, such as enhancing natural enemies, planting pest-resistant crops, adapting cultural management and using pesticides judiciously. It relies on a combination of common-sense practices.

As practised, IPM can consist of approaches to integrate two or more control techniques to manage one or more species of the same single grouping of pests, such as weeds, mites, ticks, insects, nematodes or diseases. It also can consist of approaches to integrate two or more management systems for two or more pest groupings, such as diseases and insects, or diseases, weeds, insects and nematodes.

Benbrook *et al.* (1996) view IPM systems as occurring along a continuum, which has been categorized into four levels of adoption: (i) no IPM, which corresponds to systems essentially dependent on pesticides and not using basic IPM practices like proper calibration, operation and cleaning of spray equipment, scouting for pests, and sanitation and good agronomic practice; (ii) low-level IPM, where farmers use at least the most basic IPM practices of scouting and application in accord with thresholds, avoiding or delaying resistance and secondary pest problems, optimally timing applications, and some preventive practices, such as short rotations, resistant varieties and cultivation; (iii) medium-level IPM, i.e. systems in which farmers have adopted some preventive measures, coupled with efforts to cut back on broad-spectrum pesticide use, protect beneficial organisms and assure that pesticides are applied most efficiently – includes multi-tactic approaches to limit or remove pest habitat and augment biodiversity, resistant varieties, use of cover crops and longer rotations, enhancing beneficial organisms, use of soil amendments and disease-forecasting models; and (iv) high-level, or multi-strategy biologically intensive IPM, the zone farthest along the IPM continuum, where farmers have integrated multiple preventive practices and, as a result, have become able to control pests without relying routinely on pesticides.

Integrated pest management is site specific in nature, but certain general criteria must be met for control measures to qualify as IPM practices. At a minimum, each site should have in place a management strategy, which includes prevention, avoidance, detection and suppression of pest populations, as envisioned by Dr Harold Coble of North Carolina State University, USA and the USDA IPM committee (Stall, 1999). The more biologically intensive the approach in each of these strategies, the further along the continuum the grower will be. In recent years, AWPM proponents in the USA have begun using the term areawide IPM to more accurately describe programmes currently being conducted. Strategies useful to IPM can likewise be applicable to AWPM as components of its foundation. And, of course, it is desirable that AWPM programmes be as far along the continuum as possible.

Regardless of whether IPM is being used on a farm-to-farm approach or incorporated into an areawide approach within the distinct criteria of AWPM the aim is still to maintain pest populations below damaging levels, based on proper use of the technologies available. AWPM (as does IPM) depends on the availability of adoptable, pest-specific management tools. These tools must control the pest, impact little else in the environment and not form residues on the food product, where they could be a hazard to the health of the consumer. AWPM strategies do not replace IPM concepts, but support IPM and embrace its technologies.

Technologies that can be used, depending on the situation, in AWPM approaches include:

1. Traditional biological control – the use of parasites, parasitoids, predators, pathogens, competitors and other beneficial organisms to reduce the harmful effects of pests, which may embody augmentation and conservation biological control tactics.

2. Biologically based (biorational) control – the use and application of biologically based methods (e.g. hormones, antimetabolites, feeding deterrents, repellents, pheromone and allelochemicals (semiochemicals) and other naturally produced chemicals, attracticidal compounds, traps and similar devices, autocidal methods/ sterile technology, etc.).

3. Host resistance – the use and application of pest-resistant crop cultivars and animal breeds, including genetically engineered plants and animals resistant to pests.

4. Cultural practices – the use and application of tactics such as crop rotation, intercropping, tillage approaches, cover crops or mulches, managing irrigation and drainage, fertilization, removal of crop residues and other field sanitation procedures, altering planting and harvesting schedules, and related strategies.

5. Physical and mechanical control – the use of physical and mechanical methodology, thereby exerting economic control or reducing rates of pest contamination and damage, e.g. vacuum collection, screening, trapping and other exclusion tactics, etc.

6. Chemical control – the use of broad-spectrum synthetic organic (non-naturally occurring), or analogues of, natural chemicals (e.g. pyrethroids, insect growth regulators, etc.) or inorganic chemicals for controlling animal and plant pests, including fumigation, the use of improved chemical pesticide formulations and improved pesticide application technologies (judicious use is desirable).

Models and expert systems, including predictive types and decision support systems for pest–plant/animal environmental integration, including vector–disease interaction and control agent(s) interaction are important components when available to use in order to facilitate a systems approach to maximizing plant/animal protection and environmental compatibility.

Closely related to these technologies will be an understanding and exploitation of information on the movement and dispersal of pest and beneficial species, timing of population suppression measures to coincide with low pest population densities, and optimal conditions for use of environmentally friendly technologies. The economics of the strategy are vitally important to adoption. It is essential that AWPM programmes be interfaced with multi-pest IPM systems and that systematic approaches are taken in selecting a pest(s) to be targeted for AWPM.

A number of criteria need to be considered as guidelines when implementing AWPM programmes (Kogan, 1995; Chandler and Faust, 1998; Faust and Chandler, 1998), a few of which have been mentioned above. The programme should be defined by some geographic entity that encompasses farms as well as all other non-farm components of the landscape, and should be conducted over large geographical areas with consideration of pest colonization and movement and dispersal of pest and beneficial species. The area should represent typical production settings with representative pest problems and consistent populations of the key pest(s). It is important to have assurance that the target pest(s) is amenable to control using the areawide concept over a large geographical area that may extend across county, state and, in some instances, national boundaries. Consider whether there are environmental factors that change over the area that could affect the programme. An understanding of the pest biology, ecology, genetics, behaviour, physiology, interactions with other organisms and other biological and physical characteristics of the system is critical. Is the pest genetically different in different parts of the area? What are the natural control factors? Is there a reasonable isolation of the area from other non-included infested areas such that migration into the target area or region will be minimized during the programme? What are the geographical barriers? Is there a reasonable representation of the host range (including wild relatives of the crop plant in the case of crop AWPM programmes) so that the effects of residual populations can be evaluated? What other pests exist in the ecosystem that could become important as the target/key pest(s) is managed? What are the parameters of the production system and the inputs? Has the technology been proven in smaller-scale tests?

An AWPM programme should be coordinated by groups of key participants as opposed to by just individual producers or other end-users, and the programme should involve federal (as needed), state and local extension, commodity and private grower groups, communities, agribusinesses and other stakeholders in a true partnership. Extension IPM programmes should be in place in the state or region, or planned to be developed in synchrony with the AWPM programme to ensure that multiple pest and secondary pest problems will be managed and the full impact of a combined programme will be realized. Bio-intensive, environmentally sound and economical technology must be available to the end-users and, of course, the programme should focus on reducing and maintaining a pest population at an acceptably low density, providing positive environmental benefits and food and worker safety, with a high benefit:cost ratio. Implementation of AWPM will require overall participation and compliance of growers in the area under the strategy for optimum success, as well as frequent evaluation to measure effectiveness and to assure that goals are being met. The remainder of this book will dwell in more detail on the various considerations raised here when implementing AWPM programmes.

As mentioned previously, a number of AWPM programmes in recent times have been in progress throughout the world. Earlier publications – Chandler and Faust (1998); Faust and Chandler (1998) – of the USDA's Agricultural Research Service summarized many of these programmes, and excerpts from those two publications will be included here, along with some additions to update the various activities using AWPM/IPM strategies. No great detail will be provided in this introduction of the various projects, since many that will be mentioned are already described in some detail later in this book. In any event, the various activities summarized here will serve as an indicator of the current status of AWPM.

Since the 1960s, numerous suppression programmes targeted at the pink bollworm, *Pectinophora gossypiella*, have been initiated. An areawide management programme for this pest has been in place in the San Joaquin Valley of California, USA, continuously since 1968 (Henneberry and Phillips, 1996). Most of the current pink bollworm suppression programmes that are established or under development use sterile insect releases, cotton plant destruction, mating disruption and trapping for management of the pest. The ongoing cotton boll weevil (*Anthonomus grandis*) eradication programme, which was initiated in North Carolina, South Carolina and Virginia, USA, in 1977, is another example of a successful, highly coordinated areawide management programme (Henneberry and Phillips, 1996). Suppression methods generally have included insecticides and cultural measures on in-season and over-wintering populations, use of grandlure pheromone traps to reduce weevil populations emerging in the spring, and sterile boll weevil releases.

Cotton farmers in Arkansas, USA, have voluntarily organized bollworm management communities in an attempt to suppress cotton bollworm and tobacco budworm populations areawide rather than by a field-by-field approach (Henneberry and Phillips, 1996). The aim has been to coordinate control decisions so that all cotton fields in a cotton bollworm management community are treated within a 3-day period.

In the USA two other important areawide IPM programmes have been implemented and the technologies transferred by the federal government to the affected states.

These programmes have been targeted at the gypsy moth and grasshoppers, serious pests of trees and rangeland/crops, respectively. The gypsy moth (*Lymantria dispar*) was introduced into the USA in 1869 and has defoliated thousands of acres of hardwood forests across the north-east, from Maine to North Carolina, infesting 19 states and Washington, DC. (APHIS, 2003). In 1992, the USDA's Forest Service (FS) and APHIS, along with the Department of Interior's National Park Service and eight state and university partners, embarked on a pilot project called 'Slow the Spread'. The project's goal was to slow the rate of natural spread of the gypsy moth by using IPM strategies (APHIS, 2003). In 1999, following successful completion of the pilot project, the National Gypsy Moth Slow the Spread programme was implemented along the entire 1200-mile gypsy moth frontier from North Carolina through the upper peninsula of Michigan. The programme area is located ahead of the advancing front of the gypsy moth population, and concentrates on early detection and suppression of the low-level populations along this advancing front, disrupting the natural progress of population expansion. Suppression tactics have included pheromone mating disruption, mass trapping and treatment with the microbial pesticide, *Bacillus thuringiensis* (*Bt*), diflubenzuron (except in Michigan) or a naturally occurring virus (Gypchek). The programme includes a compliance with regulations covering movement of gypsy moth host materials.

Grasshopper population outbreaks in the Great Plains and Intermountain West have occurred for many decades. In response to a grasshopper epidemic in the mid-1980s, APHIS initiated a Grasshopper Integrated Pest Management (GHIPM) Project in 1987 to develop and demonstrate new IPM technologies for transfer as a package to managers of public and private rangelands (USDA-ARS-APHIS-U.WY, 2001). APHIS had been given a congressional mandate to manage these pests on federal rangeland. APHIS had the responsibility to direct a coalition of federal agencies for the GHIPM Project. Agencies included in this project were the USDA's Agricultural Research Service, Economic Research Service, Forest Service and Extension Service (now known as the Cooperative State Research, Education and Extension Service); the US Department of the Interior's Bureau of Land Management, US Fish and Wildlife Service and National Park Service; and the US Environmental Protection Agency's Office of Pesticide Programs. Also, state departments of agriculture, land grant colleges, grazing associations and private industry joined the effort. The GHIPM demonstration project ran from 1987 to 1994 in areas of Idaho and North Dakota. Products of the programme included a *Grasshopper Integrated Pest Management User Handbook* and a CD (USDA-ARS-APHIS-U.WY, 2001). In addition to the user handbook, the CD also contains a field guide to common western grasshoppers; a section on grasshoppers (Acrididae) of Colorado: identification, biology and management;

HOPPER 4.0 and CARMA 3.3 decision support software for rangeland grasshopper management; and additional grasshopper management and GHIPM Project descriptions and information. Copies of the *Grasshopper IPM User Handbook* may be obtained from USDA, APHIS, PPQ, Operational Support Staff at 4700 River Road, Riverdale, Maryland 20737. The CD may be obtained from USDA-ARS Northern Plains Agricultural Research Laboratory, at 1500 North Central Avenue, Sidney, Montana 59270.

In 1993 USDA's Agricultural Research Service, in concert with a USDA IPM Working Group, developed a partnership framework for an AWPM initiative that would include the federal, state and private sectors as partners. On 27 September 1993, key pest management representatives from the USDA, university research and extension and several state Departments of Agriculture participated in an organizational meeting in Beltsville, Maryland. At this meeting, participants identified key pests and cropping systems for which environmentally sound pest management technologies were available for implementation on an areawide basis (Faust and Chandler, 1998). Dr Knipling played a pivotal role in the organizational meeting.

The goals of ARS's AWPM partnership initiative are: (i) to demonstrate technologies that will suppress key target pests to manageable levels using the AWPM IPM concept; (ii) increase community involvement in the initiative through educational programmes during the programme; (iii) increase economic benefits to end-users, the community and other stakeholders as a result of the programme; (iv) promote a sustainable AWPM suppression programme; and (v) introduce, transfer and promote adoption of the demonstrated pest suppression technology.

The USDA-ARS funded AWPM programme and the 5-year panel selected projects typically are structured around four key components: (i) operations (the demonstration sites); (ii) assessment (economic, sociological and environmental impacts); (iii) education (outreach and technology transfer, including training and various communication tools); and (iv) research, the results of which are intended to aid in the improvement of programme efficacy or to help circumvent obstacles to implementation (Faust and Chandler, 1998). None of the projects contains a mandatory requirement, but they do have a highly active outreach component. Extension and county agents sustain the strategies in the out-years.

The first USDA-ARS AWPM demonstration partnership project was implemented in 1994, in the north-western USA against the codling moth, *Cydia pomonella* (Calkins and Faust, 2003). Mating disruption was used to reduce the pest population while reducing the use of organophosphate insecticides. In 1995 a second project was initiated for corn rootworms (*Diabrotica* spp.) in the Midwest by using adult semiochemical insecticide bait (Chandler and Faust, 1998; Chandler, 2003). Corn rootworm populations were significantly reduced at participating sites, and new bait products were developed and evaluated for use in rootworm-infested areas. Products produced by several companies have been used in IPM wide area strategies in Hungary, Croatia, Italy and Argentina for corn rootworms (L.D. Chandler, Fort Collins, Colorado, 2006, personal communication). The ARS initiated two other AWPM IPM projects in 1996: one project was directed at insects of stored grain in Kansas and Oklahoma (Flinn *et al.*, 2003). The project used two elevator networks, one in each state, for a total of 28 grain elevators. Stored wheat was followed as it was moved from farm to the country elevator and finally to the terminal elevator, thus giving the project an areawide perspective. Fumigation using aluminium phosphide pellets, as needed, along with sampling/monitoring and decision support software, was used in the demonstration project.

The other project was directed at the leafy spurge weed (*Euphorbia esula*) and was initiated as a partnership between the ARS in Sidney, Montana; the USDA-APHIS; North and South Dakota State Universities; and Montana State University; in co-operation with the Forest Service, Cooperative States Research, Education and Extension Service (CSREES), the Bureau of Land Management, National Park Service and the state Departments of Agriculture (Anderson *et al.*, 2003). This project used biological control with emphasis on a beetle herbivore (*Apthona* sp.) of leafy spurge, and other technologies such as grazing systems, revegetation, decision aids, geographical information systems (GIS) and judicious use of herbicides, as needed.

Between 1999 and 2006, ARS initiated six additional AWPM demonstration projects, which are ongoing or just being completed, many of which are detailed in the case examples of this book. For example, in 1999 an AWPM IPM project in the Hawaiian Islands for management of tephritid fruit flies using monitoring, sanitation, male annihilation, baits, biological control and sterile male fruit flies was initiated (Mau *et al.*, 2007). The target species included Mediterranean (*Ceratitus capitata*), melon (*Bactrocera dorsalis*), oriental (*B. dorsalis*) and Malaysian fruit flies (*B. latifrons*). The overall goal is to suppress these pests below economic thresholds. Fruit flies, especially the oriental and Mediterranean, continue to show up in the continental USA, and the technologies being demonstrated in Hawaii are enhancing suppression and eradication programmes of these invasive species in the USA and elsewhere. Already, the programme has led to initiation and adoption of the AWPM tactics in Pacific Basin Areas such as French Polynesia, Fiji, Vanuatu, Guam, Cook Islands and the Northern Mariana Islands, as well as in districts in Taiwan and Queensland, Australia.

Besides the ARS partnership demonstration fruit fly project in Hawaii, a number of AWPM programmes are currently being conducted around the world that target fruit flies, largely using sterile insect techniques. These projects are in Argentina, Australia, Costa Rica, Greece, Guatemala, Mexico, Pakistan, Peru, Philippines, Portugal, Thailand and the continental USA (Hendrichs and Ortiz, 1996). A number of these programmes are coordinated mainly by grower associations and government agencies, and do prevent major economic damage to numerous fruits and vegetables.

In 2000, a project on fire ants (especially the red imported, *Solenopsis invicta*, across Florida, Mississippi, South Carolina, Texas and Oklahoma) on pastures was initiated using natural enemies (phorid fly parasites), microbial agents and attracticidal compounds (Flores and Core, 2004; Pereira, 2004; Van der Meer *et al.*, 2007).

Then, in 2001, three additional projects were implemented: (i) Russian wheat aphid (*Diuraphis noxia*) and greenbug (*Schizaphis graminum*) on wheat in the US Great Plains using customized cultural practices, pest-resistant cultivars, biological control agents and other biologically based pest control technologies (Keenan *et al.*, 2007); (ii) the *Melaleuca* weed tree (*Melaleuca quinquenervia*) in Florida using natural enemies and microbial biological control (fungus), judicious use of herbicides, mechanical (mowing) and physical (fire) control, and combinations of these tactics (Flores, 2004a; Scoles *et al.*, 2006); and (iii) the tarnished plant bug (*Lygus lineolaris*) on cotton in the delta of Mississippi and Louisiana using host destruction, host-plant resistance and remote-sensing technology, which was an extension of an ongoing ARS project (Weaver-Missick, 1999; Abel *et al.*, 2007).

In 2006, a project targeted at methyl bromide alternatives in Florida and California was initiated to assess, test and transfer an IPM wide area strategy using methyl bromide alternatives against soil pathogens, nematodes and weed pests for growers who are losing the fumigant (Schneider *et al.*, 2003) because of regulatory action. The ARS plans to continue implementing additional AWPM-IPM demonstration projects in the future as funds are released from ongoing projects being completed. For example, in October 2007 projects targeted at the Asian tiger mosquito, navel orange worm, pests and pathogens of honey bees, and weedy annual grasses of rangelands were initiated.

There have been other ARS wide area IPM projects not directly funded by the ARS AWPM initiative that have been implemented. Since 1992, ARS has led federal and state scientists in a nationally coordinated research effort to develop technologies for mitigation of the silverleaf whitefly (*Bemisia argentifolis*) problem in ornamental, vegetable, melon and fibre crops across the southern USA, and in greenhouses (De Quattro, 1997; Henneberry *et al.*, 2002). This insect has been responsible for over US$2 billion in crop loss, damage and control measures since its introduction into the USA in 1986. Areawide and community-based management approaches, covering all affected commodities, have emerged as the best strategy and have been adopted. Some crop management- and community-oriented farm practices, such as water-use patterns, proximity of alternate host crops and spatial considerations, have been implemented, resulting in whitefly population reduction. An excellent insecticide resistance management programme has been implemented to conserve a major insect growth regulator (imidocloprid) found effective under the programme. A number of other management tools have been developed and adopted by growers, including crop rotation, host-free periods, crop residue and weed destruction, host resistance and biological control (fungi, parasites and predators). Overall losses have not increased in agricultural communities where the silverleaf whitefly is a factor in crop and horticultural production and have declined in a number of cases.

In 1997 USDA-ARS implemented a 5-year wide area project against the blacklegged deer tick (*Ixodes scapularis*) in the north-east USA (Pound *et al.*, 2000a; McGraw and McBride, 2001). The project uses a device named the 'four-poster' as an alternative to eliminating deer populations or applying chemical sprays. The tick transmits the agent that causes human Lyme disease (*Borrelia burgdorferi*). The 'four-poster' consists of a bin filled with whole-kernel maize and paint rollers attached to the bin's four corners. An acaricide (amitraz) is applied to the rollers and the acaricide rubs off and kills ticks on the deer's head and neck when the animal feeds between the rollers. ARS scientists in Kerrville, Texas, developed the technology. The technique is also used in Texas against lone star ticks (*Amblyomma americanum*) (Pound *et al.*, 2000b; Flores, 2006). In 2003 scientists in Scotland began testing the '4-poster' topical applicator in that country (Flores, 2004b).

ARS scientists in New Orleans, Louisiana, USA have recently established an areawide treatment programme (Operation Full Stop) for the Formosan subterranean termite, *Coptotermes formosanus*. The project uses new termite control technologies that include monitoring/baiting technology and non-repellent termiticides

(Lax and Osbrink, 2003). The AWPM programme was established in a 15-block area of the New Orleans French Quarter with the homeowners, in the USDA-ARS campus and in southern Mississippi. The programme seems to be successful so far, and work continues to help provide long-term sustainable population control.

Several other AWPM projects, which will be covered in the case examples, include salt cedar (*Tamarix ramosissima*) in the western USA (ARS), rice insects and grain and vegetable crops in South-east Asia and grape AWPM in Italy. Other authors elsewhere have recently described areawide control or eradication efforts (Vreysen *et al.*, 2007), including the red palm weevil of coconut, the mosquito *Aedes albopictus* in Italy, mosquito control in Greece, painted apple moth in New Zealand, codling moth in British Columbia, Canada and Brazil, *Amblyomma* in the Caribbean, fruit fly in Central America, Chile, Tunisia and Sudan, mountain pine beetle in Western North America, cotton bollworm in China, tsetse fly in South Africa and Ethiopia, cactus moth in North America (including Mexico), the false codling moth in South Africa, rice stemborers in China, cotton insects in Tajikistan as well as other AWPM-related efforts.

Other potential candidate pests have been suggested, particularly during lively discussions among USDA agencies and their partners over the past decade or so. As the ARS National Program Leader assigned primary responsibility for the agency's AWPM programme initiative, the author of this introductory chapter has been privy to the many discussions and recommendations. Some of these candidates have included insects such as heliothine moths, soybean aphid, Colorado potato beetle, Asian long-horned beetle, emerald ash borer, pink hibiscus mealy bug, glassy-winged sharpshooter, European corn borer, diamondback moth, beet armyworm, cabbage looper, fall armyworm, sugarcane borer, cattle grubs and horn flies; weeds such as water hyacinth, hydrilla, Eurasian water milfoil, Old World climbing fern, German ivy, tropical and aquatic soda apple, kudzu, giant reed, hawkweeds, purple loosestrife, witch weed, knapweed, Scotch thistle, yellow star thistle, jointed goat grass, sickle pod and *Salvinia*; and pathogens/nematodes such as golden nematode, Chrysanthemum white rust, soybean cyst nematode, citrus bacterial canker, sugarcane leaf scald disease, cereal rusts, dogwood rust, late blight of potatoes, wheat scab, early blight of tomatoes and Pierce's disease of grapes (glassy-winged sharpshooter).

Not all of the pests suggested will necessarily be good candidates for AWPM/total population management, and may not fit well with the AWPM criteria, especially species that appear so sporadically that AWPM would not be justified. Some of the potential candidates mentioned are already apparently under some level of biocontrol or IPM practices. The list, of course, is probably not all-inclusive by any means.

Based on the increasing number of AWPM projects being implemented, the recent resurgence of interest in the AWPM concept and how well current practising end-users seem to have embraced the idea, the future looks good for its continued, and even accelerated, adoption and use. Organized coordination and cooperation must continue to be sought, if AWPM programmes of regional and broader geographic scope shall succeed. It is hoped that this book will contribute to the interest in AWPM and its importance to pest managers, as well as to a further understanding of what the concept has to offer.

References

Abel, C.A., Snodgrass, G.L. and Gore, J. (2007) A cultural method for the areawide control of tarnished plant bug in cotton. In: Vreysen, M.J., Robinson, A.S. and Hendrichs, J. (eds) *Area-Wide Control of Insect Pests: from Research to Field Implementation*, Springer, Dordrecht, The Netherlands (in press).

Anderson, G.L. Prosser, C.W., Wendel, L.E., Delfosse, E.S. and Faust, R.M. (2003) The ecological areawide management (TEAM) of leafy spurge program of the United States Department of Agriculture–Agricultural Research Service. *Pest Management Science* 59, 609–613.

APHIS (Animal & Plant Health Inspection Service) (2003) *Gypsy Moth: Slow the Spread Program.* Fact sheet, plant protection and quarantine, April 2003, USDA-APHIS, Riverdale, Maryland, 3 pp.

Bajwa, W.I. and Kogan, M. (2002) *Compendium of IPM Definitions (CID) – what is IPM and How is it Defined in the Worldwide Literature?* IPPC Publication No. 998, Integrated Plant Protection Center (IPPC), Oregon State University, Corvallis, Oregon, 15 pp.

Baumhover, A.H. (2002) A personal account of programs to eradicate the screw worm, *Cochliomyia hominivorax*, in the United States and Mexico with special emphasis on the Florida program. *Florida Entomologist* 85, 669–676.

Bellows, T.S. (1987) Regional management strategies in stochastic systems. *Bulletin of the Entomological Society of America* 33, 151–154.

Benbrook, C.M., Groth, E., Halloran, J.M., Hansen, M.K. and Marquardt, S. (1996) *Pest Management at the Crossroads.* Consumers Union, Yonkers, New York, 272 pp.

Bushland, R.C. (1985) Eradication program in the southwestern United States. *Miscellaneous Publications of the Entomological Society of America* 62, 12–15.

Calkins, C.O. and Faust, R.M. (2003) Overview of areawide programs and the program for suppression of codling moth in the western USA directed by the United States Department of Agriculture – Agricultural Research Service. *Pest Management Science* 59, 601–604.

Chandler, L.D. (2003) Corn rootworm areawide management program: United States Department of Agriculture – Agricultural Research Service. *Pest Management Science* 59, 605–608.

Chandler, L.D. and Faust, R.M. (1998) Overview of areawide management. *Journal of Agricultural Entomology* 15, 319–325.

Cole, T.W. and MacKeller, W.M. (1956) Cattle tick fever. In: *Animal Diseases.* US Department of Agriculture Yearbook 1956, Washington, DC, pp. 310–313.

De Quattro, J. (1997) The whitefly plan – 5 year update. *Agricultural Research* 45, 4–12.

Dickerson, W.A. and Haney, P.B. (2001) A review and discussion of regulatory issues. In: Dickerson, W.A., Brashear, A.L., Brumley, J.T., Carter, F.L. and Grefenstette, W.J. (eds) *Boll Weevil Eradication in the United States Through 1999.* The Cotton Foundation Publisher, Memphis, Tennessee, pp. 137–156.

Doutt, R.L. (1958) Vice, virtue and the vedalia. *Bulletin of the Entomological Society of America* 4, 119–123.

Faust, R.M. and Chandler, L.D. (1998) Future programs in areawide pest management. *Journal of Agricultural Entomology* 15, 371–376.

Flinn, P.W., Hagstrum, D.W., Reed, C. and Phillips, T.W. (2003) United States Department of Agriculture – Agricultural Research Service stored-grain areawide integrated pest management program. *Pest Management Science* 59, 614–618.

Flores, A. (2004a) TAMEing Melaleuca with IPM. *Agricultural Research* 52, 4–8.

Flores, A. (2004b) Tick control methods head to Scotland. *Agricultural Research* 52, 21.

Flores, A. (2006) The continuing fight against cattle ticks. *Agricultural Research* 54, 8–9.

Flores, A. and Core, J. (2004) Putting out the fire: ARS sets up regional programs to combat imported fire ants. *Agricultural Research* 52, 12–13.

Hendrichs, J. and Ortiz, G. (1996) Fruit fly (Diptera: Tephritidae) area-wide integrated management programmes in progress in the world. Paper #22- 025, abstract in *Proceedings XX International Congress of Entomology*, Florence, Italy. p. 705.

Henneberry, T.J. and Phillips, J.R. (1996) Suppression and management of cotton insect populations on an areawide basis. In: King, E.G., Phillips, J.R. and Coleman, R.J. (eds) *Cotton Insects and Mites: Characterization and Management*. The Cotton Foundation Publisher, Memphis, Tennessee, pp. 601–624.

Henneberry, T.J., Faust, R.M., Jones, W.A. and Perring, T.M. (2002) Silverleaf Whitefly: National Research, Action and Technology Transfer Plan (formerly Sweetpotato Whitefly, Strain B): Fourth Annual Review of the Second 5-year Plan and Final Report for 1992–2002, held in San Diego, California, February 10–12, 2002. US Department of Agriculture, Agricultural Research Service, June 2002, 446 pp.

Keenan, S.P., Giles, K.L., Elliott, N.C., Royer, T.A., Porter, D.R., Burgener, P.A. and Christian, D.A. (2007) Grower perspectives on areawide wheat integrated pest management in the southern US Great Plains. In: Koul, O. and Cuperus, G.W. (eds) *Ecological Based Integrated Pest Management*. CAB International, Wallingford, UK, pp. 289–314.

Klassen, W. (2003) Memorial lecture – Edward F. Knipling: titan and driving force in ecologically selective area-wide pest management. *Journal of the American Mosquito Control Association* 19, 94–103.

Knipling, E.F. (1966) Some basic principles in insect population suppression. *Bulletin of the Entomological Society of America* 12, 7–15.

Knipling, E.F. (1978) Eradication of plant pests – pro-advances in technology for insect population eradication and suppression. *Bulletin of the Entomological Society of America* 24, 44–52.

Knipling, E.F. (1979) *The Basic Principles of Insect Population Suppression and Management*. US Department of Agriculture, Agriculture Handbook no. 512, Washington, DC, 659 pp.

Knipling, E.F. (1980) Regional management of the fall armyworm – a realistic approach. *Florida Entomologist* 63, 468–480.

Knipling, E.F. and Stadelbacker, E.A. (1983) The rationale for areawide management of *Heliothis* (Lepidoptera: Noctuidae) populations. *Bulletin of the Entomological Society of America* 29, 29–37.

Kogan, M. (1982) Plant resistance in pest management. In: Metcalf, R. and Luckman, H. (eds) *Introduction to Insect Pest Management*, 2nd edition. Wiley, New York, pp. 93–134.

Kogan, M. (1995) Areawide management of major pests: is the concept applicable to the *Bemisia* complex? In: Gerling, D. and Mayer, R.T. (eds) *Bemisia: Taxonomy, Biology, Damage, Control and Management 1995*. Intercept, Andover, UK, pp. 643–657.

Lax, A.R. and Osbrink, W. (2003) United States Department of Agriculture – Agriculture Research Service research on targeted management of the Formosan subterranean termite *Coptotermes formosanus* Skiroki (Isoptera: Rhinotermitidae). *Pest Management Science* 59, 788–800.

Mau, R.F., Jang, E.B. and Vargas, R.I. (2007) The Hawaii area-wide fruit fly pest management programme: influence of partnership and a good education programme. In: Vreysen, M.J., Robinson, A.S. and Hendrichs, J. (eds) *Area-Wide Control of Insect Pests: from Research to Field Implementation*. Springer, Dordrecht, The Netherlands (in press).

McGraw, L. and McBride, J. (2001) Tick control device reduces lyme disease. *Agricultural Research*, 49, 5–7.

Myers, J.H., Savoie, A. and Randen, E. (1998) Eradication and pest management. *Annual Reviews of Entomology* 43, 471–491.

Pereira, R.M. (2004) Area-wide suppression of fire ant populations in pastures: project update. *Journal of Agricultural Urban Entomology* 20, 123–130.

Pound, J.M., Miller, J.A., George, J.E. and Le Meilleur, C.A. (2000a) The '4-poster' passive topical treatment device to apply acaricide for controlling ticks (Acari: Ixodidae) feeding on white-tailed deer. *Journal of Medical Entomology* 37, 585–594.

Pound, J.M., Miller, J.A. and George, J.E. (2000b) Efficacy of amitraz applied to white-tailed deer by the '4-poster' topical treatment device in controlling free-living lone star ticks (Acari: Ixodidae). *Journal of Medical Entomology* 37, 878–884.

Rabb, R.L. (1978) A sharp focus on insect populations and pest management from a wide-area view. *Bulletin of the Entomological Society of America* 24, 55–61.

Schneider, S.M., Rosskodf, E.N., Leesch, J.G., Chellemi, D.O., Bull, C.T. and Mazzola, M. (2003) United States Department of Agriculture–Agricultural Research Service research on alternatives to methyl bromide: pre-plant and post-harvest. *Pest Management Science* 59, 814–826.

Scoles, J.C., Pratt, P.D., Silvers, C.S., Langeland, K.A., Meisenburg, M.J., Ferriter, A.P., Gioeli, K.T. and Gray, C.J. (2006) *The Land Manager's Handbook on Integrated Pest Management of* Melaleuca quinquenervia. USDA-ARS, Fort Lauderdale, Florida, 55 pp.

Stall, W.M. (1999) IPM definition. *Journal of Vegetable Crop Production*, 4, 95–96.

USDA-APHIS (United States Department of Agriculture–Animal Plant Health Inspection Service) (1998) Eradicating screw worms from North America. *APHIS Web site* (http://www.aphis.usda.gov/OA/Screw worm.html).

USDA-ARS-APHIS-U.WY (United States Department of Agriculture–Agriculture Research Service–Animal and Plant Health Inspection Service–University of Wyoming) (2001) Grasshoppers: their biology, identification and management. In: Branson, D. and Redlin, B. (eds) (CD, June 2001). USDA-ARS, Sidney, Montana.

Van der Meer, R.K., Porter, S.D., Oi, D.H., Valles, S.M. and Pereira, R.M. (2007) Area-wide suppression of invasive fire ant populations. In: Vreysen, M.J., Robinson, A.S. and Hendrichs, J. (eds) *Area-Wide Control of Insect Pests: from Research to Field Implementation*. Springer, Dordrecht, The Netherlands (in press).

Vreysen, M.J.B., Saleh, K.M., Ali, M.Y., Abdulla, A.M., Zhu, Z.R., Juma, K.G., Dyck, A.V., Msangi, A.R., Mkonyi, P.A. and Feldmann, H.V. (2000) *Glossina austeni* (Diptera: Glossinidae) eradicated on the island of Unguja, Zanzibar, using the sterile insect technique. *Journal of Economic Entomology* 93, 123–135.

Vreysen, M.J., Robinson, A.S. and Hendrichs, J. (2007) *Area-wide Control of Insect Pests: from Research to Field Implementation*, Springer, Dordrecht, The Netherlands (in press).

Weaver-Missick, T. (1999) Banishing tarnished plant bugs from cotton, *Agricultural Research* 47, 12–14.

2 History and Ecological Basis for Areawide Pest Management

NORMAN C. ELLIOTT,[1] DAVID W. ONSTAD[2] AND MICHAEL J. BREWER[3]

[1]US Department of Agriculture, Agricultural Research Service, Plant Science Research Laboratory, Stillwater, Oklahoma, USA; [2]National Soybean Research Center, University of Illinois, Urbana, Illinois, USA; [3]Integrated Pest Management Program, Department of Entomology, Michigan State University, East Lansing, Michigan, USA

Introduction

The traditional approach to pest management is to treat the crop or commodity in a particular management unit before an economically significant infestation of the pest has developed. Determining the need to take corrective action is based on the economic threshold concept, which forms the basis of most integrated pest management programmes (Metcalf and Luckman, 1975). Areawide pest management (AWPM) can be contrasted with traditional pest management in that pest management tactics are used over a broad spatial area, often treating the whole area simultaneously to maintain the pest below economic levels or, in some cases, completely eradicate it. AWPM has potential advantages over the traditional approach. Suppression across a broad area may result in reduced reinfestation by migration from nearby unmanaged areas, and the pest management tactics employed may be more effective, particularly ecologically based tactics, when applied areawide.

A diversity of approaches exists for AWPM. The strategies used in programmes obviously must be based on the particular species that is the target of the management effort. Detailed understanding of the pest's biology and ecology, the ecological system as a whole and the pest management tactic(s) available for deployment will provide insight into the most promising avenues for effective suppression over a broad spatial area. While virtually any pest of humans or their enterprises, agricultural or otherwise, can be a target of AWPM, we will focus on AWPM of insect pests of agriculture in this chapter. The concepts are applicable to weeds and other non-insect pests, and to non-agricultural pest problems.

Dr Edward F. Knipling was among the first to formalize the idea that use of preventive approaches for managing pests on an areawide basis could be more effective and less environmentally detrimental than curative approaches, which often rely on repeated use of insecticides on individual fields (Klassen, 2003). Preventive approaches fall into two basic categories: those that make the particular management unit more resistant to pest attack and those that reduce the likelihood of attack in the first place. AWPM programmes have used both approaches, but usually rely more heavily on the second.

In most cases hosts that can serve as reservoirs for a particular pest exist outside of the managed ecosystem, from which the pest can colonize it via dispersal or migration. Knipling recognized this and considered lack of control in unmanaged ecosystems and lack of synchronous control across ecosystems as the main reasons pest populations were not being effectively controlled by management applied on an individual field basis. Knipling's (1992) principle formalized that idea when he wrote: 'Uniform suppressive pressure applied against the total population of the pest over a period of generations will achieve greater suppression than a higher level of suppression on most, but not all, of the population each generation'. Total population refers to the sum total of individuals of the species in a defined area as opposed to just those occupying a particular crop or other commodity in need of protection. Usually, the area defined for the total population is one that has the geographic integrity such that the population within it is more or less geographically isolated from other populations of the species. Knipling (1992) asserted that, in order for major advances to be made in managing many important pest problems, strategies and tactics for managing pests would need to change from the curative approaches targeting the pest on the protected crop to strategies based on suppressive measures targeting the total population preventively.

Knipling spent most of his career on eradication programmes for pests rather than on AWPM per se, but the two approaches have much in common, the major difference being the ultimate goal – areawide extermination versus suppression and maintenance at non-economic levels. Both centre on environmentally sound tactics that, when applied over a broad geographic area, can eradicate the total pest population or, in the case of AWPM, maintain it at non-economic levels. The difference between the two strategies is that eradication requires an intensive effort over a broad enough geographic area such that there is no possibility for migration by the pest into the suppression area over a reasonably long time horizon. This means that for eradication to be successful the suppression area must be effectively isolated from areas where migrants could enter and that the pest management tactics applied exert mortality much greater than the pest's reproductive capacity. These tactics must be applied within the suppression area on a frequent enough basis to maintain the population on a downward trajectory until eradication is eventually achieved. Whether or not tactics need to be applied synchronously over the entire area for the duration of the programme is arguable because some effective programmes, e.g. the boll weevil eradication programme, have focused on treating only 'hot spots' after an initial, brief phase of synchronous suppression.

A Brief History and the Development of AWPM

Klassen (2000) gives a very interesting account of the history of AWPM, which we shall recount here only briefly. Klassen notes that AWPM approaches are not new, and that early civilizations probably worked cooperatively to control pest invasions, such as those by armyworms and locusts, at scales greater than a single landholding. For example, in China an AWPM programme for the migratory locust, *Locusta migratoria maniensis*, has evolved as the result of over 3000 years of experience with periodic outbreaks along the flood plains of some major rivers. The AWPM programme for the locust now has a firm scientific basis and uses modern pest forecasting and management tools, but was initiated long before the advent of the modern scientific method, and based mainly on application of cultural practices and water management along major waterways that prior experience had shown were effective as preventive tactics to control the pest (Metcalf, 1991). Klassen (2000) discusses several additional early examples of use of the AWPM approach that grew out of practical necessity and experience with the particular pest.

Understanding of the biology and ecology of many highly mobile pests of agriculture is improving, as are the technologies for suppressing pest populations. The majority of modern-day programmes that can be classified as AWPM rely on a limited number of tactics targeted at distinctive characteristics of the pest insect's biology or ecology, most notably the sterile insect technique used in eradication programmes for the screwworm, fruit flies and other insects. Use of broad-spectrum insecticides for broad-scale suppression of mosquitoes and other pests can also be classified as an areawide approach (see Tan, 2000 for more examples). Few tactics specifically for use in AWPM exist, partially because limited emphasis has been placed on developing the AWPM concept within educational and research institutions. The expense and lack of funding available for such research has historically been a limiting factor (Linquist, 2000), but many current strategies for deploying existing host plant resistance, cultural control and biological control may be adaptable to, and benefit from, a more concerted effort in areawide deployment.

During the last decade, implementation of AWPM in the USA has been enhanced by the US Department of Agriculture, Agricultural Research Service-administered Areawide Pest Management Program (Calkins and Faust, 2003), and examples of AWPM implementation using a broader array of approaches and tactics now exist. A major goal of this book is to overview the AWPM programmes that have been implemented recently, to highlight the diversity of pests addressed by the AWPM approach and the various tactics used in the programmes.

One potential drawback to the AWPM approach is the need to coordinate the programme with stakeholders, sometimes with diverse interests. Achieving stakeholder buy-in to the potentially highly regimented and expensive programmes used in AWPM requires the pest not only to be a serious economic detriment to the commodity, but also to be consistently present at economic levels. Otherwise, the sustained sense of urgency needed to motivate stakeholders to support the programme will be difficult to achieve. Even then, it may be difficult to maintain support for the programme over the long term once the pest becomes non-significant economically and stakeholders focus on more immediately pressing issues.

For some pests it may not be feasible to undertake eradication, because these criteria cannot be met, the cost of the programme exceeds the benefits, the economic resources required are unavailable or the political will needed to initiate such a programme cannot be generated and sustained. The policy and institutional issues involved in developing and sustaining AWPM programmes are discussed in detail in later chapters of this volume. For the remainder of this chapter we will focus on ecological issues relevant to AWPM. Our objective is to highlight the role that recent advances in the study and understanding of spatial interactions in biological populations might play in optimizing AWPM approaches for pests.

Ecological Theory and Models Supporting AWPM

Optimizing an AWPM programme to maximize the suppression achieved and minimize the cost incurred would enhance the potential for success. One obvious limitation to optimal application of the total population management concept advocated by Knipling for many pests is lack of sufficient knowledge of the biology and ecology to establish the spatial extent of the total population to be managed. The total population could be as large as the geographic range of the species, or as small as a single field.

Even though AWPM was originally based on the concept of synchronous and uniform control of the total population of a pest, it has been implemented in a wide variety of ways over space and time. Management strategies may synchronize control over only part of a region infested by a pest at a time, for example by eradicating it from one edge to another; or control may be synchronized over patches and subpopulations, and not the 'total' population, when permission cannot be obtained from all public and private landholders. Lack of synchrony is exemplified by the boll weevil AWPM programme in the southern USA, where the spatial strategies for management were applied at various locations and times for a variety of logistical and political reasons. Thus, an ideal AWPM plan may only rarely be implemented in practice.

Recent advances in ecological theory have much to contribute to development of the AWPM approach. For example, Levins' (1969) classic work on metapopulation dynamics, while receiving interest predominantly from conservation biologists, was originally proposed as an explanation of the dynamics of pests existing in spatially structured populations. Since populations of most species, especially pest species, have broad spatial distributions with some degree of population subdivision, metapopulation theory may be useful for understanding population dynamics of pest species and designing effective AWPM programmes. Recent advances in landscape ecology emphasizing spatial interactions among populations (Pickett and Cadenasso, 1995) may also contribute to AWPM. Simulation modelling is obviously an important tool for understanding the dynamics of complex systems, such as populations existing in spatially heterogeneous landscapes. Next, we will focus on some spatial ecology, modelling and monitoring concepts as they relate to defining and optimizing AWPM programmes.

Temporal and Spatial Scales in AWPM Strategies

Some ecologists have expressed concern about the lack of consideration of temporal and spatial scales in ecological hypotheses (Levandowsky and White, 1977; Allen, 1989). Often, claims are made and conclusions drawn about the conditions that promote or inhibit the AWPM of arthropods without the operational temporal and spatial scales being specified. Without scales for example, we do not know whether a given concept pertains to 1 m^2 and 1 day or to 1 million km^2 and 1 year.

If management strategies are to be tested and implemented, we need to strive for more precise concepts that include general temporal and spatial scales for which the concepts are valid. Onstad (1992) evaluated this problem in epidemiology and proposed criteria for identifying appropriate scales and definitions of important terms. Criteria for identifying temporal and spatial scales should be based on consistency of observation and ecological validity. Scales must account for behaviour and longevity of the targeted pest. Temporal and spatial units must correspond, for logical reasons.

The minimum time unit for analysis of management is likely to be the generation time for the targeted arthropod. When control is synchronized for AWPM, it is likely to be synchronized by generation and not calendar time. Onstad (1992) discusses the various ways of measuring generation time for arthropods with discrete or overlapping generations. The minimum spatial unit should be the two- or three-dimensional space that is traversed on average by the targeted arthropod during a generation. The ecologically proper spatial scale depends upon a species' biology and behaviour, and must be large enough to encompass all normal movement by the average individual (Weins, 1976; Addicott *et al.*, 1987). For example, Schneider (1989) considered the temporal and spatial scales and experimental designs for field experiments needed to evaluate AWPM. Schneider (1989) concluded that two small areas could be studied for 6 years, or two larger areas could be evaluated for 2 years, to determine the feasibility of an AWPM approach.

The spatial and temporal context of the AWPM strategy should be declared. The time horizon is the term used to describe the period over which the management occurs. Every model, of course, has a time horizon for its analysis. Because of the variable nature of ecological systems in space and time, the ecological conditions on which management depends may not be valid after a certain number of arthropod generations, or when a very large number of minimum spatial units are considered together. For eradication, it is clear that the total population must be defined to include a geographic area of sufficient extent to preclude recolonization of the eradication zone over a relatively long time horizon. For areawide suppression programmes there are no similarly obvious guidelines. Obviously, the maximum spatial extent should not exceed the existing area inhabited by a species, but best estimations are needed on whether infested fringe areas that are never managed should be included within conceptual spatial boundaries of a suppression zone.

Commonly, practitioners of AWPM define boundaries of a suppression area based on a mix of biological, economic, political, sociological and other criteria (Klassen, 2000; Linquist, 2000). The 'what and where' in regard to the total population is an elusive question, but certainly should be entertained by AWPM researchers and practitioners. A metapopulation or landscape perspective may be helpful in defining ecologically optimal scales for application of AWPM programmes for particular pests. We will explore those concepts with respect to AWPM in the following sections.

Metapopulation Ecology and AWPM

Metapopulation ecology is one approach to the study of spatially structured populations. The basic assumption of Levins' (1969) metapopulation dynamics concept and subsequent refinements is that the environment is heterogeneous. This heterogeneity can be partitioned into discrete patches suitable for inhabitation by the species that are distinguishable from everything else that is considered uninhabitable. Thus, suitable habitat for a species occurs as a network of patches distributed in space and embedded in a matrix of non-suitable areas (see Fig. 2.1a). The matrix is not explicitly considered in the metapopulation approach and is assumed to be neutral in terms of its effect on population dynamics, apart from that of separating patches of suitable habitat.

Habitat patches can vary in size and quality and differ in many attributes important to a particular species. The metapopulation approach also assumes that patches are small enough that the resident populations are more or less panmictic, but far enough apart that migration between patches is an occasional event (Hanski and Gilpin, 1997).

Most insects of pest management concern are highly mobile and migratory according to the definition of Drake and Gatehouse (1994), and it might seem that the metapopulation concept would provide limited insight on effective AWPM strategies beyond that achieved by considering the total population as a panmictic unit. However, Murphy *et al.* (1990) argue that the metapopulation approach is particularly applicable for small organisms with high population growth rates that have specialized habitat requirements, traits that are exhibited by many pest species.

Migration in insects usually takes place during a single life stage, and often for only a short time; during other life stages individuals are more or less restricted to a single patch of habitat. Thus, migration is a more or less discrete event within a generation for many pest insects. Furthermore, even though agricultural crops are grown extensively in some areas, they still occur as discrete patches within a heterogeneous matrix composed of patches that may or may not be suitable for inhabitation. Byrne (Chapter 4, this volume) discusses the critical role that migration plays in the ecology of pests and in determining pest status and limitations to the effectiveness of traditional pest management programmes. Since many, if not most, agricultural and urban pests and pests of other managed ecosystems are migratory (e.g. bollworms, aphids and leaf rust), considering the role of migration is essential for determining optimal AWPM strategies.

Fig. 2.1. Comparison of the (a) metapopulation and (b) landscape approaches to population ecology.

What differentiates situations where the metapopulation concept is useful from those where an assumption of one large panmictic population is more applicable is whether the total population can be considered to be composed of discrete local breeding populations connected by migration (Hanski and Gilpin, 1997). More specifically, the pest insect's total population functions as a metapopulation when its hosts are distributed in discrete patches, subpopulations on the patches have a high probability of extinction, unoccupied patches are available to be colonized, and subpopulations do not fluctuate synchronously (Hanski and Gilpin, 1997). For many pest insects of agriculture, the first three criteria are satisfied. However, the fourth criterion is less certain, and in fact will often not be satisfied, because populations of many insects are highly dependent on climatic factors such as temperature, which are correlated over broad geographic areas.

Furthermore, the matrix is rarely without effect on population processes and can inhibit or accentuate movement of individuals among habitat patches, function as a conduit to channel movement in particular directions or serve as habitat for individuals or as a source of mortality to them (Forman, 1995). A landscape perspective that considers the composition and distribution of habitat and non-habitat patches may be insightful in designing AWPM programmes for some pests. In a landscape approach, the matrix is not considered uninhabitable, but consists of a heterogeneous mosaic of patches and other elements that interact with the species in complex ways (Fig. 2.1b). The main limitation of the landscape approach is lack of a comprehensive theoretical foundation, which limits the ability to make predictions based on well-understood theory and mathematical models (Hanski, 1998). However, this limitation does not necessarily reduce the value of the landscape approach for investigating and modelling the spatio-temporal dynamics of a particular pest species in a particular geographic setting.

Landscape Ecology and AWPM

The main difference between the metapopulation and landscape approaches to population dynamics involves the role of the matrix in the dynamics of populations on patches. In the metapopulation approach the matrix is considered to be uninhabitable but consistent in its effects on the dynamics of populations on patches. In the landscape approach the matrix can have varying effects on local populations, which can be highly influential to within-patch dynamics (Hanski, 1998). Dunning *et al.* (1992) formalized terminology for landscape effects on biological populations by identifying four classes of important ecological effects of landscapes on local populations (i.e. populations in patches): landscape complementation, landscape supplementation, source–sink dynamics and neighbourhood effects. Taylor *et al.* (1993) added the idea of landscape connectivity to the four processes defined by Dunning *et al.* (1992).

Landscape complementation occurs when an organism requires two or more non-substitutable resources that are located in different patch types. The non-substitutable resources are required by the species for different reasons, and possibly at different times during its life. The organism has to move between patch types to obtain these resources. If the patch types occur in close proximity relative to the movement ability or behaviour of the organism, then a particular patch can support a larger population than if the patches are far enough apart that resources are difficult to obtain by cross-patch movement (see Fig. 2.2a). The presence of the resources in a particular habitat patch is complemented by the close proximity of the resources in the second patch, and larger populations can be supported in the area of proximity of the patches.

Landis and Haas (1992), in a study of the European corn borer parasitoid, *Eriborus terebrans*, demonstrated landscape complementation. *Eriborus terebrans* density was greater where maize fields were adjacent to woodlots than where maize fields were adjacent to agricultural fields. The woodlots were favourable habitat for adult survival by providing a food source for adults (flowering understorey plants); maize fields provide hosts for oviposition. Both resources are essential, and their availability in close proximity promotes large local populations of *E. terebrans*.

Landscape supplementation occurs when the population in a focal patch is increased because of the close proximity of other patches that contain the same resource or one that is used for the same function, making the resource easily accessible (see Fig. 2.2b). For *Diabrotica* species beetles (*D. barberi*, *D. cristata* and *D. virgifera virgifera*), supplementation occurs in Midwestern US agricultural landscapes. These beetles are closely tied to a primary habitat – maize or prairie, depending on species – but move to secondary habitat when the relative attractiveness of food sources in the primary habitat decreases. For these species, habitat type and contrast in pollination of plant species suitable as food for the beetles are important factors influencing habitat choice. Pollen availability in habitats adjacent to maize fields or prairie supplements the beetles. Extensive use of crop and non-crop habitats within the agricultural landscape suggests that an areawide approach, focusing on both crop and non-crop habitats, would be more effective for AWPM of corn rootworms than focusing only on maize fields (Campbell and Meinke, 2006).

Source–sink population dynamics occurs when productive 'source' patches, which have more births than deaths per generation, serve as sources of emigrants that

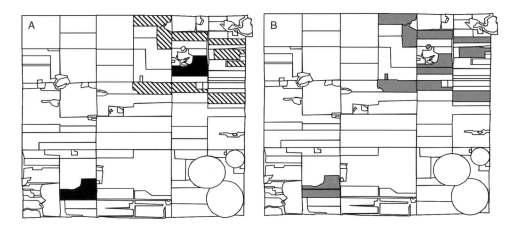

Fig. 2.2. Concepts of (a) landscape complementation and (b) supplementation in landscapes. The focal patch (in black) in the upper right area of figure (a) has a larger population than the focal patch in the lower left because of resources in nearby patches. Similarly for the corresponding patches in figure (b).

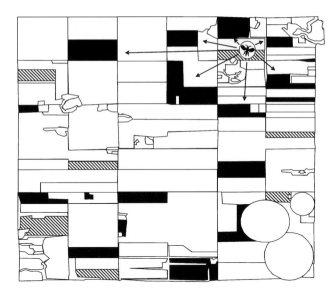

Fig. 2.3. A source/sink landscape. The source (cross-hatched) patches interact with sink (black) patches via dispersal.

migrate to less productive 'sink' patches (see Fig. 2.3). In sinks the death rate exceeds the birth rate (Pulliam, 1988; Pulliam and Danielson, 1991). Another class of patches exists, called pseudo-sinks, where the quality of the patch fluctuates between source and sink over time depending on population density, being source habitat at low density and sink habitat at high density (Watkinson and Sutherland, 1995). The concept of pseudo-sink can be extended to include habitats that fluctuate between source and sink on a seasonal basis irrespective of population density. Defined this way, the definition of pseudo-sink is useful for pest insects in agricultural settings that inhabit ephemeral habitats. The population size in a source-sink landscape can be strongly affected by the relative proportions of source and sink patches (Pulliam and Danielson, 1991).

For the Russian wheat aphid, the semi-arid landscapes of the High Plains Region of the USA can be conceptualized as a mosaic of source, sink, pseudo-sink and uninhabitable patches that changes over the course of the growing season (see Fig. 2.4). In order to maintain populations within an agricultural landscape the aphid must successfully exploit habitat patches when these are suitable and migrate to other suitable habitats when the current habitat declines in value. Both managed and unmanaged ecosystems are used. The quality of small grain fields as habitat for the Russian wheat aphid varies throughout the growing season. Patches of host cool- and warm-season grasses are sink habitat during most of the growing season, where mortality exceeds reproduction (Armstrong *et al.*, 1991).

Patches of most other land use types are uninhabitable. Population persistence is dependent on migration among habitat patches, the quality of which vary spatially and temporally. Cool-season grasses are sink or pseudo-sink habitat for Russian wheat aphids, depending on species, but their existence in the landscape is critical to Russian wheat aphid survival during summer because they represent the only suitable resource available (Brewer *et al.*, 2000). Russian wheat aphid survival over

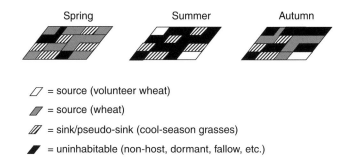

Spring   Summer   Autumn

◸ = source (volunteer wheat)

◢ = source (wheat)

◪ = sink/pseudo-sink (cool-season grasses)

◢ = uninhabitable (non-host, dormant, fallow, etc.)

Fig. 2.4. Russian wheat aphid source/sink population dynamics in Great Plains, USA, agricultural landscapes.

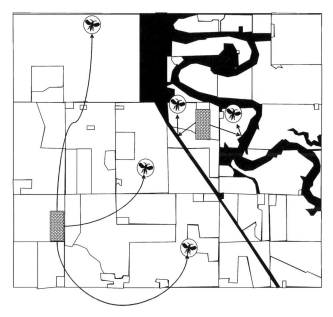

Fig. 2.5. Neighbourhood effects in landscapes; dispersal from the patch in the textured upper right area is restricted by the presence of boundaries that restrict inter-patch movement.

summer, in the absence of volunteer wheat, is therefore dependent on the extent and quality of sink and pseudo-sink habitat in the agricultural landscape. An AWPM programme for the Russian wheat aphid should be based on reducing the off-season hosts of the aphid, while simultaneously conserving natural enemies as key components (Giles *et al.*, this volume, Chapter 19).

Neighbourhood effects occur when a species in a patch is more strongly affected by the pattern of the nearby landscape than by more distant aspects of pattern. The neighbourhood concept simply formalizes the idea that an organism's ability to utilize resources in adjacent patches within a landscape can be dependent on the nature of the boundaries between patches and the shape, size and composition of immediately surrounding patches (see Fig. 2.5) (Dunning *et al.*, 1992).

As an example of neighbourhood effects, Bach (1988) demonstrated that a tomato border around squash patches inhibited movement of squash bugs from the patch to nearby patches of squash. Thus, tomato acted as an impermeable or semi-permeable boundary for movement of squash bugs between habitat patches. Wratten *et al.* (2003) demonstrated, for certain Syrphidae species of agricultural landscapes in New Zealand, that field boundaries composed of poplar hedgerows impeded movement of the syrphid flies to and from agricultural fields more than did boundaries consisting of post-and-wire fences. This study suggests that boundary structure plays a role in the functioning of spatially heterogeneous syrphid populations in agricultural landscapes by inhibiting rates of recolonization of agricultural fields following disturbance, such as insecticide application. These neighbourhood effects could alter predator–prey interactions by delaying the numerical response of the syrphids to population increases of their aphid prey (Wratten *et al.*, 2003).

Finally, connectivity, which is the degree to which the landscape facilitates or impedes movement by individuals of a species among resource patches, modifies the other landscape processes (see Fig. 2.6). Connectivity is a function of landscape structure (the composition and arrangement of elements in the landscape) and also the biology, behaviour and morphology of the particular organism (Taylor *et al.*, 1993).

Tewksbury *et al.* (2002) demonstrated that increased connectivity resulting from construction of early successional corridors in forest resulted in greater exchange of insects between patches, and also increases in plant–insect interactions (pollination). Kruess and Tscharntke (2000) demonstrated that species richness of herbivorous insects on patches of bush vetch, *Vicia sepium*, increased with an increasing area of meadows that contained bush vetch, which suggests that the insects exhibit greater

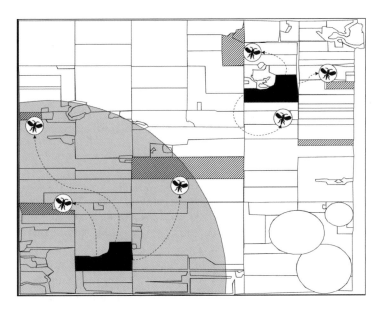

Fig. 2.6. Effects of varying connectivity (denoted by shading) on populations in heterogeneous landscapes; the high connectivity of the landscape matrix in the lower left area facilitates dispersal as compared with that in the upper right.

dispersal and patch-finding ability in a matrix with high connectivity. Not unexpectedly, species diversity decreased with increasing patch isolation.

For many pest insects, agricultural landscapes present an ever-changing mosaic of patches that exhibit classes of effects such as those defined by Dunning *et al.* (1992) and Taylor *et al.* (1993). Thorough understanding of the nature of landscape interactions for pest species and their natural enemies may aid in optimizing AWPM programmes for particular pests. For example, if a total population, in the sense defined by Knipling, is structured into distinguishable subpopulations in source and sink habitat patches, then the population can be managed without treating the subpopulations uniformly. The metapopulation and landscape approaches suggest that synchronous treating of all source populations over several generations should control, if not eradicate, the pest. Patches with sink populations may be located in natural areas or in crops that can be damaged, however, so they cannot support population growth. Thus, AWPM from a landscape perspective may be able to save treatment and environmental costs by managing only source populations, with necessary connectivity to supplementing and complementary habitat.

Concepts from metapopulation and landscape ecology may be useful for describing the dynamics of Knipling's total population in future AWPM programmes for, at least, some pest insects, and therefore may aid in developing effective AWPM strategies. Spatially explicit models are an important tool for studying populations in heterogeneous landscapes and for developing and assessing strategies and tactics for control in complex spatial systems where effects of system structure and inputs on dynamics are not always obvious. In fact, Knipling used modelling extensively in his analyses of the AWPM concept (Knipling, 1979, 1992), although the models did not explicitly consider spatial distribution. Next, we discuss the role of models and related tools in AWPM.

Ecological Modelling

How can scientists optimize the economics and efficacy of an AWPM plan given all the practical hurdles that need to be overcome? We believe that ecological theory and case studies based on modelling can help find very good, if not optimal, solutions. Modelling and theory should go hand-in-hand to help answer questions on management approaches. For example, we could ask how synchronous and how uniform management of a pest should be? Would 100% mortality on 90% of the cropland accomplish the same goal as 90% mortality on every hectare? Or 100% mortality on all cropland in 90% of the pest's generations over several years? What if the pest's phenology, genetic structure and spatial distribution vary from year to year? Models can help us explore the temporal and spatial dynamics of pests targeted by AWPM programmes.

In this section, we discuss the contribution of spatially explicit models and other tools to understanding and implementing AWPM. As with all technologies, such tools have limitations. For example, sometimes models are too specific to a particular pest or ecosystem to provide general assistance. In other cases, general models omit certain processes or conditions, thus limiting their ability for making specific recommendations. By considering a variety of approaches, we hope to derive valuable insights in developing management plans, as well as to learn what additional work may be important to further optimize the management plan.

Carrière *et al.* (2006) used GIS to account for the effects of agroecosystem hetero-geneity on dispersal and population dynamics. The results of their model allowed them to develop recommendations for AWPM that included the placement of cropland and fallow fields for better control of a regional pest. Brewster and Allen (1997) integrated a model that simulated the temporal and spatial dynamics of a pest with a digital map of heterogeneously distributed habitat. Their approach could be used to evaluate the effect of management tactics such as biological control on pests that infest multiple habitats.

Bessin *et al.* (1991) used a simulation model that considered control by host plant resistance and predation to study AWPM of *Diatraea saccharalis*. They concluded that mixtures of resistant and susceptible sugarcane deployed over broad geographic scales would help control the pest. However, their results were sensitive to assump-tions about adult dispersal. El-Sayed *et al.* (2006) reviewed the literature on empirical and modelling studies of mass trapping and drew several conclusions. They empha-sized the value of targeting low-density, isolated populations. In addition, mass trap-ping was most successful or more cost-effective for monophagous, univoltine pests with lower population growth rates over a given unit of time.

In a very interesting analysis, Byers and Castle (2005) explored the question: can the traditional pest management decision to treat individual fields asynchronously at a specified population threshold be improved by synchronously treating all fields at an average population threshold in an areawide programme? They developed a sim-ulation model of insect populations in a large set of fields that varied in exponential growth each day of a season. A portion of the insects also dispersed to adjacent fields at each time step. Byers and Castle (2005) considered a landscape with explicit spatial structure, with distances between patches determining dispersal probabilities.

In one model, populations in each field were monitored. A field was treated with insecticide if the population exceeded a threshold (asynchronous model), as performed in traditional IPM. A second model treated the entire array of fields with insecticide when the average population of all fields exceeded the same threshold (synchronous model). Byers and Castle (2005) found that the synchronous model, at all growth and dispersal rates tested, had average field populations during a season that were significantly lower (see Fig. 2.7) and required fewer treatments than the asynchronous method. Byers and Castle (2005) concluded that cooperation among growers in areawide monitoring of fields to obtain an average population estimate for use with treatment thresholds would result in significantly less insect damage and fewer insecticide treatments.

However, their conclusions may be valid for only a small portion of real situations that are likely to be encountered by farmers and their advisors. Byers and Castle (2005) allowed growth in the population at each site during each time step. This implies that each time step is a significant portion of one generation: therefore, the results are valid only for pests with overlapping generations. They also focused on pests that require control at low densities, which means that they did not incorporate density-dependent dispersal, density-dependent survival and regulation by natural enemies. Byers and Castle (2005) essentially modelled a system in which the growth of the total population is highly predictable over time. In the future, interesting extensions of the model could include variability in population growth rate over time, asynchrony in phenology of

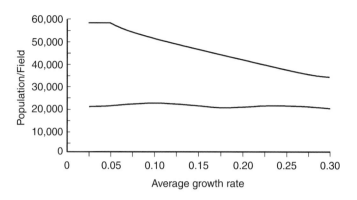

Fig. 2.7. Relationship between mean population growth rate and mean pest insect population per field per day (redrawn from Byers and Castle, 2005). The growth rate is the intrinsic rate of population increase, which is a dimensionless number; it is the natural logarithm of the ratio of population sizes at two different times.

vulnerable life stages across the landscape, sampling costs and evolution of insecticide resistance. It is possible that, with greater temporal uncertainty and temporal asynchrony, an asynchronous management strategy would be better than the synchronized areawide effort conceptualized by Byers and Castle (2005).

Theory of Host-density Thresholds

If biological control by a parasite is directed at a total population, then the concept of host-density thresholds must be considered when AWPM goals and tactics are being developed. Eradication by a host-specific parasitoid or pathogen may not be feasible for ecological reasons. Even long-term management may be difficult depending on how low the density of the targeted pest must be driven in comparison with the searching and reproductive abilities of the natural enemy in heterogeneous landscapes (Onstad and Kornkven, 1999). In the epidemiological literature (Onstad and Carruthers, 1990; Onstad, 1993; Onstad and McManus, 1996), it is generally accepted that, for a host-specific pathogen used in classical biological control, the pathogen will be extirpated at low host densities before the host is extirpated. Essentially, without continuous, multi-generation inundative releases of parasites, the parasites will drive the host density so low that the parasite will be unable to find and attack the remaining susceptible pests. This also means that efforts to use biological control in natural areas where lower densities of a pest may occur will need to compensate for the lower encounter rates between host and parasite. Furthermore, AWPM may have to rely on non-classical biological control to achieve its goals.

Insect Resistance Management

Insect resistance management can be very similar to AWPM because both depend on coordination of efforts over large regions (Onstad, 2007). Siegfried *et al.* (1998)

recognized the relationship between AWPM and insect resistance management when they warned proponents of AWPM about the increased risk of resistance evolution if areawide projects treat the landscape uniformly. They believed that some attributes of AWPM are incompatible with many conventional insect resistance management (IRM) techniques, but suggested that use of biologically based control tactics, such as behaviour-disrupting chemicals, may contribute to both AWPM and insect resistance management.

Onstad and Guse (1999) showed clearly that a regional pest population could be eradicated, if not maintained at very low levels, by the use of highly effective transgenic insecticidal crops. This was the case even when 10–30% of the landscape was planted with refugia of non-insecticidal plants. Carrière *et al.* (2003) described a real case of suppression of a regional pest using a transgenic insecticidal crop. Peck and Ellner (1997) and Peck *et al.* (1999) used spatially explict models to explore insect resistance management of a regional pest. Their conclusions were that the population growth rates and dispersal rates determine the likelihood of success of management, which are likewise relevant to AWPM over the long term.

Monitoring

How should monitoring be incorporated into an AWPM plan? Monitoring is costly and should not be performed unless absolutely necessary in AWPM. Under ideal conditions monitoring should not be performed – except following several years of synchronized, uniform control – to determine the success of the project. Certainly, monitoring should be performed at the end of the project's time horizon. In theory, one of the economic savings in AWPM is the elimination of monitoring costs in a strategy that simply inundates the environment with treatments targeting the pest on a schedule and in all areas expected to be infested. However, in practice, monitoring will probably occur. Therefore, how can ecological theory help select times and locations for monitoring?

One approach in AWPM is to trade treatments, and their cost for monitoring and its costs. In other words, use monitoring to decide when and where to treat the population to improve regional control. In situations involving metapopulations, monitoring could be used to determine the locations of source and sink patches. Then only the source subpopulations would be treated over time while sink patches were monitored to determine whether they functioned as refugia for the pest from which to recolonize source patches.

In any scenario in which the pest population density is highly variable over space and time, particularly with large areas or long periods without significant pest densities, it is possible that treatments could be optimized in their efficacy by monitoring and then treating the pest either under only low-density situations or high densities. The cotton boll weevil eradication programme in the southern USA exemplifies this approach. Monitoring is an essential component of the eradication programme, because pheromone traps are used to determine those fields to be treated with low doses of the broad-spectrum insecticide malathion. Insecticide application directed at infested fields, combined with cultural practices applied to all fields, has been successful in reducing the boll weevil to undetectable levels (Allen, this volume, Chapter 20).

Some pests may be more easily eradicated or maintained at very low densities if the management begins at low densities. This is true for mass trapping techniques (El-Sayed *et al.*, 2006). But for other pests, density at time of treatment may be irrelevant, or tactics such as biological control may be more effective at initially high densities. For example, AWPM of the Russian wheat aphid is based almost completely on preventive tactics, including conservation biological control, and density at time of implementation of AWPM is not important for this species (Giles *et al.*, Chapter 19, this volume).

Highly variable pests may not be good targets for AWPM, because monitoring of these pests (which often exist at densities below economic thresholds) is so important and control is not needed as often as for other pests. We are not capable of providing in this chapter the comprehensive discussion that the subject of monitoring deserves. Certainly, any AWPM programme that plans to use monitoring must start with an understanding of: (i) the pest's population dynamics in various habitats (Park and Tollefson, 2005); (ii) the increasing difficulty in sampling as the management programme progresses (Venette *et al.*, 2002); and (iii) the costs of extensive and intensive sampling (Nyrop *et al.*, 1986).

Conclusion

AWPM programmes, including eradication programmes, have provided some outstanding successes both in cost effectiveness and in the level and durability of control. Successful programmes were based on detailed knowledge of the biology of the target pest and on proven technologies for suppressing its populations. While some programmes have been highly successful, there are opportunities to refine approaches through the application of ecological concepts and technology, which could make AWPM applicable to a broader range of pest species and make it more economical in terms of money and time. Recent advances in ecology, particularly from the fields of metapopulation and landscape ecology, are applicable for the study of candidate species for AWPM. Application of ecological concepts, particularly when complemented by modelling and other supporting technologies, could help in defining the spatial limits of the total population from an operational viewpoint, and in optimizing AWPM programmes.

Acknowledgement

We thank Tim Johnson for producing the illustrations used in this manuscript. The views expressed herein are those of the authors only and do not necessarily reflect or represent the opinions, positions or policies of the US Department of Agriculture.

References

Addicott, J.F., Aho, J.M., Antolin, M.F., Padilla, D.K., Richardson, J.S. and Soluk, D.A. (1987) Ecological neighborhoods: scaling environmental patterns. *Oikos* 49, 340–346.

Allen, T.F.H. and Starr, T.B. (1982) *Hierarchy: Perspectives for Ecological Complexity*. University of Chicago Press, Chicago, Illinois.

Armstrong, J.S., Porter, M.R. and Peairs, F.B. (1991) Alternate hosts of the Russian wheat aphid (Homoptera: Aphididae) in northeastern Colorado. *Journal of Economic Entomology* 84, 1691–1694.

Bach, C. (1988) Effects of host patch size on herbivore density: underlying mechanisms. *Ecology* 69, 1103–1117.

Bessin, R.T., Stinner, R.E. and Reagan, T.E. (1991) Modeling the area-wide impact of sugarcane varieties and predation on sugarcane borer (Lepidoptera: Pyralidae) populations in southern Louisiana. *Environmental Entomology* 20, 252–257.

Brewer, M.J., Donahue, J.D. and Burd, J.D. (2000) Seasonal abundance of Russian wheat aphid (Homoptera: Aphididae) on non-cultivated perennial grasses. *Journal of the Kansas Entomological Society* 73, 85–95.

Brewster, C.C. and Allen, J.C. (1997) Spatiotemporal model for studying insect dynamics in large-scale cropping systems. *Environmental Entomology* 26, 473–482.

Byers, J.A. and Castle, S.J. (2005) Areawide models comparing synchronous versus asynchronous treatments for control of dispersing insect pests. *Journal of Economic Entomology* 98, 1763–1773.

Calkins, C.O. and Faust, R.J. (2003) Overview of areawide programs and the program for suppression of codling moth in the western USA directed by the United States Department of Agriculture–Agricultural Research Service. *Pest Management Science* 59, 601–604.

Campbell, L.A. and Meinke, L.J. (2006) Seasonality and adult habitat use by four *Diabrotica* species at prairie–corn interfaces. *Environmental Entomology* 35, 922–936.

Carrière, Y., Ellers-Kirk, C., Sisterson, M., Antilla, L., Whitlow, M., Dennehy, T.J. and Tabashnik, B.E. (2003) Long-term regional suppression of pink bollworm by *Bacillus thuringiensis* cotton. *Proceedings of the National Academy of Science USA* 100, 1519–1523.

Carrière, Y., Ellsworth, P.C., Dutilleul, P., Ellers-Kirk, C., Barkley, V. and Antilla, L. (2006) A GIS-based approach for areawide pest management: the scales of *Lygus hesperus* movements to cotton from alfalfa, weeds, and cotton. *Entomologia Experimentalis et Applicata* 118, 203–210.

Drake, V.A. and Gatehouse, A.G. (1994) *Insect Migration: Tracking Resources Through Space and Time*. Cambridge University Press, Cambridge, UK.

Dunning, J.B., Danielson, B.J. and Pulliam, H.R. (1992) Ecological processes that affect populations in complex landscapes. *Oikos* 65, 169–175.

El-Sayed, A.M., Suckling, D.M., Wearing, C.H. and Byers, J.A. (2006) Potential of mass trapping for long-term pest management and eradication on invasive species. *Journal of Economic Entomology* 99, 1550–1564.

Forman, R.T.T. (1995) *Land Mosaics: the Ecology of Landscapes and Regions*. Cambridge University Press, Cambridge, UK.

Hanski, I.A. (1998) Metapopulation dynamics. *Nature* 396, 41–49.

Hanski, I.A. and Gilpin, M.E. (1997) *Metapopulation Biology: Ecology, Genetics, and Evolution*. Academic Press, New York.

Klassen, W. (2000) Area-wide approaches to insect pest management: history and lessons. In: Tan, K. (ed.) *Area-wide Control of Fruit Flies and Other Insect Pests*. Penerbit Universiti Sains, Penang, Malaysia, pp. 21–38.

Klassen, W. (2003) Edward F. Knipling: titan and driving force in ecologically selective area-wide pest management. *Journal of the American Mosquito Control Association* 19, 94–103.

Knipling, E.F. (1979) *The Basic Principles of Insect Population Suppression and Management.* Agricultural Handbook 512, United States Department of Agriculture, Washington, DC.

Knipling, E.F. (1992) *Principles of Insect Parasitism Analyzed from new Perspectives: Practical Implications for Regulating Insect Populations by Biological Means.* Agricultural Handbook 693, United States Department of Agriculture, Washington, DC.

Kruess, A. and Tscharntke, T. (2000) Species richness and parasitism in a fragmented landscape: experiments and field studies with insects on *Vicia sepium. Oecologia* 122, 129–137.

Landis, D.A. and Haas, M.J. (1992) Influence of landscape structure on abundance and within field distribution of European corn borer (Lepidoptera: Pyralidae) larval parasitoids in Michigan. *Environmental Entomology* 21, 409–416.

Levandowsky, M. and White, B.S. (1977) Randomness, time scales, and the evolution of biological communities. *Evolutionary Biology* 10, 69–161.

Levins, R. (1969) Some demographic and genetic consequences of environmental heterogeneity for biological control. *Bulletin of the Entomological Society of America* 15, 237–240.

Linquist, D.A. (2000) Pest management strategies: area-wide and conventional. In: Tan, K. (ed.) *Area-wide Control of Fruit Flies and Other Insect Pests.* Penerbit Universiti Sains, Penang, Malaysia, pp. 13–19.

Loehle, C. (1987) Hypothesis testing in ecology: psychological aspects and the importance of theory maturation. *Quarterly Review of Biology* 62, 397–409.

Metcalf, R.L. (1991) Insects – man's chief competitors. In: Menn, J.J. and Steinhauer, A.L. (eds) *Progress and Perspectives for the 21st century.* Centenial National Symposium, Entomological Society of America, Lanham, Maryland.

Metcalf, R.L. and Luckman, W.H. (1975) *Introduction to Insect Pest Management.* John Wiley & Sons, New York.

Murphy, D.D., Freas, K.E. and Weiss, S.B. (1990) An environment–metapopulation approach to population viability analysis for a threatened invertebrate. *Conservation Biology* 4, 41–51.

Nyrop, J.P., Foster, R.E. and Onstad, D.W. (1986) The value of sample information in pest control decision making. *Journal of Economic Entomology* 79, 1421–1429.

O'Neill, R.V., DeAngelis, D.L., Waide, J.B. and Allen, T.F.H. (1986) *A Hierarchical Concept of Ecosystems.* Princeton University Press, Princeton, New Jersey.

Onstad, D.W. (1992) Temporal and spatial scales in epidemiological concepts. *Journal of Theoretical Biology* 158, 495–515.

Onstad, D.W. (1993) Thresholds and density dependence: the roles of pathogen and insect densities in disease dynamics. *Biological Control* 3, 353–356.

Onstad, D.W. (2007) *Insect Resistance Management: Biology, Economics, and Prediction.* Academic Press, New York.

Onstad, D.W. and Carruthers, R.I. (1990) Epizootiological models of insect diseases. *Annual Review of Entomology* 35, 399–419.

Onstad, D.W. and Guse, C.A. (1999) Economic analysis of transgenic maize and nontransgenic refuges for managing European corn borer (Lepidoptera: Pyralidae). *Journal of Economic Entomology* 92, 1256–1265.

Onstad, D.W. and Kornkven, E.A. (1999) Persistence of natural enemies of weeds and insect pests in heterogeneous environments. Chapter 19. In: Hawkins, B.A. and Cornell, H.V. (eds) *Theoretical Approaches to Biological Control.* Chapman and Hall, London, pp. 349–367.

Onstad, D.W. and McManus, M.L. (1996) Risks of host-range expansion by insect-parasitic biocontrol agents. *BioScience* 46, 430–435.

Park, Y.-L. and Tollefson, J.J. (2005) Spatial prediction of corn rootworm (Coleoptera: Chrysomelidae) adult emergence in Iowa cornfields. *Journal of Economic Entomology* 98, 121–128.

Peck, S.L. and Ellner, S.P. (1997) The effect of economic thresholds and life-history parameters on the evolution of pesticide resistance in a regional setting. *American Naturalist* 149, 43–63.

Peck, S.L., Gould, F. and Ellner, S.P. (1999) Spread of resistance in spatially extended regions of transgenic cotton: implications for management of *Heliothis virescens* (Lepidoptera: Noctuidae). *Journal of Economic Entomology* 92, 1–16.

Peters, R.H. (1988) Some general problems for ecology illustrated by food web theory. *Ecology* 69, 1673–1676.

Pickett, S.T.A. and Cadenasso, M.L. (1995) Landscape ecology: spatial heterogeneity in ecological systems. *Science* 269, 331–334.

Pulliam, H.R. (1988) Sources, sinks, and population regulation. *American Naturalist* 132, 652–661.

Pulliam, H.R. and Danielson, B.J. (1991) Sources, sinks, and population regulation: a landscape perspective on population dynamics. *American Naturalist* 137, 550–566.

Roughgarden, J., May, R.M. and Levin, S.A. (1989). *Perspectives in Ecological Theory*. Princeton University Press, Princeton, New Jersey.

Schneider, J.C. (1989) Role of movement in evaluation of area-wide insect pest management tactics. *Environmental Entomology* 18, 868–874.

Siegfried, B.D., Meinke, L.J. and Scharf, M.E. (1998) Resistance management concerns for areawide management programs. *Journal of Agricultural Entomology* 15, 359–369.

Tan, K. (2000) *Area-wide Control of Fruit Flies and Other Insect Pests*. Penerbit Universiti Sains, Penang, Malaysia.

Taylor, P.D., Fahrig, L., Henein, K. and Merriam, G. (1993) Connectivity is a vital element of landscape structure. *Oikos* 68, 571–573.

Tewksbury, J.J., Levey, D.J., Haddad, N.M., Sargent, S., Orrock, J.L., Weldon, A., Danielson, B.J., Brinkerhoff, J., Damschen, E.I. and Townsend, P. (2002) Corridors affect plants, animals, and their interactions in fragmented landscapes. *Proceedings of the National Academy of Sciences, USA* 99, 12923–12926.

Venette, R.C., Moon, R.D. and Hutchison, W.D. (2002) Strategies and statistics of sampling for rare individuals. *Annual Review of Entomology* 47, 143–174.

Watkinson, A.R. and Sutherland, W.J. (1995) Sources, sinks, and pseudo-sinks. *Journal of Animal Ecology* 64, 126–130.

Weins, J.A. (1976) Population responses to patchy environments. *Annual Review of Ecology and Systematics* 7, 81–120.

Weins, J.A. (1989) Spatial scaling in ecology. *Functional Ecology* 3, 385–397.

Wratten, S.D., Bowie, M.H., Hickman, J.M., Evans, A.M., Sedcole, J.R. and Tylianakis, J.M. (2003) Field boundaries as barriers to movement of hover flies (Diptera: Syrphidae) in cultivated land. *Oecologia* 134, 605–611.

3 Establishing Inter-agency, Multidisciplinary Areawide Pest Management Programmes

ROBERT M. NOWIERSKI[1] AND HENDRIK J. MEYER[2]

[1]USDA-CSREES, National Program Leader for Bio-Based Pest Management, Waterfront Centre, Washington, DC, USA; [2]USDA-CSREES, National Program Leader for Entomology, Waterfront Centre, Washington, DC, USA

Introduction

Pest management today usually comprises multiple complex tactics that transcend disciplines, geographical regions, climatic zones, production/management systems, production scales and economic strata. Solutions to priority problems involve research, education and extension professionals. But, equally important to project success are the users of the knowledge generated and the end-users of the 'products'. Setting a direction for the future goals, IPM has been an important activity for the 'community' of constituents who share concerns for future pest management.

Under the leadership of the USDA and land grant universities, a road map for IPM has been developed with extensive participation of diverse stakeholders. The goal of the IPM road map is to increase nationwide communication and efficiency through information exchanges among federal and non-federal IPM practitioners and service providers, including land managers, growers, structural pest managers and public and wildlife health officials. Development of the road map for the National Integrated Pest Management (IPM) Program began in February 2002, with continuous input from numerous IPM experts, practitioners and stakeholders. The road map identifies strategic directions for IPM research, implementation and measurement for pests in all major settings throughout the nation. This includes pest management for areas including agricultural, structural, ornamental, turf, museums, public and wildlife health pests, and encompasses terrestrial and aquatic invasive species.

The goal of the National IPM Program in the USA is to improve the economic benefits of adopting IPM practices and to reduce potential risks to human health and the environment caused by the pests themselves or by the use of pest management practices. Many other countries have similar programmes devoted to crop protection using the IPM approach.

Issues

Research and extension directed towards the implementation of areawide IPM includes the study of crop–pest–beneficial organism interactions (systems ecology) and interactions among pest control tactics, the impact of climate on pest management systems, the epidemiology and ecology of pests and the development of sampling protocols and predictive models for complexes of pests. Emphasis on adaptive research, the validation of IPM systems, the demonstration of new pest management approaches to end-users and regional coordination of research and extension efforts through the Regional IPM Centres and the National Plant Diagnostic Network are necessary. The area also includes work with stakeholders to identify priority needs and barriers to the implementation of IPM systems.

The peer review process ensures that competitively awarded USDA, Cooperative State Research, Education and Extension Service (CSREES) projects focus on scientifically critical areas. The Agricultural Research, Extension and Education Reform Act of 1998 process requires that formula-funded projects reflect stakeholder priorities. The competitive review process encourages innovative ideas that are likely to open new research approaches to the enhancement of US agriculture. A proven mechanism for stimulating new scientific research, the process increases the likelihood that investigations addressing important topics using well-designed and well-organized experimental plans will be funded. Each year, panels of scientific peers meet to evaluate and recommend proposals based on scientific merit, investigator qualifications and relevance of the proposed research to US agriculture.

Stakeholder input

CSREES identifies emerging issues for its IPM programmes in a variety of ways. Agency staff are active participants in IPM-related, multi-state research and extension projects that bring together agricultural scientists to address pest management issues. The annual meetings of scientists involved in these projects provide agency staff with an opportunity to keep abreast of emerging issues and needs. The advisory committees of the four regional IPM centres are another resource for the agency as it works to identify and prioritize IPM needs and issues.

Each advisory committee is a diverse group that includes agricultural producers and their representatives, private consultants, pest control operators and representatives of non-profit organizations and government agencies. Emerging issues are also identified by Pest Management Strategic Plans, which are developed for individual commodities by pest managers, research and extension experts and government regulatory staff; more than 88 have been developed and are available at http://www.pestdata.ncsu.edu/pmsp/index.cfm. CSREES also uses conferences and stakeholder forums to identify emerging issues. National IPM symposia have been held every few years since the late 1980s, and have drawn as many as 600 IPM experts from around the world to discuss new advancements and future needs.

The results of a priority-setting process provide the framework for facilitating the scientific and technological advances necessary to meet the challenges facing

US agriculture. Congress sets the budgetary framework by providing funds to ARS for intramural research programmes, and to CSREES for extramural research and extension programmes conducted primarily at land grant university partner institutions. Members of Congress also make recommendations for the scientific and programmatic administration through appropriation language and through their questions and comments during congressional hearings.

Input into the priority-setting process is sought from a variety of customers and stakeholders. The scientific community provides input through the proposals it submits each year, as well as through the proposal evaluation and funding recommendations of individual peer-review panels. Review panels for competitive programmes, federal inter-agency working groups, stakeholder workshops, the National Research Council, multi-state projects, ARS and other federal agencies involved in IPM activities are examples of important mechanisms for CSREES to identify emerging issues affecting areawide IPM development and implementation. National Program leaders attend scientific and professional meetings to keep abreast of both scientific trends that should be reflected in CSREES programmes and the coordination of priority setting with other federal agencies. National Program staff also participate in meetings with representatives of key commodity groups and other user groups to discuss current priorities, learn ways that CSREES can assist in meeting their needs and solicit comments and suggestions.

Stakeholder assessment

Although the benefits of IPM have been well documented, the extent of adoption has been limited due to several factors. A series of stakeholder workshops sponsored by the US Environmental Protection Agency (EPA) and USDA in 1992 and 1993 identified many factors constraining adoption of IPM systems, and recommended that the public and private sectors make a national commitment to overcoming these constraints (Sorensen, 1993, 1994). Major impediments included inadequate knowledge of currently available IPM tactics, a shortage of consultants and other pest management professionals to provide IPM services, the high level of management input required for implementation of some IPM systems and the lack of alternative pest control tactics for some pests. Before reforms were enacted in 1996, federal commodity programmes were other impediments to IPM adoption in cases where planting requirements restricted the ability of producers to rotate crops for pest control.

The IR-4 stakeholder process: an example

The IR-4 research planning process involves input from its many stakeholders. Most proposals for IR-4 assistance are transmitted from growers through federal and state research and extension scientists involved in high-value speciality crop pest management. IR-4 also receives proposals (called Project Clearance Requests) directly from growers and/or organizations representing a commodity. To maximize grower awareness of the programmes and their input, IR-4 personnel regularly attend grower meetings and tours. In some cases, state-level IR-4 meetings are held in which growers are invited to attend and submit Project Clearance Requests. The only groups prohibited from submitting requests are representatives of crop protection companies.

IR-4 project stakeholders are encouraged to attend the annual IR-4 Food Use and Ornamentals Workshops, where project proposals are prioritized. These workshops are critically important because IR-4 can conduct research on only 10–15% of the proposed researchable projects each year. These workshops are open forums attended by up to 200 growers, commodity organization representatives and federal and state research/extension scientists. At the workshops, every potential project is discussed in detail and its importance is considered on the basis of factors such as the availability and efficacy of alternatives, pest damage potential, performance of the proposed chemical and its compatibility with integrated pest management programmes.

The output of these meetings is a list of projects designated as having 'A', 'B' or 'C' priority or elimination from the research project list. In order to better serve the needs of growers, in 1999 IR-4 committed to a 30-month study completion policy for those projects classified as 'A' priority. Previously, most studies had taken 4–5 years to complete.

While a 100% compliance with the 30-month policy is not feasible for several reasons, it is the goal of the programme to raise the success rate from 70 to 85%. IR-4 also conducts research on as many Priority 'B' projects as possible (currently less than 25% of the total). Resources are not sufficient to allow for more 'B' or even 'C' research priority projects. Following the workshops, a National Research Planning Meeting is held to assign field and laboratory sites for the following year's research projects. About 100 food use residue projects (crop–chemical combinations) involving 700 field trials are undertaken annually, some in close cooperation and coordination with Canada's Pest Management Centre of Agriculture and Agri-Food Canada. Once projects are chosen and assigned, research protocols are drafted containing the proposed pattern of use, the number and location of field trials and instructions for the analysis of the chemical and metabolites in the commodity, as specified by the EPA. The EPA requires that this research be conducted and documented following exacting procedures outlined in the Good Laboratory Practice (GLP) guidelines.

Prioritizing needs through crop profiles and pest management strategic plans

Crop profiles (CPs) and pest management strategic plans (PMSPs) are widely recognized as a conduit for communication from growers and other IPM practitioners to regulators and granting agencies. These documents give a realistic view of crop production practices and pest management issues and strategies used in the field, and provide a forum for agricultural producers and allied professionals to set meaningful research, regulatory and educational priorities.

Strategic plans (PMSPs) are developed by growers and other stakeholders to identify the pest management needs and priorities of a particular commodity. Each plan focuses on commodity production in a particular state, region or the whole nation.

The plans take a pest-by-pest approach in identifying the current management practices (chemical and non-chemical) and those under development. Plans also state priorities for research, regulatory activity and education/training programmes needed for transition to alternative pest management practices.

Although the IPM centres have sponsored the majority of completed PMSPs to date, other agencies and groups such as EPA and grower organizations have also funded the development of these documents. The involvement of multiple organizations and facilitators makes it necessary for authors to follow a system that will ensure consistency in the content and format of all PMSPs. Completed PMSPs are hosted on the National IPM Centres web site at http://www.ipmcenters.org under CENTER PRODUCTS.

Numerous benefits may result from completing a PMSP:

- Regulators receive information on actual pest management practices and therefore will be less likely to use default assumptions in risk assessments.
- Regulators are provided with information relative to special pest management concerns (e.g. resistance management, geographical concerns).
- Stakeholders identify appropriate contact people to facilitate future communication.
- Grant seekers acquire documentation of stakeholder priority needs to support funding requests.
- Growers have available documentation to support Section 18 Emergency Exemption and Section 24(c) Special Local Needs requests.
- Commodity groups gain insight into emerging pest management issues allowing them to prioritize their research, education or other programmes they sponsor.
- Commodity representatives receive a document that can be used to convey their needs to policy makers.
- Support for IR-4 Food Use Workshop research prioritization is provided.
- Registrants may use PMSPs to identify niche markets for development of new products.
- Workshops provide a forum to discuss reduced-risk management options.

Potential changes in the 'toolbox'

Regulatory actions

EPA regulates pesticides under two major federal statutes. Under the Federal Insecticide, Fungicide, and Rodenticide Act (FIFRA), EPA registers pesticides for use in the USA and prescribes labelling and other regulatory requirements to prevent unreasonable adverse effects on human health or the environment. Under the Federal Food, Drug, and Cosmetic Act (FFDCA), EPA establishes tolerances (maximum legally permissible levels) for pesticide residues in food. For over two decades there had been efforts to update and resolve inconsistencies in the two major pesticide statutes, but consensus on necessary reforms remained elusive.

The 1996 Food Quality and Protection Act (FQPA) represented a major breakthrough, amending both major pesticide laws to establish a more consistent, protective regulatory process, grounded in sound science. The FQPA: (i) mandates a single, health-based standard for all pesticides in all foods; (ii) provides special protections for infants and children; (iii) expedites approval of safer pesticides; (iv) creates incentives for the development and maintenance of effective crop protection tools for American farmers; and (v) requires periodic re-evaluation of pesticide registrations and tolerances to ensure that the scientific data supporting pesticide registrations will remain up to date in the future.

The Montreal Protocol on Substances that Deplete the Ozone Layer is a landmark international agreement designed to protect the stratospheric ozone layer. The treaty was originally signed in 1987, and substantially amended both in 1990 and 1992. This international agreement has led to major changes for methyl bromide (MeBr), which is an odourless, colourless gas that has been used as an agricultural soil and structural fumigant to control a wide variety of pests. However, because MeBr depletes the stratospheric ozone layer and is classified as a Class I ozone-depleting substance, the amount of MeBr produced and imported in the USA was incrementally reduced until the phase-out took effect on 1 January 2005, except for the exemptions allowed by EPA. These exemptions included the quarantine and preshipment exemption to eliminate quarantine pests and the Critical Use Exemption, designed for agricultural users with no technically or economically feasible alternatives.

Environmental/Sustainability Issues

The Food Quality and Protection Act of 1996, and the resultant elimination or restrictions in the use of most broad-spectrum pesticides (especially for crops that are important foods for infants and children), have created many new challenges for areawide IPM projects. With a shift to more 'reduced risk', bio-based strategies, pest managers are finding that such systems require considerable biological and ecological knowledge of the system and a high degree of fine tuning to make them cost effective, successful and sustainable. In the tree fruit pest management programmes in the north-west, reduced risk alternatives to broad-spectrum organophosphate insecticides, while creating a safer orchard environment and enhanced opportunities for biological control, have created complex pest management systems and, in some cases, disrupted existing biological control systems.

One example of this is the disruption of the western predatory mite/phytophagous mite system in apples in the north-west. Integrated mite management, established in the late 1960s, provided stable mite control for approximately 30 years through the conservation of predatory mites by using selective rates of OP insecticides to control codling moth, thus avoiding the use of more disruptive insecticides. However, recent research has shown that higher mite populations result from the use of neonicotinoids to control codling moth as compared with the use of selective rates of Guthion in the formerly stable integrated mite management system.

Other research has shown that some of these 'reduced risk' OP alternatives may have some lethal and subtle sublethal effects (e.g. reduced fitness) on certain predator and parasitoid species. Hence, for areawide apple pest management systems in the north-west, the expectation that these 'reduced risk' products would provide stable IPM systems has yet to be realized. For these and other bio-based pest management systems in the USA (e.g. apples and tart cherries), IPM practitioners are often faced with tougher decisions and sometimes more challenging, less stable programmes.

None the less, encouraging progress is being made in the management of codling moth in western and eastern apple orchards by using mating disruption as a foundation strategy, integrated with biological and cultural control, and insecticides when necessary (Knight, this volume, Chapter 9). Recent IPM research in eastern peaches has shown that replacement reduced-risk pesticides and pest management products are more expensive, and require greater technical skill and precision. Hence, peach IPM that emphasizes more bio-based strategies will require additional research and fine tuning to enable growers to make economically and ecologically sound management decisions.

Pesticide resistance

Maintenance of our present food production and public health systems would be impossible without chemical control of agricultural and medical pests. Along with its many benefits, pest control has costs, one of the most pernicious being the evolution of pesticide resistance. Because resistance is a natural, evolutionary response of a pest population to strong selection pressure, it is a phylogenetically and geographically widespread problem that is increasing in magnitude. Resistance to insecticides, fungicides, herbicides, rodenticides and bactericides poses greater problems than ever before in agriculture and public health.

Moreover, the advent of transgenic pesticidal plants has the potential to significantly increase selection pressures for resistance relative to traditional synthetic pesticides. Therefore, resistance evolution has for the first time become a consideration in the pesticide regulatory process, both in the USA and internationally. Proactive resistance management, a requirement of current registrations, has considerable economic implications for agricultural productivity in this country and abroad.

There are two dimensions to the increase in resistance problems: the phenomenon itself and our need to respond to it. Our continued reliance on pesticides has caused the number of resistant species and populations to grow dramatically. At the same time, there is an increased awareness of this resistance problem from the regulatory community, industry and other scientists, creating an enormous demand for expert advice and information.

Like many challenges facing modern agriculture, dealing with pesticide resistance requires interdisciplinary approaches. Resistance research and management demands a threefold attack, with separate disciplines aligned along at least three separate axes.

The first axis cuts across taxonomic groups: bacteria, fungi, higher plants and arthropods. Resistance occurs in all of these, but scientists trained to specialize in one group are all too often unaware of important developments in another.

The second axis extends across levels of organization, ranging from the reductionist to more holistic and integrated ends of the continuum. Successfully dealing with resistance requires efforts at virtually all levels of biological organization, including evolution, population and molecular genetics, biochemistry, physiology and ecology, as well as contributions from studies of economics, rural sociology and other disciplines.

The third axis is the basic-applied axis. As in other areas of agricultural research, there is a premium for conducting basic research to maximize the speed and utility of its application to the problems that motivated it in the first place. Because of the extraordinary demands imposed by this interdisciplinary model, coordinated research, education and communication on resistance are of urgent importance.

Economical considerations

Building collaborations

Multi-state Research and Extension: the Hatch Act-funded multi-state research programme enables research on high-priority topics among the State Agricultural Experiment Stations (SAES) in partnership with the Cooperative State Research, Education and Extension Service (CSREES) of the US Department of Agriculture (USDA), other research institutions and agencies and with the Cooperative Extension Service (CES). In this way, technological opportunities and complex problem-solving activities, which are beyond the scope of a single SAES, can be approached in a more efficient and comprehensive way. This type of activity involves cooperative, jointly planned research employing multidisciplinary approaches. Projects are oriented toward accomplishment of specific outcomes and impacts, and based on priorities developed from stakeholder input in alignment with CSREES goals.

The very nature of the Hatch Formula Funds, allocated to each land grant university agricultural experiment station, helps ensure that collaborations will be built among institutions through the federal mandate that 20% of those funds be devoted to multi-state committee activities. These can take the form of multi-state committees (e.g. W-1185 Biological Control in Pest Management Systems of Plants), regional coordinating committees (e.g. former WRCC-66 – Biology and Control of the Russian Wheat Aphid) or rapid-response, multi-state committees to quickly bring scientists, extension specialists and state/federal entities together to address critical and emerging pest problems (e.g. NC502 for soybean aphid). The traditional multi-state committees are evaluated and, if justified, approved on a 5-year cycle, which offers adequate time to plan, coordinate and implement regional research and outreach activities.

Example one: soybean arthropod pest management projects

OVERVIEW. More soybeans are grown in the USA than anywhere else in the world. Today, farmers in more than 30 states grow soybeans, making it the second largest crop in cash sales and the number one value crop exported. In 2002, 74.31 million t of soybeans with a crop value of US$15,015 million were grown on 73.8 million acres. Soybean pest management is challenged by the simultaneous occurrence of biotic (e.g. various insects, pathogens and weeds) and abiotic (e.g. drought) stresses.

With new understandings about the physiological basis for yield loss from different stressors, an opportunity now exists to develop better strategies to address combined stressors, which are what most soybean growers experience (Higley, 1992). Additionally, the emergence of new soybean production practices, transgenic genotypes and new insect pests requires research to determine how best to manage insects and other stressors in these systems (Boethel, 2002). The potential impact of these developments on soybean profitability makes it essential that we begin addressing these new and future problems now.

Soybean growers have recently experienced increases in certain insect pest problems and the introduction of a new and potentially significant problem over the past few years. The first situation is the increase in population densities of the bean leaf beetle, *Cerotoma trifurcata* (Förster), and a corresponding rise in the incidence of bean pod mottle virus, a pathogen vectored by the beetle (Rice *et al.*, 2000). This relationship between bean leaf beetle and bean pod mottle virus, previously more common in southern states, is a relatively new occurrence in the central and northern USA. The second problem is the recent introduction of the soybean aphid, *Aphis glycines* Matsumura (Marking, 2001). Soybean growers now are facing widespread use of insecticides over potentially millions of acres of soybean in the upper Midwest. Given the native range of this insect, soybeans throughout the USA are at risk of being invaded.

In agriculture, we have seen tangible results from the landscape perspective, including areawide management of such pests as boll weevil, codling moth, Hessian fly, screwworm and gypsy moth. Significant problems face producers and scouts in soybean in the future, and at least some of these problems could be addressed by the use of remote sensing technologies. For instance, nutrient deficiencies, drought stress, insect damage, pathogen infestations and delayed maturity are all significant problems over broad geographic areas. The solutions to pest management problems in soybeans require an areawide view.

HISTORY OF PAST ACCOMPLISHMENTS.   Previous soybean entomological regional projects (e.g. S-74, S-157, S-219, S-255 and S-281, see Box 3.1) have advanced both the underlying science and the practice of pest management in soybean production.

Box 3.1.   A chronology of the multi-state arthropod soybean pest management research programmes leading to S-1010, the currently funded project.

S-74: Control Tactics and Management Strategies for Arthropod Pests of Soybeans, July, 1969–30 September 1981 (515 publications).
S-157: Tactics for Management of Soybean Pest Complexes, October 1982–30 September 1987 (338 publications).
S-219: Arthropod-induced Stresses on Soybean: Evaluation and Management, October 1987–30 September 1992 (358 publications).
S-255: Development of Sustainable IPM Strategies for Soybean Arthropod Pests, October 1992–30 September 1997 (240 publications).
S-281: Dynamic Soybean Insect Management for Emerging Agricultural Technologies and Variable Environments, October 1997–30 September 2002 (157 publications).
NC-502: Soybean Aphid: a New Pest of Soybean Production, 1 September 2000–30 September 2002.
S-1010: Dynamic Soybean Pest Management for Evolving Agricultural Technologies and Cropping Systems, 1 October 2003–present.

Collaborative, multi-state research to address the arthropod pest complex attacking soybeans in the USA began formally with the establishment of a southern region technical committee, S-74, in 1972. At that time, most of the soybean-producing states were conducting research and extension programmes that addressed control of key pests within their own states. The formation of this committee enabled a group, comprised of scientists from most of the soybean-growing states, to plan, prioritize and address key problems faced by two or more states. Even though the technical committee was administratively attached to the southern region, the membership included scientists from other regions where soybeans were grown.

Five subcommittees were established during the initial phase of this collaborative research project, with an emphasis on areas such as: (i) host plant resistance; (ii) natural control; (iii) cultural and chemical control; (iv) ecological techniques; and (v) pest management. Over the course of this initial project, significant advances were made in many areas of soybean arthropod research, an area that was in its infancy. Basic information relative to soybean pests was studied in detail; emphasis was placed on predators, parasites and diseases of soybean pests, and significant information was developed on economic thresholds for various pests, host plant resistance and the effects of various cropping systems on soybean problems.

One early suggestion from the CSRS (Cooperative State Research Service) representative was to include an agricultural economist to interject the economics of soybean pests into the group thinking to give added direction, since it would be useful in determining the economic impact of pests in relation to pesticide usage. Each successive revision of the original research project was made to address the key issues and challenges of the day.

A chronology of regional research projects is given in Box 3.1. Each project was extremely productive in terms of publications in the scientific literature. Totals are provided in the chronological listing of each multi-state project in Box 3.1; however, what is perhaps more important is that the knowledge was transferred into practice via the linkage to cooperative extension programmes in each participating state. Pest control recommendations developed by each state quickly incorporated the control strategies developed through the research effort. Pest management in soybeans moved from reliance on 'hard' pesticide usage to newer, more environmentally friendly and target-oriented pest management methods, first with the advent of organophosphates and then with development of pyrethroids and other chemical groups. Resistant plant variety development obviated the need for some pesticide use. Timing of planting and pesticide applications made control more precise for specific target pests. Biological management methods were developed and put into practice. More recently, application methodologies were developed that required lower volumes of pesticides, more accurately placed. Geographic information systems (GIS) and global positioning systems (GPS) technologies began development.

The soybean aphid, *A. glycines*, a native of Asia, was first detected in Wisconsin in 2000 and later that same year in Minnesota, Iowa, Missouri, Illinois, Indiana and Michigan. Critical Issues funding was obtained from CSREES in 2000, and a Rapid Response Multi-state Committee NC-502 (Soybean Aphid: a New Pest of Soybean Production) was formed that same year to help facilitate the development of a regional pest management effort against the aphid. This Rapid Response Multi-state

Committee merged with the current southern region project in 2002, and Hatch funding continues to date through this Multi-State Committee.

Fundamental research and IPM strategies for management of the invasive soybean aphid remain an objective of the combined project. In 2003, over 42 million acres (17 million ha) of soybean in north-central USA were infested, and over 7 million acres (2.9 million ha) were treated with insecticides to control the aphid. By 2005, the aphid had dramatically expanded its range to 22 states and was associated with millions of dollars in crop losses and management costs annually.

Additional funding for biological control and the implementation of IPM on a landscape scale was obtained in 2004, through CSREES' NRI and RAMP programmes and special grant funds. In 2005, 14 scientists from five north-central region states and the USDA received funding from the North Central Soybean Research Program to further research on classical biological control. Some of the pest monitoring protocols and predictive models developed from the soybean aphid programmes provided the framework for the development of the Pest Information Platform for Extension and Education, which focused on soybean aphid and soybean rust in 2006.

Example two: biological control in pest management systems of plants

Biological control can be defined as the deliberate use of natural enemies – predators, parasites (parasitoids) and pathogens – to suppress and maintain populations of a target pest species below that which causes economic and/or environmental damage. Biological control of arthropod pests and weeds is particularly desirable because the tactic is environmentally safe, energy self-sufficient, cost-effective, sustainable and can be readily incorporated into pest management programmes. Furthermore, in many cases benefits from the use of natural enemies accrue at no additional cost. The practice of biological control usually involves various approaches, such as: (i) the importation of exotic natural enemies (classical biological control); (ii) the conservation of resident or introduced beneficial organisms; and (iii) the mass production and periodic release of natural enemies.

In 1964, Regional Research Project W-84, 'Biological Control in Pest Management Systems of Plants', was initiated as part of an effort to coordinate biological control activities by the various agriculture experiment stations and the USDA Agricultural Research Service in the western USA. The accomplishments and benefits of W-84 from 1964 to 1989 are chronicled by Nechols et al. (1995) in a book entitled *Biological Control in the Western United States*. W-84 was one of the largest, most productive and most diverse multi-state projects concerning biological control, as evidenced by the three general chapters and 79 case histories in the book.

The present committee (W-2185) typically has 35 to 40 scientists participating on a regular basis and includes scientists from agricultural experiment stations and ARS laboratories from most of the states in the Western Region, as well as two other states outside the region (Kansas and New York) and two US territories (American Samoa and Guam). California Department of Food and Agriculture and Oregon Department of Agriculture also have been prominent participants in the project. The 237 publications (including two books, 30 book chapters and over 180 peer-reviewed articles) are testimony to the high level of productivity associated with the project. Natural enemies (predators, parasitoids, herbivores and, to a lesser extent, pathogens) of over 90 arthropod and weed pests were investigated by cooperating scientists over the course of the W-84 project.

IPM CENTRES. In 2000, the Regional IPM Centres Program was created to promote the development and implementation of IPM by facilitating collaboration across states, disciplines and purposes. The Regional IPM centres are located in each of four regions in the USA: north-central, north-eastern, southern and western. The centres serve as focal points for regional pest management information networks, collaborative team building and broad-based stakeholder participation. The result is increased coordination of IPM research, education and extension efforts and enhanced responsiveness to critical pest management challenges. Regional and national pest alerts generated by the IPM centres have provided timely and accurate information on emerging pests such as soybean rust, sudden oak death, soybean aphid, pink hibiscus mealybug, etc. In addition, the centres have played an active role in facilitating regional education and training activities relative to new invasive pests such as soybean rust and sudden oak death.

EARLY DETECTION AND RAPID RESPONSE. In 2002, the US Secretary of Agriculture established the Animal and Plant Disease and Pest Surveillance and Detection Network within CSREES. The mandate was to develop a network linking plant and animal disease diagnostic facilities across the country. In response to this, CSREES established two national networks of existing diagnostic laboratories to rapidly and accurately detect and report pathogens of national interest and provide timely information and training to state university diagnostic laboratories. The first of these is the National Plant Diagnostic Network (NPDN), which is led by five regional laboratories (Cornell, Florida, Michigan State, Kansas State and California at Davis) and one support laboratory (Texas Tech.). The mission of the NPDN is to enhance national agricultural security by quickly detecting outbreaks of pests and pathogens. To achieve this mission, a nationwide network of public agricultural institutions (land grant institutions and state departments of agriculture) was developed, which functions as a cohesive system to quickly detect and diagnose high-consequence biological pests and pathogens in agricultural and natural ecosystems.

The second of these is the National Animal Health Laboratory Network (NAHLN), which is led by 12 laboratories (University of Georgia, Texas A&M University, the University of California at Davis, the University of Wisconsin, Colorado State University, Cornell University, Rollins Laboratory in North Carolina with the Department of Agriculture and Consumer Protection, Louisiana State University, the Florida Diagnostic Laboratory with the Department of Agriculture and Consumer Protection, the University of Arizona, Washington State University and Iowa State University).

The objective of the NAHLN is to establish a national network of diagnostic laboratories to increase the nation's capability and capacity to detect foreign animal diseases. The network is a cooperative effort between two USDA agencies, CSREES and APHIS, and the American Association of Veterinary Laboratory Diagnosticians. The network is multifaceted and comprised of sets of laboratories that focus on different diseases. They use common testing methods and software platforms to process diagnostic requests and share information.

State and federal regulatory programmes, and inter-state and inter-federal agency collaborations, have played a critical role in the management of sudden oak death and soybean rust in the USA. This was accomplished, in part, by creating a functional nationwide network of public agricultural institutions with a cohesive, distributed system to quickly detect high-consequence biological pests and pathogens in agricultural and natural ecosystems. In addition to providing the means for quick identification, the NPDN also established protocols for immediate reporting to appropriate responders and decision makers.

In collaboration with CSREES' Regional Integrated Pest Management centres (IPM centres), state departments of agriculture, state plant regulatory officials and the LGU System, the NPDN system held a number of workshops and tele-conferencing sessions, which were used to train diagnosticians in the identification of sudden oak death and soybean rust pathogens. State response scenarios were conducted for each of the states involved in soybean production. In short, the NPDN allowed land grant university diagnosticians and faculty, state regulatory personnel and first detectors to efficiently communicate information, images and methods of detection throughout the system in a timely manner. National pest alert pamphlets for sudden oak death and soybean rust were also produced by IPM centres, which provided information on the distribution, life history, host range and management recommendations for sudden oak death and soybean rust.

In 2005, the Pest Information Platform for Extension and Education (PIPE) was developed in response to the soybean rust introduction in 2004. PIPE is a reporting and tracking system, developed collaboratively with the USDA Risk Management Agency, to manage pest and disease information flow via the Internet. The PIPE System provides real-time information to US crop producers, and a 'one-stop shopping' centre for timely, unbiased, national and local pest information. PIPE utilizes a reporting and tracking system for sentinel pest-monitoring plots and field observations, and includes incidence mapping, extensive coordination with extension specialists, localized suggestions for management and public and private interfaces.

The PIPE fosters good farming practices by encouraging growers to avoid unnecessary or ill-timed chemical applications, to use the proper control tactics with the proper timing to manage crop loss risk, and document practices for crop insurance purposes. The PIPE system for soybean rust saved growers millions of dollars in 2006 by providing real-time information that enabled the growers to avoid unnecessary chemical applications.

A number of grant programmes at CSREES encourage regional and national collaborations. Examples include the NRI Coordinated Agricultural Projects (CAP), the Risk Avoidance and Mitigation Program (RAMP) and Regional IPM Competitive Grants Program. CAP awards support large-scale, multi-million-dollar projects to promote collaboration, open communication and the exchange of information; reduce duplication of effort; and coordinate activities among individuals, institutions, states and regions. Project participants serve as a team that conducts targeted research in response to emerging or priority area(s) of national need. Recent CAP programmes have focused on food safety and applied plant genomics.

The goal of the RAMP programme is to enhance the development and implementation of innovative integrated pest management strategies for multi-crop food and fibre production systems, or production systems on an areawide or landscape scale. Projects typically involve multiple crops, pest species, disciplines, institutions and states; are integrated (involving research, education and extension); and emphasize a systems approach. The goal of the Regional IPM Competitive Grants Program (RIPM) is to provide knowledge and information needed for the implementation of IPM methods that: improve the economic benefits related to the adoption of IPM practices; reduce potential human health risks from pests and the use of pest management practices; and reduce unreasonable adverse environmental effects from pests and the use of pest management practices. RIPM supports projects that promote cooperative efforts across appropriate disciplines, with linkages between research and extension efforts and components of existing or emerging pest management systems. Another goal of the RIPM is to encourage collaborations among states/territories for purposes of efficiency, economy and synergy.

Inter-agency Collaborations

Federal inter-agency coordinating councils, committees and collaborations have played a critical role in addressing pest problems that threaten human/animal health, the US economy, the environment and fish and wildlife on a regional and national scale.

Invasive species

Invasive species are defined as organisms that are non-native to an ecosystem and whose introduction causes economic, social or environmental harm. Nearly every terrestrial, wetland and aquatic ecosystem in the USA has been invaded by non-native species (Lee and Chapman, 2001), with economic losses estimated at US$137 billion per year (Pimentel et al., 2000). Invasive species constitute one of the most serious economic, social and environmental threats of the 21st century.

In response to the threats posed by invasive species and the challenges to reducing their spread, the President issued Executive Order 13112 (Order) on Invasive Species (3 February 1999), which directs federal agencies to prevent the introduction of invasive species, provide for their control and minimize their impacts (see http://csrees.usda.gov/NISMP/).

This Order established the National Invasive Species Council (NISC), which is chaired by the Secretaries of Agriculture, Commerce and Interior and includes the Departments of State, Treasury, Defense, Health and Human Services, Transportation, EPA and USAID and, more recently, Homeland Security and NASA. The Order also directed the Secretary of Interior to establish a non-federal advisory committee (the Invasive Species Advisory Committee), comprised of a diverse set of stakeholders, to advise NISC on invasive species issues. The Order mandated that a National Invasive Species Management Plan (Plan) be developed through a public process and in consultation with federal agencies and stakeholders.

The first edition of the Plan was published in 2001 and included 57 action items covering areas of leadership and coordination, prevention, early detection and rapid response, control and management, restoration, international cooperation, research, information management and education and public awareness. Over the period 2006–2007, NISC revisited the Plan and came up with a reduced set of priorities that are currently being evaluated and compared to priorities identified by agencies addressing invasive species within the USDA, and by other departments.

Invasive species budget cross-cut

As called for in the National Invasive Species Management Plan, NISC agencies developed the first Invasive Species Cross-cut Budget for the fiscal year 2004. The Office of Management and Budget (OMB) encouraged NISC to develop shared (cross-cutting) goal statements, strategies and common performance measures among agencies as part of the FY04 budget process. The result was a first-of-its-kind inter-agency performance budget that facilitated the more efficient allocation of resources through enriched inter-agency cooperation. It created a starting point for more comprehensive and cooperative efforts for the FY05 to FY08 budget cycles.

The Invasive Species Cross-cut Budget is designed to: (i) encourage federal cooperation and coordination on invasive species issues that benefit from an inter-agency approach; (ii) highlight and promote inter-agency performance-based approaches to address specific invasive species issues; and (iii) provide a clear and comprehensive overview of invasive species issues and efforts across the federal government. For the FY06 Budget Cross-cut, strategic performance measures were developed for six specific invasive species initiatives: brown tree snake, emerald ash borer, leafy spurge/yellow starthistle, tamarisk, sudden oak death and Asian carp; as well as five issue- and programme-based initiatives including ballast water, prevention through education, aquatic area monitoring, early detection and rapid response and innovative control technologies.

Federal inter-agency committee for the management of noxious and exotic weeds (FICMNEW)

FICMNEW was established through a Memorandum of Understanding, which was signed by the administrators of participating agencies in 1994. This federal coordinating committee represents an unprecedented formal partnership among 16 federal agencies with direct invasive plant management and regulatory responsibilities for the USA and Territories. Through monthly meetings and other committee activities, FICMNEW members interact on important regional and national issues and share information with various public and private organizations, collaborating with the federal sector to address invasive plant issues.

FICMNEW's charter directs the Committee to coordinate information regarding the identification and extent of invasive plants in the USA and to coordinate the federal agency management of these invasive species. FICMNEW accomplishes this by developing and sharing scientific and technical information, fostering collaborative efforts among federal agencies, providing recommendations for regional and national level management of invasive plants and sponsoring technical/educational conferences and workshops concerning invasive plant species. A couple of notable publications have been produced by FICMNEW, including *Invasive Plants, Changing the Landscape of America: Fact Book* (Westbrooks, 1998) and *A National Early Detection and Rapid Response System for Invasive Plants in the United States*, which was published in 2003 (FICMNEW, 2003). FICMNEW continues to bridge the gap between federal agency plant management and scientific activities. It has been a driving force behind the national emphasis against the broader invasive species threat (see http://www. fws.gov/ficmnew/).

Technical advisory group for biological control agents of weeds (TAG)

For the past 50 years, technical review groups have assisted researchers and regulatory agencies in evaluating the safety of insect or pathogen introductions for the biological control of weeds in the USA. The original Subcommittee on Biological Control of Weeds was established in 1957, and included representatives from the Department of Interior (Bureau of Reclamation, Bureau of Land Management and Fish and Wildlife Service) and Department of Agriculture (Agricultural Research Service and Forest Service).

An informal reciprocal review of proposals for biological control of weeds began in 1962 between the USA and Canada. In 1969, the membership of the Subcommittee was expanded to include specialists in plant taxonomy, ornamentals and plant quarantine. In 1971 the Subcommittee's name was changed to Working Group, and contacts were established with Mexican officials concerning US proposals for the introduction of biocontrol of weed agents. Membership has expanded over the years to include EPA, CSREES, the Weed Science Society of America and the US Army Corps of Engineers. In 1987, the Technical Advisory Group replaced the Working Group for Biological Control Agents of Weeds.

At present, TAG is charged with recommending action to APHIS-PPQ when making a decision to issue permits, and with regard to advising researchers about the safe use of biological control agents in the environment. The expectations of TAG, and more recently the US Fish and Wildlife Service (where endangered species issues are concerned), are to engage with researchers early in the process to provide feedback on the test plant list, identify conflicts of interest and to assess the level of risk associated with the release of a particular biological control agent.

Federal inter-agency committee for the management of invasive terrestrial animals and pathogens (ITAP)

ITAP was established in 2004 to provide a forum to support technical coordination and cooperation among federal agencies on problems associated with invasive invertebrates, vertebrates and plant and animal pathogens in terrestrial ecosystems. The focus of ITAP is on invasive terrestrial vertebrates; invasive pests of human habitations; and invasive 'pests' and diseases of crops (including nursery/horticultural), domestic animals, wildlife and trees in forest, rangeland, grassland and other terrestrial ecosystems, excluding weeds and aquatic organisms. ITAP currently has seven subcommittees – focusing on invasive species issues including – Invertebrates; Vertebrates/Animal Pathogens; Plant Pathogens; Systematics; Protocols; Cross-cutting Issues; and Invasive

Species Awareness Day. One of the major accomplishments of ITAP has been the development of a report by the Systematics Subcommittee entitled *Systematics and Invasive Species: Strengthening the Federal Capacity in Systematics and Creating a Safety Net for Biosecurity*. The purpose of this document is to increase awareness of the crisis in systematics in federal agencies and the implications for US biosecurity.

Federal integrated pest management coordinating committee (FIPMCC)

The FIPMCC was created in 2002 in response to the GAO Report GAO-01-815, which concluded that the IPM initiative was missing several key management elements identified in the Government Performance and Results Act, including:

- No one is effectively in charge of federal IPM efforts.
- Coordination of IPM efforts is lacking among federal agencies and with the private sector.
- The intended results of these efforts have not been clearly articulated or prioritized.
- Methods for measuring IPM's environmental and economic results have not yet been developed.

The goal of the FIPMCC was to improve coordination of IPM activities among federal agencies and with the private sector. Also, in 2002 and preceding the formation of the FIPMCC, the development of the national road map for integrated pest management began with the goal of increasing nationwide communication and efficiency through information exchanges among federal and non-federal IPM practitioners and service providers including land managers, growers, structural pest managers, and public and wildlife health officials. Feedback for the IPM road map was obtained from over 100 individuals nationwide. Subsequent drafts of the road map were then vetted through the FIPMCC. Continuous input from numerous IPM experts, practitioners and stakeholders resulted in the current IPM road map dated 17 May 2004 (see http://northeastipm.org/whatis_ipmROAD MAP.pdf).

Pesticide Safety Education Program (PSEP)

In 1994, EPA established PSEP as a voluntary public–private partnership to reduce pesticide risk, and announced the first six PSEP partners. The USDA took responsibility for increasing adoption of IPM in US agriculture. In 1995, EPA added a Supporter category to allow organizations that train, educate or influence pesticide users to participate in PSEP and, thereby, be recognized for their contributions to reducing pesticide risk. By joining PSEP, organizations pledge that environmental stewardship is an integral part of pest management, and that they commit to working toward innovative practices that reduce risk to human health and the environment. For example, many PSEP members are adopting the use of biological pesticides or biopesticides, such as microbial pesticides, pheromones or natural compounds, which target specific pests and generally pose little or no risk to humans or the environment. In addition to promoting the use of biopesticides, PSEP advocates the adoption of integrated pest management (IPM) programmes or practices.

Web-based Extension Initiative

eXtension is an educational partnership of more than 70 land grant universities helping Americans improve their lives with access to timely, objective, research-based information and educational opportunities. Land grant colleges were founded on the ideals that higher education should be accessible to all and that the colleges should share knowledge of practical subjects with people throughout their states. eXtension will provide Internet visitors with reliable, up-to-date pest management information through online lessons.

An initial pest management-related web site provides information on the imported fire ant (see Fig. 3.1). Visitors can use the site through self-paced learning to find research-based, peer-reviewed answers to a knowledge base of commonly asked questions, which will aid in learning more about specialized areas. As with all of eXtension, the fire ant eXtension web site is being developed through the collaboration of extension professionals with expertise in this area. This site offers quality content developed in a virtual workplace by an expert team, providing the

Fig. 3.1. An initial pest management-related web site on the imported fire ant.

framework for an exemplary, trusted, electronic learning environment. By using this system, eXtension participants will gain the knowledge, skills, motivation and confidence to make their own fire ant pest management decisions.

Measurement of Results

The establishment of measurable IPM goals and the development of methods to measure progress toward achieving these goals should be appropriate to the specific IPM activity undertaken. Performance measures may be conducted on a pilot scale or on a geographic scale and scope that corresponds to an IPM programme or activity. Examples of potential performance measures follow.

Outcome: the adoption of IPM practices improves economic benefits to users

Performance measures
- In cooperation with the National Agricultural Statistics Service (NASS), design a national IPM practices adoption survey based on IPM protocols designed for specific commodities or sites within programme priorities.
- Evaluate IPM programmes on their ability to improve economic benefits using pilot studies within specific programme priority sites, and project these economic results to a regional or national basis to predict large-scale impacts using results of the practices adoption survey.
- Develop measures of public awareness of IPM.

Outcome: potential human health risks from pests and the use of pest management practices are reduced

Performance measures
- Using EPA's reduced-risk category of pesticides as the standard, document changes in pesticide use patterns over time and relate the changes to IPM practice adoption.
- Relate dietary exposure to pesticides to IPM practice adoption using the USDA Agricultural Marketing Service (AMS) Pesticide Data Program (PDP) and any other available data.
- Relate cases of the negative human health impacts caused by pest incidence (e.g. asthma cases related to cockroach infestation, insect-vectored diseases, allergic reactions to plants) to IPM practice adoption.

Outcome: unreasonable adverse environmental effects from pests and the use of pest management practices are reduced

Performance measures
- Document and relate pesticide levels in specific ground and surface water bodies, including community water supplies, to IPM practice adoption using data

from the US Geological Survey (USGS), the Natural Resource Conservation Service (NRCS) and others.

- Document and relate national indicators of natural resource health such as proportion of ground and surface water bodies with pest management-related contaminants and level of contamination to IPM practice adoption, using data from EPA and others.

- Measure the impact of IPM practice adoption on encroachment of selected invasive species in national park lands and other sites where data are available.

Areawide Pest Management Programmes

ARS areawide projects

ARS areawide pest management programmes involve coordinated research and management activities with grower participation to suppress or maintain a pest at low population levels throughout large, definable areas. This is achieved through environmentally sound, effective and economical approaches, including biological and cultural control and other sustainable agriculture practices. ARS strongly believes that IPM and areawide pest management systems, employing biologically based or pest-specific methods, can substantially substitute for, and decrease the risks from, the most hazardous chemical pesticides and simultaneously increase economic benefits for agriculture.

Corn rootworm areawide pest management project

The ARS corn rootworm areawide pest management project involves coordinated research and management efforts in Kansas, Indiana, Illinois, Iowa, Texas and South Dakota. Corn rootworm populations have been reduced by 85–95% with less than 10% of the chemicals used in previous corn rootworm control regimes. The key to the areawide corn rootworm project is to use adult attracticide baits, which were developed by ARS and are now marketed by industry. The adult baits are used in demonstration sites. This technology, together with transgenic maize, could ultimately become the management strategy of choice on the 20 million acres (8 million ha) of US cropland currently treated with insecticide for corn rootworm control. This could result in a reduction of up to 90% in the amount of soil insecticide applied to maize grown in the Midwestern USA.

TEAM Leafy Spurge project

Another example of a successful ARS areawide project was TEAM Leafy Spurge (The Ecological Areawide Management of Leafy Spurge), which was a 5-year USDA-ARS research and demonstration programme focusing on the Little Missouri River drainage system in eastern Montana and Wyoming, western North Dakota and South Dakota. The goal of this programme was to research, develop and demonstrate ecologically based integrated pest management strategies that land managers and landowners could use to achieve effective, affordable and sustainable leafy spurge control.

TEAM Leafy Spurge was funded by ARS, and managed cooperatively with the USDA Animal Plant Health Inspection Service. The project emphasized partnerships, teamwork and a cooperative approach to solving leafy spurge problems. Members of the TEAM included state and federal agencies, state cooperative extension services, land grant universities, weed managers, county and other local entities and private landowners and ranchers. The project truly utilized ecologically based, integrated weed management of leafy spurge utilizing chemical, cultural and biological control, grazing management, remote sensing and an extremely effective extension and outreach programme.

In one of the studies supported by TEAM Leafy Spurge and USDA-APHIS, flea beetles (*Aphthona lacertosa* and *A. nigriscutis*) were released at 76 sites in the vicinity of Devil's Tower, Wyoming and monitored for a 6-year period. Leafy spurge had become the dominant plant cover at each of these sites and had greatly reduced rangeland productivity. Within 3 years the beetles had reduced the average cover of leafy spurge from 60% to less than 10% at release sites (Kazmer *et al.*, 2005). The researchers found that grass cover increased from 34% to over 80% in the 6 years following flea beetle release.

CSREES Areawide Projects

Enhancing pheromone disruption project

The RAMP (Risk Avoidance and Mitigation Program) project of enhancing pheromone mating disruption programmes for lepidopterous pests in western orchards (Welter and Van Steenwyk, 2000) is well known. This project has built upon the successful areawide management project that targeted the key pest in apples and pears, the codling moth (see this volume, Chapter 9) and reduced the use of in-season organophosphate insecticides by 75%. The original project goals were to further reduce broad-spectrum pesticide use, expand the use of mating disruption using the pheromones of key insect pests and to improve opportunities for biological control of other pests in orchards. Apple and pear production systems are at risk under the 1996 Food Quality and Protection Act (FQPA) due to safety concerns and re-registration obstacles for currently used pesticides, and the fact that apples and pears comprise a significant fraction of the 'risk cup' in the diets of infants and children.

The approaches outlined in the objectives included: (i) establishment of large-scale sites to determine the difficulties and advantages of replacing broadly toxic insecticides with new selective products; (ii) evaluation and development of new, non-insecticidal – e.g. pheromones – programmes for both the primary and secondary pests; (iii) evaluation and improvement of new monitoring systems to reduce grower risk; (iv) reductions in insecticide use rates through use of feeding stimulants and baits; and (v) extension of these new programmes to new acreage, pests and crops. This project was multi-state, multi-institutional and multidisciplinary.

The research and education programmes developed by this project have reduced the use of broad-spectrum pesticides, increased farm worker safety and reduced the risk of environmental contamination. Researchers are also investigating ways to enhance biological control in the orchards, and in the process establish a low-cost, more sustainable management system. This project is expected to increase acreage under mating disruption, improve programme efficacy, reduce programme risks and reduce costs to help US agriculture compete in a global economy.

Consortium for integrated management of stored product insect pests

Another RAMP project developed a consortium for integrated management of stored product insect pests (see Ramaswamy and Subramanyam, 2000). The objectives of this project were to: (i) develop methods of pest management that reduce or eliminate the risk from pesticide residues; (ii) develop and implement information-intensive approaches to pest management based on a more complete understanding of crop and pest biology, their interactions and mutual impacts, and factors impacting the stability of pest management systems in major cropping systems; and (iii) develop outreach strategies to promote the exchange of pest management information.

Consumer demand for food free of pesticide residues, pesticide resistance in insects and the current regulatory climate have necessitated the development of effective alternatives to chemical pesticides as a means of controlling pests in stored products. This successful areawide research/extension project has developed effective management strategies for stored grain pests by using effective sampling and monitoring techniques, modelling populations, manipulating factors that create environments conducive to insect pest reproduction in storage – such as temperature and moisture – and the use of natural and alternative chemical methods to suppress insect survival.

Seeking Funds/Identifying Roles of Key Personnel

The unique mission of CSREES is to advance knowledge for agriculture, the environment, human health and well-being, and communities by supporting research, education and extension programmes in the land grant university system and other partner organizations. CSREES doesn't perform actual research, education and extension but rather helps fund it at the state and local levels and provides programme leadership in these areas. Pest management is among CSREES' targeted areas of interest and is supported through formula-based programmes (the Hatch, Smith-Lever, McIntyre-Stennis, and Evans-Allen Acts), Section 406 national competitive grant programmes, competitive special research grants, national competitive grant programmes (e.g. NRI) and inter-agency programmes (e.g. the Pesticide Safety Education Program, managed jointly by EPA and CSREES). Integrated pest management programmes supported by CSREES are detailed below.

Integrated Pest Management Programmes

Regionally focused programmes

Regional IPM centres (Centers)

Centers, through partnering with institutions and stakeholders, help facilitate the identification and prioritization of regional, multi-state IPM research, extension and education programme needs. In FY 2000, geographically based Centers were formed in the north central, north-eastern, southern and western regions to establish a national pest management information network. Centers of the future will be the focal point for team-building efforts, communication networks and stakeholder participation. Centers bring together expertise, identify needs and priorities and address a broad range of IPM research, education and outreach issues. This is a Section 406 national competitive grants programme.

Regional integrated pest management programme (RIPM)

The RIPM Program is a regionally based programme that supports development and implementation of new and modified IPM tactics and systems, the validation in production systems and the delivery of educational programmes to pest managers, advisors and producers. The programme builds stakeholder partnerships to address critical pest management needs in the region. This is a competitive special research grants programme that is managed regionally by the Centers.

Pest management alternatives programme (PMAP)

The programme goal here is to develop replacement tactics and technologies for pesticides undergoing regulatory action where there are no effective registered alternatives. This programme funds short-term development and outreach projects aimed at adaptive research and implementation of tactics that have shown promise in previous studies. The focus of the programme is primarily on developing replacements for specific tactics. The intent is to continue current programme goals and convert this programme to a component managed by the IPM Centers. This is a special research competitive grants programme.

Nationally focused programmes (discovery to implementation)

Base support to land grant universities

The underpinning of the national extramural agricultural research, education and outreach capability is accomplished through a federal/state partnership with the land grant university system. CSREES provides oversight for the federal annual base support that is provided through the Hatch, Smith-Lever, McIntyre-Stennis and Evans-Allen Acts. The federal funds are matched and multiplied by state and local resources in support of the national agricultural research, education and extension infrastructure. This is a formula-based programme.

National Research Initiative (NRI)

The NRI pest management research programme supports fundamental and mission-linked research on the biology of insects, microbes, nematodes, invasive plants and other organisms. It also supports research on the interactions among pest organisms, species of agricultural importance and their interaction with the environment. This research programme provides the foundation for the development of the next generation of IPM tools, strategies and systems. This is a national competitive grants programme.

Risk Avoidance and Mitigation Programme (RAMP)

RAMP supports the development and implementation of innovative IPM systems on an areawide or landscape-scale basis. The primary emphasis of RAMP applications should be crop productivity and profitability while addressing critical environmental quality and human health issues. RAMP applications may address major acreage crop production systems, key fruit and vegetable production systems or other agro-ecosystems where identified environmental quality or human health issues exist. The RAMP programme will fund medium-term projects that involve systems approaches. This is a Section 406 national competitive grants programme.

Crops at Risk (CAR)

The goal of the Crops at Risk programme is to create or enhance IPM practices for individual food or fibre crops grown for commercial purposes. The CAR programme will fund integrated multifunctional/multidisciplinary research, education and extension projects for crops with high-priority IPM needs as identified by stakeholders. This is a Section 406 national competitive grants programme.

Minor crop pest management (IR-4)

IR-4 is the principal public programme supporting the registration of pesticides and biological control agents for use on minor crops. This programme provides coordination, funding and scientific guidance for both field and laboratory research to develop data in support of registration packages to be submitted to EPA. IR-4 coordinates the cooperation of commodity producers, state and federal research scientists and extension specialists in identifying and prioritizing pest control needs. This is a special research competitive grants programme, with additional support from CSREES and Agricultural Research Service base funds.

Methyl Bromide Transitions Program (MBT)

This programme addresses the need to develop management technologies, systems approaches and extension delivery programmes for methyl bromide uses that may be cancelled. This is a Section 406 national competitive grants programme.

Organic Transitions Program (ORG)

The goal of this programme is the development and implementation of biologically based pest management practices that mitigate the ecological, agronomic and economic risks associated with a transition from conventional to organic agricultural production systems based on national standards. This is a Section 406 national competitive grants programme.

Extension IPM implementation

This base programme in each state and territory facilitates the development and transfer of IPM from researchers to implementation by farmers, crop consultants and other end-users. Information outreach occurs through consultations, clinics, workshops, conferences, demonstrations, field days and a wide variety of publications. This programme contributes to the scientific and extension foundation for IPM. This is a Smith-Lever 3(d) programme, with funds distributed according to a formula.

Pesticide Safety Education Program (PSEP)

The primary focus of this joint EPA/USDA programme is to provide educational programmes that support the proper use of pest management technologies. A central focus is to provide pesticide applicators with the knowledge and training needed for the safe and effective use of pesticides. Education is provided by LGU extension programmes in conjunction with state regulatory agencies that certify and license applicators. EPA provides funds (allocated on a formula basis), and CSREES manages a national programme connecting to the science education base in each state, the District of Columbia and territories.

General remarks about seeking of funding

A number of the competitive grant programmes administered by CSREES require preliminary data, strong stakeholder input, connection to crop profiles and strategic plans, and alignment with the National IPM Road Map to be competitive. For all of these competitive grant programmes the roles of key personnel must be clearly identified. In this regard, one of the most outstanding proposals submitted to the Risk Avoidance and Mitigation Program contained an appendix to the project description, with a colour-coded matrix listing subprojects by section and investigators, including the title, description, deliverables and/or preliminary data and objectives addressed.

Conclusion

Areawide approaches will no doubt continue to play a vital role in addressing regional and national pest problems. Successful programmes in the future will necessarily involve the collective efforts of many, including: (i) federal, state, commodity and stakeholder support and cooperation; (ii) inter-agency/institution collaboration and communication; (iii) research, education and extension; (iv) regulatory pragmatism; and (v) an effective system for delivering timely pest management information to growers and land managers. With the globalization of trade and travel and increasing frequency of new pest introductions, the opportunities and necessity for developing areawide collaborations have never been greater. The authors are hopeful that the information presented in this chapter and others will help contribute to the development of successful areawide projects in the future.

Acknowledgement

The views expressed herein are those of the authors only and do not necessarily reflect or represent the opinions, positions or policies of the US Department of Agriculture or its agencies, including the Cooperative State, Research, Education and Extension Service.

References

Boethel, D.J. (2002) Integrated management of soybean insects. In: Boerma, H.R. and Specht, J.E. (eds) *Soybeans: Improvement, Production, and Uses.* 3rd edn., American Society of Agronomy Monograph 16, Madison, Wisconsin, 1180 pp.

FICMNEW (2003) *A National Early Detection and Rapid Response System for Invasive Plants in the United States.* Federal Inter-agency Committee for the Management of Noxious and Exotic Weeds, Washington, DC, 24 pp. (http://www.fws.gov/ficmnew/FICMNEW_ EDRR_FINAL.pdf).

Higley, L.G. (1992) New understandings of soybean defoliation and their implications for pest management. In: Cropping, L.G., Green, M.B. and Rees, R.T. (eds) *Pest Management in Soybean.* Elsevier Applied Science, New York, pp. 56–66.

Kazmer, D., Marrs, R.W., Hunt, R., Parker-Williams, A., Boersma, J., Williams, M. and Shorma, R. (2005) *Assessing the Long-term Impact of Leafy Spurge Biological Control Agents: Conclusions from a 6-year Study.* Report for Project USDAAPHIS5179, #58-5436-1-221, 33 pp.

Lee, H. and Chapman, J.W. (2001) *Nonindigenous Species – an Emerging Issue for EPA: a Landscape in Transition: Effects of Invasive Species on Ecosystems, Human Health and EPA Goals,* Vol. 2, US EPA Office of Research and Development, Newport, Oregon, 54 pp.

Marking, S. (2001) Tiny terrors. *Soybean Digest* 61, 64–65.

Nechols, J.R., Andres, L.A., Beardsley, J.W., Goeden, R.D. and Jackson, C.G. (1995) *Biological Control in the Western United States: Accomplishments and Benefits of Regional Research Project W84 (1964–1989).* University of California, Agricultural and Natural Resources Publication #3361, Oakland, California, 356 pp.

Pimentel, D., Lach, L., Zuniga, R. and Morrison, D. (2000) Environmental and economic costs of nonindigenous species in the United States. *BioScience* 50, 53–65.

Ramaswamy, S. and Subramanyam, B. (2000) *Consortium for Integrated Management of Stored Product Insect Pests.* RAMP Project #00-51101-9674, Washington, DC.

Rice, M.E., Krell, R.K., Lam, W.F. and Pedigo, L.P. (2000) New thresholds and strategies for management of bean leaf beetles in Iowa soybean. In: *Proceedings of the Integrated Crop Management Conference,* Iowa State University Extension Services, Iowa, pp. 75–84.

Sorensen, A.A. (1993) *Constraints to the Adoption of Integrated Pest Management.* National Foundation for Integrated Pest Management Education (NFIPME), Austin, Texas, 60 pp.

Sorensen, A.A. (1994) *Proceedings of the National Integrated Pest Management Forum,* Arlington, Virginia, 17–19 June 1992. American Farmland Trust, DeKalb, Illinois, 86 pp.

Welter, S. and Van Steenwyk, R.A. (2000) *Enhancing Pheromone Mating Disruption Programs.* RAMP Project #00-51101-9673, Washington, DC.

Westbrooks, R. (1998) *Invasive Plants, Changing the Landscape of America: Fact Book.* Federal Inter-agency Committee for the Management of Noxious and Exotic Weeds (FICMNEW), Washington, DC, 109 pp.

4 Dispersal and Migration of Insects and Their Importance in Pest Management

DAVID N. BYRNE

Department of Entomology, University of Arizona, Tucson, Arizona, USA

Introduction

Insect flight is thought to have arisen from a single origin in the early Devonian, approximately 390 million years ago (Grimaldi and Engel, 2005). The evolution of this life history trait has led to much of the success of members of Insecta. Since this is particularly true of some pest insects, it follows that attempts to disrupt or exploit their dispersal could lead to control techniques that would be included in integrated pest management (IPM) programmes.

Recognition of the ability of insects to engage in migratory events did not come quickly. Although early mention of their movement was made by a number of authors for mosquitoes (Ross, 1905), butterflies (Adkin, 1925) and leafhoppers (Carter, 1927). Williams (1957) attributes the beginnings of specific considerations of insect dispersal to scientists such as Felt (1928) and Hardy and Milne (1937, 1938), these works, however, did not directly refer to purposeful movement. Hardy and Milne (1938) talked of kites, with nets attached, being towed at heights of 150 m capturing: '. . . aerial plankton . . . which consisted of small insects with weak powers of flight'.

The notion that insects could be engaged in migration was possibly not believed. This was, in part, due to biologists holding insects to standards for migration generally applied to vertebrates. For bird movement to be considered migratory, early ornithologists required that they undertake to-and-fro movement over some distance. Thomson (1926) described true bird migration as: '. . . changes in habitat, periodically recurring and alternating in direction, which tend to secure optimal environmental conditions at all times'. Some ornithologists remain as rigid. More recently Newton (2003) proclaimed: '. . . migration can be defined as a large-scale return movement of a population that occurs each year between regular breeding and wintering (or non-breeding) areas'. Although these definitions are workable for avian taxa, migration by insects does not generally involve return movements along specific routes. Adding to the problem was the fact that short-range movement was not considered by vertebrate biologists as part of the migratory process, so that most insects were specifically excluded from traditional definitions of migration (Dingle, 1996).

As recently as 2006, Dingle reported that: 'Ornithologists, and especially northern hemisphere ornithologists, have traditionally thought of migration as an annual return movement of populations between regular breeding and non-breeding grounds' (Dingle, 2006). This presents a problem of a different type, since many ornithologists seem to focus on migration as taking place at the population level. Entomologists observe movement by groups and generally consider migration to be the product of individual selection (Davis, 1980). In this way, too, insects do not seem to meet avian criteria. I guess the battle still rages on.

This is in spite of the fact that entomological definitions of migration began to separate from vertebrate definitions in the 1950s. Original interpretations, however, were partially incorrect. Williams (1957) seemed to include elements of vertebrate definitions in his definition, since he thought that insect migration had predetermined directions and distances, with changes of location, direction and distance being determined by the particular insect. Williams felt that insects had an innate sense of direction that they held on to for hours, independent of the environment. Problems relating to specific insect definitions of migration are still proving difficult. Holland *et al.* (2006) cautioned that scientists accustomed to viewing movements through the prism of return, i.e. to-and-fro trips, would probably categorize much of insect migrations as dispersal events.

It is probably best not to try to convince all vertebrate biologists that our small insects are migratory. We have our own problems within our discipline. Incorporating considerations of insect migration into IPM programmes is hampered by the fact that, until relatively recently, insect migration and dispersal have been given scant attention. We simply did not know enough about dispersal by pest insects to assist IPM practitioners in very many cases.

Also, our earlier considerations of migration dealt almost exclusively with larger insects (i.e. > 0.03 g, Byrne *et al.*, 1988) moving over relatively long distances, i.e. 20 km (Loxdale *et al.*, 1993) (e.g. Irwin and Thresh, 1988; Sappington and Showers, 1992; Halpern, 2001; Min *et al.*, 2004). This does not apply to the majority of pest species. So which insects have claimed our attention? As is often the case, most early insect migration researchers were basic scientists who were interested in the phenomenon for its own sake. Much of their attention was focused on the movement by 'charismatic megafliers',[1] such as *Danaus plexippus* (L.), the Monarch butterfly, than on the less aesthetically pleasing members of Insecta.[2]

This is not to diminish the extraordinary flight behaviour of these giant insects. Large portions of their populations do travel more than 3000 km from Canada to the Central Highlands of Mexico. Many do complete two-way trips, although multiple generations are often required to do so. Information from these studies, and other similar studies, has been extremely valuable in understanding the dispersal and migration of these insects. However, to solve the practical problems faced by our growers, we must specifically address migration by insect pests that exist, in part, in artificial systems.

Beginning in the 1960s, insect migration research began to focus more on insects of economic importance, which dispersed over relatively short distances, i.e. < 10 km

(Byrne, 1999). Emphasis, however, still remained on longer-range migration. The majority of insect migration researchers were still convinced that insects moving short distances were merely engaging in dispersal and not migratory behaviour. While attending a meeting on migration in 1986, I was good-naturedly chided when I suggested that minute insects such as *Bemisia tabaci* (Gennadius), the sweet potato whitefly, were as capable of migration as were the large lepidopterous insects that travelled from the southern tier of states northward. My reasoning was that certain whitefly species met the classical definition of insect migration, as put forth by Kennedy (1961), Southwood (1962) and Johnson (1969). Their contributions came together in Kennedy's definition of migration (1985) that could be applied to many more insects:

> Migratory behaviour is persistent and straightened-out movement effected by the animal's own locomotory exertions or by its active embarkation on a vehicle. It depends on some temporary inhibition of station keeping responses but promotes their eventual disinhibition and recurrence.

By persistent I believe Kennedy meant that dispersal was of long enough duration to carry the animal away from its original habitat to a new one. Straightened-out meant that dispersal was in a general direction, unlike foraging behaviour where the animal's track may turn back on itself. Finally, when describing inhibition to station-keeping cues, he meant that animals were ignoring those signals that they would otherwise respond to. These signals could be cues provided by plant resources or congeners. During this time the animals are often responding to sky cues while ignoring others. After a period, they are disinhibited from responding to resource cues and again react to them. It is important that Kennedy did not mention scale of movement.

Because insect migration varies so dramatically, it is impossible to arrive at a single, generally accepted definition. Regardless, with little modification, Kennedy's seems the most applicable. As I say, it does not refer to scale. Movement can take place across a distance of < 1 cm or up to 3000 km. I present a few selected case histories on insect pests moving across a variety of distances and how consideration of their migratory behaviour has, or has not, aided in their management.

Bemisia tabaci and its Parasitoids

In order to strengthen the goal of IPM, this is the best system and provides some important lessons. *Bemisia tabaci* was first found in the USA in 1894 (Russell, 1975). Russell suggested that: 'It is potentially dangerous in warmer areas of the United States . . .', indicating that it was not yet an economically important insect in 1975. In 1981, however, *B. tabaci* was found in extraordinarily high numbers in the southern tier of the USA on crops such as lettuce (*Lactuca sativa* L.), cantaloupes (*Cucumis melo* L.) (Butler and Henneberry, 1984; Byrne *et al.*, 1986) and, to a lesser extent, on cotton.

Bemisia tabaci continues to be a major problem in the 21st century. While an effective IPM plan for managing *B. tabaci*, primarily dependent on pesticides, had been developed (Ellsworth and Martinez-Carrillo, 2001; Palumbo *et al.*, 2001), the author, along with several colleagues, thought this plan could be enhanced through

Fig. 4.1. Numerous *Bemisia tabaci* moving over a field of broccoli in the Imperial Valley of California, USA, 1993 (photograph courtesy of James C. Hurt, California pest control advisor).

the inclusion of considerations of migration. This was an obvious step, because movement by *B. tabaci* can be dramatic (see Fig. 4.1). These studies were later expanded to include considerations of dispersal and migration by their parasitoids.

Whitefly flight experiments

Much of our ability to examine flight by these very small insects (*B. tabaci* weighs approximately 40 µg, while the aphelinid *Eretmocerus* spp. weighs approximately 18 µg; D.N. Byrne, unpublished data) is the result of our access to a vertical flight chamber (see Fig. 4.2). Our chamber is patterned after one used by Kennedy and Booth (1963). Blackmer and Phelan (1991) improved upon their device. With some modifications, Blackmer brought vertical flight chamber technology to the University of Arizona in 1991 for use in studying migration by *B. tabaci* (Blackmer and Byrne, 1993; Blackmer *et al.*, 1995) (see Fig. 4.2).

The Arizona chamber consists of a large, open-fronted box, 63.5 m², with a central opening in the roof into which fits a smaller open box containing a sodium vapour light (sky cue). Mounted on the side of the chamber is a 550-nm light source that simulates a vegetative cue (see Kennedy's 1985 definition of migration). Adjusting the chamber's aperture regulates draught. Digital readings from the anemometer measure wind flow from above. These data are an indirect measure of insect rate of ascent. Insects ignoring the vegetative cue while flying towards the sky cue for an extended period were identified as being migratory (Blackmer and Byrne, 1993).

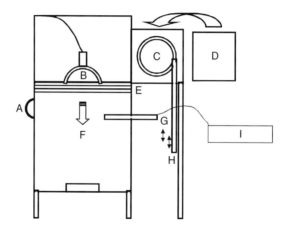

Fig. 4.2. Diagrammatic vertical section of flight chamber. A, 550 nm light source; B, sodium vapour lamp; C, aperture; D, squirrel cage fan; E, muslin screens; F, air flow direction; G, anemometer; H, handle; and I, computer (after Blackmer and Phelan, 1991; Blackmer and Byrne, 1993).

Whitefly flight laboratory studies

We used the flight chamber to determine whether flight by *B. tabaci* met generally accepted criteria for insect migration. Applying Kennedy's standards (1985), we determined that whitefly dispersal could be considered as such (Blackmer and Byrne, 1993). Their movement in the chamber was persistent: approximately 5% of the whiteflies tested flew toward the sky cue for more than 15 min. While doing so, they ignored the vegetative cue. Also, their flight was straightened out; long-duration fliers remained in the skylight column of the chamber, never landing on walls or responding to the vegetative cue for long periods (occasionally > 3 h).

Whitefly field studies

Field migration experiments corroborated laboratory findings (see Fig. 4.3). *B. tabaci* were marked and their distribution following dispersal across a 21 km² grid was noted. A significant portion of the whitefly population dispersed 5 km in a 4-h period (Byrne *et al.*, 1996), well beyond their previous habitat. Their distribution was found to be bimodal, one peak of alighting occurring at the field edge and another peak at a distance of 2.2 km. We believe that whiteflies caught near the field were engaging in foraging flight, similar to those that had flown for only short periods in the chamber, while those captured 2 km away, having ignored intervening crops, had engaged in migratory flight. The limits of our experimental design were 7 km, and we regularly caught marked whiteflies at this distance.

It was our fervent hope that such information would aid growers in making informed decisions concerning the timing of crop planting and placement. In certain isolated systems in Arizona such as the Harquehala Valley, where they grow limited

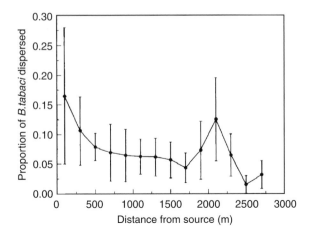

Fig. 4.3. *Bemisia tabaci* distribution following migration from a plot of cantaloupes. Those captured at the field edge were thought to have been engaging primarily in foraging flight; those caught beyond 2 km were thought to have engaged in migratory flight since they have ignored intervening vegetative cues (from Byrne *et al.*, 1996).

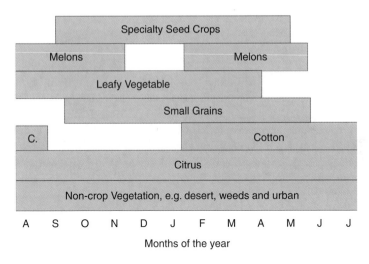

Fig. 4.4. Typical crop rotational scheme in the low desert of the US south-west. Each of these crops is a potential host for *B. tabaci*. Fields of small grains are hosts because these are weedy.

acreage of cotton in an isolated setting, these recommendations have been of value. The value of our recommendation in larger, more diverse agroecosystems was limited. The ability of *B. tabaci* to migrate at least 7 km in a single morning, combined with a patchwork cropping system (see Fig. 4.4), presents a different problem. This is largely because crop hosts are available year around. *B. tabaci* does not have to move far, 7 km appearing to be quite sufficient.

For these reasons, in the south-west, e.g. the Imperial County of California, La Paz, Maricopa and Pinal County of Arizona, *B. tabaci* migrators are herded from one rich patch (crop) to another across time. The wise use of pesticides can, in most years, keep them at manageable levels. In years in which populations are extraordinarily large (through the convergence of several biotic and abiotic factors), however, *B. tabaci* populations can exceed the capabilities of chemical control. At this point, recommendations concerning migration are even less effective. This was certainly the case in the early 1990s in Arizona, in 2005 in the Imperial Valley of California and in southern Texas in 2006.

An aside at this point can be informative, since profits and IPM are inexorably intertwined; the economic details of the vegetable markets cause supply and demand trade-offs to allow an equilibration over time. I will use lettuce, *Lactuca sativa* L., as an example. When growing conditions are favourable, e.g. lower pressure for *B. tabaci*, yields are higher and will average, for the purpose of this discussion, 1482 cartons/ha. Because of plentiful supply, however, prices can be lower, approximately US$6.75 per carton. Gross profits would be US$10,004. When *B. tabaci* populations are high, yields can be significantly lowered (988 cartons/ha). Because of limited availability, prices can increase to US$22.10 per carton. As a consequence, gross profits are US$21,835. Production costs remain the same in either situation, at US$1875.

Long-range migration

Before leaving the phenomenon of short-range migration by *B. tabaci*, the fact that this insect probably engages in longer-range movement needs to be addressed. This can completely change control dynamics across this larger scale. At present, however, evidence is circumstantial. We do know that in certain years, when the irrigation is terminated for cotton in the Mexicali Valley in the state of Baja California, Mexico, large populations of *B. tabaci* appear in the Coachella Valley of California after a 2-week delay (see Fig. 4.5). The distance between these two points is approximately 100 km. During these same years, large numbers of whitefly cause visual problems for individuals flying above 366 m (John Jessen, personal communication).

Parasitoid flight

We were undaunted in the face of others' concerns; most have discounted biological in non-protected agroecosystems (Hoelmer, 1996; Bellamy *et al.*, 2004). We continued to believe that an IPM system for *B. tabaci* would be more efficient if agents of biological control could be included. After all, the wasp *Eretmocerus eremicus* had been effective in field cage studies (Simmons and Minkenberg, 1994). The first step in determining the feasibility of the use of biological control in outdoor situations was to examine the flight behaviour of a commercially available parasitoid that has been deployed against *B. tabaci*. We examined the flight characteristics of *E. eremicus* in the flight chamber and found them capable of sustained flight, i.e. some flew in excess of

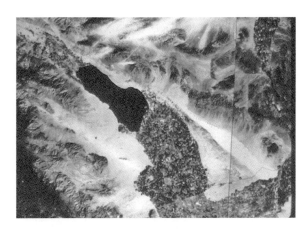

Fig. 4.5. Aerial view of Coachella Valley, California, USA, in the north-west; the Salton Sea and the Imperial Valley mid-lower part of image; and the Mexicali Valley across the international border. *Bemisia tabaci* is thought to leave the Mexicali Valley when cotton irrigation water is terminated and subsequently move into the Coachella, a distance of > 100 km.

Table 4.1. Flight duration (min) for the whitefly parasitoid *Eretmocerus eremicus*, as determined in a vertical flight chamber (total flown = 92). Numbers with the same superscript are not significantly different according to a Tukey-Kramer multiple comparison test ($P = 0.05$).

	Non-mated	Mated
Females	34.4[a]	10.2[b]
Males	6.7[b]	0.7[c]

2 h. We also found mating status and gender to have a significant effect on flight duration (see Table 4.1).

Field trials were necessary, and a feasibility study was conducted in an open cantaloupe crop. Unfortunately, we found the ability of *E. eremicus* to reduce *B. tabaci* populations in cantaloupe plots to be extremely limited (Bellamy *et al.*, 2004) (see Fig. 4.6). Even after releasing the equivalent of 12,350,000 parasitoids/h (five release dates of 247,000 wasps/release in each of four 0.36 ha replicated plots), whitefly populations did not significantly decline. Recommended rates were 24,700/ha at the time (Arbico Organics, Tucson, Arizona, USA, http://store.arbico-organics.com/ and Koppert Biological Systems, the Netherlands, http://www.koppert.nl/e005.shtml).

We feel certain this failure was, in part, attributable to the low relative dispersal distances of *E. eremicus* as compared with those of *B. tabaci*. This would explain the lack of control in a relatively large patch. We also speculated that *E. eremicus* females were leaving the area in search of whitefly hosts. But perhaps, most importantly, immigrating pest populations overwhelmed any suppression of the resident whitefly population through these augmentative releases. Bellamy and Byrne (2001) predicted

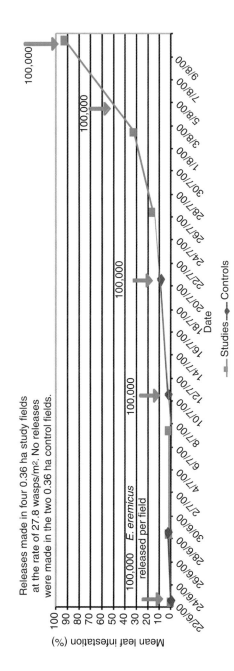

Fig. 4.6. Effect on *Bemisia tabaci* populations of the insertion (↓) of the equivalent of 1,000,000,000 *Eretmocerus eremicus* per release into a cantaloupe field (from Bellamy *et al.*, 2004).

that 95% of captured *E. eremicus* would be found within 75 m of plot centres while, as stated, Byrne *et al.* (1996) found *B. tabaci* capable of moving 4 km during the same time period. A second contributing factor was massive immigration by whiteflies into plots. Collier and Van Steenwyk (2004) argued that parasitoid emigration was a potential problem, as well as pest immigration.

As an additional measure of success, rates of parasitism were monitored over a 52-day period. Disappointingly, mean rates of parasitism did not increase with time nor did results differ from control plots. Hence, the use of *E. eremicus* alone is not an efficient means of reducing whitefly populations in melon crops in south-western USA.

Implications for biological control in greenhouses

Work on whitefly parasitoids continues because we believe that, given enough under-standing of migration and its contributing factors, we could make this strategy fit in greenhouses. We are now involved in a series of experiments concerning the use of whitefly parasitoids in closed systems. This, of course, now seems obvious given that immigration was found to be one of the two major reasons for the failure of parasitoids in open-field settings. Greenhouses should have been one of the first arenas we investigated. The literature is replete with examples where the use of parasitoids has been helpful in managing whitefly populations (Parrella *et al.*, 1991; Heinz, 1996; Hoddle *et al.*, 1997; Hoddle and Van Driesche, 1999; Van Driesche *et al.*, 2001; Simmons *et al.*, 2002; Qiu *et al.*, 2004; Stansly *et al.*, 2005).

Locally, we are fortunate that greenhouse crop production is an expanding industry in Arizona. EuroFresh Farms®, with facilities near Bonita, Arizona, controls more than 80% of US tomato greenhouse production, in a 160-ha facility. Their operation is nearly pesticide free (one application for *Aculops lycorsici* Masse, tomato rust mite, is sometimes necessary early in the season). Otherwise, this organization relies exclusively on *E. eremicus* and *Encarsia formosa* Gahan to control whitefly popula-tions. We have examined three other species as a means of comparison (see Table 4.2). We again collected data concerning wasp flight behaviour using our verti-cal flight chamber in the laboratory (see Table 4.2). Using the criterion of flight pro-pensity (percentage that flew), we determined that the two *Eretmocerus* species had the greatest potential for controlling whiteflies in greenhouses. This is because they may spread throughout the crop more effectively than *Encarsia* spp. We now believe this characteristic may not always be desirable, however. Our goal is to continue testing parasitoid wasps in the laboratory in order to select the best candidates for greenhouse experiments. Several seem promising (see Table 4.2).

Our next step in this process is to determine, in a systematic manner, which endemic whitefly parasitoids are available in Arizona. Observations in Yuma County indicate that *Encarsia sophia*, *Eretmocerus ethiopia* and *Eretmocerus hayati* are plentiful. We suspect that other wasps that are worthy of examination are also present, but in lower numbers. We hope to find the most effective parasitoid to use in Arizona greenhouses under different sets of circumstances. For example, when pest popula-tions are localized it may be best to use a whitefly parasitoid that has poor migratory

Table 4.2. Summary statistics from vertical flight chamber studies of 1-day-old females of five tested parasitoid species.

Species	Flight propensity (%)	Estimated flight distance (m)	Estimated flight distance range (m)
En. formosa	4.8	55.8	NA
En. pergandiella	8.1	62.3 ± 25.7	28.8 ± 138.3
En. sophia	10.9	36.5 ± 9.5	19.8 ± 52.9
Er. nr. *emiratus*	66.7	31.1 ± 14.3	2.8 ± 109.9
Er. eremicus	23.6	38.1 ± 24.35	5.1 ± 280.5

Means are presented as ± SEM. *En.*, *Encarsia*; *Er.*, *Eretmocerus*; NA, not available (from M.A. Asplen, unpublished data).

ability. When pest populations are widely distributed, however, a parasitoid that readily migrates might be best. Knowledge of migratory ability can be invaluable when selecting the best parasitoid, or combination of parasitoids, to deploy.

Other Case Histories

Myzus persicae and potatoes in north-western USA

An outstanding, but under-publicized, IPM programme involving potatoes, *Solanum tuberosum* L., potato leafroll virus and *M. persicae*, the green peach aphid, was initiated in north-western USA in the early 1970s. It actually began in 1926 with the use of potato seed pieces that were certified as being pathogen-free. Certain growers readily accepted it, but it was not until the 1960s, following a high incidence of potato leafroll virus in commercial potato fields, that the use of certified seed was broadly accepted. Prior to that, it was common for growers to hold back smaller potatoes from the previous crop for that year's seed (G.W. Bishop, personal communication). These undersized 'single drop' potatoes were often of smaller size because they were infected with the pathogen. With the use of disease-free seed, levels of pre-season inoculum were reduced to manageable levels. These steps alone, however, were not enough.

In the northern tier of states *M. persicae* is holocyclic, with winter populations of eggs being primarily maintained on peach, *Prunus persica* (L.). Beginning in the spring, adult migrants move to a series of herbaceous hosts, including potatoes. This continues through the summer, with aphids returning to woody perennial hosts for the winter. Scientists in Idaho decided that one way to reduce aphid populations was by eliminating or ameliorating their overwintering host. There are three major potato-growing areas in this state: one found west of Boise, one centred on Twin Falls in the south-central portion of Idaho and one in the eastern part of the state near Aberdeen.

It was evident from the start that, because of commercial plantings, there were too many peach trees in the western part of Idaho for this plan to be effective. In the other two growing regions, however, peach trees were grown only as ornamentals or as part of backyard gardens. A scouting programme identified every peach tree in central and eastern Idaho. They (Sandvol *et al.*, 1977) found 1334 trees near Twin Falls and 372 trees near Aberdeen. After having *P. persica* declared a noxious weed in these areas, owners were offered two options: they could have their tree replaced by a non-host ornamental or they could agree to have their tree(s) treated every year with a systemic insecticide in order to eliminate spring migrants. This programme was in place for more than 15 years until other methods could be found to manage *M. persicae*. This was a system developed near the early part of the IPM era, where an intimate understanding of pest migratory patterns was not critical in developing a pest control system.

The implementation of this programme was met with enthusiasm, and there are many who feel that it was successful in reducing the number of spring migrants that eventually moved into potato crops. Unfortunately, these IPM successes are almost impossible to quantify and, as in this case, remain undocumented. Regardless, shortly after federal monetary support became no longer available, enthusiasm waned and the programme faded. North-western growers have returned to managing *M. persicae* on a field-by-field basis and have abandoned community-wide efforts. One major concern centred on cost effectiveness: leaders of the programme spent money treating all peach trees in the two growing areas. One wonders what would have happened if early efforts had been made to define the effective dispersal distances of *M. persicae* in the north-west in order to make informed decisions so that only those that were significant sources of *M. persicae* would have been treated. A more economically efficient programme might have been more palatable to growers, and upfront considerations of aphid dispersal might well have been included in such a system.

Agrotis ipsilon (black cutworm) migration

One of the best-studied cases of migration by an insect pest involves *Agrotis ipsilon* (Hufnagel), the black or greasy cutworm. This insect is found in 49 states of the USA, and in many other parts of the Old and New World. It feeds on a wide range of field and garden crops, with a special preference for maize and tobacco. Other known hosts include asparagus, bean, beet, cabbage, castor bean, cotton, grape, lettuce, groundnut, pepper, potato, radish, spinach, squash, strawberry and tomato. Black cutworms are among the most destructive of all cutworms. C.V. Reilly reported that the greasy cutworm was destroying seedling maize in 1868 along the Mississippi and Missouri rivers – it has been with us for a while. The larvae sever plants from roots near the soil line; usually no other feeding damage is present. Many larvae move from plant to plant on successive nights, while some stay to feed on the roots and underground stems of cut plants.

Importantly for this discussion, *A. ipsilon* has prodigious powers of dispersal. One of the first reports was by Bishara (1932), who reported them as moving from hotter to cooler climates in Egypt and Israel. There were later reports of the same behaviour. Jia (1985) marked *A. ipsilon* with fluorescent material at 23° N in South-east

China in early and late March. By early April he had captured a few at 38° N and one at 42° N. These constitute distances of 1344 and 1818 km, respectively. Showers *et al.* (1989) reared *A. ipsilon* on diet containing a red dye. These were released in Crowley, Louisiana on 4–11 June and were recaptured 1142 km north near Elkhart, Iowa on 13 June. Similarly reared moths were released on 17–21 July near Rock Rapids, 1266 km away.

Also, the construction of wind trajectories to trace previous paths of air parcels provided information on possible sources (Showers, 1997). Examining migrating moths that had been naturally marked with pollen found only in southern states substantiated these assessments. *A. ipsilon* has been found in Iowa and Minnesota with pollen from *Pithecelobium* and *Calliandra* (both *Mimosaceae*) (Hendrix and Showers, 1992). Since these plants are found along the Rio Grande River, this provides empirical evidence of a spring/summer northward migration.

For all this extensive study of *A. ipsilon* migration, there is no documented evidence that it was ever of any direct use in IPM. I don't know of any direct use in IPM beyond a general explanatory knowledge base.

Peronospora tabacina (tobacco blue mould): dispersal and IPM

The science of aerobiology has contributed greatly to an IPM system involving fungal migration (e.g. Stukenbrock *et al.*, 2006) by *Peronospora tabacina* Adam (*Peronosporaceae*), blue mould of tobacco and *Nicotiana tabacum*, commercial tobacco (Main *et al.*, 2001; Main, 2003). This pathogen thrives well during periods of cool, wet weather. *P. tabacina* spreads rapidly northward when conditions are wet. This is often across great distances, e.g. from Cuba to Canada (Moss and Main, 1988). Each year tobacco in the USA is exposed to asexual, windborne sporangiospores originating from inoculum sources on commercial winter tobacco and to wild *Nicotiana* species in the tropical zones south of the 30th parallel. The fungus is not known to overwinter in the more temperate zones, so inoculum must be reintroduced each year into the USA. Having entered the USA, the inoculum produced in one tobacco-growing area can quickly spread northward on wind currents to other, distant production areas.

The long-range atmospheric fungal migration of *P. tabacina* sporangiospores has been extensively studied in an attempt to predict blue mould epidemics. Considerable effort has been expended to identify primary inoculum sources and pathways of moving spore masses. Scientists at North Carolina State University, using the weather prediction systems of the National Oceanic and Atmospheric Administration that forecast weather 48 h into the future, produce prognostications concerning the northern distributions of the pathogen. These forecasts are an outstanding example of basic science being used to provide practical information to growers concerning the movement of pest species (Aylor and Taylor, 1983).

The idea of developing a reporting system for growers concerning the epidemiology of *P. tabacina* was generated at North Carolina State University in 1981 (Moss and Main, 1988). The *P. tabacina* prediction system was made available to all growers and the industry beginning in 1995 (Main and Keever, 1999). The system is dependent on timely and well-documented disease reports provided by growers, extension specialists and industry representatives who identify suspected cases of *P. tabacina* along the eastern seaboard. Reports are posted daily, then North Carolina State University personnel assess these reports. The reporting network data and diagnosis of spore transport in the atmosphere are then used to calculate trajectory models that predict spore movement (Draxler, 1992). A trajectory weather section describes the recent past, present and future weather conditions at the source and along the anticipated pathway for northerly movement of *P. tabacina*. Growers use this information for making decisions about when to make prophylactic applications of fungicides. Isard and Gage (2001) acknowledge that this system represents a model for IPM users involved with continent-wide pest movement.

Pectinophora gossypiella (pink bollworm): resistance and migration from refuges

Pectinophora gossypiella (Saunders), the pink bollworm, a native of Asia, had reached the cotton belt in the southern USA by the 1920s. It was a major pest in southwestern cotton until the early part of the 1990s, when a variety of methods involving cultural, biological and chemical techniques were used to manage this insect pest. These insects spend the winter as pupae after entering the soil in the late summer/early autumn. They emerge as adults in the spring, perhaps in advance of susceptible cotton being available. As a result, control experts recommended planting short-season cotton varieties that would not be mature at the time of early *P. gossypiella* emergence in the early spring. This cotton could also be harvested before *P. gossypiella* was ready to pupate. Pesticides were recommended at the pinhead square stage of cotton development (the time when flowers have fallen and bolls are just forming). Additional pesticide treatments were recommended based on levels of *P. gossypiella* infestations.

Growers and university personnel were at odds because the former group wanted to treat when levels were at 5% of checked bolls. Extension employees were certain that levels could be three times that high. Strategies of the two ranks were also at contretemps because growers wanted to leave the crop in for longer periods to ensure maximum yield. To do so required several additional applications of pesticides. In 1990 the average cotton field was receiving seven applications for *P. gossypiella* alone.

Fortunately, in the late 1990s, *Bt* cotton was introduced. Through genetic engineering, the gene used by *Bacillus thuringiensis* (*Bt*) to produce a protein crystal toxic to lepidopterous pests could be inserted into the cotton genome, so that the plant produced its own *Bt* toxin. Cotton plants expressing these modified genes provided control of several pests, including *P. gossypiella*. In 1998, *Bt* cotton accounted for over one-quarter of US harvested cotton acreage.

Because the toxin was continually being expressed in the plant, and nearly all members of the pest population were exposed, resistance was of great concern (Gould, 2000). In 1996 the Environmental Protection Agency (USEPA, 2001), working with the agrochemical industry, took steps to slow or prevent the development of resistance. They did so by requiring cotton growers to plant a portion of their crop to a refuge of non-*Bt* varieties. In these refuges the pests could feed on plants that did not contain *Bt* toxin, and could thereby maintain *Bt*-susceptible alleles. The theory was that susceptible insects produced in the refugia would mate with the toxin-resistant animals that had survived the *Bt*, and so dilute the alleles for resistance and prolong the pest population's susceptibility. With modifications, this IPM strategy is successfully in place today. Resistance to *Bt* cotton has not been developed to date (Sisterson *et al.*, 2005). This is in spite of the fact that approximately 80% of the cotton in the low desert cotton-growing areas of Arizona is *Bt* cotton. Nationwide, the average is about 60% (T.J. Dennehy, personal communication).

One of the principal reasons why this strategy has succeeded has been the science that led to the proper placement of these refuges, here determined by analysis of the patterns observed for the dispersal/migrational habits of *P. gossypiella*.

Using a series of statistical models, Sisterson *et al.* (2005) examined the relationship between the temporal and spatial placement of *Bt* cotton fields and refuges of non-*Bt* cotton on the development of resistance. Dispersal of *P. gossypiella* between these two crops increases chances that resistant adults mate with susceptible adults from refuges. Experimental results indicated that interactions among the relative abundance and distribution of refuges and *Bt* cotton fields could alter the effects of movement on resistance development. With fixed field locations and all *Bt* cotton fields adjacent to at least one refuge, resistance evolved most slowly with low movement. However, low movement and fixed field locations favoured rapid resistance evolution when some *Bt* crop fields were isolated from refuges. Information of this sort would have resulted in preventative language in initial instructions, so that non-*Bt* cotton would not be planted more than 200 km away, a practice used by some growers early in the programme (T.J. Dennehy, personal communication).

The results of this experiment demonstrated that fixing field locations and distributing refuges uniformly to ensure that *Bt* crop fields were not isolated from refuges could delay resistance. While the conclusions of these models provided valuable information, they are based on certain assumptions about how far *P. gossypiella* disperses between these crops. More accurate models will be produced when more precise information is made available about dispersal distances. Y. Carrière is currently generating these data (personal communication). In this situation it was imperative that a system be put in place at the beginning of the programme, so delay in the generation of information concerning placement distance is clearly understandable. It is my feeling, however, that inclusion of considerations of migration at the beginning would have been invaluable.

Conclusions

I have offered a small sample of case histories where scientists have tried to incorporate elements of insect migration into integrated pest management programmes. There have been some startling successes. As an example, I have described C.E. Main's Cooperative Extension programme concerning the advancement of *P. tabacina*, tobacco blue mould, from subtropical regions of the New World up the eastern US seaboard. I have provided examples where, in spite of not being considered initially, migrational elements were later successfully incorporated into IPM program-mes, e.g. movement of *P. gossypiella* out of refuges into *Bt* cotton where they mated with resistant individuals. I have discussed instances where migration-related infor-mation is soon to become part of IPM strategies, but has not yet done so. Unfortu-nately, the largest groups of programmes are those where the migration and dispersal of pest organisms have been elegantly and extensively studied, but the information has never, and probably never will, be incorporated into practical control systems.

Charlie Main's network of projecting tobacco blue mould dispersal seems to be exemplary of how to incorporate considerations of migration into IPM programmes. He is a practical researcher whose initial motivation was to solve problems faced by growers. He had a basic understanding of the role of aerobiology in the movement of fungal pathogens (see http://www.ces.ncsu.edu/depts/pp/bluemold/). He built on that knowledge to develop a forecasting system for *P. tabacina* spread that serves as a basis for growers making decisions about when to apply fungicides. The key to Main's success is that dispersal was a primary focus from the programme's beginning.

I have mentioned a programme that acknowledged the importance of migration and dispersal from the onset, but initially had more pressing problems to solve. Sisterson *et al.* (2005) were concerned with preventing the development of pesticide resistance by *P. gossypiella*. Along with other scientists, their focus was to delay this potential problem, which was of immediate concern. As part of this effort they were concerned with refuges for susceptible pink bollworm populations. Models were con-structed that made assumptions about the dispersal capabilities of this moth. They are now in the process of generating data that will refine their predictions by deter-mining the insect's migratory ability.

In many cases, promising experiments have been completed concerning migra-tion of pest insects. Unfortunately, little of this information has directly contributed to IPM programmes. For example, a great deal is known about dispersal by *A. ipsilon*: so much so that one might believe that this information would be at the core of an IPM system. Apparently this is not the case, as no recommendations exist that allow growers to know in advance when this moth will arrive in their area. Irwin (1999) presents an outstanding theoretical model for the spread of soybean mosaic virus based, in part, on the work of Halbert *et al.* (1981). Work on utilizing this information as part of an IPM programme seems to have halted, however, at least temporarily, as scientists have moved to other projects.

I find myself in a similar situation. The projects that I have been working on with a number of colleagues (e.g. Byrne *et al.*, 1986, 1988, 1996; Byrne and von Bretzel, 1987; Byrne and Houck, 1990; Blackmer and Byrne, 1993; Isaacs and Byrne, 1998; Veenstra and Byrne, 1999; Bellamy *et al.*, 2004) concern migration and dispersal by *B. tabaci* and its parasitoids. This is an example of an answer looking for a question. We have learned many things about whiteflies – e.g. which trap provides the best monitoring system for movement, the relationship between wing loading and other flight mechanics, that *B. tabaci* is truly migratory and how far it migrates in the field, a great deal about flight physiology, and the impact of whitefly parasitoids in open agriculture. The question is, can this information lead to improvement of an IPM programme for whiteflies? After more than 20 years of research my response has to be, not yet.

I was hopeful that dispersal distance information would allow growers to make informed decisions concerning crop planting and placement. Apparently this is not possible, due to the extremely high reproductive potential of whiteflies on crops that are in close proximity to one another. We know they can move up to 7 km in 4 h, and probably much farther, in a system where vegetable and cotton fields are regularly interspersed. In this situation immigration concerns trump all others. We have found that highly touted biological control agents have no impact on *B. tabaci* in open agriculture. We feel this is, in part, attributable to differences in dispersal capability between the pest and its parasitoids. So, we know why biological control does not work under the conditions tested, but this does not provide a solution to the problem. We have also started to investigate the relationship between reproductive timing and parasitoid dispersal. This is all very interesting scientifically, but does not come close to offering a practical solution to growers.

These may represent the growing pains experienced by any group of scientists who begin an investigation of short-range migration by insects where little is known initially. We have explored blind alleys, partially because we were prejudiced by previous experiences. I helped lead the effort by insisting that basic science could eventually provide practical solutions. I still believe this to be true, and there having been certain advances made in our basic understanding of migration by very small insects, but I have as yet been unable to contribute to an IPM programme for *B. tabaci*.

This effort has not been for naught, however. We have now found a situation where what we have learned about dispersal by whiteflies and their parasitoids can be put to practical use. This involves the growing of vegetables in a 400-acre greenhouse system. EuroFresh® Farms (http://www.eurofresh.com/) began growing tomatoes and cucumbers under conditions of protected agriculture in Arizona in 1990. From the beginning, they have used biological control as the mainstay of their management system. We believe we can help them improve their management programme. We are now in the process of examining a variety of endemic and imported parasitoid wasps in our flight chamber. We have found their flight characteristics to be dramatically different from one another (M.A. Asplen, unpublished data). Considering the laboratory data we have collected to date, we expect that some of these aphelinid wasps will be more suitable under certain conditions than will others. Our expectations are that this will be dictated by whitefly population density and distribution. We are eagerly waiting preliminary testing of our hypotheses in commercial greenhouses. Perhaps now we will be able to offer practical information to the greenhouse industry in Arizona.

It is likely that my inability to provide information that is of immediate value to growers was because I was focused on the wrong arena. I was working in traditional agricultural production of vegetables. Most strategies, other than the judicious use of pesticides, are economically unfeasible (imagine the cost of so many *E. eremicus*!) and do not provide levels of whitefly population suppression that would be acceptable to growers. Realizing biological control would not be practical in open agriculture because of immigration, I should have looked toward management of greenhouse pests earlier.

I am convinced that the key to incorporating migration into IPM programmes is to maintain that as a primary goal, carefully considering alternative opportunities. While scientists repeatedly state that understanding migration is paramount to the success of IPM programmes (e.g. Byrne *et al.*, 1997; Aylor and Irwin, 1999), we too often fail to meet this challenge for many reasons: programmes are not sustained, we lose sight of our goal of providing information of practical value or economically our solutions are unfeasible. It is imperative that we maintain contact with the end-users of our scientific efforts, the growers.

Notes

[1]My thanks to Cleveland Amory.
[2]This is not universally true, but merely an expression of the author's own prejudice.

References

Adkin, R. (1925) Dispersal of butterflies and other insects. *Nature* 116, 467.

Aylor, D.E. and Irwin, M.E. (1999) Aerial dispersal of pests and pathogens: implications for integrated pest management – preface. *Agricultural and Forest Meteorology* 97, 233–234.

Aylor, D.E. and Taylor, G.S. (1983) Escape of *Peronospora tabacina* spores from a field of diseased tobacco plants. *Phytopathology* 73, 525–529.

Bellamy, D.E. and Byrne, D.N. (2001) Effects of gender and mating status on self-directed dispersal by the whitefly parasitoid *Eretmocerus eremicus*. *Ecological Entomology* 26, 571–577.

Bellamy, D.E., Asplen, M.K. and Byrne, D.N. (2004) Impact of *Eretmocerus eremicus* (Hymenoptera: Aphelinidae) on open-field *Bemisia tabaci* (Hemiptera: Aleyrodidae) populations. *Biological Control* 29, 227–234.

Bishara, I.E. (1932) The greasy cutworm (Agrotis ipsilon née ypsilon Rott.) in Egypt. *Egyptian Agricultural Technical Science Bulletin* 114, 1–58.

Blackmer, J.L. and Phelan, P.L. (1991) Behavior of *Carpophilus hemipterus* in a vertical flight chamber – transition from phototactic to vegetative orientation. *Entomologia Experimentalis et Applicata* 58, 137–148.

Blackmer, J.L. and Byrne, D.N. (1993) Flight behaviour of *Bemisia tabaci* in a vertical flight chamber: effect of time of day, sex, age and host quality. *Physiological Entomology* 18, 223–232.

Blackmer, J.L., Byrne, D.N. and Tu, Z. (1995) Behavioral, morphological, and physiological traits associated with migratory *Bemisia tabaci* (Homoptera, Aleyrodidae). *Journal of Insect Behavior* 8, 251–267.

Butler, G.D. and Henneberry, T.J. (1984) *Bemisia tabaci*: effect of cotton leaf pubescence on abundance. *Southwestern Entomologist* 9, 91–94.

Byrne, D.N. (1999) Migration and dispersal by the sweet potato whitefly, *Bemisia tabaci*. *Agricultural and Forest Meteorology* 97, 309–316.

Byrne, D.N. and Houck, M.A. (1990) Morphometric identification of wing polymorphism in *Bemisia tabaci* (Homoptera: Aleyrodidae). *Annals of the Entomological Society of America* 83, 487–493.

Byrne, D.N. and von Bretzel, P.K. (1987) Similarity in flight activity rhythms in coexisting species of Aleyrodidae, *Bemisia tabaci* and *Trialeurodes abutilonea*. *Entomologia Experimentalis et Applicata* 43, 215–219.

Byrne, D.N., von Bretzel, P.K. and Hoffman, C.J. (1986) Impact of trap design and placement when monitoring for the bandedwinged whitefly and the sweetpotato whitefly (Homoptera: Aleyrodidae). *Environmental Entomology* 15, 300–304.

Byrne, D.N., Buchmann, S.L. and Spangler, H.G. (1988) Relationship between wing loading, wingbeat frequency and body mass in homopterous insects. *Journal of Experimental Biology* 135, 9–23.

Byrne, D.N., Rathman, R.J., Orum, T.V. and Palumbo, J.C. (1996) Localized migration and dispersal by the sweet potato whitefly, *Bemisia tabaci*. *Oecologia* 105, 320–328.

Carter, W. (1927) Extension of the range of *Eutettix tenellis* Baker and curly top of sugar beet. *Journal of Economic Entomology* 20, 714–717.

Collier, T. and Van Steenwyk, R. (2004) A critical evaluation of augmentative biological control. *Biological Control* 31, 245–256.

Davis, M.A. (1980) Why are most insects short fliers? *Evolutionary Theory* 5, 103–111.

Dingle, H. (1996) *Migration, the Biology of Life on the Move*. Oxford University Press, New York.

Dingle, H. (2006) Animal migration: is there a common migratory syndrome? *Journal of Ornithology* 147, 212–220.

Draxler, R.R. (1992) *Hybrid Single-particle Lagrangian Integrated Trajectories (HY-SPLIT)*. Version 3.0 – user's guide and model description. NOAA Technical Memo. ERL ARL-195.

Ellsworth, P.C. and Martinez-Carrillo, J.L. (2001) IPM for *Bemisia tabaci*: a case study from North America. *Crop Protection* 20, 853–869.

Felt, E.P. (1928) Dispersal of insects by air currents. *Bulletin of the New York State Museum* 274, 59–129.

Gould, F. (2000) Testing *Bt* refuge strategies in the field. *Nature Biotechnology* 18, 266–267.

Grimaldi, D. and Engel, M.S. (2005) *Evolution of the Insects*. Cambridge University Press, Cambridge, UK.

Halbert, S.E., Irwin, M.E. and Goodman, R.M. (1981) Alate aphid (Homoptera, Aphididae) species and their relative importance as field vectors of soybean mosaic virus. *Annals of Applied Biology* 97, 1–9.

Halpern, S. (2001) *Four Wings and a Prayer*. Pantheon Books, New York.

Hardy, A.C. and Milne, P.S. (1937) Insect drift over the North Sea, 1936. *Nature* 139, 520.

Hardy, A.C. and Milne, P.S. (1938) Aerial drift of insects. *Nature* 141, 602.

Heinz, K.M. (1996) Predators and parasitoids as biological control agents of *Bemisia* in greenhouses. In: Gerling, D. and Mayer, R.T. (eds) *Bemisia: 1995 Taxonomy, Biology, Control and Management*. Intercept Ltd, Andover, UK, pp. 435–449.

Hendrix, W.H. and Showers, W.B. (1992) Tracing black cutworm and armyworm (Lepidoptera: Noctuidae) northward migration using *Pithecelobium* and *Calliandra* pollen. *Environmental Entomology* 21, 1092–1096.

Hoddle, M.S. and Van Driesche, R. (1999) Evaluation of *Eretmocerus eremicus* and *Encarsia formosa* (Hymenoptera: Aphelinidae) Beltsville strain in commercial greenhouses for biological control of *Bemisia argentifolii* (Homoptera: Aleyrodidae) on colored poinsettia plants. *Florida Entomologist* 82, 556–569.

Hoddle, M., Van Driesche, R. and Sanderson, J. (1997) Biological control of *Bemisia argentifolii* (Homoptera: Aleyrodidae) on poinsettia with inundative releases of *Encarsia formosa* (Hymenoptera: Aphelinidae): are higher release rates necessarily better? *Biological Control* 10, 166–179.

Hoelmer, K.A. (1996) Whitefly parasitoids: can they control field populations of *Bemisia*. In: Gerling, D. and Mayer, R.T. (eds) *Bemisia: 1995 Taxonomy, Biology, Control and Management*. Intercept Ltd, Andover, UK, pp. 451–476.

Holland, R.A., Wikelski, M. and Wilcove, D.S. (2006) How and why do insects migrate? *Science* 313, 794–796.

Irwin, M.E. (1999) Implications of movement in developing and deploying integrated pest management strategies. *Agricultural and Forest Meteorology* 97, 235–248.

Irwin, M.E. and Thresh, J.M. (1988) Long-range aerial dispersal of cereal aphids as virus vectors in North America. *Philosophical Transactions of the Royal Society of London* 321, 421–446.

Isaacs, R. and Byrne, D.N. (1998) Aerial distribution, flight behaviour and eggload: their inter-relationship during dispersal by the sweetpotato whitefly. *Journal of Animal Ecology* 67, 741–750.

Isard, S.A. and Gage, S.H. (2001) A decision support system for managing the blue mold disease of tobacco: a case study. In: Isard, S.A. and Gage S.A. (eds) *Flow of Life in the Atmosphere: an Airscape Approach to Understanding Invasive Organisms*. Michigan State University Press, East Lansing, Michigan, pp. 143–162.

Jia, P. (1985) A brief report of long-distance migration and mark-recapture of cutworms, *Agrotis ipsilon* (erroneously *ypsilon*) Rott. *China Plant Protection* 11, 20.

Johnson, C.J. (1969) *Migration and Dispersal of Insects by Flight*. Methuen and Company Limited, London.

Kennedy, J.S. (1961) Turning point in the study of insect migration. *Nature* 189, 785–791.

Kennedy, J.S. (1985) Migration, behavioural and ecological. In: Rankin, M.A. (ed.) *Migration: Mechanisms and Adaptive Significance*. Contributions in Marine Science Suppl., Vol. 27, Marine Science Institute, University of Texas, Port Arkansas, Texas, pp. 7–26.

Kennedy, J.S. and Booth, C.O. (1963) Free flight of aphids in the laboratory. *Journal of Experimental Biology* 40, 67–85.

Loxdale, H.D., Hardie, J., Halbert, S., Foottit, R., Kidd, N.A.C. and Carter, C.I. (1993) The relative importance of short- and long-range movement of flying aphids. *Biological Reviews* 68, 291–311.

Main, C.E. (2003) Aerobiological, ecological, and health linkages. *Environment International* 29, 347–349.

Main, C.E. and Keever, Z.T. (1999) *Forecasting Transport of Spores and Transport of Tobacco Blue Mold* (http://www.ces.ncsu.edu/depts/pp/bluemold/, accessed 24 June 1999).

Main, C.E., Keever, T., Holmes, G.J. and David, J.M. (2001) *Forecasting Long-range Transport of Downy Mildew Spores and Plant Disease Epidemics*. APSnet (http://www.apsnet.org/online/feature/forecast/).

Min, K.J., Jones, N., Borst, D.W. and Rankin, M.A. (2004) Increased juvenile hormone levels after long-duration flight in the grasshopper, *Melanoplus sanguinipes*. *Journal of Insect Physiology* 50, 531–537.

Moss, M.A. and Main, C.E. (1988) The effect of temperature on sporulation and viability of isolates of *Peronospora tabacina* collected in the United States. *Phytopathology* 78, 110–114.

Newton, I. (2003) Geographical patterns in bird migration. In: Berthold, P., Gwinner, E. and Sonnenschein, E. (eds) *Avian Migration*. Springer, New York, pp. 211–224.

Palumbo, J.C., Horowitz, A.R. and Prabhaker, N. (2001) Insecticidal control and resistance management for *Bemisia tabaci*. *Crop Protection* 20, 739–765.

Parrella, M.P., Paine, T.D., Bethke, J.A., Robb, K.L. and Hall, J. (1991) Evaluation of *Encarsia formosa* (Hymenoptera: Aphelinidae) for biological control of sweet-potato whitefly (Homoptera: Aleyrodidae) on poinsettia. *Environmental Entomology* 20, 713–719.

Qiu Y.T., van Lenteren, J.C., Drost, Y.C. and Posthuma-Doodeman, C.J.A. (2004) Life-history parameters of *Encarsia formosa*, *Eretmocerus eremicus* and *Er. mundus*, aphelinid parasitoids of *Bemisia argentifolii* (Hemiptera: Aleyrodidae). *European Journal of Entomology* 101, 83–94.

Ross, R. (1905) An address on the logical basis of the sanitary policy of mosquito reduction. *British Medical Journal* 1, 1024–1029.

Russell, L.M. (1975) Collection records of *Bemisia tabaci* (Gennadius) in the United States. (Hemiptera: Aleyrodidae). *USDA Cooperative Economic Insect Report* 25, 229–230.

Sandvol, L., Byrne, E.A., Byrne, D.N. and Homan, H.W. (1977) Green peach aphid. *Idaho Agricultural Experiment Station Current Information Series* 399, 1–2.

Sappington, T.W. and Showers, W.B. (1992) Reproductive maturity, mating status, and long duration flight behavior of *Agrotis ipsilon* (Lepidoptera: Noctuidae) and the conceptual misuse of the oogenesis flight syndrome by entomologists. *Environmental Entomology* 21, 677–688.

Showers, W.B. (1997) Migratory ecology of the black cutworm. *Annual Review of Entomology* 42, 393–425.

Showers, W.B., Whitford, F., Smelser, R.B., Keaster, A.J., Robinson, J.F., Lopez, J.D. and Taylor, S.E. (1989) Direct evidence for meteorologically driven long-range dispersal of an economically important moth. *Ecology* 70, 987–992.

Simmons, A.M., Abd-Rabou, S. and McCutcheon, G.S. (2002) Incidence of parasitoids and parasitism of *Bemisia tabaci* (Homoptera: Aleyrodidae) in numerous crops. *Environmental Entomology* 31, 1030–1036.

Simmons, G.S. and Minkenberg, O.P.J.M. (1994) Field cage evaluation of augmentative biological control of *Bemisia argentifolii* (Homoptera: Aleyrodidae) in Southern California cotton with the parasitoid *Eretmocerus* nr. *californicus* (Hymenoptera: Aphelinidae). *Environmental Entomology* 23, 1552–1557.

Sisterson, M.S., Carrière, Y., Dennehy, T.J. and Tabashnik, B.E. (2005) Evolution of resistance to transgenic crops: Interactions between insect movement and field distribution. *Journal of Economic Entomology* 98, 1751–1762.

Southwood, T.R.E. (1962) Migration of terrestrial arthropods in relation to habitats. *Biological Reviews* 37, 171–214.

Stansly, P.A., Calvo, F.J. and Urbaneja, A. (2005) Augmentative biological control of *Bemisia tabaci* biotype 'Q' in Spanish greenhouse pepper production using *Eretmocerus* spp. *Crop Protection* 24, 829–835.

Stukenbrock, E.H., Banke, S. and McDonald, B.A. (2006) Global migration patterns in the fungal wheat pathogen *Phaeosphaeria nodorum*. *Molecular Ecology* 15, 2895–2904.

Thomson, A.L. (1926) *Problems of Bird Migration*. Houghton Mifflin Company, Boston, Massachusetts.

US Environmental Protection Agency (USEPA) (2001) Biopesticides registration action document – *Bacillus thuringiensis* plant-incorporated protectants (http://www.epa.gov/pesticides/biopesticides/pips/bt_brad2/4-irm.pdf).

Van Driesche, R.M., Hoddle, M.A., Roy, S. and Sanderson, J.P. (2001) Effect of parasitoid release pattern on whitefly (Homoptera: Aleyrodidae) control in commercial poinsettia. *Florida Entomologist* 84, 63–69.

Veenstra, K.H. and Byrne, D.N. (1999) Does dispersal affect the reproductive physiology of the sweet potato whitefly, *Bemisia tabaci? Physiological Entomology* 24, 72–75.

Williams, C.B. (1957) Insect migration. *Annual Review of Entomology* 2, 163–180.

5

A Landscape Perspective in Managing Vegetation for Beneficial Plant–Pest– Natural Enemy Interactions: a Foundation for Areawide Pest Management

MICHAEL J. BREWER,[1] TAKUJI NOMA[1] AND NORMAN C. ELLIOTT[2]

[1]Integrated Pest Management Program, Department of Entomology, Michigan State University, East Lansing, Michigan, USA
[2]US Department of Agriculture, Agricultural Research Service, Stillwater, Oklahoma, USA

Introduction

In the USA, Europe and increasingly in other regions, cropping systems designed for high production output are significant features of the landscape. Deployment of mechanized and high-input cropping systems over the last 50 years has resulted in substantial transformation and fragmentation of major grassland, shrubland and woodland systems throughout the world. These cropping systems are typically less diverse in species composition, structure and ecological functioning than those found in the original plant community (Altieri, 2004). Decreases in plant diversity of agroecosystems (i.e. the crops themselves and surrounding remnants of the original plant system) have negatively affected ecosystem functions (Freemark, 2005). For agriculture, declines in agroecosystem diversity can result in increased crop herbivory and decreased beneficial organisms that feed on pests (Letourneau, 1998; Altieri, 2004).

Agricultural plant diversification is advocated as a remediation method to reverse these pest management challenges associated with modern cropping systems (Banks, 2000; Benton *et al.*, 2003; Altieri, 2004; Schmidt *et al.*, 2004), adding to other efforts to restore disturbed areas to their original plant community (Freemark, 2005). Mechanistically, this approach is based in part on outcomes of vegetation-driven plant–herbivore–natural enemy interactions predicted from the resource concentration, enemies, associational resistance and plant apparency hypotheses (Root, 1973; Banks, 2000; Altieri, 2004).

A landscape perspective can help refine vegetative-based management approaches to pest management locally within, or adjacent to, agricultural fields of interest (e.g. Vorley and Wratten, 1987; Murphy *et al.*, 1996; Banks, 2000). More recently, landscape ecologists have assessed the health of ecosystem services involving insects across vegetative conditions that extend to the neighbourhood and broader landscape (e.g. Kruess and Tscharntke, 1994; Marino and Landis, 1996; Duelli, 1997; Elliott *et al.*, 1998b; Fahrig and Jonsen, 1998; Thies *et al.*, 2005). In the young field of landscape ecology, studies of the effects of landscape elements on arthropod natural enemies of pest insects have come from predominantly forested regions that have been fragmented to various degrees by forest harvesting and other human activities (e.g. Roland and Taylor, 1997), while some studies consider crop–woodland landscapes (e.g. Menalled *et al.*, 1999) and, much less commonly, crop–grassland/shrubland landscapes (e.g. Elliott *et al.*, 1998a). Adding a landscape perspective provides ecosystem context in which plant–pest–natural enemy (PPNE) interactions must function, with more regional effects possibly impeding, enhancing or not affecting outcomes of species interactions at the lower organizational level of individual fields (Noss, 1990).

In a companion chapter, Byrne (this volume, Chapter 4) focused on dispersal and migration of insects and their importance in understanding the dynamics of pest spread across agroecosystems. We broaden the discussion to consider the relationships of PPNE interactions to vegetation, ranging from vegetation within agricultural fields (e.g. Nentwig, 1989), adjacent boundaries and fields (e.g. Vorley and Wratten, 1987; Dennis *et al.*, 2000) and in the broader landscape (e.g. Marino and Landis, 1996; Elliott *et al.*, 1998b; Thies *et al.*, 2005). Here, we pay special attention to examples from crop-grassland/shrubland landscapes to complement previous reviews and perspectives on crop–woodland and –forest landscapes (Roland and Taylor, 1997; Menalled *et al.*, 1999). Others have considered a landscape perspective to areawide application of mating disruption techniques (i.e. pheromone and sterile-male techniques) (Jones and Casagrande, 2000). A review of landscape characteristics and principles applicable to management of pests is provided.

We also introduce landscape analysis approaches for characterizing and assessing landscape composition, structure and scale of vegetation relevant to PPNE interactions. Throughout, we use a case example on regulation of cereal aphids by natural enemies in North America, and supplemental examples of similar PPNE systems in Europe. We propose that understanding of the role of vegetation in PPNE interactions increases with a landscape perspective and positions practitioners to best apply vegetation-based approaches in pest management, both locally and areawide.

A Landscape Perspective to Improve Pest Management

Composition of landscape elements

A landscape perspective of pest management considers PPNE interactions within the context of landscape elements, emphasizing vegetation as a principal element affecting these interactions (Banks, 2000; Altieri, 2004). Species lists of herbivores (key pests and possibly those causing incidental herbivory), their natural enemies, their host plants and the biological resources available for pests and natural enemies, along with their physiological and behavioural traits, are relevant to considering how biological control functions (Letourneau, 1998). Likewise, characteristics of crop cultivars, non-crop plants and pest variants in plant virulence are relevant in assessing how vegetation can reduce pest feeding and damage (Banks, 2000). One friendly addition to these details is the characterization of relevant abiotic conditions. Plants, herbivores and natural enemies, especially in temperate systems, must function within the temperature and moisture ranges of the region of interest and may provide clues to development and reproduction of pests and natural enemies (Jervis, 2005).

Species lists, species characterization and abiotic conditions allow initial cataloguing of PPNE interactions relevant to pest management. Yet estimations of presence and intensity of species, their traits and abiotic conditions in field studies are labour and knowledge intensive. For field assessment, key questions are: what compositional details are essential to gather and what surrogate measurements are reliable to gauge beneficial PPNE functioning (Duelli, 1997)? The extensive agricultural ecology literature provides a foundation to make judicious selection of key species, biological resources and abiotic conditions for measurement. After initial surveys and consultation of the literature, representatives of key taxa may be selected to measure abundance as an indicator of health of PPNE interactions. None the less, the diversity of pest and beneficial organism fauna and their interactions present challenges to predicting pest management outcomes (Sheehan, 1986).

Structure of landscape elements

Inclusion of vegetation structure can greatly help in understanding pest regulation. Structure of the ecosystem is delineated by the arrangement of land elements (e.g. hills, waterways, soil types, roadways) and the managed and unmanaged biological elements (e.g. arrangement of agricultural fields and borders, and non-crop patches and corridors). Standardized landscape metrics can be used to characterize patch size, spatial arrangement of vegetation patches and corridors, and the degree of saturation and mixing of vegetation types across a landscape (Elliott *et al.*, 1988a). Temporal patterns of cultivation of managed plants and the growth period of unmanaged plants are also relevant to understanding PPNE interactions, especially in temperate climates (Wissinger, 1997; Barbosa, 1998). Structural and temporal details have been used in qualitative assessments of PPNE interactions (e.g. Vorley and Wratten, 1987; Cowgill *et al.*, 1993; Murphy *et al.*, 1996; Ahern and Brewer, 2002), but their use in a quantitative assessment has been a more recent development.

Methodology to calculate landscape metrics from mapping products and insert them within an analytical framework to assess their relationship to PPNE interactions is improving as remote sensing and geographic information system (GIS) tools are applied. The process of gathering remotely sensed imagery from appropriate sensors for vegetation classification and classification into pertinent vegetation layers has been used to study PPNE interactions (Elliott *et al.*, 1998a). Imagery is obtained from various sources: satellite sensors, low-altitude aerial photography and ground survey products (Jensen, 2000). Land cover classification is facilitated by GIS software (e.g. IMAGINE, ERDAS, Inc., Atlanta, Georgia. USA); and landscape metrics can be calculated from the classified land cover products using landscape pattern analysis programs, such as FRAGSTATS (McGarigal and Marks, 1995).

As an example of quantification of composition and structure applied to pest management, farm- and regional-scale plant diversity in the wheat-growing west-central region of the Great Plains of North America was hypothesized to affect the abundance of two hymenopteran parasitoids of the key damaging aphid of wheat (*Diuraphis noxia*, the Russian wheat aphid) (Brewer *et al.*, 2005; Noma *et al.*, 2005). The range of vegetation variation in composition, grain size and spatial distribution was captured in a classified Landsat image of south-eastern Wyoming, western Nebraska and north-eastern Colorado, USA, based on methodology of Elliott *et al.* (1998a). A thematic map of the 14,000 km^2 study area was prepared from satellite imagery taken by Landsat 7 ETM + Scanner (US Geological Survey, Reston, Virginia) in June and August 2000. The images from the two dates were combined into 14 layers of spectral data, which were used to classify various ground cover types using IMAGINE (Elliott *et al.*, 1998a). The final classified map product depicted 12 land cover classes: wheat, fallow, sunflower, lucerne, maize, millet, other vegetation, Conservation Reserve Program grassland (Mitchell, 1988), other grasslands principally used for cattle grazing, riparian areas, urban areas and water (see Fig. 5.1). The thematic map had a 76% overall accuracy (ranging by 50–100% accuracy for each land cover class) in matching randomly selected classified pixels with ground-truth data (see Table 5.1).

Vegetation Diversity Gradient

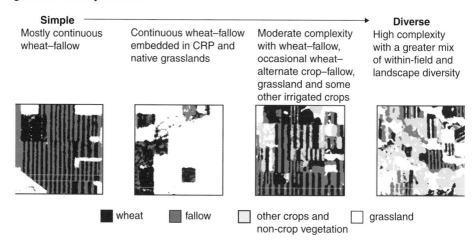

Fig. 5.1. Vegetation variation in composition, grain size and spatial distribution was captured in 5 × 5 km sections of landscapes detected in a classified Landsat image of the wheat-growing west-central region of the Great Plains of North America, based on the methodology of Elliott *et al.* (1998a). The original 12 land cover classifications were reclassified into four principal classes for depiction here.

Table 5.1. Percentages of various land cover types in heterogeneous and homogeneous vegetation regions surrounding wheat production farms of the wheat-growing west-central region of the Great Plains of North America.

| Land cover | Regional-scale diversity (%) | | Accuracy[a] |
	Heterogeneous	Homogeneous	
Grass-based vegetation	52.93 (3.00)[g]	74.38 (3.98)[h]	
Wheat	12.56 (2.40)	18.62 (4.41)	77.78
CRP grassland[b]	19.09 (1.70)	22.76 (4.93)	79.55
Other grassland[c]	21.27 (5.44)	33.99 (8.45)	76.79
Non-grass vegetation	32.60 (1.84)[g]	7.95 (1.42)[h]	
Sunflower	8.98 (4.15)	1.26 (0.60)	50.00
Lucerne	6.39 (1.72)	0.76 (0.36)	95.45
Maize	1.52 (1.07)	0.29 (0.21)	68.75
Millet	0.21 (0.16)	0.11 (0.06)	100.00
Other vegetation[d]	11.64 (1.45)	2.92 (0.68)	52.17
Riparian area[e]	3.87 (1.52)	2.61 (0.69)	100.00
Grass/non-grass ratio	1.68 (0.15)[g]	15.75 (5.95)[h]	
Other land cover[f]	14.46 (3.04)	17.67 (4.31)	77.08

The numbers are a mean percentage of the total patch areas (standard error) occupied by each land cover type within 25 km^2 circular areas represented in the region ($n = 8$ for each landscape type). Different letters within a row indicate a significant difference at $\alpha = 0.05$.
[a]Accuracy of classification in matching randomly selected classified pixels on the thematic map, based on ground-truth surveys.
[b]Grasslands managed for wildlife conservation, as sponsored by the US Department of Agriculture Conservation Reserve Program (Mitchell, 1988).
[c]Grasslands managed principally for cattle grazing.
[d]Patches consisting largely of unclassified crop or weedy vegetation.
[e]Patches consisting largely of shrubs and trees along a body of water.
[f]A combination of non-vegetation land cover types (fallow, urban and water) that was excluded from analysis.

The land cover classification was used to select a vegetation gradient that represented extant farm-scale plant diversity that was being managed by wheat farmers and regional-scale diversity that was affected by acreage allotments to farmers participating in a conservation programme of the US Department of Agriculture (i.e. Conservation Reserve Program (Mitchell, 1988)). In simple farms, the crop rotation was a series of spatially alternating wheat and fallow strips of 30–60 m in width (wheat–fallow). In diverse farms, the cropping area consisted of a series of wheat, alternative spring-sown crop and fallow strips (wheat–alternate crop–fallow) (see Fig. 5.1). The regional-scale landscapes were selected to represent relatively heterogeneous or homogeneous regions in which farm sites were nested based on degree of grass-based vegetation.

We used 25 km^2 circular regions (5.6 km diameter) with the farm as proximate centre to evaluate the regional vegetation. Within each circle, total patch areas for each vegetation class on the thematic map were quantified using FRAGSTATS (McGarigal and Marks, 1995). Based on the thematic map, the patch area surrounding the farm sites was classified as either heterogeneous or homogeneous in regional-scale vegetation diversity (see Fig. 5.1, Table 5.1). The homogeneous regions consisted of relatively large areas of grass-based vegetation (combination of wheat, Conservation Reserve Program grasslands and other grasslands) and small areas of non-grass vegetation (combination of other agriculture and riparian areas), while the heterogeneous regions consisted of relatively small grass-based vegetation and large non-grass vegetation (see Table 5.1). Through this process we categorized four combinations of farm-scale and regional-scale diversity in the 14,000 km^2 study area, as linked to two scales appropriate to farm-level management and regional agricultural programme management (see Fig. 5.1).

As a note of caution, quality of classification of land cover (e.g. plant species, non-crop land management type, crop type) varies considerably across mapping products. In our example the classifications were derived from satellite imagery. We acknowledge that the mechanics of classification into pertinent vegetation layers can be laborious and require specialized computing, software and human resources (Elliott *et al.*, 1998a). Standardized mapping products with refined crop and vegetation data layer information are welcome tools, and are becoming more widely available (USDA, 2007). We anticipate that high-quality, standardized cropland data layer products will facilitate the broader use of structure of vegetation in understanding PPNE interactions. Low-altitude photography has also been used, and its finer grain may be useful in identifying vegetation corridors and other landscape features that are not easily differentiated from satellite imagery (Jensen, 2000).

Function of landscape elements

Plant–pest–natural enemy interactions can be affected by structural and temporal patterns of vegetation in the ecosystem. In general, the composition, grain size and spatial and temporal arrangement of the vegetation – and, potentially, other elements that comprise a landscape – may play important roles in determining an organism's population size. For agricultural pests, studies have shown that parasitism and predation rates on pest insects tend to be higher and crop damage lower in structurally diverse agricultural landscapes than in simplified landscapes (Menalled *et al.*, 1999; Altieri, 2004), although this is not a certain outcome across systems (Sheehan, 1986). This relationship is probably not the result of landscape diversity per se, but rather depends on whether specific requisites of natural enemies, as well as pests, are more or less likely to be present and accessible in a diverse spatial mosaic of habitats than in a landscape with few habitat types that are accessible to natural enemies (Menalled *et al.*, 1999). This assumption is consistent with hierarchy theory, in which higher organizational levels (i.e. composition and structure of vegetation of an area) constrain the interactions at lower levels (i.e. specific PPNE interactions in a cropped field) (Noss, 1990). From a landscape perspective, the metrics of key and surrogate elements may serve as important indicators of the health of PPNE interactions.

From a practitioner's viewpoint, if only a few compositional elements are key to pest management in a simple landscape structure (e.g. a widely planted monoculture that dominates the landscape), then understanding of key mechanistic functions driving the interaction may be sufficient to evaluate approaches to pest management. For example, when the organisms have limited or highly preferred biological resources to use along with limited mobility, mechanistic approaches may reveal opportunities to manage a few key vegetation elements in order to improve ecological functioning of biological control agents.

A landscape perspective is still relevant, particularly if movement between resources is needed (Wissinger, 1997), and a qualitative assessment may be completely satisfactory. Through use of baffled water traps to detect parasitoid movement, Vorley and Wratten (1987) found that barley and early-sown wheat adjacent to late-sown wheat served as a significant source of hymenopteran parasitoids (*Aphidius* spp.) to control potentially damaging levels of cereal aphids populating late-sown wheat. Also in England, Cowgill *et al.* (1993) found that flowering, non-crop plants next to cereal crops increased the abundance of adults and eggs of the syrphid *Episyrphus balteatus* (Degeer) in winter wheat. In the USA, Ahern and Brewer (2002) found that addition of spring-sown sunflower into a strip rotation of winter wheat and fallow increased the abundance of several hymenopteran parasitoids (Braconidae and Aphelinidae) that attack the key cereal aphid, *Diuraphis noxia* (Mordvilko). And, as an example outside the cereal aphid system, Murphy *et al.* (1996) found that early-season abundance of an egg parasitoid, *Anagrus epos* (Girault), of the grape leafhopper increased twofold when prune trees were near vineyards. The landscape features of proximity, prevailing wind direction and seasonality of biological resources were key compositional, physiological and behavioural attributes of these studies.

When multiple compositional, physiological and behavioural attributes are relevant (e.g. broad host ranges, multiple biological resources, high mobility, varied abiotic conditions), both composition and structure of vegetation elements may affect PPNE interactions. It is this situation where a landscape quantitative assessment may be most valuable in assessing the relative importance of vegetation structure to PPNE interactions. If there are common relevant features or surrogates to a larger relevant group of compositional elements that can be classified, landscape analysis techniques may help assign (at least in sign if not in intensity) probable pest management-based outcomes. For example, both Thies *et al.* (2005) in Germany and Menalled *et al.* (1999) in the USA found that complex landscapes were associated with higher parasitism of herbivores in agricultural lands than in agricultural lands nested in simpler regional vegetation.

It is the potential for multifactor interactions that makes a solely mechanistic approach to devising vegetation management recommendations prone to difficulties in assessing interactions experimentally and prone to unexpected consequences. As an example, lack of improvement of pest management services or even undesirable outcomes, such as increased pest pressure, are possible if the addition of vegetation elements benefits the pest organisms and overshadows benefits to plant or natural enemy regulation of the pests. This concern may be particularly relevant when managing polyphagous and mobile pests with specialized and less mobile natural enemies. In England, Vorley and Wratten (1987) recognized that the benefit of early-sown grains to increase parasitoids must be balanced against the potential increased risk of barley yellow dwarf virus in the cereal-based system.

In Germany, Thies *et al.* (2005) noted that the increases in cereal aphid parasitism in complex landscapes were offset by higher aphid colonization in the same complex landscapes.

Scale and pattern of landscape elements

The effects of spatial scale and temporal patterns of landscape elements on PPNE interactions are more recently appreciated topics in conservation biological control (Letourneau, 1998) and the deployment of cropping system strategies (Helenius, 1997; Benton *et al.*, 2003). As noted above, functioning of PPNE interaction may be associated with the structure and composition of within-field vegetation (e.g. Nentwig, 1989; Ahern and Brewer, 2002), adjacent agricultural fields and field borders (Vorley and Wratten, 1987; Cowgill *et al.*, 1993), and the broader regional structure of remnants of the original plant community and managed cropped and noncropped areas (Marino and Landis, 1996; Elliott *et al.*, 1998b; Letourneau, 1998).

In addition, the seasonal nature of natural and managed vegetation (temporal vegetation structure), especially in temperate zones, may have a strong effect on PPNE interactions (Wissinger, 1997; Barbosa, 1998). In addition, one or more spatial scales from highly local arrangements of specific plant species to more regional arrangement of general vegetation classes (e.g. mixes of plant communities and cropped fields) may affect the functioning of PPNE interactions. More defined scale and magnitude of landscape effects are most probably related to the organism's habitat and foraging characteristics (Dunning *et al.*, 1992; Fahrig and Merriam, 1994). It follows that species with different needs and behaviours will be affected differently by the scale in landscape structure brought about by natural processes, such as disturbance and succession, or by humans, such as cropping system deployment and implementation of vegetation-based farm practices.

In our North American cereal aphid example (see Fig. 5.2, Table 5.2), two dominant parasitoids, *Aphelinus albipodus* Hayat & Fatima (Hymenoptera: Aphelinidae) and *Lysiphlebus testaceipes* (Cresson) (Hymenoptera: Braconidae, Aphidiinae), differ in physiological and behavioural attributes (see Table 5.2).

These biological characteristics may be useful in deriving hypotheses on the relative responsiveness of parasitoids to changes in farm- and regional-scale plant diversity found throughout the wheat production area of this region (see Fig. 5.1). Based on these attributes, a reasonable hypothesis is that *L. testaceipes* would be more sensitive to neighbourhood and regional plant diversity because of its mobility, large host aphid range and adult food requirements. The responsiveness of *A. albipodus*, a representative *Aphelinus* sp., to plant diversity may not extend beyond the farm-scale strategy to add a spring-sown grain to the traditional wheat–fallow strip rotation. *Aphelinus* spp. are less mobile, have a smaller host aphid range and can feed on aphid hosts more effectively than *L. testaceipes*. Alternatively, these differences may not be sufficiently large to show differential responses between the species to the two agricultural landscape scales of interest: farm-scale plant diversity managed by wheat farmers and regional-scale diversity affected by acreage allotments to farmers participating in a federally sponsored conservation programme. The dilemma for pest managers is that they work with a diverse fauna, both in composition (number of species) and in the

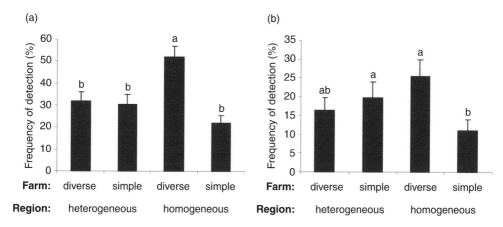

Fig. 5.2. Two experimental designs for analysing the effects of scale heterogeneity on populations: (a) factorial design (2 × 2 factor design shown here) allowing estimates of relative and joint contributions of local (farm) and regional (areawide) spatial scales of special interest; and (b) regression approach (special-interest local zone surrounded by differing regional conditions shown here) allowing estimates of scale most relevant to ecological functioning of organisms of interest in the local zone.

Table 5.2. The range of physiological and behavioural attributes of two hymenopteran parasitoids (*Lysiphlebus testaceipes* and *Aphelinus* sp.) that prey upon *Diuraphis noxia* and other aphids in the west-central region of the North American Great Plains.

Biological characteristic	Parasitoid	
	L. testaceipes	*Aphelinus* sp.
Mobility	Moderate mobility – flight is common[a]	Low mobility – mostly by walking[b]
Response to attractants	Prey and plant volatiles[c]	Less known, less indication of response to volatiles[d]
Host range	More host species[e, f]	Fewer host species[f, g]
Adult food sources	No aphid host feeding, aphid honeydew[h]	Aphid host feeding, aphid honeydew[i]

[a]Fernandes *et al.* (1997); [b]Mason and Hopper (1997); [c]Schuster and Starks (1974); [d]De Farias and Hopper (1997); [e]Pike *et al.* (2000); [f]Kaiser *et al.* (2007); [g]Elliott *et al.* (1999); [h]Quicke (1997); [i]Boyle and Barrows (1978).

variety of physiological needs and behaviours, and a diverse agricultural landscape. This diversity nevertheless provides opportunities to optimize management approaches, locally and areawide.

The composition and quality of a plant community across a landscape both change seasonally with plant phenology and cultivation practices, especially in the temperate agricultural regions of the world. In our North American Great Plains cereal aphid example, winter wheat strips are mature or harvested during summer, resulting in a greatly reduced function as habitats of cereal aphids and aphid parasitoids (Brewer *et al.*, 2005). During this time period, spring-sown crops and some non-crop plants in grasslands are available, some of which harbour aphids known to be used by parasitoids of *D. noxia* (Donahue *et al.*, 2000; Brewer *et al.*, 2005). In contrast, wheat strips harbour aphids in the spring, when spring-sown crop plants are not available (Brewer *et al.*, 2005). Thus, quality and relative suitability of wheat strips and other vegetation as habitats of aphid and aphid parasitoids change as seasons progress. Increasing cereal aphid parasitoids early in the season by planting early-sown cereals adjacent to late-sown cereals (Vorley and Wratten, 1987) is another example where temporal patterns in landscape elements may play important roles in PPNE interactions.

Analytical Approaches to Discerning Local and Regional Landscape Effects

The use of landscape analysis methods in discerning the relevance of scale and pattern of vegetation provides great opportunity to transition to a more quantitative assessment for planning cropping system deployment and adoption of vegetation-based farm practices, both locally and areawide (Elliott *et al.*, 1998b; Thies *et al.*, 2005). In our North American Great Plains cereal aphid example, the classification of farm-scale diversity and regional-scale diversity reflected the extant variation in wheat production (crop strip rotation) and regional agricultural land use in the west-central Great Plains of North America. The large study region allowed consideration of scale effects through a factorial design. For this study, scales are nested in each other: two types of crop strip rotations used on farms are nested in regional agricultural land use that we categorized in two classes. All possible combinations of the levels within each scale of interest were considered in this 2×2 (farm-scale diversity $\times$ regional-scale diversity) factorial (see Fig. 5.3a). The factorial structure was appropriate in assessing the joint effects of the two landscape scales on the abundance of the two primary parasitoids of the key wheat-damaging aphid.

Our farm scale was within the range of scales found by Vorley and Wrattn (1987) and Thies *et al.* (2005) to be significant to cereal aphid parasitoid functioning. Our regional scale was designed to capture surrounding vegetation patches which are typically perceived in US land survey units of 1.6×1.6 km squares (sections) to 9.6×9.6 km squares (townships) for land use planning (such as the Conservation Reserve Program) and gathering of agricultural land use statistics (Elliott *et al.*, 1998a; USDA, 2005). The farm-scale vegetation diversity (evaluated by the type of wheat-based crop system used) had a greater effect on parasitoid prevalence than the regional-scale vegetation diversity (see Fig. 5.3a). The findings were consistent with those of Thies *et al.* (2005), who determined that landscape structure at the spatial scale of 0.5–2.0 km (approximates our farm scale) had the most significant influence on cereal aphid parasitoids.

In addition, the farm- and regional-scale factorial design of plant diversity revealed that parasitoid abundance in homogeneous areas especially benefited from

(a)   (b)

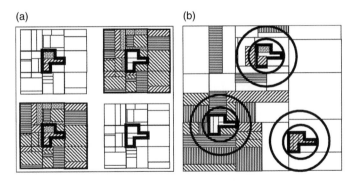

Fig. 5.3 Prevalence of two cereal aphid parasitoids, *Aphelinus albipodus* (a) and *Lysiphlebus testaceipes* (b), sampled in the wheat production region of the west-central Great Plains of North America, August 2001, using the methodology of Noma *et al.* (2005). There was a significant interaction on prevalence of the parasitoids (a, $\chi^2 = 1.07$, df = 1, $P = 0.0009$; b, $\chi^2 = 6.40$, df = 1, $P = 0.01$) across a farm (simple and diverse farms) and regional (homogeneous and heterogeneous areas) diversity gradient (see Fig. 5.1 and Table 5.1 for further details). Different letters indicate significant differences between means. Error bars represent standard errors of means. No aphids were found in wheat strips after harvest by August, while aphids (*Aphis helianthi*) were present in the spring-sown sunflower strips at the densities of 0.4–1.8 aphids per plant.

addition of farm-scale plant diversity in the form of adding a spring-sown crop into the wheat–fallow strip rotation (see Fig. 5.3a). These vegetation diversity–parasitoid interactions found in August 2001 were similar between the two parasitoid species (*A. albipodus* and *L. testaceipes*) (see Fig. 5.3a). The results supported the theory that similar spatial scales were relevant for the prevalence of both parasitoids, representative of the two major families of aphid parasitoids, and both parasitoids benefited by plant diversity in similar manners.

An alternative to this factorial approach is a regression approach, which has been used in several studies of landscape effects on populations or communities (e.g. Elliott *et al.*, 1998b; Thies *et al.*, 2005). In the regression method, spatial extent is varied using hierarchically increasing spatial extent, where landscape composition and pattern variables are measured for each experimental unit at each extent (see Fig. 5.3b). The ecological variable of interest is regressed against landscape variables measured at each particular spatial extent to determine the scale accounting for the greatest amount of variation.

For example, landscape effects on insect parasitism were assessed at multiple spatial scales in a temperate forestland modified by intensive agriculture (Schmidt *et al.*, 2004; Thies *et al.*, 2005). The landscape sector encompassing each insect sampling site was quantified for five land use types (arable land, grassland, forest, hedgerow and garden land/settlement), using digitized thematic maps. The land use types were quantified within concentric circles of seven spatial scales ranging from 0.5–6.0 km diameter, with the sampling site as the centre point. A multiple regression was performed for each spatial scale in which the response variables (aphid density and aphid parasitism rate) were regressed against predictive factors, including a landscape variable (percentage of arable land). By comparing *F*-statistics and levels of significance among the seven spatial scales analysed, the scale associated with the highest explanatory power in the regression was determined.

The advantage of using a classification gradient in a complete factorial design is its ability to show relative contributions and interactions (joint effects) between scales that are relevant to agricultural management interests. Using the North American Great Plains cereal aphid example, a more complex wheat-based system that included a spring-sown annual crop was a good strategy to promote parasitoids, and the approach was especially important in the more grass-based homogeneous regions of the study area (see Fig. 5.3).

The implication from a cropping system perspective is that farmers, especially in highly homogeneous vegetation areas, can enhance parasitoids by diversifying their wheat strip crop system. Schmidt *et al.* (2005) also utilized a factorial design to differentiate effects of local management and wider landscape context on ground-dwelling farmland spiders. In contrast, the regression approach has benefits in finer-scale discrimination of the functioning of PPNE interactions. Thies *et al.* (2005) concluded that smaller spatial scales were more relevant for cereal aphid parasitoids (0.5–2.0 km) as compared with spatial scales relevant for dispersal of cereal aphids (up to 6.0 km) within the spatial scales studied. Not surprisingly, the major finding of the two analytical approaches in these studies of cereal aphids was consistent, with the differences reflecting the intent of the studies. For both, local vegetation, whether actively managed by a farmer or extant, is the scale most closely associated with parasitoid abundance.

Practitioner Support in Areawide Application of Vegetation-based Management

In practice, within-field and near-field manipulation of vegetation has benefited from understanding of PPNE interactions (Powell, 1986; Barbosa, 1998). Regionally, extant vegetation may be associated with differing risks to pests, because of different levels of pest management service related to different levels of plant diversity found across the agroecosystem (Marino and Landis, 1996; Elliott *et al.*, 1998b; Thies *et al.*, 2005). Conceptually, this information is useful in encouraging farmer adoption of land management practices and regional land use planning that will be most likely to preserve and enhance pest management services, as well as to reverse the trend of biodiversity loss in major agricultural zones of the world.

From a practitioner perspective, pest managers are being encouraged through incentives mechanisms (Casey *et al.*, 1999) and challenged through regulatory mechanisms (Johnson and Bailey, 1999) to adopt ecologically and vegetation-based IPM practices. The European Union, Canada and the USA, among others, have begun to institute conservation policies affecting growers (Casey *et al.*, 1999; Anon., 2006; Hoard and Brewer, 2006). Both financial and technical assistance through governmental conservation programmes are available to growers to encourage adoption of specific IPM practices on farms that are linked to conservation of natural resources and ecosystem services (Anon., 2006; Hoard and Brewer, 2006). More detailed regional planning efforts to optimize ecosystem services, including pest management, are less structured in governmental programmes, but the potential impact of widespread grower participation in such programmes, such as the impact on agriculture and wildlife conservation of the Conservation Reserve Program in the Great Plains of North America (Mitchell, 1988), cannot be understated.

Summary of Value of a Landscape Perspective to Pest Management

Understanding the role of vegetation may facilitate on-farm, vegetation-based recommendations to improve pest management, assessment of benefits of regional plant diversity to pest management, or both. The former, specific vegetation-based recommendations for grower adoption, certainly have on-farm value. The latter has obvious implications for areawide pest management, either accumulating the effects of local vegetation structure in and around agricultural fields or in a synergistic or detrimental fashion where regional plant diversity constrains the interactions at lower organizational levels (Noss, 1990).

For areawide pest management application, local vegetation management recommendations applied regionally may show simple additive improvements to pest management, or the regional vegetation composition and structure may further enhance (or impede) beneficial PPNE interactions. Neutral or enhanced benefits serve areawide pest management, although the potential for capturing pest management enhancements areawide is of special interest to planning cropping system deployment strategies and adoption of vegetation-based farm practices. In our cereal aphid examples, the local farm-scale effect of vegetation management was clear in work from Germany, the UK and North America. And, in the case of the North American example, farm-scale crop diversification had special appeal in areas where the vegetation was regionally homogeneous.

The dilemma for pest managers interested in areawide pest management and vegetation-based management approaches is that opportunities and complexity are probably highest when there are available a diverse fauna and diverse agricultural landscape. This diversity begs the question of how we may use composition and structure of extant managed and unmanaged lands to support pest management services; and how additional farm- or regional-scale management shifts can further benefit pest management. A landscape perspective, and a trend toward more quantitative analytical methods and more readily accessible land cover products, may become increasingly valuable as conservation and other societal interests encourage practitioners to use vegetation management as a tool to manage pests, locally and areawide.

Acknowledgements

We appreciate the financial support of the USDA CSREES National Research Initiative, Biologically based Pest Management Program (grants 2000- 02559 and 2002- 04573).

We thank the University of Wyoming, grower cooperators and students (M. DeWine, S. Grabowski, K. Hoff, A. Kelsey and S. Yan) for support and assistance in carrying out field studies in the wheat-growing west-central region of the Great Plains of North America. We thank Tim Johnson for producing Fig. 5.2. The views expressed herein are those of the authors only and do not necessarily reflect or represent the opinions, positions or policies of the US Department of Agriculture.

References

Ahern, R.G. and Brewer, M.J. (2002) Effect of different wheat production systems on the presence of two parasitoids (Hymenoptera: Aphelinidae, Braconidae) of the Russian wheat aphid in the North American Great Plains. *Agriculture, Ecosystems and Environment* 92, 201–210.

Altieri, M.A. (2004) *Biodiversity and Pest Management in Agroecosystems.* Food Products Press, New York, 236 pp.

Anon. (2006) *Project Eligibility Guidelines for Environmental Cost-share Programs Available to Farmers through the Canada-Ontario Environmental Farm Plan.* Agriculture and Agri-Food Canada, Guelph, Ontario, Canada, 23 pp.

Banks, J.E. (2000) Natural vegetation in agroecosystems: pattern and scale of heterogeneity. In: Ekbom, B., Irwin, M. and Robert, Y. (eds) *Interchanges of Insects between Agricultural and Surrounding Landscapes.* Kluwer Press, Dordrecht, Netherlands, pp. 215–229.

Barbosa, P. (1998) *Conservation Biological Control.* Academic Press, San Diego, California, 396 pp.

Benton, T.G., Vickery, J.A. and Wilson, J.D. (2003) Farmland biodiversity: is habitat heterogeneity the key? *Trends in Ecology and Evolution* 18, 182–188.

Boyle, H. and Barrows, E.M. (1978) Oviposition and host feeding behavior of *Aphelinus asychis* (Hymenoptera: Chalcidoidea: Aphelinidae) on *Schizaphis graminum* (Homoptera: Aphididae) and some reactions of aphids to this parasite. *Proceedings of the Entomological Society of Washington* 80, 441–455.

Brewer, M.J., Noma, T. and Elliott, N.C. (2005) Hymenopteran parasitoids and dipteran predators of the invasive aphid *Diuraphis noxia* after enemy introductions: temporal variation and implication for future aphid invasions. *Biological Control* 33, 315–323.

Casey, F., Schmits, A., Swinton, S. and Zilberman, D. (1999) *Flexible Incentives for the Adoption of Environmental Technologies in Agriculture.* Kluwer Academic Publishers, Norwell, Massachusetts, 370 pp.

Cowgill, S.E., Wratten, S.D. and Sotherton, N.W. (1993) The effects of weeds on the numbers of hoverfly (Diptera: Syrphidae) adults and the distribution and composition of their eggs in winter wheat. *Annals of Applied Biology* 123, 499–515.

De Farias, A.M.I. and Hopper, K.R. (1997) Responses of female *Aphelinus asychis* (Hymenoptera: Aphelinidae) and *Aphidius matricariae* (Hymenoptera: Aphidiidae) to host and plant-host odors. *Environmental Entomology* 26, 989–994.

Dennis, P., Fry, G.L.A. and Anderson, A. (2000) The impact of field boundary habitats on the diversity and abundance of natural enemies in cereals. In: Ekbom, B., Irwin, M. E. and Robert, Y. (eds) *Interchanges of Insects Between Agricultural and Surrounding Landscapes.* Kluwer Academic Publishers, Dordrecht, The Netherlands, pp. 195–214.

Donahue, J.D., Brewer, M.J. and Burd, J.D. (2000) Relative suitability of crested wheatgrass and other perennial grass hosts for the Russian wheat aphid (Homoptera: Aphididae). *Journal of Economic Entomology* 93, 323–330.

Duelli, P. (1997) Biodiversity evaluation in agricultural landscapes: an approach at two different scales. *Agriculture Ecosystems and Environment* 62, 81–91.

Dunning, J.B., Danielson, B.J. and Pulliam, H.R. (1992) Ecological processes that affect populations in complex landscapes. *Oikos* 65, 169–175.

Elliott, N.C., Hein, G.L., Carter, M.R., Burd, J.D., Holtzer, T.O., Armstrong, J.S. and Waits, D.A. (1998a) Russian wheat aphid (Homoptera: Aphididae) ecology and modeling in Great Plains agricultural landscapes. In: Quisenberry, S.S. and Peairs, F.B. (eds) *Response Model for an Introduced Pest – the Russian Wheat Aphid.* Proceedings Thomas Say Publications in Entomology, Entomological Society of America, Lanham, Maryland, pp. 31–64.

Elliott, N.C., Kieckhefer, R.W., Lee, J. and French, B.W. (1998b) Influence of within-field and landscape factors on aphid predator populations in wheat. *Landscape Ecology* 14, 239–252.

Elliott, N.C., Lee, J.H. and Kindler, S.D. (1999) Parasitism of serveral aphid species by *Aphelinus asychis* (Walker) and *Aphelinus albipodus* Hayat and Fatima. *Southwestern Entomologist* 24, 5–12.

Fahrig, L. and Jonsen, I. (1998) Effect of habitat patch characteristics on abundance and diversity of insects in an agricultural landscape. *Ecosystems* 1, 197–205.

Fahrig, L. and Merriam, G. (1994) Conservation of fragmented populations. *Conservation Biology* 8, 50–59.

Fernandes, O.A., Wright, R.J., Baumgarten, K.H. and Mayo, Z.B. (1997) Use of rubidium to label *Lysiphlebus testaceipes* (Hymenoptera: Braconidae), a parasitoid of greenbugs (Homoptera: Aphididae), for dispersal studies. *Environmental Entomology* 26, 1167–1172.

Freemark, K. (2005) Farmlands for farming and nature. In: Wiens, J.A. and Moss, M.R. (eds) *Issues and Perspectives in Landscape Ecology.* Cambridge University Press, Cambridge, UK, pp. 193–200.

Helenius, J. (1997) Spatial scales in ecological pest management (EPM): importance of regional crop rotations. *Biological Agriculture and Horticulture* 15, 163–170.

Hoard, R.J. and Brewer, M.J. (2006) Adoption of pest, nutrient, and conservation vegetation management using financial incentives provided by a US Department of Agriculture Conservation Program. *HortTechnology* 16, 306–311.

Jensen, J.R. (2000) *Remote Sensing of the Environment: an Earth Resource Perspective.* Prince Hall, Upper Saddle River, New Jersey, 544 pp.

Jervis, M.A. (2005) *Insects as Natural Enemies.* Springer, Dordrecht, Netherlands, 748 pp.

Johnson, S.L. and Bailey, J.E. (1999) Food Quality Protection Act 1996: major changes to the Federal Food, Drug, and Cosmetic Act; the Federal Insecticide, Fungicide, and Rodenticide Act and impacts of the changes to pesticide regulatory decisions. In: Ragsdale, N.N. and Seiber, J.N. (eds) *Pesticides: Managing Risks and Optimizing Benefits.* American Chemistry Society, Washington, DC, pp. 8–15.

Jones, O.T. and Casagrande, E.D. (2000) The use of semiochemical-based devices and formulations in area-wide programmes: a commercial perspective. In: Tan, K. (ed.) *Area-wide Control of Fruit Flies and Other Insect Pests.* Penerbit Universiti Sains Malaysia, Penang, Malaysia, pp. 285–294.

Kaiser, M.E., Noma, T., Brewer, M.J., Pike, K.S., Vockeroth, J.R. and Gaimari, S.D. (2007) Hymenopteran parasitoids and dipteran predators found using soybean aphid after its Midwestern United States invasion. *Annals of the Entomological Society of America* 100, 196–205.

Kruess, A. and Tscharntke, T. (1994) Habitat fragmentation, species loss, and biological control. *Science* 264, 1581–1584.

Letourneau, D.K. (1998) Conservation biology: lessons for conserving natural enemies. In: Barbosa, P. (ed.) *Conservation Biological Control.* Academic Press, San Diego, California, pp. 9–38.

Marino, P.C. and Landis, D.A. (1996) Effect of landscape structure on parasitoid diversity and parasitism in agroecosystems. *Ecological Applications* 6, 276–284.

Mason, P.G. and Hopper, K.R. (1997) Temperature dependence in locomotion of the parasitoid *Aphelinus asychis* (Hymenoptera: Aphelinidae) from geographical regions with different climates. *Environmental Entomology* 26, 1416–1423.

McGarigal, K. and Marks, B.J. (1995) FRAGSTATS: spatial pattern analysis program for quantifying landscape structure. USDA Forest Service General Technical Report, PNW-351, Fort Collins, Colorado, 134 pp.

Menalled, F.D., Marino, P.C., Gage, S.H. and Landis, D.A. (1999) Does agricultural landscape structure affect parasitism and parasitoid diversity? *Ecological Applications* 9, 634–641.

Mitchell, J.E. (1988) *Impacts of the Conservation Reserve Program in the Great Plains: Symposium Proceedings.* USDA Forest Service General Technical Report RM-158, Rocky Mountain Forest and Range Experiment Station, Fort Collins, Colorado, 134 pp.

Murphy, B.C., Rosenheim, J.A. and Granett, J. (1996) Habitat diversification for improving biological control: abundance of *Anagrus epos* (Hymenoptera: Mymaridae) in grape vineyards. *Environmental Entomology* 25, 495–504.

Nentwig, W. (1989) Augmentation of beneficial arthropods by strip-management II: successional strips in a winter wheat field. *Journal of Plant Diseases and Protection* 96, 89–99.

Noma, T., Brewer, M.J., Pike, K.S. and Gaimari, S.D. (2005) Hymenopteran parasitoids and dipteran predators of *Diuraphis noxia* in the west-central Great Plains of North America: species records and geographic range. *BioControl* 50, 97–111.

Noss, R.F. (1990) Indicators for monitoring biodiversity: a hierarchical approach. *Conservation Biology* 4, 355–364.

Pike, K.S., Starý, P., Miller, T., Graf, G., Allison, D., Boydston, L. and Miller, R. (2000) Aphid parasitoids (Hymenoptera: Braconidae: Aphidiinae) of North-west USA. *Proceedings of the Entomological Society of Washington* 102, 688–740.

Powell, W. (1986) Enhancing Parasitoid Activity in Crops. In: Waage, J. and Greathead, D. (eds) *Insect Parasitoids.* Academic Press, London, pp. 319–340.

Quicke, D.L.J. (1997) *Parasitic Wasps.* Chapman & Hall, London, 470 pp.

Roland, J. and Taylor, P.D. (1997) Insect parasitoid species respond to forest structure at different spatial scales. *Nature* 386, 710–713.

Root, R.B. (1973) Organization of a plant–arthropod association in simple and diverse habitats: the fauna of collards (*Brassica oleracea*). *Ecological Monographs* 43, 95–124.

Schmidt, M.H., Thewes, U., Thies, C. and Tscharntke, T. (2004) Aphid suppression by natural enemies in mulched cereals. *Entomologia Eperimentalis et Applicata* 113, 87–93.

Schmidt, M.H., Roschewitz, I., Thies, C. and Tscharntke, T. (2005) Differential effects of landscape and management on diversity and density of ground-dwelling farmland spiders. *Journal of Applied Ecology* 42, 281–287.

Schuster, D.J. and Starks, K.J. (1974) Response of *Lysiphlebus testaceipes* in an olfactometer to a host and a non-host insect and to plants. *Environmental Entomology* 3, 1034–1035.

Sheehan, W. (1986) Response by specialist and generalist natural enemies to agroecosystem diversification: a selective review. *Environmental Entomology* 15, 456–461.

Thies, C., Roschewitz, I. and Tscharntke, T. (2005) The landscape context of cereal aphid–parasitoid interactions. *Proceedings of the Royal Society. B, Biological Sciences* 272, 203–210.

USDA (2005) Agricultural statistics database. US Department of Agriculture, Beltsville, Maryland (http://www.nass.usda.gov/Census_of_Agriculture/index.asp).

USDA (2007) Cropland data layer. US Department of Agriculture, Beltsville, Maryland (http://www.nass.usda.gov/research/Cropland/SARS1a.htm).

Vorley, V.T. and Wratten, S.D. (1987) Migration of parasitoids (Hymenoptera: Braconidae) of cereal aphids (Hemiptera: Aphididae) between grassland, early-sown cereals and late-sown cereals in southern England. *Bulletin of Entomological Research* 77, 555–568.

Wissinger, S.A. (1997) Cyclic colonization in predictably ephemeral habitats: a template for biological control in annual crop systems. *Biological Control* 10, 4–15.

6 Social and Economic Aspects of Areawide Pest Management

SEAN P. KEENAN[1] AND PAUL A. BURGENER[2]

[1]Department of Entomology and Plant Pathology, Oklahoma State University, Stillwater, Oklahoma, USA
[2]Department of Agricultural Economics, Panhandle Research and Extension Center, University of Nebraska-Lincoln, Scottsbluff, Nebraska, USA.

Introduction

Areawide pest management (AWPM) programmes build upon past achievements in agricultural innovation, expanding the implementation of integrated pest management (IPM) practices to larger geographical scales (Kipling, 1980; Kogan, 1998). Implementation on a broad geographical scale means that social, institutional and financial capital must be dedicated to the task:

> Social, political, and economic factors must come together with science before an areawide program can succeed. In addition, scientific challenges include defining the appropriate geographical area, selecting the control approaches to test and combine, and addressing the different life cycles of the target pest as well as secondary pests.
>
> (Faust, 2001)

Because AWPM programmes have typically relied upon voluntary adoption, pest management practices must demonstrate economic advantage to farmers over their existing practices. Adoption will also be facilitated if AWPM practices have low complexity, ease of trial adoption, rapidly observable results and high compatibility with other aspects of farm management (see Rogers, 2003).

This chapter explores demonstration elements from the US Department of Agriculture, Agricultural Research Service's demonstration programme for cereal aphid AWPM. We explore elements of the demonstration programme as agricultural innovations. We discuss potential adoption of these elements by wheat producers on the Great Plains and implications of the programme outcomes for other wheat-growing regions of the world. We begin with a history of research on the adoption of agricultural innovations, which provides the context for evaluating the cereal aphid programme from the perspective of farmer adoption. Chapter 19, this volume, by

Kristopher Giles *et al.* provides a summary and assessment of the research and technological developments of the cereal aphid AWPM programme.

Adoption of Agricultural Innovations

The social and economic challenges confronting AWPM are the same as those confronted by promoters of past agricultural innovations. An innovation may be any idea, practice or object whose adoption is 'new' to a group of potential adopters (Rogers, 2003). That is, application of the innovation can be 'new' even if the innovation itself is not new. Innovation is a social process that occurs when there is increasing interest (public awareness/discussion) in some form of technology and a concerted effort to encourage adoption (new programmes, new organizations, etc.). So, for example, while aphid-resistant wheat varieties have been available to wheat producers for a significant period, the promotion of resistant cultivars as part of a comprehensive AWPM programme can be innovative.

First published in 1962, *Diffusion of Innovations* by Everett Rogers brought together ideas developed by rural sociologists in an effort to characterize and improve the diffusion of agricultural innovations. An influential study that launched this effort was an effort to promote the adoption of hybrid seed maize in Iowa (Ryan and Gross, 1943). Core aspects of the innovation–diffusion model developed through an 'invisible college' of rural sociologists interested in assisting cooperative extension with the diffusion of agricultural innovations (see North Central Rural Sociology Committee, 1955; Fliegel with Korsching, 2001; Rogers, 2003). The history of this literature was summarized in *Diffusion Research in Rural Sociology* by Frederick Fliegel (first published in 1993 by Greenwood Press and then in 2001 by the Social Ecology Press, with an additional chapter by Peter Korsching).

Attention of rural sociologists in the USA turned toward the international context in the 1960s (Rogers, 2003). The innovation–diffusion concept proved useful in describing how new technologies spread in developing nations. The title of the second edition of Rogers' book, *Communication of Innovations: a Cross-Cultural Approach*, reflected this international perspective (Rogers with Shoemaker, 1971). As diffusion research became global, a broad research literature developed around the problem of distinguishing when innovations were more likely to succeed. Subjects of study included education, nutrition, family planning, health and medicine. Following this burst of interest in international topics, the study of diffusion became more closely associated with the developing fields of mass communication and marketing (Rogers, 2003).

Fundamental concepts of innovation diffusion

In *Diffusion of Innovations*, Rogers (2003) summarizes four main elements that are useful for distinguishing successful from unsuccessful innovations. The first is concerned with *characteristics of the innovation* itself, which make it more or less attractive to potential adopters. The other three elements are concerned with the social context of diffusion – these are the *communication process*, the *temporal process* and the *social networks* of diffusion.

Characteristics of the innovation

The study of innovation diffusion begins with the characteristics of the innovation. Rogers (2003) summarized five characteristics of innovations:

- The concept of *relative advantage* is meant to encompass social, economic and technical attributes of an innovation, but the concern is with the *experience* of early adopters and the *perception* of potential adopters who observe the experience of early adopters.
- *Compatibility* is likewise meant to encompass socio-economic attributes as well as technical compatibility with other practices. To what extent is the innovation compatible with existing practices that will not change with adoption of the innovation?
- *Trial adoption* refers to the degree to which an innovation may be tested by a potential adopter on a limited basis prior to adopting it fully.
- *Observable results* refer to the degree to which favourable results of adopting the innovation may be seen early in the adoption process.
- *Complexity* means that innovations will be less likely to diffuse rapidly if they are technically complex, difficult to integrate with other practices or require extensive learning or practice to use. As with the other four characteristics, complexity is concerned with the perceptions/experiences of potential adopters as well as the technological aspects of complexity.

The communication process

Regarding communication channels, some potential adopters will learn about a given innovation through mass media channels, while others will learn about it through interpersonal channels (Rogers, 2003). Additional information about the innovation (technical aspects, testimonials, meetings, new organizations, etc.) may be obtained through either or both of these channels. Naturally, mass media channels have the potential to reach the largest number of people quickly, while interpersonal channels may have greater influence on the adoption decision, particularly for individuals who have little trust and/or less access to mass media channels. When implementing programmes like AWPM, it is important to bear in mind that individuals will differ in terms of both their access to and their preference for communication channels.

The temporal process

In terms of the temporal process, some innovations are quickly adopted while others require a significant period before the innovation achieves 'take-off' (innovations that never 'take off' are characterized as 'failed innovations' after interest in them wanes). Graphically represented, the cumulative percentage of persons adopting a successful innovation over time will be represented by some form of an S-shaped curve (see Fig. 6.1). The take-off stage is closely related to the social networks of diffusion, as discussed below.

Related to the temporal process, innovation researchers have summarized characteristics of the *innovation–decision process* and characteristics describing the *relative innovativeness* of potential adopters (North Central Rural Sociology Committee 1955; Rogers 2003). With respect to the decision process, there is a logical progression of

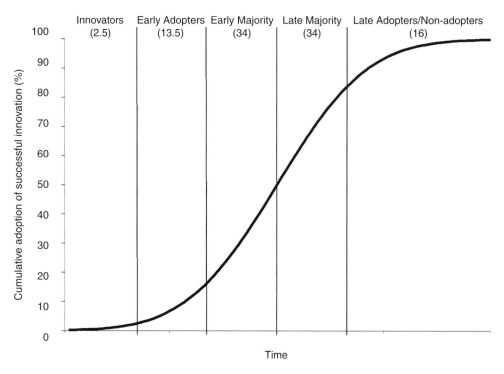

Fig. 6.1. Diffusion curve and adopter categories (adapted from similar illustrations by North Central Rural Sociology Committee 1955; Dent, 1999; Rogers, 2003; Fuchs 2007).

events that influences the rate of innovation adoption. Potential adopters must first become aware of an innovation, form a positive or negative attitude toward it, make a decision to adopt or not adopt, then implement its use and, finally, evaluate the results. Bennett (1977) represented a similar decision process with the acronym, KASA: Knowledge, Attitude, Skills and Aspirations. Regardless of the temporal sequence or rapidity with which potential adopters acquire these attributes, all four are necessary for innovation adoption.

The *relative innovativeness* of potential adopters means that early adopters may be qualitatively different from later adopters of innovations, as illustrated in Fig. 6.1. In general, earlier adopters have higher education, more access/use of mass media communication channels and greater technical competence than later adopters. Rogers (2003) summarized five categories of potential adopters in terms of their relative innovativeness: innovators, early adopters, early majority, late majority and late adopters, or laggards (see also North Central Rural Sociology Committee, 1955). Individuals in these categories are presumed to share qualitative attributes that dispose them to be either earlier or later adopters of innovations; hence, the categories are related to zones of the S-shaped adoption curve.

As summarized in Table 6.1, innovators are viewed by peers as *venturesome*; by nature they are a small minority of the group of potential adopters. They are willing to experiment with new innovations and thus serve as gatekeepers for innovations –

Table 6.1. Adopter categories in terms of how individuals in these categories are viewed by peers (potential adopters) and roles they play in the innovation diffusion process (adapted from information in North Central Rural Sociology Committee, 1955; Rogers, 2003).

Adopter category	Peer view (reputation)	Role in diffusion
Innovator	Venturesome	Gatekeeper
Early adopter	Respected	Community opinion leader
Early majority	Deliberate	Local adoption leader
Late majority	Sceptical	Acceptance
Late adopter/non-adopter	Traditional	Confirmation, preservation

they are the first to see success with beneficial innovations, but few will follow their lead. Of those who will, many are community leaders that have broad social ties and keep abreast of developments in their field/industry. Because they tend to be well known and *respected*, community leaders can facilitate the 'take-off' stage of an innovation (in Fig. 6.1, take-off occurs when the rate of adoption first increases to its highest rate, or at the beginning of the steepest part of the curve).

Early majority adopters are locally significant leaders who are more *deliberate* in their practices and decisions as compared with innovators and community leaders. They pay close attention to community leaders and have many local ties as well (where local may refer to geographical and/or social network proximity). Consequently, these individuals play a key role in the successful diffusion of innovations.

Late majority adopters are similar to early majority adopters except that they are more *sceptical* and have fewer social ties. What distinguishes the late majority is that they adopt an innovation at a time when it is transforming from an innovation to an accepted (normative) practice. Late adopters and non-adopters are individuals who, for various reasons, are either resistant to an innovation or do not perceive it to be useful to their situation. They are viewed by peers as *traditional*, or dedicated to older ways of doing things.

Social networks

It is apparent from these characteristics that social ties between potential adopters can have a significant influence on the success or failure of innovation diffusion. Besides the interrelations of potential adopters, other characteristics of social networks may influence the relative success of innovation diffusion. Rogers discussed the importance of communication network characteristics, opinion leadership, social ties (links) and the point of critical mass (take-off) as influences on the rate of innovation diffusion. Rogers used the example of the Cooperative Extension System as an illustration of a successful innovation–diffusion network (Rogers, 2003). The Extension System illustrates successful technology transfer, combined use of mass media and interpersonal communication channels, and strategies for overcoming *heterophily* – differing degrees and types of technical competence – between change agents and potential adopters.

Alternatives to (or expansions on) the innovation diffusion model

Some scholars have encouraged extension leaders to adopt a different framework than the innovation diffusion model, favouring other models of social networks such as social learning theory and actor–network theory (see Coughenour and Chamala, 2000; Coughenour, 2003; Leeuwis and van den Ban, 2004). Leeuwis and van den Ban argued that the new model for extension should be one of facilitation and communication (social learning) rather than technology transfer of singular innovations. The innovation diffusion model categorized potential adopters with the assumption that everyone is, or needs to be, moving in the same direction. In practice, extension professionals understand that innovation occurs through unplanned change, informal networking and conflict. Thus, Leeuwis and van den Ban (2004) argued, designers of extension should build programmes that help farmers develop and reinvent technologies and social relationships instead of simply adopting uniform technological innovations from university-sponsored research. Consistent with this view, Coughenour (2003) observed that the development of conservation tillage in Kentucky involved broad changes in farming practices and a cooperative reinvention process that required the participation of broader social networks encompassing private companies, farmers' organizations and cooperative extension.

Diffusion of IPM

Sociologists and extension professionals have applied the concepts of innovation diffusion to the implementation of IPM technologies (Buttel *et al.*, 1990; Ridgley and Brush, 1992; Bechinski, 1994; Cuperus and Berberet, 1994; Nowak *et al.*, 1996; Cuperus *et al.*, 2000). Fuchs (2007) described the importance of change agents and 'reinvention' of IPM innovations for commercial agriculture. IPM has been similar to other forms of system-level agricultural change in that change agents have included a broader range of participants than just extension professionals. IPM has involved scientists from governmental, non-governmental and for-profit organizations. The high level of technical competence of these change agents suggests that a challenge of IPM is a high degree of heterophily with potential adopters (i.e. greater technical competence of change agents versus potential adopters – the farmers). Fuchs notes that IPM programmes have tried to overcome this by involving extension professionals in adaptive research programmes; this facilitated ongoing reinvention efforts and greater collaboration with the end users.

Petrzelka *et al.* (1997) identified a range of challenges in implementing an integrated crop management programme in Iowa, particularly the difficulty of illustrating successful results and profitability advantages early enough in the programme to maintain producer interest. Petrzelka *et al.* (1997) also discussed the importance of producers' trust in programme proponents as an important factor in successful programme implementation. Similarly, Baumgärtner *et al.* (2007) described how institutional structures and adaptive management are important to the design and implementation of IPM programmes (see also Dent, 1995; Kogan *et al.*, 1999; Baumgärtner *et al.*, 2003).

Following in this vein, areawide pest management programmes supported by the US Department of Agriculture, Agricultural Research Service (USDA-ARS) made significant efforts to use a cooperative, team-building approach. Essential features of areawide pest management are implementation of control tactics over large geographical areas, coordination (development of social networks) among diverse organizations within these geographical areas and a focus on reducing pest populations to an acceptably low density (Chandler and Faust, 1998). Many of the chapters of this book discuss the relative successes of areawide programmes in their efforts to involve change agents and agricultural producers in the implementation of AWPM.

Winter Wheat and Areawide Pest Management for Cereal Aphids

Wheat remains a key food grain throughout the world. Wheat production can be found in all of the agricultural production regions, with major production areas located in the semi-arid regions of Asia, Europe, North America, South America, Africa and Australia. World wheat production is near 600 million t on an annual basis. The USA contributes nearly 10% of this production, approximately 60 million t annually. Of the USA production, nearly half, or 25 million t, is hard winter wheat, which is primarily produced in the Great Plains states of Texas, Oklahoma, Kansas, Colorado, Wyoming and Nebraska. Much of this production is on millions of dryland production acres that produce less than 60 bushels per acre annually, and a large percentage is in a wheat–fallow system that splits these low yields into production on a semi-annual basis.

Though each of these production areas has its own specific insect pest concerns, aphid pests can be found in all of these critical production areas. For winter wheat producers in the Great Plains of the USA, the Russian wheat aphid, *Diuraphis noxia* (Mordvilko) and the greenbug, *Schizaphis graminum* (Rondani) are the major aphid pests. The Russian wheat aphid (RWA) has caused in excess of US$1.2 billion in losses to the wheat and barley industries since its appearance in 1986. Annual greenbug losses have been estimated as high as US$400 million, depending on the year. Presently, the control of RWA and greenbug is nearly all through chemical insecticides, and losses from annual infestations of these pests can be attributed in a large part to the cost of insecticide control. For many winter wheat producers in the Great Plains, the cost of treatment may be excessive. These dryland wheat producers base their profitability on low-cost and low-input production systems. Therefore, wheat producers need to use alternative IPM strategies to control insects across a wide area.

In autumn 2001, USDA-ARS initiated a 5-year areawide demonstration programme for suppression of RWA and greenbug. A cooperative research team was assembled from five universities – the University of Nebraska, Colorado State University, Kansas State University, Oklahoma State University and Texas A & M University. The research team worked with USDA-ARS to establish cooperative relationships with wheat producers and field demonstration sites.

The area of concern for RWA and greenbug is vast, encompassing the majority of the area of the US Great Plains where winter wheat is grown. The RWA and greenbug areas depicted in Fig. 6.2 span portions of six states and stretch

National Atlas of the United States

Fig. 6.2. Map showing the three zones of the cereal aphid areawide programme; the dashed ellipse indicates areas of Russian wheat aphid (RWA) concern and the solid ellipse indicates areas of greenbug concern.

approximately 600 miles (375 km) north–south and 400 miles (250 km) east–west. The areawide programme focused on working with a series of demonstration field sites and a small group of participating wheat producers recruited from within the three zones identified in Fig. 6.2: a northern area (Zone 1), where RWA is the primary insect pest; a south-western area (Zone 2), which is concerned about both RWA and greenbug; and a south-eastern area (Zone 3), where the greenbug is the major aphid pest.

Farm operator participation

A total of 141 producers participated in the project for the entire 4-year demonstration phase of the programme. As noted in Table 6.2 each zone was well represented, with 45 growers in Zone 1, 42 in Zone 2 and 54 in Zone 3. The average age of the producers was 48.9 years in 2003, with little difference across the three zones. Participating growers closely reflect the average age of farmers in the region, being slightly younger than the 52 years of average age for all farmers. The youngest producer in the project was 22 years of age, while the oldest was 76.

Table 6.2.   Characteristics of farm operators participating in the areawide programme, 2003.

	Programme zone			
	1	2	3	All zones
Number of participating operators	45	42	54	141
Operator's age/education (years)				
Minimum age	22	27	31	22
Maximum age	76	76	69	76
Average age	49.5	48.1	49.1	48.9
Average education	14.5	14.7	14.9	14.7
Number of years as a farm operator				
Minimum	3	5	12	3
Maximum	55	55	50	55
Average	25.9	24.2	26.9	25.8
Average that farm has been in family (years)	79.5	56.0	77.0	71.5
Portion of farm labour hired (%)	18.7	26.2	22.0	22.2
Farm acreage (sum for all operators, 1000s acres)				
Dry cropland	150.1	126.5	96.2	372.8
Irrigated cropland	8.2	30.8	4.6	43.5
Pasture	65.0	58.0	50.8	173.9
Conservation Reserve Program (CRP)	21.8	26.0	4.1	51.9
Farmland crop shared (%)	37.2	42.9	47.5	42.8
Portion of farmland cash leased (%)	9.4	9.2	26.1	15.6
Average head (1000) of cattle per year (sum for all operators)	13.8	10.4	22.2	46.4

Education and farm experience were also similar across each of the zones. The education level averaged 14.7 years for all 141 growers. Producers involved in the programme averaged 25.8 years of experience, ranging from 3 years to 55 years. Some of these farms had been in the same family for more than 75 years. The farms were family-based operations with less than 25% of the labour being hired on average.

The farms in this programme managed 372,800 acres (151,000 ha) of dryland crops, 43,500 acres (17,600 ha) of irrigated crops, 173,900 acres (70,400 ha) of pasture and included 52,000 acres (21,100 ha) of Conservation Reserve Program (CRP) participation. Over 50% of the farmland acres were leased, not unlike the general farm population in the region. Of the leased acres, the amount share leased is nearly 75%.

The demographics of the producers in this project were similar to the averages for the region, providing a representative group for acquiring information about farming practices in the region. Based on these characteristics and information gleaned from interactions in focus groups, it was evident that several of the key programme participants were early innovators and community leaders, while most of the remaining producers could be classified as early majority adopters. These individuals were helpful in evaluating programme elements and, at a later point in time, would be critical in increasing the rate of adoption by peers.

Annual cost-of-production interviews with participating wheat growers provided information on farm operating costs and revenues. Focus groups with producers at the beginning and end of the demonstration programme were a way of initiating relationships with producers while learning about their farming history and decision making (Keenan *et al.*, 2007a, b). The plan for the demonstration phase was to observe growers' practices without significant intervention in their farming practices. However, the programme did have some interventions. Operators with demonstration fields were provided with an aphid-resistant seed variety where appropriate to their location. This allowed the areawide research team to evaluate the effectiveness of the resistant variety. Also, focus groups provided an opportunity for operators to learn from one another, and educational materials (newsletters, information on field scouting methods) did provide operators with information about the programme elements.

The strategy of the demonstration was to enhance the effectiveness of biological control with diversified cropping and, where appropriate, the use of cultivars resistant to RWA or greenbug. In this context, increased use of simplified field scouting methods by farm operators would help reduce use of insecticide treatments; field scouting would also help farm operators to monitor the effectiveness of biological control. Additionally, the programme was an opportunity to advance remote sensing and information technology (IT) applications for areawide pest management implementation.

Aphid-resistant cultivars

In general, RWA-resistant varieties are most adapted for use in eastern Colorado, with many of the varieties developed through Colorado State University. The greenbug-resistant variety, TAM 110, is most adapted for use in the Texas Panhandle. Programme participants reflected these characteristics in the use of these wheat varieties. Table 6.3 summarizes the use of resistant wheat varieties among programme participants. Some producers in Zones 1 and 2 had been using RWA-resistant varieties since these varieties first became available. In Zone 1 (mostly in Northern Colorado), 14.7–19.2% of annual wheat acres planted by programme participants were a RWA-resistant variety. Use was more common among Zone 2 producers (mostly in south-eastern Colorado), where between 19.1 and 25.2% of programme participants' annual wheat acres were in a RWA-resistant variety.

Resistant cultivars of wheat seed have been used for the past decade in the RWA areas of Colorado, Kansas, Wyoming and Nebraska. These cultivars have helped farmers produce winter wheat in the region without having to treat with chemical pesticides. The genetic resistance was bred into several cultivars that have allowed producers to use the resistant technology in most of the production areas across the region. These resistant cultivars had significant success until an additional RWA biotype was discovered in the region that is not affected by the resistance in the existing cultivars.

Table 6.3. Acres of all wheat varieties (summed for 141 programme participants) and percentage of acres in Russian wheat aphid (RWA) and greenbug-resistant varieties by programme zone and year.

	Crop year			
	2002	2003	2004	2005
Zone 1				
Sum of wheat acreage[a]	56,015	56,669	54,453	63,253
RWA resistant (%)[b]	19.2	18.3	14.7	18.2
Greenbug resistant (%)[c]	2.5	7.8	9.6	7.2
Zone 2				
Sum of wheat acreage[a]	62,404	78,788	65,236	73,298
RWA resistant (%)[b]	24.7	25.2	19.1	21.2
Greenbug resistant (%)[d]	13.9	18.5	26.4	13.6
Zone 3				
Sum of wheat acreage[a]	65,789	67,745	71,900	72,145
RWA resistant (%)	–	–	–	–
Greenbug resistant (%)	–	–	–	–

[a]Summed acreage for known wheat varieties for all 141 programme-participating producers.
[b]Russian wheat aphid-resistant wheat varieties were: Halt, Prairie Red, Prowers 99, Yumar, Ankor and Stanton.
[c]Wheat varieties Above and AP502CL.
[d]Wheat variety TAM 110.

New research is under way to provide additional resistant cultivars that will have resistance to all of the different RWA biotypes. While producers used this technology, it was not intended for all of the cereal area on the farm. It was expected that farmers would use resistant cultivars on a portion of their acres, the most susceptible to RWA attack, and use other non-resistant varieties on the remainder of the acres. With the discovery of new RWA biotypes, the sowing of resistant cultivars has been reduced, but not eliminated. Producers in high RWA pressure areas continue to use these cultivars to reduce the presence of the initial biotype, with the understanding that recently discovered biotypes will remain in the wheat. If resistant cultivars can reduce pressure to levels that are below the economic damage threshold for treatment, there is a positive response from the use of resistant cultivars.

The proportion of Zone 2 wheat acres that were planted with TAM 110 varied between 13.6 and 26.4% of the acres planted by programme participants. In focus groups, several producers indicated that they liked the greenbug resistance trait of TAM 110, but many indicated that traits for disease resistance, drought resistance, yield potential and forage potential were bigger considerations in their variety selection decision.

In addition to TAM 110, the varieties Above and AP502CL are greenbug resistant. TAM 110 was used by some of the Zone 2 producers (primarily among those in the Panhandle region of Texas). Above and AP502CL were mostly grown by Zone 1 producers, with the proportion of acres in these varieties varying from 2.5 to

9.6% annually among all programme participants. However, focus group discussions suggested that producers were more likely to be growing these varieties for weed management benefits or for sale as seed wheat rather than for the benefits of greenbug resistance. TAM 110 is widely used to assist producers in managing greenbug pressures, but this cultivar has its own set of drawbacks. In recent years, there has been significant rust pressure in the southern wheat-growing areas, forcing producers to manage for multiple pest pressures in the same region. TAM 110 is susceptible to rust, which forces wheat producers to take a decision on the risk factors between rust pressure and greenbug pressure.

Plant breeders continue to work on solutions to these problems, while attempting to maintain yield and quality characteristics necessary for new cultivars to be accepted by farmers. Aphid-resistant cultivars are generally not adapted to Zone 3, and none of the programme participants in that zone indicated growing resistant varieties.

Field scouting

Field scouting is critical to the successful control of insect pests in these areas. While field scouting may be critical, many producers do not spend an adequate amount of time and effort on this management strategy. Although the need for field scouting can be easily quantified for producers, the critical times for scouting are also very busy times for many producers and the scouting gets pre-empted by other critical crop production tasks. There are crop consultants in the area that could be hired to complete this task, but the cost is high for these services and wheat is a low-cost, low-input system, as noted previously. Another factor that limits the amount of scouting done by wheat producers is the size of their farms. Many wheat farmers produce more than 2000 acres (800 ha) of wheat each year. The size of the farm limits the ability of the farmer to adequately scout all of the acres for insect, disease and weed pests.

A simplified method of field scouting has been recently modified to incorporate natural enemy identification (Elliott *et al.*, 2004; Royer *et al.*, 2005a, b). This system, referred to as *Glance 'N' Go*, has made a significant effort to increase the rate of adoption by farmers by improving upon the characteristics of field scouting as an agricultural innovation: relative advantage, compatibility, trial adoption, observable results and complexity (see Cuperus and Berberet, 1994).

Table 6.4 summarizes dryland wheat field scouting practices indicated by areawide programme participants at the beginning of the programme. Overall, 29.8% indicated that they did not practise any field scouting of dryland wheat, and another 29.1% relied on a private crop consultant or other crop advisor (including cooperative extension educators) to scout wheat. Of those who did their own field scouting, 36.9% indicated that they had scouted irregularly or infrequently (e.g. only when they had heard about an aphid outbreak in their area), and only 4.3% indicated that they had scouted at regular intervals for preventive purposes. By project zone, a slightly higher percentage of programme participants in Zone 3 indicated that they had scouted regularly, 7.4%, compared with Zones 2 (2.4%) and 1 (2.2%). Zone 2 producers were the most likely to use a crop consultant or crop advisor

Table 6.4. Field scouting methods as indicated by wheat producers by areawide programme zone (2003).

Field scouting carried out by:	Percentages within programme zones			
	1	2	3	All zones
Crop consultant or crop advisor	15.6	42.9	29.6	29.1
Self, infrequently or irregularly	35.6	23.8	48.2	36.9
Self, regular interval	2.2	2.4	7.4	4.3
None	46.7	31.0	14.8	29.8
Percentage totals	100.0	100.0	100.0	100.0
Total number of producers	45	42	54	141

(42.9%). Zone 1 producers were most likely to do no scouting at all (46.7%) or to scout infrequently or irregularly (35.6%).

Focus groups with programme participants helped to reveal some of the characteristics of producers who had frequently scouted. In the case of Zone 3, many of the producers who had scouted at regular intervals were concentrated in an area of more frequent greenbug outbreaks. Some of these producers were also more likely to have smaller acreages of wheat and to be intensive farm managers (attention to detail). They were also in an area where cooperative extension educators had made significant efforts to inform producers of greenbug problems and IPM methods – including field scouting. Focus groups also suggested that programme participants in Zone 3 were more familiar with the *Glance 'N' Go* field scouting system at the end of the demonstration programme than they were at the initiation of the programme. In particular, more participants indicated that they had scouted for the presence of beneficial insects as well as aphids since they had become aware of the *Glance N' Go* system. In the second-round focus groups with programme participants in Zones 1 and 2, most had become aware of the discovery of the Russian wheat aphid biotype 2 during the programme, and some had increased their field scouting efforts as a result.

Diversified dryland cropping

Recently, Great Plains dryland crop production systems have moved toward less tillage and more intensive cropping. Increases in acres of dryland maize, grain sorghum, sunflowers, proso millet, cotton and other alternative crops reinforce this observation. As traditional wheat producers look for options to increase profits, lower risk and mitigate pest losses, they have looked to the potential for additional crop diversity. The nature of these new production systems has made it necessary to move toward limited tillage in conjunction with the move to diversified cropping systems. Diversifying crops in the rotation minimized annual yield variability (Anderson *et al.*, 1999). This statement, while simple and short, may be the key to producers considering changes in the Great Plains production system. If long-term yield variability can be reduced, the profitability from year to year will also increase.

The major pest problem for monoculture wheat systems in the Great Plains is the presence of winter annual grasses (downy brome, jointed goatgrass, feral rye) in the winter wheat crop. Diverse cropping systems can effectively control winter annual grasses in winter wheat systems, allowing wheat producers to deliver a higher-yielding crop that meets quality guidelines (Daugovish *et al.*, 1999). Typically, producers make the move toward crop diversity to control weed or insect pests in the system more often than to increase profits. Pest issues may be so severe that the only option is to move to another crop. In these cases, profitability may have suffered significantly enough that there is increased profitability from diversity by default.

Several recent studies have looked at the profitability of diversified crop rotations across the Great Plains. Dhuyvetter *et al.* (1996) determined that profitability of diverse systems with a crop grown on 67–75% of the acres was higher than in the traditional wheat–fallow system in eight of the nine locations in studies from North Dakota to Texas. In this study, tillage systems were also evaluated for profitability in different cropping systems. For the wheat–fallow system, no-till systems were never more profitable than either conventional or reduced tillage systems. However, in the more intensive systems, no-till or reduced tillage was always more profitable than the conventional tillage system. Given these results, the change to diverse cropping systems seems to be most successful when combined with a change in tillage systems. Kaan *et al.* (2002) showed that diverse systems in Eastern Colorado were more profitable than wheat–fallow over several years. These results were based on a set of studies at two sites in Colorado that represented two of the zones in the areawide project.

When diverse production systems are adopted to assist with control of either insect or weed pests, producers may not need the system to show significantly high profit levels. The farm may actually be better off if the diversified system is equally as profitable as the monoculture system, while providing pest management benefits to the entire farm. Another potential benefit is the opportunity to reduce risk in these highly risky areas. Production of several crops will allow the farmer to produce crops that enter into different markets, grow during different seasons and utilize different sets of resources. Hail and drought are key weather risks throughout this area and, by growing different crops, a producer may be able to spread the risk of both hail and drought. Markets for different crops and types of crops may not move in similar directions on a yearly basis, allowing the farm to capture profits in one market in a year when another market may be soft.

For evaluating the relative advantage of crop diversity, Table 6.5 summarizes annual averages from 4 years of net returns to land and management (in US dollars) for producers participating in the AWPM demonstration programme. (In the context of this project, net return to land and management is defined as return prior to any charges for land or management for the farm producer.) Differences in per-acre returns by zone illustrate the advantages of climate and rainfall from the north-west to south-east regions of the central US Great Plains. The overall average net return for Zone 1 producers was US$23.35 per acre (0.4 ha) compared with US$39.18 for Zone 2 and US$70.78 for Zone 3 producers. This is related to the acres farmed, illustrated earlier in Table 6.2, producers in Zones 1 and 2 typically farm larger acreages than producers in Zone 3, somewhat levelling the differences in overall economic returns to the whole farm.

Table 6.5. Dryland crop diversity and net return per acre summary by programme zone and crop categories.

Average returns to land and management (US$), 2002–2005 by project zone and crop	Crop diversity[a]			
	Low	Medium	High	All producers
Zone 1 (all dryland crops)	24.58	23.51	22.15	23.35
Wheat–fallow	24.17	21.51	24.40	23.18
Lucerne	157.15	108.84	–	124.94
Other hay, forage and silage crops	8.67	31.47	21.53	21.08
Other dryland crops	21.32	22.67	20.27	21.53
Zone 2 (all dryland crops)	29.13	33.34	58.00	39.18
Wheat–fallow	24.65	20.99	35.52 [b]	26.19
Other hay, forage and silage crops	32.29	54.72	99.91	64.27
Cotton	–	112.15	112.32	112.25
Other dryland crops	52.86	48.04	63.07	53.78
Zone 3 (all dryland crops)	70.42	58.62	87.19	70.78
Wheat–fallow	48.53	33.76	63.48b	47.74
Lucerne	233.81	250.91	340.00	258.56
Other hay, forage and silage crops	48.59	68.11	65.72	57.60
Cotton	35.63	55.47	158.68	69.59
Other dryland crops	43.33	22.89	69.00[b]	43.75

[a]The crop diversity variable ranks producers based on percentage of cultivated acres in crops other than wheat, lucerne or other hay and forage crops for the period 2002–2005. Low diversity = 0–10% (35 out of 141 producers); medium diversity = 11–30% (67 producers); and high diversity = > 30% of cultivated acreage (39 producers).
[b]The average for high-diversity operations was significantly greater than the average for medium-diversity categories based on one-way analysis of variance and LSD post hoc comparisons ($P < 0.05$).

Evident in Table 6.5, lucerne and cotton are the most profitable crops on a per-acre basis. Lucerne, however, is typically maintained as a stand for 3–5 years (hence, lucerne is not typically rotated with winter wheat or other crops on an annual basis). While lucerne and cotton are the most profitable crops, these are grown in selected locations where they grow well and where there are established markets or processing facilities (cotton gins). Other hay, forage and silage crops are also pre-sented separately in Table 6.5, because many producers have at least some cultivated acreage in these crops to provide food for livestock. Thus, these crops may or may not increase the overall crop diversity of a given farm operation.

The simplest (least diverse) dryland cropping system in the programme area is either continuously planted winter wheat or a wheat–fallow rotation. The crop diversity of farm operations is represented by three categories in Table 6.5, ranking operations as low, medium or high crop diversity. The least diverse farm operations had 10% or less of cultivated dryland acres in a crop other than wheat, fallow, lucerne, or other hay, forage, or silage crops. The most diverse had 30% or more of their cultivated dryland in some other crop. Crops rotated with winter wheat on an annual or semi-annual basis varied by project zone.

In Zone 1, 45 producers participating in the areawide programme collectively averaged about 55,000 acres (22,000 ha) of wheat and 53,000 acres (21,000 ha) of fallow per year, in the period 2002–2005. The three most common crops rotated annually or semi-annually with wheat and fallow were proso millet (16,000 acres (6400 ha)), sunflower (9000 acres (3600 ha)) and maize (5000 acres (2000 ha)). Dryland lucerne was produced by only three programme participants, with an average of only 86 acres (35 ha) annually; other hay, forage and silage crops, however, accounted for about 3000 acres (1200 ha) annually. The much larger acreage of wheat and fallow compared with the other crops is a reflection of the prevalence of the wheat–fallow dryland farming system in Zone 1.

The figures in Table 6.5 do not indicate a profit advantage for more diversified farm operations in Zone 1. Overall net returns were slightly higher among the least diverse, US$24.58, compared with US$22.15 among the most diverse. Lucerne appears influential in this difference, but few operators (only three out of 45 producers) in Zone 1 had lucerne. Medium-diversity operators had the greatest returns from other hay, forage and silage crops, US$31.47, while the low-diversity operators averaged much lower, at US$8.67. Returns from other dryland crops were quite similar among all Zone 1 producers. None of the averages observed for Zone 1 were statistically significant based on one-way analyses of variance and LSD post hoc comparison tests.

The wheat–fallow system is as prevalent in Zone 2 as it is in Zone 1. In Zone 2, 42 producers in the areawide programme collectively averaged 63,000 acres (25,000 ha) of wheat and 40,000 acres (16,000 ha) of fallow annually during the period 2002–2005. Among the other dryland crops that were grown in annual or semi-annual rotations with winter wheat in Zone 2 were grain sorghum (18,000 acres (7200 ha)), cotton (3000 acres (1200 ha)), sunflower (1000 acres (400 ha)) and maize (840 acres (340 ha)). Hay, forage and silage crops accounted for about 2000 dryland acres (800 ha) among Zone 2 producers.

Table 6.5 does indicate higher average returns overall for high crop-diversity farm operations in Zone 2: the figure for high-diversity operations is US$58.00, contrasted with US$33.34 among medium-diversity operations and US$29.13 among low-diversity operations. This result was partly due to significantly higher average returns from wheat and fallow acres among the higher-diversity operations. The average returns from wheat and fallow for high-diversity operations, US$35.52, were significantly greater than the average for medium-diversity operations, US$20.99 (based on one-way analysis of variance and LSD post hoc comparisons). However, the difference between high diversity, Us$35.52, and low diversity, US$24.65, was not statistically significant. This result was due to high variability in net returns among producers within both categories – high-diversity and low-diversity operations. Zone 2 medium-diversity operations averaged about the same return per acre for cotton, US$112.15, as the high-diversity operations, US$112.32, but none of the low-diversity operations produced dryland cotton.

In Zone 3, continuous wheat (without a fallow period) is the norm due to higher rainfall as compared with the other two zones. Collectively, 54 producers in Zone 3 of the areawide programme averaged 70,000 acres (28,000 ha) of wheat annually and only 237 acres (91 ha) of fallow. Leading crops grown in rotation with wheat among these producers were grain sorghum (7000 acres (2800 ha)), soybeans (4000 acres (1600 ha)), cotton (3000 acres (1200 ha)) and maize (3000 (1200 ha) acres). Lucerne (5000 acres (2000 ha)) and other hay, forage and silage crops (4000 acres (1600 ha)) also accounted for a substantial portion of dryland crop acreage among Zone 3 producers.

Continuous wheat production is known to exacerbate grassy weed problems, which is one probable reason that the more diversified operations in Zone 3 averaged significantly higher average returns from wheat: US$63.48 per acre among high-diversity operations compared with US$33.76 for medium-diversity operations. However, many low-diversity operations in Zone 3 also had above-average returns from wheat acres (group average = US$70.42), resulting in a non-significant difference in returns from wheat comparing low- and high-diversity operations in Zone 3.

The trend was comparable for other dryland crops, where high-diversity operations also averaged significantly higher returns (US$69.00) compared with medium-diversity operations (US$22.89), but not significantly higher than the average for low-diversity (US$43.33) operations, again due to high within-group variability. The average returns from cotton for high-diversity operations appear advantageous (US$158.68) compared with medium- (US$55.47) and low-diversity (US$35.63); however, only nine out of 54 producers (and only two high-diversity operators) grew cotton, resulting in high standard errors for the observed averages. Results observed for lucerne followed a similar pattern.

Conclusions

The AWPM producer group was diverse, in terms of production systems and geographic location. Within the group there will be significant differences in the level of adoption of each programme element and corresponding levels of success with the adopted elements.

As many have in the past, producers will probably adopt resistant cultivars as they are made available, although the initial use will be on a trial basis, as it is with most technology adoption by farmers. These initial trials will make it easy for the producers to evaluate and determine the compatibility with the individual farm characteristics and insect pressures. During the focus groups, producers mentioned the need for resistance to insects other than aphids. If resistant cultivars for other insects are made available, it can be assumed that producers will sow these varieties on a trial basis for evaluation. The location and regularity of aphid problems will be critical to the long-term and widespread adoption of resistant cultivars, both in the USA and in other areas of the world.

Development of simplified field scouting systems that are adaptable to different geographic locations and a variety of insect pests will have the potential for initial adoption, with increased use over time. Decreasing the complexity and the time requirements for field scouting has and will continue to enhance the acceptability of this technology. At the present time a simplified, quick field scouting process (*Glance 'N' Go*) is applicable only to greenbug management. A similar process would be widely accepted across the entire programme region if made available. This technology could also prove highly useful in other wheat-growing areas of the world, particularly in locations of smaller-acreage fields and farms.

The adoption of crop diversification will be more difficult to apply on a widespread basis in the arid environment of the US Great Plains. Regardless of where it occurs, crop diversification is complex because it involves broader changes in farm operation, goals, personal and financial characteristics of operators, and potential resistance from landlords and agricultural lenders. Crop selection encompasses personal, technical, financial and economic factors (Makeham and Malcolm, 1993; Corselius *et al.*, 2003). Adoption of even relatively simple crop diversity can require changes in tillage system, marketing management and an investment in machinery to be successful. This limits the amount of trial adoption that producers are willing to entertain. Instead, farmers are more likely to undertake extensive reading and research before making the change to crop diversity and fewer tillage operations, then shift the entire farm to the new system.

In the short term there will be difficulties associated with this change, which will have the producer questioning the decision. This lack of observable results is a key challenge faced in the adoption of crop diversity on an areawide basis. Several cost-of-production analyses have pointed out the advantages, or at least a lack of disadvantages, to the adoption of crop diversity form and economic perspective. Delivery of this information in conjunction with information of the insect, weed and disease management benefits will be critical to the continued areawide adoption of crop diversity.

To date, the Cooperative Extension System has played the major role in the diffusion of IPM technologies for dryland winter wheat. Along with continued interaction with farm producers who are innovators and community leaders, USDA-ARS should continue to find ways to coordinate with CES research and extension professionals to achieve adoption of AWPM programme elements. CES agricultural extension agents/educators remain in the best position to act as change agents by determining the applicability of programme elements and information needs of producers in local areas. CES agents/educators already alert growers to potential aphid outbreaks through multiple communication channels: radio, newsletters, e-mail alerts and personal communication. They also already assist growers with cultivar selection, field scouting and crop diversification decisions. CES agents/educators could benefit from AWPM information technology (IT) advancements in their efforts to communicate effectively with producers. Thus, the research community can continue to provide technologies appropriate to producers by working through traditional cooperative extension channels.

References

Anderson, R.L, Bowman, R.A., Nielsen, D.C., Vigil, M.F., Aiken, R.M. and Benjamin, J.G. (1999) Alternative crop rotations for the central Great Plains. *Journal of Production Agriculture* 12, 95–99.
Baumgärtner, J., Getachew, T., Melaku, G., Sciarretta, A., Shifa, B. and Trematerra, P. (2003) Cases for adaptive ecological systems management. *Redia* LXXXVI, 165–172.

Baumgärtner, J., Pala, A.O. and Trematerra, P. (2007) Sociology in integrated pest management. In: Koul, O. and Cuperus, G.W. (eds) *Ecologically Based Integrated Pest Management*. CAB International, Wallingford, UK, pp. 154–179.

Bechinski, E.J. (1994) Designing and delivering in-the-field scouting programmes. In: Pedigo, L.P. and Buntin, G.D. (eds) *Handbook of Sampling Methods of Arthropods in Agriculture*. CRC Press, Boca Raton, Florida, pp. 683–706.

Bennett, C.F. (1977) *Analyzing Impacts of Extension Programs*. Publication ESC-575, Extension Service, US Department of Agriculture, Washington, DC.

Buttel, F.H., Gillespie, G.W., Jr. and Power, A. (1990) Sociological aspects of agricultural sustainability in the United States: a New York case study. In: Edwards, C.A. (ed.) *Sustainable Agricultural Systems*. Soil and Water Conservation Society, Ankeny, Iowa, pp. 515–532.

Chandler, L.D. and Faust, R.M. (1998) Overview of areawide management of insects. *Journal of Agricultural Entomology* 15, 319–325.

Corselius, K.L., Simmons, S.R. and Flora, C.B. (2003) Farmer perspectives on cropping systems diversification in northwestern Minnesota. *Agriculture and Human Values*, 20, 371–383.

Coughenour, C.M. (2003) Innovating conservation agriculture: the case of no-till cropping. *Rural Sociology* 68, 278–304.

Coughenour, C.M. and Chamala, S. (2000) *Conservation Tillage and Cropping Innovation: Constructing the New Culture of Agriculture*. Iowa State University Press, Ames, Iowa.

Cuperus, G.W. and Berberet, R.C. (1994) Training specialists in sampling procedures. In: Pedigo, L.P. and Buntin, G.D. (eds) *Handbook of Sampling Methods of Arthropods in Agriculture*. CRC Press, Boca Raton, Florida, pp. 669–681.

Cuperus, G.W., Mulder, P.G. and Royer, T.A. (2000) Implementation of ecologically based IPM. In: Rechcigl, J.E. and Rechcigl, N.A. (eds) *Insect Pest Management: Techniques of Environmental Protection*. Lewis Publishers, Boca Raton, Florida, pp. 171–204.

Daugovish, O., Lyon, D.J. and Baltensperger, D.D. (1999) Cropping systems to control winter annual grasses in winter wheat (*Triticum aestivum*). *Weed Technology* 13, 120–126.

Dent, D. (1995) Programme planning and management. In: Dent, D. (ed.) *Integrated Pest Management*. Chapman & Hall, London, pp. 120–151.

Dhuyvetter, K.C., Thompson, C.R., Norwood, C.A. and Halvorson, A.D. (1996) Economics of dryland cropping systems in the Great Plains: a review. *Journal of Production Agriculture* 9, 216–222.

Elliott, N.C., Royer, T.A., Giles, K.L., Kindler, S.D., Porter, D.R., Elliott, D.T. and Waits, D.A. (2004) A web-based decision support system for managing greenbugs in wheat. *Crop Management* (online) doi:10.1094/CM-2004- 1006- 01-MG.

Faust, R. (2001) Forum – invasive species and areawide pest management: what we have learned. *Agricultural Research* (http://www.ars.usda.gov/is/AR/archive/nov01/form1101. htm).

Fliegel, F.C. with Korsching, P.F. (2001) *Diffusion Research in Rural Sociology: the Record and Prospects for the Future*. Social Ecology Press, Middleton, Wisconsin.

Fuchs, T.W. (2007) Diffusion of IPM programmes in commercial agriculture: concepts and constraints. In: Koul, O. and Cuperus, G.W. (eds) *Ecologically Based Integrated Pest Management*. CAB International, Wallingford, UK, pp. 432–444.

Kaan, D.A., O'Brien, D.M., Burgener, P.A., Peterson, G.A. and Westfall, D.G. (2002) *An Economic Evaluation of Alternative Crop Rotations Compared to Wheat–Fallow in North-eastern Colorado*. Colorado State University, Agricultural Experiment Station, Technical Bulletin TB02-1.

Keenan, S.P., Giles, K.L., Burgener, P.A., Christian, D.A. and Elliott, N.C. (2007a) Collaborating with wheat producers in demonstrating areawide integrated pest management. *Journal of Extension* 45 (http://www.joe.org/joe/2007february/a7.shtml).

Keenan, S.P., Giles, K.L., Elliott, N.C., Royer, T.A., Porter, D.R., Burgener, P.A., and Christian, D.A. (2007b) Grower perspectives on areawide wheat integrated pest management in the southern US Great Plains. In: Koul, O. and Cuperus, G.W. (eds) *Ecologically Based Integrated Pest Management*. CAB International, Wallingford, UK, pp. 289–314.

Kipling, E.F. (1980) Areawide pest suppression and other innovative concepts to cope with our more important insect pest problems. In: *Minutes of the 54th Annual Meeting of the National Plant Board*, Sacramento, California, pp. 68–97.

Kogan, M. (1998) Integrated pest management: historical perspectives and contemporary developments. *Annual Review of Entomology* 43, 243–270.

Kogan, M., Croft, B.A. and Sutherst, R.F. (1999) Applications of ecology for integrated pest management. In: Huffaker, C.B. and Gutierrez, A.P. (eds) *Ecological Entomology*. John Wiley & Sons, New York, pp. 681–728.

Leeuwis, C. and van den Ban, A. (2004) *Communication for Rural Innovation: Rethinking Agricultural Extension*. 3rd edn., Blackwell Science, Oxford, UK.

Makeham, J.P. and Malcolm, L.R. (1993) *The Farming Game Now*. Cambridge University Press, Cambridge, UK.

North Central Rural Sociology Committee (1955) *How Farm People Accept New Ideas*. North Central Regional Publication No. 1 of the Agricultural Extension Services (Reprinted in 1981 as Special Report No. 15, Iowa State University Cooperative Extension (http://www. soc.iastate.edu/extension/publications/SP15.pdf, accessed 26 March 2007)).

Nowak, P., Padgett, S. and Hoban, T.J. (1996) Practical considerations in assessing barriers to IPM adoption. *Proceedings of the Third National IPM Symposium/Workshop*. US Department of Agriculture, Economic Research Service, MP-1542, Washington, DC.

Petrzelka, P., Padgitt, S. and Connelly, K. (1997) Teaching old dogs survival tricks: a case study in promoting integrated crop management. *Journal of Production Agriculture* 10, 596–602.

Ridgley, A. and Brush, S.B. (1992) Social factors and selective technology adoption: the case of integrated pest management. *Human Organization*, 51, 367–378.

Rogers, E.M. (2003) *Diffusion of Innovations* 5th edn., The Free Press, New York.

Rogers, E.M. with Shoemaker, F.F. (1971) *Communication of Innovations: a Cross-cultural Approach*. The Free Press, New York.

Royer, T.A., Giles, K.L. and Elliott, N.C. (2005a) *Glance 'n Go Sampling for Greenbugs in Winter Wheat*. Spring edn., Oklahoma Cooperative Extension Service, Oklahoma State University Extension Facts, L-306, Stillwater, Oklahoma.

Royer, T.A., Giles, K.L. and Elliott, N.C. (2005b) *Glance 'n Go Sampling for Greenbugs in Winter Wheat*. Fall edition, Oklahoma Cooperative Extension Service, Oklahoma State University Extension Facts, L-307, Stillwater, Oklahoma.

Ryan, B. and Gross, N.C. (1943) The diffusion of hybrid seed corn in two Iowa communities. *Rural Sociology* 8, 15–24.

7 Environmental Monitoring in Areawide Pest Management

J.D. CARLSON[1] AND ALBERT SUTHERLAND[2]

[1]Oklahoma State University, Biosystems and Agricultural Engineering, Oklahoma, USA
[2]Oklahoma State University, Biosystems and Agricultural Engineering, National Weather Center, Norman, Oklahoma, USA

Introduction

The environment plays a crucial role in agriculture, as it affects not only crops and livestock but also their associated insect and disease pests. Accordingly, weather measurements play an essential role in integrated pest management. With knowledge of how environmental conditions affect a particular insect or disease pest, one can better time scouting activities and pesticide applications to maximize pest control. Use of weather-based decision support models can often result in reduced pesticide usage, which in turn minimizes environmental contamination and pesticide residues in foods, reduces costs and slows the development of resistance in target pests (WMO, 1988; Gillespie, 1994; McFarland and Strand, 1994).

In contrast to stand-alone environmental monitoring stations, automated weather station networks in conjunction with weather-based models can provide a regional view of biological pest development. Weather station networks, especially if coupled with efficient information dissemination, also maximize the numbers of growers and others in the agricultural community who can participate and use such information in their IPM decisions.

This chapter focuses on environmental monitoring in areawide pest management (AWPM). Topics to be discussed include the environmental factors important in pest management systems, along with the development and importance of environmental monitoring networks to provide the necessary information needed for such systems. The Oklahoma Mesonet is highlighted here, not only as a state-of-the-art environmental monitoring network but also for the weather-based decision support products it offers. The chapter concludes by looking at the Internet as a data conduit, effective educational programmes for growers and crop consultants, and future innovations in AWPM.

Environmental Factors that Regulate Pest Management Systems

As with the crops and livestock they infest, insect and disease pests are influenced by environmental conditions. Knowledge of how the weather influences a particular pest provides the rationale for choosing suitable environmental measurements and designing appropriate schemes for pest management (Gillespie, 1994).

With respect to insects, temperature (air, and sometimes soil) is the primary influence, although other factors such as light, wind, moisture and rainfall may be important for various life stages. A typical insect pest passes through stages from egg to larva to adult, with possibly a rest period (diapause) in climates with a cold season. Temperature is the controlling factor on the rate of development and survival at each stage, controlling the number of generations, the arrival of a particular damaging growth stage and the density of populations (Gillespie, 1994).

Accordingly, a number of temperature-based models has been developed for use in IPM. Many of these models use thermal units or 'degree-days', which are used to estimate the amount of energy available for the development of a particular insect stage. Typically using daily maximum and minimum temperatures to calculate the average daily temperature (T), these degree-day models use a lower temperature development threshold (T_L) and, sometimes, an upper temperature threshold (T_H). The assumption is that development occurs if the daily average temperature lies between T_L and T_H. Daily degree-days are usually accumulated from a particular start date or 'biofix' point; when the accumulated degree-day total reaches a particular value or range, the specific damaging growth stage of the insect pest is assumed to occur. Excellent reviews of the use of thermal units and degree-days in insect modelling are given in McFarland et al. (1991). A study showing the influence of degree-days, not only on insect pests but also on the crops they infest and how that affects strategies for planting date, is given in Carlson and Gage (1989).

Some examples of degree-day-based insect pest models used in Oklahoma include the weevil and pecan nut casebearer (Carlson and Sutherland, 2006). The lucerne weevil (*Hypera postica*) model uses 8.9°C (48°F) as its low temperature developmental threshold and uses a start date of 1 January. The pecan nut casebearer (*Acrobasis nuxvorella*) model has 3.3°C (38°F) as its T_L and, depending on location, uses site-specific start dates ranging from 21 March to 24 April. A study describing the development of this model for use in Oklahoma is given in Grantham et al. (2002).

Relative humidity has a definite impact on insect biology. Studies have shown that relative humidity can impact insect distribution (Platt et al., 1958), development (Guarneri et al., 2002; Duale, 2005; Perez-Mendoza and Weaver, 2006), habitat selection (Lorenzo and Lazzari, 1999) and mortality (Pelletier, 1995). While relative humidity affects insects in many ways, relative humidity is a rarely used insect model parameter. Perez-Mendoza noted that air temperature and population were the main factors affecting wheat stem sawfly development, overshadowing changes in relative humidity.

Other forms of moisture such as rainfall can also have an affect on insect pests. Rainfall may inhibit the flight of adults and possibly contribute to mortality in adults and newborn larvae, especially when accompanied by strong winds (Gillespie, 1994). Rainfall has also been shown to be a factor in the emergence of some insects from the ground, such as the pecan weevil (Raney et al., 1970; Mulder et al., 2004).

Wind speed and direction are important on two geographic scales – locally and regionally (long-range transport). Insect adults use winds to disperse from over-wintering locations and to travel throughout their adult lives to find favourable host plants for food and egg laying. This travel can occur on a local scale, but also can cover long distances and several days if upper-level winds and temperatures are suitable. One case study involving the long-range migration of potato leafhopper into Michigan from the Gulf region is presented in Carlson *et al.* (1992). Looking at upper-level winds, temperatures, precipitation and using trajectory analysis, the authors showed that the sudden arrival of potato leafhopper in Michigan during the spring of 1989 from potato leafhopper source regions could have resulted from 24–36 h of continuous flight or from two to three successive night-only flights. It is hypothesized that the potato leafhopper adults use the warm temperatures and favourable winds in the nocturnal jet stream to fly at levels 500–1500 m above ground. This is consistent with radar observations of migrating insects (Drake, 1985; Westbrook *et al.*, 1991) as well as other insect trajectory analyses (Domino *et al.*, 1983; McCorcle and Fast, 1989).

With respect to plant diseases, it is usually necessary to consider at least temperature and moisture as major factors in development, although winds also play a role through local dispersal and long-range transport of spores. A generalized life cycle begins with the arrival of a potentially infectious body (e.g. a spore) at the host plant, followed by infection and colonization of the plant, a period of incubation and finally the development of new propagules to be dispersed by wind or water (Gillespie, 1994).

With respect to moisture, some models rely on precipitation while others use 'leaf wetness', which is either calculated or obtained through leaf wetness sensors (Duan and Gillespie, 1987). At the onset of the Oklahoma Mesonet of automated weather stations in the early to mid-1990s (see section below), leaf wetness sensors were installed, but attendant problems led to the use of relative humidity as a surrogate in disease models.

Currently, disease models on the Oklahoma Mesonet use combinations of temperature and relative humidity to calculate 'infection hours' – periods of time deemed suitable for infection of the host plant (Carlson and Sutherland, 2006). For example, the pecan scab model uses temperatures $\geq 21.2°C$ (70°F) and relative humidity $\geq 90\%$ to calculate infection hours (Driever *et al.*, 1998). The model checks for the number of accumulated infection hours during the unprotected portion of the preceding 14 days (a fungicide application is assumed to protect for 2 weeks). Depending on the disease susceptibility of the pecan cultivar, a spray is recommended if infection hour totals exceed 10 h for susceptible varieties, 20 h for moderately susceptible or 30 h for resistant or native pecans.

Winds also play an important role, dispersing infectious bodies such as spores both locally and long distances, the latter via long-range transport. Like insects, long-range dispersal of biotic agents (e.g. spores) via winds in the upper levels of the planetary boundary layer has been well documented (Hirst *et al.*, 1967; Aylor *et al.*, 1982; Davis and Main, 1986).

Traditionally, pest models, whether for insects or diseases, run on current/past weather data from either manually observed measurements or automated weather stations. They are often described as 'forecast' models, despite the fact that no forecast information has been included. Yet these models do provide a 'forecast' of pest activity, because they provide indications of insect or disease development prior to when they could be noted with 'typical field scouting'.

The opportunity exists to add weather forecasts to insect and disease pest models, which will provide more lead-time for making pest management decisions. Using forecast information for disease modelling is more problematic, however, as forecasts of relative humidity, precipitation or leaf wetness (calculated) are much less accurate (and more site specific) than temperature forecasts.

Of all the weather parameters, forecasts of air temperature are the most accurate. This is true for both short-range forecasts for up to 5 days in the future and longer 30- and 90-day forecasts. In an analysis of long-range air temperature and precipitation forecast, Jeanne Schneider, USDA Grazinglands Climatologist, has shown how long-range forecasting accurately predicts air temperature trends above and below normal (Schneider *et al.*, 2005). The weather forecasts she analysed were from the National Weather Service's Climate Prediction Center.

In their present form, the National Weather Service's long-range forecasts issued by the Climate Prediction Center would be difficult to add to a model, because they do not predict probable temperatures. Instead, they provide a map of the probability of air temperatures or precipitation being in one of three categories, above normal, near normal or below normal, as shown in Fig. 7.1.

What is possible is to incorporate short-range weather forecasts into insect and disease models. Numerical models run by the US National Weather Service, for

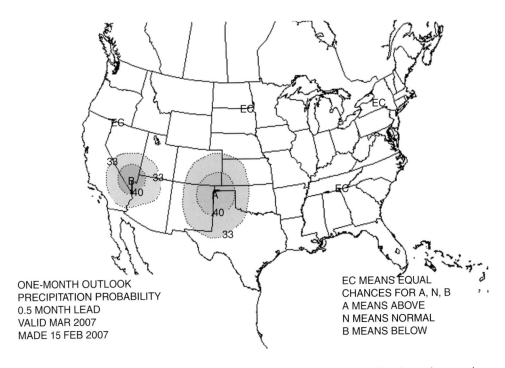

ONE-MONTH OUTLOOK
PRECIPITATION PROBABILITY
0.5 MONTH LEAD
VALID MAR 2007
MADE 15 FEB 2007

EC MEANS EQUAL
CHANCES FOR A, N, B
A MEANS ABOVE
N MEANS NORMAL
B MEANS BELOW

Fig. 7.1. One-month precipitation outlook for March 2007 from the national weather service climate prediction centre.

example, produce a 84-h digital output of various weather variables in 1-h increments (e.g. North American Model (NAM)). With interpretive software, the hourly values of air temperature and other required variables such as relative humidity can be extracted from National Weather Service files and incorporated into insect and disease IPM models. This has been done for the Oklahoma Mesonet spinach white rust model (Jabrezemski and Sutherland, 2006). The bold portion of the solid line in the graph in Fig. 7.2 represents the forecast of spinach white rust infection hours calculated from NAM forecast values.

Forecasting is being added to more and more online insect pest models. One example, in North America, is the online phenology and degree-day models produced by the Integrated Plant Protection Center at Oregon State University, and the Pacific North-west Coalition (http://pnwpest.org/cgi-bin/ddmodel.pl).

Weather forecasts are also being used in disease models: examples include fruit and vegetable disease models available to New Zealand farmers from HortPlus (www.hortplus.com), pest models created by Dacom Plant Service, Netherlands (http://www.dacom.nl) and nut and vegetable disease models operated by the Oklahoma Mesonet, USA (Carlson and Sutherland, 2006).

Environmental Monitoring Networks

History

Today we take automated tasks for granted. When it comes to weather data collection, automation is a relatively recent technology. Even routine, hand-collected weather data are a relatively new event. Early efforts to organize weather data collection began in the UK in the 1820s and 1830s (Walker, 1993). In the USA, the first instrumented weather observations were taken as early as 1715 (Fiebrich, 2007), but it wasn't until the 1800s that attempts began to collect data from broad geographical regions. In 1844, the Smithsonian Institution in collaboration with the US government was collecting weather observations from observers across the country. The invention of the telegraph was the technology advance that allowed weather information from many locations to be collected for analysis at a single location.

Advances in technology have driven meteorological observation and observation systems. The first major technological advance was a thermometer that could be used outside of the laboratory; then came the telegraph; next were advances in radars, with the first system set up to monitor storms in Texas, Oklahoma, Kansas and Nebraska in 1947 (Anon., NOAA personnel, 1947). The first system of automated weather sensors was a set of weather buoys set up in the early 1940s near Toronto, Canada (Fiebrich, 2007).

During the 1960s, the US National Weather Service was in the early stage of adapting computer technology to transfer, ingest, store and retrieve weather data (Fenix, 2006). In 1969, the US National Weather Service designed the Remote Automatic Meteorological Observing System (RAMOS) to collect weather data for aviation use (Fiebrich, 2007). A second system, Aviation Automated Weather Observing System (AV-AWOS), came about in the late 1970s. The late 1980s was a time of

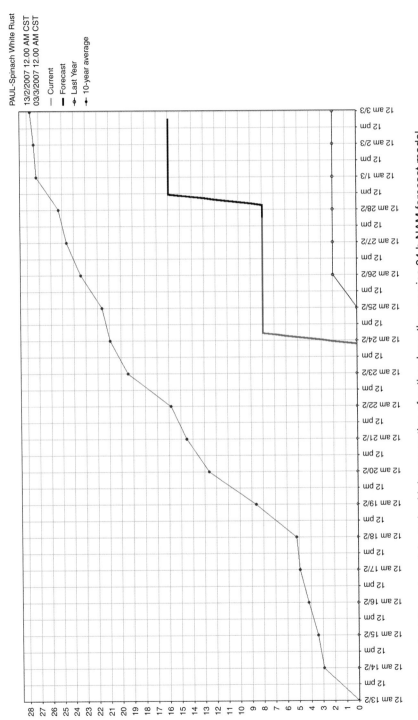

Fig. 7.2. Spinach white rust model graph with incorporation of national weather service 84 h NAM forecast model.

increased deployment of automated weather stations by the US government (Fiebrich, 2007).

Around the same time, non-federal automated weather stations and networks were being developed and deployed across the USA and Canada (Meyer and Hubbard, 1992). Fuelled by the need for more specific meteorological data for agricultural and other purposes, the state and private sector became involved. Spurred on by the need for real-time and near-real-time data, data not routinely reported by federal sources (e.g. solar radiation) and the greater spatial density of stations, there was a dramatic growth in the number of such stations and networks in the 1980s. This has continued onward through the 1990s (e.g. the Oklahoma Mesonet became operational in 1994) and into this decade. The survey conducted by Meyer and Hubbard (1992) showed 831 automated weather stations and 150 mobile stations across the USA and Canada. With respect to 'networks', 51% of them had five stations or fewer; 35% had between six and 20 and 14% had more than 20.

One of the more prominent non-federal networks is CIMIS, the California Irrigation Management Information System, consisting of over 120 automated weather stations throughout California. The IPM programme at University of California-Davis uses these stations to offer a wide range of IPM products, including crop-specific degree-day models, insect models and disease models (http://www.ipm.ucdavis. edu). Another state of the art network, the Oklahoma Mesonet, is discussed in following sections.

It is interesting to note that automated weather systems became a reality only after computer technology had been developed to make automatic data collection, transmission and ingest possible. Data collection, one of the most basic functions of meteorology, was being automated 20 years after deployment of the first operational radars and at the same time as satellites. The first geostationary satellite, GOES 1-M, was launched by the USA in 1975 (http://www.oso.noaa.gov/history/operational.htm).

Areawide networks – advantages and disadvantages

Regional weather systems offer a wide view of the environment and environmental changes. The weather data they provide can be used to create large-scale maps that indicate where early-season pest activity is more likely. Collecting data from multiple sites provides a broad view of environmental changes. It provides a perspective on the environmental changes in nearby locations and an indication of how the current season is progressing in relation to climate norms. An example can be seen in the Oklahoma map of lucerne weevil degree-day units accumulated between 1 January and 26 March 2006 in Fig. 7.3. In the southern part of the state the degree-day units had reached a high of 694, while in the colder region of the western Panhandle only 290 degree-day units had accumulated. Such information can be used by lucerne producers in southern parts of the state to begin scouting programmes as warranted by the earlier spring heating in their region of the state. This model helps southern growers avoid the mistake of waiting too long before scouting lucerne fields. The model helps northern growers from wasting labour resources by scouting too early.

The advantage of areawide weather-based systems can also be their weakness. A single station has the advantage of more accurately reflecting the crop microclimate.

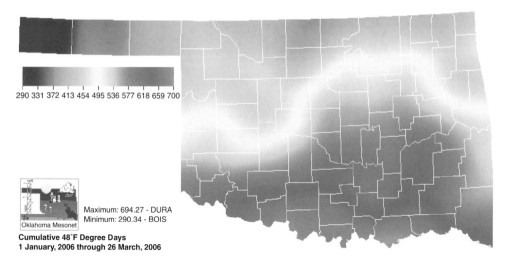

290 331 372 413 454 495 536 577 618 659 700

Maximum: 694.27 - DURA
Minimum: 290.34 - BOIS
Oklahoma Mesonet

Cumulative 48°F Degree Days
1 January, 2006 through 26 March, 2006

Fig. 7.3.   Map from the lucerne weevil model of cumulative base 48°F (9°C) degree-days as of 26 March 2006.

The Encyclopedia Britannica Online defines microclimate as any climatic condition in a relatively small area, within a few metres or less above and below the Earth's surface and within canopies of vegetation (http://www.britannica.com/eb/article-9052497). Individual weather stations are an excellent way to monitor local field crop conditions. The tower locations of automated weather monitoring networks may not adequately reflect individual field situations (Shock *et al.*, 2004). In these cases, it may be necessary to install a weather station that can monitor field environmental conditions. The need for in-field weather monitoring is more critical for diseases, since disease outbreak is highly dependent on moisture (e.g. leaf wetness, relative humidity, precipitation).

The value of a regional weather system comes with the long-term data collection, management, dissemination and cost. In order to reap the benefits of an in-field weather station it needs to be moved as seasonal crops are rotated between fields. Moving a weather monitoring station terminates the data that could build to create a climate record. Automated weather monitoring stations that are part of a weather system are located for quality of weather data and site permanence. Permanent stations provide long-term data that build over time to become a climate record. Regional weather networks are typically managed by weather experts and qualified technicians. Their expertise is important in maintaining the quality of the data through instrument calibration and maintenance. A grower-oriented system is less likely to be maintained. If moved, the grower may not have the expertise to properly site and set up the station in the new location. Using data from non-calibrated sensors as input to a model introduces new errors into the model output, resulting in errors in pest prediction.

Another advantage of weather networks over singly owned stations is access and dissemination of the data. In the case of on-farm stations, the producer owns the weather data and may choose not to make it available to others. Even if the data owner is willing to share the data, these may not be in a proprietary format that can be accessed by others. On the other hand, weather networks can efficiently disseminate data via the Internet to maximize data access by agricultural producers and professionals. Thus IPM implementation can be greater when areawide networks are used, in contrast to single station sites.

Installation and maintenance costs have been obstacles to establishing regional weather network systems. The cost advantage of installing regional systems is that the funding can be distributed between more people. Reliable regional weather monitoring systems can be used for weather forecasting, agriculture, construction, insurance, emergency management and education, as well as by the public. These systems greatly enhance the meteorological communities' ability to forecast and monitor weather events. What first appears as an expensive undertaking is actually a valuable environmental monitoring infrastructure with a high benefit–cost return.

If each farmer in a region installed their own weather station, the cost of data could be substantially higher than a regional system. Typical orchard systems can range from US$500 to 6000 (Travis, 2006). The data might or might not be shared, depending on the attitude of the individual and local technology capabilities. Data from the individually purchased weather stations will have different sensor manufacturers and/or models. Sensors on these systems may or may not be calibrated by the manufacturer. In the field, there would be no quality assurance testing and only the rare farmer would conduct regular sensor calibration.

Funding and support for environmental monitoring systems

Weather data collection requires stable funding sources. For agriculture, more than just the current data are of value. Long-term data collection provides a climate record that also has great value. Climate data provide a prediction of what to expect in any given location. With the warming climate (IPCC, 2007), long-term weather data are critical in plotting how a specific location's weather patterns are changing.

Traditionally, people involved in agricultural production undertook weather data collection. Advances in computer technology and weather sensors have allowed automated environmental monitoring systems to be installed to augment human data collection.

A wide variety of environmental monitoring networks has been installed across the globe. Those countries with more scientific expertise and available resources have installed the most sophisticated systems. The political climate and weather expertise of countries across the globe has had a big impact on funding and support for environmental monitoring systems. The developed countries, such as the USA, those within the European Union (EU), China, India, Japan, Canada, Australia and New Zealand have the most extensive and sophisticated systems. Other countries are adding meteorological data collection and expertise as resources allow. The World Meteorological Organization (WMO) is working to help less-developed countries establish weather data collection systems and enhance government meteorological agencies.

Countries that are leaders in environmental monitoring systems are those who have actively pursued applications of weather data. An example of commitment to securing weather data has been investment in weather satellites. By the end of 2005 the following countries had operational weather satellites: the USA (GOES-10, GOES-12, NOAA-17 and NOAA-1), the EU (Meteosat-5, Meteosat-7 and Meteosat-8), China (FY-2 C and FY-1 D), Japan (MTSAT-1R), the Russian Federation (METEOR-3 M N1) and India (Kalpana-1) (WMO, 2005).

Weather monitoring systems require a commitment from a broad group of supporters. The Oklahoma Mesonet was started with an initial investment of US$2.1 million in 1993. As of 2006, the initial investment is close to the annual budget needed for system maintenance, data collection, quality assurance and data distribution. Additional funding from federal and state grants provides the support needed to test new ideas and explore promising areas of research.

Funds for operation of the Oklahoma Mesonet have come primarily from state funding. The value in this is that network decisions are made by Oklahoma-based leadership. These administrators have local knowledge of the data needs and unique environmental characteristics. Local management has been key to creating products that best serve Oklahoma agriculture.

The funding model in Europe is quite different from that in North America. European countries do not provide weather data for free: instead, data access is fee based. This has created solid funding for weather data within each country. The downside of this funding approach is that data are hard to share between countries, inhibiting regional data use and products.

The Oklahoma Mesonet, a State-of-the-art Environmental Monitoring Network: the Network

In 1987, meteorologists from the University of Oklahoma joined with scientists from Oklahoma State University to design and seek funding for an advanced automated weather monitoring system. Their vision was to create a uniform system of towers across the state of Oklahoma. All sites would have identical calibrated sensors, the data would be quality assured and data would be automatically transmitted to a central site for storage and dissemination. The system would be known as the Oklahoma Mesonet. The attention to system uniformity and data quality has made the Oklahoma Mesonet one of the premier weather monitoring systems in the world. Its unique characteristics were formally recognized through a special citation from the American Meteorological Society in 2005.

One of the original founders, entomologist Gerrit Cuperus, was instrumental in creating integrated pest management products. Dr Cuperus also challenged those on the Oklahoma Mesonet steering committee to explore avenues to serve a variety of weather data and product users.

Operational since 1994, the Oklahoma Mesonet is a statewide, automated weather station network operated jointly by the University of Oklahoma and Oklahoma State University (Elliott *et al.*, 1994; Brock *et al.*, 1995; McPherson *et al.*, 2007). With an annual budget of US$2.2 million, Mesonet obtains about 81% of its funding from the state of Oklahoma. Each Mesonet station consists of an automated instrument tower (10 m in height) along with other sensors within a 10×10 m plot of land. A typical Mesonet weather station is shown in Fig. 7.4.

A wide array of weather and soil variables are measured. Weather variables include air temperature (1.5 and 9.0 m), relative humidity (1.5 m), solar radiation, wind speed (1.5 and 10 m), wind direction (10 m), rainfall and barometric pressure. Soil variables include soil temperature at various depths (5 and 10 cm under bare soil; 5, 10 and 30 cm under sod cover) and soil moisture at various depths under sod cover (5, 25 and 60 cm).

The Oklahoma Mesonet is a *mesoscale* network with respect to both space and time. With respect to space, the network in 2007 consisted of 116 sites with an average spacing of 30 km (19 miles). The tower locations are shown in Fig. 7.5.

Fig. 7.4. Oklahoma Mesonet 10 m automated weather station tower at Stillwater, Oklahoma.

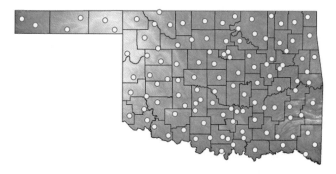

Fig. 7.5. Location of automated weather station sites in the Oklahoma Mesonet.

Note that there is at least one station in every county; some counties have three or four stations.

With respect to time, the reporting of weather and soil observations by the Mesonet also falls within the mesoscale range. For this purpose, the Oklahoma Mesonet uses an already existing statewide telecommunications network, the Oklahoma Law Enforcement Telecommunications Network (OLETS). Observations are sent by radio signal to a nearby OLETS tower (or by a repeater to an OLETS tower). Weather observations are sent every 5 min, while soil temperatures are sent every 15 min and soil moisture measurements every 30 min.

For clientele in agriculture and natural resources, one of the major benefits of the Oklahoma Mesonet is the spatial density of the network. With the average station spacing of 30 km, growers and others are typically within 15 km (9 miles) of a Mesonet tower and are able to use essentially local weather information. Prior to Mesonet, all that was available in near-real-time was information from synoptic scale networks.

The other major benefit is the real-time availability of the data. Not only are weather and soil observations made and sent in time intervals ranging from 5 to 30 min, but these observations are made available on the Internet within minutes of being received at the Oklahoma Climatological Survey in Norman, Oklahoma. Examples where such timeliness is important include prescribed burning and wildfires. This temporal and spatial density is useful for a wide range of other applications, such as weather and hydrological forecasting.

Since 1994, the Oklahoma Mesonet has become an increasingly used source for weather-based decision-support products for the agricultural and natural resources community. During the first 2 years, such products were offered to this community via a bulletin board service, whereby files were downloaded over telephone lines and then viewed using Mesonet-developed software. With the increasing availability of the Internet to clientele, this method of product distribution began to be replaced by the World Wide Web starting with the first agricultural web site in March 1996, which featured the Oklahoma Fire Danger Model. The advent in 1996 of this Internet product and others was particularly well timed, coinciding with the National Weather Service's cancellation of its agricultural and private fire weather services.

Products for agriculture and natural resources are currently offered on the 'Oklahoma Agweather' web site (http://agweather.mesonet.org). The home page as of 2007 is shown in Fig. 7.6. Note that there are menu icons not only for weather and soil products, but also for specific commodities and markets.

Many of the products rely on 'WXSCOPE' plug-in software developed by the Oklahoma Climatological Survey (OCS). This software allows the creation of speciality maps, graphs, tables and other products on the user's local computer (only data are downloaded from OCS). With respect to map products, the software allows the viewer to zoom in and out, add county and other geographical boundary overlays and animate the maps. An example of a contour map for relative humidity is shown below in Fig. 7.7. It shows the location of a 'dry line', which is a sharp discontinuity of relative humidity. Relative humidity values ahead of (to the east) of the dry line are 60% or higher, while values behind the dry line fall to values lower than 15%. The day illustrated was one of extreme fire danger in the very dry and windy air behind the dry line.

Fig. 7.6. The 'Oklahoma Agweather' home page.

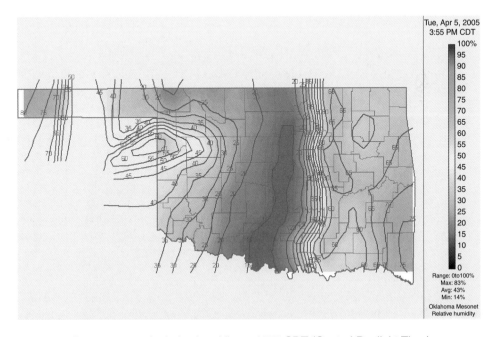

Fig. 7.7. Contour map of relative humidity at 1555 CDT (Central Daylight Time) on 5 April 2005 (produced by the WXSCOPE Plugin software).

Decision-support products

A variety of decision-support products is available on the Oklahoma Agweather site. They fall generally into three main areas: (i) weather and soil products based on current, recent or historical data; (ii) weather-based models for specific applications in agriculture and natural resources; and (iii) forecasts.

Weather and soil products

A wealth of products, most using the plug-in software, are available, for various weather and soil variables measured directly by or calculated through use of Mesonet data.

With respect to weather data, standard synoptic maps are available, as are speciality maps for temperature, humidity (relative humidity and dew point), wind speed and direction, rainfall, atmospheric pressure, solar radiation and lapse rate conditions near the surface. Through use of the plug-in software, many of these maps permit zooming, geographical overlays and animation. An example of a colorized contour map for solar radiation is shown in Fig. 7.8. The zooming feature has been used to focus in on north-east Oklahoma.

Meteograms, which are graphs/charts of various weather variables from a point in the past to the current time, are also available for specific Mesonet sites. Figure 7.9 shows a 24 h meteogram for the Jay station. Air temperature and dew point are shown on the first chart, while wind information is shown on the second.

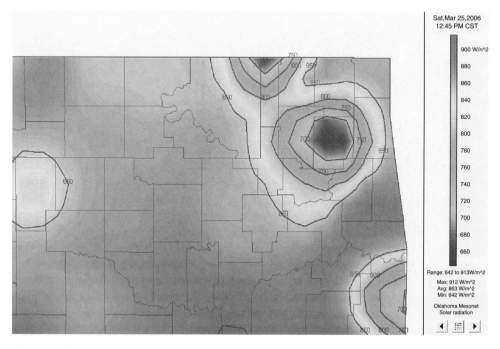

Fig. 7.8. Contour map of solar radiation at 1255 CST (Central Standard Time) on 25 March 2006.

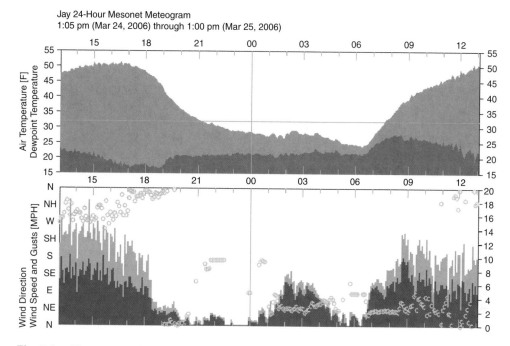

Jay 24-Hour Mesonet Meteogram
1:05 pm (Mar 24, 2006) through 1:00 pm (Mar 25, 2006)

Fig. 7.9. Meteogram of temperature and wind conditions over the previous 24 h at Jay, Oklahoma.

Other types of products include tables. Examples include cumulative rainfall and short- and tall-grass reference evapotranspiration for specific Mesonet sites over various time periods (e.g. previous 7–90 days).

Soil products include maps for soil temperature and soil moisture at various depths. With respect to soil temperature, maps showing current values at 5, 10 and 30 cm are available as are maps showing averages over the past 1, 3 and 7 days. With respect to soil moisture, maps of 'fractional water index' are available showing the fractional amount of saturation at that level (0 = no water, 1 = complete saturation). Figure 7.10 shows a contour map of fractional water index at 60 cm depth.

Other Mesonet data products include monthly station site summaries showing both daily and monthly statistics for the site (e.g. daily/monthly averages and extremes for such variables as temperature, and daily/monthly totals and extremes for rainfall). Historical Mesonet weather and soil data, in digital format, are also available for research studies and other purposes. While not based on Mesonet data, it should be mentioned that the user also has access to satellite and NEXRAD radar imagery, the latter of which uses the plug-in software, thus allowing zooming and animation. Figure 7.11 shows a thunderstorm line firing up just ahead of the dry line depicted in Fig. 7.7.

Weather-based models for specific applications

The Oklahoma Agweather web site features a variety of weather-based models for agriculture and natural resources. While these could be categorized in a number of

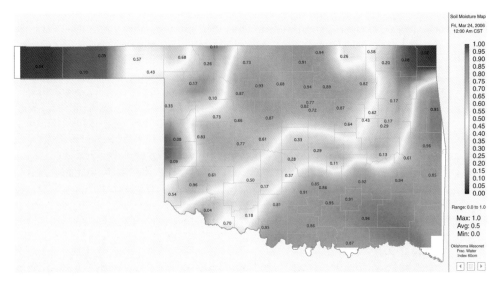

Fig. 7.10. Soil moisture map showing fractional water index at 60 cm depth as of 24 March 2006.

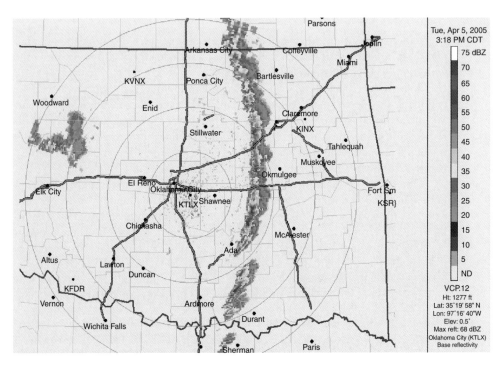

Fig. 7.11. Formation of thunderstorms ahead of the dry line at 1518 CDT on 5 April 2005.

different ways, one illustrative way is based on how many Mesonet weather variables are used in the particular model.

A number of models are based solely on temperature and, in particular, on cumulative degree-days from a specific start date and using a specific temperature threshold for phenological development. As of 2007, two Oklahoma Mesonet insect management models fall into this group, the lucerne weevil and pecan nut casebearer models. These models use degree-day accumulation to predict a particular damaging growth stage of the insect pest, thus fostering optimal timing of scouting activities and insecticide application. The models are updated once daily. The pecan nut casebearer model is based on base 38°F (3.3°C) degree-days from site-specific start dates ranging from 21 March to 24 April. The lucerne weevil model is based on base 48°F (8.9°C) degree-days from 1 January (see Fig. 7.3). Aside from the two insect models, there are also degree-day calculators for various crops (e.g. maize, soybeans, cotton), where the user can enter a specific start and end date for the degree-day accumulations.

A number of models are based on temperature and relative humidity. As of 2007, there are four disease management models for peanut leafspot, pecan scab, watermelon anthracnose and spinach white rust. Scientists at Oklahoma State University have developed these models through field research. The models are based on cumulative 'infection hours', which are periods when certain temperature and relative humidity constraints have been met. If the grower has sprayed a fungicide during the current crop season, the date of the last application is also taken into account (see Fig. 7.12). Two of the models, pecan scab and spinach white rust, have an 84 h forecast integrated into the model to estimate infection hours over the coming 3 days.

Some models are calculated from more than two weather variables. In this category are included models for lesser grain borer flight, cattle heat/cold stress, evapotranspiration and irrigation scheduling, atmospheric dispersion and fire danger. The atmospheric dispersion model is relevant to IPM, as a guide for applying pesticides during optimal dispersion conditions. As of 2007, these models will use only Oklahoma Mesonet data to calculate model output (i.e. there is no predictive component). In the years ahead, 3-day weather forecasts will be integrated into all Oklahoma Mesonet agricultural and natural resource models, as has been done for two of the disease management models.

Forecast products

The Oklahoma Mesonet, being a weather and soil data reporting system, does not make forecasts. However, forecasts are especially important to those in agriculture and natural resource management, so the Oklahoma Agweather web site includes links to various forecast products from the US National Weather Service (NWS). Aside from links to NWS forecast products, there is a stand-alone forecast product for use within the agriculture and natural resources community. This product uses 60-h MOS (Model Output Statistics) forecast output from the NGM (Nested Grid Model) for locations within and surrounding Oklahoma (NWS, 1992). Tabular as well as graphical (maps) output is available. The tables are specific to the MOS locations and include forecasts for various weather variables in 3 h increments throughout the 60 h forecast period. Output from the Oklahoma Dispersion Model has been added as well. These forecasts are updated twice daily.

Fig. 7.12. Spinach white rust decision support web page.

The Internet as a Data Conduit

The Internet has made it possible instantly to deliver specific data to anyone with Internet access. While television and radio offer the same outreach, they differ markedly from the Internet in data delivery. Television and radio broadcast professionals determine what is of interest to the greatest number of viewers or listeners. The Internet, on the other hand, is dramatically different in that the end-user selects what they are interested in. Not only does the user select the product or information of interest, they also choose how much time they spend viewing a product or information set.

Computers have made it possible to rapidly collect, test, catalogue and transmit weather data. An example of this can be found with the Oklahoma Mesonet. Weather data are collected from over 116 locations spread across 181,186 km^2 (69,956 square miles) in the state of Oklahoma. These stations transmit weather data every 5 min to a central location in Norman, Oklahoma. Once on computers in Norman, all data are tested by quality assurance software to determine whether they pass quality control checks. The data are then used to update over 66,000 products and web pages maintained by the Oklahoma Climatological Survey. The most recent weather and product data are viewable via the Internet in 6–7 min from the time of data collection.

The Internet makes it possible to deliver updated weather information to tens of thousands of individuals within minutes. This completely dwarfs the number of people who could be reached in the past via telegraph, telephone and fax technology. Radio and television can send out data just as rapidly to mass audiences, but only the data or information the station selects for broadcast.

When technology is available and can aid farmers, they make use of it. The vast majority of young farmers and ranchers in the USA have gained access to Internet communications technology. As of 2004, in the 12th annual American Farm Bureau survey of young farmers, which included 342 producers from 45 USA states between the ages of 18 and 35, 92.4% indicated they had a computer at home or on the farm and 88.3% had Internet access (Thornton and Lipton, 2004). Five years earlier, in 1999, only 52.2% of the surveyed young farmers and ranchers had Internet access. In the 13th survey, computer use rose slightly from 92.4% in 2004 to 94% in 2005 (Keller and Taylor, 2005).

One shortcoming with the Internet is that not all farm households have Internet access. Thus, these households are not able to access Oklahoma Mesonet data or products. The good news is that the vast majority of farmers consider having Internet access critical to their farm enterprise. Not only is the Internet important for weather information, it provides access to market prices, market news, equipment location, product information, crop cultural information, etc.

In Oklahoma, rural phone cooperatives and major telephone companies have made it a priority to bring high-speed Internet access (DSL or cable Internet) to rural communities (Pioneer Telephone Cooperative, AT&T). This has helped to increase the quality of Internet service and the number of Oklahoma farms with Internet access.

Even with Internet access at the farm headquarters, access has not been available in the field. In 2006 and 2007, Internet access via cellphones is slowly becoming more common, but access is limited to large municipalities and narrow corridors along interstate highways in Oklahoma and the rest of the USA. Long-range wireless Internet has been talked about for many years, but at the beginning of 2007 this technology, too, is not yet in rural locations.

One of the technology options that has not been pursued by the Oklahoma Mesonet is the use of mobile telephones to transmit pest information. This would be a valuable service to agricultural professionals and producers: cellphone use by agricultural professionals and producers in the USA is high. In the 2004 American Farm Bureau's young farmers and ranchers survey, 89.7% reported that they had used a cellphone for farm operations (Keller and Taylor, 2005).

Worldwide mobile telephone use has surpassed that of fixed telephone lines. As of 2004, the International Telecommunications Union reported that there were 77 mobile telephone subscribers and 54 fixed telephone lines per 100 inhabitants in developed countries (ITU, 2006). In developing countries, in the same report, there were 28 mobile telephone subscribers and 18 fixed telephone lines per 100 inhabitants. Individual countries' figures can vary significantly from such surveys, one example being the estimated 2 million cellphone subscribers and only 20,000 fixed-line telephones reported in 2007 in Afghanistan (Foster, 2007).

This wide and rapid proliferation of cellphone technology and its ability to receive voice and text make it a valuable communications tool in reaching agricultural professionals and producers. The cellphone makes it possible to provide simple weather and pest data in a timely manner wherever needed.

Grower and Crop Consultant Educational Programmes

With new and innovative weather-based models, what are the ways that can be used to encourage farmer use of these models to better time pest management actions and reduce pesticide use? Weather-based models must be decision-support tools that maintain or increase crop yield and/or quality, while minimizing crop inputs. Crop producers need tools that increase their profit margin. Once a model has been accepted as having economic value, growers and crop consultants need to be taught how to interpret model data and use the model.

Successful farmers and crop consultants know from experience when crops are in good health or need attention. Weather-based models do not override human knowledge and experience. Weather-based computer models quantify this knowledge and experience. Quantification allows the farmer to make more precise management decisions and record crop information. This new level of decision support creates a more precise measure of pest activity and management. It provides a better approach to assess the economic value of pest management strategies.

Sometimes from a university or government perspective, agricultural producers and professionals seem slow to adopt new technology. It is not out of slowness, but out of caution that farmers are hesitant to implement new technology. Each farmer lives and works with the risk of crop failure. In underdeveloped nations, when crops fail, the farm family goes hungry. In developed nations, the family may not go without food, but the farmer may be forced to quit farming and liquidate farm assets. With such a constant risk facing them, farmers want verification that new technologies will be profitable before they risk money, time and crop value.

The accuracy of weather-based models is a consideration in their use. Growers must have confidence that model errors will not cause crop losses. Models need to be conservative estimators of pest activity. Model errors that predict insect activity when there is none, or higher activity than is present in the field, can be tolerated. Growers cannot tolerate models that predict no activity when damaging pests are present.

When using models, it is important for users to remember that models are only tools. Models have limitations and are not perfect predictors. Agricultural professionals and producers can use models to improve the effectiveness of their scouting programmes. The models can alert them to times when they need to be more vigilant of pest outbreaks. Weather-based computer models do not replace the need to inspect plants during field visits.

The Oklahoma Mesonet has an outreach programme that works in cooperation with Oklahoma State University Cooperative Extension specialists and county educators. To manage the agricultural outreach programme there is one faculty and one professional, as of 2007. Their duties involve conducting educational meetings, holding product demonstrations at in-state farm shows, production of a monthly newsletter, creation of educational materials in a variety of formats and one-on-one contacts. These educational methods and materials are the resources individual agricultural professionals and producers have available to learn how to interpret and use Oklahoma Mesonet pest models.

In 2003, there was a major revision of the Oklahoma Agweather web site and implementation of detailed web statistics. In 2003, the average number of unique visitors per month was 3691. The average number of unique visitors per month had increased by 13% in 2004 to 4173. In 2005 this had increased by 16% to 4856 (average number of unique visitors per month); in 2006, the increase was 29% to 6298. With an estimated potential Oklahoma agricultural audience with Internet access of 100,000, this leaves a lot of room for growth in the use of the Oklahoma Agweather web site.

In the USA, farmers have extensive access to free weather data and weather-based decision support models. While much of the data and model output is free, they must invest personal time to learn how to use and interpret pest models, or hire a qualified consultant. For growers with limited exposure to the Internet and computers, learning how to interpret and use Internet-based pest models can be a daunting task. US growers with large farm operations often secure access to a crop consultant. As of 2007 in the USA and Canada, there were close to 14,000 Certified Crop Advisors (http://www.agronomy.org/cca/). Many Certified Crop Advisors work for agricultural service businesses, who provide crop consulting services to farmers who purchase fertilizer and plant protection products from them. A smaller percentage of Certified Crop Advisors are hired directly by the farmer.

Weather data access in Europe is fee based. In Switzerland, farmers and agricultural professionals must pay to access data, agricultural forecasts, pest models, radar and phenology bulletins from the government weather service, MeteoSwiss (http://www.meteoswiss.ch/web/en/services.html). A number of companies provide weather data access and pest models, such as Dacom Plant Service. Like Certified Crop Advisors in the USA, European weather service companies act as advisors to reduce the time the farmer must spend learning decision-support model functions and how properly to interpret model output.

Future Innovations

The future in AWPM will revolve around accuracy, automation, alliances and advances.

Weather forecasting will continue to become more accurate as technology advances. Forecasting accuracy has consistently tracked technology changes. Faster computers allow more data to be assessed in shorter timespans. Satellites provide a global view of weather patterns and weather events. The blending of advances in these technologies continues to advance the science of weather forecasting.

Automated weather monitoring networks will continue to expand: new automated stations are being installed daily across the globe and data from these networks are more accurate and consistent. Consistent high-quality data from these networks will also improve forecast accuracy.

Alliances will and have already been created to add weather as a component of pest management services. The alliance between Syngenta Corporation and Damcom Plant Services and Syngenta Corporation and Accuweather, Inc. are examples of this. Syngenta Corporation manufactures and sells pest control products. To provide more grower education and service they have contracted with Damco Plant Services to provide pest models and with Accuweather, Inc. for weather data. In 2007 in the

USA, this is being provided to agricultural producers and professionals at no cost via the Syngenta FarmAssist web site (http://www.farmassist.com). On Prince Edward Island, Canada, Atlantic AgriTech, Inc., Damcom Plant Services and Agriculture and Agri-Food Canada participated in a joint project to forecast European corn borer during the 2006 and 2007 potato production seasons.

As more is learned about insect and disease biology and the role weather plays in their development and behaviour, new pest models will be developed and older ones improved. Advances in satellite and radar technology will provide new and better areawide monitoring tools. These improvements will develop along with denser data networks and more accurate forecasts to create dependable, accurate pest prediction models.

Summary

Environmental monitoring is a relatively recent tool that brings new information to integrated pest management. When regional environmental monitoring networks are established, the weather data collected can be used to model insect activity and behaviour, as well as disease development, over the area covered by the network. The most important data collected from a regional network for pest management decisions are air temperature and relative humidity. Ironically, in an age of computers, satellites and radar, the building of regional environmental networks to collect these two basic weather parameters is in its infancy.

The Oklahoma Mesonet has been a trendsetting environmental network in tower distribution, weather data collection, data quality assurance, product development and outreach educational activities. Forecast output from the National Weather Service is being added to Oklahoma Mesonet data to create integrated pest management decision support models that provide additional lead-time for agricultural producers and professionals.

Many industries, of which agriculture is only one, can benefit from regional environmental network weather data. It takes a diverse group to gain the political support necessary to fund environmental network installation and ongoing maintenance. Data from regional networks have day-to-day value and, over time, can be used to build a climate record.

The future will see more sharing of data and more alliances between private and public sector entities to better serve the public. AWPM and traditional pest management can be enhanced through alliances between the various entities involved in agricultural pest management.

References

Anon., NOAA personnel (1947) *NOAA History – Stories and Tales of the Weather Service: Radar Detection*. National Oceanic and Atmospheric Administration (http://www.history.noaa.gov/stories_tales/radar_detect.html).

Aylor, D.E., Taylor, G.S. and Raynor, G.S. (1982) Long-range transport of tobacco blue mold spores. *Agricultural Meteorology* 27, 217–232.

Brock, F.V., Crawford, K.C., Elliott, R.L., Cuperus, G.W., Stadler, S.J., Johnson, H.L. and Eilts, M.D. (1995) The Oklahoma Mesonet: a technical overview. *Journal of Atmospheric and Oceanic Technology* 12, 5–19.

Carlson, J.D. and Gage, S.H. (1989) Influence of temperature upon crop and insect pest phonologies for field corn and the role of planting date upon their interrelationships. *Agricultural and Forest Meteorology* 45, 313–324.

Carlson, J.D. and Sutherland, A.J. (2006) The Oklahoma Mesonet: decision-support products for agriculture and natural resources using a mesoscale automated weather station network. *Conference CD, 27th Conference on Agricultural and Forest Meteorology*, American Meteorological Society, 22–25 May 2006, San Diego, California, Paper 1.3, 8 pp.

Carlson, J.D., Whalon, M.E., Landis, D.A. and Gage, S.H. (1992) Springtime weather patterns coincident with long-distance migration of potato leafhopper into Michigan. *Agricultural and Forest Meteorology* 59, 183–206.

Davis, J.M. and Main, C.E. (1986) Applying atmospheric trajectory analysis to problems in epidemiology. *Plant Disease* 70, 490–497.

Domino, R.P., Showers, W.B., Taylor, S.E. and Shaw, R.H. (1983) Spring weather pattern associated with suspected black cutworm moth (Lepidoptera: Noctuidae) introduction into Iowa. *Environmental Entomology* 12, 1863–1871.

Drake, V.A. (1985) Radar observations of moths migrating in a nocturnal low-level jet. *Ecological Entomology* 10, 259–265.

Driever, G.F., Smith, M.W., Carlson, J.D., Duthie, J.A. and vonBroembsen, S.L. (1998) Field evaluations for a weather-based model for scheduling fungicide applications to control pecan scab. *1998 American Pathology Society Meeting*, Las Vegas, Nevada, 8–12 November 1998.

Duale, A.H. (2005) Effect of temperature and relative humidity on the biology of the stem borer parasitoid *Pediobius furvus* (Gahan) (Hymenoptera: Eulophidae) for the management of stem borers. *Environmental Entomology* 34, 1–5.

Duan, R.X. and Gillespie, T.J. (1987) A comparison of cylindrical and flat-plate sensors for surface wetness duration. *Agricultural and Forest Meteorology* 40, 61–70.

Elliott, R.L., Brock, F.V., Stone, M.L. and Harp, S.L. (1994) Configuration decisions for an automated weather station network. *Applied Engineering in Agriculture* 10, 45–51.

Fenix, J.L.R. (2006) *The National Weather Service Gateway: a History in Communications Technology Evolution*. National Oceanic and Atmospheric Administration, History web site, Tools of the Trade, Weather Prediction and Detection, Gateway History (http://www.history. noaa.gov/stories_tales/gateway.html).

Fiebrich, C. (2007) History of weather observations across the United States. Transitioning the historical climate archives to data from newly automated sites – maintaining continuity in the temperature climate record. PhD thesis, The University of Oklahoma, Norman, Oklahoma, Chapter 2.

Foster, M. (2007) *Cell Phones Vital in Developing World*. Associated Press, News Release, 27 January 2007 (Kansas.com, The Wichita Eagle, http://www.kansas.com/mld/kansas/business/ 16563247.htm).

Gillespie, T.J. (1994) Pest and disease relationships. In: Griffiths, J.F. (ed.) *Handbook of Agricultural Meteorology*, Oxford University Press, New York, pp. 203–209.

Grantham, R.A., Mulder, P.G., Cuperus, G.W. and Carlson, J.D. (2002) Evaluation of pecan nut casebearer *Acrobasis nuxvorella* (Lepidoptera: Pyralidae) prediction models using pheromone trapping. *Environmental Entomology* 31, 1062–1070.

Guarneri, A.A., Lazzeri, C., Diotaiuti, L. and Lorenzo, M.G. (2002) The effect of relative humidity on the behaviour and development of *Triatoma brasiliensis*. *Physiological Entomology* 27, 142–147.

Hirst, J.M., Stedman, O.J. and Hogg, W.H. (1967) Long-distance spore transport: methods of measurement, vertical spore profiles and the detection of immigrant spores. *Journal of General Microbiology* 48, 329–355.

IPCC (Intergovernmental Panel on Climate Change) (2007) *Climate Change 2007: the Physical Science Basis: Summary for Policymakers.* Intergovernmental Panel on Climate Change, Working Group I, 5 February 2007 (http://www.ipcc.ch/SPM2feb07.pdf).

ITU (International Telecommunications Union) (2006) *World Telecommunication/CT Development Report 2006: Measuring ICT for Social and Economic Development.* 8th edn., International Telecommunications Union (specialized agency of the United Nations) (http://www.itu.int/pub/D-IND-WTDR-2006/en).

Jabrzemski, R. and Sutherland, A.J. (2006) An innovative approach to weather-based decision-support for agricultural models. *American Meteorological Society International Conference on Interactive Information Processing Systems for Meteorology, Oceanography, and Hydrology, 86th American Meteorological Society Annual Meeting,* 1 February 2006, Atlanta, Georgia, Paper 12.7.

Keller, R. and Taylor, T. (2005) *Land Availability Concerns Young Farmer/Ranchers.* American Farm Bureau Federation, News Release, 18 March 2005 (http://www.fb.org/index.php?fuseaction = newsroom.fbnarchive).

Lorenzo, M.G. and Lazzari, C.R. (1999) Temperature and relative humidity affect the selection of shelters by *Triatoma infestans,* vector of chagas disease. *Acta Tropica* 72, 241–249.

McCorcle, M.D. and Fast, J.D. (1989) Prediction of pest distribution in the corn belt: a meteorological analysis. *Preprints, 19th Conference on Agricultural and Forest Meteorology and 9th Conference on Biometeorology and Aerobiology,* 7–10 March 1989, American Meteorological Society, Boston, Massachusetts, pp. 298–301.

McFarland, M.J. and Strand, J.F. (1994) Weather-wise planning in farm management. In: Griffiths, J.F. (ed.) *Handbook of Agricultural Meteorology.* Oxford University Press, New York, pp. 264–272.

McFarland, M.J., McCann, I.R. and Kline, K.S. (1991) Synthesis and measurement of temperature for insect models. In: Goodenough, J.L. (ed.) *Insect Modeling.* American Society of Agricultural Engineering monograph, St Joseph, Michigan.

McPherson, R.A., Fiebrich, C., Crawford, K.C., Elliott, R.L., Kilby, J.R., Grimsley, D.L., Martinez, J.E., Basara, J.B., Illston, B.G., Morris, D.A., Kloesel, K.A., Stadler, S.J., Melvin, A.D., Sutherland, A.J., Shrivastava, H., Carlson, J.D., Wolfinbarger, J.M., Bostic, J.P. and Demko, D.B. (2007) Statewide monitoring of the mesoscale environment: a technical update on the Oklahoma Mesonet. *Journal of Atmospheric and Oceanic Technology* 24, 301–321.

Meyer, S.J. and Hubbard, K.G. (1992) Nonfederal automated weather stations and networks in the United States and Canada: a preliminary survey. *Bulletin American Meteorological Society* 73, 449–457.

Mulder, P.G., McCraw, B.D., Reid, W. and Grantham, R.A. (2004) *Monitoring Adult Weevil Populations in Pecan and Fruit Trees in Oklahoma.* Oklahoma State University, Oklahoma Cooperative Extension Service, Norman, Oklahoma, Fact Sheet F-7190, 8 pp.

NWS (US National Weather Service) (1992) *NGM-Based MOS Guidance – the FOUS14/FWC Message.* Technical Procedures Bulletin 408, USA Department of Commerce, Washington, DC, 16 pp.

Pelletier, Y. (1995) Effects of temperature and relative humidity on water loss by the Colorado potato beetle, *Leptinotarsa decmlineata* (Say). *Journal of Insect Physiology* 41, 235–239.

Perez-Mendoza, J. and Weaver, D.K. (2006) Temperature and relative humidity effects on postdiapause larval development and adult emergence in three populations of wheat stem sawfly (Hymenoptera: Cephidea). *Physiological Ecology* 35, 1222–1231.

Platt, R.B., Love, G.J. and Williams, E.L. (1958) A positive correlation between relative humidity and the distribution and abundance of *Aedes vexans. Ecology* 39, 167–169.

Raney, H., Eikenbary, R.D. and Flora, N. (1970) Population density of the pecan weevil under 'Stuart' pecan trees. *Journal of Economic Entomology* 63, 697–700.

Schneider, J.M., Barbrecht, J.D. and Steiner, J.L. (2005) Seasonal climate forecasts: summary of current opportunities across the contiguous United States. *Presentation, American Society of Agronomy-Crop Science Society of America-Soil Science Society of America International Annual Meeting,* 6–10 November 2005, Salt Lake City, Utah.

Shock, C.A., Shock, C.C., Saunders, L.D., Jensen, L.B., Locke, K., James, S.R., Carlson, H.L., Kirby, D.W. and Charlton, B.A. (2004) Evaluation of the Wallin model for regional predictions of potato late blight in semi-arid regions of Oregon. *American Journal of Potato Research* 81, 88.

Thorton, M. and Lipton, D. (2004) *Young Farmer and Rancher Optimism Hits Survey High.* American Farm Bureau Federation, News Release, 18 March 2004 (http://www.fb.org/index. php?fuseaction=newsroom.fbnarchive).

Travis, J.W. (2006) *Collecting Weather Information in Orchards, Part 1: Cultural Information.* Pennsylvania Tree Fruit Production Guide, 2006–2007 edn., Pennsylvania State University (http://tfpg.cas.psu.edu/39.htm).

Walker, J.M. (1993) The meteorological societies of London. *Weather* 48, 364–372.

Westbrook, J.K., Wolf, W.W., Adams, S.D., Rogers, C.E., Pair, S.D. and Raulston, J.R. (1991) Observations and analyses of atmospheric transport of noctuids from northeastern New Mexico and south-central Texas. *10th Conference on Biometeorology and Aerobiology and Special Session on Hydrometeorology,* 10–13 September 1991, American Meteorological Society, Boston, Massachusetts, pp. 100–103.

WMO (World Meteorological Organization) (1988) *Agrometeorological aspects of operational crop protection.* WMO Technical Note No. 192, WMO, Geneva, Switzerland.

WMO (World Meteorological Organization) (2005) *Annual Report 2005: Weather, Climate, Water and Sustainable Development.* WMO-No.1000 : 4, WMO, Geneva, Switzerland.

8 The Role of Databases in Areawide Pest Management

VASILE CATANA,[1] NORMAN ELLIOTT,[2] KRIS GILES,[1]
MUSTAFA MIRIK,[3] DAVID PORTER,[2] GARY HEIN,[4]
FRANK PEAIRS[5] AND JERRY MICHELS[3]

[1]Oklahoma State University, Department of Entomology and Plant Pathology, Stillwater, Oklahoma, USA
[2]US Department of Agriculture, Agricultural Research Service, Plant Science Research Laboratory, Stillwater, Oklahoma, USA
[3]Texas A&M University, Agricultural Experiment Station, Bushland, Texas, USA
[4]Department of Entomology, University of Nebraska Panhandle R&E Center, Scottsbluff, Nebraska, USA
[5]Department of Bioagricultural Sciences and Pest Management, Colorado State University, Fort Collins, Colorado, USA

Introduction

The simplest definition of the term 'database' is given in Webster's dictionary as 'a comprehensive collection of related data organized for convenient access, generally in a computer' (Random House, 1996). This term appeared in the late 1960s because of the evolution of computer software and the need to distinguish the specialized computer systems for the storage and manipulation of data, called database management systems (DBMS) (Neufeld and Cornog, 1986). Today, the acronym 'DBMS' is universally understood within Information Technology (IT), just like the acronym '*Bt*' for '*Bacillus thuringiensis*' is in the field of biological pest control. At the present time there are numerous DBMS products available on the market. The most popular are Oracle©, dBase©, DB2©, MS SQL Server© and Access©. Access is a part of the Microsoft Office product and can be considered as a prototype of DBMS with limited functionality. These products vary in price and capacity, and therefore the budgetary constraints and the requirements of a particular database application determine their utility.

The evolution of database products has been rapid, reflecting advances in the theory of databases during the last 35–40 years. Beginning with simple data files with direct access, these database products now include very sophisticated file systems with complex interrelationships. More recently, there has been a series of new database applications named Relational Database Management Systems (RDBMS).

Their development was a product of the advancement in IT, which forced DBMS to adapt. One of these advancements was the creation of distributed computer systems using local or wide-area networks (LAN/WAN) at the end of the 1980s and early 1990s (Date, 2003). These networks stimulated development of new methods for remote database connection, and the improvements of client/server technologies when databases are organized on a computer server separated from those remotely accessed and used to enter data. The uses of DBMS can be very diverse, but this chapter focuses on the application of DBMS in the field of biology, more specifically in entomology and integrated pest management (IPM).

The first uses of databases in biology were mainly to share knowledge about a particular subject. An elegant example of this type of application is deciphering the genetic code of a particular organism. This type of application has been typically addressed by collaboration of research teams working at different geographic locations (King, 2004). For example, scientists communicating or collaborating through a database application announced in April 2003 that the human genome (DNA sequence) was completely decoded (Human Genome Project, 2003). The potential uses of DBMS in plant protection are diverse, and the benefits of using DBMS are not limited to information dissemination as suggested in some publications (e.g. Xia *et al.*, 2002). In fact, DBMS can be used to address many problems in biology, from research questions like deciphering the genetic code to building and maintaining national biodiversity inventories.

Database and IPM

Today, the acronym IPM is universally accepted to mean integrated pest management. Although there are many definitions of IPM, most stress the role of information and a systems approach to decision making in pest management (Kogan, 1998). In principle, IPM is an ecologically sound strategy that relies on natural mortality factors such as natural enemies, weather, crop management and pesticides when needed to control pests (Kogan, 1998). Thus, many IPM programmes seek to integrate non-chemical control tactics as much as possible for pest management. The US Department of Agriculture and the Environmental Protection Agency work to decrease pesticide application by demonstrating a variety of alternative control techniques to producers in order to minimize pesticide use. These techniques include biological control, genetic resistance, tillage, crop rotation and a wide variety of cultural control methods and other techniques (Dent, 1995).

Two often-ignored components of IPM implementation are database organization and data analysis for purposes of facilitating management decisions. The definition of IPM implies that managing pests is a complex process including many internal and external factors. The chosen strategy is, ideally, based on detailed knowledge of the current state of the agricultural system, future behaviour of the system and the options available for pest control. Appropriately integrating all of this information for a complex agricultural system is often beyond the capability of IPM practitioners. Clearly, control decisions in many IPM systems are data intensive, requiring the use of DBMS and decision-making tools, such as expert systems.

There are many examples of database implementation in IPM, a few of which are discussed in this chapter. At the beginning of the 1980s in the Netherlands, a disease and pest management system was implemented for winter wheat based on information collected in producer fields (Zadoks, 1981). The system was so efficient that other European countries developed their own version of the Dutch project (e.g. Germany and Norway). Folkedal and Breving (2004) give a detailed description of the wheat database implemented in Norway. In the middle of the 1980s in Byelorussia (ex-USSR), a mainframe-oriented database for Colorado potato beetle (*Leptinotarsa decemlineata* Say) was developed to optimize insecticide use in large areas.

During the 1960s in the UK, a network of 16 suction traps (Rothamsted Insect Survey) was activated mainly for aphid monitoring (Tatchell, 1991; Knight *et al.*, 1992). The data collected by the Rothamsted Insect Survey were used to describe fundamental factors affecting the dynamics of aphid populations and to provide information on aphid control for growers and other pest managers. Presently, there is a series of similar suction trap networks throughout Western and Central Europe. In 1999, various European countries decided to organize the EXAMINE project (EXploitation of Aphid Monitoring systems IN Europe). The information provided by the EXAMINE network has been very useful in helping to define the role of aphids, not only as pests but also as vectors of plant disease (EXAMINE, 2000).

Over the last 20–25 years, computer applications have focused mainly on simulation models and decision-support systems for pest management, and did not make progress toward extensive use of databases. For example, Legg and Bennett (1992, 1993) developed a computer-based decision support system to aid managers with Russian wheat aphid control in the western Great Plains of the USA, and implemented the system on hand-held computers used in the field. The system has components for selecting an appropriate economic injury level or threshold for automated decision making through a sequential sampling scheme, and for providing management support in cases where sampling continues to a user-defined maximum number without having reached a control decision. In another example, Mann *et al.* (1986) developed an advisory system for cereal aphid management. The system was essentially based on simulation models for aphid population dynamics.

Later systems developed along similar lines but included the creation of a database as a constituent part of the computer-based advisory system (Zintzaras and Tsitsipis, 2003; Elliott *et al.*, 2004). Perini and Susi (2004) developed a 'systemic' approach to build decision-support systems (DSS) used in agriculture by the advisory service for pest management in Italy. Murali *et al.* (1999) described the impact of a DSS product – 'PC – Plant Protection' on agriculture in Denmark after it had been commercialized in 1993. This system is distributed by the Danish Agricultural Advisory Centre and has been well accepted by growers because of increased profits flowing from the reliable and economically valid recommendations provided by the system.

As a rule, insect pests have very large geographic distributions; therefore both research and management may be more efficient when accomplished by multiple teams located throughout the dispersal range of the pest (Taylor, 1977; Elliott *et al.*, 1998). Implementation of a database can be very powerful if several teams simultaneously work on a common project at geographically separated locations. Area-wide pest management (AWPM) programmes typically are of this type (Faust and

Chandler, 1998), often with both state and federal research teams working in a coordinated fashion on specific demonstration sites at several geographic locations. After the establishment of the project goals, the entomologists and specialists in other relevant disciplines typically formulate tasks to standardize a common data collection protocol and to implement it at all locations. The protocol can set the standards for management practices, and can include the same sampling and monitoring methods for a project. When all the participants in an AWPM project follow the same protocol, with the evaluation of the data performed on a project-wide basis, the programmes are particularly amenable to, and in need of a database where information can be entered, analysed and summarized project wide.

Other IT Tools for Plant Protection

A common database in AWPM projects guarantees that compatible and concise data are collected from different geographic locations. One obvious, straightforward and useful addition to data collected at several geographic locations is to include the longitude and latitude using a Global Position System (GPS) receiver (Nemenyi *et al.*, 2003). Today, there are relatively inexpensive hardware and software products that can combine GPS with computers (Ohio Geospatial Program, 2003). These products determine the coordinates of the sample location with high accuracy and are very easy to use in field conditions. Moreover, the use of these tools can reduce the work involved in database development because of the electronically collected data from the field.

When the coordinates of the sample locations are built into the structure of an AWPM database, a bridge can be developed between the collected information and software that facilitates spatial data analysis. For example, Geographic Information Systems (GISs) can be used to visualize and preliminarily analyse the spatially referenced data. Zintzaras and Tsitsipis (2003) describe this type of application. Their information system includes aphid data from suction and yellow traps. It also has a Windows-based application that creates contour maps representing the spatial expansion of particular aphid species.

We argue that, in AWPM projects, a project-wide database should be considered as a starting point so that monitoring and management activities can be undertaken on a project-wide basis without excessive data recoding and compilation of data. Data recoding and compilation create inherent problems (errors and compatibility) when research teams operate independently and maintain separate databases, even if attempts are made to standardize data collection for all sites. With a database, data from all sites are entered into the database in a consistent format and are accessible for all participants to permit comparative analysis simultaneously. At the outset of such a project, it is important to develop and implement consistent sampling and monitoring methods to be used by all teams involved in the project. The use of hand-held computers for entering data in the field is helpful in this respect, because data entry templates can be constructed using software packages such as Farmworks Sitemate© at the beginning of the project. The templates ensure that all individuals who collect and enter data use the same format. As the data accumulate over time, the value of the database increases substantially for addressing questions of both a planned and unanticipated nature. This has been an important issue proved by projects such as EXAMINE, because continuously collected data over time and space have been invaluable for addressing research questions such as detecting and predicting the impact of changes in land use on the dynamics of aphids considered as key pests in many terrestrial agroecosystems (EXAMINE, 2000).

Today the majority, if not all, DBMSs permit application through web-based development. In this context, the work of Sivertsen (2005) is relevant. Sivertsen (2005) developed a database to combine an agrometeorological system with simulation models for making measurements and prognoses of crop development over time and for monitoring and predicting crop diseases and other pests. A web site containing real-time information on pest problems, densities and density fluctuations over time could be a valuable resource to alert growers and other pest managers of the pest status or need for pest monitoring in their area. The potential application of this approach is illustrated in Fig. 8.1.

A permanently maintained AWPM database linked to web pages can form the basis of a pest outbreak risk warning system for pest management. Proper use of regional or areawide pest databases with Internet technology has the potential to provide a valuable service to growers, with relatively little work required from persons involved in sampling and monitoring pest populations in agriculture, above and beyond their current workloads. While there is considerable effort required to create a database application for region-wide pests, once such a system is developed, tested and revised it can be extremely useful for real-time updating and visualizing of pest conditions over broad geographic areas.

When considered on an individual basis, where sampling may be limited to a few fields, the value of the data for assessing regional pest status is limited. But, when many individuals each make a small effort to collect data and combine it in a coordinated fashion, the database has the capability for triggering pest-scouting activities in a timely manner throughout a region. When sample points are automatically georeferenced using a hand-held computer with an attached ground GPS receiver

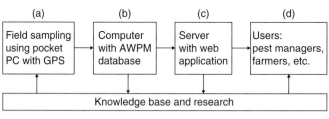

Fig. 8.1.   Flow of data from field to end-users: (a) data collection in the field using a palm-top computer equipped with a Global Position System unit; (b) data transformation from palm-top computer into an areawide pest management database; (c) connection between a database and an Internet application providing field data or descriptive statistics for end-users (pest managers, growers, scientists, administrators and other interested personnel); and (d) interpretation of data for pest management by end-users.

while collecting data, spatio-temporal analysis and display of the data can be easily accomplished. Such an analysis of spatially referenced population data is an obvious and logical activity to aid in understanding the spatial distribution and dynamics of the pest populations in relation to concomitant biotic and abiotic environmental factors.

Today, there are specialized statistical software packages such as the STATISTICAL ANALYSIS SYSTEM© (SAS Inc., Cary, North Carolina, USA) and SURFER© (Golden Software Inc., Golden, Colorado, USA) that facilitate construction of surfaces displaying pest density in space. Each surface can be considered analogous to a solution of a diffusion equation if we assume that the distribution of the pest species in an area follows the laws of thermodynamics. This subject falls outside the scope of this chapter, but it is notable that numerous papers dealing with mathematically described arthropod spatio-temporal population dynamics through partial differential equations have recently been published (e.g. Yamamura, 2002; Bianchi, 2003; Filipe and Maule, 2003; Eason *et al.*, 2004).

The information, accumulated in a well-constructed, comprehensive database, has potential utility long after the AWPM research programme ends. For example, data on the population dynamics of species collected over broad geographic areas and then compiled during a study can be a source of information for different space–time studies. It can be a database used to develop simulation models of pest population dynamics or to explore the utility of various pest management strategies. Once a database is constructed it requires little effort to maintain, and it has a much longer lifespan than the duration of a particular project.

An Example of Database Development

This section describes one way to develop an inexpensive database application for a spatially distributed AWPM project, but the reader should be aware that there are many other approaches and tools to accomplish the development of such a database (e.g. Isard *et al.*, 2006). In our example, database development requires only a dedicated computer with moderate capabilities and a link to the Internet. The computer should have either Windows 2000 Advanced Server or Windows 2003 Server operating system installed. We developed the DBMS using Oracle 9i RDBMS©. The Oracle 9i RDBMS©, with the choice of an operating system, can be downloaded at http://www.oracle.com. Oracle 9i kit installation has two distinct parts: the server and the client components. The server component is used to generate and install an instance of the Oracle database on a separate computer. The client component has to be installed on all computers that will be connected to the future database. It contains two distinct components: the Oracle Net Manager (ONM) and the library of classes that generate Oracle objects for database connection. The installation process of server and client components is very well documented and there are numerous sites on the Internet providing information about Oracle 9i (Oracle 9i, 2003).

During the database creation process, the developer will be prompted by the Database Configuration Assistant to give a unique Global Database Name to the future database. This unique name, with the server IP address and its port number, will be used by the ONM to establish a remote connection database via the Internet from a distant computer using TCP/IP protocol. The ONM is a very powerful tool that provides enterprise-wide connectivity solutions in a distributed, heterogeneous computing environment. It reduces the complexity of network configuration and management, maximizes performance and improves network diagnostic capabilities. It also enables a network session from a client application to an Oracle database server. Once a network session is established, the ONM acts as a data courier for the client application and the database server. The ONM is responsible for establishing and maintaining the connection between the client application and database server, as well as exchanging messages between them. The ONM can perform these functions because it is located on every computer in the network (Oracle9i, 2002; see Fig. 8.2).

The ONM is an alternative to the Microsoft ODBC (Open Database Connectivity) technology. From our point of view, ODBC is more complicated and less stable. One more reason to use the Oracle 9i RDBMS is that inside of Oracle products exists another tool: Oracle Objects for OLE, called 'OO4O', which is a COM (Component Object Model) component that facilitates software development. The OO4O can easily be integrated in Microsoft Visual Studio 6.0© and permits software development using Visual Basic or Visual C++. The OO4O can also be used from Microsoft Visual Basic for Application© (VBA), which is a tool of each Microsoft Office component.

For example, from an Excel table using VBA and OO4O, the user can easily establish a remote connection with an AWPM database. The code presented in Appendix A is an example of such a VBA application. It is presented in the OO4O documentation. The first 'Set' operator will start the 'Oracle In Process Server', which provides an interface between our application and the Oracle database. The second 'Set' operator will connect our Excel table to the 'Exampledb' database, where 'scott' is a user name and 'tiger' is his password. When an Oracle 9i database is generated, the 'Exampledb' database and the 'scott' user are created automatically.

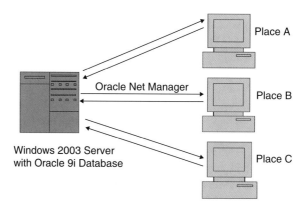

Fig. 8.2. A stable link between a Windows Server 2003 with an Oracle 9i database and a series of computers with different platforms at remote locations using Oracle Net Manager.

The third 'Set' operator will create a set of records and will select all the rows from the 'emp' table and assign the result to the global EmpDynaset variable. The parameter in this operator is an expression written in Structured Query Language (SQL), and it specifies what information will be withdrawn from the database. The next operators will put the contents of the EmpDynaset variable on 'Datasheet' worksheet of our Excel table.

Once the structure of a common AWPM database is finished, the respective tables can be generated inside the database using SQL. Appendix B contains an example of such a table creation. This example can be executed directly on the server or remotely from an application. Table organization should diminish the volume of stored information. For example, the repeated information should be stored in separate tables and be linked to other tables by common keys.

Once the database development is finished, we can start to develop a software application to facilitate remote data input. The steps described here are widely applied today between programmers. However, because this field of knowledge progresses rapidly it is likely that this approach will soon evolve.

As a rule there are several programming languages inside Microsoft Visual Studio (MVS): Visual Basic, Visual C++ and Java. The most recent version of MVS (MVS .NET) has one more language: C#. Another feature of this version is that all these languages have in their background the same hierarchy of classes; the applications developed in one language can be used in another. This work is done by the Microsoft .NET version 2.0 work frame.

The OO4O within MVS is used exactly in the same manner as in the example for VBA from Appendix A. It is a good idea to develop a common application for all participants in an AWPM project. It can be easily performed, for example, in Visual Basic. Once the application is developed it can be uploaded to an ftp site with necessary install instructions for all users. Such an application should be able to read Excel tables that will contain sampling data from all places involved in the AWPM project. It also should have the capability to check the format of all data before they are added to the database. For security reasons it is a good idea to protect the connection to the AWPM database by a user name and password. For this, it is enough to organize a separate user table inside the AWPM database. The creation of such a table is presented in Appendix B.

Once the AWPM database has enough information, and we need to share it with other people, it is a good idea to develop a web application that will let users visualize and download data. Such a web application can be developed within MVS. The latest version of MVS .NET has several facilities for this purpose. In particular, it allows development of a web application using the Active Server Pages (ASP) technology. ASP is Microsoft's product that enables Internet pages to be dynamically created using HTML, scripts and reusable ActiveX server components. ASP is a component of the Internet Information Server (IIS) that runs on Windows NT platforms. When a browser requests an ASP page from a server, the server generates the web page with HTML code and gives it to the browser. ASP.NET is the next version of ASP. ASP.NET allows programmers to use a fully featured programming language such as C# or VB.NET to build web applications rapidly. The ASP.NET is mostly used to access data from a database, and those data are then built into the returned web page 'on the fly'.

Our AWPM project web site will be used as an example of ASP technology: it is located at http://199.133.145.58/GPIPMWebApplication/WebAWPMPForm.aspx and is linked to an Oracle 9i database on a Windows 2003 Server. The structure of our database corresponds to the field-collected data and it is schematically represented in Fig. 8.3. Each rectangle stands for a separate table that corresponds to a distinct type of sampling; a common key links the tables. The AWPM project for wheat was conducted in the Great Plains of the USA from 2002 to 2006. This region was chosen because cereal crops, primarily rain-fed winter wheat, dominate it. Growers across the region use a range of agronomic practices for wheat production. In the south, growers commonly use winter wheat fields as pasture for cattle from November to February, whereas in the north wheat is used solely for grain production. Wheat efficiently uses the soil moisture accumulated during the winter, which is important in the Western Great Plains where annual precipitation is < 400 mm.

The major pests of wheat and other cereals, including barley and sorghum, in the Great Plains are several aphid species. Two of these aphids, the greenbug (GB), *Schizaphis graminum* (Rondani) and the Russian wheat aphid (RWA), *Diuraphis noxia* (Mordvilko), are common economically important pests. The RWA was introduced into the USA in 1986 from Mexico (Halbert and Stoetzel, 1998). Depending upon local climatic conditions, only one of these species is important, while both are important in others. The economic importance of these aphids was discussed in Morrison and Peairs (1998).

There are several objectives of the AWPM project, the most important of which are to: (i) estimate two distinct agricultural methods – with crop rotation and without it (non-traditional and traditional farming); (ii) improve the monitoring of aphid mega-populations on a large scale; (iii) determine the role of the parasites and predators as regulators of the aphid density; and (iv) investigate environmental conditions that cause aphid outbreaks. All these, together, should improve the potential for effective IPM through the development of forecast models and other predictive tools.

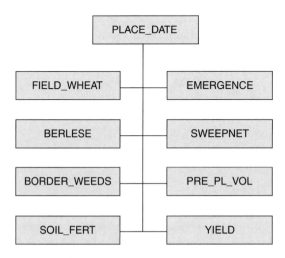

Fig. 8.3. The structure of Oracle 9i database; each rectangle represents a table, and tables are linked by a common key.

Our AWPM project involves universities in Colorado, Kansas, Nebraska, Oklahoma and Texas; these states represent the western and southern Great Plains. Up to six rainfed farms were selected in each state and a series of observations were made bi-weekly or monthly on wheat fields on each farm, depending on the insect abundance. There are 23 study farms, and the location of each farm is represented on our web site map by a shaded rectangle (see Fig. 8.4). The user can choose a farm by clicking on the corresponding rectangle and, on the next web page, will be prompted to choose one of several sampling methods.

Most of the fields sampled were rectangular, and size varied from approximately 100 to 130 ha. Each field was split up into 25 equal-sized grid cells, and one sample was established at the centre of each cell. Coordinates of each sample were taken with a GPS receiver to determine its location. The borders of each field were digitized using GIS software (FarmWorks Inc., Hamilton, Indiana, USA) installed on a pocket personal computer (PC). The pocket PC is equipped with a GPS unit, and the location of each sampling point was determined with high precision.

There were four different insect samples taken in the fields: two were made at each point and the remaining two were brought to the laboratory. We developed a simple template on a pocket PC to facilitate entry of each sample in the field. The first field sample was a visual count, and it consisted of examining four tillers in a

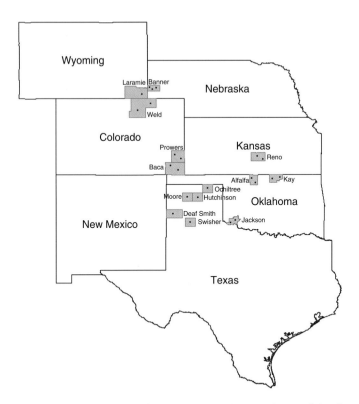

Fig. 8.4. Location of the sampling fields in the south-central part of the Great Plains, USA.

radius of 1 m around the main sampling point. Three groups of objects (insects) (aphids, aphid mummies and predators) were counted. The counts were directly entered in a template with 58 columns on the palm computer. The columns of this table were common for all sampling fields, and the table structure corresponded to the common protocol established by agreement among all research teams involved in the AWPM project. There were three distinct aphid species sampled: the greenbug, Russian wheat aphid and bird cherry oat aphid, and if there were other species they were entered in the column 'Other Aphids'. As a rule, this group consisted of the rice root aphid (*Rhopalosiphum rufiabdominalis*), the corn leaf aphid (*R. maidis*) and the English grain aphid (*Sitobion avenae*). Black and gold mummies were counted separately to distinguish two distinct groups of parasitoids.

The other visual sample made in the field is an aphid predator count in two-row-foot of wheat at each sample location. In this predator group, important species were individually distinguished, e.g. *Hippodamia convergens*, *Coccinella septempunctata*, *Coleomegilla maculata* and *H. sinuata*, but some predators that corresponded to the same genus (family, tribe, etc.) were combined as a single entry to reduce the volume of collected information. Other predators recorded included spiders, nabids, carabids, staphylinids, syrphids, *Geocoris*, *Orius*, *Scymnus* and green and brown lacewings. We also sampled other important pests of wheat such as pest mites, fall armyworm, armyworm, army cutworm and the 15 most important weeds.

There were three more sampling methods used to determine more accurately and precisely the density of categories of insects at each sampling point. Sweep net sampling was performed to estimate populations of highly mobile predators, such as lady beetles, that are difficult to count accurately by the visual count method. Berlese sampling was done to determine more precisely aphid density. Emergence canister sampling was performed to obtain more accurate estimation of parasitism. Data from each of these three sampling methods have their individual table within the database. Along with these samples, weather data (precipitation and temperature) were collected throughout the growing season at 15 min intervals for all fields using Watchdog Data loggers, model 425iR (Spectrum Technology Inc., East-Plainfield, Illinois, USA).

Using C# from Microsoft Visual Studio.NET, we developed an Internet application and then made a link between our AWPM database and our web site (see above) that facilitated visualization of the data. When a visitor logs on to the web site, they can choose a field on the map and will then be prompted to select a date for one of the eight sample types: field (visualization), Berlese, emergence canister, sweepnet, border weeds, pre-plant volunteer wheat, soil fertility and yield. After choosing the sampling date from a combo box, the visitor can simply click on the corresponding button and the data set selected will be visible on their browser in the form of a table. The data can be also downloaded in the form of an Excel table by clicking the corresponding button.

If the visitor chooses a data set collected in one of the fields that contain three categories of data: aphids, parasitoids and predators, they will also have the capability to examine three-dimensional surfaces for each category. If the selected data are not all equal to zero, each surface will describe the spatial distribution of abundance of the selected aphid, parasitoid or predator for the chosen date. The surfaces are automatically generated on our web server by 'SURFER 8' software using kriging

(Surfer, 2006). If there is enough information to calculate a linear variogram, kriging will generate a surface that has a smooth form. All three figures show spatial allocations of the insect populations in the selected field for the chosen date.

The web application described above is a prototype for a larger web infrastructure with online analytical processing. Figures 8.5 and 8.6 represent the total aphid densities, including GB and BCOA, collected on 24 March 2003 and on 2 April 2003 in wheat field No. 1 located in Jackson County, Oklahoma. The aphid density on the first date was much higher than on the second date. It can be easily analysed by accessing our online database and selecting the field from 23 fields. It can be seen on the first date that parasitoid density was very low and total predator density was zero. The total parasitoid and predator densities are represented in the Figs 8.7 and 8.8, respectively, for 2 April 2003. It can be seen that they were high, but at the same

OK_JACKSON_1_Field_3_24_2003 Total of Aphids

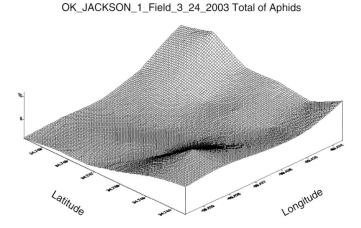

Fig. 8.5. The total aphid density in field No.1 located in Jackson County, Oklahoma, on 24 March 2003.

OK_JACKSON_1_Field_4_2_2003 Total of Aphids

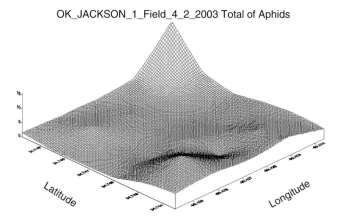

Fig. 8.6. The total aphid density in field No.1 located in Jackson County, Oklahoma, on 2 April 2003.

OK_JACKSON_1_Field_4_2_2003 Total of Parasitoids

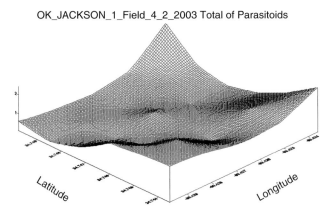

Fig. 8.7. The total parasitoid density in field No.1 located in Jackson County, Oklahoma, on 2 April 2003.

OK_JACKSON_1_Field_4_2_2003 Total of Predators

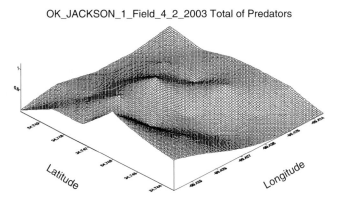

Fig. 8.8. The total predator density in field No.1 located in Jackson County, Oklahoma, on 2 April 2003.

time the aphid density was low (see Fig. 8.6). This suggests that the high aphid density on 24 March 2003 was followed by an increase in predators and parasitoids, which had controlled the aphid population by 2 April 2003.

The example above shows some new possibilities of our approach in interpreting insect population evolution in time and space. Presently, there are numerous theoretical publications explaining spatial and temporal population dynamics by diffusion equations. These theoretical conclusions are not validated by real data because it is difficult to bring them together. The AWPM database can do it. First, it is possible to determine whether the insect population's behaviour is similar to a physical system that follows the thermodynamic law like heat in a solid body. Using the AWPM database this can be done at smaller (field) or larger (landscape or region) scale. However, this is not the only possibility of the AWPM database in analysing insect population dynamics. We can also estimate the climatic influence on this process, or the influence of different farming methods on pest abundance, etc.

Conclusions

Today, the DBMS can easily be implemented in pest management programmes such as AWPM projects. A database application achieves full functionality and IT elements only if it is properly organized. For this purpose, careful preliminary coordination should occur between all teams involved in the project, so that database structure satisfies anticipated needs for data collection, storage, manipulation and analysis. At the outset of a project, efforts should be focused on unifying the data-gathering methods and protocols, and the structure of the information to be collected during the project. The next step is to analyse and develop the structure of the future database. This work should include the requirements and the right IT tools to develop a database. Presently, there are numerous commercially available tools for the purpose of database development, but they differ markedly in their performance and price, and the right choice can be a key to successful database development.

The database concept can contribute to future progress in IPM, because it forms an information source for IPM programme development and implementation. One exciting and interesting research avenue is the development of forecasting methods to predict pest population dynamics in space and time. The database might contain, for example, the data to assess the relationship between environmental conditions and the occurrence of pest outbreaks. Furthermore, it can provide the data to evaluate the roles of predators and parasitoids as pest population regulators. This is an important point, because implementation of a spatially distributed database provides the ability to study the dynamics of insect populations and their interactions using a spatio-temporal approach in a way not possible within the framework of a typical research project.

References

Bianchi, F.J.J.A. (2003) Usefulness of spatially explicit population models in conservation biological control: an example. Landscape management for functional biodiversity. *IOBC wprs Bulletin* 26, 13–18.

Date, C.J. (2003) *An Introduction to Database Systems*. 8th edn., Addison Wesley, New York.

Dent, D. (1995) *Integrated Pest Management*. Chapman & Hall, London.

Eason, A., Tim, U.S. and Wang, X. (2004) Integrated modeling environment for statewide assessment of groundwater vulnerability from pesticide use in agriculture. *Pest Management Science* 60, 739–745.

Elliott, N.C., Hein, G.L., Carter, M.R., Burd, J.D., Holtzer, T.O., Armstrong, J.S. and Waits, D.A. (1998) Russian wheat aphid (Homoptera: Aphididae) ecology and modeling in Great Plains agricultural landscapes. In: Quisenberry, S.S. and Peairs, F.B. (eds) *Response Model for an Introduced Pest – the Russian Wheat Aphid*. Entomological Society of America, Lanham, Maryland, pp. 31–65.

Elliott, N.C., Royer, T.A., Giles, K.L., Kindler, S.D., Porter, D.R., Elliott, D.T. and Waits, D.A. (2004) A web-based decision support system for managing greenbugs in wheat. *Crop Management* doi: 10.1094/CM-2004-09XX-01-MG.

EXAMINE (2000) http://www.rothamsted.bbsrc.ac.uk/examine/

Faust, R.M. and Chandler, L.D. (1998) Future programs in areawide pest management. *Journal of Agricultural Entomology* 15, 371–376.

Filipe, J.A.N. and Maule, M.M. (2003) Analytical methods for predicting the behavior of population models with general spatial interactions. *Mathematical Biosciences* 183, 15–35.

Folkedal, A. and Breving, C. (2004) VIPS – a web-based decision support system for crop protection in Norway. In: Thysen, I. and Hocevar, A. (eds) *Online Agrometeorological Applications with Decision Support on the Farm Level. COST Action 718: Meteorological Applications for Agriculture.* DINA Research Report No. 109, pp. 18–27.

Halbert, S.E. and Stoetzel, M.B. (1998) Historical overview of the Russian wheat aphid (Homoptera: Aphididae). In: Quisenberry, S.S. and Peairs, F.B. (eds) *Response Model for an Introduced Pest – the Russian Wheat Aphid.* Entomological Society of America, Lanham, Maryland, pp. 12–30.

Human Genome Project (2003) http://www.ornl.gov/sci/techresources/Human_Genome/home.shtml

Isard, S.A., Russo, J.M. and DeWolf, E.D. (2006) The establishment of a national pest information platform for extension and education. *Plant Health Progress*, doi 10.1094/ PHP-2006- 0915- 01-RV.

King, G.J. (2004) Bioinformatics: harvesting information for plant and crop science. *Seminars in Cell and Developmental Biology* 15, 721–731.

Knight, J.D., Tatchell, G.M., Norton, G.A. and Harrington, R. (1992) FLYPAST: an information management system for the Rothamsted aphid database to aid pest control research and advice. *Crop Protection* 11, 419–426.

Kogan, M. (1998) Integrated pest management: historical perspectives and contemporary development. *Annual Review of Entomology* 43, 243–270.

Legg, D.E. and Bennett, L.E. (1992) A mobile workstation for use in integrated pest management program on the Russian wheat aphid. In: Morrison, W.P. (ed.) *Proceedings of the 5th Russian Wheat Aphid Conference*, 26–28 January 1992, Fort Worth, Texas, pp. 66–69.

Legg, D.E. and Bennett, L.E. (1993) Development of a transportable computing system for on-site management of the Russian wheat aphid in wheat. *Trends in Agricultural Science and Entomology* 1, 31–39.

Mann, B.P., Wratten, S.D. and Watt, A.D. (1986) A computer-based advisory system for cereal aphid control. *Computers and Electronics in Agriculture* 1, 263–270.

Morrison, W.P. and Peairs F.B. (1998) Response model concept and economic impact. In: Quinsenberry, S.S. and Peiars, F.B. (eds) *Response Model for an Introduced Pest – the Russian Wheat Aphid.* Entomological Society of America, Lanham, Maryland, pp. 1–11.

Murali, N.S., Secher, B.J.M., Rydahl, P. and Andreasen, F.M. (1999) Application of information technology in plant protection in Denmark: from vision to reality. *Computers and Electronics in Agriculture* 22, 109–115.

Nemenyi, M., Mesterhazi, P.A., Pecze, Z. and Stepan, Z. (2003) The role of GIS and GPS in precision farming. *Computers and Electronics in Agriculture* 40, 45–55.

Neufeld, M.L. and Cornog, M. (1986) Database history: from dinosaurs to compact discs. *Journal of the American Society for Information Science* 37, 183–190.

Ohio Geospatial Program (2003) http://geospatial.osu.edu/resources/hand-heldgps.html

Oracle9i (2002) *Net Services Administrator's Guide* (http://www.oracle.com/pls/db92/db92.docindex).

Oracle9i (2003) *Database Release Notes* (http://download-east.oracle.com/docs/html/B10924_01/toc.htm).

Perini, A. and Susi, A. (2004) Developing a decision support system for integrated production in agriculture. *Environmental Modeling and Software* 19, 821–829.

SAS Institute (1990) *SAS/STAT User's Guide, Version 6.* SAS Institute, Cary, North Carolina.

Sivertsen, T.H. (2005) Implementation of a general documentation system for web-based administration and use of historical series of meteorological and biological data. *Physics and Chemistry of the Earth* 30, 217–222.

Surfer (2006) http://www.goldensoftware.com/products/surfer/surfer.shtml

Random House Webster's Unabridged Electronic Dictionary, 2nd edn., Copyright© 1996.

Tatchell, G.M. (1991) Monitoring and forecasting aphid problems. In: Peters, D.C., Webster, J.A. and Chlouber, C.S. (eds) *Aphid–Plant Interactions: Populations to Molecules*. Oklahoma Agricultural Experiment Station, MP–132, Norman, Oklahoma, pp. 215–231.

Taylor, L.R. (1977) Migration and the spatial dynamics of an aphid, *Myzus persicae*. *Journal of Animal Ecology* 46, 411–423.

Xia, Y., Stinner, R.E. and Chu, P.-C. (2002) Database integration with the web for biologists to share data and information. *Electronic Journal of Biotechnology* 5, 154–161.

Yamamura, K. (2002) Dispersal distance of heterogeneous populations. *Population Ecology* 44, 93–101.

Zadoks, J.C. (1981) EPI-PRE: a disease and pest management system for winter wheat developed in the Netherlands. *EPPO Bulletin* 11, 365–369.

Zintzaras, E. and Tsitsipis, J.A. (2003) Centaur database: an information system for aphid surveys and pest control. *Journal of Applied Entomology* 127, 534–539.

Appendix A

An example of VBA application that demonstrates how Oracle Objects for OLE can be used to access Oracle 9i database.

```
Sub Get_Data()
    'Create and initialize the necessary objects
    Dim OraSession As Object
    Dim OraDatabase As Object
    Dim EmpDynaset As Object
    Dim ColNames As Object

    Set OraSession = CreateObject('OracleInProcServer.XOraSession')
    Set OraDatabase = OraSession.OpenDatabase('Exampledb', 'scott/tiger', 0&)
    Set EmpDynaset = OraDatabase.DbCreateDynaset('select * from emp', 0&)

    'Using field array, ie. ColNames('ename').value, is significantly faster than using
        'field lookup, ie. EmpDynaset.fields('ename').value
    Set ColNames = EmpDynaset.Fields
    'Place column headings on sheet
    For icols = 1 To ColNames.Count
        Worksheets('DataSheet').Cells(1, icols).Value = ColNames(icols – 1).Name
    Next
    'Place data on sheet using CopyToClipboard
    EmpDynaset.CopyToClipboard –1
    Sheets('DataSheet').Select
    Range('A2').Select
    ActiveSheet.Paste
End Sub.
*
```

Appendix B

Creation of 'Users' table inside of 'Exampledb' database using SQL language.
CREATE TABLE 'EXAMPLEDB'.'USERS'
('USER_NAME' VARCHAR2(25),
'PASSWORD' VARCHAR2(25),
PRIMARY KEY ('USER_NAME') ENABLE);

9   Codling Moth Areawide Integrated Pest Management

Alan L. Knight

Yakima Agricultural Research Laboratory, Agricultural Research Service, Wapato, Washington, USA

Introduction

Codling moth (CM) is an insidious pest, tunnelling to the core of valuable commodities that are typically marketed with exceptional quality standards for appearance, firmness and sweetness. While there is no mention in the Bible of whether the apple that Eve gave to Adam graded 'Extra fancy', it is likely that if this fruit had been infested with CM, the human race would not be as anxious about returning to a pest-ridden garden of paradise. Nevertheless, since Noah allowed two adult CM to disembark from his boat, the distribution of this pest has closely followed man's cultivation of its hosts around the world (Shel'deshova, 1967). Historically, commercial plantings of both apple (*Malus domestica* Borkhausen) and pear (*Pyrus communis* L.) have been heavily sprayed with seasonal programmes of broad-spectrum insecticides as part of the management of CM (Barnes, 1959; Madsen and Morgan, 1970). These intensive and indiscriminate management practices have not only defined the efficacy of control for this key pest, but also the population dynamics of a suite of secondary pests and their associated natural enemies, and the occurrence of several negative spill-over effects related to the environment and human safety (Prokopy and Croft, 1994).

Growers in the USA were offered in the early 1990s a new integrated pest management programme (IPM) for their orchards that hinged on the adoption of sex pheromones for mating disruption (MD) of CM, an intensive monitoring programme and the judicious use of more selective pesticides (Barnes *et al.*, 1992). Initial testing of this integrated approach, when applied to individual small orchards with low CM pressure, was mostly successful (Howell and Britt, 1994; Knight, 1995a; Gut and Brunner, 1998). Yet, some growers experienced higher levels of CM damage. The cost of the new IPM programme was higher than most growers' current management programmes because many growers applied only a few seasonal sprays for CM, and subsequent reductions in the use of pesticides to manage secondary pests were minimal (Knight, 1995a; Williamson *et al.*, 1996). Secondly, new pest problems developed in many orchards that required the application of additional sprays, further disrupting the implementation of IPM and raising the new programme's overall cost (Knight, 1995a; Gut and Brunner, 1998). Thirdly, the performance of the early dispensers was poor as the sex pheromones were not well protected and their emission rates were not adequate late in the season (Brown *et al.*, 1992; Knight *et al.*, 1995). Perhaps not surprisingly, a significant proportion of growers initially adopting the use of MD dropped out of the IPM programme prior to 1995 (Howell and Britt, 1994).

An areawide pest management (AWPM) approach was proposed as a potentially more effective strategy that could improve both the performance of sex pheromones and biological control (BC) (Kogan, 1995). AWPM accepts that pests and their natural enemies do not recognize individual orchards' boundaries and that effective management requires a coordinated, regionally focused project (Knipling, 1979). Essentially, the programme was conceived as a 'pyramid scheme', where more and more small growers situated in the centre of an ever-expanding project would benefit as all potential sources of CM impacting their orchards would be treated with MD and intensively monitored, and that the expanding area coming under a more selective management programme would harbour a significant increase in populations of natural enemies and their contribution to BC would also increase.

Demonstration of this concept was initiated in 1995 in a multi-institutional programme created by a close collaboration of university and governmental researchers in Washington, Oregon and California, with primary funding provided by the US Department of Agriculture's (USDA) Agricultural Research Service (ARS) (Kogan, 1994). The 5-year CAMP (CM Areawide Management Program) was the first of the areawide programmes initiated by USDA. The goal of this programme was to implement, assess, research and educate the industry about promising new IPM technologies. CAMP was highly successful in fuelling the rapid adoption of a new paradigm in orchard pest management that resulted in significant reductions in fruit injury using nearly 80% less broad-spectrum insecticides.

Constraints in Developing Areawide CM Management

Management of CM is difficult due to a number of operational, biological and ecological factors. CM is well adapted to the temperate climate zones, can have one to four generations per year and overwinters as a mature larva hibernating in protected bark crevices (Riedl and Croft, 1978). Both sexes are winged and they can disperse widely between managed and unmanaged hosts (White *et al.*, 1973; Knight *et al.*, 1995). Unmanaged sites can include backyard fruit trees, municipal plantings of crab apples, trees surviving at old homesteads and along pasture fencerows and in poorly managed orchards.

Female moths emerge with mature oocytes and have to mate only once to lay a full complement of fertile eggs (Howell, 1991). Females can deposit 50–100 eggs that are laid individually on or adjacent (< 15 cm) to fruits (Jackson, 1979). This oviposition strategy minimizes predation and larval competition for fruits, while maximizing the proportion of fruits that are attacked. Levels of fruit injury can rise rapidly between generations, and unmanaged orchards can experience over 80% fruit injury (Myburgh, 1980). Neonate larvae do not generally feed before entering the fruit, which significantly reduces the effectiveness of many of the selective insecticides that require ingestion (Croft and Riedl, 1991). Larvae tunnel through the flesh of the fruits to feed on the seeds, rendering fruit infested by even a single larva worthless.

Natural control of CM due to predation or parasitism of eggs and larvae is low in unmanaged sites (Falcon and Huber, 1991) or in orchards under MD (Knight *et al.*, 1997), and natural regulation of CM populations is more strongly influenced by density-dependent factors, such as crop load and available overwintering sites (Ferro *et al.*, 1974).

Effective control of CM requires the repeated applications of cover sprays to maintain an effective toxic residue during the season (Barnes, 1959). Unfortunately, CM has evolved resistance to every class of insecticides applied by growers, from the early use of lead arsenate (Hough, 1928) and DDT (Cutwright, 1954) to the more recently developed insect growth regulators (Sauphanor and Bouvier, 1995) and granulosis virus (Fritsch *et al.*, 2005). The organophosphate (OP) insecticide, azinphosmethyl, has been the primary insecticide used in the USA for CM since the mid-1960s and, surprisingly, resistance was not detected until 1989 (Varela *et al.*, 1993). Resistance mechanisms in CM have included a number of physiological pathways, including altered target sites and amplified detoxification enzymes (Reyes *et al.*, 2007). Cross-resistance among classes of insecticide has apparently reduced the effectiveness of new classes of insecticide, even before they had become widely adopted (Sauphanor and Bouvier, 1995; Dunley and Welter, 2000).

Codling moth pressures in Washington State orchards by the early 1990s had increased significantly, with seasonal moth catches nearly tripling and spray applications doubling from the mid-1980s for many growers (Howell and Britt, 1994). In addition, in response to elevated levels of resistance, many growers increased their application rate of azinphosmethyl and tightened spray intervals. The use of both methyl parathion and chlorpyrifos for CM increased precipitously during this period, causing serious impacts on both pollinators and BC of secondary pests (Gut *et al.*, 1992). Coincidentally, the Alar 'scare' (plant growth hormone) in 1989, the cancellation of phosphamidon (aphicide) in 1990 and the withdrawal of cyhexatin (miticide) in 1992 were harbingers for the eventual loss of even more pesticides registered for use in pome fruits. In particular, the future of the nine OP insecticides registered for tree fruit in 1995 seemed dim.

In 1996 the Food Quality Protection Act was passed, and the Environmental Protection Agency (EPA) was ordered to undertake a reassessment of all pesticide tolerances using a new risk standard of 'a reasonable certainty of no harm'. A new quantitative approach was adopted that considered both the aggregate exposure (all exposure pathways) and the cumulative exposure (all chemicals with the same modes of action considered together), adding up to a tenfold safety factor to protect children. Use restrictions of certain insecticides were tightened by extending worker re-entry periods and preharvest spray intervals and by reducing the total amount of material that could be used per year. By 2006 EPA had reviewed all 9600 pesticide tolerances and revoked 3200, modified 1200 and left 5237 unchanged (Willett, 2006). Currently, there are only four OP insecticides registered in tree fruits in the USA, and a complete phase-out for azinphosmethyl is now scheduled for 2012.

Quarantine security and phytosanitary requirements have serious impacts on the international marketing of apples and pears (Hansen and Johnson, 2007). For example, Japan requires a postharvest quarantine treatment (probit-9 efficacy) of US cherries with methyl bromide to disinfest fruit for codling moth (MAFF-Japan, 1950). Other countries have strict tolerances for the incidence of pests such as CM, which can lead to rejection of shipments and the eventual shutdown in the market. Postharvest treatments are a significant cost added to the marketing of these fruits, and can have serious impacts on the quality of the treated commodity and risks to human health and environmental degradation (Hansen and Johnson, 2007). These strict international tolerances for CM force growers to integrate a system of various biological and operational production and postharvest factors that can provide near-quarantine security levels of pest-free produce prior to shipment (Jang and Moffitt, 1994).

Disruption of the natural control of secondary pests in pome fruits by the sprays applied for CM can contribute to the use of additional sprays and subsequent development of resistance in species, such as aphids, leafhoppers and mites (Croft and Bode, 1983). Most noteworthy has been the history of repeated development of resistance to new classes of pesticides by pear psylla, *Cacopsylla pyricola* (Van de Baan and Croft, 1991) and the tetranychid mites, *Tetranychus urticae* and *Panonychus ulmi* (Croft, 1979). The application of additional sprays to manage these OP-resistant pests can cause secondary outbreaks of pests. For example, the western tentiform leaf miner (WTLM), *Phyllonorycter elmaella*, developed resistance to azinphosmethyl in the early 1980s, and BC of this pest by the eulophid, *Pnigalio flavipes*, is disrupted by summer use of chlorpyrifos and methyl parathion (Barrett, 1988). The subsequent use of the carbamate, oxamyl, to control leaf miners disrupts integrated mite control, forcing growers to apply costly miticides (Hoyt, 1983).

Conversely, reductions in the use of broad-spectrum sprays can release populations of other pests from chemical control. When broad-spectrum sprays were reduced in MD orchards, minor problems with sporadic pests, such as true bugs, increased (Gut and Brunner, 1998). Of greater concern, however, were the outbreaks of the tortricid leafrollers, *Pandemis pyrusana* and *Choristoneura rosaceana*, which caused significant levels of fruit injury (Knight, 1995a; Gut and Brunner, 1998). Infestations of orchards by leafrollers can occur from importation of infested nursery stock and the immigration of adult moths from unsprayed, non-bearing blocks of apple, and from cherry orchards after mid-summer harvest (Knight, 2001). While a number of parasitoids can attack these leafroller species, parasitism levels are typically low in conventional orchards (Brunner and Beers, 1990). The use of OP insecticides in the spring and summer for these pests further destabilizes secondary pest populations (Beers *et al.*, 1998) and has selected for resistance in some populations (Smirle *et al.*, 1998). Selective control of leafrollers is possible with *Bacillus thuringiensis* (*Bt*) but the level of control is variable (Brunner, 1994). MD for leafrollers is an added expense, and preliminary trials were only marginally effective due to the relatively high population density of these polyphagous pests and their dispersal capabilities (Gut *et al.*, 1992).

Studies prior to CAMP had demonstrated that BC of secondary pests was not significantly improved when growers reduced their use of broad-spectrum sprays for CM (Howell and Britt, 1994; Knight, 1995a; Gut and Brunner, 1998). For example, the population densities of generalist aphid predators and an egg parasitoid of white apple leafhopper (WALH), *Typhlocyba pomaria*, were higher, but pest levels were unaffected in sex pheromone-treated orchards (Knight, 1995a; Gut and Brunner, 1998). However, the full potential of BC was difficult to assess in these studies as growers only marginally reduced their use of pesticides for secondary pests (Knight, 1995a). The often marginal effectiveness of the available selective insecticides (*Bt* for leafrollers, soaps for aphids and oil for CM eggs) was a significant constraint in 1995.

Organic tree fruit production increased dramatically, from 1988 when the certification programme began in Washington State with 11 growers farming 40 ha, to 1990 with 100 growers farming 800 ha. A small survey of organic apple-growing practices in 1990 found that orchards were treated, on average, with 14 botanical and microbial sprays per season for CM, yet many orchards still suffered high levels of fruit injury (Knight, 1994). A high proportion of organic growers adopted MD after 1991, but this approach was often ineffective when used as the sole tactic to manage high population densities of CM (Trimble, 1995).

One major constraint common in implementation of IPM technologies has been the establishment of effective systems of information delivery to participants (Travis and Rajotte, 1995). Traditionally, growers obtain information from a large variety of sources, including university extension activities (meetings, publications and field demonstrations), fieldmen, consultants, agricultural supply companies, packing houses and cooperatives. However, acquiring real-time information concerning pest populations in surrounding orchards is more difficult. While some growers may know something about the pest pressures in the adjacent surrounding orchards, they are unlikely to be able to assess pest pressures impacting them from more distant sources.

A number of farm operational factors have negative impact on the management of CM. Many orchards (60% in the Yakima Valley; Howell and Maitlen, 1987) are irrigated with overhead sprinklers. Overhead irrigation can wash off spray residues, forcing growers to apply higher rates of insecticides and to spray more frequently (Howell and Maitlan, 1987). Similarly, the use of evaporative cooling to reduce sunburn can remove residues and require similar increased spray use in orchards (Williams, 1993). Storage and transport of bins can introduce CM infestations into clean orchards (Newcomber, 1936; Proverbs and Newton, 1975).

Asynchronous emergence of adult moths from bin piles can create unexpected periods of CM activity (Higbee *et al.*, 2001). This problem is heightened because bins are introduced into orchards at variable time periods during the season, and the strong temperature gradient that exists from the inside to the outside of large bin piles can extend the moth's emergence period (Higbee *et al.*, 2001). Finally, attaining complete coverage and fruit protection in large, three-dimensional tree canopies is difficult (Byers *et al.*, 1984). The deposition patterns achieved by growers vary widely, and tractor speed, nozzle type, water volume and air velocity all have a significant impact on these (Howell and Maitlen, 1987).

Codling moth has always been a greater problem on orchard borders rather than in the interior (Madsen and Vakenti, 1973; Gut and Brunner, 1998). Moths, both immigrating into and emigrating out of orchards, pool along the borders resulting in higher densities of injured fruits (Knight, 2007). In addition, spray coverage can be poor along the edges of orchards. The distribution of sex pheromone along borders is also reduced by higher wind speed and turbulence (Milli *et al.*, 1997).

Monitoring CM with sex pheromone-baited traps is widely practised by growers (Riedl *et al.*, 1986). A number of factors influence moth catches, including a rapid degradation of the sex pheromone within lures (Knight and Christianson, 1999). Thresholds of moth catches have been used as recommendation for the application of insecticides, but the frequent occurrence of 'false negatives', where traps fail to detect local infestations, is a particular problem for MD orchards (Knight and Light, 2005). Recommendations for monitoring CM in MD orchards have suggested that growers should use a higher density of traps and replace lures more frequently, practices that increase monitoring costs (Gut and Brunner, 1996).

Mating disruption prior to CAMP was used on 3800 ha in Washington State in 1994 (see Fig. 9.1). The original Isomate-C dispenser was a clear polyethylene tube characterized by a rapid reduction in its emission rate and increased degradation of the pheromone when placed in full sunlight (Brown *et al.*, 1992). A number of factors were known to influence the performance of MD for codling moth, such as dispenser characteristics, pest population density and an orchard's isolation and topography (Charmillot, 1990). For example, MD failed in only one site, Y6, in the 3-year Yakima Valley study (Knight, 1995b). This orchard had a number of poor characteristics that lessened the likely success of MD: a high initial population of CM, a 6% slope, a high proportion of missing trees and an uneven orchard canopy (tree heights ranged from 2 to 5 m). Specific studies addressing the optimal use of the Isomate-C dispenser for CM MD – such as characterizing the seasonal dispenser emission rate, the influence of dispenser density and positioning within the canopy and the role of sex pheromone blend – were not sufficiently characterized until 1995 (Knight, 1995b; Knight *et al.*, 1995; Weissling and Knight, 1995).

The cost of Isomate-C dispensers in 1991 started at US$326/ha and application costs, depending on the method used by growers, varied from US$27 to 69/ha (Knight, 1995a). This initial cost of MD was equivalent to four applications of azinphosmethyl, but most growers in Washington State were applying more than

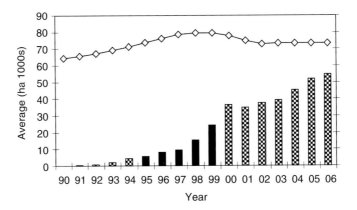

Fig. 9.1. The acreage of total apple and pear production (◊) and acreage treated with sex pheromones (vertical bars) for codling moth in Washington State, 1990–2006. The acreage treated during the 5-year codling moth areawide management programme (CAMP) is shown by solid bars.

three sprays per season (Knight, 1995a). An economic break-even analysis found that the cost of MD would have to decline by 30–73% to be equivalent to the growers' then current spray programmes (Williamson *et al.*, 1996). Furthermore, this analysis did not include the grower's increased cost of monitoring of MD orchards. Perhaps, in response to both its higher cost and variable effectiveness, over one-third of growers surveyed from 1991 to 1994 had stopped using this new technology (Howell and Britt, 1994).

Factors Available for Successful AWPM Implementation

There are many constraints that impact the management of CM and the various secondary pests in tree fruits. However, there are also a number of very important factors that were in place in the early 1990s that facilitated the development of effective AWPM programmes. CM has a narrow host range, and in the major tree fruit growing regions in the western USA there are not a large number of unmanaged orchards or large sources of CM outside of commercial orchards (Barnes, 1991). In general, CM populations are maintained at very low levels in commercial orchards and most growers in Washington State prior to 1995 applied no more than two sprays per season (see Table 9.1). Pest boards are funded in nearly every county to deal with the presence of pest problems emanating from unmanaged sites, and are usually mandated either to spray orchards or to remove trees at the owner's expense.

Table 9.1. Survey of insecticide usage for management of codling moth and leafrollers in Washington State, USA, apple from summaries of agricultural chemical usage in 1991, 1995, 1999 and 2005 (from NASS reports).

	Mean no. of sprays (percentage of area treated)			
Insecticide	1991	1995	1999	2005
Azinphosmethyl	2.8 (90)	3.3 (94)	2.3 (78)	1.8 (72)
Chlorpyrifos[a]	1.4 (65)	1.3 (80)	1.3 (65)	1.1 (58)
Phosmet	2.1 (9)	2.4 (2)	2.0 (7)	1.4 (15)
Ethyl parathion	1.0 (32)	—[c]	—[c]	—[c]
Methyl parathion[a]	1.5 (28)	1.2 (19)	1.1 (5)	—[c]
Bacillus thuringiensis[b]	NA	2.2 (21)	2.0 (19)	1.3 (18)
Spinosad[b]	NA	NA	1.4 (39)	1.3 (62)
Acetamiprid	NA	NA	NA	1.2 (41)
Thiacloprid	NA	NA	NA	1.1 (2)
Novaluron	NA	NA	NA	1.2 (12)
Lambda-cyhalothrin	NA	NA	NR	1.7 (3)
Granulosis virus	NA	NR	NR	1.5 (9)

NA, not available; NR, no records.
[a]These insecticides are applied for control of either codling moth or leafrollers.
[b]These insecticides are applied for control of leafrollers.
[c]No longer registered for use in apple.

Experience developed for the use of CM MD during the 3-year transition programmes conducted in the Yakima and Wenatchee Valleys in Washington State (Brunner *et al.*, 1992), the first coordinated areawide programme developed by five pear growers in Randall Island, California, in response to OP resistance (Varela *et al.*, 1993) and the use of MD on a large, contiguous apple block (400 ha) by a single grower in north-central Washington (Knight, 1992) were key events that provided the industry with an important early assessment of the potential outcome of adopting MD. Risk assessment indices were developed to assess the probable success of MD based on orchards' previous and current levels of fruit injury and moth catches in sex pheromone-baited traps (Gut and Brunner, 1996). Four risk categories (very low, low, moderate and high) were created, each with associated guidelines for the supplemental use of insecticides and suggested dispenser application rates (Gut and Brunner, 1996). Recommendations for the monitoring and managing of secondary pests in MD orchards were also outlined (Gut *et al.*, 1995).

Despite the detection of incipient levels of resistance to azinphosmethyl in some orchards, the majority of CM populations monitored remained susceptible (Varela *et al.*, 1993; Knight *et al.*, 1994). The availability of both methyl parathion and chlorpyrifos, which exhibited negative cross-resistances with azinphosmethyl (Dunley and Welter, 2000), allowed growers to use other effective materials if needed. The existing integrated mite management programme present in most orchards was largely created by the development of resistance by the phytoseiid mites to azinphosmethyl (Hoyt, 1969). This was also true for the effective BC of the western tentiform leaf miner by *P. flavipes* (Beers *et al.*, 1993).

Several new selective approaches were developed to manage leafrollers in the early 1990s. *Bt* sprays could be optimized if the first spray was applied at the maximum rate, delayed until petal fall and applied only with a forecast of extended warm weather (Knight, 1997). Studies found that significant BC of leafrollers could develop in orchards with selective programmes (Brunner, 1992). Demonstrations that a generic, partial sex pheromone blend could be effective for several of the suite of leafroller species attacking a number of horticultural crops in western USA increased the likelihood of commercial development (Knight, 1996; Knight *et al.*, 1998; Knight and Turner, 1999). The first testing of MD in the USA for leafrollers in larger (16 ha) replicated orchard blocks demonstrated the effectiveness of this selective approach (Knight *et al.*, 1998).

Several improvements were made with sex pheromone-based products early in the 1990s. A new Isomate-C Plus dispenser was developed that significantly reduced the degradation of the pheromone and had an improved seasonal emission profile (Gut *et al.*, 1995). Competition among several small companies registering MD dispensers for CM caused significant reductions in the retail price of dispensers. The application cost of MD declined as growers developed more efficient methods of applying dispensers. For example, the total cost of Isomate-C dispensers and their application dropped US$100/ha from 1991 to 1993 (Knight, 1995a). In particular, the cost of MD dropped most dramatically for growers who cut their application rate of dispensers. Advantages – such as no re-entry waiting periods, compatibility with overhead irrigation, lowered risk of incidence of insecticide resistance, improved worker safety and no container disposal – all combined to generate growing interest in this technology (Brunner *et al.*, 1992).

A few studies demonstrated that CM could be managed in organic orchards successfully using MD. Successful organic production was demonstrated in Canada by removal of injured fruits during the season, banding of trees to remove overwintering larvae and the use of sex pheromones (Judd *et al.*, 1997). Gut and Brunner (1998) were able to clean up an infested organic orchard using two applications of MD dispensers and 16 supplemental sprays of ryania and *Bt*. In subsequent years, CM was effectively managed using only MD in this orchard.

Management of the problematic orchard borders was achieved with a number of approaches. Typically, growers sprayed borders of MD orchards with insecticide sprays (Knight, 1995a; Gut and Brunner, 1998). A few growers applied additional dispensers on borders or extended the pheromone-treated area to include adjoining blocks of hosts or non-hosts, windbreaks or fencerows (Gut and Brunner, 1998). Treating larger, contiguous areas reduced the relative size of orchards' borders relative to the total area, and subsequently the importance of these areas.

Several institutional factors were present prior to CAMP that strongly benefited the development of AWPM. Thresholds and sampling plans were developed for nearly all secondary pests (Beers *et al.*, 1993). A predictive phenology model to time the first cover spray to coincide with the start of CM egg hatch had been validated and was widely used (Brunner *et al.*, 1982). Weather monitoring networks, such as Washington State's Public Agricultural Weather System (PAWS), were established and provided input for a number of insect, disease and plant models that growers could readily access. The various land grant university extension services were relatively well-funded, gathering and disseminating information for growers through publications, workshops, field days and via telephone, fax, mail and the fledgling Internet. In addition, private consultants and fieldmen representing chemical supply companies, cooperatives and packing houses provided monitoring services and made informed management recommendations for growers.

The tree fruit industry in the western USA has a history of providing generous support for pest management-related research (Ing, 1999). The first research project funded by the Washington Tree Fruit Research Commission (WTFRC) for CM MD (US$6000) was granted in 1991, along with a budget of US$100,000 for entomological research (Ing, 1992). Levels of funding by WTFRC for entomological research increased by US$100,000 each year prior to CAMP, with over US$200,000 granted for CM and leafroller MD research alone in 1994 (Ing, 1995).

The Structure of CAMP

The AWPM programme was developed as a partnership of federal agencies, university researchers and extension personnel, state departments of agriculture and the private sector, including growers, commodity groups and other stakeholders (Calkins and Faust, 2003). CAMP was constructed as a coordinated programme requiring active grower involvement to apply environmentally sound, effective and economical approaches over large, contiguous areas of tree fruit production (Calkins, 1998). The objectives of the programme focused primarily on entomology and did not incorporate new approaches for either horticultural or postharvest control.

The specific objectives were to:

- Reduce the use of neurotoxic insecticides by 80%.
- Demonstrate that MD worked better when applied over large areas, partially through the need for less pheromone and lower costs.
- Develop companion technologies to supplement MD that have a lower cost.
- Increase the role of BC in managing pest populations.
- Develop an effective areawide monitoring programme and establish the use of thresholds.
- Improve worker safety.
- Improve public perception that fruit production is safe for consumers (Calkins et al., 2000).

The expected benefits of this programme were that the AWPM programme would be as effective as, or better than, conventional programmes for CM, reduce the need for sprays for other pests and reduce the use of broad-spectrum insecticides. These benefits would be reached through coordination and optimization of growers' actions, including expanded monitoring, adoption of action thresholds and the use of selective and efficacious tactics (Calkins et al., 2000). A transition to this new programme for growers was eased by providing a direct subsidy to growers within the pilot projects for 3 years (US$125/ha), as well as providing funding for each project to hire a manager and supplies. The intention of the project was that government involvement would end after 5 years and AWPM would be maintained through a sustainable framework created by farmers, consultants and local organizations (Calkins and Faust, 2003).

Five pilot projects were established in the western USA, with one project in California and Oregon and three in Washington State. These five sites were selected to encompass a broad geographical area and included a range of climatic conditions, fruit varieties and cultural practices, as well as differences in pest management practices. A few basic criteria were established to identify the suitability of each site: (i) a typical fruit production area in the region with consistent pest pressure from CM and other pests; (ii) producers within the project would be willing to cooperate and share costs; and (iii) each group would have the ability to construct a local organizational structure to support and continue the use of AWPM (Calkins and Faust, 2003).

Similar management tactics were applied to orchards in all projects (Calkins, 1999). Orchards were treated with the full rate of sex pheromone dispensers (Isomate-C+ was the sole product used in all but one project), and growers were encouraged to apply one spray of azinphosmethyl to lower the initial population density of CM. All orchards were monitored with a high density of traps (one per ha) baited with high-load sex pheromone lures, and placed in the upper third of the canopy. Moth catch thresholds were established to recommend the use of supplemental sprays during the second moth generation. Secondary pests and natural enemies were closely monitored in a proportion of blocks, and the supplemental use of insecticides for secondary pests was based on the use of accepted thresholds (Calkins, 2003). Comparison orchards outside of the projects were selected based on their similarity in cultivars and horticultural practices, but these conventionally managed orchards were not treated with MD.

The organization of CAMP was structured into four subsections: administration, implementation, research and education. Dr Calkins, the research leader at the

ARS laboratory in Wapato, Washington, was the project's overall administrative leader and controlled the funding provided to the various pilot projects, research projects and education and extension outreach efforts. An Areawide CM Industry Advisory Committee comprised of a cross-section of industry leaders met with the administrators, project coordinators and researchers to review progress at each project site and discuss the related ongoing research concerning control of CM and other orchard pests. This group then reported their findings to other committees, such as the Washington State Horticultural Association Agriculture Chemical Committee, within the industry of each state. ARS administrators from the beginning emphasized the importance of bringing growers into the process early.

Implementation efforts in each of the five original CAMP sites were managed by scientists at the universities and at ARS. Funding provided to each of the pilot projects ranged from US$50,000 to 185,000 per year depending on the size and specific needs of the project. The first year of the projects was the most difficult due to some concern by growers that they were being forced, even if by peer pressure, to join a government-mandated programme. Before becoming active participants, growers had to be assured that they had control of the project. The use of the 3-year subsidy appeared to have been a very effective enticement for growers to join the programme.

CAMP projects

Howard Flat, Washington, was characterized as an isolated, typical tree fruit production area (90% apple) in north-central Washington, with flat topography. A preliminary coordinated study began in 1994 using MD on 125 ha. The funding provided by CAMP increased the size of the project to 486 ha, with 176 blocks farmed by 34 growers. The Howard Flat Management Board was formed with five fruit industry fieldmen (individuals who worked for local packing houses and agricultural chemical distribution companies) and three local growers. A Technical Advisory Committee was created with a group of applied entomologists (university and government researchers), and weekly breakfast meetings were held from the start to the end of each growing season. A project coordinator was hired to handle the daily activities, such as orchard monitoring and data summation. Monitoring information from all blocks was disseminated through weekly meetings, postings at a centrally located kiosk within the project, an electronic bulletin board and a monthly newsletter.

Parker Heights was considered to be a challenging area (190 ha), characterized by mixed-crop production (80:20 apple and pear) and situated across a hilly terrain. Pome fruit orchards were interspersed among 60 ha of stone fruit (cherry and peach), creating an extensive array of MD-treated borders in the project. An ARS employee served as the site coordinator and managed the project, along with a steering committee of two growers and three fruit industry fieldmen.

Oroville was a unique site consisting of 154 ha situated on the Canadian border on either side of Lake Osoyoos. Thirteen growers farmed 65 orchard blocks with 90% apple production. All orchards received weekly releases of sterilized CM adults provided by the Canadian Sterile Insect Release Program in Osoyoos, British Columbia (Bloem and Bloem, 2000). ARS hired a project coordinator, released the moths, monitored the orchards and maintained an office on site where growers could access project information and discuss issues with the coordinator's staff.

Medford, situated in southern Oregon, was characterized by a flat topography and 90% of its 121 ha were planted in 13 cultivars of pear and farmed by seven growers. The project was organized by Oregon State University personnel and began in 1994 on 30 ha, using a selective programme based on MD and repeated applications of horticultural oil for CM. The project had the same coordinator for all 5 years, and bi-weekly meetings were held with all participants during each growing season.

Randall Island was the first coordinated areawide project for CM and was started by five Bartlett pear growers on 308 ha in 1993, with support from the University of California in Berkeley, California. The project focused on the use of MD to combat the development of high levels of OP resistance (five- to eightfold) in local CM populations. Initially, growers used two dispenser applications and evaluated the use of rotations of methyl parathion and azinphosmethyl in combination with MD to manage CM and OP resistance.

CAMP funded 17 one-year sites from 1997 to 1999 (see Table 9.2). Each project received US$40,000 to hire a project manager and purchase basic supplies, such as traps and lures. The criteria used to select these new sites were that they be comprised of > 160 ha of contiguous orchards, used MD and be farmed by at least five growers. Selection preference was given to sites that: (i) had some prior experience with MD; (ii) were farmed mostly by small growers; (iii) had a unique feature that would help extend the fruit industry's knowledge and adoption of the areawide control approach – such as pest complex, pest pressure or the site's topography or location; and (iv) could demonstrate a strong likelihood of continuing the project after the 1-year support ended (Calkins, 1999).

Support provided for research

CAMP provided nearly US$1 million to support various research projects that:

- Addressed improved monitoring of CM and leafrollers.
- Quantified the atmospheric sex pheromone concentration within orchards and the release rates of lures and dispensers.
- Improved monitoring for stink bugs.
- Enhanced BC of CM, leafrollers and the tarnished plant bug, *Lygus lineolaris*, through classical and augmentative releases of parasitoids.
- Assessed the importance of biodiversity and the population dynamics of selected BC predator species in orchards.
- Characterized the impact of seasonal spray programmes of horticultural oil and kaolin particle films on pest management and plant growth.
- Tested new technologies for MD of CM and leafrollers.

Proposed projects were submitted on a yearly basis to CAMP and evaluated by a panel of ARS scientists, with recommendations provided by representatives of WTFRC and the California Pear Advisory Board. The goals of this collaborative project were to avoid duplicative funding by WTFRC and/or the Pear Pest Management

Table 9.2. General characteristics of the project sites established during the CAMP programme.

Year (s)	Location	Crop mix[a]	Hectares	No. of growers
1995–1999	Randall Island, CA	0:100	308	5
1995–1999	Medford, OR	20:80	121–202[b]	5–7[b]
1995–1999	Parker, WA	80:20	190–224[b]	7–11[b]
1995–1999	Howard's Flat, WA	90:10	486–688[b]	36–57[b]
1995–1999	Oroville, WA	95:5	154–526[b]	66–09[b]
1997	Progressive Flat, WA	100:0	243	25
1997	Brewster, WA	96:4	1902	41
1997	Manson, WA	98:2	410	68
1997	Wapato, WA	90:10	364	18
1997	Mendocino, CA	0:100	223	10
1998	Chelan, WA	100:0	263	11
1998	E. Wenatchee, WA	95:5	202	12
1998	Quincy, WA	100:0	283	7
1998	Bench Road, WA	95:5	506	9
1998	Moxee, WA	100:0	271	6
1998	Lower Roza, WA	90:10	688	23
1998	Rogers Mesa, CO	100:0	243	17
1999	Milton Freewater, OR	100:0	422	20
1999	Entiat Valley, WA	50:50	565	32
1999	Highland, WA	95:5	690	24
1999	West Valley, WA	92:8	338	12
1999	Lake County, CA	0:100	202	9
1995–1999	Total (1999)		9763	533

[a]Ratio of apple to pear within the project.
[b]Figures increased between 1995 and 1999.

Research Fund in California and to maximize the overall impact of the dollars spent in supporting research. The impact of the CAMP programme was clearly expanded by the acquisition of additional funding to support both research and implementation by the various participants in the project.

Centralized project coordination

One necessary structural component envisioned to develop an AWPM programme was the establishment of a central authority to run the project (Kogan, 1995). However, due to the large geographical size and number of participants, a centralized authority for CAMP proved to be cumbersome and largely unnecessary to coordinate the activities at each site. Several useful tools were developed at Oregon State University that aided the project, including: (i) a generalized bibliographic database for CM; (ii) an online weather data and degree-day web site supporting 102 sites from Oregon, Washington and Idaho, with 20 insect and disease models and crop models; (iii) GIS tools to develop maps of each of the sites and to summarize moth catches and levels of fruit injury among orchards; and (iv) grower satisfaction surveys in 1995 and 1996.

Orchard monitoring

One major impetus for growers to adopt MD for CM was the belief that achieving significant reductions in the use of broad-spectrum insecticides for this key pest would allow them also to reduce their use of sprays for secondary pests. Thus, CAMP spent nearly US$200,000 each year monitoring pests and natural enemies in comparison blocks under areawide MD versus conventional OP-based programmes (Beers *et al.*, 1998). Researchers at Washington State University developed the standardized protocols for monitoring, including sampling plans and data sheets for both apple and pear. Data were collected from a subsample of orchards (4 ha blocks) in each of the five original pilot project sites from 1996 to 1999 (see Table 9.3). In addition, data were collected from apple plots within four large, contiguous orchard sites in Washington State managed by individual growers (GRABS – Growers Resource Acquisition Baseline Study). Unfortunately, an additional objective of this project to 'assess the effect of MD versus conventional management on spray practices for secondary pests and its economic significance' was not completed.

Twelve different types of samples were collected in apple and pear during the project (Beers *et al.*, 1998). The eight samples collected from apple included: (i) the

Table 9.3. Summary of secondary pest sampling conducted in selected orchards (4 ha plots) treated with sex pheromone within areawide projects and compared with similar (tree age, cultivar and training system) conventional orchards.

Site/state	Crop	Contiguous hectares of MD	No. of blocks sampled (no. of comparison blocks)			
			1996	1997	1998	1999
Oroville/WA	Apple	526	12(6)	5(5)	5(5)	5(5)
Howard Flat/WA	Apple	344	12(8)	12(8)	12(8)	12(8)
Parker/WA	Apple	179	14–16(8)	15–17(8)	15–25(8)	13–28(8)
Parker/WA	Pear	45	9–20(4–8)	9–10(5–9)	9–14(4–6)	15–21(6–9)
Medford/OR	Pear	202	3(3)	4(4)	4(4)	4(4)
Randall Island/CA	Pear	308	9–11(2)	8(8)	8(8)	8 (8–11)
Vantage/WA[a]	Apple	202	4(4)	4(4)	4(4)	4(4)
Brewster/WA[a]	Apple	486	4(4)	4(4)	4(4)	4(4)
Bridgeport/WA[a]	Apple	202	4(4)	4(4)	4(4)	4(4)
Pateros/WA[a]	Apple	233	4(4)	4(4)	4(4)	4(4)

[a]GRABS (Grower Resource Acquisition Baseline Study) sites were large, contiguous areas of apple owned by a single grower under MD.

density of overwintering eggs and levels of parasitism for WALH; (ii) WALH nymphal densities during the first and second generations; (iii) aphid population (*Aphis* spp.) and natural enemy counts at five time periods during the season; (iv) leaf samples of WTLM mines and parasitism rates during the second and third generations; (v) *Campylomma* nymph counts during petal fall; (vi) leafroller larval counts for the overwintering and summer generations; (vii) mite binomial and leaf brushing samples; and (viii) fruit damage samples at mid-season and preharvest. The four types of pear samples collected were: (i) pre-bloom cluster samples for mites and psylla; (ii) post-bloom leaf brushing for mites and psylla; (iii) limb taps for psylla and generalist predators; and (iv) fruit damage counts at mid-season and preharvest.

Extension and education activities

CAMP was extremely active in the collection and distribution of information concerning the various aspects of pest management under the new AWPM programme. An Extension Program Coordinator was hired in 1996 and stationed with Washington State University Extension in Wenatchee, Washington. A similar 2-year extension position was funded in Yakima, Washington, during the last 2 years of the project. Extension personnel were also active in promoting MD for CM in both Oregon and California, though no new positions were created.

The Program Coordinator published 31 issues of the *Areawide IPM Update Newsletter* from 1996 to 1998. This newsletter provided a comprehensive review of research findings and information concerning each of the CAMP sites. Summer IPM tours and winter workshops were held to present the latest findings on the use of MD and the progress of the areawide programmes. The Coordinator produced an informational booklet on using MD and guidelines to establish new areawide projects (Alway, 1998a). Guides were also produced specifically for pear pest identification and monitoring (VanBuskirk *et al.*, 1998; Bush *et al.*, 1999).

Outcome of the CM AWPM Programme

CAMP was considered a great success by the industry because most growers were able to reduce their use of OP insecticides and their levels of CM injury without a noticeable increase in production costs (Calkins, 1999). This generalized result was sufficient to create a 'buzz' and promote a more rapid rate of adoption of MD, starting in 1998 (see Fig. 9.1). However, the specific components of pest management used in each project varied and a clear interpretation of the project's overall results is more nuanced. A number of additional factors affecting the economics of tree fruit production probably impacted the increased rate of adoption of MD that occurred after the conclusion of the project. Of particular importance was the development of OP resistance and the anticipation of use restrictions implemented by EPA for certain OP insecticides (Willett, 2006).

Levels of CM injury were generally higher in orchards during 1994 prior to the start of CAMP, and declined strongly during the 5-year project (see Fig. 9.2a).

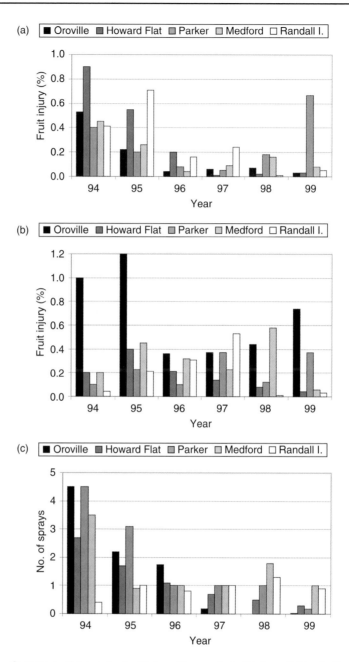

Fig. 9.2. Summary of the extent of (a) codling moth fruit injury; (b) leafroller fruit injury; and (c) organophosphate spraying applied per season for codling moth in the five original CAMP sites prior to the start of the project (1994) and during the 5-year project (1995–1999).

Injury levels in the first year of CAMP declined in all sites except Randall Island, where 1995 was actually the third year of the project. The ability of growers to reduce their use of supplemental OP sprays at this site during the project's first 3 years (1993–1995) was limited by a continued high pest pressure and elevated levels of OP resistance. CM populations in the later years of the project were significantly reduced by the rotation of methyl parathion and azinphosmethyl during the season and the occasional postharvest use of chlorpyrifos.

The significant increase in CM injury in the last year of the Parker project was also noteworthy (see Fig. 9.2a). The project coordinator hypothesized that this spike was the result of growers sharply reducing their dispenser density from 1998 after the cost subsidy was dropped by ARS (Higbee and Calkins, 2000). An apparent poor correlation of moth counts in traps with pest pressure 'false negatives' in some orchards in this hilly terrain allowed growers to forego sprays, and high levels of fruit injury occurred in 1999.

Levels of fruit injury by leafrollers were generally higher than CM injury in all of the pilot projects (see Figs 9.2a, b). Interestingly, growers in the Oroville site were not aware of leafrollers being present in their orchards prior to 1994. Levels of leafroller injury increased in all projects in the first year of CAMP (Fig. 9.2b). Injury levels were variable between years in each of the sites, with populations gradually increasing in Oroville, declining in Howard Flat, variable in Parker and declining in the last 2 years of the projects in Medford and Randall Island (see Fig. 9.2b).

The use of OP sprays targeted specifically for CM declined by nearly 75% in orchards within the five projects compared with either the levels used prior to the project in 1994 or the comparison blocks surveyed during the project (see Fig. 9.2c). The use of sprays declined most precipitously in Oroville, probably due to the supplemental control provided by the SIT programme. Use of OP sprays also declined sharply in Howard Flat and Parker; however, as previously mentioned, the sharp reduction in sprays applied in Parker was correlated with an equally sharp increase in CM injury (see Fig. 9.2a). The use of one seasonal OP application remained consistent in both the Randall Island and Medford pear sites.

In the first year of the project some growers did not spray for leafrollers because of the high cost of MD for CM (Beers et al., 1998). However, in general, leafrollers were a problem in most orchards inside and outside the projects and all growers increased their use of both chlorpyrifos against delayed dormant and methyl parathion during the summer, as well as the use of Bt for spring and summer leafroller control (see Table 9.1). Unfortunately, the spray records for materials applied for other secondary pests such as aphids were not summarized from each site.

Sampling of secondary pest and natural enemy populations found a few significant differences between apple and pear blocks in the CAMP versus conventional blocks (see Table 9.4). WALH population densities were lower in CAMP than in conventional blocks only in 1997 (early-summer nymphs) and 1999 (overwintering eggs); however, parasitism levels of the overwintering eggs by the mymarid, *Anagrus epos*, were higher in all years. No significant differences were found in either the densities of green aphids or their assemblage of generalist predators (coccinellids, syrphids, lacewings, mirids and cecidomyiids) during the study. The density of WTLM was lower and parasitism levels by *P. flavipes* were higher in some samples in 1998 and 1999. The density of phytophagous mites was lower in CAMP only in 1998, but the density of predator mites was higher during the last 3 years of the project. In pear, phytophagous mite densities were lower in CAMP only in 1996 and predatory mites were higher only in 1998. The mid-summer density of pear psylla was lower in CAMP from 1996 to 1998 (see Table 9.4).

The assessment of fruit injury from 1996 to 1999 found that CM injury was lower in apple CAMP blocks from 1996 to 1998 (see Table 9.5). Similarly, fruit injury by leafrollers was also lower from 1996 to 1998. Levels of fruit injury by the western flower thrips, *Frankliniella occidentalis*, were higher in CAMP in 1997 than in conventional blocks, but *Campylomma* injury levels were lower in CAMP in 1996. Tarnished plant bug injury was higher in CAMP than in conventional blocks in 1997 but lower in 1999. No significant differences in injury were found between treatments due to the cutworm, *Lacanobia subjuncta*, the stink bug, *Euschistus conspersus* or green aphids (see Table 9.5). Overall, levels of fruit injury were lower in CAMP than in conventional apple orchards in both 1998 and 1999. Total pear injury was lower in CAMP in both 1997 and 1998, primarily due to lower levels of injury from tarnish plant bug and, in particular, pear psylla in these years (see Table 9.6). No difference was found for levels of injury in pear due to CM, leafroller, *L. subjuncta*, the grape mealybug, *Pseudococcus maritimus*, or the stink bug, *E. conspersus*.

Table 9.4. Summary of secondary pest and natural enemy sampling in apple and pear blocks where significant differences were found in the areawide sex pheromone-treated blocks versus those in the comparison conventional blocks, 1996–1999.

		Significant difference in arthropod population densities in areawide MD versus the comparison conventional orchards			
Crop/sample timing	Sample	1996	1997	1998	1999
Apple/overwintering	Leafhopper eggs				Lower
Apple/early summer	Leafhopper nymphs		Lower		
Apple/mid-summer	Leafroller larvae		Lower	Lower	
Apple/mid-summer	Mines per leaf			Lower	Lower
Apple/late summer	Mines per leaf				Lower
Apple/mid-summer	Mites per leaf			Lower	
Apple/late summer	Mites per leaf			Lower	
Pear/pre-bloom	Pear psylla eggs per cluster			Lower	
Pear/mid-summer	Mites per leaf	Lower			
Pear/mid-summer	Pear psylla per leaf	Lower	Lower	Lower	
Apple/overwintering	Parasitized leafhopper eggs	Higher	Higher	Higher	Higher
Apple/mid-summer	Parasitism of leaf miners			Higher	Higher
Apple/late summer	Parasitism of leaf miners				Higher
Apple/mid-summer	Predatory mites per leaf		Higher	Higher	Higher
Pear/mid-summer	Predatory mites per leaf			Higher	

Table 9.5. Summary of percentage fruit injury in the apple blocks treated with sex pheromone (MD) within areawide projects versus conventional blocks (Conv) outwith the Oroville, Howard Flat, Parker, Vantage, Brewster, Pateros and Chelan areawide codling moth (CM) projects.

Treatment	Year	CM	Leafroller	Cutworm	Thrips	Campylomma	Lygus	Stink bug	Aphids	Total
MD	1996	0.17[a]	0.15[a]	–	0.08[a]	0.01[a]	0.14[a]	–	0.17[a]	0.72[a]
Conv	1996	0.34[b]	0.25[b]	–	0.03[a]	0.05[b]	0.09[a]	–	0.09[a]	0.85[a]
MD	1997	0.05[a]	0.05[a]	–	0.11[b]	0.07[a]	0.15[b]	–	0.01[a]	0.44[a]
Conv	1997	0.15[b]	0.18[b]	–	0.02[a]	0.07[a]	0.03[a]	–	0.02[a]	0.47[a]
MD	1998	0.09[a]	0.10[a]	0.10[a]	0.24[a]	0.05[a]	0.06[a]	0.06[a]	0.02[a]	0.72[a]
Conv	1998	0.60[b]	0.29[b]	0.10[a]	0.01[a]	0.06[a]	0.06[a]	0.08[a]	0.04[a]	1.24[b]
MD	1999	0.32[a]	0.47[a]	0.20[a]	0.20[a]	0.11[a]	0.13[a]	0.20[a]	0.02[a]	1.65[a]
Conv	1999	0.59[a]	.92[a]	0.23[a]	0.19[a]	0.22[a]	0.39[b]	0.36[a]	0.05[a]	2.95[b]

Means within year followed by a different superscript letter were significantly different, $P < 0.05$.

Table 9.6. Summary of fruit injury in the pear blocks treated with sex pheromone (MD) within areawide projects versus conventional blocks (Conv) outwith the Parker, Medford and Randall Island areawide codling moth (CM) projects.

Treatment	Year	CM	Leafroller	Cutworm	Psylla	Mealybug	Lygus	Stink bug	Total
MD	1996[c]	0.19[a]	0.26[a]	—	0.02[a]	0.0[a]	0.26[a]	–[d]	0.73[a]
Conv	1996[c]	0.02[a]	0.14[a]	—	0.09[a]	0.0[a]	0.62[a]	–[d]	0.87[a]
MD	1997	0.08[a]	0.31[a]	—	0.52[a]	0.03[a]	0.48[a]	0.03[a]	1.45[a]
Conv	1997	0.05[a]	0.03[a]	—	1.38[b]	0.02[a]	0.64[a]	0.02[a]	2.14[b]
MD	1998	0.32[a]	0.12[a]	0.02[a]	1.42[a]	0.00[a]	0.15[b]	0.01[a]	2.04[a]
Conv	1998	0.05[a]	0.02[a]	0.02[a]	4.94[b]	0.00[a]	0.05[a]	0.01[a]	5.09[b]
MD	1999	0.08[a]	0.13[a]	0.02[a]	0.56[a]	0.05[a]	0.12[a]	0.00[a]	0.96[a]
Conv	1999	0.09[a]	0.11[a]	0.04[a]	0.53[a]	0.01[a]	0.10[a]	0.02[a]	0.90[a]

Means within year followed by a different superscript letter (a, b) were significantly different, $P < 0.05$.

[c]Injury by pear psylla, mealybug and sucking bugs was collected only in Medford in 1996.

[d]All sucking bug fruit injury was grouped together as Lygus injury in 1986.

Evaluations of the 17 one-year areawide projects found that all growers considered the project to be a success (Coop *et al.*, 1999). In general, growers in these projects used reduced rates of dispensers, except along orchard borders or in problem hot spots. Low-pressure sites were able to use none to one cover sprays, while mid- and high-pressure orchards used one to two sprays for the first generation and a reduced number of sprays in the second generation, mostly along borders and in hot spots. These spray programmes contrasted with the previous use of the three to five cover sprays that many growers had applied prior to the projects. In general, no CM injury was reported in these sites or, in some cases, injury was confined to a few small areas within each project, usually known hot spots.

The major secondary pest problems experienced in these projects were either from true bugs, such as the western boxelder bug, *Leptocoris rubrolineatus*, or stink bugs and the noctuid, *Lacanobia subjuncta*. Leafrollers were a concern for growers in a number of projects, as moth counts in traps were generally very high during the season. Injury levels from leafrollers, however, were low to moderate in only a few orchards. The large (1902 ha) Brewster project included a 200 ha contiguous area that was treated with MD for both CM and leafrollers (Dual MD) for 3 years (Knight *et al.*, 2001). Leafroller injury in the Dual MD project was reduced by 41%, and growers used approximately one fewer *Bt* spray compared with the orchards in the regular MD programme for CM. Despite these promising results leafroller injury was considered too high, and growers dropped the use of MD and switched to newly registered insecticides, i.e. spinosad (see Table 9.1).

Economic Evaluation of the CM AWPM Programme

Growers have consistently cited the high cost of adopting MD for CM as the major disincentive for them in adopting this new technology (Coop *et al.*, 1999). Prior to CAMP, an analysis found that costs had increased by US$740/ha for pear growers in Sacramento County, California who made two dispenser applications per season (Weddle, 1994). Similarly, Williamson *et al.* (1996) found MD was US$188/ha more expensive than conventional management in Washington State apple orchards. Other economic impacts that influenced grower adoption of MD included the effectiveness of MD, the difficulty in applying dispensers, the uncertainty associated with monitoring CM, the need for increased attention in orchards (US$80–120/ha to hire a consultant or for scouting) and the increased risk of secondary pest problems (Weddle, 1993).

Proponents have hypothesized from the earliest adoption of MD that growers would experience reduced injury levels in treated orchards, partially because the evolution of OP resistance was curtailing the effectiveness of the current programme in conventionally managed sites (Weddle, 1994); and that enhanced BC would allow growers to reduce their spray bill in the second and subsequent years (Connor and Higbee, 1999). Williamson *et al.* (1996) suggested that the benefits of this programme would outweigh its costs when dispensers cost less, growers could further reduce their spray use and the proportion of fruit packed could increase in MD orchards. This optimistic analysis, however, suggested that the benefits of adopting MD would be reduced by the continued need for growers to spray orchards with other pesticides or crop amendments, such as calcium chloride.

The economic analysis following the 5-year CAMP was similar (Connor and Higbee, 1999). MD could only be cost-effective in a straight benefit–cost analysis if growers reduced their dispenser densities by half in the second and subsequent years and could maintain a reduced spray programme. Unfortunately, this analysis did not include the costs of two 'down side' outcomes: the higher cost of monitoring and the higher risk of new pest problems developing in MD orchards. Again, the potential impact of MD on BC was mentioned but not included in the analysis (Connor and Higbee, 1999).

The limited value of these analyses was highlighted by: (i) their failure to include the benefits accrued to growers using MD who developed experience with an alternative technology prior to the loss of OP insecticides; (ii) growers' greater ability to schedule workers to re-enter unsprayed orchards and complete other essential operations, i.e. thinning, irrigation, etc.; and (iii) growers' reduced liability when the potential for worker exposure to OP residues was eliminated.

Interestingly, two analyses found that the cost of implementing a MD-based programme in pear was less expensive than the then-current conventional programmes. Growers in the Medford project reduced their pesticide use by US$520/ha as compared with conventional growers (VanBuskirk *et al.*, 1999). Total operating costs of production for pear orchards using the MD aerosol 'puffer' for Lake County, California were marginally higher in the first year of the project, but 3% lower in the later years (Elkins *et al.*, 2005).

Growers' Responses to the CM AWPM Programme

Twenty-two areawide projects, including 533 growers farming 9763 ha, were organized within CAMP (see Table 9.2). Nearly all growers responding to a survey were very pleased with the results obtained with AWPM and the organizational structure of the CAMP projects (Coop *et al.*, 1999). This enthusiasm was reflected in the growth in both the size and number of participants in four of the five original sites that occurred during the 5-year period (see Table 9.2). In addition, new groups of growers were eager to join the 17 one-year projects (Calkins, 1999).

The structure of these projects varied in terms of both the information provided to the growers and the actions requested by the participants. Generally, all projects created a structure that allowed groups of growers to meet and discuss mutual problems and interests. Usually, there were highly knowledgeable advisors associated with the projects to help solve problems and answer bug-related questions. Projects implemented intensive monitoring programmes for CM and often a select group of secondary pests, and these data were summarized and shared among members. The projects allowed growers to manage their pests effectively through scouting, use of action thresholds and reliance on selective integrated tactics. The outcome of nearly all of these projects was to corral a group of growers together and transform them all into practitioners of IPM through greater knowledge and the use of research-based programmes. The successes of the projects were in direct proportion to the intensity of the grower involvement in the group (Knight, 1999).

The number one factor indicating the potential for success has been the group's efforts to clean up the problem orchards in their area. Every successful project included the previously non-participating (NP) grower. For example, the successful grassroots movement at Howard Flat eventually convinced 34 of the 36 growers to join the project. In Oroville, all but one grower joined. NP growers had a number of reasons for not joining the projects: some were against government programmes of any kind; others did not want to work closely with their neighbours either because of past grievances or due to a fear that they would be criticized or lose their management independence. Many growers were initially sceptical of the efficacy of the programme or felt that it was too expensive. And, perhaps the number one reason why pockets of CM existed initially in each of the projects, was because some farmers did not farm full-time and because events in their personal life prevented them from effectively focusing their management skills (Knight, 1999). Areawide problems with CM were dramatically reduced once these growers joined with their neighbours.

Conversely, the major reasons why growers joined the project were because they saw that it was not a top-down government programme; growers in the projects actively reached out to educate and persuade others to join; and, after the first year, the programme was demonstrated to be working and other growers realized they could save money and avoid some of the pesticide-related headaches by joining the project.

Unexpected Outcomes of the CM AWPM Programme

Immediately following the first registration of CM MD in the USA, growers expressed their concern that MD was too expensive (Knight, 1995a). The initial implementation of CAMP required that growers receive a subsidy of US$125/ha (Kogan, 1995). From the beginning, growers adopting MD seemed to push the limits of this technology by stretching the established orchard risk categories to allow them to reduce their dispenser density (Gut and Brunner, 1996). For example, by 1997 only 55% of Washington State growers were using the full rate of pheromone (Alway, 1997), and this had declined further to 27% by 1998 (Alway, 1998b). The potential negative impact of reducing dispenser density on CM management could be observed in the Parker CAMP site (see Fig. 9.2a). While growers reduced their use of MD, they also continued to apply broad-spectrum insecticides for CM on more than 80% of orchards (Alway, 1998b). This grower-developed programme did not allow CAMP to quite achieve its goal of reducing OP use by 80% (see Fig. 9.2c).

The relative importance of BC in the CAMP projects was similar to the earlier results found in the 3-year transition studies conducted in individual orchards (Knight, 1995a; Gut and Brunner, 1998). In both studies, growers saw an increase in BC but few significant reductions in most secondary pest populations (Beers *et al.*, 1998). The one exception was the significant reductions that occurred in pear psylla populations in the Parker and Medford pear projects (see Table 9.6). Also, similar to findings from earlier studies, certain pests became more important in MD orchards

(Brunner, 1999). For example, the grape mealybug increased gradually in the Parker site in both apple and pear (Higbee and Calkins, 2000). Leafrollers became a much more important problem requiring specific sprays in all sites.

Sucking bugs, such as stink bugs and tarnish plant bug, were major problems in some orchards, particularly along their borders. The noctuid, *L. subjuncta*, was a new pest for growers that caused serious problems in some blocks (Landolt, 1998). Nevertheless, other potential pests – such as the lesser appleworm, *Grapholitha prunivora*, which had become a pest in commercial apple orchards in the eastern USA (Krawczyk and Johnson, 1996) and is known to be present in wild hosts such as hawthorn and native plum in the Pacific north-west (Brown, 1953) – did not become a problem.

Apple and pear production is ultimately not driven by CM management success but by the economics of farm management. By 1999 the economics of tree fruit production were very poor for growers, and this led to both an overall reduction in the area under production and a noticeable slowing in the adoption rate of MD (see Fig. 9.1). In particular, some growers with older varieties such as Red Delicious went out of business, as did many smaller operations. Other orchards were replanted with wine grapes, cherry or stone fruit. The continued sprawl of towns and cities has forced the conversion of many orchards into rural housing developments. Orchards owned by absentee investment groups and speculators were managed with a thin array of low-cost inputs. These financial conditions led to neglect of some orchards, and CM population levels skyrocketed in some districts. Pest control boards became largely ineffective due to the abundance of problem sites. In many former CAMP sites it took about 3 years for pest problems to build to the levels existing prior to the project.

The ARS-funded project effectively brought together personnel from government, industry and several universities. Tremendous successes were reached in the implementation of AWPM projects, in new research discoveries and in outreach efforts to educate the industry. Not unexpectedly, a few problems occurred with the functioning of such an independent group of experts. The group did not easily adopt the AWPM tenet of having a centralized structure for programme coordination and data collection and dissemination (Kogan, 1995), as some project leaders were hesitant to share information. Control of research funding by ARS created some dissension among researchers when their projects were not fully funded. The unity of the group appeared to break apart by the end of the project, and a summary report was never completed.

Future Prospects for the CM AWPM Programme

CAMP was an extremely well-funded project, well received by growers and the industry and was clearly influential in the shift that has occurred in tree fruit pest management away from OP-based programmes. CAMP demonstrated that MD could substitute for some use of insecticides, but also emphasized that insecticides are still needed to maintain pest populations at low levels. The use of MD for CM has continued to grow since CAMP ended, and comprises nearly 75% of the production area in Washington State (see Fig. 9.1). Today, with such a large proportion of orchards under MD, there are many contiguous areas treated with MD but, in general, these growers do not work together.

Developing and maintaining a coordinated approach is more difficult than having all growers in a region adopt a similar technology. For example, pear growers in Lake County adopted the use of an areawide grid of aerosol puffers for MD (Shorey and Gerber, 1996). The University of California Extension, with some support from CAMP, ran a demonstration programme for 3 years in this area, and this has been smoothly adopted by the local private consultants (PCAs). Today, these pear growers are on autopilot for control of CM, while the management of secondary pests varies widely among orchards (Elkins, 2002).

Few coordinated AWPM programmes exist today. Government support was necessary in organizing the 22 CAMP projects and it appears that in most regions there is not sufficient organizational structure for growers to maintain their own projects without government funding. Various factors caused projects to dissolve following the CAMP programme. Many projects were fragile, consisting of growers expressing a stereotypical American Wild-West 'go-it-alone' mentality. Projects such as Rogers Mesa in western Colorado were abandoned as growers switched production from apple to stone fruits. Howard Flat lost a large proportion of its tree fruit production due to poor economics and the steady usurpation of orchards by rural real estate development. Some projects were able to function for more than one year with CAMP funding or by obtaining additional government funding or working within funded research projects. Having a few large, contiguous blocks of orchards monitored by one PCA has allowed several large areas to continue under a centralized stewardship.

Other projects, where growers worked with multiple PCAs or sent their fruit to several cooperatives or packing houses, have tended to dissolve. Some such as Brewster were able to exist for a few years due to a combination of factors, such as stretching their use of CAMP funds, obtaining additional government funding, working with a government-funded research project and by forming a non-profit organization that could allocate participants a fee to fund a centralized monitoring and data dissemination programme. However, this project ended after 5 years due to poor farm economics and a lack of a strong and unified grower commitment to the project.

Today, only two of the original 22 CAMP projects remain: Ukiah Valley in California and Milton Freewater in Oregon. Ukiah Valley started in 1996 with a grant from EPA and then extended the 1-year CAMP funding in 1997 to fund a 3-year project headed by University of California Extension personnel. Pear growers farming 536 ha formed the non-profit Ukiah Pear Grower Association and continue to allocate growers a fee to hire a trap checker who monitors orchards and distributes information to all participants. Unfortunately, the cohesiveness of the project is threatened by both the reduced problems in managing CM and the emergence of new pests. The Milton Freewater growers organized themselves through the Blue Mountain Horticultural Society in 1998, and have maintained a coordinated project on nearly 1000 ha in north-eastern Oregon. Interestingly, not all growers use MD in the project. Growers are assessed at US$45/ha to fund a monitoring programme. Data are e-mailed to all growers and the various warehouses in the district and are posted on several bulletin boards within the project site. General information is exchanged, and the group's cohesion is maintained at weekly meetings held at the local extension office.

In summary, 7 years after the end of CAMP there remains a general lack of local coordination between growers' pest management activities, but there has been an exponential increase in the knowledge of how to implement MD (Brunner *et al.*, 2007). CAMP was followed by other, well-funded USDA projects, which achieved further improvements in MD and tested alternative, selective tactics to replace the use of OPs (Initiative for Future Agricultural and Food Systems, Risk Avoidance and Mitigation for Major Food Crops Systems, American Farmland Trust and Sustainable Agriculture Research and Education). Unfortunately, CM remains the number one pest problem for most growers and management programmes have become more expensive as they shifted from OPs to MD supported by a diversity of supplemental spray programmes using several new insecticides (see Table 9.1). The evolution of resistance to these new insecticides and their negative impacts on BC continue to be key concerns in implementing sustainable IPM programmes (Brunner *et al.*, 2005).

Epilogue

Several factors have contributed to the success of the CAMP programme (Coop *et al.*, 1999). These can be grouped into two categories: (i) operational – the availability of several effective and selective tactics for both the key and secondary pests backed by technical support; and (ii) organizational – well-funded, coordinated programmes directly involving growers, researchers, industry leaders and governmental administrators. The lesson learned from the CAMP programme is that pest management is similar to rocket science and requires attention, experience and skill to be effective (Knight, 1999). Dissemination of knowledge and coordination of actions by individual growers have been shown to improve pest management, and offer tremendous benefits to society. Future efforts should focus on how similar, grower-based organizations can be developed and sustained.

References

Alway, T. (1997) Mating disruption use in Washington State: grower survey report. *Areawide IPM Update* 2, 1–3.

Alway, T. (1998a) *Codling Moth Mating Disruption and Establishing a Pheromone-based Codling Moth Management Site in the Pacific Northwest.* Washington State University Cooperative Extension, Wenatchee, Washington.

Alway, T. (1998b) Codling moth mating disruption use in Washington State. *Areawide IPM Update* 3, 1–2.

Barnes, M.M. (1959) Deciduous fruit insects and their control. *Annual Review of Entomology* 4, 343–362.

Barnes, M.M. (1991) Codling moth occurrence, host race formation, and damage. In: van der Geest, L.P.S. and Evenhuis, H.H. (eds) *Tortricid Pests: their Biology, Natural Enemies and Control.* Elsevier, Amsterdam, pp. 313–327.

Barnes, M.M., Millar, J.G., Kirsch, P.A. and Hawks, D.C. (1992) Codling moth (Lepidoptera: Tortricidae) control by dissemination of synthetic female sex pheromone. *Journal of Economic Entomology* 85, 1274–1277.

Barrett, B.A. (1988) The population dynamics of *Pnigalio flavipes* (Hymenoptera: Eulophidae), the major parasitoid of *Phyllonorycter elmaella* (Lepidoptera: Gracillariidae) in central Washington apple orchards. PhD dissertation, Washington State University, Pullman, Washington, 173 pp.

Beers, E.H., Brunner, J.F., Willett, M.J. and Warner, G.M. (1993) *Orchard Pest Management.* Good Fruit Grower, Yakima, Washington.

Beers, E.H., Himmel, P.D., Dunley, J.E., Brunner, J.F., Knight, A., Higbee, B., Hilton, R., VanBuskirk, P. and Welter, S. (1998) Secondary pest abundance in different management systems. *Washington Horticultural Association Proceedings* 94, 121–127.

Bloem, K.A. and Bloem, S. (2000) SIT for codling moth eradication in British Columbia, Canada. In: Tan, K.H. (ed.) *Areawide Control of Fruit Flies and Other Insect Pests.* Penerbit University Sains Malaysia, Pulau Pinang, Malaysia, pp. 207–214.

Brown, D.F., Knight, A.L., Howell, J.F., Sell, C.R., Krysan, J.L. and Weiss M. (1992) Emission characteristics of a polyethylene pheromone dispenser for mating disruption of codling moth (Lepidoptera: Tortricidae). *Journal of Economic Entomology* 85, 910–917.

Brown, E.E. (1953) Life cycle of the lesser apple worm in northeastern Oregon. *Journal of Economic Entomology* 46, 1631–1641.

Brunner, J.F. (1992) Leafroller biocontrol: promising new parasites discovered. *Washington Horticultural Association Proceedings* 88, 285–288.

Brunner, J.F. (1994) Using Bt products as tools in pest control. *Good Fruit Grower* 45, 34–38.

Brunner, J.F. (1999) New pests: a challenge for areawide programs. *Washington Horticultural Association Proceedings* 95, 154–158.

Brunner, J.F. and Beers, E.H. (1990) *Pandemis and Obliquebanded Leafrollers.* Washington State University Extension Bulletin 1582, Washington State University, Pullman, Washington.

Brunner, J.F., Hoyt, S.C. and Wright, M.A. (1982) *Codling Moth Control – a New Tool for Timing Sprays.* Washington State University Extension Bulletin 1072, Pullman, Washington.

Brunner, J., Gut, L. and Knight, A. (1992) Transition of apple and pear orchards to a pheromone-based pest management system. *Washington Horticultural Association Proceedings* 88, 169–175.

Brunner, J.F., Beers, E., Doerr, M. and Granger, K. (2005) *Managing Apple Pests without Organophosphates.* Good Fruit Grower, Yakima, Washington.

Brunner, J.F., Dunley, J.E., Beers, E.H. and Jones V.P. (2007) Building a multi-tactic biologically intensive pest management system for Washington orchards. In: Felsot, A.S. and Racke, K.D. (eds) *Crop Protection Products for Organic Agriculture: Environment, Health, and Efficacy Assessment.* American Chemical Society, Washington DC, pp. 131–143.

Bush, M., Higbee, B., Peterson, B., Simone, N. and Weddle, P. (1999) *Pear Insect Field Guide.* Good Fruit Grower, Yakima, Washington.

Byers, R.E., Lyons, C.G., Yoder, K.S., Horsburgh, R.L., Barden, J.A. and Donohue, S.J. (1984) Effects of apple tree size and canopy density on spray chemical deposit. *Horticultural Science* 19, 93–94.

Calkins, C.O. (1998) Review of the codling moth areawide suppression program in the western United States. *Journal of Agricultural Entomology* 15, 327–333.

Calkins, C. (1999) Codling moth areawide management: a successful 5-year program. *Washington Horticultural Association Proceedings* 95, 151–153.

Calkins, C.O. (2003) Areawide program for suppression of codling moth in the western United States. *IOBC wprs Bulletin* 26, 21–25.

Calkins, C.O. and Faust, R.J. (2003) Overview of areawide programs and the program for suppression of codling moth in the western USA directed by the United States Department of Agriculture-Agricultural Research Service. *Pest Management Science* 59, 601–604.

Calkins, C.O., Knight, A.L., Richardson, G. and Bloem, K.A. (2000) Area-wide population suppression of codling moth. In: Tan, K.H. (ed.) *Areawide Control of Fruit Flies and Other Insect Pests.* Penerbit University Sains Malaysia, Pulau Pinang, Malaysia, pp. 215–219.

Charmillot, P.J. (1990) Mating disruption technique to control codling moth in western Switzerland. In: Ridgway, R. Silverstein, R.M. and Inscoe, N. (eds) *Behavior-modifying Chemicals for Insect Management: Applications of Pheromones and other Attractants.* Marcel Dekker, Inc., New York, pp. 165–182.

Connor, J.D. and Higbee, B. (1999) The economics of mating disruption based on pest control in central Washington apples. *Washington Horticultural Association Proceedings* 95, 161–166.

Coop, L., Kogan, M. and Bajwa, W. (1999) Areawide codling moth: extending the principles and lessons learned outside the project and to other commodities. *Washington Horticultural Association Proceedings* 95, 176–183.

Croft, B.A. (1979) Management of apple arthropod pests and natural enemies relative to developed resistance. *Environmental Entomology* 8, 583–586.

Croft, B.A. and Bode, W.M. (1983). Tactics for deciduous fruit IPM. In: Croft, B.A. and Hoyt, S.C. (eds) *Integrated Management of Insect Pests of Pome and Stone Fruits.* John Wiley and Sons, New York, pp 219–270.

Croft, B.A. and Riedl, H.W. (1991) Chemical control and resistance to pesticides of the codling moth. In: van der Geest, L.P.S. and Evenhuis, H.H. (eds) *Tortricid Pests: their Biology, Natural Enemies and Control.* Elsevier, Amsterdam, pp. 371–387.

Cutwright, C.R. (1954) A codling moth population resistant to DDT. *Journal of Economic Entomology* 47, 189–190.

Dunley, J.E. and Welter, S.C. (2000) Correlated insecticide cross-resistance in azinphosmethyl-resistant codling moth (Lepidoptera: Tortricidae). *Journal of Economic Entomology* 93, 955–962.

Elkins, R.B. (2002) *Areawide Implementation of Mating Disruption in Pears Using Puffers.* California Department of Pesticide Regulation, Agreement No. 00-0198 S, Sacramento, California.

Elkins, R.B., Klonsky, K.M. and DeMoura, R.L. (2005) Cost of production for transitioning from conventional codling moth control to aerosol-released mating disruption ('puffers') in pears. *Acta Horticulturae* 671, 559–563.

Falcon, L.A. and Huber, J. (1991) Biological control of the codling moth. In: van der Geest, L.P.S. and Evenhuis, H.H. (eds) *Tortricid Pests: their Biology, Natural Enemies and Control.* Elsevier, Amsterdam, pp. 355–369.

Ferro, D.N., Sluss, R.R. and Harwood, R.F. (1974) Changes in the population dynamics of the codling moth, *Laspeyresia pomonella*, after release from insecticide pressures. *Environmental Entomology* 3, 686–690.

Fritsch, E., Undorf-Spahn, K., Kienzle, J., Zebitz, C.P.W. and Huber, J. (2005) Apfelwicker granulovirus: erst hinweise auf unterschiede in der empfindlichkeit lokaler apfelwickler populationen. *Nachrichtenblatt Deutsche Pflanzenschutzdienst* 57, 29–34.

Gut, L.J. and Brunner, J.F. (1996) *Implementing Codling Moth Mating Disruption in Washington Pome Fruit Orchards.* Tree Fruit Research and Extension Center Information Series, No.1, Washington State University, Pullman, Washington.

Gut, L.J. and Brunner, J.F. (1998) Pheromone-based management of codling moth (Lepidoptera: Tortricidae) in Washington apple orchards. *Journal of Agricultural Entomology* 15, 387–405.

Gut, L.J., Brunner, J.F. and Knight, A. (1992) Mating disruption as a control for codling moth and leafrollers. *Good Fruit Grower* 43, 56–60.

Gut, L., Brunner, J.F., and Knight A. (1995) Implementation of pheromone-based pest management programs. *Washington Horticultural Association Proceedings* 91, 227–234.

Hansen, J.D. and Johnson, J.A. (2007). Introduction. In: Hansen, J.D. and Johnson, J.A (eds) *Heat Treatments for Postharvest Pest Control*. CAB International, Wallingford, UK, pp. 1–26.

Higbee, B.S. and Calkins, C.O. (2000) Five year summary of the Parker Heights CAMP project. *Proceedings Western Orchard Pest and Disease Management Conference* 74, 15–16.

Higbee, B.S., Calkins, C.O. and Temple, C.A. (2001) Overwintering of codling moth (Lepidoptera: Tortricidae) larvae in apple harvest bins and subsequent moth emergence. *Journal of Economic Entomology* 94, 1511–1517.

Hough, W.S. (1928) Relative resistance to arsenical poisoning of two codling moth strains. *Journal of Economic Entomology* 22, 325–329.

Howell, J.F. (1991) Reproductive biology. In: van der Geest, L.P.S. and Evenhuis, H.H. (eds) *Tortricid Pests: their Biology, Natural Enemies and Ccontrol*. Elsevier, Amsterdam, pp. 157–174.

Howell, J.F. and Maitlen, J.C. (1987) Accelerated decay of residual azinphosmethyl and phosmet by sprinkler irrigation above trees and its effect on control of codling moth based on laboratory bioassays as estimated by laboratory simulation of insecticide deposits. *Journal of Agricultural Entomology* 4, 281–288.

Howell, J. F. and Britt, R. (1994) IPM changes associated with using mating disruption to control codling moth in commercial apple production. *Washington Horticultural Association Proceedings* 89, 258–264.

Hoyt, S.C. (1969) Integrated chemical control of insects and biological control of mites on apple in Washington. *Journal of Economic Entomology* 62, 74–86.

Hoyt, S. (1983) Biology and control of the western tentiform leafminer. *Proceedings of the Washington State Horticultural Association* 79, 115–118.

Ing, G. (1992) *Apple 1992*. Washington Tree Fruit Research Commission, Wenatchee, Washington.

Ing, G. (1995) *Apple Entomological Research Review*. Washington State Tree Fruit Research commission, Wenatchee, Washington.

Ing, G. (1999) Almost thirty years of grower codling moth research funding. *Washington Horticultural Association Proceedings* 95, 184.

Jackson, D.M. (1979) Codling moth egg distribution on unmanaged apple trees. *Annals of the Entomological Society of America* 72, 361–368.

Jang, E.B. and Moffitt, H.R. (1994) Systems approaches to achieving quarantine security. In: Sharp, J.L. and Hallman, G.J. (eds) *Quarantine Treatments for Pests of Food Plants*. Oxford & IBH Publishing Co., New Delhi, India, pp. 225–237.

Judd, G.J.R., Gardiner, M.G.T. and Thomson, D.R. (1997) Control of codling moth in organically managed apple orchards by combining pheromone-mediated mating disruption, post-harvest fruit removal and tree banding. *Entomologia Experimentalis et Applicata* 83, 137–146.

Knight, A. (1992) New dimensions in mating disruption of codling moth. *Washington Horticultural Association Proceedings* 88, 166–168.

Knight, A. (1994) Insect pest and natural enemy populations in paired organic and conventional apple orchards in the Yakima Valley, Washington. *Journal of the Entomological Society of British Columbia* 91, 27–36.

Knight, A. (1995a) The impact of codling moth (Lepidoptera: Tortricidae) mating disruption on apple pest management in Yakima Valley, Washington. *Journal of the Entomological Society of British Columbia* 92, 29–38.

Knight, A.L. (1995b) Evaluating pheromone emission rate and blend in disrupting sexual communication of codling moth (Lepidoptera: Tortricidae). *Environmental Entomology* 24, 1396–1403.

Knight, A.L. (1996) Sexual biology and mating disruption of orange tortrix, *Argyrotaenia citrana* (Lepidoptera: Tortricidae). *Journal of the Entomological Society of British Columbia* 93, 111–120.

Knight, A. (1997) Optimizing the use of Bts for leafroller control. *Good Fruit Grower* 48, 47–49.

Knight, A.L. (1999) Spatial versus sociological-based pest management: biological and economic constraints. *Proceedings Washington Horticultural Association* 95, 167–168.

Knight, A.L. (2001) Monitoring the seasonal population density of *Pandemis pyrusana* (Lepidoptera: Tortricidae) within a diverse fruit crop production area in the Yakima Valley, WA. *Journal of the Entomological Society of British Columbia* 98, 217–225.

Knight, A.L. (2007) Influence of within-orchard trap placement on catch of codling moth (Lepidoptera: Tortricidae) in sex pheromone-treated orchards. *Environmental Entomology* 36, 1485–1493.

Knight, A. and Christianson, B. (1999) Using traps and lures in pheromone-treated orchards. *Good Fruit Grower* 50, 45–51.

Knight, A.L. and Light, D.M. (2005) Developing action thresholds for codling moth (Lepidoptera: Tortricidae) with pear ester- and codlemone-baited traps in apple orchards treated with sex pheromone mating disruption. *The Canadian Entomologist* 137, 739–747.

Knight, A.L. and Turner, J.E. (1999) Mating disruption of *Pandemis* spp. (Lepidoptera: Tortricidae). *Environmental Entomology* 28, 81–87.

Knight, A.L., Brunner, J.F. and Alston, D. (1994) Survey of azinphosmethyl resistance in codling moth (Lepidoptera: Tortricidae) in Washington and Utah. *Journal of Economic Entomology* 87, 285–292.

Knight, A.L., Howell, J.F., McDonough, L.M. and Weiss, M. (1995) Mating disruption of codling moth (Lepidoptera: Tortricidae) with polyethylene tube dispensers: determining emission rates and the distribution of fruit injuries. *Journal of Agricultural Entomology* 12, 85–100.

Knight, A.L., Turner, J.E. and Brachula, B. (1997) Predation of codling moth (Lepidoptera: Tortricidae) in mating disrupted and conventional orchards in Washington. *Journal of the Entomological Society of British Columbia* 94, 67–74.

Knight, A.L., Thomson, D.R. and Cockfield, S.D. (1998) Developing mating disruption of obliquebanded leafroller (Lepidoptera: Tortricidae) in Washington State. *Environmental Entomology* 27, 1080–1088.

Knight, A., Christianson, B., Cockfield, S. and Dunley, J. (2001) Areawide management of codling moth and leafrollers. *Good Fruit Grower* 52, 31–33.

Knipling, E.F. (1979) *The Basic Principles of Insect Population Suppression and Management.* Agricultural Handbook No. 512, USDA-ARS, Washington, DC.

Kogan, M. (ed.) (1994) *Areawide Management of the Codling Moth, Implementation of a Comprehensive IPMPprogram for Pome Fruit Crops in the Western US.* Integrated Plant Protection Center, Oregon State University, Corvallis, Oregon.

Kogan, M. (1995) Areawide management of major pests: is the concept applicable to the *Bemisia* complex? In: Gerling, D. and Mayer, R.T. (eds) *Bemisia 1995: Taxonomy, Biology, Damage, Control and Management.* Intercept Ltd, Andover, UK, pp. 643–657.

Krawczyk, G. and Johnson, J.W. (1996) Occurrence and pest status of lesser appleworm *Grapholita prunivora* Walsh in Michigan apple orchards. *IOBC wprs Bulletin* 19, 356–357.

Landolt, P.J. (1998) *Lacanobia subjuncta* (Lepidoptera: Noctuidae) on tree fruits in the Pacific Northwest. *Pan-Pacific Entomology* 74, 32–38.

Madsen, H.F. and Morgan, C.V.G. (1970) Pome fruit pests and their control. *Annual Review of Entomology* 15, 295–320.

Madsen, H.F. and Vakenti, J.M. (1973) Codling moth: use of codlemone-baited traps and visual detection of entries to determine need of sprays. *Environmental Entomology* 2, 677–679.

MAFF-Japan (1950) Plant protection law enforcement regulation. In: *Ministerial Ordinance No. 848, Article 3, Annexed Table.* Ministry of Agriculture, Forestry and Fisheries, Kyoto, Japan, pp. 27–33.

Milli, R., Koch, U.T. and de Kramer, J.J. (1997) EAG measurement of pheromone distribution in apple orchards treated for mating disruption of *Cydia pomonella*. *Entomologia Experimentalis et Applicata* 82, 289–297.

Myburgh, A.C. (1980) Infestation potential of the codling moth. *Deciduous Fruit Grower* 10, 368–377.

Newcomber, E.J. (1936) Orchard sanitation for control of codling moth. *Washington State Horticultural Association Proceedings* 31, 140–141.

Prokopy, R.J. and Croft, B.A. (1994) Apple insect pest management. In: Metcalf, R.L. and Luckmann, W.H. (eds) *Introduction to Pest Management.* John Wiley and Sons, Inc., New York, pp. 543–585.

Proverbs, M.D. and Newton, J.R. (1975) Codling moth control by sterile insect release: importation of fruit and fruit containers as a source of reinfestation. *Journal of the Entomological Society of British Columbia* 72, 6–9.

Reyes, M., Franck, P., Charmillot, P.J.,Ioratti, C., Olivares, J., Pasqualini, E. and Sauphanor, B. (2008) Diversity of insecticide resistance mechanisms in European populations of the codling moth, *Cydia pomonella. Pesticide Science* (In press).

Riedl, H. and Croft, B.A. (1978) The effects of photoperiod and effective temperatures on the seasonal phenology of the codling moth (Lepidoptera: Tortricidae). *The Canadian Entomologist* 110, 455–470.

Riedl, H., Howell, J.F., McNally, P.S. and Westigard, P.H.. (1986) *Codling Moth Management: Use and Standardization of Pheromone Trapping Systems.* University of California Division of Agriculture and Natural Resources, Bulletin 1918, Oakland, California.

Sauphanor, B. and Bouvier, J.C. (1995) Cross-resistance between benzoylureas and benzoylhydrazines in the codling moth, *Cydia pomonella* L. *Pesticide Science* 45, 369–375.

Shel'deshova, G.G. (1967) Ecological factors determining the distribution of the codling moth, *Laspeyresia pomonella* L. (Lepidoptera: Tortricidae) in the northern and southern hemispheres. *Entomological Review* 46, 349–361.

Shorey, H.H. and Gerber, R.G. (1996) Use of puffers for disruption of sex pheromone communication of codling moths (Lepidoptera: Tortricidae) in walnut orchards. *Environmental Entomology* 25, 1398–1400.

Smirle, M.J., Vincent, C., Zurowski, C.L. and Rancourt, B. (1998) Azinphosmethyl resistance in the obliquebdnaded leafroller, *Choristoneura rosaceana*: reversion in the absence of selection and relationship to detoxification enzyme activity. *Pesticide Biochemistry Physiology* 61, 183–189.

Travis, J.W. and Rajotte, E.G. (1995) Implementing IPM through new technologies and the non-agricultural community. *Journal of Agricultural Entomology* 12, 219–227.

Trimble, R.M. (1995) Mating disruption for controlling the codling moth, *Cydia pomonella* (L.) (Lepidoptera: Tortricidae) in organic apple production in southwestern Ontario. *The Canadian Entomologist* 127, 493–505.

VanBuskirk, P., Hilton, R., Simone, N. and Alway, T. (1998) *Orchard Pest Monitoring Guide for Pear: a Resource Book for the Pacific Northwest.* Good Fruit Grower, Yakima, Washington.

VanBuskirk, P., Hilton, R., Westigard, P. and Naumes, L. (1999) Orchard monitoring – the foundation of an areawide program. *Washington Horticultural Association Proceedings* 95, 159–160.

Van de Baan, H.E. and Croft, B.A. (1991) Resistance to insecticides in winter and summer form of *Psylla pyricola. Pesticide Science* 32, 225–233.

Varela, L.G., Welter, S.C., Jones, V.P., Brunner, J.F. and Riedl, H. (1993) Monitoring and characterization of insecticide resistance in codling moth (Lepidoptera: Tortricidae) in four western states. *Journal of Economic Entomology* 86, 1–10.

Weddle, P.W. (1993) *Barriers to the Implementation of Mating Disruption for Control of Codling Moth.* Weddle, Hansen & Associates, Placerville, California.

Weddle, P.W. (1994) *Management of Codling Moth in Bartlett Pears in California: a Preliminary analysis of the Relative Costs of Insecticide- and Pheromone-based IPM Strategies.* Weddle, Hansen & Associates, Placerville, California.

Weissling, T.J. and Knight, A.L. (1995) Vertical distribution of codling moth adults in pheromone-treated and untreated plots. *Entomologia Experimentalis et Applicata* 77, 271–275.

White, L.D., Hutt, R.B. and Butt, B.A. (1973) Field dispersal of laboratory-reared fertile female codling moth and population suppression by release of sterile males. *Environmental Entomology* 2, 66–69.

Willett, M. (2006) The 1996 Food Quality Protection Act after a decade. *Washington Horticultural Association Proceedings* 103, 162–165.

Williams, K.M. (1993) Use of evaporative cooling for enhancing apple fruit quality. *Washington Horticultural Association Proceedings* 89, 97–99.

Willamson, E.R., Folwell, R.J., Knight, A. and Howell, J.F. (1996) Economics of employing pheromones for mating disruption of the codling moth, *Carpocapsa pomonella. Crop Protection* 15, 473–477.

10 Corn Rootworm Areawide Pest Management in the Midwestern USA

LAURENCE D. CHANDLER,[1] JAMES R. COPPEDGE,[2]
C. RICHARD EDWARDS,[3] JON J. TOLLEFSON,[4]
GERALD R. WILDE[5] AND ROBERT M. FAUST[6]

[1]USDA-ARS-NPA, Fort Collins, Colorado, USA
[2]USDA-ARS-SPA, College Station, Texas, USA
[3]Department of Entomology, Purdue University, West Lafayette, Indiana, USA
[4]Department of Entomology, Iowa State University, Ames, Iowa, USA
[5]Department of Entomology, Kansas State University, Manhattan, Kansas, USA
[6]US Department of Agriculture, Agricultural Research Service, Beltsville, Maryland, USA

Introduction

For many maize (*Zea mays*) producers across the Midwestern USA, as well as in parts of the northern and southern plains, the corn rootworm complex (*Diabrotica* spp.; Coleoptera: Chrysomelidae) has represented one of the greatest challenges to efficient, quality maize-grain production over the past 50–60 years. Three species of corn rootworm are particularly troublesome: the western corn rootworm, *Diabrotica virgifera virgifera*, the northern corn rootworm, *D. barberi* and the Mexican corn rootworm, *D. virgifera zeae*, all have consistently been important economic pests of maize (Metcalf, 1986b). Prior to the introduction of transgenic maize varieties designed for corn rootworm management, from 8–10 million ha were treated with soil-applied insecticides to protect maize roots from larval feeding, with no absolute guarantee that the plants would be protected.

Corn rootworms damage maize primarily through larvae feeding on the roots. Severe feeding can result in lodged or stunted plants due to substantial root loss (Chandler, 2003). Over the years prior to the mid-1990s yield losses varied annually, but when the value of those losses was combined with the cost of control, these averaged at around US$1 billion per year (Chandler *et al.*, 1998). Adult corn rootworms can also inflict damage to maize plants through feeding on silk and pollen. Large numbers of rootworms feeding on silk can substantially reduce pollination, which interferes with kernel set and reduces yield.

During the late 1980s and 1990s several changes occurred related to corn rootworm biology and ecology, which increased the difficulty of managing this pest. Soil insecticides, which had been used extensively to manage corn rootworm larvae (Chandler *et al.*, 2000), came under increased scrutiny for their perceived negative environmental impact. Although there are probably no published data supporting and/or refuting the environmental effects of insecticides applied directly for rootworm control, public concern over the practice helped focus the need for development of alternative control methods to manage this insect complex. The single most important concern related to insecticide application to the soil was the known fact that many insecticides were used without knowledge of need – e.g. did the number of potential larvae within a maize field warrant the use of the insecticide to protect the crop? (Sutter *et al.*, 1991; Gray *et al.*, 1992). Additionally, soil insecticides did not always reduce corn rootworm populations and thus did not serve as reliable management tools for reducing pest densities that would attack maize roots in subsequent growing seasons.

To determine the need for insecticide applications it is best to base the decision on an insect-sampling regime that targets a critical stage of an insect's life cycle. For corn rootworms, sampling the immature stages could provide the best information on the need to intervene to protect maize roots. However, sampling for eggs and larvae of this insect complex is difficult, expensive and, perhaps, unreliable. In many instances it was just simpler, as well as more economical, for growers to apply an insecticide without knowledge of pest density. Prophylactic insecticide application served as a simple insurance policy that was generally cheap and reliable.

In addition to soil insecticides, foliar insecticides have been used to reduce adult corn rootworm populations, in theory limiting the number of females laying eggs that would potentially hatch the following year. However, in the 1960s, western corn rootworm beetle populations developed resistance to foliar applied insecticides in the chlorinated hydrocarbon class (Ball and Weekman, 1962), and later to foliar-applied carbamates and organophosphates in Nebraska during the mid-1990s (Meinke *et al.*, 1998). In parts of Nebraska significant levels of resistance to microencapsulated methyl-parathion and carbaryl, two widely used foliar insecticides, were detected (Meinke *et al.*, 1998; Siegfried *et al.*, 1998; Wright *et al.*, 2000). Although most instances of resistance to these compounds have been reported in adult populations, evidence also suggests that larvae of these same resistant adult populations may be somewhat tolerant to certain soil insecticides (Wright *et al.*, 2000).

Resistance to the chlorinated hydrocarbons in the 1960s forced many growers to start alternating maize with other crops on an annual basis. Corn rootworm larvae do not travel great distances to seek maize roots (Short and Luedtke, 1970), and they cannot feed and survive on roots of numerous alternative crop species commonly grown in many Midwestern production areas (e.g. soybean, lucerne, wheat, etc.). By rotating crops it was assumed that one could significantly disrupt the corn rootworm life cycle and thus reduce the insecticide use needed to manage rootworm populations and minimize damage in maize.

Insects do tend to adapt to perturbations forced upon their populations. Although rotating crops appeared to work effectively for many years, during the 1980s numerous populations of northern corn rootworms in parts of the Midwest developed an extended diapause in response to the selection pressure applied by annual crop rotation of large areas of crop land. This biological phenomenon of extended diapause allowed the insects to survive as eggs in the soil for more than one growing season (Krysan *et al.*, 1986). Thus, rotation of crops did not completely solve northern corn rootworm management, nor did it reduce the insecticide use problems.

More recently, as a response to crop rotation during the 1990s, western corn rootworms in Illinois/Indiana began to oviposit in soybean (*Glycine max*) (Edwards *et al.*, 1996). When these eggs hatched the following season, where maize had been planted after the soybeans, the maize suffered substantial economic damage. This phenomenon has spread to several surrounding states (e.g. Ohio, Michigan and perhaps Iowa) and has created numerous dilemmas for maize production specialists. A similar phenomenon was observed with Mexican corn rootworms laying eggs in grain sorghum (*Sorghum bicolor*) in Texas during the 1990s (J. R. Coppedge, personal communication). This probably occurred as a direct response to annual rotation between maize and grain sorghum.

As management of the corn rootworm complex became more challenging in the 1980s to mid-1990s, it was apparent that alternative control tactics were needed to effectively limit larval feeding damage to maize roots and to provide an economical alternative to the traditional soil and/or foliar insecticides. Thus, during this critical timespan, scientists at USDA-ARS and cooperating land grant universities initiated research to identify new management tools to effectively combat the economic threat posed by this insect complex (Chandler, 2003). The remainder of this chapter will discuss development and evaluation of a novel insect behavioural-based technology, designed to manage corn rootworm adult populations and limit oviposition, thus reducing the number of larvae available to feed on and damage maize roots early in the growing season. We will also discuss implications for use of such a technology in areawide pest management (AWPM).

Technology Development

Beginning in the 1980s and continuing into the early to mid-1990s, collaborating scientists began to address development of a new corn rootworm management technology that took advantage of the corn rootworm beetle's natural attraction to plants in the family *Cucurbitaceae*. Many early biological/ecological studies had previously identified the strong attraction of beetles to various compounds, including flowers, from plants in the melon family. Several plant- and flower-produced chemicals, or their analogues, were identified that attracted beetles and/or served as feeding stimulants (Metcalf *et al.*, 1982, 1987; Ladd *et al.*, 1983; Metcalf, 1986a; Lampman and Metcalf, 1987; Lampman *et al.*, 1987; Metcalf and Lampman, 1989; Lance and Elliott, 1991; Sutter and Lance, 1991). These compounds appeared to mimic the odours of floral structures and attracted both insect sexes.

Conceptually, use of these compounds in a novel formulation designed to attract beetles to a single point and to stimulate their willingness to feed on a bait containing a low dose of a highly toxic insecticide seemed to hold promise as an environmentally friendly management tool (Chandler, 2003). With this foundation, scientists worked for approximately 4–5 years (ending in the early 1990s) to develop a semiochemical insecticide bait that would prove to be effective in killing a high percentage of beetles in maize fields where it was applied (Lance, 1988; Lance and Sutter, 1991, 1992, 1993; Sutter and Lance, 1991; Weissling and Meinke, 1991a, b; Sutter and Hesler, 1993; Sutter *et al.*, 1998).

The bait was composed of natural cucurbitacins, a common insecticide used as a toxicant, and a non-toxic, edible carrier. Cucurbitacins are bitter-tasting tetracyclic triterpenoids that stimulate feeding in rootworm adults and repel non-rootworm insects. Cucurbitacins are found in most plants within the *Cucurbitaceae*, and were found in high concentrations in roots of the wild buffalo gourd, *Cucurbita foetidissima*, as well as in fruit of several melon species (Chandler, 2003). Meinke (1995) provided evidence of potential benefits from this behaviour-based approach. These included: (i) effective beetle control obtainable with small amounts of insecticide per hectare; (ii) few adverse effects on non-target organisms because the baits must be ingested; and (iii) human exposure to insecticides and potential environmental contamination were reduced because of the small amount of toxin in the bait.

During the initial research and development stage, scientists evaluated numerous bait compounds. After several experimental trials, two compounds, SLAM® (Microflo Co. and BASF Corp.) and Compel® (Ecogen, Inc.), were chosen to be more fully tested in large-scale field trials (Sutter *et al.*, 1998). These baits were composed of dried, powdered root of the buffalo gourd, carbaryl and a carrier. SLAM® was produced as a microsphere and included minute amounts of carbaryl as the toxicant. The microsphere breaks down into a suspension in water and can easily be sprayed (Chandler, 1998). Compel® was developed as a flowable compound where all ingredients, including carbaryl, were combined with a sticky carrier prior to being flung out from a spinning cone attached to the wing of an aircraft. Both baits adhere to plant surfaces and stimulate rootworm adult feeding, which exposes the insect to efficacious levels of toxicant (Chandler *et al.*, 2000). Both compounds were found to be highly efficacious against corn rootworm beetles, using about 95–98% less insecticide per ha than traditional foliar applications of carbaryl.

Following extensive testing, SLAM® proved easier to use and thus was selected as a preferred product for additional evaluation and possible use in AWPM programmes (Sutter *et al.*, 1998). Numerous field applications to rootworm-infested maize were made with SLAM®, with most studies clearly showing that the bait was effective and highly efficacious against the adults, could reduce adult corn rootworm abundance during the critical oviposition period and could easily be applied via conventional aerial or ground applicators (Hoffmann *et al.*, 1996a, b, 1998; Chandler and Sutter, 1997; Chandler, 1998). Additionally, SLAM® was found generally to have no significant negative impact on beneficial arthropods (Ellsbury *et al.*, 1996a, b; Chandler and Sutter, 1997; Hoffmann *et al.*, 2000). Thus the bait appeared to be both effective in reducing corn rootworm beetle populations and environmentally friendly.

Additional bait products were developed in the mid- to late 1990s as part of the natural evolution of the existing products, the discovery of other important sources of high concentrations of cucurbitacin and numerous user concerns over the manufacturing and quality control of SLAM®. These concerns eventually led to the withdrawal of SLAM® as a commercially available product in 1998/1999. These new products included Invite® (Florida Food Products, Inc.) and CideTrak® CRW (Trece, Inc.). Unlike SLAM®, both Invite® and CideTrak® CRW were not formulated with insecticide during manufacturing (Chandler, 2003). Similar to Compel®, insecticides of choice were added to these compounds at the time of mixing for application. Both new products were efficacious against corn rootworm beetles and demonstrated to be viable alternatives to SLAM®.

As with any insecticide-based insect management tool, timing of application of a toxicant is critical to its effectiveness. In theory, semiochemical baits are applied to maize plants when adult corn rootworms are most abundant and preferably at or before the females start to oviposit. Beetle numbers and life stage identification are important factors needed to time an application.

To determine effectively the number of corn rootworm beetles in a field, several techniques were evaluated in conjunction with bait applications. For optimal and economical decisions to support bait applications, the unbaited Pherocon® AM (yellow sticky) trap manufactured by Trece, Inc. was found to be the best for monitoring beetle populations. It compared favourably to traditional visual counts of the number of adults per maize plant (Chandler, 2003). A cumulative catch of four to seven beetles per trap per day over 7 consecutive days, when gravid females were present, was determined to be an effective threshold to initiate bait applications in maize. Placement of adequate numbers of traps in maize fields, beginning at the time of adult emergence from the pupal stage, allows pest management specialists effectively to time bait applications. Monitoring subsequent adult populations must continue for 6 weeks to determine whether additional applications are needed. Economics of trapping (trap costs and labour involved in placing and collecting traps) is the limiting factor for widespread adoption of yellow sticky traps as monitoring tools.

Formal AWPM Programme Implementation

As development of the semiochemical insecticide baits continued in the early 1990s the USDA-ARS initiated an Areawide Pest Management Program (APMP) in response to the USDA Pest Management (IPM) Initiative (USDA, 1993, 1994). The goal of this initiative was to implement IPM over at least 75% of the US crop acreage by 2000. In 1993, ARS, various university research and extension personnel and personnel from several state Departments of Agriculture met in Washington, DC to identify key pests and cropping systems for which environmentally sound pest management technologies were available for implementation on an areawide basis (Faust and Chandler, 1998). One of the projects selected was the management of adult corn rootworms within the maize pest management system in the Midwest using a semiochemical insecticide bait.

In 1995, the USDA-ARS-sponsored corn rootworm AWPM programme was initiated. An ad hoc committee comprising individuals from ARS, USDA-CSREES, USDA-ERS and the Extension Service was formed to guide the development and implementation of the programme. The committee organized a stakeholder meeting in St Louis, Missouri during this initial year to gather input from interested parties and to determine the feasibility of implementation. After this meeting several additional meetings were held to seek further input and to develop a team of research and extension personnel who would roll out the programme across the US maize belt.

From these meetings a conceptual plan was developed and the programme fully implemented in the spring of 1997 (Chandler, 2003).

The programme was divided into four distinct phases: (i) site development and information gathering – 1996; (ii) implementation – 1997; (iii) programme continuation, 1998–2001; and (iv) final assessment and technology transfer – 2002 (Chandler *et al.*, 2000). The mission of the programme was the successful establishment and implementation of an areawide demonstration programme that: (i) is the result of a partnership of growers, private consultants, applicators and suppliers, research and extension personnel and local, state and federal agencies who have a stake in the development and adoption of improved crop management technologies; and (ii) clearly demonstrates the advantages of enhanced grower profits, reduced risks, enhanced environmental compatibility and superiority of IPM approaches compared with current pest control approaches (Chandler *et al.*, 2000).

The established goals of the programme were to: (i) demonstrate an AWPM concept for the control of corn rootworm and other pests of maize such that voluntary adoption would occur throughout all maize production regions; and (ii) develop a partnership of federal, state, local and private interests that would be involved in the programme from conception to adoption (Chandler *et al.*, 2000).

ARS provided approximately US$550,000 for Phase I funding in fiscal year 1996, with approximately US$1.6 million provided annually from 1997 to 2001. Funds were distributed to cooperators using established ARS intra-agency fund transfers and specific cooperative agreements with cooperating universities. At the time of programme initiation, the potential advantages of conducting an AWPM programme for corn rootworm included: (i) consistency of control using standardized management strategies across a wide geographic area; (ii) reduced insect pest movement among fields; (iii) cost effectiveness compared with a field-by-field management approach; and (iv) reduced pest populations within a defined area (USDA-ARS/IDEA, 2004).

Site Selection and Programme Execution

Five corn rootworm AWPM sites, with associated non-areawide-treated control fields, were identified for use in the programme (Chandler *et al.*, 2000; Chandler, 2003). Each site was selected for its uniqueness and ability to adequately reflect various growing conditions and corn rootworm challenges faced across the US maize production areas. Five sites were selected:

Site A, the Illinois/Indiana site located in eastern Iroquois County, Illinois and western Benton County and Newton County, Indiana was approximately 41 km^2 in size with a total of 45 cooperating growers and approximately 4600 ha of maize and soybean in more than 160 fields. This site was within the heart of the region experiencing significant western corn rootworm behavioural changes resulting in oviposition in soybean. The faculty from Purdue University managed the site, with assistance from colleagues at the University of Illinois.

Site B, the Iowa site located in Clinton County, Iowa was approximately 41 km^2 in size, with a total of 40 cooperating growers and more than 2500 ha of maize in over 100 fields. This site targeted both northern and western corn rootworm in a primarily continuous maize production area. Some soybean was grown in rotation along with pasture crops. The faculty from Iowa State University managed the site.

Site C, the Kansas site located in Republic County, Kansas was approximately 41 km^2 in size, with 36 cooperating growers and more than 1700 ha of maize in over 90 fields. This site targeted western corn rootworm in a furrow-irrigated, continuous maize production area. The faculty from Kansas State University managed the site.

Site D, the Texas site located in Bell County, Texas was approximately 21 km^2 in size, with eight cooperating growers and approximately 800 ha of maize and 300 ha of grain sorghum. This site targeted Mexican corn rootworm in a primarily dryland maize/grain sorghum production system and was initiated a year earlier (in 1996) than the other sites. The site was managed by ARS personnel from College Station, Texas, in cooperation with the faculty from the Texas Agricultural Extension Service and Texas A&M University.

Site E, the South Dakota site located in Brookings County, South Dakota. The site was approximately 41 km^2 in size, with 20 cooperating growers and approximately 1400 ha of maize in over 50 fields. The site targeted northern and western corn rootworm in a primary maize/soybean rotation area. Some continuous maize was grown under centre-pivot irrigation systems. The site was managed by ARS personnel from Brookings, South Dakota in cooperation with the faculty from South Dakota State University.

Corn rootworm monitoring

Each AWPM site team was charged with monitoring corn rootworm beetle populations using traditional sampling methodologies (Chandler, 2003). Pherocon® AM (yellow sticky) traps were selected as the primary tool for monitoring beetle populations in maize and for initiating semiochemical insecticide-bait applications in maize at Sites B, C and E. Pherocon® AM traps were also used to monitor beetles and initiate bait applications to soybean in Site A. A minimum of six yellow sticky traps equally distributed throughout a field, regardless of size, were required for monitoring (Chandler *et al.*, 1999). Trapping began as soon as adults began to emerge and continued through a large portion of the remaining growing season. Hein and Tollefson (1985) had previously reported on the effectiveness of Pherocon® AM traps as a scouting tool for predicting damage by larvae the following year based on beetle counts the previous year.

Adult counts from maize plants were used to trigger bait applications to maize at Sites A and D. A cumulative count of from four to seven beetles per yellow sticky trap per day over 7 consecutive days – when gravid female rootworms were present – was used as the threshold to apply bait in maize. Bait applications in soybean were made when two beetles per trap per day over 7 days had been captured. Maize plant counts triggered bait applications when counts of 0.5 beetles per plant or more were observed and when gravid females were present. Baits were reapplied as necessary if corn rootworm populations returned to the above treatment thresholds (Chandler *et al.*, 2000).

Baits and applications

SLAM® was the original bait of choice for the programme. Each site used SLAM® exclusively from 1997 to 1999 and parts of 2000. At that time, SLAM® was no longer manufactured and each site had a choice of using Invite® or CideTrak® CRW combined with carbaryl through the remainder of the programme. Applications of bait were made using aircraft at all sites and following methodology described by Hoffmann *et al.* (1996a, b). Baits were applied in 3.785 l of water/ha at rates ranging from 53–106 g of carbaryl/ha.

Effectiveness assessment

In addition to the trap and whole plant counts conducted before and following each bait application, maize root damage ratings using the Iowa 1–6 scale (Hills and Peters, 1971) were conducted in each year to determine the effectiveness of the programme in protecting maize roots (Chandler, 2003). Higher root ratings indicated greater feeding damage, with ratings of three or over considered economic. Root injury was also assessed in control fields outwith the management areas where growers had applied insecticides to the soil at planting for larval control and left an area that had not been treated with an insecticide. These comparisons contrasted with the AWPM approach of controls applied by individual farmers. Additionally, some sites used beetle-emergence cages (1 m^2 in size) in the management area and control fields to assist in monitoring the sex ratios of the corn rootworm population and in determining the success of baits in reducing rootworm densities in maize fields 1 year after treatment compared with conventional practices.

Supporting Research

In addition to the bait application and population monitoring portion of the programme, four additional supporting research activities were conducted to assist in final assessment of the AWPM approach:

Insecticide resistance management

Entomologists at the University of Nebraska, in cooperation with ARS personnel in Brookings, South Dakota, monitored corn rootworm adult susceptibility to carbaryl, related insecticides and cucurbitacin at all areawide sites throughout the duration of the programme. Associated studies were conducted to determine factors responsible for development of carbaryl and methyl parathion resistance in various western corn rootworm populations, as well as to determine the biochemical and molecular mechanisms involved in resistance (Chandler, 2003).

Economic assessment

Agricultural economists at Purdue University collected data and provided vital information for producers on the economic impact of the programme related to farm profits. The data determined the type of maize production system most likely to benefit from an AWPM approach (Chandler, 2003).

Sociological assessment

Rural sociologists from Iowa State University conducted on-site surveys and assessments to determine barriers and opportunities for successful implementation of the programme. Business models were identified that could be used to transfer the technology and sustain it within the private sector (Chandler, 2003).

Non-target arthropod assessment

Entomologists from South Dakota State University, in cooperation with ARS personnel in Brookings, South Dakota and faculty/staff of Iowa State University, conducted population density assessments of non-target arthropods following SLAM® applications in sites B and E.

Results and Conclusions

As with any large-scale IPM programme, the results of the corn rootworm AWPM programme varied greatly from site to site. The measures of success with a programme of this type should be viewed not only in the ability to manage a difficult insect pest, but also in the amount of new knowledge generated from the actions of a large and diverse team of scientists, stakeholders and end-users. Listed below are summary findings from each of the five management sites:

Site A: the site root ratings conducted in 2001 showed a significant reduction in corn rootworm larval feeding damage within the areawide site compared with fields in the companion control area. Ratings averaged 1.73 and 3.26 (Iowa 1–6 root rating scale) in the management site and control area, respectively. The difference in root ratings indicated that areawide suppression had had an effect on rootworm populations in this particular year, but similar results were not seen in other years. Substantial problems occurred related to efficacy of the semiochemical baits, which resulted in an increased number of applications in each year. Questions on whether this approach is economical in this area still exist (C.R. Edwards, 2001, personal communication; Chandler, 2003; Gerber, 2004).

Site B: over the life of the programme, western corn rootworm larval feeding damage within the management site was less than that observed in control areas where no soil insecticides had been applied. Root ratings averaged 2.2, 2.6, 1.4 and 1.9 (Iowa 1–6 root rating scale) in 1998, 1999, 2000 and 2001, respectively, in the

AWPM site compared with 3.5, 3.6, 2.5 and 2.0 during the same years in the control fields. During the same period corn rootworm beetles produced in the managed area averaged 232,000, 447,000, 235,000 and 514,000/ha, respectively, compared with 909,000, 1,161,000, 420,000 and 1,773,000/ha in the control fields (Tollefson, 1998; J.J. Tollefson, 2001, personal communication; Chandler, 2003).

Despite the substantial reduction in beetle numbers, there are still questions as to whether the remaining numbers pose an economic threat to maize producers in the following year. In general, the number of bait applications needed to manage the pest remained stable in each year. If 80% of the 12,693 ha of maize protected by the adult corn rootworm management programme over 5 years had received the commonly used 229 g of organophosphorus insecticide/ha, 10,154 ha would have been treated with 232,527 g of active ingredient. This represents 14,226,850 g of actual toxicant rather than the 665,417 g of carbaryl that were applied to the Iowa managed area during the study, more than a 20-fold reduction in insecticide load in the environment.

Site C: following implementation of the programme in 1998, the percentage of maize hectarage above treatment thresholds requiring bait applications within the management site had been reduced from 51 to 15% by 2001. The percentage of hectarage above treatment thresholds in the control (unmanaged) areas ranged from 74.5 to 87.5 over the same period. An average of only 33.5% of maize fields in the managed area exceeded economic beetle populations over the programme compared with 82.8% in the control (unmanaged) fields (Wilde *et al.*, 1998; G.E. Wilde, 2001, personal communication; Chandler, 2003; USDA-ARS/IDEA, 2004). It appears that a continuous maize production system in irrigated portions of the western Midwest is the most likely target for successful implementation of the AWPM concept against western corn rootworms.

Site D: large numbers of Mexican corn rootworms surviving in grain sorghum, as well as in maize, resulted in a substantial challenge for consistent areawide management results. However, over the life of the programme, the percentage of maize hectarage requiring bait applications to manage the insect was reduced from 90 to 18%. These results would indicate that the programme was successful (Lingren, 1999; J.R. Coppedge, 2001, personal communication; USDA-ARS/IDEA, 2004). However, the main challenge to maintaining a successful programme in typical Texas maize production areas is related to the economics of dryland agriculture. Droughts reduce yields and thus growers are reluctant to maintain intensive pest management programmes, which cut into their profit margins.

Site E: large numbers of northern corn rootworms within the management site rendered the use of semiochemical baits extremely challenging. This particular species of rootworm does not respond as well to the feeding stimulants/attractants within the baits, unlike western corn rootworms. Thus the management site did not see true reductions in the number of necessary bait applications from the start until the termination of the programme. The baits were as effective as traditional soil insecticides in maintaining reduced amounts of larval root feeding, but the difficulties encountered in managing northern corn rootworm make this particular maize production area a challenge for areawide implementation (L.D. Chandler, 2001, personal communication; USDA-ARS/IDEA, 2004).

Despite the numerous challenges encountered in implementing an AWPM programme against the corn rootworm complex, this programme resulted in substantial positive findings that will have some impact on corn rootworm management for many years to come. The use of baits, although overall no more successful in reducing root feeding damage than soil insecticides, did significantly reduce the total amount of insecticide applied to the areawide management sites. As previously mentioned, in the entire Iowa site, the bait application programme reduced the total amount of insecticide active ingredient used by as much as 20-fold compared with conventional soil insecticides, if all hectares were treated as required. Similar results were achieved in all sites by the entire AWPM programme (USDA-ARS/IDEA, 2004). The environmental savings achieved because of the programme have positive consequences for the entire USA.

In addition to the above environmental benefits, several other products were developed during the life of the programme that benefit producers (USDA-ARS/IDEA, 2004):

- Refined aerial bait application methods: aerial application protocols developed by scientists in the programme are now consistently used to optimize bait effectiveness.
- New traps: traps were developed by scientists in cooperation with Trece, Inc., to monitor corn rootworm beetle emergence and to trigger bait applications; the traps are now commercially available.
- Biological model: a new model was developed by scientists at Iowa State University that allows beetle emergence from the soil to be predicted with more precision; growers now need only monitor maize fields during a 4-week period of the growing season rather than the 8–10 weeks previously required (Nowatzki *et al.*, 2002; Park and Tollefson, 2006).
- Rapid identification of rootworm insecticide resistance; University of Nebraska and ARS researchers developed a vial bioassay technique that more quickly and easily assesses the insecticide resistance status of field-collected western corn rootworms. This tool can be used to make informed selections of insecticides when treatments are required.
- Pherocon® AM (yellow sticky) traps were used to develop economic injury level (EIL) and economic thresholds (ET) for western corn rootworm captures in soybean in Indiana and Illinois. The EIL was set at seven beetles captured per trap per day and the ET was determined to be five beetles captured per trap per day (Gerber, 2004).
- Two new insecticide baits were developed, CideTrak CRW® and Invite®, which remain available from Trece, Inc. and Florida Food Products, Inc, respectively (Chandler, 2003). Pingel *et al.* (2001) assisted this effort by evaluating techniques to improve residual activity of the baits, and Schroder *et al.* (1998, 2001) assisted in development of new cucurbitacin sources resulting in development of Invite®. These baits provided growers with the additional freedom of selecting and using an insecticide of their choice within the mixtures.

As the programme progressed, much was learned about the economic and sociological benefits of conducting an AWPM programme across maize production areas of the USA. Studies conducted by Purdue University economists indicated that the Kansas areawide site showed the greatest potential to benefit economically from the adoption of AWPM of rootworms. This conclusion was based on a study of net present value analyses over an 8-year planning horizon (Quan, 1999). Kansas growers can more readily afford the costs of scouting and bait applications, and results of the analyses indicated they would see an increase in their on-farm income.

Sociologists at Iowa State University developed three business models that could continue to sustain a corn rootworm AWPM programme after scientists and other interested parties have stopped the research and demonstration component of the programme (Padgitt, personal communication, 2004; USDA-ARS/IDEA, 2004):

- Private supplier model – private business management as a service to individual growers, at an economic advantage compared to traditional costs (preferred by participants in the corn rootworm AWPM programme).
- Collective enterprise model – growers organized as a cooperative/non-profit enterprise to provide AWPM.
- Special-use model – special tax revenues are collected from a geographical area or district designated by an authoritative body to cover the costs of AWPM.

At the end of the programme, many questions remained concerning the viability of the AWPM concept for corn rootworm. The onset and subsequent expanded use of genetically modified maize varieties targeted at corn rootworm has reduced the concerns over traditional corn rootworm management techniques (soil insecticides, foliar insecticides and crop rotations), and has thus negatively affected the probability of using AWPM techniques across the maize belt. Currently variations of AWPM for corn rootworm continue to be used in portions of western Kansas and in selected pockets in southern and central Texas. Additionally, it is apparent that AWPM-supported activities have achieved several scientific successes during the life of the USDA-ARS-sponsored programme. Some important conclusions and/or findings from the programme include:

- Combining several techniques such as baits, genetically modified maize varieties, insecticides and efficient crop scouting techniques is crucial for achieving true IPM success and minimizing insecticide resistance development in adult corn rootworms.
- Carbaryl susceptibility of western corn rootworm populations was reduced in areas with repeated SLAM® applications over a 4-year period (Zhu *et al.*, 2001).
- The efficacy of feeding stimulants or baits, such as Invite®, in combination with certain insecticides may be compromised by previously identified resistance and by insecticides that antagonize the feeding stimulation of the cucurbitacin bait, with both inoxacarb and fipronil providing effective alternatives to carbaryl within the bait (Parimi *et al.*, 2003).
- Use of semiochemical-based baits did not significantly affect population densities of non-target arthropods when used as prescribed in an area management programme (Boetel *et al.*, 2005).
- AWPM of adult beetles may result in fewer fields to scout and reduced insecticide use – lowering both economical and environmental costs/concerns.
- Crop rotation remains an economically viable control strategy unless the area has a high number of western corn rootworm beetles laying eggs in both maize and soybean, or a high number of northern corn rootworms exhibiting extended diapause traits.

- Successful adoption of AWPM strategies will require innovative marketing and the willingness of growers to work together for a common goal. Making growers aware of the potential advantages of AWPM, increasing the use of IPM techniques and providing opportunities for enterprising agribusiness suppliers will encourage adoption of new and unique rootworm management options.

Acknowledgements

The authors would like to express their sincere appreciation to the numerous individuals who have worked on and supported the corn rootworm AWPM concept since the early 1990s. In particular, we would like to acknowledge the following individuals (in alphabetical order) for their contributions to various aspects of the programme: Mr Mike Athanas, Mr Dave Beck, Ms Amber Beckler, Dr Robert Behle, Dr Ed Berry (retired), Mr Larry Bledsoe, Dr Mark Boetel, Dr Wayne Buhler, Dr Larry Buschman, Dr Jesse Cocke (retired), Dr Stan Daberkow, Dr Mike Ellsbury, Dr Wade French, Dr Billy Fuller, Dr Corey Gerber, Dr Mike Gray, Dr Leslie Hammack (retired), Ms Anetra Harbor, Ms Deb Hartman, Dr Randy Higgins (deceased), Dr Clint Hoffmann, Ms. Denise Hovland, Mr Aaron Howell, Dr Gretchen Jones, Mr Nimesh Kadakia, Dr Buddy Kirk (retired), Dr Dave Lance, Dr Les Lewis, Dr Scott Lingren, Dr Marshall Martin, Dr Michael McGuire, Ms. Sarah McKenzie, Dr Lance Meinke, Mr Tara Mishra, Dr Tim Nowatzki, Mr Mike O'Neil, Dr Eldon Ortman (retired), Dr Steve Padgitt, Dr Srinivas Parimi, Dr Yung Lak Park, Dr Peggy Petrzelka, Dr Randy Pingel, Mr Peter Quan, Dr Walt Riedell, Dr Mike Scharf, Dr Robert Schroder (retired), Dr Roxanne Shufran, Dr Blair Siegfried, Dr Phil Sloderbeck, Ms Christina Welch Stair, Dr Jenny Stebbing, Dr Kevin Steffey, Dr Gerald Sutter (retired), Dr Jeff Whitworth, Dr Bob Wright, Dr X. Zhou and Dr K.Y. Zhu.

Additionally, the programme members thank members of Private Industry representing BASF Corp., MicroFlo Co., Ecogen, Inc., Trece, Inc. and Florida Foods International for their valuable assistance. Finally, the programme would not have been possible without the assistance of the more than 150 farmers and consultants who participated by allowing access to their farms for conduct of the studies.

References

Ball, H.J. and Weekman, G.T. (1962) Insecticide resistance in the adult western corn rootworm in Nebraska. *Journal of Economic Entomology* 55, 439–441.

Boetel, M.A., Fuller, B.W., Chandler, L.D., Tollefson, J.J., McManus, B.L., Kadakia, N.D. and Mishra, T.A. (2005) Nontarget arthropod abundance in areawide-managed corn habitats treated with semiochemical-based bait insecticide for corn rootworm (Coleoptera: Chrysomelidae) control. *Journal of Economic Entomology* 98, 1957–1968.

Chandler, L.D. (1998) Comparison of insecticide-bait aerial application methods for management of corn rootworm (Coleoptera: Chrysomelidae). *Southwestern Entomologist* 23, 147–159.

Chandler, L.D. (2003) Corn rootworm areawide management program: United States Department of Agriculture-Agricultural Research Service. _Pest Management Science_ 59, 605–608.

Chandler, L.D. and Sutter, G.R. (1997) High clearance sprayer methods for application of corn rootworm (Coleoptera: Chrysomelidae) semiochemical-based baits. _Southwestern Entomologist_ 22, 167–178.

Chandler, L.D., Woodson, W.D. and Ellsbury, M.M. (1998) Corn rootworm IPM: implementation and information management with GIS. In: Wilkinson, D. (ed.) _Proceedings of the American Seed Trade Association 52nd Annual Corn and Soybean Research Conference_, Chicago, Illinois, pp. 129–143.

Chandler, L.D., Ellsbury, M.M. and Woodson, W.D. (1999) _Area-wide Management Zones for Insects_. PPI Site-specific management guidelines, SSMG-19, 4 pp.

Chandler, L.D., Coppedge, J.R., Edwards, C.R., Tollefson, J.J. and Wilde, G.E. (2000) Corn rootworm area-wide management across the United States. In: Tan, K.H. (ed.) _Area-Wide Control of Fruit Flies and Other Insect Pests_. Penerbit Universiti Sains Malaysia, Penang, Malaysia, pp. 159–168.

Edwards, C.R., Bledsoe, L.W. and Obermyer, J.L. (1996) The dramatic shift of the western corn rootworm, _Diabrotica virgifera virgifera_ LeConte (Coleoptera: Chrysomelidae), to maize in rotation with soybeans in Indiana. In: Anon. (ed.) _Proceedings of the XX International Congress of Entomology_, Florence, Italy, paper #15- 082, pp. 469.

Ellsbury, M.M., Gaggero, J.M. and Johnson, T.B. (1996a) Mortality of lady beetle adults and lacewing larvae following exposure to semiochemical-based corn rootworm baits. _Arthropod Management Tests_ 21, 402–403.

Ellsbury, M.M., Gaggero, J.M. and Johnson, T.B. (1996b) Survival of ground beetles (Carabidae) after feeding on western corn rootworm adults killed by semiochemical-based baits. _Arthropod Management Tests_ 21, 403–404.

Faust, R.M., and Chandler, L.D. (1998) Future programs in areawide pest management. _Journal of Agricultural Entomology_ 15, 371–376.

Gerber, C.K. (2004) Testing the areawide pest management concept on western corn rootworm (_Diabrotica virgifera virgifera_ LeConte) in northwestern Indiana and east central Illinois. PhD Dissertation, Purdue University, West Lafayette, Indiana.

Gray, M.E., Felsot, A.S., Steffey, K.L. and Levine, E. (1992) Planting time application of soil insecticides and western corn rootworm (Coleoptera: Chrysomelidae) emergence implications for long-term management programs. _Journal of Economic Entomology_ 85, 544–553.

Hein, G.L. and Tollefson, J.J. (1985) Use of Pherocon® AM trap as a scouting tool for predicting damage by corn rootworm (Coleoptera: Chrysomelidae) larvae. _Journal of Economic Entomology_ 78, 200–203.

Hills, T.M. and Peters, D.C. (1971) A method of evaluating post-plant insecticide treatments for control of western corn rootworm larvae. _Journal of Economic Entomology_ 64, 764–765.

Hoffmann, W.C. (1996) Applying areawide control to corn rootworms. _Ag-Pilot International_ December 1996, pp. 64–65.

Hoffmann, W.C., Lingren, P.S., Coppedge, J.R. and Kirk, I.W. (1996a) Aerial application of SLAM®. _Ag-Pilot International_ March 1996, 18–20.

Hoffmann, W.C., Lingren, P.S., Coppedge, J.R. and Kirk, I.W. (1996b) Corn rootworm control with a semiochemical-based insecticide. In: _1996 NAAA/ASAE Joint Technical Session_, Paper No. AA96-007. ASAE, St Joseph, Michigan.

Hoffmann, W.C., Lingren, P.S., Coppedge, J.R. and Kirk, I.W. (1998) Application parameter effects on efficacy of a semiochemical-based insecticide. _Applied Engineering in Agriculture_ 14, 459–463.

Hoffmann, W.C., Coppedge, J.R., Lingren, P.S. and Chandler, L.D. (2000) Impact of a semiochemical insecticide (SLAM®) on abundance of beneficial insects in corn. *Southwestern Entomologist* 25, 31–38.

Krysan, J.L., Foster, D.E., Branson, T.F., Ostlie, K.R. and Cranshaw, W.S. (1986) Two years before the hatch: rootworms adapt to crop rotation. *Bulletin of the Entomological Society of America* 32, 250–253.

Ladd, T.L., Stinner, B.R. and Krueger, H.H. (1983) Eugenol, a new attractant for the northern corn rootworm (Coleoptera: Chrysomelidae). *Journal of Economic Entomology* 76, 1049–1051.

Lampman, R.L. and Metcalf, R.L. (1987) Multicomponent kairomonal lures for southern and western corn rootworms (Coleoptera: Chrysomelidae) *Diabrotica* spp. *Journal of Economic Entomology* 80, 1137–1142.

Lampman, R.L., Metcalf, R.L. and Anderson, J.F. (1987) Semiochemical attractants of *Diabrotica undecimpunctata howardi* Barber, the southern corn rootworm and *Diabrotica virgifera virgifera* LeConte, the western corn rootworm (Coleoptera: Chrysomelidae). *Journal of Chemical Ecology* 13, 959–975,

Lance, D.R. (1988) Potential of 8-methyl-2-decyl propanoate and plant-derived volatiles for attracting corn rootworm beetles (Coleoptera: Chrysomelidae) to toxic baits. *Journal of Economic Entomology* 81, 1359–1362.

Lance, D.R. and Elliott, N.C. (1991) Seasonal responses of corn rootworm beetles (Coleoptera: Chryomelidae) to non-pheromonal attractants. *Journal of Entomological Science* 26, 188–196.

Lance, D.R. and Sutter, G.R. (1990) Field-cage and laboratory evaluations of semiochemical-based baits for managing western corn rootworm beetles (Coleoptera: Chrysomelidae). *Journal of Economic Entomology* 83, 1085–1090.

Lance, D.R. and Sutter, G.R. (1991) Semiochemical-based toxic baits for *Diabrotica virgifera virgifera* (Coleoptera: Chrysomelidae): some effects of particle size, location, and attractant content. *Journal of Economic Entomology* 84, 1861–1868.

Lance, D.R. and Sutter, G.R. (1992) Field tests of a semiochemical-based toxic bait for suppression of corn rootworm beetles (Coleoptera: Chrysomelidae). *Journal of Economic Entomology* 85, 967–973.

Lingren, P.S. (1999) Management of Mexican corn rootworm, *Diabrotica virgifera zeae* Krysan and Smith (Coleoptera: Chrysomelidae), through areawide adult suppression and crop rotation. PhD Dissertation, Texas A&M University, College Station, Texas.

Meinke, L.J. (1995) *Adult Corn Rootworm Management.* University of Nebraska Bulletin MP63-C, Nebraska.

Meinke, L.J., Siegfried, B.D., Wright, R.J. and Chandler, L.D. (1998) Adult susceptibility of Nebraska western corn rootworm (Coleoptera: Chrysomelidae) populations to selected insecticides. *Journal of Economic Entomology* 91, 594–600.

Metcalf, R.L. (1986a) Coevolutionary adaptions of rootworm bettles (Coleoptera: Chrysomelidae) to cucurbitacins. *Journal of Chemical Ecology* 12, 1109–1124.

Metcalf, R.L. (1986b) Foreword. In: Krysan, J.L. and Miller, T.A. (eds) *Methods for the Study of Pest* Diabrotica. Springer-Verlag, New York, pp. vii–xv.

Metcalf, R.L. and Lampman, R.L. (1989) Estragole analogues as attractants for corn rootworm (Coleoptera: Chrysomelidae). *Journal of Economic Entomology* 82, 123–129.

Metcalf, R.L., Ferguson, J.E., Lampman, R. and Anderson, J.F. (1987) Dry cucurbitacin-containing baits for controlling diabroticite beetles (Coleoptera: Chrysomelidae). *Journal of Economic Entomology* 80, 870–875.

Metcalf, R.L., Rhodes, A.M., Metcalf, R.A., Ferguson, J., Metcalf, E.R. and Lu, P.Y. (1982) Cucurbitacin contents and diabroticite (Coleoptera: Chrysomelidae) feeding upon *Cucurbita* spp. *Environmental Entomology* 11, 931–937.

Nowatzki, T.M., Tollefson, J.J. and Calvin, D.D. (2002) Development and validation of models for predicting the seasonal emergence of corn rootworm (Coleoptera: Chrysomelidae) beetles in Iowa. *Environmental Entomology* 31, 864–873.

Parimi, S., Meinke, L.J., Nowatzki, T.M., Chandler, L.D., French, B.W. and Siegfried, B.D. (2003) Toxicity of insecticide-bait mixtures to insecticide resistant and susceptible western corn rootworms (Coleoptera: Chrysomelidae). *Crop Protection* 22, 781–786.

Park, Y.L. and Tollefson, J.J. (2006) Development and economic evaluation of spatial sampling plans for corn rootworm *Diabrotica* spp. (Coleoptera, Chrysomelidae) adults. *Journal of Applied Entomology* 130, 337–342.

Pingel, R.L., Behle, R.W., McGuire, M.R. and Sasha, B.S. (2001) Improvement of the residual activity of a cucurbitacin-based adult corn rootworm (Coleoptera: Chrysomelidae) insecticide. *Journal of Entomological Science* 36, 416–425.

Quan, P.U. (1999) Corn rootworm control: economic evaluation of an areawide pest management approach. M.S. thesis, Purdue University, West Lafayette, Indiana.

Schroder, R.F.W., DeMilo, A.B., Lee, C.J. and Martin, P.A.W. (1998) Evaluation of water-soluble bait for corn rootworm (Coleoptera: Chrysomelidae) control. *Journal of Entomological Science* 33, 355–364.

Schroder, R.F.W., Martin, P.A.W. and Athanas, M.M. (2001) Effect of phloxine B-cucurbitacin bait on diabroticite beetles (Coleoptera: Chrysomelidae). *Journal of Economic Entomology* 94, 892–897.

Short, D.E. and Luedtke, R.J. (1970) Larval migration of the western corn rootworm. *Journal of Economic Entomology* 63, 325–326.

Siegfried, B.D., Meinke, L.J. and Scharf, M.E. (1998) Resistance management concerns for areawide management programs. *Journal of Agricultural Entomology* 15, 359–369.

Sutter, G.R. and Hesler, L.S. (1993) New biorational and environmentally safe technology for management of insect pests in maize production systems, In: Granholm, N.H. (ed.) *Proceedings of Stress Symposia: Mechanisms, Responses, Management.* College of Agricultural and Biological Sciences, South Dakota State University, Brookings, South Dakota, pp. 303–306.

Sutter, G.R. and Lance, D.R. (1991) New strategies for reducing insecticide use in the corn belt. In: Rice, B.J. (ed.) *Sustainable Agriculture Research and Education in the Field – 1990.* Board of Agriculture National Research Council, National Academy Press, Washington, DC, pp. 231–249.

Sutter, G.R., Branson, T.F., Fisher, J.R. and Elliott, N.C. (1991) Effects of insecticides on survival, development, fecundity, and sex ratio in controlled infestations of western corn rootworm (Coleoptera: Chrysomelidae). *Journal of Economic Entomology* 84, 1905–1912.

Sutter, G.R., Lance, D.R., Meinke, L.J., Frana, J.E., Metcalf, R.L., Levine, E. and Gaggero, J.M. (1998) *Managing Corn Rootworms with a Granular Semiochemical-based Bait.* USDA-ARS Publication #143, Washington, DC, 40 pp.

Tollefson, J.J. (1998) Rootworm areawide management in Iowa. *Journal of Agricultural Entomology* 15, 351–357.

USDA (1993) *Three Agency Release, Presidential Announcement Regarding IPM Adoption.* Office of Communications, 23 June 1993, USDA, Washington, DC.

USDA (1994) *USDA Announces National Plan to Increase use of Integrated Pest Management.* Release Number 0943.94; and Backgrounder: USDA's IPM Initiative, Release Number 0942.94, Office of Communications, USDA, Washington, DC.

USDA-ARS/IDEA (2004) Controlling corn rootworm – the billion $$$ insect; research provides new strategies. *IDEA 22* July 2004, 4 pp.

Weissling, T.J. and Meinke, L.J. (1991a) Potential of starch encapsulated semiochemical-insecticide formulations for adult corn rootworm (Coleoptera: Chrysomelidae) control. *Journal of Economic Entomology* 84, 601–609.

Weissling, T.J. and Meinke, L.J. (1991b) Semiochemical-insecticide bait placement and vertical distribution of corn rootworm (Coleoptera: Chrysomelidae) adults: implications for management. *Environmental Entomology* 20, 945–952.

Wilde, G.E., Whitworth, R.J., Shufran, R.A., Zhu, K.Y., Sloderbeck, P.E., Higgins, R.A. and Buschman, L.L. (1998) Rootworm areawide management project in Kansas. *Journal of Agricultural Entomology* 15, 335–349.

Wright, R.J., Scharf, M.E., Meinke, L.J., Zhou, X., Siegfried, B.D. and Chandler, L.D. (2000) Larval susceptibility of an insecticide-resistant western corn rootworm (Coleoptera: Chrysomelidae) population to soil insecticides: laboratory bioassays, assays of detoxification enzymes, and field performance. *Journal of Economic Entomology* 93, 7–13.

Zhu, K.Y., Wilde, G.E., Higgins, R.A., Sloderbeck, P.E., Buschman, L.L., Shufran, R.A., Whitworth, R.J., Starkey, S.R. and He, F. (2001) Evidence of evolving carbaryl resistance in western corn rootworm (Coleoptera: Chrysomelidae) in areawide-managed cornfields in north central Kansas. *Journal of Economic Entomology* 94, 929–934.

11 Grape Areawide Pest Management in Italy

CLAUDIO IORIATTI,[1] ANDREA LUCCHI[2] AND
BRUNO BAGNOLI[3]

[1]Istituto Agrario di San Michele a/A, San Michele a/A (Trento), Italy
[2]Dip. C.D.S.L. Sez. Entomologia Agraria, Università di Pisa, Pisa, Italy
[3]CRA-Istituto Sperimentale per la Zoologia Agraria, Florence, Italy

Introduction: Description of the Problem and Need for an Areawide Pest Management Approach

Pest significance in Italian vineyards

Vine growing in Italy covers approximately 868,000 ha, 90% of which is allocated to the production of wine grapes and about 10% to table grapes. The economic share of wine production within the entire agricultural sector is 7.2%. All regions of Italy are involved in one way or another in vine growing, which has historically been primarily a visually appealing feature of the agricultural landscapes of Italy and an attraction for tourists. Hence, the health of vineyards plays a very important environmental, social and economic role in addition to its more basic role as the bedrock of an agricultural commodity.

Among the biotic adversities of the vine, pathogens are, broadly speaking, the most significant threat and, for over a century, two diseases – downy mildew and powdery mildew – have yearly required the adoption of repeated measures everywhere. As to pests, the grapevine moth, *Lobesia botrana* (Lepidoptera: Tortricidae), is still regarded as the key insect. It affects all the national vine-growing regions and, in many of these, this species often reaches such population densities as to require containment measures. The grape berry moth, *Eupoecilia ambiguella* (Lepidoptera: Tortricidae), is present, on the other hand, mostly in northern Italy, where it is locally distributed and only exceptionally reaches a worrying population density (Ioriatti *et al.*, 2005).

The many changes that have occurred since World War II have substantially altered the complex of the Arthropoda associated with viticulture, which has been augmented by new and important exotic species, such as the North American Auchenorrhyncha, *Metcalfa pruinosa* (Flatidae) and, especially, *Scaphoideus titanus* (Cicadellidae), the fearsome carrier of the phytoplasma of flavescence dorée (FD).

Currently, in Italy, the insects associated with the vine that can cause significant infestations, no matter how local or occasional, amount to some dozens of species.

They include, in addition to the ones listed above, the thysanopterans *Drepanothrips reuteri* and *Frankliniella occidentalis* (Thripidae); the hemipterans *Empoasca vitis*, *Zygina rhamni* (Cicadellidae), *Daktulosphaira vitifoliae* (Phylloxeridae), *Planococcus* spp., *Parthenolecanium corni*, *Pulvinaria vitis* and *Targionia vitis* (Coccoidea); the lepidopterans *Theresimima ampelophaga* (Zygaenidae), *Argyrotaenia ljungiana* (Tortricidae), *Cryptoblabes gnidiella*, *Ephestia* spp. (Pyralidae), *Noctua pronuba*, *N. fimbriata* (Noctuidae) and *Peribatodes rhomboidaria* (Geometridae); the dipterans *Janetiella oenophila* (Cecidomyiidae) and *Drosophila* spp. (Drosophilidae); the coleopterans *Tropinota squalida* (Scarabaeidae), *Altica ampelophaga* (Chrysomelidae), *Amphicerus bimaculatus*, *Sinoxylon* spp. (Bostrychidae), *Vesperus* spp. (Cerambycidae) and *Otiorrhynchus* spp. (Curculionidae); and some hymenopteran Vespidae (Bagnoli, 1990; Brunelli *et al.*, 1993; Lucchi, 1997a).

In the last few years, Sicily and Sardinia have experienced a recrudescence of infestations of the leafhopper, *Jacobiasca lybica* (Cicadellidae), which, unless appropriately fought, can cause substantial economic damage (Mazzoni *et al.*, 2003). The leafhopper, typically thermophyle, could in future be a threat also for other southern vine-growing regions, due to the gradual increase in the mean temperature of the Mediterranean areas.

As regards mites, the most interesting species are the eriophyid *Calepitrimerus vitis* and the tetranychids *Eotetranychus carpini* and *Panonychus ulmi*. These typical secondary pests have been a problem in the past due to the abuse of synthetic phytosanitary products, such as dithiocarbamides, carbamic esters and organophosphates. The adoption of the principles of integrated pest management and the increased knowledge of natural antagonists have put the problem in perspective and now, at least broadly speaking, the problem seems to be manageable with more sophisticated growing techniques and the enhancement of the population of Phytoseiidae (*Kampimodromus aberrans*, *Typhlodromus exhilaratus*, *T. pyri*, etc.), predator mites of great ecological value, recognized as the new frontier in the natural control of Tetranychidae and Eriophyidae (Duso *et al.*, 2003).

Over the last few decades, vine defence against arthropod pests has made huge progress nearly everywhere, and today it is increasingly complying with IPM principles, often appropriately and effectively translated into regionwide guidelines.

The case of the grape phylloxera, *Daktulosphaira vitifoliae*, which has been excellently controlled since the first decade of the 20th century by grafting European vines on to American rootstock, can be regarded as the epitome of areawide pest management (AWPM) in Italy. Such an approach has recently found its first modern application in Trentino South-Tyrol (provinces of Trento and Bolzano), through the large-scale adoption of the mating disruption method for the containment of vine moths.

Current Management Systems and Approaches for the Vine Moth

Eupoecilia ambiguella and *Lobesia botrana* are both polyphagous and polyvoltine, developing on the vine over two or three full generations, the first on the flowers (anthophagous) and the others on the grapes (carpophagous). Hibernation takes place at the pupal stage under the rhitidome of the stump. *L. botrana* prevails in hot/dry regions, while *E. ambiguella* prefers the colder and more humid climates of

Central Europe, is locally spread, especially in the north, and is less harmful than *Lobesia*.

Apart from a few exceptions of poorly fruiting cultivars, the damage due to vine moths basically depends on the larval populations of the carpophagous generations. As to the growing of winemaking grapes, the damage results from the susceptibility of damaged bunches to rot rather than from the direct loss of product from insect feeding. Rot mostly affects vines with compact infructescence, which are more sensitive to fungal and bacterial infections. Usually the second generation of the moths is the most harmful one for early-ripening varieties, while the third generation is the most important for late-ripening vines. In Trentino-South Tyrol, the third generation has no economic relevance since it mostly affects the still-green bunches on secondary shoots, which have no economic value.

For compact-bunch cultivars like Sangiovese, Chardonnay, Pinot and Sauvignon, an empirical tolerance threshold of 5–10% of egg-infested bunches for the second and third generations has been found to be effective. Such a threshold can be raised to over 20% for loosely bunched vines, which are less sensitive to rots, such as Merlot and Cabernet.

Currently, the methods for direct moth control include the use of traditional neurotoxic insecticides (chlorpyriphos methyl, chlorpyriphos ethyl, fenitrothion), new-generation neurotoxic insecticides (indoxacarb and spinosad), chitin synthesis inhibitors (flufenoxuron and lufenuron), moulting accelerating compounds (tebufenozide and methoxyfenozide) and *Bacillus thuringiensis kurstaki* and *B. t. aizawai*.

Since the 1990s, the mating disruption method has become widespread, partly because it is also applicable to 'organic farming' systems.

The experience gained over the last decade in some important vine-growing regions of Europe and Italy has actually been encouraging for industry researchers and operators as regards the possibility of obtaining positive results in moth control uniquely through the use of such a method (Stockel *et al.*, 1994; Charmillot and Pasquier, 2000; Varner *et al.*, 2001a; Carlos *et al.*, 2005; Ioriatti *et al.*, 2005; Lucchi and Bagnoli, 2007).

Nowadays, the area of pheromone-treated vineyards in Europe is estimated to cover about 100,000 ha (60,000 in Germany, 12,000 in Italy, 10,000 in France, 12,000 in Spain, 6500 in Switzerland, 1000 in Austria, 500 in Portugal and about 1000 in the Eastern European countries of the Czech Republic, Slovakia and Hungary).

Limitations of current management approaches

Moth control through the application of insecticides stands out for its remarkable flexibility in terms of the choice of, if and when to act on the different vineyards on the farm. While products containing *B. thuringiensis*, undoubtedly having a high ecological value, have phytosanitary limits chiefly due to their poor persistence, many of the active ingredients currently used show highly critical points due to their acute toxicity for vertebrates, the risk of toxic residues in foods and in the environment, their poor selectivity towards auxiliaries and, more generally, their poor eco-compatibility.

For these reasons, as in the USA (Food Quality Protection Act 1996), provisions were also issued in Europe to reduce the use of such substances (EU 349/2002).

The problem of the toxicity of pesticides is even more deeply felt in highly urbanized productive settings and/or tourist resorts.

Conditions for successful application of AWPM

Some important vine-growing regions of Italy seem to be particularly suitable for the application of AWPM. The most significant conditions include:

- Substantial damage caused by wide geographical diffusion of and great difficulty in controlling vine moths.
- Local presence of associations (wine growers' cooperatives, producers' consortia, professional and trade organizations) and public and private bodies that can provide technical assistance and control over the productive chain.
- Presence of schools, universities and research centres specializing in vine growing and winemaking that can help improve and transfer knowledge to industry specialists and operators.

Such conditions are prerequisites for developing a coordinated, synergic defence management system according to AWPM standards.

Description of the AWPM Programme and Approaches

Moth control in terms of AWPM had its earliest version in the practice of the 'mandatory larvae collection' of vine moths according to the law passed on 30 April 1870 in the Trentino-South Tyrol region (Mach, 1890). This compulsory control measure was justified by the higher efficacy of a concerted action given the mobility of the insects.

As mentioned before, in the same region and for the same phytophages, the conditions for a modern AWPM approach have been met one century later through the application of the pheromone mating disruption technique.

AWPM technologies and approaches

As is well known, pheromone mating disruption (MD) implies the use of an insect's sex pheromone to prevent or delay the mating of insects. In viticulture, this technique is currently based on the use of reservoir-like dispensers available in different formulations: Basf twin-ampulle Rak, Shin-Etsu twist-tie ropes and twin-type Isonet and Suterra Check-Mate membranes.

Such dispensers may contain either or both active ingredients of *L. botrana* and *E. ambiguella* which, in this case, are the main components of the two natural pheromone blends, E7,Z9-12:Ac for the first species and Z9-12:Ac for the second. Apart from the active ingredient, which is the same for all kinds of dispensers, there is great variation in other features, such as specific loading, physical and chemical material features, dispenser size and shape, and thickness of walls, leading to varying performance in terms of release rates, duration and final efficacy.

In areas where damage by third-generation *L. botrana* may occur, dispensers must ensure pheromone release from the end of March to the end of September. In Trento province, an average release of at least 23 ± 8 mg/ha/h is usually sufficient to ensure an efficient control of *L. botrana*.

The necessity of controlling both grapevine moths in that area led to the creation of dual-purpose dispensers. The industrial products are Basf Rak 1 + 2 ampullae and Shin-Etsu twin tubes, Isonet LE. In recent years Shin-Etsu has introduced the Isonet Lplus, a compromise solution that carries within a twist-type single-rope dispenser the pheromone standard needed to control the three *L. botrana* flights, mixed with a small amount of pheromone for MD control of *E. ambiguella* in areas where it occurs at low population levels (Varner *et al.*, 2001a, b). This dispenser releases, for the length of the season, both species' pheromones but in different amounts per unit time. A release rate of 2.3 ± 0.5 mg/ha/h has been verified as sufficient for the prevention of *E. ambiguella* outbreaks in the larger part of Trento province. Isonet LE was recently registered (March 2007) and Isonet Lplus is still pending registration in Italy, but both have been granted a special use exception so far.

Mating disruption is typically a preventive method for which the best application period is before the first flight of the year (Charmillot, 1992). Most of the early 1990s failures recorded in Italy and abroad have as their main reason the application aimed at the second flight (Arias *et al.*, 1992; Bagnoli *et al.*, 1993; Cravedi, 1993, 1995; Bagnoli and Goggioli, 1996).

MD protocols for viticulture traditionally specify 500 dispensers per ha, evenly distributed and each covering an area of 20 m². Depending on vine spacing it may be worth considering application on alternate rows, thus reducing fieldwork.

Although dispenser placement plays an important role in the MD results, not much information is available to optimize it in relation to vineyard agronomical features. Protocols provide only generic recommendations on the minimum area to be treated, buffer zones and border treatment. In addition, it would be helpful to suggest differentiated distributions in relation to slope and row directions as they affect the loss of released pheromone. In Trento province, as well as in other hilly regions, specific experiences show that an uneven dispenser distribution of 70% dosage in the 30% upper side of the vineyard slope area, with the remaining 30% evenly applied to the rest of the area, could be more effective.

Regardless of the type, it is advisable that dispensers are fixed to the vine shoot at the height of the future clusters, where they are protected from direct sun radiation and high temperatures. Estimated application time varies between 1.5–3.0 h/ha per worker depending on the vineyard topographical conditions.

The MD efficacy evaluation is carried out by assessing the presence of the pest by concerted field scouting. Counting male catches in pheromone traps placed in treated areas is the easiest method of indirect efficacy evaluation, but its accuracy is not precise enough (Charmillot, 1992; Ioriatti *et al.*, 2005). As a consequence, the more time-consuming evaluation of infested bunches is normally considered more appropriate for the purpose.

The first European MD tests in vineyards were conducted in the late 1970s in France and Switzerland for *L. botrana* and *E. ambiguella*, respectively (Roehrich *et al.*, 1977;

Rauscher and Arn, 1979), and 10 years later in Italy (Vita *et al.*, 1985; Ioriatti and Vita, 1990) to control *L. botrana*. This pilot experiment conducted in the Trento province on a small scale (8 ha) and with handmade dispensers provided unsatisfactory results. Soon after, further experiences carried out with industrial dispensers in the same area confirmed the need to apply on wide areas to achieve success with the method.

In 2006, the vine-growing area subjected to mating disruption in Trentino-South Tyrol was 9500 ha, accounting for approximately 92% of the whole area of Italy treated with sex pheromones against vine moth.

The adoption of MD in that region has been favoured by specific socio-economic, agronomic and target species bio-ethological factors. First, the scientific activity and extension service of the Agricultural Institute of San Michele a/A (IASMA) created a foundation that demonstrated the potential effectiveness of the MD technique. Secondly, these activities were promoted by an active cooperative extension organization that is well-respected within this region. In particular, work carried out by Cantine Mezzacorona (Mezzocorona, Trento) was particularly successful in overcoming some of the operational obstacles that have limited the adoption of MD in the past due to the average small size of individual farms. In fact, the vine-growing area of the region is extremely fragmented, being divided into small grape fields (frequently < 1 ha), usually spread over wide areas.

Within this context, a successful application of MD was only possible through an AWPM approach enabling operators to overcome this limitation. Decisive for the success was the role of the growers' associations and their involvement in the AWPM project. Since the early 1990s the Cantine Mezzacorona has promoted awareness campaigns, encouraged the use of MD and provided the necessary financial and organizational support to associated members (Varner and Ioriatti, 1992; Varner *et al.*, 2002). Moreover, the active collaboration of the chemical industry with the technical and scientific institutions and the local cooperative system has been a key factor contributing to the rapid adoption of MD in this area.

Development and Implementation of the AWPM Programme for the Control of Vine Moth

The vine moth areawide suppression programme in the province of Trento was initiated in 1998, when the local government decided to support the application of MD as an areawide approach. Partners in the programme included scientists from IASMA, consultants of the advisory service (ex ESAT) as well as the growers operating as members of the wineries cooperatives. Objectives of the programme were:

- Replacement of organophosphate treatments by the use of non-pesticidal systems.
- Application of MD on a large scale to take advantage of the increased efficiency of the method.
- Reinforcement of the biological control of mites.
- Improvement in workers' safety.
- Improvement in the professional skill of growers and technicians.
- Reduction of public exposure to insecticides.

- Improvement in the perception that grapes and wine are produced with high consumer safety standards in mind.
- Promotion of the image of a 'clean' agriculture that can integrate instead of hampering the value of the natural environment in the tourist market.

This step was the consequence of 10 years of research carried out at IASMA in close collaboration with the advisory service and the main pheromone-producing companies. Preliminary activities permitted the setting up of the methods and the realization that three main factors were involved in a successful approach: the population density of the target pest, the quality of the dispensers and the areawide approach.

Population density of the target pest

The first larval generation feeds on flowers and does not cause significant damage when infested bunches are less than 40% if chemical control is performed on the summer generations. The threshold is far below this when MD is in place. If there are more than 5% of inflorescences occupied by larvae, a chemical treatment is needed in the second generation. According to the load of the main cultivars occurring in Trento province, this threshold corresponds to about 5000–7000 larvae/ha. These data are in the range of those reported by Von Feldhege *et al.* (1995) as being the maximal level of population density of moths for an effective application of MD.

Quality of the dispensers

The close cooperation with pheromone companies has allowed the verification of the suitability of commercial dispensers for local viticulture. As mentioned above, from the 1990s cooperation with Shin-Etsu has permitted the development of dispensers for the control of both vine moths, *L. botrana* and *E. ambiguella* (Isonet L in 1995, Isonet LE in 1998 and Isonet Lplus in 2000).

Areawide approach

Since the pilot experimental trials carried out in Trento province in collaboration with the Cantine Mezzacorona, the proposed technology, when applied over a sufficient area (> 50 ha), has been able to control vine moth at the same level or even better than the traditional method based on insecticide applications. The efficacy of MD increased according to the treated area – 8 ha in 1988, with a progressive increase to 232 ha in 1997.

These satisfactory results pushed the local government to promote an AWPM project supporting part of the costs of the dispensers. This constituted a first favourable condition for the growers' agreement to the project. Consultants and cooperatives worked together in promoting the MD and in organizing the field application of the dispensers. Initially, the treated areas were selected by the consultants as those closest to the residential areas and those most uniform in size and shape. Rapidly, MD became the standard technology for the control of vine moth in the region.

From 1998 to 2001 the grant from the local government ranged from 0 to 50% of the cost of the dispensers, according to the year and the type of dispenser considered. The cost of dispensers was supported 100% for certified organic farming. In the same period the treated area increased from 700 to 5600 ha and, as a consequence, the outlay for the programme increased from €35,000 to 110,000.

Since 2002 the system of subsidy for the improvement and spread of environmentally friendly actions has changed: growers owning a minimum of land (about 0.2 ha) and applying MD for at least 5 years would receive the difference in cost between the more expensive MD and the cheaper chemical control, estimated at €75 and 130/ha, respectively, for the control of only *L. botrana* and for the combined control of the two pests. Nevertheless the total outlay for the programme has not changed, and the investment of the local government was confirmed at about €110,000/year. This has promoted the further spread of the method which, 5 years after the beginning of the AWPM project, had been adopted over about 90% of the entire grape-growing area (see Fig. 11.1).

Development and Implementation of Education and Technology Transfer Programmes

As mentioned above, the expansion of MD in the Trento province was mediated by the cooperatives. Since the beginning of the project the dispensers have been bought by the cooperatives and distributed to the members under the supervision of the local IASMA consultant. The consultant usually calls the growers during the winter to

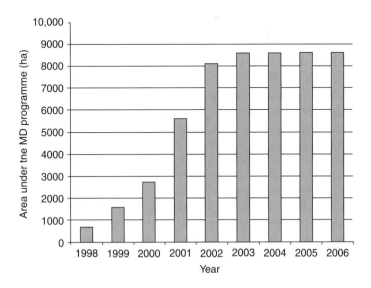

Fig. 11.1. Development of areawide application of mating disruption (MD) against vine moth in Trento province, Italy, 1998–2006 (from Ioriatti *et al.*, 2007).

explain the purpose of the programme and to present the previous year's results. Close to the application period the growers are called again by the advisor to receive the dispensers and the instructions for the correct application in the field. Extra dispensers are provided to the growers whose fields are located on the perimeter or in the upwind part of the treated area.

Selected fields are chosen for the scouting activity according to their position (perimeter) and/or the known presence of high insect population (hot spots). Monitoring traps baited with the lure of the two vine moths are placed in those vineyards and checked weekly by the local advisor. Captures in the traps are used as an alerting tool; no curative treatments are decided on this information.

The efficacy assessment on the first generation is performed when the larvae are finally forming nests on the flower clusters. According to the infestation level, treatments against the second generation are decided using the threshold established by Charmillot *et al.* (1997) and confirmed for the Trento province by Varner *et al.* (2001b). At the beginning of the second generation, scouting activity is performed in selected fields by trained people in order to assess the presence of eggs on the grape berries. The dispenser performance in relation to the third-flight adults is easily evaluated by checking the eggs laid on the green clusters of the secondary shoots (late August–early September). The results are then illustrated to the growers during the following winter meetings.

Compatibility of the AWPM programme with crop management and control of co-occurring pests

In the province of Trento, the mating disruption method proved to be fully compatible with the management of the varying growing techniques. In fact, the dispensers are used at a time of year when vine growing and winemaking labour is at its lowest. At the same time, the presence of the dispenser in the vineyard does not hinder any growing process. In addition, the dispensers do not necessarily have to be removed at the end of the season and, in any case, they can be removed either during the winter pruning or during the new installation. In this respect, note that dispensers made of biodegradable materials are under development (Ioriatti *et al.*, 2005).

As mentioned above, the phytophagous arthropods associated with the vine amount to a few dozen. Partly depending on the level of equilibrium existing in the agroecosystem, such species can occasionally or locally cause more or less harmful infestations that are managed with measures that are defined on a case-by-case basis and comply, increasingly, with IPM standards. For *Metcalfa pruinosa* and *Scaphoideus titanus*, in some Italian regions the AW approach has been based on the classic biological control against the planthopper and the mandatory chemical control against the leafhopper. In the province of Trento, such an approach properly matched the AWPM moth control system.

The case of *Metcalfa pruinosa*: AWPM by biological control

Appearing in Europe in the late 1970s in the province of Treviso (Zangheri and Donadini, 1980; Dlabola, 1981), the North American homopteran, *Metcalfa pruinosa* (Flatidae), quickly spread in virtually all of the Italian territory and into some neighbouring European countries (Wilson and Lucchi, 2007), favoured by its remarkable polyphagy and the lack of specific and effective natural antagonists.

On the vine, the insect can cause damage by spoiling the plant's organs with the abundant waxy secretions produced by its juveniles, the subtraction of sap and the release of honeydew, on which a sooty mould is then formed (Bagnoli and Lucchi, 2000). Some of its typical bio-ethological and physiological traits make *M. pruinosa* unsuitable for chemical control. In addition, the new apicultural interest for *Metcalfa* in the places where it settles, with summer production of substantial amounts of honeydew that could turn out to be contaminated, clashes with the use of insecticidal molecules (Lucchi, 1997b). For these reasons and because of the exotic nature of the pest, the chosen approach was that of using a North American antagonist, *Neodryinus typhlocybae* (Ashmead) (Hymenoptera: Dryinidae) (Girolami and Camporese, 1994; Mazzon *et al.* 2003).

Since the 1990s, this insect has been spread throughout central and northern Italy under agreements between public facilities (universities, research centres, regional and provincial bodies) and/or private bodies (Bioplanet of Cesena), and has proved to be perfectly adaptable to the Italian climate and surprisingly effective in containing the populations of *Metcalfa* in the short–medium term. So far, the number of Italian 'releases' of this entomophagous insect has been estimated at about 800.

In Trentino, 12 releases have been carried out since 1998 in six sites, using about 100 individuals of the parasitoid per site (sex ratio 1:1).

To summarize, the AWPM scheme for the biological control of *Metcalfa pruinosa* with *Neodryinus typhlocybae* in the province of Trento has involved:

- Breeding of the antagonist on the natural host, and collection and storage of cocoons.
- Preparation of release packages containing cocoons of the parasitoid grown on a vegetal support.
- Location, across the vine-growing region, of suitable sites of release having natural vegetation adjoining the vineyard and a high population of the host.
- Choice of the best time to place the packages in the natural area, usually in the first week of June, when the host population is mostly composed of individuals at the III, IV and V pre-imaginal stage.
- Checking whether the population has settled.

The experience gained in Trentino (Angeli *et al.*, 2005) and in other Italian regions (Lucchi and Wilson, 2003) in the last decade has highlighted the fact that *N. typhlocybae* can offer a long-term solution for control of the planthopper. The parasitoid has not only steadily settled in the sites of release, but it has progressively colonized the surrounding areas, some of which were very distant, thus resulting in a remarkable reduction in the flatid populations, even in the short-term applications.

The case of *Scaphoideus titanus*: AWPM by mandatory control

As is well known, *Scaphoideus titanus* is the specific vector of the phytoplasma of flavescence dorée, *Candidatus phytoplasma vitis* (IRPCM, 2004), a disease subject to quarantine for which, in Italy, the Ministry of Agricultural and Forestry Policies issued a law in 2000 for the mandatory control, setting forth the measures to be taken in the presence of FD and/or its vector.

Scaphoideus titanus is a species of North American origin that is present in Croatia, France, Portugal, Serbia, Spain, Slovenia, Switzerland and Hungary. In Italy it was found for the first time in 1964, and so far its presence has been recorded all over central and northern Italy and in two southern regions, where it is assumed to have been carried recently by anthropic diffusion through the use of egg-infested propagation materials (Lessio and Alma, 2004).

In its region of origin, *S. titanus* is a polyphagous species, while in Europe it is closely dependent on the vine, on which it completes just one generation per year and hibernates at the egg stage (Mazzoni *et al.*, 2005).

Recent researches suggest that the leafhopper might be subjected to both a natural control, through the use of specially introduced specific antagonists (Nusillard *et al.*, 2003), and a biotechnological control of the MD type, through manipulation of the acoustic communication between the two sexes (Lucchi *et al.*, 2004). However, at the moment, *S. titanus* can be effectively controlled only through the use of insecticides.

The most appropriate time for applying the treatment is the month of June, and in any case before the adult appearance, in order to control the juveniles that are more sensitive to the insecticide and less dangerous than the adults with regard to spreading of the disease.

The range of usable insecticides includes both traditional and new-generation chitin inhibitors and neurotoxic agents.

Since the harmfulness of *S. titanus* comes only from its ability to transmit FD phytoplasma, the control of this species will clearly have to be attuned to the risk of disease spread. In other words, proper control of the leafhopper must be based on a careful geographical monitoring of FD and its vector. Such monitoring has been operating for years in several Italian regions, and relies on cooperation between the regional phytosanitary services, public and private research centres and technical advisory centres. It constitutes a fundamental tool for the management of the control of *S. titanus* in terms of AWPM.

In the province of Trento, the leafhopper has been present since 1988, while some cases of FD were found in the southern part of the region in 2000. The monitoring of the two organisms throughout the regional territory showed, in 2003, the contemporary presence of the phytoplasma and of its vector in some limited areas. Since then, insecticidal treatments have been planned and applied on approximately 10% of the province's vine-growing area.

The proper application of these procedures (monitoring of both insect and disease) was crucial in limiting the area that had to be compulsorily subjected to the insecticidal treatment without endangering the sustainability of MD.

Evaluation of the AWPM Programme

Effectiveness of the AWPM programme in controlling the target pest(s)

Trento province has provided the most significant evidence in Italy that the effectiveness of MD increases with repeated applications and with the increase in the size of the treated area. This can be corroborated by the experience of the viticultural area of Mezzocorona, in which from 1997 to 2001 the area subjected to MD rose from 170 to 1100 ha, i.e. the whole vine-growing area. Over the 5 years of application, the mean infestation at the end of the second generation in the pheromone-covered area dropped from 1.2 to 0.2% (Varner *et al.*, 2001a, b) versus a mean attack rate in the traditionally defended areas of about 5–8% of infested bunches (Ioriatti *et al.*, 1993).

Before MD had been applied, 60% of the vine growers were controlling vine moths with two treatments per year, 31% treated only once and 9% did not treat at all (Ioriatti *et al.*, 1993). Given that chlorpyrifos methyl was the most widely applied insecticide, it can be estimated that 17 t of commercial formulation (Reldan, recommended field dose: 2 kg/ha) were annually applied in the vineyards. At present, insecticide applications are no longer needed for grape moth control in most of the grape-growing areas of Trento province. Sporadic chemical treatments are needed only in small and isolated vineyards during the more favourable years for pest development. The risk of negative side-effects of the pesticides on beneficial organisms, as well as on human beings, is avoided entirely.

Unintended negative and positive consequences of the AWPM programme

The presence of phytophagous species different from the target ones is a problem concerning both initial and ongoing phases in the MD application decision-making process. In a given agroecosystem, the presence of noxious insects requiring the use of conventional insecticides could reduce, from both an ecological and agronomical point of view, the innovative weight of the method, lowering interest in its adoption. On the other hand, repeated application of MD may trigger occurrence of secondary pests no longer controlled by conventional chemical treatments (Neumann, 1990). Local outbreaks of the tortricid *Argyrotaenia ljungiana* and coccids *Parthenolecanium corni* and *Pulvinaria vitis* in MD areas of Trento province were reported by Varner *et al.* (2001b), and recently confirmed by Barrios Sanromà *et al.* (2006) in Spain.

Nevertheless, the last 10 years of experience have shown that high selectivity of this method, in the medium to long term, leads to a range of positive effects on entomophagous populations allowing a natural regulation of secondary pests. In other words, the hypothesis could be a flow chart such as: MD application → spray reduction → occurrence of secondary pests → increase of natural enemies populations → natural control of secondary pests, as confirmed by a number of studies in Italy (Varner *et al.*, 2001b) and abroad (Delbac *et al.*, 1996a, b; Stockel *et al.*, 1997; Schirra and Louis, 1998; Koclu *et al.*, 2005).

The impact of MD on non-target Lepidoptera within vineyards has not yet been fully investigated. For instance, in traps of *L. botrana* and *E. ambiguella* placed within MD vine-growing areas of Tuscany, non-target species such as *Idaea biselata*, *I. filicata*, *I. straminata* (Geometridae) and *Lobesia bicinctana* (Tortricidae) are no longer captured (Ioriatti *et al.*, 2005). For this reason, further ecological studies on the interference of the pheromone on both the biodiversity and the potential co-victims of entomophagous insects associated with vine moth populations deserve to be carried out.

Costs and Benefits of the AWPM Programme

The cost of mating disruption is generally higher than the specific conventional chemical control of vine moth. This difference is larger when an OP is applied (50% additional cost) and smaller when an IGR is used (30% a.c.). The gap is even higher when double dispensers are required.

As mentioned above, in the AWPM project promoted in the Trento province the extra cost was covered by contribution. Note, however, that in 2001, the year in which the public subsidy for buying the dispensers for *L. botrana* was temporarily stopped, the area covered by MD rose from 2750 to 5600 ha, proving that the vine growers firmly believed in this technology.

One advantage in using AWPM was a reduction of the cost per unit area. In fact the growers, as the programme progressed, controlled the pest with a lower number of dispensers (by 20–50%) because of the reduced pressure of the pest population and by taking advantage of the incompletely exhausted dispensers of the previous year (Anfora *et al.*, 2005).

Such costs were further reduced by the introduction of Isonet Lplus dispensers, which cost just 2% more than Isonet L (for *L. botrana*), but 28% less than Isonet LE. In any case, once fully implemented as an AWPM strategy, MD was more cost effective than conventional control.

The areawide application of MD mostly benefited the smaller farms, especially the organic ones which, as is known, have fewer moth control systems.

Sometimes the vine growers used participation in the project as a factor to increase the commercial value of their product. The perception that grapes and wine are produced with high consumer safety standards in mind contributes to spreading the image of environmentally safe agriculture that can enhance, instead of decrease, the value of natural environments for tourism.

The risk of developing resistance from the widespread application of a control system for a given species, as happens in the AWPM approach reported by Siegfried *et al.* (1998), does not exist in the cases we mentioned, since no reports of true resistance to MD are yet available. Although genetic heterogeneity of *L. botrana* may affect responsiveness to pheromones and promote the selection of less susceptible moths, the argument is still only speculation. Moreover, resistance caused by sex-linked mutations is even less probable because it would have to occur in both male and female moths, which would involve the production of a mutant sex pheromone by females and the coincident mutation of antennal pheromone receptors in males.

The mandatory control of *S. titanus* through one or two applications of insecticides per year – either synthetic (buprofezin, thiametoxan, chlorpyrifos methyl, etc.) or natural (pyrethrins, rotenone, etc.) and, what is more, on just 10% of the AW area – does not seem to have created any problem yet in terms of onset of resistant strains. The biological control of *M. pruinosa* with its specific parasitoid *N. typhlocybae* does not involve, by its very nature, any risk of resistance.

Sociological evaluation of the AWPM programme

When, 20 years ago, the first field trials with MD started in the province of Trento, we could not forecast such a great success. The benefit of MD technology is not only valuable in terms of higher efficacy as compared with chemical control, the greatest consequence of the widespread application of MD is the improvement of the quality of life for growers and the public living around the growing areas. The generalized application of MD removes much of the pesticide exposure risk to workers, both during treatments and for general work in the vineyard.

The elimination of insecticidal treatments in summer, when environmental sensitivity is particularly high due to the increase in outdoor activities, has reduced the conflict between the agricultural and the non-agricultural world. This becomes extremely important in a region like Trentino-South Tyrol, where tourism is a key industry.

Summary and Future Directions

In Italian viticulture, the most important AWPM approach is represented by the use of pheromone mating disruption (MD) for the control of vine moth, adopted in recent years in the province of Trento (northern Italy).

In that area the adoption of MD has been favoured by specific socio-economic, agronomic and target species bio-ethological factors. Partners in the programme included scientists of IASMA, consultants of the advisory service and the growers operating as members of the wineries cooperatives.

The real vine moth AWPM programme was initiated in 1998 with the support of the local government. In the first 3 years public contribution partly covered the cost of the dispensers, and the pheromone-treated area increased from 700 to 2750 ha. In 2002, even when the public subsidy for buying the dispensers for *L. botrana* had temporarily been stopped, the area subjected to MD rose from 2750 to 5600 ha, proving that the vine growers firmly believed in this technology.

At present, insecticide applications are no longer needed for grape moth control in most of the grape-growing areas of Trento province. Sporadic chemical treatments are needed only in small and isolated vineyards during the more favourable years for the pest's development.

The areawide application of MD mostly benefited the smaller farms, especially the organic ones which, as is known, have fewer moth control systems. Sometimes they used participation in the scheme as a factor to increase the commercial value of their product.

The classic biological control and mandatory chemical control carried out, respectively, against the co-occurring pests *M. pruinosa* and *S. titanus* matched the AWPM moth control system. As regards *S. titanus*, the careful monitoring of insect and disease was crucial in limiting the area that had to be compulsorily subjected to the insecticidal treatment without endangering the sustainability of MD.

In 2006, the area covered by MD in Trentino-South Tyrol amounted to 9500 ha, accounting for approximately 92% of the whole vine-growing area of Italy treated with sex pheromones against vine moth. The benefit of the areawide approach was valuable not only in terms of higher efficacy and medium-term cost reduction as compared with that of chemical control: the greatest consequence of the widespread application of MD in Trento province was the improvement of the quality of life for the growers and the public living around the growing areas.

Unfortunately, the size of the other Italian vine-growing areas treated with MD is still limited despite the extensive research and the testing activities carried out in the past 20 years by both public and private institutions. Factors limiting the adoption of MD in those areas are not related to its efficacy, which has been proved to be competitive with chemical control, nor to its impact on the management of other pests, but relate mainly to sociocultural and economical conditions existing in the particular vine-growing area: (i) growers are more strongly impacted by the perceived higher costs; (ii) they are not likely to assume the risks initially associated with this technology; (iii) they show little interest in innovative methods; and (iv) they are generally lacking a structured and tight organization such as the cooperatives, which were the prime factor in the implementation of AWPM in the province of Trento.

Clearly, new investment in basic and fundamental research is a critical factor that will allow AWPM to expand in viticulture. In this respect, much attention needs to be paid to novel pheromone application systems (false trail following, aerosol puffers, microencapsulated pheromones, sprayables, auto-confusion, attract and kill) and to the identification of plant volatiles as possible cues for ovipositing females.

Nevertheless, in Italy we believe that MD, if applied with the right materials and protocols, can represent from this present time the main means for the fulfilment of AWPM projects in viticulture. In addition, in valuable viticultural areas marked by a high population density of *L. botrana* or by peculiar phytosanitary problems (e.g. presence of phytoplasma vectors), MD could be adopted as a grape protection platform on which to insert, in case of need, integrative insecticide treatments suitable for both conventional and organic viticulture. Such a strategy seems, at the moment, the solution which best marries the need for an effective and selective control of the key pest, *L. botrana*, with the need to oppose, case by case, attacks by other harmful species.

In this process, all the initiatives addressed to ensure a vineyard control management for entire areas (growers cooperatives, spin-off systems, etc.) are welcome.

Acknowledgments

The success of this experience would not have been possible without the support from Luisa Mattedi, Enzo Mescalchin, Mauro Varner and Vitterio Veronelli. The authors are also indebted to the consultants of the advisory service in viticulture (IASMA-CAT) for providing their professional skills to the growers in applying mating disruption techniques and in the valuable field scouting activity.

References

Anfora, G., Tasin, M., Bäckman, A.C., De Cristofaro, A., Witzgall P. and Ioriatti, C. (2005) Attractiveness of year-old polyethylene Isonet sex pheromone dispensers for *Lobesia botrana*. *Entomologia Experimentalis et Applicata* 117, 201–207.

Angeli, G., Maines, R., Fanti, M., Forti, D., Sofia, M., Baldessari, M., Tomasi, C., Sandri, O., Delaiti. L., Ioriatti, C., and Girolami, V. (2005) Biological control of *Metcalfa pruinosa* with *Neodryinus typhlocybae*: establishment and diffusion of the parasitoid in Trentino Alto Adige (Italy). *IOBC/wprs Bulletin* 28 (7), 271–274.

Arias, A., Bueno M., Nieto, J., Valenzuela, M., Perez, A., Cuenda, B., Gallego, F., Alamada, A. and Castillo, M.A. (1992) Essais de confusion sexuelle de *Lobesia botrana* Den. & Schiff. pendant 1989 et 1990 dans Tierra De Barros (Espagne). *Bulletin OILB/srop* 15 (2), 18.

Bagnoli, B. (1990) Incidenza delle infestazioni da artropodi e difesa dei vigneti in Toscana. *La Difesa delle Piante* 3–4, 89–112.

Bagnoli, B. and Goggioli, D. (1996) Application of mating disruption technique to control the grape moth *Lobesia botrana* (Den. and Schiff.) in Tuscany. *Proceedings of the XX International Congress of Entomology*, Florence, Italy, abstract (poster) 15–194, p. 497.

Bagnoli, B. and Lucchi, A. (2000) Dannosità e misure di controllo integrato. In: Lucchi, A. (ed.) *La Metcalfa negli Ecosistemi Italiani*. ARSIA, Regione Toscana, pp. 65–88.

Bagnoli, B., Goggioli, D. and Righini, M. (1993) Prove di lotta con il metodo della confusione sessuale contro *Lobesia botrana* (Den. e Schiff.) nella zona del Chianti. *Redia* 76, 375–390.

Barrios Sanromà, G., Moret, V.D. and Aybar, J.R. (2006) Control de polilla del racimo (*Lobesia botrana*) en viñedos de Cataluña mediante la técnica de la confusión sexual. La experiencia de Raimat. *Phytoma España* 183, 23.

Brunelli, A., Borgo, M., Bagnoli, B. and Cravedi, P. (1993) Strategie di difesa integrata per le uve da vino. In: Bisiach, M. and Bagnoli, B. (eds) *Proceedings of the meeting 'Lotta integrata in viticoltura'*, ISPaVe, Rome, pp. 129–159.

Carlos, C., Costa, J., Gaspar, C., Domingos, J., Alves, F. and Torres, L. (2005) Mating disruption to control grapevine moth, *Lobesia botrana* (Den. & Schiff.) in Porto wine region: a three-year study. *IOBC/wprs Bulletin* 28 (7), 283–287.

Charmillot, P.-J. (1992) Mating disruption technique to control grape and wine moths: general considerations. *IOBC/wprs Bulletin* 15 (5), 113–116.

Charmillot, P.-J. and Pasquier, D. (2000) Lutte par confusion contre les vers de la grappe: succès et problèmes rencontres. *IOBC/wprs Bulletin* 23 (4), 145–147.

Charmillot, P.-J., Pasquier, D., Schmid, A., Emery, S., de Montmollin, A., Desbaillet, C., Perrottet, M., Bolay, J.M. and Zuber, M. (1997) Lutte par confusione contre les vers de la grappe eudémis et cochylis en Suisse. *Revue Suisse Viticulture Arboriculture Horticulture* 5, 291–299.

Cravedi, P. (1993) Confusione sessuale nel controllo delle tignole della vite. In: Bisiach, M. and Bagnoli, B. (eds) *Proceedings of the meeting 'Lotta integrata in viticoltura'*. ISPaVe, Rome, pp. 91–102.

Cravedi, P. (1995) I feromoni nella difesa integrata dei vigneti di uva da vino. *L'Informatore Agrario* 13, 59–61.

Delbac, L., Fos, A., Lecharpentier, P. and Stockel, J. (1996a) Confusion sexuelle contre l'Eudemis. Impact sur la cicadelle verte dans le vignoble bordelais. *Phytoma, la Défense des Végétaux* 488, 36–39.

Delbac, L., Lecharpentier, P., Fos, A. and Stockel, J. (1996b) La confusion sexuelle contre l'Eudemis. Vers un equilibre biologique de l'acarofaune du vignoble. *Phytoma, la Défense des Végétaux* 484, 43–47.

Dlabola, J. (1981) *Metcalfa pruinosa* (Say, 1830), eine schädliche nordamerikanische flatide als erstfund in der palaearktis (Insecta : Homoptera, Auchenorrhyncha). *Faunistiche Abhandlungen* 8, 91–94.

Duso, C., Malagnini, V., Drago, A., Pozzebon, A., Galberto, G., Castagnoli, M. and de Lillo, E. (2003) The colonization of Phytoseiid mites (Acari: Phytoseiidae) in a vineyard and the surrounding hedgerows. *IOBC/wprs Bulletin* 26 (4), 37–42.

Girolami, V. and Camporese, P. (1994) Prima moltiplicazione in Europa di *Neodryinus typhlocybae* (Ashmead) (Hymenoptera: Dryinidae) su *Metcalfa pruinosa* (Say) (Homoptera: Flatidae). *Proceedings of the XVII Congresso Nazionale Italiano di Entomologia*, Udine, Italy, 13–18 June 1994, pp. 655–658.

Ioriatti, C. and Vita, G. (1990) Résultats préliminaires d'un essai de lutte par confusion sexuelle contre le vers de la grappe (*L. botrana* Schiff.) dans un vignoble du Trentin. *IOBC/wprs Bulletin* 13 (7), 80–84.

Ioriatti, C., Angeli, G., Delaiti, L., Delaiti, M. and Mattedi, L. (1993) Un solo intervento mirato. *Terra Trentina* 39 (7), 24–28.

Ioriatti, C., Bagnoli, B., Lucchi, A. and Veronelli, V. (2005) Vine moth control by mating disruption in Italy: results and future prospects. *Redia* 87, 117–128.

Ioriatti, C., Mattedi, L., Meschalchin, E. and Varner M. (2007) 20 años de esperiencia en la aplicación de feromonas para el control de polilla del racimo (*Lobesia botrana*) en viñedos del Trentino Alto Adige (Italia). *Ias Jornadas Internacionales sobre Feromonas y su uso en Agricultura*, Murcia, Spain, 21–22 November 2006. Consejeria de Agricultura y Agua, Comunidad Autónoma de la Region de Murcia, Spain,, pp. 73–79.

IRPCM Phytoplasma/Spiroplasma Working Team-Phytoplasma Taxonomy Group (2004) Description of the genus *Candidatus Phytoplasma*, a taxon for the wall-less, non-helical prokariotes that colonize plant phloem and insects. *International Journal of Systemic Evolution and Microbiology* 54, 1243–1255.

Koclu, T., Altindisli, F.O. and Ozsemerci, F. (2005) The parasitoids of the European grapevine moth (*Lobesia botrana* Den. & Schiff.) and predators in the mating disruption-treated vineyards in Turkey. *IOBC/wprs Bulletin* 28 (7), 293–297.

Lessio, F. and Alma, A. (2004) Seasonal and daily movement of *Scaphoideus titanus* Ball (Homoptera: Cicadellidae). *Environmental Entomology* 33, 1689–1694.

Lucchi, A. (1997a) Intense defogliazioni causate da Altica della vite (*Altica ampelophaga*). *L'Informatore Agrario* 6, 81–83.

Lucchi, A. (1997b) *Metcalfa pruinosa* and honey production in Italy. *American Bee Journal* 137, 532–535.

Lucchi, A., and Bagnoli, B. (2007) Seis años de interrupción del acoplamiento (confusión sexual) para el control de la polilla europea de la vid, en Toscana. *Ias Jornadas Internacionales sobre Feromonas y su uso en Agricultura*, Murcia, Spain, 21–22 November 2006. Consejeria de Agricultura y Agua, Comunidad Autónoma de la Region de Murcia, pp. 53–59.

Lucchi, A. and Wilson, S.W. (2003) Notes on the dryinid parasitoids of planthoppers (Hymenoptera: Dryinidae; Hemiptera: Flatidae, Issidae). *Journal of the Kansas Entomological Society* 76, 34–37.

Lucchi, A., Mazzoni, V., Presern, J. and Virant-Doberlet, M. (2004) Mating behaviour of *Scaphoideus titanus* Ball (Hemiptera: Cicadellidae). *Proceedings of the 3rd European Hemiptera Congress*, Saint Petersburg, Russia, 8–11 June 2004, pp. 49–50.

Mach, E. (1890) *Misure per Combattere la Tortrice o Tignola dell'Uva* (Caròl, Cajòl, Bissòl). Consiglio Provinciale daAgricoltura, Innsbruck, Austria, 16 April 1890, 10 pp.

Mazzon, L., Lucchi, A., Girolami, V. and Santini, L. (2003) Investigation on voltinism of *Neodryinus typhlocybae* (Ashmead) (Hymenoptera: Dryinidae) in natural context. *Frustula Entomologica n.s.* 24, 9–19.

Mazzoni, V., Lucchi, A., Varner, M., Mattedi, L., Bacchi, G. and Bagnoli, B. (2003) First remarks on the leafhopper population in a vine-growing area of South-Western Sicily. *IOBC/wprs Bulletin* 26 (8), 227–231.

Mazzoni, V., Alma, A. and Lucchi, A. (2005) Cicaline dell'agroecosistema vigneto e loro interazioni con la vite nella trasmissione di fitoplasmi. In: Bertaccini, A. and Braccini, P. (eds) *Flavescenza Dorata e altri Giallumi delle Vite in Toscana e in Italia*. Quaderno ARSIA 3/05, LCD srl, Florence, Italy, pp. 55–73.

Neumann, U. (1990) Commercial development: mating disruption of the grape berry moth. In: Ridgway, R.E., Silverstein, R.M. and Inscoe, M.I. (eds) *Behavior-modifying Chemicals for Insect Management: Applications of Pheromones and Other Attractants*. Marcel Dekker, Inc., New York, pp. 539–546.

Nusillard, B., Malausa, J.-C., Giuge, L. and Millot, P. (2003) Assessment of a two-year study of the natural enemy fauna of *Scaphoideus titanus* Ball in its North American native area. *IOBC/wprs Bulletin* 26 (8), 237–240.

Rauscher, S. and Arn, H. (1979) Mating suppression in tethered females of *Eupoecilia ambiguella* by evaporation of (Z)-9-dodecenyl acetate in the field. *Entomologia Experimentalis et Applicata* 25, 16–20.

Roehrich, R., Carles, J.P. and Tresor, C. (1977) Essai préliminaire de protection du vignoble contre *Lobesia botrana Schiff.* au moyen de la phéromone sexuelle de synthèse (méthode de la confusion). *Revue de Zoologie Agricole et de Pathologie Végétale* 76, 25–36.

Schirra, K.J. and Louis, F. (1998) Occurrence of beneficial organisms in pheromone-treated vineyards. *IOBC/wprs Bulletin* 21 (2), 67–69.

Siegfried, B.D., Meinke, L.J. and Scharf, M.E. (1998) Resistance management concerns for areawide management programs. *Journal of Agricultural Entomology* 15, 359–369.

Stockel, J., Schmitz, V., Lecharpentier, P., Roehrich, R., Torres-Vila, L.M. and Neumann, U. (1994) La confusion sexuelle chez l'Eudémis *Lobesia botrana* (Lepidoptera Tortricidae). Bilan de 5 années d'expérimentation dans un vignoble bordelais. *Agronomie* 14, 71–82.

Stockel, J., Lecharpentier, P., Fos, A. and Delbac, L. (1997) Effets de la confusion sexuelle contre l'Eudémis *Lobesia botrana* sur les populations d'autre ravageurs et d'auxiliaires dans le vignoble Bordelais. *IOBC/wprs Bulletin* 20 (1), 89–94.

Varner, M. and Ioriatti, C. (1992) Mating disruption of *Lobesia botrana* in Trentino (Italy): organization of the growers and first results. *IOBC/wprs Bulletin* 15 (5), 121–124.

Varner, M., Lucin R., Mattedi L. and Forno, F. (2001a) Experience with mating disruption technique to control grape berry moth, *Lobesia botrana*, in Trentino. *IOBC/wprs Bulletin* 24 (2), 81–88.

Varner, M., Mattedi, L., Rizzi, C. and Mescalchin, E. (2001b) I feromoni nella difesa della vite. Esperienze in provincia di Trento. *Informatore Fitopatologico* 10, 23–29.

Varner, M., Mattedi, L., Forno, F. and Lucin, R. (2002) Twelve years of practical experience using mating disruption against *Lobesia botrana* and *Eupoecilia ambiguella* in the vineyards of 'Cantine Mezzacorona' located in the piana Rotaliana Valley. IOBC, *Working Goup Meeting 'Pheromones and Other Semiochemicals in Integrated Production'*, Erice, Italy, 22–25 September 2002, Scientific Programme and Abstracts, p. 5.

Vita, G., Caffarelli, V. and Pettenello, M. (1985) Esperienze di lotta integrata in un comprensorio viticolo del Lazio. *Proceedings of the XIV Congresso Nazionale Italiano di Entomologia*, Palermo, Italy, 28 May–1 June, pp. 891–895.

Von Feldhege, M., Louis, F. and Schmutterer, H. (1995) Untersuchungen über falterabundanzen des bekreuzten traubenwicklers *Lobesia botrana Schiff.* im Weinbau. *Anzeiger Schädlingskunde Pflanzenschutz Umweltschutz* 68, 85–91.

Wilson, W.S. and Lucchi, A. (2007) Feeding activity of the flatid planthopper *Metcalfa pruinosa* (Hemiptera: Fulgoroidea). *Journal of the Kansas Entomological Society* 80, 175–178.

Zangheri, S. and Donadini, P. (1980) Comparsa nel Veneto di un omottero neartico: *Metcalfa pruinosa* (Say) (Homoptera, Flatidae). *Redia* 63, 301–305.

12 Stored-grain Insect Areawide Pest Management

David W. Hagstrum,[1] Paul W. Flinn,[2]
Carl R. Reed[3] and Thomas W. Phillips[4]

[1]USDA-ARS Grain Marketing and Production Research Center, Manhattan, Kansas, USA (retired)
[2]USDA-ARS Grain Marketing and Production Research Center, Manhattan, Kansas, USA
[3]Department of Grain Science and Industry, Kansas State University, Kansas, USA
[4]Department of Entomology and Plant Pathology, Oklahoma State University, Stillwater, Oklahoma, USA

Introduction

Over 67 million t of wheat (2.5 billion bushels) are moved by truck, railcar or barge through the grain marketing system in the USA during a year (Hagstrum and Heid, 1988; http://www.ers.usda.gov/Data/Wheat/YBtable04.asp). In the USA, commercial grain storage facilities are called elevators (Reed, 2006).

Typically, wheat is transported from the farm first to the country elevators, then to terminal elevators and finally to mill or export elevators. Elevators received their name from the bucket elevator that conveys the grain to the top of the building, from where it can be distributed to different bins. Country (local) elevators tend to receive grain from a smaller geographical area and generally have less grain storage and handling capacity than do terminal elevators. Because the terminal, mill and export elevators receive grain from large geographical areas, stored-grain insect pests are managed most effectively by areawide pest management (AWPM) practices that target local infestations before insects are transported with the grain to larger elevators (Flinn et al., 2003a, b, 2007a).

Wheat is harvested in the USA from June in the southern part of the wheat-growing region until August in the northern part (Hagstrum and Heid, 1988). Grain temperature, grain moisture and storage time are the main factors determining the risk of economic losses from insect infestation. Wheat harvested in June in southern regions is generally at higher risk because grain temperature is suitable for insect movement and reproduction for a longer time than is wheat harvested in August in northern regions. Currently, management of insect pests in the wheat marketing system is usually sufficiently effective to slow insect population growth and prevent insect populations from increasing exponentially. During the 1977 and 1978 storage seasons in the USA, the average insect pest population in the wheat marketing system increased from 0.15 insects/kg (four insects per bushel) in June to 0.44 insects/kg (12 insects per bushel) in October, and then decreased as grain cooled in autumn and winter (Hagstrum and Heid, 1988).

The profits and losses from merchandizing grain are greater and more apparent to managers of grain businesses than the costs and losses from insect pests in the elevator. Perhaps, for the same reason, grain merchandizing is generally given the highest priority and the greatest attention, and pest management and other aspects of grain management receive less attention. Any advanced insect pest management programme that is introduced into this corporate culture must be cost-effective, minimize the risk of insect problems and require minimal attention.

Elevator Operation

The operating practices of elevators need to be considered when developing an AWPM programme. At elevators, grain is sampled, weighed and stored. Often, it also is dried, segregated, blended, aerated and/or fumigated. The bucket elevator moves the grain from the dump pit below ground level to the top of the building or elevator leg. At small elevators, a distributor directs grain to the storage bins. At larger elevators, belt, drag or screw conveyors must move the grain laterally from bucket elevator to the storage bins. Most elevators have more than one dump pit and bucket elevator. Bins in most elevator buildings are unloaded through a discharge spout at the bottom of each bin on to a reclaim conveyor that transports it laterally to the bucket elevator. The distributor, located just beneath the top of the bucket elevator, directs the grain to another storage bin, a lateral conveyor, a weighing bin or a load-out bin. Load-out bins are small bins located over the truck dump. Storage bins not located in the elevator building may have an overhead load-out spout on the external wall for unloading.

Elevators often store several different types of grain. Grain of the same type usually is segregated by moisture content, protein content or test weight (bulk density). To facilitate this segregation, elevators frequently have many bins, often of different capacities (see Table 12.1). Terminal elevators may have over 100 bins (Elevators 9, 11 and 12) and some large terminal elevators 1000 or more. Records related to grain quality typically are kept on a large diagram of the elevator's bin layout, referred to as a bin board. Information including grain quality characteristics, volume of grain in the bin and fill and fumigation dates is recorded on this bin board. Fumigation is the treatment of grain by a toxic gas, usually phosphine, to kill insect pests (Hagstrum *et al.*, 1999).

Often, grain from several bins is blended to consistently meet grain quality requirements or to upgrade grain quality. Grain from one to ten bins (see Table 12.2) is often blended together by metering grain from each bin on to a reclaim conveyor. Generally, grain from fewer bins is blended at country elevators than at terminal elevators.

Table 12.1. Numbers of concrete bins of different storage capacities at 12 elevators in Kansas.

Elevator	Storage capacity (thousands of bushels) per bin								Total	Total capacity at elevator (bushels)
	1–2	2–5	5–10	10–20	20–30	30–40	40–50	69		
1	1	6	3	4	8	0	0	0	22	264,158
2	1	3	0	8	3	0	2	0	17	277,404
3	5	12	2	6	0	0	0	2	27	323,432
4	3	17	1	20	0	0	0	0	41	401,459
5	1	19	6	0	14	0	0	0	40	517,708
6	1	1	10	5	1	0	0	0	26	539,986
7	1	6	18	1	22	0	8	0	49	815,086
8	1	8	21	0	2	7	1	0	51	976,092
9	12	12	54	6	48	0	0	0	132	1,646,774
10	1	5	18	8	29	0	17	0	78	1,801,929
11	2	42	42	29	34	6	0	0	155	2,075,390
12	0	9	42	12	4	22	44	0	133	3,035,777
Total	29	140	217	99	165	35	84	2		

Table 12.2. Number of bins used to make wheat blends at Kansas elevators.

| Number of bins | Internal movement | | | | Outbound movement | |
| | Terminal elevator | | Country elevator | | All elevators | |
	Frequency	% of total	Frequency	% of total	Frequency	% of total
1	63	26.8	94	78.3	9	11.7
2	51	21.7	19	15.8	17	22.1
3	50	21.3	2	1.7	16	20.8
4	36	15.3	5	4.2	11	14.3
5	24	10.2			10	13.0
6	6	2.6			3	3.9
7	4	1.7			2	2.6
8	0				4	5.2
9	1	0.4			1	1.3
10					4	5.2
Total	235		120		77	

Economic Losses Attributed to Insects

Insects are objectionable by their mere presence in a food commodity, but also produce kernel damages, and may leave webbing or an undesirable odour in the grain. In the grain industry in the USA, wheat millers are very sensitive to the kernel damage caused by the internal-feeding insect species, i.e. the lesser grain borer, *Rhyzopertha dominica*, and the rice weevil, *Sitophilus oryzae*. The larvae of these species develop within individual kernels, consuming grain material and leaving cast exoskeletons and other insect filth inside a kernel of grain. When the adult emerges from a kernel of wheat, a distinctive orifice is formed. Kernels with this injury are called insect-damaged kernels (IDK) by industry and regulatory personnel. Each adult insect is associated with one IDK, and the density of IDK increases over time if the insect population is not suppressed. Even if the insects are killed by fumigation, the IDK levels remain the same.

Internally infested kernels lead to insect filth in the flour made from the wheat. Microscopic fragments of insect exoskeletons in flour are considered animal filth, and are limited by a bakers' contract specifications and by government regulators. To minimize the insect filth in the flour and to ensure that the miller can mill flour with fewer than 15 insect fragments per 50 g, flour millers limit the density of IDK in wheat through contract specifications and receiving norms. Typical limits are very strict, often five or fewer IDK per 100 g of grain, or about 0.14% or less by weight. The regulator's (Food and Drug Administration – FDA) actionable limit is 75 fragments per 50 g of flour. Similarly, the official grain-grading agency (Federal Grain Inspection Service (FGIS) of the Grain Inspection, Packers and Stockyards Administration – GIPSA) may condemn wheat containing more than 32 IDK per 100 g as being unfit for human consumption, and allowing it to be used only for animal feed or non-food industrial uses.

Insect-related costs in grain elevator operations in the USA are of two types (Hagstrum *et al.*, 1999). The first is penalties levied for insect presence and/or insect damage in grain by the purchaser or receiver. The second is the cost of managing insects to prevent the price penalties. In many cases, the cost of managing insects may be much larger than the cost of damage or penalties.

Penalties often take the form of a price discount, i.e. a lower price per unit of grain when grain containing live insects or insect-damaged kernels is received. Less apparent, but just as real, are the additional costs of locating an alternative market and of transporting the grain to that market when a shipment of grain is rejected for insect-related reasons at a destination. In many cases the costs are not apparent, such as when grain cannot be shipped to the preferred, higher-price market because of insect presence or damage, or when a supplier is quoted a lower price than a competitor because of his reputation for infested and/or damaged grain. Researchers have no way of quantifying several of these types of penalties or of separating the insect-related penalties from penalties related to other grain quality factors. For example, wheat that is desirable for its high bulk density and protein content would be less likely to be refused or discounted if an insect were found in the sample than would grain of lower quality in which the same type of insect was found. Similarly, it is difficult to quantify the value of the storage space occupied by insect-damaged grain as it waits to be blended into better-quality grain. In many cases, this space otherwise would generate storage revenue.

At the 'farm-gate' end of the grain marketing system, such as when the farmer delivers wheat to a country or terminal elevator, or a flour mill, or when wheat moves from country to terminal elevators, samples are taken and examined before the grain is accepted. Each receiver has acceptance criteria and discount scales for various quality factors, including the presence of live adult insects and the kernel damages caused by the insects. Nearer to the final consumer, the receiver is more likely to accept the official inspection certificate instead of investing in sampling and grading. A competent shipper may be able to avoid the price discounts while shipping large quantities of damaged grain by keeping the level of damage from affecting the grade. Thus, the costs to the shipper of the insect-related damage may be only those related to the cost of holding and blending the damaged grain. This common practice is the mechanism by which the grain trade in the USA deals with most IDK.

The cost of insect-control practices is similarly difficult to separate from the costs of activities that would be performed even if insects were not present. Where wheat is harvested wet, it must be dried and cooled to prevent damage by moulds. It must be transported to storage facilities and blended whether insects are present or not. Thus, the portion of the cost for these activities that should be assigned to insect pest management is arbitrary. Where wheat is stored in bins without aeration capability, it is turned – moved from one bin to another – to cool it. The fumigation is accomplished during the turn, but the cost of the fumigant is minor compared with the costs of electricity, manpower, capital depreciation and grain shrink (loss in volume of grain resulting from handling).

The grain sampling rates of 0.5 kg of wheat per 2000 to 3000 bushels used in the USA are adequate for characterizing the grain quality, but they are too low to provide much sensitivity to insect presence. This low sampling rate sometimes results in large numbers of insects being transported in the grain without detection. A case study by the authors at the Kansas terminal elevator illustrates the commercial impact of different sampling rates and methods. Researchers sampled for insects in wheat outbound to a 100-car train by taking a total of 31.7 kg of grain per railcar. The average insect density observed was 2.1 insects/kg. At the same time, official grain inspectors examined 1 kg of sample per railcar, using samples collected by automatic diverter samplers. They found sufficient insects to declare that 14 of the 100 railcars contained infested wheat. The shipper called for a re-inspection, which was performed by manually probing the grain surface in each suspect railcar. Based on this second examination, all lots that had been graded 'infested' were deemed to be 'not infested', so certificates without the 'infested' designation were issued. If the grain contained the same insect density when it came to rest in the railcars as it had in the elevator basement where the researchers' samples were collected, nearly 19 million insects were shipped in that unit train without a single grade certificate indicating the presence of insects. Some grain receivers rely on the grading certificate only in making their purchase, and would not have been advised of the insect presence.

Insect Population Trends

Since the grain-handling industry began adopting phosphine fumigation more than 50 years ago, managers in the US grain-handling firms generally have not looked to science to provide solutions to their insect pest management problems. This may in part be because previous scientific studies did not provide the information needed to improve pest management. Financial support from government and industry has been sufficient to establish research programmes on stored-product insect problems at only a few land-grant universities. Recently, researchers have provided information on insect populations at typical grain elevators in the USA that can improve pest management.

Sampling with a vacuum probe at elevators in Kansas and Oklahoma has shown that insect populations increased from June to October, then levelled off and declined as autumn and winter temperatures cooled the grain (see Fig. 12.1).

The primary insect pests of stored wheat are *Cryptoletes ferrugineus* (the rusty grain beetle), *R. dominica*, *S. oryzae*, *Tribolium castaneum*, (the red flour beetle) and the sawtoothed grain beetle, *Oryzaephilus surinamensis*. A computer simulation model correctly predicted that, with an immigration rate of 0.35 *R. dominica* per ton of wheat per day, *R. dominica* would increase from 0.1 to 3.5 insects per kg of wheat from 20 September to 14 December (see Fig. 12.2). Populations decreased in March, April and May primarily due to low grain temperatures. The immigration rate for a new model for elevators was 50% higher than that for the old model predicting insect population growth on farms. The immigration rates at these elevators were probably higher than normal and rates that are ten- or 100-fold lower are probably common.

Insects generally do not infest wheat in the field in Kansas or Oklahoma. However, some infested grain that has been stored on farms may be delivered to elevators along with newly harvested grain. In Kansas, probe trap catches indicate that insects present at elevators infest grain soon after it is stored in bins (Reed *et al.*, 2001). Insects first infest grain at the surface, and insect densities decrease with the depth

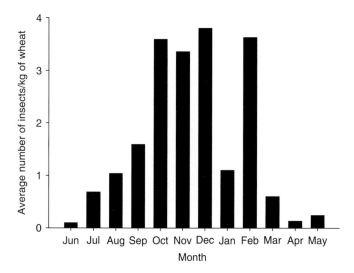

Fig. 12.1. Seasonal trends for average number of insects/kg sample of wheat for grain stored at elevators in Kansas and Oklahoma.

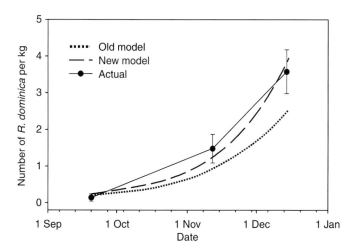

Fig. 12.2. Growth rate of *R. dominica* in Kansas and predictions of computer simuiation models (from Flinn *et al.*, 2004).

below the grain surface (see Fig. 12.3). *R. dominica* tended to move down during the grain storage period. This depth distribution pattern may be altered when grain is moved or when the grain from two or more bins is blended.

If the depth distribution of the insect populations is known, the insect density in the grain can be estimated at any point in the bin discharge. Wheat in discharge spouts has been shown frequently to contain a higher insect density than the bulk of

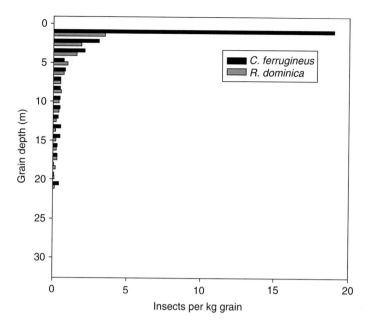

Fig. 12.3. Depth distribution of two species of insect pests in concrete bins at elevators in Kansas.

the grain, so the first grain discharged is likely to have a high insect count. If the grain surface is more densely infested than the rest of the mass, the point in the discharge when the majority of the insects are loaded out depends on whether the grain flow is of the funnel-flow, mass-flow or transitional type. The bin dimensions and geometry (Reed, 2006) influence this flow pattern.

In tall bins, mass flow, with every kernel moving towards the discharge spout at the same velocity, occurs until the distance between the grain surface and bin floor is 1.5 to 3.5 times the bin diameter. The grain near the bottom is discharged first. The flow pattern converts to funnel flow when the distance between the grain surface and bin floor is 1.5 to 3.5 times the bin diameter, and then only the grain directly above the discharge spout and a thin layer from the grain surface are in motion.

The species composition of insects in grain at elevators in Kansas was studied by taking several types of grain samples. Discharge spout (DS) samples were taken by opening the slide gate at the bottom of the bin and allowing about 10 kg of grain to fall on the stationary belt. A 3.9 l (3 kg) sample was taken from this grain. When a pile of residual grain was found outside bins or in an empty bin, a 3.9 l sample was collected. If the pile was smaller than 3.9 l, the entire pile was taken as a sample of grain residue. Grain samples of 3.9 l were collected with a vacuum probe as the probe was pushed through a 1.2 m layer of grain. For truck samples, 3.9 l of grain were removed from the grain stream as the grain was dumped from the truck.

The species composition in grain residues was found to be quite different from that in the wheat stored in bins (see Table 12.3). *R. dominica* density was low in grain

Table 12.3. Species composition of insect pests in various types of grain samples from Kansas elevators (from Reed et al., 2003; Arthur et al., 2006).

Type of sample, location or time grain sampled[a]	Percentage of each insect pest species				
	R. dominica	S. oryzae	C. ferrugineus	T. castaneum	O. surinamensis
DS samples after 89 days, bins not cleaned	1.1	47.8	14.6	35.0	1.5
Residual grain outwith bins	1.4	46.7	36.7	13.8	1.3
Maize residues from empty bins	2.1	25.2	55.3	17.4	0.1
DS samples after 49 days, bins not cleaned	5.0	50.4	8.7	35.6	0.3
DS samples January–June	8.1	4.5	78.7	1.1	7.7
Grain residues from empty bins	9.0	32.4	45.0	10.6	3.0
DS samples June–December	9.3	14.3	53.9	15.6	6.7
Residual wheat from empty bins	14.0	32.1	40.7	8.2	5.0
DS samples after 89 days, clean bins	19.1	32.7	32.7	4.4	11.1
DS samples after 49 days, clean bins	22.6	60.3	14.4	2.1	0
Vacuum probe samples January–June	29.3	5.7	38.6	18.7	7.7
Vacuum probe samples June–December	35.3	7.9	38.7	6.3	11.3
Vacuum probe samples 0–3.7 m depth	44.2	1.0	35.7	19.0	0.1
Truck samples for grain from elevators	46.0	7.3	23.1	19.8	3.8
Truck samples for grain from farms	52.4	0.4	8.8	36.0	2.4
Vacuum probe samples at 4–12 m depth	83.5	1.1	7.0	7.7	0.2

[a] See text for explanation of the types of samples, location or time grain sampled.

residues outwith bins, grain residues in empty bins and the first grain passing through the discharge spout, but high in wheat stored in bins. *S. oryzae* populations were low in the wheat stored in bins, but high in grain residues outwith bins, grain residue in empty bins and the first grain passing through the discharge spout. Insect densities in the more accessible grain residues inside empty bins and outside the bins, and the first grain passing through the discharge spout, were ten times higher than those in the wheat stored in bins (Reed *et al.*, 2003; Arthur *et al.*, 2006). The large numbers of insects in the first grain discharged from the discharge spout may be the result of insects from grain residues infesting grain stored near the bottom of the bin.

Cleaning the empty bins before refilling resulted in a lower insect density in the discharge spout sample from new grain stored in these bins for a period of up to 3 months. Insect densities in grain residues outside bins were higher in samples collected at the ground or subterranean level (discharge spouts, residues in empty bins and spills in the basement or tunnel) than at the top of the elevator, where much of the grain being conveyed is new grain being loaded into bins. Investigators concluded that routine sanitation practices, including prompt clean-up of spills, thorough cleaning of empty bins, and periodic flushing of discharge spouts, should greatly reduce the resident population of stored-grain insects at elevators.

The numbers of *R. dominica* and *T. castaneum* in wheat delivered from farm to country or terminal elevators were higher and the numbers of other species were lower than in wheat delivered from country elevators to terminal elevators. The overall mean insect density in wheat delivered from farms (4.18 ± 1.38 SE, $n = 909$) was higher than the overall mean insect density in wheat delivered from a country elevator (0.50 ± 0.06 SE, $n = 4554$). Higher mean insect densities in farm-stored wheat may in part be a result of the grain being stored in smaller bins where grain is more accessible to insects. In both cases, 80% of the wheat samples did not have insects. In bins that had received wheat at harvest time, the wheat between 12 m below the surface and 1 m up from the bottom was generally inaccessible to insects until the grain was moved. During several months of storage, this grain had a lower insect density than grain closer to the top and bottom surfaces. In wheat stored in elevator bins in Kansas and Oklahoma, *C. ferrugineus* was more prominent in the top 3.7 m, while *R. dominica* was more prominent at 4–12 m.

Natural Enemies of Stored-grain Insects

Hymenopteran insect parasitoids of pests were found in the wheat stored in bins at 13 out of 16 elevators in Kansas (Reed *et al.*, 2003). *Cephalonomia waterstoni* and its host, *C. ferrugineus*, were most prevalent in grain samples from wheat stored in bins or grain residues in empty bins, while *Anisopteromalus calandrae* and its host, *S. oryzae*, were most prevalent in samples from grain residues found outwith bins (Arthur *et al.*, 2006; Table 12.4). *Theocolax elegans*, a parasitoid of *R. dominica*, and *Habrobracon hebetor* Say, a parasitoid of the Indianmeal moth, *Plodia interpunctella*, were found in much smaller numbers.

Parasitoids were found in the grain residues outwith bins at all nine elevators from which grain residues samples were collected, and they were found in 1.6–9.4%

Table 12.4. Species composition of beneficial insects in various types of grain samples from elevators in Kansas.

Location of sample	Percentage of each beneficial insect species			
	H. hebetor	*A. calandrae*	*T. elegans*	*C. waterstoni*
Grain stored in bin	10.39	14.77	1.17	73.67
Grain residue from empty bin	7.40	12.50	1.00	79.10
Grain residue from outwith bin	0	89.93	3.90	6.17

of the grain samples taken at any one elevator. Because of the small size of parasitoids and their tendency to leave grain samples, their prevalence was probably underestimated. However, the numbers and prevalence of parasitoids indicate that they are important in reducing the numbers of insect pests at elevators.

Current Insect Pest Management Practices at Elevators

Aeration, sanitation and fumigation are the primary methods used to manage stored-grain insects at elevators in the USA (Hagstrum *et al.*, 1999). Treating grain with a residual insecticide or inert dust as a protectant is less common at elevators (Subramanyam, 2003) than on farms. In a 1997 survey based upon 1956 responses to a questionnaire from elevator managers in 14 states, representing 82% of the wheat produced in the USA, malathion was used on 1.5% of the wheat, chlorpyrifos-methyl was used on 1.4% and inert dust was used on 0.2%.

One potential disadvantage of protectants is that grain may be treated more than once, resulting in insecticide residues exceeding tolerances of 8 ppm for malathion and 6 ppm for chlorpyrifos-methyl. Pesticide data programme residue analysis of 1563 wheat samples from 29 states between 1995 and 1997 showed that 0.002–7.600 ppm of malathion were found on 68–71% of samples, that 0.002–3.300 ppm chlorpyrifos-methyl residues were found on 52–73% of samples and that 40–48% of samples had residues of both insecticides. These residues must be the results of protectants being applied on farms before wheat is delivered to elevator.

The cost of aeration systems is often justified as a means of managing grain moisture, especially in autumn crops. In the Kansas/Oklahoma study, 29% of the concrete bins and 44% of the steel bins at elevators in Kansas, and 18% of the concrete bins and 50% of the steel bins at elevators in Oklahoma were equipped for aeration. Automatic aeration controllers can be used to cool wheat during the summer and autumn by running fans only when air temperature is appropriate for the cooling of grain.

Aeration was tested as a way of managing insects in stored wheat. In a Kansas study in a large upright bin equipped with 20-HP, positive-pressure aeration fan located near the bin floor, a 5-HP fan extracting air on the roof and an aeration controller, two aeration fronts were moved through the grain cooling it from 28.9 to 18.3°C by the end of October. The principal insect pest species cannot survive and reproduce at this temperature. The cost of aeration was US$188 per bin, or US$0.004 per bushel.

Sanitation programmes remove residual grain or grain dust so that insect populations cannot reproduce in them. Sanitation alone cannot eliminate insect populations, but it does reduce populations and may improve the effectiveness of other pest management methods. Inspection of elevators in Kansas has shown that spilled grain residues usually consist of only a few bushels per elevator and that these usually were cleaned up in less than a week. However, the grain residues were sometimes swept into the nearest bin without killing the insects infesting them.

Fumigation of grain with phosphine to kill insect pests is most effective when a lethal dose is maintained throughout the commodity and storage structure for the period required to kill the insects (normally 3–5 days). To maintain this lethal concentration for several days, the storage structure must be sealed or fumigant must be added from pressurized cylinders to replace the fumigant that leaks out of the structure. Studies showed that the fumigation was often carried out on a calendar schedule instead of when it was needed or when it would have been most effective. Our AWPM project found that wheat was fumigated throughout the storage period in Kansas (see Fig. 12.4).

However, only a small portion of the wheat stored in the nearly 486,000 t (18 million bushels) of storage capacity at 13 elevators was generally fumigated each month. Some of the wheat stored for a long time may be fumigated two or three times. The peak number of fumigations in September was a result of managers fumigating wheat before the autumn harvest of maize, grain sorghum and soybeans. In a 1997 nationwide survey, elevator managers in the USA indicated that phosphine was

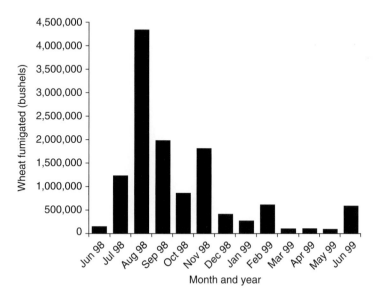

Fig. 12.4. Total bushels of wheat at 13 elevators in Kansas fumigated each month, June 1998–June 1999.

used on only 11.5% of wheat (Subramanyam, 2003). Phosphine fumigation costs US$0.33 to 0.43/ton when probed into the grain stored in a steel bin, and US$0.75 to 1.10/ton when fumigant was added to the grain stored in concrete bins by moving the grain to another bin (Hagstrum and Subramanyam, 2006). The cost of moving grain is US$0.48 to 0.67/ton.

Phosphine fumigants leave no pesticide residue on grain, so insects from other bins and grain residues can infest the grain as soon as the fumigant has dissipated and when the ambient air temperature is high enough for insect movement. Also, because of the difficulty of sealing bins, all of the insects are not killed during a fumigation of a typical elevator bin. The densities of the major insect pest species in wheat stored at two elevators in Kansas, where all of the bins were fumigated, were reduced by 97–99%. In Oklahoma, no insects were detected after fumigation in eight out of 11 bins, but insects were detected in the other three bins (Phillips *et al.*, 2001). These insect pest populations will often recover after fumigation.

Maintaining fumigant concentrations is most difficult near the grain surface, where insect densities are highest (see Fig. 12.3). Studies (Reed, 2006; Flinn *et al.*, 2007c) have shown that the optimal method of determining where to apply the fumigant when grain is turned from one bin to another is based on the difference between grain temperature and outside air temperature. When the post-turn grain temperature and the air temperature during the fumigation period are similar, distributing the fumigant evenly throughout the full depth of grain provided the greatest chance of success. When the grain was significantly warmer than the average outside air temperature during the fumigation period, applying the fumigant to grain loaded into the bottom half of the bin was most successful because air currents moved the fumigant towards the top of the grain mass.

Fumigating only the bins in which grain has high levels of insect infestation will minimize the use of pesticides, the cost of pest management and the rate at which insects become resistant to pesticides. Minimizing the use of pesticides in the USA is important because insects have already started to become resistant (Zettler and Cuperus, 1990). Timing of fumigation is very important. Fumigating too early can result in insect populations recovering before grain cools in the autumn and winter, and a second fumigation will be needed. Fumigating too late can result in substantial economic losses attributable to insects. Killing insects resistant to pesticides by aerating to cool the grain in the autumn and winter will also slow the development of resistance.

Areawide Pest Management

Areawide pest management (AWPM) reduces the overall insect pest levels over time by managing them over a wide area. Stored-grain insect pests are very mobile and can quickly reinfest grain that has been disinfested (Hagstrum and Subramanyam, 2006). An example of a successful AWPM programme is the one implemented for the central bulk grain handling organization in Queensland, Australia following two storage seasons (1989–1991) during which approximately 60% of storages were infested (Bridgeman and Collins, 1994).

The widespread use of the residual insecticides bioresmethrin and fenitrothion as grain protectants in Australia had led to complacency among managers, and sanitation had gradually declined over several seasons because of heavy reliance upon these residual insecticides. According to Bridgeman and Collins, the development of insecticide resistance, inadequate sealing of structures for fumigation and the practice of receiving or shipping grain from a storage facility before fumigation had been completed also contributed to poor insect pest management.

The project addressed these issues by developing written standards for sanitation, providing training and conducting periodic sanitation audits. Amorphous silica was used instead of residual insecticides to treat storage facilities and grain was cooled by aeration. Additional sealing was carried out to increase the effectiveness of the fumigation. Farmers and private grain traders were encouraged to use pest management methods other than residual insecticides and to deliver residue-free grain. An insect-trapping programme was developed to provide early warning of insect problems so that the timing of insect pest management could be optimized.

As a result of the programme, the percentage of storage facilities that were infested was reduced from 60 to 16. The percentage of storages in which residual insecticide residues were detected decreased from 90 to 30, and the cost of insect pest management was reduced from A$1.50 to 0.60/ton (approximately 37 bushels) of wheat.

During the 2000–2001 and 2001–2002 crop years, all of the wheat at 28 elevators in Kansas and Oklahoma was sampled for insects with a vacuum probe at a sampling rate of 0.07–0.13 kg/ton (Flinn *et al.*, 2003a, b, 2007a). Insect numbers in grain samples collected by vacuum probe from the top 12 m (40 feet) of grain were highly correlated with the insect numbers in grain samples taken as the bin was unloaded ($r^2 = 0.79$) (Reed *et al.*, 2001). Also, 96% of *C. ferrugineus* and 94% of *R. dominica* were found in the top 12 m of the grain stored in bins (see Fig. 12.3). Thus, the vacuum probe provided a convenient and reliable method of routinely sampling the grain for insects without having to move the grain.

Vacuum probe sampling of grain stored in bins at elevators for insects cost US$0.0092 per bushel of grain at a sampling rate of 0.07–0.13 kg/ton (B. Adam, 2006, personal communication; Adams *et al.*, 2006). This includes the labour cost to set up and take down the sampling equipment (2.5 h), taking grain samples (2 ± 1 min per sample), sieving insects from the grain sample (0.4 ± 0.3 min per sample) and counting the numbers of each insect species in the grain sample (3.01 ± 0.93 min per sample). This also includes the purchase of a US$8000 vacuum probe sampler with an expected lifespan of 10 years. If the cost of the vacuum probe is not included, sampling costs US$0.002/bushel.

Decision support software (STORED GRAIN ADVISOR PRO, http://ars.usda.gov/npa/gmprc/bru/sga, Flinn *et al.*, 2007b) was developed for wheat stored at elevators. This program uses information on insect density from vacuum probe grain samples to make decisions for each bin at an elevator. It also uses a computer simulation model to forecast the insect-related risks based upon the current estimates of insect density, grain temperature and grain moisture. Risk analysis is presented graphically to the elevator manager as a bin layout diagram. The manager is also given a printed report with insect pest management recommendations and economic analysis for each bin.

For the 533 bins at 28 elevators in Kansas and Oklahoma sampled to a depth of 12 m every 6 weeks by vacuum probe, decision support software was used to predict which bins were at low risk of economic losses attributable to insects (< 2 insects/kg predicted in 1–2 months), moderate risk (2–10 insects/kg predicted in 1–2 months) or high risk (> 2 insects/kg when sampled and 10 insects/kg predicted in 1–2 months). For bins with high risk, fumigation was recommended, and for the bins with low or moderate risk, sampling again in 6 weeks was recommended. STORED GRAIN ADVISOR PRO failed to predict when grain was at a high risk for only two bins, and in both cases the insect density was high only near the grain surface, suggesting recent insect immigration. Probe traps could be used to detect these re-infestations near the surface (Toews *et al.*, 2005).

Sampling of the 533 bins discussed above showed that only a small portion of the bins at each elevator needed to be fumigated at any one time (see Table 12.5). The percentage of bins needing fumigation increased from 1.7% in June to 19% in October, and then tended to decrease from November to May. This reduction was perhaps the result of heavily infested wheat being fumigated by elevator managers who detect insects by monitoring grain temperature for evidence of heating.

Because the percentage of bins at an elevator that needed to be fumigated varied from 0 to 60 and averaged 10, sampling to locate infested bins was cost effective. Early fumigation of the grain stored in these few bins may reduce the overall insect infestation at elevators. The remaining bins at an elevator may not need to be fumigated at all, because fumigating the few bins that had high insect densities earlier should reduce insect immigration into the other bins and thus prevent insect density in these bins from reaching unacceptable levels. Survival of the natural enemies in the bins that are not fumigated may also reduce insect populations. Thus, risk analysis software can improve pest management by predicting when insect pest populations will reach economic injury levels and by reducing the amount of grain fumigated. The areawide, sampling-based pest management approach can be used cost-effectively at a single elevator (each bin representing a field and all of the bins representing a large geographical area) as well as throughout the wheat marketing system.

Fumigating the grain in all of the bins at an elevator storing 19,048 t (700,000 bushels) of wheat would cost US$14,000. However, when the elevator manager knows that only three out of 30 of these bins are likely to have insect densities of ≥ 2 insects/kg during the following 2 months, fumigating these three bins would cost only US$1400 and the cost of fumigation would thus be reduced by US$12,600.

Table 12.5. Number of bins at each elevator in which wheat required fumigation (data from 28 elevators in Kansas and Oklahoma).

Number out of 100 bins	Frequency (%)
0–10	71
11–20	12
21–40	13
41–60	4
>60	0

Commercial Scouting and Consulting

With vacuum probe sampling of all bins and the decision support software discussed above, a private consulting company has provided scouting services to more than 70 elevators in Kansas, Oklahoma and Nebraska during the past 5 years (Flinn *et al.*, 2007b). The sampling programme has improved insect pest management by ensuring that fumigation is done at the time when it will be most effective. Scouting may have helped to reduce the average incidence of insect-damaged kernels by as much as 24%. The average number of IDK was 2.5 per 100 g of wheat during the first year of scouting and 1.9 per 100 g of wheat during the second year.

Initially, managers often did not follow the recommendations of scouting reports, but after receiving several reports many of the managers started following the recommendations. Managers have used the improved grain quality information from the scouting report for the full bin depth to better merchandize their grain. Information about current insect infestation levels and forecasts of future insect infestation levels allow elevator managers first to sell the grain that is most likely to need fumigation if it would have been kept for another month or more.

Conclusion

When wheat is mixed with wheat from other locations as it moves through the grain-marketing system insect infestation can be spread to larger quantities of wheat, increasing the overall cost of insect pest management. In Kansas and Oklahoma, insect infestations are currently managed primarily by calendar-based fumigation of all of the wheat at an elevator. Grain is not sampled to determine the most effective time for fumigation. Insufficient sealing and poor timing of the fumigations reduce the cost-effectiveness of fumigation. The low sampling rates used for grain inspection result in large numbers of insects being shipped with the wheat, thus spreading the insect infestation throughout the marketing system.

Calendar-based fumigation of all of the grain at an elevator is not cost-effective because usually only a percentage of the bins require fumigation at any one time. Delaying the fumigation, when possible, until the autumn has several benefits: (i) the grain can be cooled with aeration after an autumn fumigation, which decreases subsequent population growth; (ii) insect immigration rates into grain bins in the autumn and winter are lower than during the summer; and (iii) the necessity for a second fumigation is greatly reduced because of (i) and (ii). Areawide, sampling-based pest management at each elevator or across the grain-marketing system using decision-support software can minimize the cost and maximize the effectiveness of insect pest management. In addition, it should reduce the risk of economic losses due to insects, the amount of wheat that is fumigated and the frequency of fumigation.

References

Adams, B.D., Phillips. T.W. and Flinn, P.W. (2006) The economics of IPM in stored grain: why don't more grain handlers use IPM? In: Lorini, I., Bacaltchuk, B., Beckel, H., Deckers, D.,

Sundfeld, E., dos Santos, J.P., Biagi, J.D., Celaro, J.C., Faroni, L.R.D'A., Bortolini, L. de O.F., Sartori, M.R., Elias, M.C., Guedes, R.N., da Fonseca, R.G. and Scussel, V.M. (eds) *Proceedings 9th International Working Conference on Stored Product Protection*, Brazilian Post-harvest Association, Campinas, Brazil, pp. 3–12.

Arthur, F.H., Hagstrum, D.W., Flinn, P.W., Reed, C.R. and Phillips, T.W. (2006) Insect populations in grain residues associated with commercial Kansas grain elevators. *Journal of Stored Product Research* 42, 226–239.

Bridgeman, B.W. and Collins, P.J. (1994) Integrated pest management in the GRAINCO, Queensland Australia, storage system. In: Highley, E., Wright, E.J., Banks, H.J. and Champ, B.R. (eds) *Proceedings of the 6th International Working Conference on Stored-Product Protection*, CAB International, Wallingford, UK, pp. 910–914.

Flinn, P.W., Hagstrum, D.W., Reed, C. and Phillips, T.W. (2003a) United States Department of Agriculture-Agricultural Research Service stored-grain areawide integrated pest management program. *Pest Management Science* 59, 614–618.

Flinn, P.W., Hagstrum, D.W., Reed, C. and Phillips, T.W. (2003b) Areawide integrated pest management program for commercial grain stores. In: Credland, P.F., Armitage, D.M., Bell, C.H., Cogan, P.M. and Highley, E. (eds) *Advances in Stored Product Protection*. CAB International, Wallingford, UK, pp. 99–102.

Flinn, P.W., Hagstrum, D.W., Reed, C. and Phillips, T.W. (2004) Simulation model of *Rhyzopertha dominica* population dynamics in concrete grain bins. *Journal of Stored Product Research* 40, 39–45.

Flinn, P.W., Hagstrum, D.W., Reed, C. and Phillips, T.W. (2007a) Area-wide IPM for commercial wheat storage. In: Vreysen, M.J.B., Robinson, A.S. and Hendrichs, J. (eds) *Area-wide Control of Insect Pests: From Research to Field Implementation*, Springer, Dordrecht, Netherlands, pp. 239–246.

Flinn, P.W., Hagstrum, D.W., Reed, C. and Phillips, T.W. (2007b) Stored Grain Advisor Pro: decision support system for insect management in commercial grain elevators. *Journal of Stored Product Research* 43, 375–383.

Flinn, P.W, Reed, C., Hagstrum, D.W. and Phillips, T.W. (2007c) Seasonal and spatial changes in commercial elevator bins: implications for phosphine fumigation. *International Miller* 2006–2007, 36–40.

Hagstrum, D.W. and Heid Jr., W.G. (1988) US wheat marketing system: an insect ecosystem. *Bulletin of the Entomological Society of America* 34, 33–36.

Hagstrum, D.W. and Subramanyam, Bh. (2006) *Fundamentals of Stored-Product Entomology*. AACC International, St Paul, Minnesota.

Hagstrum, D.W., Reed, C. and Kenkel, P. (1999) Management of stored wheat insect pests in the USA. *Integrated Pest Management Review* 4, 127–142.

Phillips, T.W., Doud, C.W., Toews, M.D., Reed, C., Hagstrum, D.W. and Flinn, P.W. (2001) Trapping and sampling stored-product insects before and after commercial fumigation treatments. In: Donahaye, E.J., Navarro, S, and Leesch, J. (eds) *Proceedings of the International Conference on Controlled Atmosphere and Fumigation in Stored Products*, Executive Printing Services, Clovis, California, pp. 685–696.

Reed, C.R. (2006) *Managing Stored Grain to Preserve Quality and Value*. AACC International, St Paul, Minnesota.

Reed, C., Hagstrum, D.W., Flinn, P.W. and Phillips, T.W. (2001) Use of sampling information for timing fumigations at grain elevators. In: Donahaye, E.J., Navarro, S. and Leesch, J. (eds) *Proceedings of the International Conference on Controlled Atmosphere and Fumigation in Stored Products*, Executive Printing Services, Clovis, California, pp. 699–705.

Reed, C.R., Hagstrum, D.W., Flinn, P.W. and Allen, R.F. (2003) Wheat in bins and discharge spouts, and grain residues on floors of empty bins in concrete grain elevators as habitats for stored-grain beetles and their natural enemies. *Journal of Economic Entomology* 96, 996–1004.

Subramanyam, Bh. (2003) Pesticide residue an important issue. *Milling Journal* 4, 42–44.

Toews, M.D., Phillips, T.W. and Payton, M.E. (2005) Estimating populations of grain beetles using probe traps in wheat-filled concrete silos. *Environmental Entomology* 34, 712–718.

Zettler, J.L. and Cuperus, G.W. (1990) Pesticide resistance in *Tribolium castaneum* (Coleoptera: Tenebrionidae) and *Rhyzopertha dominica* (Coleoptera: Bostrichidae) in wheat. *Journal of Economic Entomology* 83, 1677–1681.

13 Aphid Alert: How it Came to be, What it Achieved and Why it Proved Unsustainable

EDWARD B. RADCLIFFE,[1] DAVID W. RAGSDALE,[1] ROBERT A. SURANYI,[2] CHRISTINA D. DIFONZO[3] AND ERIN E. HLADILEK[4]

[1]*Department of Entomology, University of Minnesota, St Paul, Minnesota, USA*
[2]*MGK® McLaughlin Gormley King Company, Minneapolis, Minnesota, USA*
[3]*Department of Entomology, Michigan State University, East Lansing, Michigan, USA*
[4]*Department of Entomology, University of Kentucky, Lexington, Kentucky, USA*

Introduction

'Aphid Alert' was the name used to identify a series of research and outreach initiatives undertaken from 1992 to 2003, and in some instances since, to address potato virus problems in seed potato production in the Northern Great Plains (NGP) of the USA, in particular north-western Minnesota and eastern North Dakota. Aphid Alert was adopted from the name of a pest management advisory newsletter sent to Minnesota and North Dakota seed potato growers in 1994, and again from 1998 to 2003.

The name found popular acceptance and was applied, even retroactively, to a series of related research/outreach activities. This chapter will focus primarily on the areawide aphid-trapping network operated by the University of Minnesota from 1992 to 1994, and again from 1998 to 2003. Data presented here on potato seed lot rejections due to potato viruses were provided by the Minnesota Department of Agriculture Seed Potato Certification Program (courtesy of Willem Schrage, potato programme supervisor). Data presented here on aphids (reported as numbers or percentages of total captures) are from the subset of traps that were located in the NGP portion of the network (see Fig. 13.1).

Virus Management in Seed Potato Production

Access to high quality, disease-free seed potatoes has been described as 'the single most important integrated pest management practice available to potato growers'

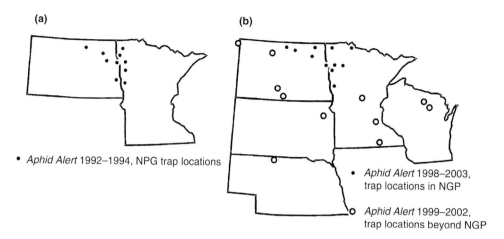

Fig. 13.1. Aphid Alert trap locations, 1992–1994 and 1998–2003.

(Gutbrod and Mosley, 2001) and is essential for successful commercial production. For almost a century, state-regulated seed potato certification programmes have been the primary mechanism for ensuring the cultivar integrity and tuber health of US seed potatoes (Rieman, 1956; Franc, 2001). Seed potato lots can be downgraded or rejected for recertification for a myriad of defects, but aphid-transmitted potato viruses far exceed all others.

Nearly all seed potato certification programmes use a limited-generation production system. Typical production systems permit field increase for five to eight generations. In modern practice, seed potato increase is initiated with tissue culture-derived seedlings tested by enzyme-linked immunosorbent assay (ELISA) to assure freedom from viruses. State seed certification programme personnel inspect seed potato increase fields periodically during the growing season, and representative samples of harvested tubers are indexed in a winter grow-out for virus or other defects. Tolerances for recertification are stringent for all generations (typically ranging from 0.0 to 1.0% total virus) and are usually relaxed incrementally with successive generations of increase.

Potato leafroll virus (PLRV) and potato virus Y (PVY) are both aphid transmitted (for a discussion of vector biology see Robert and Bourdin, 2001; Radcliffe and Ragsdale, 2002). Transmission of PLRV is persistent and circulative in the body of the vector (Nault, 1997). PLRV can be acquired and transmitted only by aphid species that phloem feed on potato, but not all potato-colonizing species transmit PLRV. *Myzus persicae*, green peach aphid, is the most cosmopolitan, abundant and efficient vector of PLRV (Ragsdale *et al.*, 2001). All other aphid-transmitted potato viruses are non-persistent and borne on the insect's mouthparts (stylets). Many aphid species, including species that do colonize potato, are capable of transmitting PVY. *M. persicae* is the most efficient vector, but other potential vector species are often much more abundant and thus of greater importance in PVY spread in particular locations or years.

Association of the spread of aphid-transmitted potato viruses with aphid flight activity is well documented (Boiteau and Parry, 1985; Sigvald, 1989, 1992; Halbert *et al.*, 1990; Pickup and Brewer, 1994). Therefore, trapping networks designed to monitor activity of aphid vectors have been used as decision tools for management of viruses in seed potatoes. In the early development of aphid-trapping networks in seed potato production, the focus was on *M. persicae* (Hille Ris Lambers, 1972). Once researchers recognized the importance of other less efficient but more abundant virus vectors to PVY spread (van Harten, 1983; Harrington *et al.*, 1986; Sigvald, 1987; Harrington and Gibson, 1989; Heimbach *et al.*, 1998), those operating aphid-trapping networks began routinely monitoring these species also. Over the past 50 years, aphid-trapping networks have existed, at least temporarily, in many countries. The oldest and most extensive of these is the EXAMINE (EXploitation of Aphid Monitoring systems IN Europe) network, which presently operates more than 70 suction traps (after the Rothamsted design), 12.2 m tall, in 19 European countries (Harrington and EXAMINE Consortium, 2007).

Generally, aphid-trapping networks are intended to monitor flight of vector species on a regional basis. At any particular location, e.g. an individual farm, flight activity may not be detected because aphid populations are low or the trap site not representative. Other limitations are that the traps may not be monitored frequently enough and that expertise in aphid taxonomy is required to identify the captured aphids to species. However, traps can be effective in detecting sudden influxes of winged aphids into seed potatoes from other crop or weed hosts in the immediate vicinity. This information can be used to time application of foliar aphidicide or crop oil, or in early vine kill where tuber development and yield permit.

Seed Potato Production in Minnesota and North Dakota

Minnesota and North Dakota ranks as the third largest potato-growing region (180,000 ha) in the USA, producing over 2 million t of potatoes per year (USDA/NASS, 2007). Seed potatoes represent an important component, 15–20%, of the on-farm value of this production. Minnesota and North Dakota have had the reputation of being especially favourable for seed potato increase, in part because of a 'northern vigour' imparted by growing conditions, but mostly because historically the region seemed relatively free of aphid-transmitted potato viruses.

In 2006, Minnesota and North Dakota produced 21.7% (12,300 ha) of the seed potatoes certified in the USA (NPC, 2006). However, as recently as 1990, Minnesota and North Dakota produced 31.4% (22,500 ha) of US certified seed potatoes (Slack, 1993). The number of Minnesota and North Dakota farming operations growing seed potatoes has declined to less than half the number that did so in 1990. Many of those who quit were second- and third-generation seed potato growers. The major contributing factor in their decision to quit growing seed was the persistent occurrence of aphid-transmitted viruses, especially PVY. Frequent seed lot rejections make production economically unsustainable, since investment in early-generation seed production cannot usually be recouped before at least three generations of increase.

Aphid/Potato Virus Research at the University of Minnesota

Research on insect-transmitted plant diseases has a long tradition at the University of Minnesota. Entomologist A.A. Granovsky was hired to collaborate with pathologist

J.D. Leach in research on insect-transmitted plant pathogens, and they first co-taught a course on the subject in 1931. In later years, Granovsky became an expert aphid taxonomist and assembled a large reference collection, still maintained at the University of Minnesota. That collection was to prove invaluable in the implementation of Aphid Alert, being used by project personnel to hone skills in aphid identification and as a reference for confirming identifications of less common aphid species.

Much of the early University of Minnesota research on potato viruses (i.e. prior to Aphid Alert) was focused on PLRV and *M. persicae*. With the emergence of PVY as a major concern to the Minnesota/North Dakota seed potato industry in the mid-1980s, attention shifted to PVY. Our first large-scale field experiments on PVY were carried out at Rosemount, Minnesota in 1991. This location, far from any seed potato production, was selected because of seed grower concerns that research using PVY inoculum sources might present a risk to nearby seed potato production. Insecticide efficacy trials were conducted, with all foliar sprays applied by helicopter to avoid risk of mechanical spread of virus by machinery moving through the plots. None of the aphidicides tested in 1991, all products then commonly used on potato, reduced PVY spread. However, in other experiments we found no evidence of mechanical transmission of PVY in seed handling or from machinery moving through the field during cultural operations, thereby allaying that grower concern (Banttari, 1993). Lack of mechanical transmission of PVY has since been confirmed in New Brunswick (Sturz *et al.*, 2000).

Most of our post-1991 aphid/virus research was done in collaboration with the Minnesota/North Dakota potato industry, on grower-owned seed farms or university experiment stations near NGP seed potato production. Significant funding for this research was provided by: (i) the Red River Valley Potato Growers Association (a commodity organization representing NGP potato growers and now known as the Northern Plains Potato Growers Association); (ii) the Minnesota Area II Potato Promotion and Research Council; (iii) Minnesota Certified Seed Potato Growers Association; and (iv) the North Dakota Seed Potato Growers Association. Minnesota Agricultural Experiment Station and competitive grants provided additional funding from USDA/CSREES, NC-IPM. Cooperators, among others, in this effort included the Minnesota Extension Service-IPM Program, plant pathologists at North Dakota State University, the state seed potato certification programmes in Minnesota and North Dakota and a number of leading seed potato growers in Minnesota and North Dakota.

Aphid Alert, 1992–1994

Knowledge of which vector species are present, their abundance and comparative efficiency in transmitting PVY, is necessary to design area-specific management practices to limit spread of the virus. The Aphid Alert trapping network was established to provide this information for the NGP seed potato industry. The first iteration of the Aphid Alert network, operated from 1992 to 1994, was primarily research driven and only secondarily intended for providing seed producers with current-season pest management advisories. Trapping was done each year on five to eight owner-operated seed potato farms and one or two university experiment stations (see Fig. 13.1a).

Traps consisted of green and yellow ceramic tiles (Dal-Tile, Dallas, Texas) placed individually in 1.4 l plastic containers (Servin Saver, Rubbermaid, Wooster, Ohio) partially filled with a 50:50 mixture of technical grade propylene glycol and water (DiFonzo *et al.*, 1997). Four traps, two green and two yellow, were used at each location. The green tiles were intended to mimic foliage and provide an unbiased measure of aphid landing rates (Irwin, 1980), whereas yellow was selected because it is attractive to certain aphid species, especially *M. persicae* (Eastop, 1955). Traps were emptied weekly and the collected aphids counted and identified to species, but species identification of other than *M. persicae* was not completed until after the growing season.

To determine the phenology of PVY spread, healthy, potted, indicator plants (equal numbers of *Physalis pubescens* (= *floridana*) and potato) were exposed on bait boards to aphid landing for 1-week intervals at each aphid trapping site (DiFonzo *et al.*, 1997). The bait boards were 1.9 cm plywood with an area of 1.2 m^2, painted either yellow to attract *M. persicae* or brown to mimic soil. One board of each colour was placed at each site. Eight indicator plants were arranged in a circle on each board, with a PVY-infected potato plant placed in the centre. We tested the hypothesis that aphids would land by chance on the infected plant, acquire PVY and then transmit the virus to the adjacent indicator plants.

After exposure, the indicator plants were moved to an aphid-proof screen cage, held for 6 weeks and then tested for PVY by ELISA. This identified the time frame during which PVY transmission occurred most frequently. Comparison of these data with aphid captures in the tile traps was used to infer association of PVY spread with the abundance of specific vector species. For example, across the 3 years, 89% of PVY transmissions to indicator plants occurred between 8 July and 19 August, suggesting that cereal aphids – e.g. bird cherry-oat aphid, *Rhopalosiphum padi*; corn leaf aphid, *R. maidis*; greenbug, *Schizaphis graminum*; and English grain aphid, *Sitobion avenae* – were important PVY vectors in the NGP.

Aphid Alert newsletter

A 1994 USDA/CSREES North Central IPM (NC-IPM) grant funded a University of Minnesota 'research/demonstration' project on use of crop borders to reduce PVY spread in seed potatoes (DiFonzo *et al.*, 1996). A component of this project was distribution of a printed newsletter, *Aphid Alert*, mailed weekly from mid-summer to harvest to all Minnesota and North Dakota seed potato-farming operations. This newsletter was originally envisioned as a vehicle to promote use of crop borders as a means of limiting PVY spread. However, it was also used to provide 'real-time' summaries of aphid capture data from the aphid-trapping network and report other observations, e.g. updates on the status of cereal aphids, thus alerting seed potato growers that these potential PVY vectors were about to leave their cereal hosts and move into adjacent crops.

After the 1994 growing season, with the initial research objectives largely accomplished, the Aphid Alert trapping network was discontinued. Happily for NGP

seed potato growers, 1993 was a year of exceptionally low vector pressure (as indicated by aphid captures in the Aphid Alert trapping network) and, that year, potato
seed lot rejections were well within historic norms. Unfortunately, this respite proved
to be temporary and, by 1997, potato seed lot rejection levels were again considered
disastrous, this time with both PVY and PLRV at epidemic levels. Seed potato growers lobbied the University of Minnesota to reactivate the Aphid Alert network and
the Minnesota Legislature to provide funding to support research on potato viruses.

Aphid Alert, 1998–2004

In spring 1998, the Minnesota State Legislature authorized the Rapid Agricultural
Response Initiative, providing flexible funding to enable University of Minnesota
researchers to respond to emerging urgent issues that affected Minnesota's agriculture and natural resource-based industries. An initial allocation of US$1.5 million
was provided for designated projects in 1998, and a recurring allocation of US$1 million
was created in 1999. One of the first 'Rapid Response' projects funded, 1998–2001,
was to develop approaches for managing aphid-transmitted viruses in seed potatoes
(Suranyi *et al.*, 1999). Additional projects with Aphid Alert in their title, or that were
specifically represented as complementary when proposed, were funded by the Red
River Valley Potato Growers Association, the Minnesota Area II Potato Promotion
and Research Council and several competitive grants. Sponsors of one or more of
these complementary grants included USDA/ARS, NC-IPM, USEPA and the
Rapid Agricultural Response Fund. Total extramural funding for Aphid Alert,
1992–1994 and 1998–2004, was in the range of US$1.5 million. Research units par-
ticipating in Aphid Alert 1998–2003 included all that had participated in Aphid
Alert 1992–1994.

   In 1998, the aphid-trapping network was re-established, this time using low-volume (2.4 m³/min) suction traps and green tile traps. Most locations had one suction
trap and two green tile pan traps, but from 1998 to 2002 some locations had only pan
traps. The suction traps were miniaturized versions (2.3 m tall) of the 8.5 m model
designed to monitor wheat aphid, *Diuraphis noxia*, in western USA (Allison and Pike,
1988).

   The project began with traps at 12 locations throughout Minnesota and North
Dakota in 1998 and, eventually, in 2001 it was introduced in five states and 26 locations. Nebraska, South Dakota and Wisconsin were included from 2000–2002, and a
location in Montana was added in 2002. In the final year of the project, 2003, traps
were operated at eight Minnesota and North Dakota locations only. A parallel
aphid-trapping network was instituted in Manitoba, Canada in 2003, and that network has continued to operate (Manitoba Agriculture, Food and Rural Initiatives,
2007).

   Traps of both networks were emptied weekly and within 2 days the aphids were
identified to species, counted and the data reported to seed potato growers via a
renewed *Aphid Alert* newsletter, this time also published on the Internet (http://
www.ipmworld.umn.edu/alert.htm), and distributed by e-mail to over 800 persons
worldwide. The electronic *Aphid Alert* newsletter provided seed potato growers with

near real-time information on vector abundance. Information presented in Aphid Alert was used by growers in making pest management decisions, e.g. in timing the application of aphidicide or crop oil or to 'vine-kill' early. Every edition of the newsletter contained information, and often short articles on some aspect of vector/virus biology, ecology and management. Information on the management of other potato pests was included when appropriate. For many growers, and even crop consultants and seed certification personnel, much of this information was both new and of immediate practicality. One grower self-reported keeping every issue of *Aphid Alert* (over 75 in all were produced) for permanent reference.

Aphid captures

More than 57,000 aphids representing 41 species or species complexes were captured during the 9 years of Aphid Alert trapping. Aphid abundance in the NGP, as measured by cumulative captures per trap, varied widely from year to year in both total numbers and species composition. Overall, 95% of the aphids captured were identified to species. About 90% of the aphids identified belonged to 16 species reported in the literature as being capable of acquiring and transmitting PVY.

Three potato-colonizing species, *M. persicae*, *Macrosiphum euphorbiae* (potato aphid) and *Aphis nasturtii* (buckthorn aphid), were collected regularly, but most years represented less than 5% of total captures. *Myzus persicae* constituted from 0.2% (2001) to 13.1% (1999) of total aphid captures, but exceeded 2% of total captures only in the years 1998–2000. Cumulative captures of *M. persicae* ranged from a high of 687 in 1999 to lows of 7 (0.4% of total captures in 1993) and 5 (0.04% of total captures in 2001) (see Fig. 13.1).

Abundant among the species captured that do not colonize potato were sunflower aphid, *Aphis helianthi*; turnip aphid, *Lipaphis pseudobrassicae* (= *erysimi*); *R. maidis*; several common cereal aphids including *R. padi*, *S. avenae*, *S. graminum* and soybean aphid, *Aphis glycines*. The latter species, a recent introduction to North America (Venette and Ragsdale, 2004) and an efficient vector of PVY (Davis *et al.*, 2005), was first detected in NGP trap captures in 2001, but has been abundant in the NGP since. Most of the potato non-colonizers were aphids associated with annual crops common to the NGP including small-grain cereals, maize, canola, soybean and sunflower. However, a few, including *Capitophorus* spp., *Hayhurstia atriplicus* and mealy plum aphid, *Hyalopterus pruni*, preferentially colonize various broadleaf weeds and grasses.

Virus Trends in Minnesota and North Dakota Seed Potatoes

For at least 30 years prior to the mid-1980s, Minnesota seed potato lots were seldom rejected for recertification because of PVY (Robinson, 1978). It is possible that the prevalence of PVX in those years made it much easier to identify and rogue infected plants. Multiple infections of PVX and PVY (i.e. strain PVY^O, the only variant of PVY known to occur in North America prior to 1990 (Singh, 1992)) tend to be expressed as 'severe mosaic'.

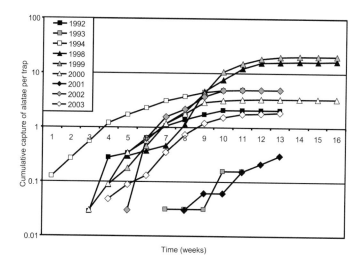

Fig. 13.2. Cumulative capture of *M. persicae* per trap in NGP, 1992–1994 and 1998–2003; week 8 ends on or before 8 August.

PVY first emerged as a major concern for NGP seed potato producers in 1988. That year, 8.2% of seed lots entered into the Minnesota Seed Potato Certification Program winter grow-out were above tolerance for PVY, with nearly twice that percentage in 1989 and 1990 (see Fig. 13.3). The PVY epidemic reached its initial zenith in 1991, when a previously unprecedented 32.1% of Minnesota seed potato lots were not eligible for recertification because of severe mosaic (a classification that supposedly distinguishes PVY from other foliage-mottling viruses, e.g. potato virus S, that tend to express as 'mild mosaics'). The situation was somewhat better in 1992, when Minnesota seed lot rejections due to PVY dropped to 19.1%. This phase of the epidemic effectively ended in 1993, a year in which the abundance of all vector species was exceptionally low.

However, PVY quickly rebounded, again reaching catastrophic proportions across the region in 1997, this time with PLRV also increasing to epidemic proportions (see Fig. 13.4). That year, 19.3% of Minnesota seed lots were above threshold for PVY and 23.7% were above the threshold for PLRV. Unfortunately, this was but a harbinger of what was to come. The following year, 52.2% of Minnesota seed potato lots were above tolerance for PVY and 31.1% above tolerance for PLRV. On average from 1988 to 2006, 29.1% of all seed lots entered into the annual Minnesota Seed Potato Certification Program winter trials exceeded tolerance for PVY, and from 1997 to 2000, 28.8% of seed lots also exceeded tolerance for PLRV. However, seed potato lots with appreciable PLRV usually also had PVY sufficient to cause rejection even if PLRV had not been present. The PLRV epidemic ended with the 2001 season (a year with low *M. persicae* abundance), but PVY has remained at high prevalence to the present. Since 1997, Minnesota seed lot rejections due to PVY have ranged from 28.3% (2001) to 61.8% (2004).

Prevalence of PVY in the seed potato production system has been persistent and particularly severe in Minnesota, but this problem is not unique to Minnesota and

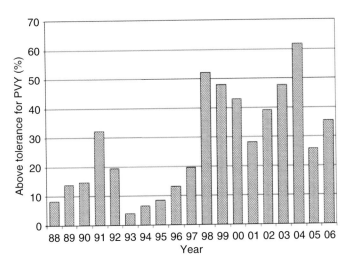

Fig. 13.3. Percentage of Minnesota seed potato lots above tolerance for PVY in winter grow-out, 1988–2006.

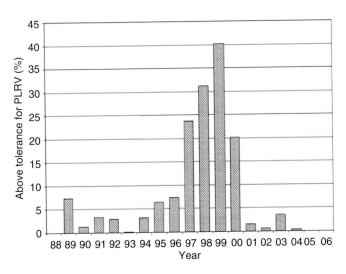

Fig. 13.4. Percentage of Minnesota seed potato lots above tolerance for PLRV in winter grow-out, 1988–2006.

North Dakota. Over the past 20 years, North American seed certification programmes have proved increasingly unsuccessful in purging PVY from state and national seed potato production systems. Many factors are suspected as contributing to this problem, including the emergence of new PVY strains. Since 2004, PVY^N (the 'tobacco veinal necrotic strain' of PVY) and PVY$^{N:O}$ recombinants have largely replaced PVY^O as the dominant PVY strain in North America (Crosslin *et al.*, 2002; Piche *et al.*, 2004; Davis, 2006). The effects of PVY^N and PVY$^{N:O}$ are generally mild, or even undetectable, in most potato cultivars, thus reliance on visual indexing for purposes of seed certification has become problematic (Singh and Singh, 1997; Sturz *et al.*, 1997; Singh *et al.*, 1999, 2003). This has made roguing difficult if not impossible for most growers, and tends to compromise the reliability of current-season virus readings and visual indexing of winter grow-outs by seed potato certification programmes (Davis, 2006). Moreover, it is now recognized that late-season infection with PLRV can go undetected in winter grow-outs, particularly so if indexing is terminated early. In the 2007 Minnesota winter grow-out (2006 crop), no seed lots were rejected for PLRV, but our serological assays showed the virus to be prevalent (data unpublished).

Molecular testing (polymerase chain reaction (PCR)) is the only method that can provide absolute assurance that a potato seedling is free of virus. Serological testing is a more realistic alternative, but even that is time-consuming and expensive compared with visual virus indexing. Serological testing of foliage from all the plants grown in a state seed potato certification programme winter grow-out would be a formidable task, and is perhaps impractical because of the handling time required to collect and process such a high volume of samples. Moreover, present seed certification regulations specify visual indexing as the standard to be used. Also, there could be an economic disadvantage to being the first certification agency to use more sensitive testing methods, especially in situations where there are two states growing seed in the same landscape, as is the case in the NGP.

Other factors suggested as contributing to the present PVY problem include:

- The popularity of certain essentially asymptomatic cultivars (Souza-Dias and Slack, 1987; Russo *et al.* 1999; Mollov and Thill, 2004).
- The introduction of a new vector (*A. glycines*) that has changed the dynamics of PVY epidemiology in the NGP by contributing massive aphid flights in early summer when potatoes are most susceptible to infection (Davis *et al.*, 2005).
- Changing cropping systems (e.g. expansion of canola and soybean production in the NGP).
- Changing pesticide use patterns; for example, the Colorado potato beetle, *Leptinotarsa decemlineata*, developed resistance to all classes of insecticides in common use in the NGP in the 1980s and early 1990s, leading to intensive spray schedules that tended to flare *M. persicae* populations; and emergence of more virulent strains of the potato late blight pathogen, *Phytophthora infestans* DeBary, necessitated greatly increased use of protective fungicides that in turn interfered with aphid entomopathogens, thus favouring *M. persicae* survival (Lagnaoui and Radcliffe, 1998).
- Perhaps even global warming (Davis, 2006). The only element that could end this epidemic would be the development of certain and efficient means of eliminating PVY inoculum from the seed production system and immediate landscape.

In the NGP there are no known perennial hosts of PLRV, and thus potato is the only source of PLRV inoculum. However, there is limited isolation of seed potato production from commercial (ware) production. That, and the fact that seed lots passing summer inspections can be sold for ware production even if they are not tested in a winter grow-out or exceeded tolerance for virus in the grow-out, tends to ensure the presence of virus inoculum in the vicinity of seed production.

The Rise and Fall of Aphid Alert

A review of the origin and eventual demise of Aphid Alert may be instructive to others contemplating development and implementation of areawide pest management programmes. The primary impetus for initiation of this programme of research and outreach came from the potato industry. The support was broad, and initially included not only seed potato growers and state seed certification programmes, but also ware producers and processors. Seed potato production, per se, represents a comparatively small sector of the overall NGP potato industry, whether measured by number of growers, hectares, tuber yield or dollar sales, but all recognized the essential importance of clean seed. There was an imperative for the initiation of the Aphid Alert programme because a seed production system that had seemed relatively secure suddenly presented extreme economic risk.

Potato production is a highly specialized and technically sophisticated form of agriculture. Potato growers tend to be innovators and, not surprisingly, strong supporters of scientific research. The Aphid Alert could not have been implemented on the scale that it was without the substantial funding provided by the various potato grower organizations and the on-site cooperation of participating seed potato growers. Indeed, we had many more invitations to locate traps on farms than we could accept because of the impracticality of our handling a greater number of samples. However, while these groups and individuals provided financial support for the research we considered part of the overall effort to address the problem of PVY, and continued to support virus-related research, grower perception of Aphid Alert tended to be narrower than that. For them, Aphid Alert was the aphid-trapping network and weekly newsletters. Clearly, seed potato growers valued this service. A survey of seed growers at the end of the 1999 growing season indicated that 78% of respondents ($n = 35$) used information presented in the newsletter to make pest management decisions (Radcliffe and Ragsdale, 2002).

Over the 9 years that the Aphid Alert trapping network was in operation, grower organizations and the Minnesota Legislature's Rapid Response Initiative provided more than two-thirds of the funding that directly supported that activity and newsletter. We were able to leverage this support to advantage in competing for extramural grants that permitted accomplishment of a wide range of complementary research studies.

These included research on:

- The use of crop borders to limit PVY spread (DiFonzo *et al.*, 1996).
- The impact of potato fungicides on *M. persicae* populations (Ruano-Rossil *et al.*, 2001).
- The use of crop oils to limit spread of PVY (Suranyi, 2000).
- The role of landscape ecology and aphid behaviour on the dynamics of vector dispersal and virus spread (Carroll, 2005).
- Site-specific targeting of foliar insecticide applications for *M. persicae* control (Carroll *et al.*, 2008).
- The development of meteorological models using the duration of low-level jets to predict *M. persicae* abundance and virus spread (Zhu *et al.*, 2006).
- The transmission efficiencies of vector species (Davis *et al.*, 2006).

- The virus strain presence in Minnesota (Davis *et al.*, 2006).
- The reliability of various methodologies of virus detection for purposes of seed certification (Davis *et al.*, 2006).

In many respects Aphid Alert was a success. Seed potato growers tended to be strongly supportive and many were quick to adopt new control technologies, e.g. scouting for aphids, using insecticides more selectively, using crop borders, applying crop oils, earlier planting and vine kill and targeting application of insecticides – i.e. a single spray width applied by aircraft to field margins bordering fallowed land (e.g. headlands) for control of newly colonizing *M. persicae* (Carroll *et al.*, 2008). The latter provided excellent aphid control while reducing use of foliar aphidicides by over 90%. There seemed to be consensus within the potato industry that the annual benefits of adopting these practices greatly exceeded the research investment (Agricultural Utilization Research Institute, 2002). The trapping network also provided data essential for accomplishing several of the research studies, particularly the third and fourth items on the above list.

So why did the Aphid Alert trapping network prove unsustainable and could its demise have been prevented? Before addressing those questions we must ask: what were the failures or shortcomings of Aphid Alert, and could these have been corrected? The most obvious failure of Aphid Alert is that the current PVY epidemic has persisted in the NGP since 1994. In contrast, the PLRV epidemic of 1997–2000 ran its course and ended. It is obvious that the PVY problem will not be solved quickly, and perhaps will remain intractable so long as growers are obliged to rely on current technologies.

Farmers still tend to think of aphidicides as their primary defence against current-season spread of insect-transmitted viruses, but such use has proved to be of inconsistent benefit (Perring *et al.*, 1999). Movement of aphid-transmitted viruses into clean potato fields from outside inoculum sources is almost exclusively by winged aphids (Boiteau, 1997). Insecticides are seldom of any benefit in preventing the spread of either non-persistent viruses (e.g. PVY) (Perring *et al.*, 1994; Ragsdale *et al.*, 1994) or persistent viruses (e.g. PLRV) by winged aphids already capable of transmission (Hanafi *et al.*, 1995). Movement of PVY and other non-persistently transmitted viruses is thought to be almost exclusively by winged aphids because their transmission requires no latent period in the vector, and ability to transmit is lost in the first few feeding probes following acquisition (Ragsdale *et al.* 1994).

In contrast to PVY, within-field spread of PLRV is often by apterae walking from plant to plant (Hanafi *et al.*, 1989). Insecticides targeted against aphid vectors, either as systemics at planting or as foliar sprays when needed, tend to be effective in preventing spread of PLRV from within-field sources because the time lag between acquisition of the virus and onset of ability of the vector to transmit is sufficient for the insecticide to have effect (Woodford *et al.*, 1988; Hanafi *et al.*, 1989; Flanders *et al.*, 1991; DiFonzo *et al.*, 1995; Boiteau and Singh, 1999).

Weekly application of crop oil can provide considerable protection against the spread of non-persistent viruses (Boiteau and Singh, 1982; Secor *et al.*, 2004). Thus, it is useful to know when aphids begin flying, even for PVY. However, since species composition and abundance of aphid populations tends to vary greatly between locations and years (DiFonzo *et al.*, 1997, Suranyi, 2000), a more site-specific monitoring of vector populations may be required to be useful as a decision tool in PVY control.

Breeding for aphid and virus resistance offers some promise (e.g. Novy *et al.*, 2002; Davis, 2006), but will take many years to accomplish by conventional plant breeding. Cultivars have been genetically transformed to express resistance to both PLRV and PVY (Brown *et al.*, 1995; Berger and German, 2001). While this novel technology proved far more effective in reducing virus spread than presently available tactics, these cultivars have been withdrawn from the market because they were not accepted for fear of possible public backlash against the technology (Thornton, 2003).

One major shortcoming of the Aphid Alert project was that the service component, i.e. operation of the network, identifying the insects and reporting the trap results to growers on a near real-time basis, fell upon personnel whose professional responsibility was primarily or exclusively research. While undergraduate technicians were used in assisting with some of the more routine aspects of this work, most aspects of establishing and maintaining the network and all of the aphid identification was assumed by graduate students. This worked because the graduate students had dissertation projects that necessitated obtaining aphid capture data. However, as a service function, operation of an aphid-trapping network was not an appropriate activity for a research university.

The benefits to the seed potato industry might have justified the annual investment required, but implementation was not simple. Realistically, this was not an activity a grower organization would want to assume, even if they were prepared to finance the operation. An assessment of ~US$7.50/ha would be adequate to support an aphid-trapping network in the NGP if all Minnesota and North Dakota seed potato growers contributed. The most appropriate agency to operate such a network might be the State Department of Agriculture, possibly as an activity of the potato seed certification programme.

Appreciable seed rejections in the NGP due to PLRV have not occurred since 2000. This made the service of aphid-trapping network less of an imperative for seed potato growers. However, considerable PLRV inoculum is still present in the NGP seed potato production system. A year of high *M. persicae* abundance could easily produce a PLRV epidemic comparable to that experienced in the NGP from 1997 to 2000. The seeming intractability of the PVY problem made it inevitable that the grower organizations would eventually want to reprioritize their research investments. Although clean seed is an essential requirement for potato production, the industry faces many other challenges and the dollars available to these associations to support research are mostly from assessments paid by growers producing for processing or fresh market. The fact that seed potato production in this region crosses state, and indeed international, boundaries presents a further complication in the financing and coordinating of an areawide aphid-trapping network.

In spite of the importance of PVY in seed production ware producers have not been unduly impacted, as there is still sufficient certified seed being produced. However, the presence of new PVY strains, especially the recent introduction of PVY[NTN] (Singh *et al.*, 2003), could make PVY very much of an industry problem; PVY[NTN] ('tuber necrosis subgroup') infection can cause serious tuber defects (potato tuber necrotic ringspot disease, PTNRD) and breakdown in storage. Effects of PVY[NTN] are so devastating on tuber health that the disease tends to be self-eliminating. Most NGP seed potato growers would probably like to see restoration of the Aphid Alert network, but to be cost-effective such a network would need to serve a broader group of commodities, e.g. soybean growers and producers of small-grain cereals.

References

Agricultural Utilization Research Institute (2002) Seed money: Aphid Alert enriches potato producers in Northern Minnesota. *Ag Innovation News: the Newspaper of the Agricultural Utilization Research Institute* 11(3), 12.

Allison, D. and Pike, K.S. (1988) An inexpensive suction trap and its use in an aphid monitoring network. *Journal of Agricultural Entomology* 5, 103–107.

Banttari, E.E. (1993) Potato virus Y research, Verticillium wilt tests. *Valley Potato Grower* 59, 8–10.

Berger, P.H. and German, T.L. (2001) Biotechnology and resistance to potato viruses. In: Loebenstein, G., Berger, P.H., Brunt, A.A. and Lawson, R.H. (eds) *Virus and Virus-like Diseases of Potatoes and Production of Seed-potatoes.* Kluwer Academic Publishers, Dordrecht, Netherlands, pp. 341–363.

Boiteau, G. (1997) Comparative propensity for dispersal of apterous and alate morphs of three potato-colonizing aphid species. *Canadian Journal of Zoology* 75, 1396–1403.

Boiteau, G. and Parry, R.H. (1985) Monitoring of inflights of green peach aphids, *Myzus persicae* (Sulzer), in New Brunswick potato fields by yellow pans from 1974 to 1983: results and degree-day simulation. *American Potato Journal* 62, 489–496.

Boiteau, G. and R.P. Singh. (1982) Evaluation of mineral oil sprays for reduction of virus Y spread in potatoes. *American Potato Journal* 59, 253–262.

Boiteau, G. and Singh, R.P. (1999) Field assessment of imidacloprid to reduce the spread of PVY^O and PLRV in potato. *American Journal of Potato Research* 76, 31–36.

Brown, C.R., Smith, O.P., Damsteegt, V.D., Yang, C.-P. Fox L. and Thomas, P.E. (1995) Suppression of PLRV titer in transgenic Russet Burbank and Ranger Russet. *American Potato Journal* 72, 589–597.

Carroll, M.W. (2005) Spatial distribution of the green peach aphid, *Myzus persicae* (Sulzer), and management applications in seed potato. PhD dissertation, University of Minnesota, St Paul, Minnesota.

Carroll, M.W., Radcliffe, E.B., MacRae, I.V., Ragsdale, D.W., Olson, K.D. and Badibanga, T. (2008) Border application to reduce insecticide use in seed potato production: biological, economic, and managerial analysis. *American Journal of Potato Research* (submitted).

Crosslin, J.M., Hamm, P.B., Eastwell, K.C., Thornton, R.E., Brown, C.R., Corsini, D.L., Shiel, P.J. and Berger, P.H. (2002) First report of the necrotic strain of potato virus Y (PVY^N) potyvirus on potatoes in the northwestern United States. *Plant Disease* 86, 1177.

Davis, J.A. (2006) Identifying and mapping novel mechanisms of host plant resistance to aphids and viruses in diverse potato populations. PhD dissertation, University of Minnesota, St Paul, Minnesota.

Davis, J.A., Radcliffe, E.B. and Ragsdale, D.W. (2005) Soybean aphid, *Aphis glycines* Matsumura, a new vector of potato virus Y in potato. *American Journal of Potato Research* 82, 197–201.

Davis, J.A., Radcliffe, E.B. and Ragsdale, D.W. (2006) Effects of high and fluctuating temperatures on *Myzus persicae* (Hemiptera: Aphididae). *Environmental Entomology* 35, 1461–1468.

DiFonzo, C.D., Ragsdale D.W. and Radcliffe, E.B. (1995) Potato leafroll virus spread in differentially resistant potato cultivars under varying aphid densities. *American Potato Journal* 72, 119–132.

DiFonzo, C.D., Ragsdale, D.W., Radcliffe, E.B., Gudmestad, N.C. and Secor, G.A. (1996) Crop borders reduce potato virus Y incidence in seed potato. *Annals of Applied Biology* 129, 289–302.

DiFonzo, C.D., Ragsdale, D.W., Radcliffe, E.B., Gudmestad, N.C. and Secor, G.A. (1997) Seasonal abundance of aphid vectors of potato virus Y in the Red River Valley of Minnesota and North Dakota. *Journal of Economic Entomology* 90, 824–831.

Eastop, V.F. (1955) Selection of aphid species by different kinds of insect traps. *Nature (London)* 176, 936.

Flanders, K.L., Radcliffe, E.B. and Ragsdale, D.W. (1991) Potato leafroll virus spread in relation to densities of green peach aphid (Homoptera: Aphididae): implications for management thresholds for Minnesota seed potatoes. *Journal of Economic Entomology* 84, 1028–1036.

Franc, G.D. (2001) Seed certification as a virus management tool. In: Loebenstein, G., Berger, P.H., Brunt, A.A. and Lawson R.H. (eds) *Virus and Virus-like Diseases of Potatoes and Production of Seed-potatoes*. Kluwer Academic Publishers, Dordrecht, Netherlands, pp. 407–420.

Gutbrod, O.A. and Mosley, A.R. (2001) Common seed potato certification schemes. In: Loebenstein, G., Berger, P.H., Brunt, A.A. and Lawson, R.H. (eds) *Virus and Virus-like Diseases of Potatoes and Production of Seed-potatoes*. Kluwer Academic Publishers, Dordrecht, Netherlands, pp. 421–438.

Halbert, S., Connelly, J. and Sandvol, L. (1990) Suction trapping of aphids in western North America (emphasis on Idaho). *Acta Phytopathologica et Entomologica Hungarica* 25, 411–422.

Hanafi, A., Radcliffe, E.B. and Ragsdale, D.W. (1989) Spread and control of potato leafroll virus in Minnesota. *Journal of Economic Entomology* 82, 1201–1206.

Hanafi, A., Radcliffe, E.B. and Ragsdale, D.W. (1995) Spread and control of potato leafroll virus in the Souss Valley of Morocco. *Crop Protection* 14, 145–153.

Harrington, R. and EXAMINE Consortium (2007) *EXploitation of Aphid Monitoring Systems IN Europe* (http://www.rothamsted.ac.uk/examine/).

Harrington, R. and Gibson, R.W. (1989) Transmission of potato virus Y by aphids trapped in potato crops in southern England. *Potato Research* 32, 167–174.

Harrington, R., Katis, N. and Gibson, R.W. (1986) Field assessment of the relative importance of different aphid species in the transmission of potato virus Y. *Potato Research* 29, 67–76.

Heimbach, U., Thieme, T., Weidemann, H.-L. and Thieme, R. (1998) Transmission of potato virus Y by aphid species which do not colonise potatoes. In: Nieto, J.M., Nafría, H. and Dixon, A.F.G. (eds) *Aphids in Natural and Managed Ecosystems*. Universidad de León, León, Spain, pp. 555–559.

Hille Ris Lambers, D. (1972) Aphids: their life cycles and their role as virus vectors. In: de Bokx, J.A (ed.) *Viruses of Potatoes and Seed-Potato Production*. Pudoc, Wageningen, Netherlands, pp. 36–56.

Irwin, M.E. (1980) Sampling aphids in soybean fields. In: Kogan, M. and Herzog, D. (eds) *Sampling Methods in Soybean Entomology*. Springer-Verlag, New York, pp. 239–259.

Lagnaoui, A. and Radcliffe, E.B. (1998) Potato fungicides interfere with entomopathogenic fungi impacting population dynamics of green peach aphid. *American Journal of Potato Research* 75, 19–25.

Manitoba Agriculture, Food and Rural Initiatives (2007) *Insects, Insect Updates, Aphid Monitoring on Potato* (http://www.gov.mb.ca/agriculture/crops/insects/index.html).

Mollov, D.S. and Thill, C.A. (2004) Evidence of potato virus Y asymptomatic clones in diploid and tetraploid potato breeding populations. *American Journal of Potato Research* 81, 317–326.

Nault, L.R. (1997) Arthropod transmission of plant viruses: a new synthesis. *Annals of Entomological Society of America* 90, 521–541.

Novy, R.G., Nasruddin, A., Ragsdale, D.W. and Radcliffe, E.B. (2002) Genetic resistances to potato leafroll virus, potato virus Y, and green peach aphid in progeny of *Solanum tuberosum*. *American Journal of Potato Research* 79, 9–18.

Perring, T.M., Gruenhagen N.M. and Farrar, C.A. (1999) Management of plant viral diseases through chemical control of insect vectors. *Annual Review of Entomology* 44, 457–481.

Piche, L.M., Singh, R.P., Nie, X. and Gudmestad, N.C. (2004) Diversity among Potato virus Y isolates obtained from potatoes grown in the United States. *Phytopathology* 94, 1368–1375.

Pickup, J. and Brewer, A.M. (1994) The use of aphid suction-trap data in forecasting the incidence of potato leafroll virus in Scottish seed potatoes. *Proceedings Brighton Crop Protecion Conference: Pests and Diseases* 1, 351–358.

Radcliffe, E.B. and Ragsdale, D.W. (2002) Invited review. Aphid-transmitted potato viruses: the importance of understanding vector biology. *American Journal of Potato Research* 79, 353–386.

Ragsdale, D.W., Radcliffe, E.B., DiFonzo, C.D. and Connelly, M.S. (1994) Action thresholds for an aphid vector of potato leafroll virus. In: Zehnder, G.W., Powelson, M.L., Jansson, R.K. and Raman, K.V. (eds) *Advances in Potato Pest Biology and Management.* American Phytopathology Society, St Paul, Minnesota, pp. 99–110.

Ragsdale, D.W., Radcliffe, E.B. and DiFonzo, C.D. (2001) Epidemiology and field control of PVY and PLRV. In: Lobenstein, G., Berger, P.H., Brunt, A.A. and Lawson, R.H. (eds) *Virus and Virus-like Diseases of Potatoes and Production of Seed-potatoes.* Kluwer Academic Publishers, Dordrecht, Netherlands, pp. 237–270.

Rieman, G.H. (1956) Early history of seed certification in North America, 1913–1922. *Potato Handbook, Seed Certification Issue.* New Brunswick and Potato Association of America, Orono, Maine, pp. 6–10.

Robert, Y. and Bourdin, D. (2001) Aphid transmission of potato viruses. In: Lobenstein, G., Berger, P.H., Brunt, A.A. and Lawson, R.H. (eds) *Virus and Virus-like Diseases of Potatoes and Production of Seed-potatoes.* Kluwer Academic Publishers, Dordrecht, Netherlands, pp. 195–225.

Robinson, D.P. (1978) Red River Valley insect survey, 1976. MSc Thesis, Part II, University of Minnesota, St Paul, Minnesota.

Ruano-Rossil, J.M., Radcliffe, E.B. and Ragsdale, D.W. (2004) Disruption of entomopathogenic fungi of green peach aphid, *Myzus persicae* (Sulzer), by fungicides used to control potato late blight. In: Simon, J.C., Dedryver, C.A., Rispe, C. and Hullé, M. (eds) *Aphids in a New Millennium.* INRA, Versailles, France, pp. 365–370.

Russo, P., Miller, L., Singh, R.P. and Slack, S.A. (1999) Comparison of PLRV and PVY detection in potato seed samples tested by Florida winter field inspection and RT-PCR. *American Journal of Potato Research* 76, 313–316.

Secor, G.A., Gudmestad, N.C. and Suranyi, R.A. (2004) Fungicide compatibility with Aphoil for late blight control. *American Journal of Potato Research* 81, 87–88 (abstr.).

Sigvald, R. (1987) Aphid migration and the importance of some aphid species as vectors of potato virus Y^O (PVY^O) in Sweden. *Potato Research* 30, 267–283.

Sigvald, R. (1989) Relationship between aphid occurrence and spread of potato virus Y^O (PVY^O) in field experiments in southern Sweden. *Journal of Applied Entomology* 108, 35–43.

Sigvald, R. (1992) Progress in aphid forecasting systems. *Netherland Journal of Plant Pathology* 98 (Suppl. 2), 55–62.

Singh, R.P. (1992) Incidence of the tobacco veinal necrotic strain of potato virus Y (PVY^N) in Canada in 1990 and 1991 and scientific basis for the eradication of the disease. *Canadian Plant Disease Survey* 72, 113–119.

Singh, M. and Singh, R.P. (1997) Potato virus Y detection: sensitivity of RT-PCR depends on the size of fragment amplified. *Canadian Journal of Plant Pathology* 19, 149–155.

Singh, M., Singh, R.P. and Moore, L. (1999) Evaluation of NASH and RT-PCR for the detection of PVY in dormant tubers and its comparison with visual symptoms and ELISA in plants. *American Journal of Potato Research* 76, 61–66.

Singh, R.P., McLaren, D.L. Nie, X. and Singh, M. (2003) Possible escape of a recombinant isolate of Potato virus Y by serological indexing and methods of its detection. *Plant Disease* 87, 679–685.

Slack, S.A. (1992) A look at potato leafroll and PVY: past, present, future. *Valley Potato Grower* 87, 679–685.

Souza-Dias, J.A.C. and Slack, S.A. (1987) Relation of potato leafroll virus concentration in potatoes to virus concentration in aphids. *American Potato Journal* 64, 459 (abstr.).

Sturz, A.V., Diamond, J.F. and Stewart, J.G. (1997) Evaluation of mosaic expression as an indirect measure of the incidence of PVY^O in potato cv. Shepody. *Canadian Journal of Plant Pathology* 19, 145–148.

Sturz, A.V., Stewart, J.G., McRae, K.B., Diamond, J.F., Lu, X. and Singh, R.P. (2000) Assessment of the importance of seed cutting, in-season cultivation, and the passage of row equipment in the spread of PVY^O. *Canadian Journal of Plant Pathology* 22, 166–173.

Suranyi, R.A. (2000) Management of potato leafroll virus in Minnesota. PhD dissertation, University of Minnesota, St Paul, Minnesota.

Suranyi, R., Radcliffe, T., Ragsdale, D., MacRae, I. and Lockhart, B. (1999) *Aphid Alert: research/outreach initiative addressing potato virus problems in the northern Midwest USA*. In: *World Potato Congress* (http://www.potatocongress.org/sub.cfm?source=146).

Thornton, M. (2003) The rise and fall of newleaf potatoes. In: Ristow, S., Rosa, E.A. and Burke, M.J. (eds) *Biotechnology: Science and Society at a Crossroad*. National Agricultural Biotechnology, Council Boyce Thompson Institute, Ithaca, New York, pp. 235–243.

USDA/NASS (2007) National Agricultural Statistics Service (http://www.nass.usda.gov/QuickStats/index2.jsp).

van Harten, A. (1983) The relation between aphid flights and the spread of potato virus Y^N (PVY^N) in the Netherlands. *Potato Research* 26, 1–15.

Venette, R.C. and Ragsdale, D.W. (2004) Assessing the invasion by soybean aphid (Homoptera: Aphididae): where will it end? *Annals of the Entomological Society of America* 97, 219–226.

Woodford, J.A.T., Gordon, S.C. and Foster, G.N. (1988) Side-band application of systemic granular pesticides for the control of aphids and potato leafroll virus. *Crop Protection* 7, 96–105.

Zhu, M., Radcliffe, E.B., Ragsdale D.W., MacRae, I.V. and Seeley, M.W. (2006) Low-level jet streams associated with spring aphid migration and current season spread of potato viruses in the US northern Great Plains. *Agriculture Forest Meteorology* 38, 192–202.

14 Areawide Suppression of Fire Ants

M.D. Aubuchon and R.K. Vander Meer

USDA-ARS, Center for Medical and Veterinary Entomology, Gainesville, Florida, USA

Introduction

Significance of the pest management problem

The imported fire ants, *Solenopsis invicta* and *S. richteri*, were inadvertently introduced into the USA in the early 1900s and currently inhabit over 129 million ha in Puerto Rico and 12 southern states, from Texas to Virginia (Callcott and Collins, 1996; USDA-APHIS map). Imported fire ants have also become established in isolated sites in California, Arizona, New Mexico and Maryland. Strict quarantine procedures have limited the spread of this pest (Lockley and Collins, 1990), but eventually populations will expand westward in increasing numbers in New Mexico, Arizona and California. They will also move upward along the Pacific coast, southward into Mexico and the Caribbean and northward in Oklahoma, Arkansas and Tennessee and along the eastern seaboard into Maryland and possibly Delaware (Korzukhin *et al.*, 2001).

Mature monogyne (single queen) fire ant colonies contain 100,000 to 250,000 workers (Tschinkel, 1988, 1993) and reach infestation rates of over 130 mounds/ha. In the last few decades, polygyne fire ant colonies (multi-queen colonies) appear to be proliferating in the southern states. With polygyne populations, the number of mounds may reach over 500/ha (Porter *et al.*, 1991; Porter, 1992), resulting in interconnected super-colonies because of the lack of territoriality among polygyne colonies (Morel *et al.*, 1990; Vinson, 1997). Control is difficult because more queens must be killed.

Imported fire ants destroy many ground-inhabiting arthropods and other small animals (Vinson and Greenberg, 1986; Porter and Savignano, 1990; Jusino-Atresino and Phillips, 1994; Wojcik, 1994; Forys *et al.*, 1997; Allen *et al.*, 1998; Williams *et al.*, 2003). Because fire ants are highly aggressive when their nests are disturbed, this often results in painful stings to humans and their pets. Between 30 and 60% of the people in the infested areas are stung each year, with hypersensitivity occurring in 1% or more of those people (deShazo *et al.*, 1990, 1999; deShazo and Williams, 1995), suggesting that over 200,000 persons per year may require medical treatment.

Imported fire ants adversely affect yields of several important agricultural crops (Adams, 1986; Lofgren, 1986). Reductions in soybean yields are associated with the ant feeding on germinating seeds and roots of surviving plants, thus lowering plant density and causing estimated annual crop losses of over US$100 million (Adams, 1986; Thompson and Jones, 1996; Shatters and Vander Meer, 2000). Other affected crops include maize, potatoes, aubergine and okra. Studies have demonstrated that imported fire ants can seriously damage young citrus trees (Adams, 1986) by feeding on bark, flowers, newly set fruit and other plant tissue. Tree replacement in established groves (average of five replants/ha) costs US$145.57/ha/year (Adams, 1986; Lofgren, 1986). Imported fire ants will also kill chicks and injure young livestock. In a survey of Texas cattle producers, an estimated US$67 million per year in losses was due to fire ants (Barr and Drees, 1996). Total economic losses (cost of control and damage) in the USA are estimated at nearly US$6 billion per year (Pereira *et al.*, 2002).

Description of current management systems and approaches

Several mound drenches have been developed for fire ant control, but are impractical on a scale other than for residential use. The most effective and environmentally responsible method of control is the use of toxic baits because the fire ant has a very effective foraging and resource distribution system that gets the bait/active ingredient to the target. Fire ant bait is typically composed of a vegetable oil phagostimulant that also acts as a solvent for an oil soluble toxicant. This solution is then absorbed on to a defatted maize grit granule that will absorb 20–30% oil and still maintain flowability. The bait is spread on the ground and the foraging ants find it and bring it back to nest mates. The toxicant must exhibit delayed toxicity to give the foraging workers time to distribute the oil/toxicant to all colony members.

Limitations of current management approaches

Although there are several commercial toxic baits available for imported fire ants, these baits are expensive and many are not registered for large acreage. Most fire ant active ingredients have adverse effects on the environment. Toxic bait development and EPA registration efforts by the chemical industry have primarily focused on the lucrative urban market, and thus few companies have pursued registration of baits for use in agricultural settings. Even when available, toxic baits are expensive and require continuous reapplication because of the rapid reinfestation of treated areas. The non-specific nature of the active ingredients adversely impacts non-target native ant species, as well as the environment. Altogether, chemical treatment strategies alone are not a viable option for large tracts of land such as rangeland and pastures. In addition, with increasing emphasis on quarantine expressed by APHIS, and with mandates from the Department of Defense, the Environmental Protection Agency and the public to reduce risks associated with pesticides, there is a need for a different fire ant strategy.

Anticipated benefits of AWPM

Recent USDA research has led to the availability of self-sustaining biological control agents and, along with effective toxic baits, has provided tools for development of an integrated pest management (IPM) system for suppression of fire ant populations. The advent of these fire ant control tools has led to an ARS headquarters-funded Areawide Pest Management Project (AWPM), the goal of which was to maintain low fire ant populations with reduced need for bait toxicants by using available self-sustaining fire ant biological control agents in conjunction with bait toxicants.

Anticipated benefits are manifold:

- Spread of self-sustaining biological control agents will help restore the ecological balance between the imported fire ant and native fauna.
- Areawide management technology, especially biological control agents, will be transferred to state and federal agencies, as well as to state and private land managers.
- Sustained fire ant population reduction will be achieved.
- Farm workers will be able to work in a safer environment.
- Fire ant economics will be better understood.
- Developed methodology will be transferred to a variety of end-users via web site development and other educational media.
- Pesticide risk will be reduced.

Description of the AWPM Programme and Approaches

AWPM management technologies and approaches

The Imported Fire Ant and Household Insects Research Unit (IFAHIRU) and cooperators have created the single most successful control programme for fire ants to date. Development, assembly and refinement of a complex array of control techniques have resulted in the first AWPM programme for fire ants. The IFAHIRU developed the fire ant bait toxicant concept that has been most effective and environmentally friendly, while simultaneously discovering, importing and releasing fire ant-specific, self-sustaining biological control agents such as microsporidian pathogens and phorid fly (*Pseudacteon*) parasites.

Although the flies cause direct fire ant mortality, they also reduce foraging and mating flight activities, resulting in weakened fire ant colonies and reduced reproductive potential. The microsporidian pathogen stresses infected colonies, resulting in reduced colony lifespan and rendering colony members more susceptible to bait toxicants. Establishment of some of these self-sustaining biocontrol agents was critical to development of an integrated management plan for control of fire ants using an IPM approach (combination of bait toxicants and biological control agents).

Compatibility of the fire ant AWPM programme with other pest management or land improvement practices

Fire ant baits are generally considered also to have an effect on non-target ant species but not the general arthropod diversity; therefore, baits are mainly neutral in terms of other pest or land improvement practices. It is possible that baits could be admixed with certain types of fertilizer and co-distributed. This process would decrease the overall cost for the use of baits for fire ant control. The self-sustaining biological control agents have been demonstrated to be very specific to the fire ant genus and often species specific; therefore, they are not expected to have negative or positive effects on non-fire ant pest control or on land improvement practices. In contrast, non-fire ant pest treatments may have a negative effect on fire ant populations. Also, land improvement practices may negatively affect fire ant populations, e.g. liquid ammonia fertilizer and controlled burning.

Development and implementation of the AWPM programme

Cooperators

ARS has expertise in parasite and pathogen biological control, as well as molecular biology and chemical ecology. ARS scientific expertise was supplemented with:

- An agricultural economist from Texas A&M University.
- APHIS involvement in large-scale phorid fly rearing and in assisting and advising on the use of aerial bait treatments.
- The education component was directed by a University of Florida extension specialist who developed a web site, educational brochures, videos and other presentation materials. These information tools were used to educate stakeholders, e.g. extension specialists, high-value property owners, local government and the public, about the AWPM programme.
- Each of the AWPM project's cooperators were charged with the task of developing the within-state infrastructure needed to carry out the complex assessments required for execution of the programme and evaluation of programme success.
- Environmental impact was assessed using ARS and state cooperator expertise.
- ARS directed a portion of their research effort toward specific problems associated with the AWPM project.

All of the above contributed to the successful demonstration of the first continuous AWPM programme for fire ants in five US states, representing diverse ecological conditions and over a multiple-year period.

Development and implementation of education and technology transfer programmes

Education programme

The educational component provided extensive positive outreach to our partners and customers, as exemplified by the following:

- A programme web site was created, and updated continuously with new information.
- Videos describing the fire ant microsporidian disease and phorid flies were produced and distributed via the web site and on CD (over 1000 were distributed; included in ARS Congressional Budget hearings package).

- Programme brochures were produced and distributed by direct mailing, insertion in trade magazines and to the public at state agricultural fairs and public presentations (40,000 to 50,000 distributed).

Public interest has been enormous – 42,288 distinct visits to the web site in 2005 and 58,387 in 2006. Part of a video describing the parasitic phorid fly was the subject of an article by a nationally syndicated columnist. This article caused such a huge number of requests to the web site that the server crashed. The areawide web site will continue to be maintained and updated with progress in the newly established 'high value' demonstration sites.

Technology transfer

Phorid fly parasite rearing is complex, labour intensive and not likely to be taken on by private industry. Thus, APHIS provided funding to transfer the ARS-developed phorid fly rearing technology to the Florida Division of Plant Industry (DPI), in Gainesville, Florida. Similar technology was also transferred to the University of Texas, Austin; Louisiana State University, Baton Rouge; and the ARS, Biological Control of Pests Research Unit, Stoneville, Mississippi. The technology transferred included mass rearing of phorid flies, methods of releasing and establishing phorid fly parasites and numerous requests to release flies in the USA. Development of methods to mass rear the phorid fly parasites was essential to the success of the AWPM programme. An unintended consequence of the rearing technology was participation of numerous additional cooperating institutions in phorid fly releases.

The ARS developed novel methods for infecting fire ant colonies with the microsporidian pathogen, *Thelohania solenopsae*. These methods were crucial in facilitating the spread of the disease in fire ant populations in the AWPM programme. In addition, the AWPM project also promoted inoculation and spread of a microsporidian pathogen by university and state department of agriculture cooperators in five other fire ant-infested states. Currently, the technology is being used among high-value properties (e.g. parks, golf courses, hunting clubs, natural areas, military facilities) where fire ant control is highly desirable. These sites are being used to demonstrate that biological control, in combination with toxic bait applications, can be used in many different situations to provide safe, effective, economical fire ant control. Other researchers have adopted these methods for infecting colonies throughout the range of introduced fire ants.

ARS scientists developed a simple and reliable method for estimating fire ant population densities by utilizing a food lure and establishing an action threshold for treatment. Cooperators adopted this method after ARS demonstrated strong correlations between the new method and the previously used mound count and population index methods. The food lure method reduces the time needed to estimate populations by at least 50%, requires no specialized training and is easily transferable, thus simplifying the implementation of fire ant integrated pest management (IPM).

The research component of the project responded to the need for a rapid, sensitive method for detecting the presence of the microsporidian pathogen with an easy-to-use PCR method that was transferred to our project partners, as well as to fire ant researchers worldwide. Additional fire ant biological control agents from South America are currently in our quarantine facility, undergoing the extensive testing required for obtaining permission for their release in the USA.

Evaluation of the AWPM Programme

Effectiveness of the AWPM programme at controlling fire ant populations

The AWPM programme has had significant impact. Fire ant population levels have been suppressed below target thresholds in all demonstration sites in pastures. For the first time, fire ant control has been maintained at more than 80% over a total area of about 8896 ha for 4–5 years. These properties are now serving as examples for neighbouring property owners, and have provided for a continuing expansion of interest in fire ant IPM in different regions of the USA. Further examples of impact are listed as follows:

- In Florida, fire ant reduction has averaged 88% where the IPM approach was used, as compared with only 71% where fire ants were controlled only by chemical pesticides. In Texas, plots with high phorid fly populations were correlated with lower fire ant populations.
- Sustainable biological control agents were successfully released into all five states where the AWPM programme was implemented, and in dozens of other locations throughout the infested area in cooperation with APHIS and cooperators in each state (see Table 14.1).
- *Pseudacteon tricuspis*, the first species of phorid fly released, is currently well established in eight states: Florida, Alabama, Georgia, Mississippi, Louisiana, South Carolina, Texas and Arkansas (see Table 14.1).
- Two biotypes of *P. curvatus* have been established in the USA. The first biotype is established on black imported fire ants in Mississippi, Alabama and Tennessee; a

Table 14.1. Total area currently occupied by phorid flies, and the human population impacted. Five hundred thousand phorid decapitating flies (*Pseudacteon tricuspis* and *P. curvatus*) were released at the 83 sites in 12 states.

State	Release sites		Total area impacted (km²)	Human population in impacted area
	P. tricuspis	*P. curvatus*		
Areawide states:				
Florida	6	10	92,324	13,420,532
Mississippi	2	2	33,249	1,085,755
Oklahoma	6	3	4,023	48,198
South Carolina	5	1	1,959	334,609
Texas	19	2	8,819	953,408
Other states (7)	21	7	120,968	4,211,527
Total	59	25	261,342	20,054,000

second biotype is established in Florida, South Carolina, Texas and Oklahoma on red imported fire ants (see Table 14.1).

- The total area impacted by phorid fly parasites is > 260,000 km², an area comprising around 20 million people. We anticipate that, over the next 4–5 years, the flies will expand their range to over 1,200,000 km².

- *Thelohania solenopsae*, a microsporidian pathogen that debilitates fire ant queens and eventually kills the colony, is established and spreading in Florida, Texas, South Carolina and Oklahoma – e.g. 60% increase in Florida's IPM site and natural spread from 0–12% infected colonies in the bait toxicant-only site.

- The AWPM project has helped promote inoculations of *T. solenopsae* by university and state department of agriculture cooperators in ten infested states. During the AWPM project it has been documented that the pathogen has become widespread in multiple-queen fire ant populations, where it may be prevalent in well over 155,000 km², with infection rates averaging about 51%.

- Phorid flies and the *T. solenopsae* parasite have reduced fire ant populations by at least 1 and 33%, respectively. These reductions have translated into tens of millions of dollars saved for those in impacted areas.

- Farm worker safety has been significantly improved due to reduced exposure to fire ants.

- There have been fewer mechanical and electrical equipment repairs due to fewer fire ants and fewer mounds.

Unintended positive consequences of the AWPM programme

Efforts of the Areawide Suppression of Imported Fire Ants programme have led to several unintended positive results. *Pseudacteon tricuspsis*, the first species of phorid fly released, is currently well established in Alabama, Georgia and Louisiana, in addition to five participating areawide states (see Table 14.1). One biotype of *P. curvatus* has been established on black imported fire ants in Tennessee and Alabama, as well as in Mississippi. A second biotype, *P. curvatus*, is established in Florida, South Carolina, Texas and Oklahoma on red imported fire ants (see Table 14.1). As multiple species of phorid flies spread beyond the confines of areawide field sites, they provide an added benefit for people living within these areas. The presence and expansion of phorid flies also helps the native and endangered species that have been adversely affected by fire ant aggression and environmental domination.

Economic evaluation of costs and benefits of the AWPM programme

Economic surveys were prepared by an agricultural economic team from the Texas A&M University and sent to the farmers involved in the demonstration sites, as well as to the researchers in each state. These surveys assessed the impact of the fire ant pests on farm activities, as well as assessing the costs and benefits of the AWPM programme. These surveys are being analysed, and the data obtained so far have been used to estimate the economic impact of fire ants on both US agriculture and individual states. Texas and Florida represent approximately 50% of the estimated impact of fire ants in the USA, with the remaining 50% divided among all other infested states, including California. Although California initiated an eradication programme against fire ants, the estimated impact for California assumes that the infestation survives.

Prospects for the long-term sustainability of the AWPM programme

This AWPM project has enabled USDA and its cooperators to implement IPM of fire ants over large areas, over a sustained length of time and in diverse areas of the USA. A significant part of fire ant IPM has been the dissemination of self-sustaining parasites and pathogens in the infested areas. For the most part these biocontrol agents have become established and spread as anticipated or at an even greater rate and population density. In South America fire ant populations are five to ten times lower than in the USA, without the use of pesticides.

If the introduction of natural enemies of the fire ant reduces their population to one-half of what it is in South America, then reductions in the USA would be in the order of 40–45%, significantly reducing pesticide use for fire ant control and diminishing both the human impact of fire ants and their negative effects on agriculture and the environment. Results with biocontrol agents are not dramatic, but they are very encouraging for the long-term future (10–20 years), as additional biocontrol agents are released.

Ongoing and new research initiatives in biological control, bait improvement, biologically based control and new methods of fire ant detection and/or population assessment will continue to be highlighted on the areawide web site. In addition, we will maintain close contact with our demonstration site partners to provide consultation, and transfer new technology as it develops.

Summary and Future Directions

The areawide Suppression of Imported Fire Ants Project has entered the last 2 years of its expected duration. A new protocol has been developed to expand the project from the initial demonstration sites to other, smaller, sites in areas under different land use. Current sites were all established on improved, grazed pastures under cattle production. New demonstration sites were established on 'high value' properties where fire ant control is highly desirable and represents a high economic, environmental and/or aesthetic value (e.g. parks, poultry farms, hunting clubs, natural areas, military facilities, urban horticulture, etc.).

The objective is to expand the AWPM concept to other customers besides cattle farmers and to demonstrate that the concept of using biological controls in combination with toxic bait applications can be used in many different situations. This will apply what has been learned from the large-scale AWPM programme on pastures to properties and owners that have a high probability of continuing the fire ant IPM programme after project funding expires. It is expected that these properties will serve as examples for neighbouring property owners, and thus create a knowledge base on fire ant management and biological control that will provide for continuing expansion of interest in fire ant IPM in different regions in the USA.

References

Adams, C.T. (1986) Agricultural and medical impact of the imported fire ants. In: Lofgren, C.S. and Vander Meer, R.K. (eds) *Fire Ants and Leaf-cutting Ants: Biology and Management.* Westview Press, Boulder, Colorado, pp. 48–57.

Allen, C.R., Lutz, R.S. and Demaris, S. (1998) Ecological effects of the invasive nonindigenous ant, *Solenopsis invicta*, on native vertebrates: the wheels on the bus. *Transcriptions of the North American Wildlife and Natural Resources Conference* 63, 56–65.

Barr, C.L. and Drees, B.M. (1996) *Texas Cattle Producer's Survey: Impact of Red Imported Fire Ants on the Texas Cattle Industry Final Report.* Texas Agricultural Extension Service, Texas A&M University, College Station, Texas.

Callcott, A.-M.A. and Collins, H.L. (1996) Invasion and range expansion of imported fire ants (Hymenoptera: Formicidae) in North America from 1918–1995. *Florida Entomologist* 79, 240–251.

deShazo, R.D., Butcher, B.T. and Banks, W.A. (1990) Reactions to the stings of the imported fire ant. *New England Journal of Medicine* 323, 462–466.

deShazo, R.D. and Williams, D.F. (1995) Multiple fire ant stings indoors. *Southern Medical Journal* 88, 712–715.

deShazo, R.D., Williams, D.F. and Moak, E.S. (1999) Fire ant attacks on residents in health care facilities: a report of two cases. *Annals of Internal Medicine* 131, 424–429.

Forys, E.A., Allen, C.R. and Wojcik, D.P. (1997) *The Potential for Negative Impacts by Red Imported Fire Ants (*Solenopsis invicta*) on Listed Herpetofauna, Mammals, and Invertebrates in the Lower Florida Keys.* Florida Game and Fresh Water Fish Commission, Tallahassee, Florida.

Jusino-Atresino, R. and Phillips Jr., S.A. (1994) Impact of red imported fire ants on the ant fauna of central Texas. In: Williams, D.F. (ed.) *Exotic Ants. Biology, Impact, and Control of Introduced Species.* Westview Press, Boulder, Colorado, pp. 259–268.

Korzukhin, M.D., Porter, S.D., Thompson, L.C. and Wiley, S. (2001) Modeling temperature-dependent range limits for the fire ant *Solenopsis invicta* (Hymenoptera: Formicidae) in the United States. *Environmental Entomology* 30, 645–655.

Lockley, T.C. and Collins, H.L. (1990) Imported fire ant quarantine in the United States of America: past, present, and future. *Journal of the Mississippi Academy of Science* 35, 23–26.

Lofgren, C.S. (1986) The economic importance and control of imported fire ants in the United States. In: Vinson, S.B. (ed.) *Economic Impact and Control of Social Insects.* Praeger, New York, pp. 227–256.

Morel, L., Vander Meer, R.K. and Lofgren, C.S. (1990) Comparison of nestmate recognition between monogyne and polygyne populations of *Solenopsis invicta* (Hymenoptera: Formicidae). *Annals of the Entomological Society of America* 83, 642–647.

Pereira, R.M., Williams, D.F., Becnel, J.J. and Oi, D.H. (2002) Yellow head disease caused by a newly discovered *Mattesia* sp. in populations of the red imported fire ant, *Solenopsis invicta. Journal of Invertebrate Pathology* 81, 45–48.

Porter, S.D. (1992) Frequency and distribution of polygyne fire ants (Hymenoptera: Formicidae) in Florida. *Florida Entomologist* 75, 248–257.

Porter, S.D. and Savignano, D.A. (1990) Invasion of polygyne fire ants decimates native ants and disrupts arthropod community. *Ecology* 71, 2095–2106.

Porter, S.D., Bhatkar, A.K., Mulder, R., Vinson, S.B. and Clair, D.J. (1991) Distribution and density of polygyne fire ants (Hymenoptera: Formicidae) in Texas. *Journal of Economic Entomology* 84, 866–874.

Shatters, R.G.J. and Vander Meer, R.K. (2000) Characterizing the interaction between fire ants (Hymenoptera: Formicidae) and developing soybean plants. *Journal of Economic Entomology* 93, 1680–1687.

Thompson, L. and Jones, D. (1996) Expanding the Arkansas IFA farm survey over the south. In: Collins, H. (ed) *Proceedings 1996 Imported Fire Ant Research Conference*, USDA, APHIS, PPQ, IFA Lab, Gulfport, MS, New Orleans, Louisiana, pp. 81–83

Tschinkel, W.R. (1988) Colony growth and the ontogeny of worker polymorphism in the fire ant, *Solenopsis invicta*. *Behavioral Ecology and Sociobiology* 22, 103–115.

Tschinkel, W.R. (1993) Sociometry and sociogenesis of colonies of the fire ant *Solenopsis invicta* during one annual cycle. *Ecological Monographs* 64, 425–457.

Vinson, S.B. (1997) Invasion of the red imported fire ant (Hymenoptera: Formicidae): spread, biology, and impact. *American Entomologist* 43, 23–39.

Vinson, S.B. and Greenberg, L. (1986) The biology, physiology, and ecology of imported fire ants. In: Vinson, S.B. (ed.) *Economic Impact and Control of Social Insects*. Praeger, New York, pp. 193–226.

Williams, D.F., Oi, D.H., Porter, S.D., Pereira, R.M. and Briano, J.A. (2003) Biological control of imported fire ants (Hymenoptera: Formicidae). *American Entomologist* 49, 150–163.

Wojcik, D.P. (1994) Impact of the red imported fire ant on native ant species in Florida. In: Williams, D.F. (ed.) *Exotic Ants: Biology, Impact, and Control of Introduced Species*. Westview Press, Boulder, Colorado, pp. 269–281.

15 Salt Cedar Areawide Pest Management in the Western USA

RAYMOND I. CARRUTHERS,[1] C. JACK DELOACH,[2] JOHN C. HERR,[1] GERALD L. ANDERSON[3] AND ALLEN E. KNUTSON[4]

[1]USDA-ARS, Western Regional Research Center, Albany, California, USA
[2]USDA-ARS, Grassland, Soil and Water Research Laboratory, Temple, Texas, USA
[3]USDA-ARS, Northern Plains Agricultural Research Center, Sidney, Montana, USA
[4]Texas A&M University and Texas Agricultural Experiment Station, Dallas, Texas, USA

Introduction: Description of the Problem and Need for an Areawide Pest Management Approach

Salt cedar (*Tamarix* spp. (*Tamaricaceae: Tamaricales*)) is a group of exotic shrubs to small trees that have invaded many riparian areas and lake shores across western North America. Of the 54 species known worldwide (Baum, 1967, 1968), ten species of salt cedars have been introduced into the USA (Crins, 1989), primarily from their countries of origin across Europe and Central Asia. They are also native in Africa and the Indian subcontinent of Asia. No species from the entire family *Tamaricaceae* are native to North America, and only a restricted group of six species of more distantly related plants (*Frankenia: Frankeniaceae*) in the entire Order *Tamaricales* are known to exist in the USA.

Introduction of salt cedars began in the early to mid-1800s (first noted in 1823), when these species were used extensively for wind and water erosion control along railroads and waterways. Such use continued well into the mid- to late 1900s through plantings supported by federal, state and local governments and private land owners/managers. In many areas where they were planted salt cedars naturalized, became well established and spread throughout riparian areas of the west. It is now estimated that these exotic invasive shrubs infest more than 800,000 ha of highly valued riparian land from the central Great Plains to the Pacific coast, and from northern Mexico to the Canadian border.

Recent DNA studies (Gaskin and Schaal, 2003) indicate that *T. ramosissima* Ledeb. and *T. chinensis* Lour. and their hybrids are the most widespread and damaging species in the western USA, along with *T. parviflora* D.C. in California and Nevada, and *T. canariensis* Willd. (and hybrids) in some south-western areas of the country, especially along the Gulf Coast of Texas and Louisiana. They are all deciduous, deep-rooted, woody shrubs to small trees, which bloom in the spring and/or summer with pink to whitish flowers and are vegetated with foliage of juniper (cedar)-like bracts (see Plate 1).

Salt cedars have many characteristics that enable them to invade and occupy riparian areas and adjacent upland sites. They produce copious amounts of small windblown or waterborne seed throughout the spring and summer seasons at times when competing native plants are not typically reproductive. They also reproduce vegetatively, which helps them spread and re-establish following floods that periodically scour western waterways (Everitt, 1980). Salt cedars are deep-rooted, facultative phreatophytes that can use groundwater, soil moisture or surface waters. Thus, once established, they can occupy areas further from the stream banks and may consume more water across a flood plain than shallow-rooted native phreatophytes (Smith *et al.*, 1998). Salt cedars further have the advantage of being facultative halophytes that can use saline groundwater by excreting excess salts through leaf glands. This excretion results in increased salt levels in adjacent soils that may be highly limiting for other, less salt-tolerant, plant species. The resulting high soil salinity inhibits many native competitors and often leads to extensive monocultures of salt cedar (see Plate 1a, b).

Salt cedars are also tolerant of fire, drought, inundation, livestock or wildlife browsing (although herbivory is quite limited in North America), and thus have reproduced and spread widely with few natural controls. North American native insects and other wildlife did not evolve with salt cedar and thus rarely use it as a food resource, except that many pollinating adult insects (most produced as immatures on nearby native plants) visit salt cedar flowers for pollen and nectar (DeLoach and Tracy, 1997; DeLoach *et al.*, 2000).

Since native insects do not feed on the vegetation, roots, boles, flowers or seeds of salt cedar, they exert no noticeable level of natural control. Salt cedars are further tolerant to mechanical damage, and readily resprout from underground lateral buds after heavy scouring or other above-ground physical injury. The lack of herbivory, coupled with salt cedar's innate ability to withstand adversity and regrow under harsh conditions, has further led to its high densities and an expanding range over the past few decades.

In addition, salt cedars also interact synergistically with human-induced ecosystem changes in ways that further increase their competitive advantages over native and other beneficial plants (DeLoach, 1991; DeLoach *et al.*, 2000). Most significantly, changes in the timing of water flow through damming of streams in the spring and release of water in the summer, have decreased synchrony of spring-seeding native plants such as willows and cottonwoods, and have benefited summer-seeding salt cedars.

Significance of the Pest Management Problem

The invasion of riparian ecosystems by exotic salt cedars has caused one of the worst ecological disasters in the history of the western USA (Brown *et al.*, 1989; Lovich and DeGouvenain, 1998). Salt cedars displace native riparian plant communities, degrade wildlife habitats (including that of many declining or endangered species), use great quantities of scarce groundwater, increase soil salinity and wildfire frequency, and interfere with recreational usage of natural areas. These invasive shrubs increase bank aggradation, narrow and deepen stream channels and alter water temperature and quality. Salt cedars damage the habitat of many aquatic invertebrates, fish and riparian animals by eliminating backwaters and open sand and gravel bars, and by changing riffle and bank structure. They often create an impenetrable thicket that can exclude large wildlife and livestock from scarce water resources in arid grazing areas. The negative aspects of the salt cedar invasion have alarmed many ecologists and environmentalists (Sala *et al.*, 1996; Smith *et al.*, 1998; Shaforth *et al.*, 2005), water users, ranchers, park and wildlife managers and recreationalists, who are now requesting and/or demanding its control.

Salt cedar also has some limited beneficial value, mostly for controlling streambank erosion (for which it was originally introduced), to a lesser degree as an ornamental shrub and, occasionally, as a maintenance plant for honeybees in some locations. A few birds, including the endangered south-western subspecies of the willow flycatcher (*Empidonax trailii* Audubon *extimus* Phillips), the white-winged dove (*Zenaida asiatica* (L.)) and other miscellaneous animals use it for cover, as nesting sites and may feed on the pollinating insects found on its flowers. Despite these limited benefits, the USDI Fish and Wildlife Service (FWS) and other environmentally oriented groups have become aware of the damage caused by salt cedar, and now support the AWPM of this exotic plant through use of biological control and other integrated practices.

This was not always the case, but in-depth risk analyses (DeLoach, 1991; DeLoach and Tracy, 1997; DeLoach *et al.*, 2000), including environmental and economic analyses (Brown *et al.*, 1989; Zavaleta, 2000) have demonstrated that the damage caused by salt cedar far outweighs its few beneficial attributes, and thus helped alter the minds of conservation-oriented natural resource managers. It is now well accepted that salt cedars often competitively displace native plant communities, degrade wildlife habitat and contribute to the population decline of many species of birds, fishes, mammals and reptiles, including some 40 threatened or endangered species (see DeLoach and Tracy, 1997).

Most recently, the critical nature of the drought in the south-western USA has threatened the water supplies of municipalities and of irrigated agriculture, and caused default of water agreements between states and of the water treaty between the USA and Mexico. Salt cedar thickets typically use 4–5 acre feet of water per year, that in the present drought severely reduces the water available for agricultural irrigation, and municipal and environmental use. This has further engendered more political and monetary support for AWPM of salt cedar, using biological control as the central technology for controlling existing infestations and limiting spread of this invasive shrub.

Description of Current Management Systems and Approaches

In the absence of insect natural enemies and disease-causing organisms, salt cedar grows very aggressively and is highly competitive with native vegetation, especially in areas where the natural hydrology has been altered in ways that limit stream flow during spring months when native cottonwoods and willows are naturally seeding. Therefore, many of the nation's most productive and diverse ecological regions are being negatively affected by the invasion of this exotic invasive plant. Local, state and federal land and waterway managers have been fighting salt cedar in an expensive and losing battle using one or more traditional tools, such as the broadcast application of chemical herbicides, burning, physical removal via hand-cutting and stump treatment, or by bulldozing areas to remove both above and below ground salt cedar biomass.

Each of these methods has its strong and weak points but, overall, these treatments are extremely expensive, difficult to implement in wildlands or riparian areas and are highly disruptive to non-target flora and fauna in and around the affected habitats. Furthermore, extensive limitations exist due to state and federal regulations on both chemical and physical control methods in or adjacent to sensitive waterways. When used in an integrated manner, however, and linked with both areawide know-how and appropriate revegetation technology, these control methods can be effective and economical in many circumstances. If further linked with new biologically based methods that may be used as the keystone technology of an AWPM programme, they can become highly effective and sustainable.

The effectiveness and use of herbicides against salt cedars were reviewed by Sisneros (1991), who reported that Arsenal® (chemical name: imazapyr) used as an aerial spray alone or mixed with Rodeo® (glyphosate) to reduce costs, and Garlon® 4 (chemical name: triclopyr) as a cut-stump or stem-slash treatment, all provide good control. Arsenal® is a very broad-spectrum herbicide (except for legumes), and the label lists 176 species that it controls: 55 species of grasses, including salt grass, which is a common pasture grass in saline areas where salt cedar grows, 75 broadleaf weeds, 13 vines and brambles and 34 brush and tree species, including some very valuable native plants like cottonwoods and willows (BASF, 2004).

Although these chemicals are potentially effective, both chemicals are expensive and also kill many native plants, thus these controls are unsatisfactory in most natural areas of mixed vegetation where the objective is to kill the invading weed and preserve the beneficial and native plant species.

Less disruptive methods of chemical control involve the direct application of chemicals such as Garlon® 4 or Pathfinder® II (both triclopyr-based products) to the cut stumps of salt cedars, where it is absorbed into the opened cambium layer of the plant. Triclopyr is a synthetic auxin, a naturally occurring plant hormone that acts in plants to promote rapid growth and excessive cell division. Control of undesirable salt cedar from triclopyr treatments is essentially due to an auxin overdose that causes uncontrolled growth of the plant and eventual death at appropriate dosages. Both the cutting and application of the chemical to salt cedar stumps are labour intensive and thus costly. These methods, however, are much less disruptive to the background flora and fauna than broadcast application of a herbicide; however, the skeletons of cut salt cedars can cause increased fire danger if not physically removed from the treatment areas.

Fire, itself, has also been used extensively to help control salt cedar in many areas, as it is one of the few cost-effective ways of thinning extensive stands of salt cedar. The salt cedar tree typically begins regrowing within a month or so after a fire,

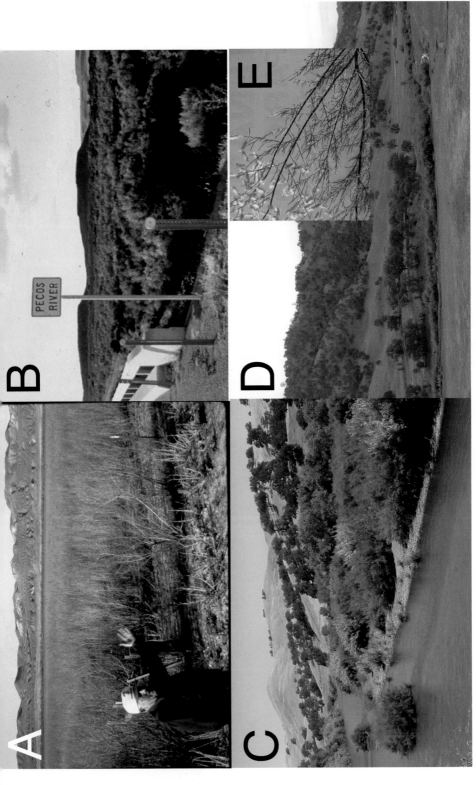

Plate 1. (A) An extensive infestation of saltcedar (*Tamarix ramosissima*) near Lovelock, Nevada (1997). This dormant saltcedar represents several thousand acres infested during a single flood event on the Humboldt River that spread seed nearly uniformly across the area. (B) A similar infestation of *T. ramosissima* along the Pecos River in Texas showing the dense foliage during the summer season. (C) *Tamarix parviflora* infesting Cache Creek, California, note the saltcedar growing directly in the water channel where they trap sediments, block the main channel and thus induce flooding. (D) *T. parviflora* in bloom along Bear Creek in California during early spring. (E) *T. ramosissima* blooms are variable in colour and occur throughout the summer season.

2.

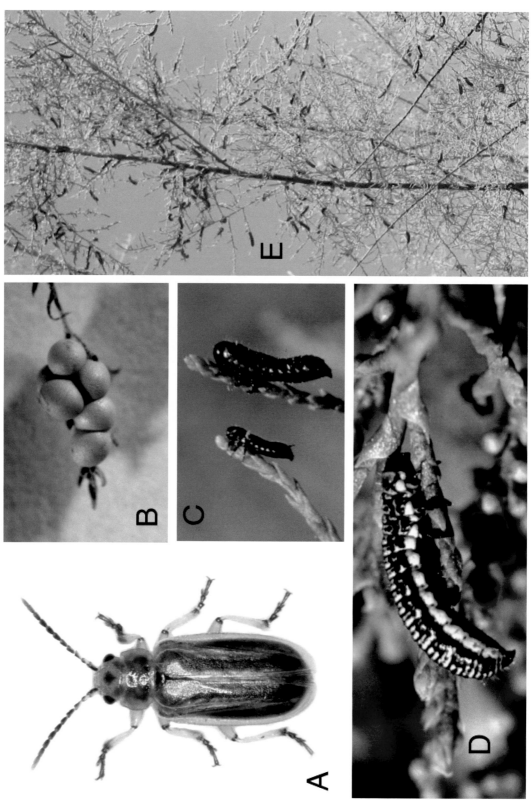

Plate 2. Life stages of *Diorhabda elongata* showing adults, eggs, 1st, 2nd and 3rd instars (A–D, respectively) and a common density found in the Areawide Pest Management

3.

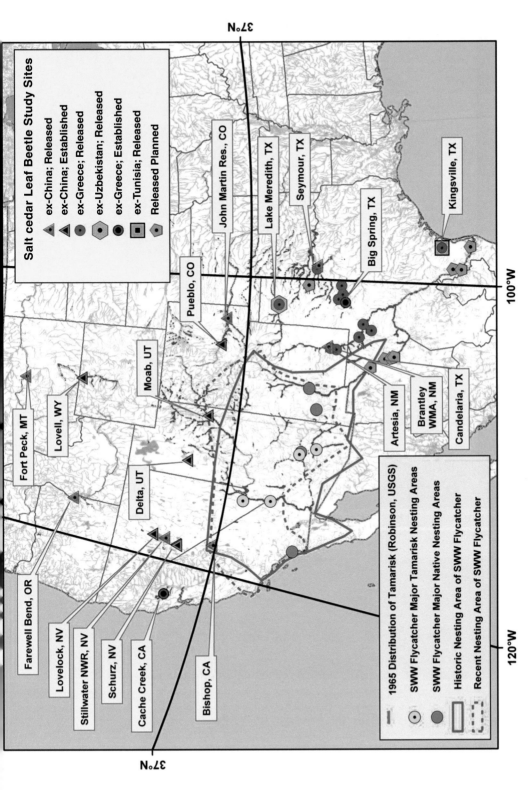

Plate 3. Study sites, including areas where the AWPM of saltcedar has been investigated. Note that *Diorhabda elongata* beetles from northern Asia worked best in more northern release sites and similarly *D. elongata* beetles from Greece have been released and are more effective in the southern USA.

Salt cedar Leaf Beetle Study Sites

- ex-China; Released
- ex-China; Established
- ex-Greece; Released
- ex-Uzbekistan; Released
- ex-Greece; Established
- ex-Tunisia; Released
- Released Planned

Farewell Bend, OR
Lovelock, NV
Stillwater NWR, NV
Schurz, NV
Cache Creek, CA
Bishop, CA
Fort Peck, MT
Lovell, WY
Delta, UT
Moab, UT
Pueblo, CO
John Martin Res., CO
Lake Meredith, TX
Seymour, TX
Big Spring, TX
Artesia, NM
Brantley WMA, NM
Candelaria, TX
Kingsville, TX

37°N
37°N
120°W
100°W

- 1965 Distribution of Tamarisk (Robinson, USGS)
- SWW Flycatcher Major Tamarisk Nesting Areas
- SWW Flycatcher Major Native Nesting Areas
- Historic Nesting Area of SWW Flycatcher
- Recent Nesting Area of SWW Flycatcher

4.

5.

Plate 4. The centre of the *Diorhabda elongata* release site (area of brown vegetation) is readily evident near the looping dirt road (A and B). Outside of the release area, saltcedar plants are lush and green (C), while within the release area (D) the saltcedar have been totally stripped of green foliage by *D. elongata* adults and larvae.

Plate 5. (A) Heavy first-year defoliation caused by *Diorhabda elongata* at the Shurz, Nevada, release site. Often, the plants dieback more severely than caused by the defoliation alone, leaving much dried leaf material on the plants. (B) Intermixed with heavily defoliated saltcedar, many native species flourish due to decreased competition from the saltcedar. No feeding has been noted on any non-target species at any of the release sites (note healthy green cottonwoods, willows, sagebrush and other species intermixed with the defoliated saltcedar (A, B)). (C) Adult *D. elongata* populations at these sites were very high. (D) Larval populations at the same sites were at tremendous levels for multiple seasons causing repeated defoliations over several seasons.

Plate 6. (A) In early spring, *Tamarix parviflora* blooms heavily with dark pink flowers prior to leafout. (B) Such blooming patterns aid in visual detection of this species from the air. (C) Aerial photographs of *T. parviflora* have been spatially georectified and digitally analysed using both colour and texturing techniques. (D) This allowed the development of a mosaic map showing the distribution and density of saltcedar along the watercourse, helping local land managers to plan AWPM of this invasive plant.

7.

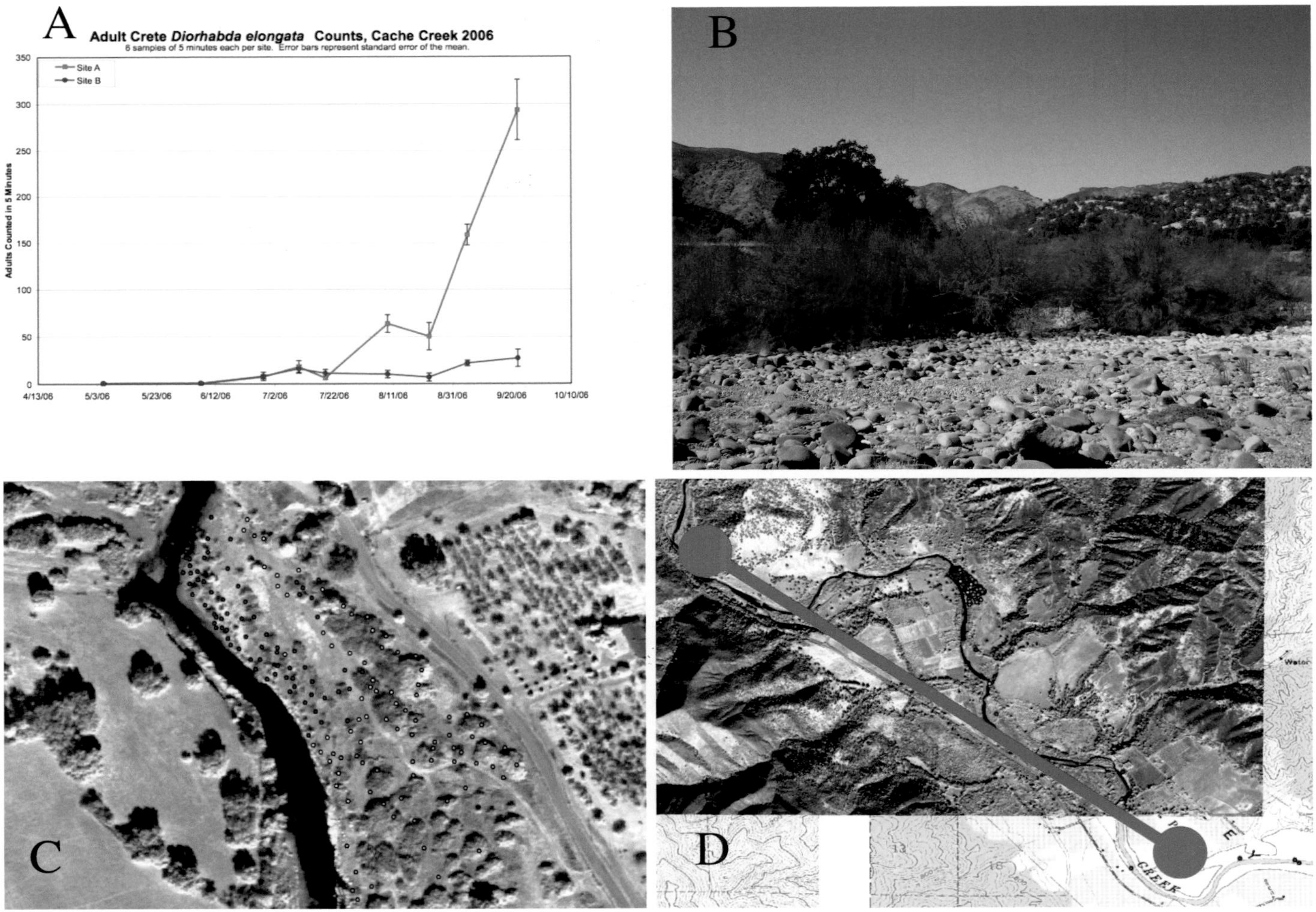

Plate 7. (A) Infestations start slowing right after the beetles are released, but grow exponentially as there are few natural enemies to limit their populations as seen in this graph depicting adult counts at the Cache Creek release area where beetles from Crete, Greece were released. (B) Once *Diorhabda elongata* numbers increase to a critical threshold level, heavy defoliation is seen on impacted plants near the release areas. (C) This heavy defoliation induces adult beetle movement that the AWPM project tracks first on the ground using GPS and vegetation sampling. (D) However, the beetles and the associated defoliation quickly spread, as in this Cache Creek release site where within a single season beetles were detected over an 8 km (5 mile) stretch of habitat.

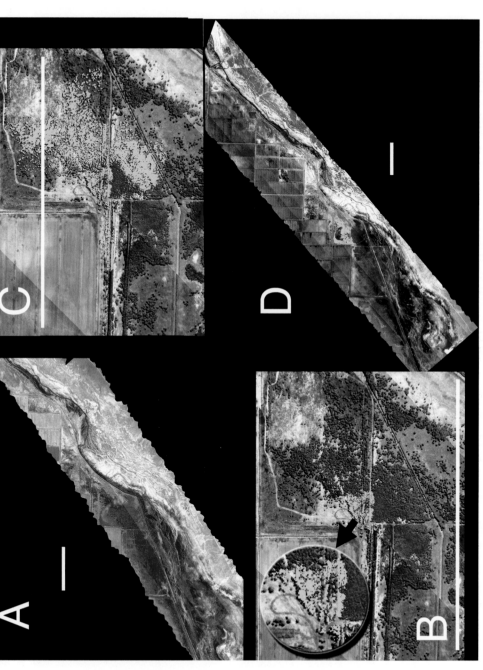

Plate 8. (A) An infrared image shows healthy saltcedar plants (red coloration depicts healthy vegetation) across the Lovelock, Nevada, release site in July 2002 (the white bar in each panel represents ca. 1 km). (B) Note the yellow artificially coloured area (0.35 ha), where the *Diorhabda elongata* beetles totally stripped the foliage by the end of August 2002. Note the loop road that can be seen next to the release site in Plate 4 (A and B). These hyperspectral images allow detailed assessments to be made over wide areas to track beetle impact through time. (C) In July 2003, the defoliated area had significantly expanded to over 3.4 ha of actual canopy and significantly more ground area. (D) By September 2003, 76.7 ha of canopy had been totally defoliated, and repeated defoliation occurred within the entire test area impacted in previous beetle generations.

9.

Plant Defoliation by Year and Beetle Generation

2002	canopy/ground area	2004	canopy/ground area
July	0.0/0.0 ha	July	607.0/810.0 ha
Sept	0.3/1.0 ha	Sept	1500.0/2025.0 ha
2003		**2005**	
July	3.2/4.3 ha		all salt cedar within
Sept	76.6/200 ha		26,300 ha area

Plate 9. Beetle damage to saltcedar at Lovelock, Nevada, was tremendous by 2004, impacting thousands of hectares and was beginning to cause saltcedar mortality around the original release site. By 2005, beetles were found > 150 km from the original release area and most saltcedars in northern Nevada were impacted to some level with defoliation continuing to increase at many sites.

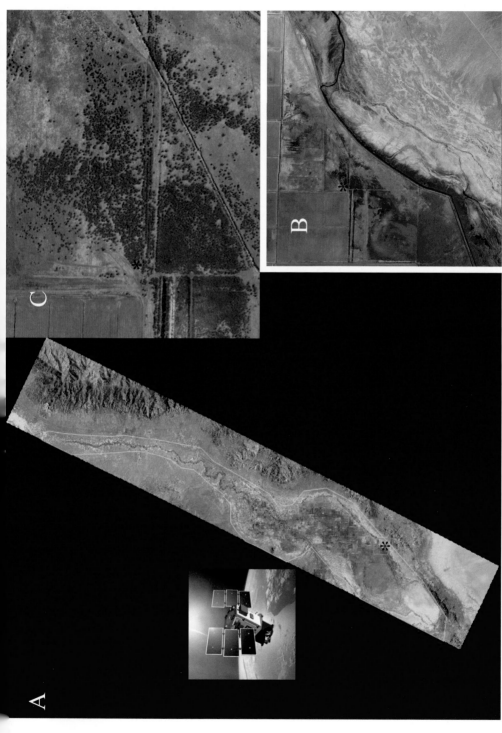

Plate 10. Neither ground nor aerial surveys proved adequate to assess the areas of defoliation caused by *Diorhabda elongata* across northern Nevada and Utah test sites. Alternatively, Quickbird satellite imagery was used to help assess saltcedar defoliation along the Humboldt River (An * indicates the original release site in all three images). (A) An approximate 80-mile swath representing multispectral Quickbird data collected to aid in defoliation assessment. (B) Shows a satellite depiction of the original Lovelock release site, again with the pinkish red coloration depicting green vegetation. (C) A maximum zoom image of the release site showing individual dying saltcedar with no signs of live vegetation. Imagery supplied by Digital Globe Corporation.

but such prescribed fires typically thin salt cedar thickets enough to allow worker entry into an area where follow-up herbicidal applications can be made on plant regrowth. The USDI has used this approach in managing salt cedar in heavily infested areas of California (West, 2007; http://www.invasivespeciesinfo.gov/docs/news/workshopSep96/west.html); however, it is typically very detrimental to native woody species such as cottonwoods and willows in the same or adjacent areas. One hundred-foot (30 m) buffer areas are recommended between treated zones and areas that are to be protected from the prescribed burn.

Bulldozers and other heavy equipment have also been used in attempts to control salt cedar, but since this plant easily resprouts from root fragments, it alone is typically not adequate to control this shrub unless repeated treatments or combined treatments with other control measures are used. Physical treatment of this type is also usually highly disruptive to local habitats and is the least economical and least desired method of control. In some circumstances, however, it has been one of the only ways to destroy heavily infested sites where fire is not an option, thus allowing land managers open access to areas that were previously unpenetrable for other ground-based control actions.

Limitations of Current Management Approaches

Although traditional approaches to salt cedar control may be successful in the short term they do not provide permanent control of the problem, as salt cedar often grows back or re-invades from surrounding areas. Such treatment also affects the quality of the treated habitat, which is typically degraded by the control action due to unwanted non-target impacts on native flora and fauna. Furthermore, most areas of infestation are spread along linear waterways or throughout a network of complex riverine tributaries within different-sized watersheds. Such watersheds may range from the extensive Colorado and Rio Grande River systems to localized areas such as isolated desert springs.

Due to the nature of most western riparian corridors, salt cedar infestations often cross many different political and ecological boundaries. Thus control programmes are not often implemented holistically, as the logistics of coordinating many land-owners and managers across such geographic and political boundaries are often difficult, if not impossible. Thus localized implementation of any such control efforts usually has only short-term effects unless conducted cooperatively, by using water-shed level practices to synchronize management actions and to avoid re-invasion.

When targeting small, isolated areas of infestation, labour-intensive cutting and stump treatment has been most effective. This approach has been successfully used by the National Park Service locally to eradicate salt cedar from restricted areas (see http://www.nature.nps.gov/biology/ipm/manual/exweeds1.cfm). On a larger scale, other management groups have targeted entire tributaries, such as major sections of the Pecos River in New Mexico and Texas, with broadcast chemical applications. On the Pecos River near Artesia, New Mexico, *c.* 2000 ha of salt cedar were treated from 2001 to 2004, with 85–90% kill.

Along the Pecos River, Texas from 1999 to 2004, 5169 ha of salt cedar along 434 km of the river and its tributaries have been treated, at a cost of US$2.5 million

(US$407/ha), with 85–90% kill (Hart *et al.*, 2000, 2005; Hart, 2004, 2006), and the present cost is now US$580/ha. Such areawide chemical application techniques often require several follow-up treatments to kill salt cedar regrowth and, when successful, are unfortunately equally devastating to most of the other vegetation in the area, as the chemicals used for salt cedar control have little or no selectivity. More environmentally sensitive cut-stump treatments require much more physical labour and often cost between US$2000 and US$3000/ha (J. Watson, California, 2006, personal communication).

Such areawide chemical application programmes have achieved the organization of landowners and management groups within a watershed to obtain more complete control of salt cedar in order to limit re-invasion potential. However, planning and implementation is difficult and not always successful, as many landowners often do not want to invest the resources necessary to conduct such work, nor do they agree easily on the combined technology or the economics of the control effort. Effective follow-up is even more difficult, as once the primary control effort passes, sustained enthusiasm seems hard to maintain, especially as continuing investments are often required. Even when systematic watershed management practices are followed and the majority of the land is treated, re-invasion may occur across waterways through seed transport or other methods of movement. In these situations, parallel AWPM using biological control can significantly aid in lowering salt cedar densities by limiting regrowth and spread.

Furthermore, without adequate revegetation methods used in direct follow-up to such a scorched-earth policy, severe erosion problems can quickly destroy the rapid gains made by eliminating salt cedar, since other covering vegetation is also lost. Such rapid removal of all the vegetation from an area and the associated ecological disturbance may result in other noxious and invasive weeds repopulating treated areas rather than the desired beneficial vegetation. More gradual and sustained means of reducing and/or eliminating salt cedar significantly improve the long-term healing and sustainability of any such management practice, as they allow a more gradual transition from an infested to a restored state.

Use of AWPM with biological control as the central and unifying technology provides such an approach, as biological control typically requires 3–5 years of repeated defoliation to cause salt cedar mortality. Although biological control agent feeding may cause rapid leaf loss and limit water use very quickly, actual mortality is caused by the combined impact of the beetles, a declining photosynthetic capability of the weed and competition from other plants in the area, rather than by chemical toxicity of herbicides. Thus salt cedar mortality is often caused secondarily through competition with other plants in the local community that respond to increasing water and light availability, and thus are released by the decline in salt cedar vigour to repopulate areas where biological control has been used.

Anticipated Benefits of AWPM

We expect the areawide application of biological control of salt cedar to reduce, gradually (over a period of 3–5 years) and permanently (through longer periods of time) the abundance of salt cedar to below the level of economic or environmental damage, but not to eradicate it. In this situation, both salt cedar and the insects used for biological control would remain at fluctuating low population levels, and thus the agent or agents would always remain present at some level to rebound and control regrowth or re-invasion of windblown or waterborne seeds, if needed. Biological control can work alone or in combination with other methods of salt cedar control, depending upon the local needs of the land and waterways managers. In some areas, rapid knock-down of salt cedar is desired to facilitate flood control, fire management or other purposes, and then the biological control agents can be used to control regrowth or, if control over a longer period of time is desired, insects can be used alone or with follow-up treatments using other control methods.

Under most conditions, we expect the native plant communities to re-establish naturally in areas where depth to water table and soil salinity are not too great. This should improve wildlife habitat and allow the recovery of many species of birds and fish and some mammals and reptiles, including several threatened and endangered species. Control of salt cedar is expected to increase the amount and quality of water available for irrigated agriculture and municipal and environmental use, and to help fulfil the interstate water rights agreements and between the USA and Mexico.

Control is also expected to increase recreational usage of parks and wildland areas, to reduce wildfires and to allow the gradual reduction of salinity levels of surface soils in presently infested areas. Large-scale revegetation projects are also under development by the USDI Bureau of Reclamation for areas where natural revegetation may be insufficient (Lair and Wynn, 2002). In some areas where salt cedar has formed extensive monocultures for many decades, where the natural hydrologic cycles have been interrupted, soil salinity increased or water tables lowered, natural revegetation is unlikely and will require additional interventions.

Since biological control is a gradual process rather than a quick-acting control method such as chemical and physical control, and also since it is target selective, we feel that it holds the best potential for allowing various habitat restoration processes to be developed and implemented gradually, where other controls would not be easily compatible with recovery efforts.

Description of the AWPM Programme and Approaches

The low beneficial value of salt cedar, its lack of closely related plants in the Western hemisphere and the large number of host-specific and damaging natural enemies that attack it within its native distribution in the Old World, make salt cedar an almost ideal invasive weed for an AWPM programme using biological control as its keystone technology. Over the past decade, biological control of salt cedar has seen a major research effort within the USDA-ARS and its cooperators. A consortium of scientists and land managers has recently developed and field-tested the use of this new technology in several western states, including California (CA), Colorado (CO), Montana (MT), New Mexico (NM), Nevada (NV), Oregon (OR), Texas (TX), Utah (UT) and Wyoming (WY) (see DeLoach et al., 2004).

The programme began with surveys for potential natural enemies carried out in Italy, Israel, Iran, India, Pakistan (Gerling and Kugler, 1973; Habib and Hassan, 1982) and Turkey (Pemberton and Hoover, 1980), together with extensive ecosystem studies in the former Soviet Union (Mityaev, 1958; Kovalev, 1995) and by our team in China and Kazakhstan (Li and Ming, 2001–2002; Mityaev and Jashenko, 2007).

This combined team effort revealed over 300 specific and highly damaging insect species that are now being considered as potential biological control agents for salt cedar. Research began at Temple, TX, in 1986, with a thorough review of the literature and an analysis assessing the benefits and risks of such a programme (DeLoach *et al.*, 2004). Overseas testing of control agents was then begun under quarantine condition at Temple, TX, in 1992 (see DeLoach *et al.*, 1996) and in Albany, CA, in 1998 (see Lewis *et al.*, 2003a).

AWPM management technologies and approaches

The leaf beetle, *Diorhabda elongata deserticola* Chen (1961), was selected as the first potential introduced biological control agent of salt cedar, since it was observed to severely defoliate this target plant in research sites in China, Kazakhstan and other Eurasian locations. Defoliation was typically heavy as the beetles attacked in mass, causing vegetative damage that induced severe dieback of most branches on affected plants. In these native areas, most defoliated salt cedar resprout from the base of the plant late in the season, with the next generation of insects either continuing to feed on these same plants or moving on to other adjacent plants, allowing the impacted stand to recover.

Diorhabda elongata populations in these areas were limited by their own suite of natural enemies, most notably a tachinid larval parasite (*Erynniopsis antennata* (Rondani)) and a microsporidian pathogen (*Nosema* spp.), and thus rarely reached extremely high population densities in native areas (a factor that does not hold in North America once top-down population regulators were removed from the system in quarantine).

In Eurasia, an ecological balance seems to have been established between low densities of the native salt cedars, the many diverse natural enemies that use the plant and their associated parasites and predators in areas of origin. The leaf beetle, however, seemed very appropriate as a potential biological control agent of salt cedar, as it caused high levels of defoliation (for short times in limited areas), had a high reproductive rate and further exhibited extensive dispersal capabilities in nature. Thus, in March 1994, a petition was submitted to USDA Animal and Plant Health Inspection Service (APHIS) Technical Advisory Group on Biological Control of Weeds (TAG) in 1998, allowing for the caged release of the leaf beetle *D. elongata* from China and Kazakhstan into several western US states. This was the beginning of the research and developmental phases of the AWPM programme for salt cedar control in the USA.

Thus, *D. elongata* was selected as the first of ten priority agents for consideration as a manipulated biological control agent to be used in North America, and will be the only insect discussed here (see DeLoach *et al.*, 1996 for a discussion of other potential biological control agents). Parallel studies on *D. elongata* biology, host specificity and efficacy were outlined and implemented by our internal USDA-ARS AWPM team and affiliated members of the Saltcedar Biological Control Consortium, formed in 1998 (see section on Technology Transfer and Education). Through joint efforts of this consortium and core project scientists, funding for a coordinated research and developmental programme was acquired through competitive selection of a USDA-CSREES-IFAFS grant. Without such funding, an AWPM approach and the regional testing and development would not have been possible (Carruthers *et al.*, 2000; Carruthers, 2003, 2004).

The biology of *D. elongata* – from both Fukang, Xinjiang Autonomous Region, China and from Chilik (120 km ENE of Almaty), Kazakhstan (see Plate 2a–e) – was assessed by our team in Kazakhstan (Mityaev and Jashenko, 1999–2005), in China (Li and Ming, 2001–2002) and at Temple, TX, Albany, CA and the other various release sites in the USA (see subsequent sections). Both adults and larvae of *D. elongata* feed on the foliage of salt cedar, and the large larvae also de-bark small twigs causing the distal foliage to die. The adults overwinter and the larvae pupate under litter and in the soil beneath the trees. In the laboratory, an average female oviposits approximately 200 eggs over a 12-day period, but can live up to several months (1–3 months) in the field.

Lewis *et al.* (2003b) measured the duration of each life stage, calculated an optimal net reproductive rate (R_o) of 88.2, time per generation (T) of 39.9 days and the rate of increase, showing that the population can double each 6.2 days. Further temperature-dependent developmental and survival characteristics were measured and used to construct predictive models to estimate phenology, survival and reproduction under both insectary and field conditions (Herrera *et al.*, 2005). Field cage studies conducted in multiple sites revealed a range of population increases by location, averaging approximately 30-fold per generation (DeLoach *et al.*, 2004).

The synchronization of the beetle life stages with normal salt cedar leaf-out enables the overwintering beetle adults to emerge from April to early May, depending upon heat accumulation patterns in the local area and, after a short preovipositional period, to lay eggs from early April to late May. First-generation larvae are typically present from May through June, pupate in the leaf litter/soil interface under the salt cedar and then emerge as first-generation adults in late June–mid-July. In most areas north of the 38° parallel, where the daylength is sufficient, the first-generation adults reproduce actively and initiate a second generation of larvae that follow the same general cycle as the first generation.

In all areas the beetles end the season as adults, emerging over a period from August to October, depending upon local conditions. The second-generation adults feed for a short time, rarely oviposit and then aggregate in the soil/litter interface where they overwinter in high-density masses, often of thousands of beetles or more. In warmer locations with earlier spring warm-up, such as California and southern Texas, beetles can emerge as early as late March and may have up to four or five generations within a growing season, as they also stay active much later into the autumn.

Heavy population densities, especially of large larvae and adults, can produce severe defoliation in one or more generations, each year. In the more southern areas, where the salt cedar growing season is longer, some strains of beetles may complete four or more generations in a single season. The number of generations of beetles is dependent upon their temperature-dependent developmental rates (Herrera *et al.*, 2005) and their diapause induction characteristics, which are affected by a combination of temperature and photoperiod (Lewis *et al.* 2003b; Bean *et al.*, 2007a, b).

The initial populations of beetles from western China and Kazakhstan were not well adapted to the more southern latitudes of the south-western USA such as California, New Mexico and Texas. For that reason, additional biotypes (potentially different species, although not yet documented in the taxonomic literature) were collected from more southern latitudes in Eurasia and North Africa, and were then used successfully in the south-western areas of the USA (see Carruthers *et al.*, 2006, 2007; DeLoach *et al.*, 2007a, b). Signs now indicate that potential selection among the original stocks from China and Kazakhstan may be occurring, and thus may become adapted in terms of diapause induction to more southern latitudes (Thompson *et al.*, 2007).

Extensive host range testing overseas, in conjunction with quarantine tests conducted at Temple, TX and Albany, CA, indicated that *D. elongata* is very host-specific in its feeding patterns and is virtually restricted to feeding on invasive salt cedars in North America, with the exception of potentially feeding on a group of herbaceous to shrubby plants in the family *Frankeniaceae*, Order *Tamaricales*.

The *D. elongata* Fukang/Chilik ecotype was extensively tested for its ability to develop, reproduce and complete its entire life cycle on 84 different test plant accessions, including six species and 26 accessions of *Tamarix*, four species of the somewhat related and native *Frankenia* and 71 species of more distantly related plants, habitat associates, agricultural crops and ornamental plants in 24 families in more than 24 tests over 14 years (DeLoach *et al.*, 2003; Lewis *et al.*, 2003a; Dudley and Kazmer, 2005; Milbrath and DeLoach, 2006a, b; Herr *et al.*, 2007).

The most at-risk species were found to be *Frankenia salina* (Molina) I.M. Johnston, *F. jamesii* Torrey and the formerly endangered *F. johnstonii* Correll. Extensive additional testing of *Frankenia* was conducted under APHIS permits in the laboratory, greenhouse, field cages and, eventually, in the open field. These tests demonstrated that *D. elongata* from both Fukang, China and from Chilik, Kazakhstan, were safe to release in the field. Similar laboratory testing was conducted on six other strains of beetles collected at later dates, but only strains from China, Kazakhstan and Greece received the full battery of tests that warranted field release into areas where *F. salina*, the most at-risk species, was present (Herr *et al.*, 2007).

Open-field tests conducted on the most vulnerable species indicate that, on *Frankenia* spp., we expect only occasional attraction to, or feeding and oviposition on, the plants, if they grow adjacent to *Tamarix*. Although possible, we do not expect the beetles to develop self-sustaining populations on *Frankenia*, nor do we expect *Frankenia* to be a sustaining host plant in nature (see Milbrath and DeLoach, 2006a, b; Herr *et al.*, 2007). In parallel with all beetle releases, forced selection studies have been and are continuing to assess potential host shifts in the laboratory, and additional field testing and monitoring of *F. salina* is also being conducted, and will continue for several years to assess any possible unwanted negative impacts with time; however, none are anticipated.

Compatibility of the AWPM programme with crop management and management of co-occurring pests and native species

Currently, salt cedar is controlled in most locations as a single, invasive species; however, it can co-occur with a number of other invasive weeds in complexes that differ with geographical location. These species vary across the western USA, but include problematic weeds such as giant reed (*Arundo donax* L.), tall whitetop (*Lepidium latifolium* L.), yellow starthistle (*Centaurea solstitialis* L.), Russian knapweed (*Acroptilon repens* (L.) D.C.) and Russian olive (*Elaeagnus angustifolia* L.). Individual land management teams currently deal with these species as follow-up problems after first controlling for salt cedar to increase access to infested riparian corridors. This control is typically conducted chemically or through physical removal of associated weed species. No integrated multiple species pest control efforts have been developed or implemented to control these assemblages of invasive weeds; however, USDA-ARS has initiated new biological control efforts for most of these species and, hopefully, new natural enemies will be incorporated into more holistic integrated weed management programmes, once they become available.

However, additional effort has been directed toward making AWPM of salt cedar compatible with the management of other beneficial species, including native willows and cottonwoods, which represent natural vegetation that is desired in many riparian areas, along with an assemblage of other native and beneficial plant and animal species. In fact, much effort is currently oriented toward revegetation technologies (Lair and Wynn, 2002), and is discussed in a previous section. One ongoing issue of compatibility, however, still exists with the use of AWPM for biological control and the management of an endangered native bird.

The listing of the south-western willow flycatcher (*Empidonax trailii* subsp. *extimus* Phillips (SWWFC)) as federally endangered, in March 1995, required consultation with FWS and the preparation of a Biological Assessment, which was submitted to FWS Region 2 (Albuquerque, NM) in October 1997. This analysis revealed that the flycatcher used salt cedar extensively for nesting habitat in some areas of Arizona, but little in adjacent states, and that other potentially harmful effects of salt cedar reduced reproductive success of the flycatcher to half of that in its native willow habitat (DeLoach and Tracy, 1997; DeLoach *et al.*, 2000).

Due to potential impacts of rapidly reducing salt cedar canopies through extensive defoliation and the questionable impact to nesting sites of an endangered subspecies of bird (the SWWFC), the project was allowed to proceed only slowly, in phases and in locations over 200 miles from SWWFC nesting sites. The US FWS required that initial beetle releases be conducted within cages and then gradually scaled up to open-field releases once experience had been gained on the beetle's biology and impact under North American conditions. Thus, the AWPM effort followed both timelines and monitoring requirements set through negotiations between the regulatory agencies (US FWS, USDA-APHIS and various states), USDA-ARS and the Saltcedar Biological Control Consortium.

To further assess this situation, a research proposal to FWS was submitted on 28 August 1998 (DeLoach and Gould, 1998). It specified a research phase in which:

- *Diorhabda elongata* could be released into secure field cages at ten specified sites in different climatic zones in CA, CO, NV, TX, UT and WY, all more than

200 miles from where the SWWFC nests in salt cedar. The beetles were to be carefully monitored in the cages for 1 year to determine their overwintering ability, mortality factors, rate of increase and damage to salt cedar and non-target plants in the cages.

- The beetles then could be released into the open field over a 2-year period, during which the degree and rapidity of control, rate of natural dispersal and effects on native plant and wildlife communities would be monitored.

After this combined 3-year test period, FWS, ARS and APHIS would review the research results and determine the conditions under which an Areawide Implementation Phase could be conducted. A Letter of Concurrence was issued by FWS on 28 December 1998 (revised 3 June 1999) and an Environmental Assessment was prepared by USDA-APHIS on 18 March 1999. APHIS issued a Finding of No Significant Impact (FONSI) on 7 July, and permits to release in field cages during July 1999.

By taking this approach, it significantly slowed the development of AWPM of salt cedar, but it has also provided much experience with salt cedar control in areas away from the SWWFC, and thus scientists working in the areas where this bird nests in salt cedar are better prepared to consider using biological control as a management option in the future. Currently, no such programme is planned for the heart of the endangered SWWFC habitat in AZ, but programme expansion into parts of NM and southern CA will soon require that this issue be addressed.

Development and Implementation of the AWPM Programme

Experimental releases and results in field cages: July 1999–May 2001

Diorhabda elongata desericola from Fukang, China, were released into field cages during July and August 1999 at eight sites (see Plate 3) on a privately owned ranch near Seymour, TX; on Bureau of Reclamation land near Pueblo, CO; National Park Service lands near Lovell, WY; Paiute Indian tribal lands near Schurz, NV; a privately owned farm near Lovelock, NV; Los Angeles County Water District lands near Bishop, CA; and Hunter-Liggett Military Base, near Lockwood, CA. Beetles from Chilik, Kazakhstan were released on Bureau of Land Management land near Delta, UT.

During the spring of 2000, beetles from Fukang also were released into cages at Stillwater National Wildlife Refuge near Fallon, NV and on private land at Cache Creek near Woodland, CA. These beetles successfully overwintered in the cages at the eight most northerly sites, although only weakly so at Stillwater and Cache Creek. They failed to overwinter at the two most southerly sites, at Seymour, TX and Hunter-Liggett, CA. At the six sites where strong overwintering occurred (Pueblo, CO, Lovell, WY, Delta, UT, Lovelock and Schurz, NV and Bishop, CA), the beetles increased to large numbers during the summer and completely defoliated the plants inside the cages during both 1999 and 2000. The two generations of larvae during June and August produced extensive damage to the caged salt cedar, such that additional cages had to be established over fresh plants where beetles were transferred to preserve the outdoor colonies. After the failure to overwinter at the most southern sites, the beetles were restocked but failed to reproduce, entered diapause in early July, and again failed to overwinter.

Field observations and experiments in outdoor cages indicated that the most probable cause of the failure to overwinter and reproduce was the short summer daylengths at latitudes below 38° (see Lewis *et al.*, 2003b; Bean *et al.*, 2007a, b), which induced premature diapause. Daylength near the origin of these beetles at Fukang (44°17′ N) and Chilik (43°33′ N latitude) attains a maximum of 15.5 h; maximum daylength at more southerly locations in the USA was only around 14 h.

In follow-up laboratory studies, Bean *et al.* (2007a, b) determined that *D. elongata* from Fukang required at least 14.75 h daylength to avoid entering overwintering diapause. Thus, in the most southerly release areas, the beetles often began diapausing in early July without adequate winter reserves to survive the extended time in dormancy (Lewis *et al.*, 2003b).

These findings led to further overseas exploration and the acquisition of additional beetle strains from a wider latitudinal gradient, from North Africa, through Europe and into Western Asia, to supplement existing populations of *D. elongata* from Kazakhstan and China. Thus, the programme proceeded as planned in the more northerly areas toward open-field releases of beetles, and worked to assess potential new germplasm for use in more southerly locations.

Currently, beetles from Crete (Greece) are the primary insect natural enemy being successfully used below the 38° parallel; however, investigations are continuing on other potential beetle populations, which will be used as needed where the currently released beetles seem ecologically unsuitable. See DeLoach *et al.* (2007a, b) for an up-to-date assessment of the different ecotypes of beetles currently being used in this programme throughout the country.

Releases and results in the open field in northern areas: May 2001–late summer 2006

The results of the releases into field cages and of the additional test results of *D. elongata deserticola* were submitted to USDA-APHIS and the US FWS on 25 August 2000, requesting releases into the open field. These permits were granted and adult beetles were released into the open field beginning May 2001 at the six sites where the beetles had successfully overwintered in cages. Additional releases were made during the remainder of the year as beetles were produced in nursery cages. Altogether, approximately 27,000 adults and larvae were released at Lovell, WY; 6900 adults plus many larvae at Pueblo, CO; 15,000 at Delta, UT; 3500 at Schurz and 1400 at Lovelock, NV; 4400 larvae and 2000 adults at Bishop, CA; and 498 adults at Seymour, TX. Upon release, the beetles typically dispersed rapidly and were difficult to observe in large numbers in or around the release area.

At most sites, only a few to moderate numbers of eggs, larvae and adults were found in the field throughout the remainder of the summer of 2001, until late August or early September, when no more insects were found; and thus we assumed that the adults had entered overwintering diapause. In the autumn of 2001, the most severe damage was observed at Pueblo, CO, where the beetles defoliated approximately two-thirds of a rather large tree about 10 m from where they had been released. Similar low densities of beetles were found during the spring and early summer of 2002, and were detectable only over an area of *c.* 50–100 m in radius from their release points.

Then, when large larvae of the second generation developed in mid-August, extensive damage was found at some sites. The most spectacular impact was in Lovelock, NV: tremendous populations of large and some medium-sized larvae were found on 13 August that rapidly defoliated salt cedar trees near the original release site. On 28 August, the larvae had destroyed 95–98% of the foliage of all trees within an area 100 m in diameter (< 1 ha), centred at the release cage (see Plate 4). Heavy feeding, but not total defoliation, had occurred in an additional concentric ring 50 m wide outside the core of the affected area (DeLoach *et al.*, 2004).

By the end of the third growing season in late August 2003, the Fukang/Chilik ecotype of *D. elongata* had begun a rapid and dramatic defoliation of salt cedar at five of the seven release sites north of the 38th parallel. At Lovelock, NV, the beetles had defoliated *c.* 1 ha of a dense stand of salt cedar in early September 2002 (total area of land infested by salt cedar, not canopy cover of the trees), which had increased to 4.3 ha by early July 2003, and to *c.* 200 ha by early September 2003, along a 5 km reach of the Humboldt River (Carruthers *et al.*, 2005). By September 2003, several plants had resprouted profusely from the base and occasionally from the upper branches, but enough beetles had remained in the stand to totally defoliate this regrowth. At Schurz (see Plate 5), the beetles had defoliated *c.* 12 ha along the Walker River.

By the end of the third growing season after release (late June 2004), defoliation by the Fukang/Chilik beetles at five northern sites had increased by three- to fivefold over the amount in August 2003, to an estimated 600 ha at Lovelock and 120 ha at Schurz. Defoliation increased exponentially in most release areas, both in terms of the insect numbers, their linear spread and the area of salt cedar impacted. By the summer of 2006 many thousands of acres had been totally defoliated at most of the release sites. Most significantly, as of late 2006, trees at the centre of the release sites were beginning to die. For example, approximately 65% of the trees defoliated in 2002 at the site in Lovelock (see Plate 4) were completely dead (T.L. Dudley, California, 2006, personal communication), and we expect to see increasing whole-tree death in wider areas in 2007.

Although salt cedar mortality took several years, the repeatedly high levels of defoliation (95–100% for each generation) have reduced normal plant respiration and transpiration significantly each season, thus reducing salt cedar water usage to very low levels, and virtually eliminating seed production in most defoliated areas. This defoliation has also opened the canopy, allowing other plants to begin increasing in growth and number. Bird populations have increased at some sites, including Lovelock, NV, due to the increased food supply consisting largely of high populations of *D. elongata* beetles (Longland *et al.*, 2007). Additional releases of the Fukang/Chilik beetles have now been made in several different states, both as research test sites and in an AWPM implementation effort conducted by USDA-APHIS, various state Departments of Agriculture, weed control districts and private landowners. Nursery sites have been set up in several western states (ten sites approved in each of the participating states) where establishment has been achieved at many locations; however, quantitative results from this effort are not yet available.

Based on further laboratory and field assessments, *D. elongata* beetles from Greece were chosen as being the most likely strain of beetles to work effectively in more southern climates, and thus have been fully safety tested and released into both TX and CA, where they have successful overwintered and are now causing similar levels of defoliation to those seen in the early release years in the more northerly areas (see Carruthers *et al.*, 2006, 2007; DeLoach, 2007a, b). The areas impacted by these beetles are still small (< 50 ha); however, population expansion is following a similar trajectory to the increases seen with the northern beetles released in 2001 (more details provided in a subsequent section). Further expansion of the associated acreage and the resulting impact on salt cedar is expected in these southern locations during the spring and summer of 2007.

Development and implementation of education and technology transfer programmes

Cooperator and customer education and technology transfer were considered and planned for very early in the development of the AWPM programme for salt cedar. In fact, outreach efforts began prior to any North American field experimentation on the project. This was done in a variety of ways, but most importantly through the formalization of the Saltcedar Biological Control Consortium (SBCC). The SBCC was organized in November 1998 to provide coordination between agencies and to obtain input, guidance and oversight in the research programme from users and environmental organizations, and to solicit political and monetary support in furthering the project (Stenquist, 2000).

The group has met annually since then, and now has representatives from *c.* 50 federal and state agencies, universities and private user and environmental groups. The consortium was instrumental in gaining user support, developing and implementing operational plans and served as a means of linking critical groups into functional units that moved the wider-scale project forward through complicated partnerships with wide-ranging programme goals. It also allowed diverse input from all interested parties, providing dialogue and a means for rapid negotiations of any critical or controversial issues.

Only through the activities of the consortium were roadblocks removed regarding release of biological control agents because they had the potential to affect the endangered SWWFC. Such openness of operation and willingness of all parties holistically to address the salt cedar problem allowed a successful path to open. This and other similar situations typically required concessions from all sides but, through time, such discussions built trust and accelerated programme success. For example, the consortium explicitly developed a detailed monitoring plan that allowed the US FWS to issues its concurrence with both field-cage and open-field releases of the first salt cedar biological control agents. Without this linkage, the programme might never have met these first critical milestones. Additionally, several other outreach activities, such as symposia, extension brochures, joint agency press releases and educational information, were developed and released using the consortium as a clearing house.

Maintaining a team independent of any one group or governmental agency proved beneficial, as it depolarized many potential problems and cross-agency conflicts. Subsequently sister groups, such as the Tamarix Coalition in Grand Junction, CO and the Rio Grande Institute in western TX, were organized as local extensions or partners with the original consortium, and have worked effectively to accelerate local projects using the salt cedar AWPM technologies. Most recently, a new TX, NM and Mexico section of the SBCC was formed to provide guidance and coordination for the special circumstances encountered in this southern area. Standing committees of these consortia include subcommittees such as: Science and Research, Wildlife and Environment, Federal, State, Private Liaison, US–Mexico Relations, Native American Relations and Public Education/Information (DeLoach *et al.*, 2005). The SW Consortium group, in particular, was instrumental in obtaining approval from the Mexican Government in early 2007 to expand the AWPM of salt cedar into the Rio Grande River and northern Mexico. For more complete information on organization and activities see the web site at http://bc4weeds.tamu.edu/weeds/rangeland/saltcedar-bcc-2005.pdf

Evaluation of the AWPM Programme

Effectiveness of the AWPM programme at controlling the target pest

The leaf beetle, *D. elongata*, has now been tested at several locations, where it has established reproductive populations, increased dramatically in numbers and spread extensively across salt cedar-infested areas where the beetles have caused extensive defoliation of salt cedar for multiple seasons (see Plates 4 and 5). In these test locations, the leaf beetles are significantly impacting salt cedar growth and development, while no non-target plants have been negatively affected (see examples in Plate 5 and in DeLoach *et al.*, 2005). Such qualitative statements are true, but not often convincing to the scientific community or the regulatory agencies responsible for permitting the use of biological control technologies.

Thus, to quantitatively assess the effectiveness of this programme and to ensure environmental safeguards, a comprehensive monitoring programme was developed through the SBCC, approved by the USDA-APHIS and US FWS permitting units, and implemented by our project personnel. Ground sampling of beetle populations and their impact on target salt cedar and adjacent beneficial species was conducted at all of the release sites, along with various assessments targeting revegetation of beneficial plants and assessment of impacts on local and migratory wildlife. This was accomplished through an intensive monitoring programme that was designed by the SBCC and was carried out as required by the FWS Letter of Concurrence and the APHIS permits, as described in the Research Proposal to FWS of 28 August 1998 (Gould *et al.*, 2000). This included both monitoring of the insects and plants in the initial field cages, and then open-field assessments conducted both pre- and post-release.

The basic open-field monitoring plan specified a 10 ha sampling circle centred at the beetle release point, with 100 permanently marked sentinel salt cedar trees with 25 trees in the inner 1 ha circle, 25 trees in a middle 2 ha concentric ring and

50 trees in an outer 7 ha ring, these trees being divided equally and at random within each quadrant of each ring.

Diorhabda elongata and other insect populations, percentage defoliation produced, plant growth and condition were measured periodically on four marked, 40 cm-long branch terminals on each salt cedar tree. Once annually the vegetation was sampled by measuring tree height and diameter, estimating percentage healthy vegetation, yellowing or dead branches and estimating the distance, species identification and size of the three nearest neighbour trees. Understorey vegetation was also measured in two 1 m² quadrats, one under and one outside the tree canopy, in which percentage cover of all species and of litter and bare soil was estimated. This sampling plan was appropriate initially, and shortly after release, but became inadequate because the beetles rapidly overflowed this area and completely defoliated the entire sampling circle at most release sites. A modification such as the inclusion of longer transects (up to several km) was initiated using sentinel trees where similar data were collected.

Additionally, timed adult beetle counts were used to follow low-level beetle populations and, later, a beetle pheromone and a salt cedar extract were formulated (Cossé *et al.*, 2005) that were very attractive to *D. elongata* beetles from China and Kazakhstan over distances of 10–20 m. These baits were incorporated into a trapping system and have been effectively used in monitoring the dispersal of the beetles at Lovelock, NV and other locations.

All of these techniques aided local site managers in assessing overwintering success, beetle movement and eventual spread into new areas. The results from this monitoring were extensive and varied from test site to test site, but have been summarized by DeLoach *et al.* (2004, 2005, 2007a, b) and various other unpublished consortium reports. Here, we will summarize the impact and results of this programme primarily through the use of remote sensing and GIS visuals later in this section, as these techniques give a more comprehensive impression of the beetle spread, defoliation and impact.

In addition to the vegetation- and beetle-monitoring methods, a number of wildlife species were also assessed. For example, a bird-monitoring plan combined two riparian sampling areas that were selected, one with near-monotypic salt cedar and one with nearly pure native vegetation. In every area ten permanent point-count areas were located, each 100 m in diameter and separated from each other by 100 m. Three times annually during the breeding season, the numbers of each bird species seen or heard are counted from the centre point of each circle. This allowed a direct comparison between salt cedar and native vegetation, and also of populations in salt cedar before and after biological control, with the native vegetation plots used as a statistical control.

Results from such monitoring assessments showed that, initially, salt cedar had fewer numbers and less diversity of native birds inhabiting areas of invasion than adjacent areas populated with native plant species. As defoliation increased, birds moved back into the areas where AWPM using *D. elongata* was under way, where they used the insects as a food resource. With time, we expect other native plant populations to increase in beetle areas, and the relative amount of salt cedar should further provide more diversity of habitat, feeding and nesting space for many bird species (see DeLoach *et al.*, 2005; Longland *et al.*, 2007 for more details). Procedures also have been developed for monitoring butterflies, other insects, small mammals and reptiles, but will not be discussed here (see DeLoach *et al.*, 2005).

Monitoring through remote sensing

In the first 2 years following beetle release, ground-based sampling proved adequate; however, the scale of beetle population growth, spread and impact quickly made ground-based field sampling difficult, expensive and thus impractical. Previous studies (Everitt and DeLoach, 1990) have documented the effective use of remote sensing for the assessment of salt cedar infestations. In support of the overall project monitoring and assessment efforts, remote sensing was conducted to characterize salt cedar infestations, to follow beetle establishment, impact and spread and, eventually, to document the return of beneficial vegetation into areas where salt cedar has been controlled.

A comprehensive assessment programme has used a combination of colour aerial photography, airborne hyperspectral imagery and multispectral satellite data to assess salt cedar populations prior to and following beetle release, and to follow beetle defoliation patterns from the original release sites across wide areas of impact. The coupling of colour and texture analysis was first used to identify, classify and map invasive *T. parviflora* during spring bloom, along a 50-mile (80-km) segment of Cache Creek in central CA, with an estimated 90% accuracy (Ge *et al.*, 2006).

An ongoing effort is further evaluating the use of hyperspectral imagery during mid-summer to identify vegetative salt cedar (without blooms) interspersed with other background native vegetation in this and other infested areas. Additional use of hyperspectral image assessment and GIS mapping has allowed biological control specialists to track and evaluate beetle performance at many pilot release sites where the beetles have spread across thousands of acres. A combination of aerial and ground sampling has clearly documented the success and safety of this project in multiple study areas. Now, however, due to the magnitude of the impact, aerial or satellite reconnaissance seems to be the only practical method of assessing the full impact, as tens of thousands of hectares have now been positively impacted by this AWPS programme.

Based on field results of the AWPM programme (documented primarily with remote sensing assessments), several state and federal agencies are highly enthusiastic about AWPM based on biological control, and have initiated efforts to use this technology over even yet wider areas. Thus remote sensing is expected to play a major role in the continued assessment of this beneficial invasive species management project.

Here, we will present remote sensing results from only a few of over a dozen specific research sites examined. Examples have been provided to cover the many different methods that have been used, alongside detailed assessments of aerial photography from Cache Creek, CA of salt cedar density and hyperspectral imagery from Lovelock and Schurz, NV to demonstrate beetle establishment, spread and defoliation impact on salt cedar populations.

However, these three types of remote sensing technology, along with videography and satellite imagery, have been used to assess salt cedar infestations and biological control impacts at most of our research locations in all of the states where the salt cedar leaf beetle has been released. Individual approaches varied at each site, depending upon the local biological situation and the personnel and equipment available in each area to conduct the work (see Carruthers *et al.*, 2005).

Aerial photography

Cache Creek is a small river that is located north and west of Woodland, CA (38°40′02″N latitude and 128°45′30″W longitude), and is the location of one of our Cretian beetle release sites. The stream course is highly infested with *T. parviflora* intermixed with a diverse combination of native flora, including many species of willows, cottonwood and other common riparian plant species (see Plate 1c). This area is also infested with other exotic plant species of interest to our research team, including *Arundo donax*, *Lepidium latifolium* and *Centaurea solstitialis*, which are also being assessed using remote sensing. Cache Creek is of primary interest to us, as *T. parviflora* blooms in distinctive purplish-red blossoms (see Plate 6a) early in the spring prior to leaf-out of salt cedar and most other riparian vegetation. Therefore, this salt cedar is easily identified visually on the ground and from the air (see Plate 6b), making it possible for us to separate it from other vegetation at this time of the year.

During mid-season, however, it is difficult to separate *T. parviflora* from other green vegetation, as the plants are often intertwined and hard to see due to visual barriers caused by adjacent plant canopies. This area was chosen as one of our primary study sites because an extensive historical record of aerial photography exists for this location and thus provided a means of assessing salt cedar expansion through time. Currently, the salt cedar infestation on Cache Creek runs along approximately 80 km of river channel, much of which is hard to access on the ground, and is thus very difficult to map and manage. The use of remote sensing to assess and develop comprehensive distributional maps did allow local land managers to implement salt cedar removal, and allowed our team to better assess biological control impacts.

A total of 42 natural colour aerial photographs was taken, at a scale of 1:12,000, along Cache Creek in April 2002. At the time of photography, the salt cedar was in full bloom and was purplish-red in colour, making them distinctive from other associated riparian vegetation and the physical background in the study area (see Plate 6b). For analysis, the photographs were scanned at a 1-foot resolution (1000 dpi) in full colour using three (blue, green and red) channels. Once processed for analysis, images were orthorectified and georeferenced to 1 m resolution digital orthoquads from USGS, using a second-order polynomial function. Each individual digital image was then evaluated for salt cedar, classified and then mosaiced to provide an areawide map of the salt cedar infestation.

Based on the colour similarity and relationships between various types of native vegetation and the invasive species, vegetation was initially divided into eight types: *T. parviflora*, evergreen trees, non-evergreen trees, shrubs, crops, bare fields (including agriculture and rangeland), water bodies (including wetlands) and rocks and roads. Colour patterns alone were not adequate to separate and classify the salt cedar from the other categories of habitat, and thus more complex methods of analysis were required. To distinguish salt cedar from associated vegetation and background required the combined use of colour and texture analysis to provide an adequate measure of salt cedar density (see Ge *et al.*, 2006).

Since colour alone did not accurately separate salt cedar from the surrounding habitats (even though it was easily recognizable using the human eye due to large-scale pattern recognition), more complex pattern recognition and texture analysis was used as an added component in the analysis. Using this approach, the overall average separability of salt cedar from other habitat types was significantly improved and habitat classification errors were minimized, with vegetation covers comparing reasonably to actual observed plant distributions (see Plate 6c).

Comparisons made between this automated salt cedar recognition process from the photographs resulted in a classification providing 90% accuracy through actual field validation (Ge *et al.*, 2006). These classified images were then mosaiced across the study area in order to provide comprehensive infestation maps and to estimate the invasive salt cedar cover in a 40-km test area of the Cache Creek drainage. The total area of salt cedar infestation for this single 40-km section of stream was estimated at 3.98 km^2. These maps (see Plate 6d) are now being used by local land managers in their salt cedar control programmes and as a basis for comparison against more detailed hyperspectral analyses being conducted.

Plate 7 shows aerial imagery from our Cache Creek test site, along with actual ground-based counts of adult beetle populations showing the exponentially increasing beetle population density. These data were collected across the study area as depicted on the GIS map showing the impacted trees on the ground. A further time sequence of remote sensed data collected via Quickbird satellite using multispectral imagery (scheduled but not collected) will provide a time series of defoliation events over the next few years that should allow more accurate and widespread assessment of beetle dispersion and impact, salt cedar decline and the beneficial response of native vegetation in the area.

Hyperspectral remote sensing to assess beetle density and impact

To best demonstrate the use of remote sensing for assessment of beetle defoliation and impact, we concentrated on the use of hyperspectral imagery that was collected at multiple insect releases sites, including our test area in Lovelock, Nevada (40°1′20″N latitude and 118°31′24″W longitude). The Lovelock site is adjacent to the Lovelock and Oreana Valleys that are along the Humboldt River, just upstream from the Humboldt Sink. The Lovelock Valley is one of Nevada's primary agricultural areas, where lucerne is grown mainly for seed production.

Salt cedar heavily infests the Humboldt River and side irrigation channels in this area, where salt cedar extends upstream for approximately 166 km. Salt cedar is especially damaging in this drainage, as it both reduces available water for agriculture and blocks channel flow during times of heavy runoff. Local agricultural producers have been fighting salt cedar all along the Humboldt River Valley for the past two decades, and have been losing the battle. Salt cedar leaf beetle releases in this area have been extremely successful (see Plates 4 and 5). The beetles have highly defoliated the salt cedar and have spread > 150 km from the release site over the past 5 years. In fact, due to the extensive dispersal and impact of the AWPM programme, beetle impact can now only easily be evaluated using aerial or satellite reconnaissance.

Hyperspectal aerial imagery was acquired over study areas using a CASI II imaging system on 2 July 2002, 29 August 2002, 18 July and 10 September 2003 and 18 September 2004. In the last two seasons (2005 and 2006), the area of impact was so large that it was only practical to use satellite imagery to assess the area of defoliation. Thus, Quickbird satellite imagery was acquired to assess beetle damage. To our knowledge, this is the only biological control programme that has ever used such technology to assess natural enemy spread and programme impact.

The CASI II is a line scanner, covering 545 nm, between 400 and 1000 nm, that recorded the calibrated radiance values that were used in this analysis. Spectral and spatial coverage were changed over the course of the study to best accommodate biological control damage assessment, but ranged between 1 and 2 m pixel resolutions on the ground. Both were adequate to assess beetle defoliation; however, as the programme increased in impact we had to decrease resolution to increase the area being covered to circumscribe the defoliation within the available time of data collection. Spectral resolution also varied between years, ranging between 32 and 48 bins (400–1000 nm) and covering between 10 and approximately 3170 ha, as necessary, to encompass the target area where beetles were active.

Geographic coordinates for each line of the CASI images were recorded in real time using a differential global positioning system (GIS) coupled with the system's inertial momentum unit. Image to image registration was performed to compensate for much of the spatial registration error that confounds temporal difference analysis. Images were then transformed to the Normalized Difference Vegetation Index (NDVI), and change detection analysis was used to determine differences collected over the period of study. Areas showing change were photo-interpreted and matched to ground observations on beetle occurrence to ensure that the change was the result of *D. elongata* feeding damage to salt cedar. All areas of salt cedar that had been defoliated were subsequently masked (in yellow) as a region of interest (ROI), and the impacted canopy area was quantified (in ha, see Plate 8).

Analysis for the 2 July 2002 imagery indicated a total imaged area of approximately 1697 ha, which was the most infested with monotypic salt cedar (see Plate 8a). During this sample period we could see no visible salt cedar damage. Significant defoliation of salt cedar was first identified by ground crews, just prior to the August 2002 flight, where the damaged plants clearly show in the infrared area of the spectra. Change detection between the July 2002 and August 2002 image pairs (see Plate 8a and b, respectively) indicated that the biological control agents had defoliated approximately 0.35 ha (canopy area and *c.* 1 ha in ground area) over the 2-month period.

By July of 2003 the total amount of salt cedar defoliated had increased to 3.4 ha of canopy (4.3 ha of ground area, see Plate 8c) and, in the interval between the 18 July 2003 and 10 September 2003 flights (62 days), *D. elongata* had defoliated approximately 76.7 ha of canopy (*c.* 200 ha of ground area, see Plate 8). Between the initial flight of 2 July 2002 and the 10 September 2003 flight, the defoliation had progressed approximately 2.8 km from the original release site and produced an impact span of approximately 5.2 km (south-west to north-east). The beetles themselves had actually spread much further along the Humboldt drainage, but defoliation levels were not detectable at these low densities.

By 2004 the area of impact had grown so large (see Plate 9) that it could no longer be easily assessed using fixed-winged aircraft, and thus satellite imagery was used to assess beetle spread and impact. Although we are still working out the details of this assessment technology, we have estimated that every salt cedar plant within a 27,500 ha area around the Lovelock release site was highly defoliated by the beetle. Unfortunately, we do not have pre-release imagery on this same scale, so we are not able to conduct a more comprehensive change detection analysis. However, totally defoliated and dying salt cedar plants are readily identifiable in the Quickbird imagery (see Plate 10) at the centre of the original release site.

Results indicate an exponential rate of beetle population growth, spread and salt cedar defoliation at each of our successful release sites. A total of nine other sites were monitored using similar techniques to those used in Lovelock, NV (not reported here). Many of these other areas have more typical riparian vegetation, with a mixture of salt cedar and other plants. Defoliation rates in these areas indicate that impacts to salt cedar are somewhat slower across most of the plant's range, but that insect establishment has occurred, that the beetle populations are building and that salt cedar is being negatively impacted at almost all of the test locations (see DeLoach *et al.*, 2007a).

Unintended negative and positive consequences of the AWPM programme

The largest unexpected consequence of this biologically based AWPM programme was that its effectiveness and spread proceeded almost too quickly. The establishment of beetles at most of our test sites was rapid, with exponential increases in beetle numbers and tremendous spread of the agent throughout areas adjacent to the targeted watersheds. In some areas the insects and associated defoliation has literally spread hundreds of kilometres and impacted tens of thousands of hectares of salt cedar within 6 years following a one-time point release of beetles. The spread and impact of the beetle highly surpassed our expectations and overwhelmed our short-term monitoring capabilities.

However, no unintended negative impacts have been seen throughout the course of this programme, only extensive and beneficial defoliation of salt cedar. We believe that this is primarily due to the extensive safety and efficacy testing that was conducted in advance of the release of this insect into the environment. This rapid impact has been a very positive attribute of the AWPM programme using *D. elongata*; however, its rapid impact means that land managers now need quickly to follow up with activities to keep other invasive species from filling opening niches.

Researchers and implementation teams, however, are working hard to conduct accelerated revegetation research and implementation efforts in preparation for the wide areas that may soon have significantly declining populations of salt cedar (Lair and Wynn, 2002). The consequences of such declines are overwhelmingly positive, however, long-term monitoring and assessment by land management agencies, such as the Bureau of Land Management and the Bureau of Reclamation, are now extremely important. USDA-APHIS as the lead organization in the areawide implementation of this effort must directly interface with both public and private land and waterway managers to continually assess and take action where needed. Such activities have been planned and are now being implemented across release areas to identify and quantify environmental damage if it occurs.

Economic evaluation of the AWPM programme

The economic benefits of the AWPM effort to control salt cedar are potentially very great although not yet quantified, as several questions remain as to the natural balance that will be achieved between the leaf beetle and the salt cedar populations across different parts of the USA. Clearly, different evaluation metrics are appropriate as we move from totally environmental values to those involving human uses of land and water resources for agriculture and metropolitan uses of water. Both the short-term impact of AWPM using biological control and its long-term sustainability will need to be accounted for in any such assessment. Economists from the University of California Davis have been contracted to conduct such an assessment (Carruthers *et al.*, 2000), but longer-term impact assessments are required prior to their analysis having any real meaning.

Clearly, however, now that the research and developmental stages of the salt cedar AWPM programme using *D. elongata* have been developed, implementation is very inexpensive and requires only a minimum of effort to establish beetles, develop nursery and release sites to multiple beetle numbers, organize and implement redistribution mechanisms and deploy resources to conduct follow-up monitoring as needed to assess and enact revegetation efforts.

Prospects for the long-term sustainability of the AWPM programme

The prospect for long-term sustainability of this programme is very high, as it has been in many similar biological programmes for control of weeds over the years. However, it is unknown what the long-term future actually holds for the dynamic interactions between salt cedar and *D. elongata*. Currently, all field populations of this beetle seem to be free of host-specific natural enemies, as those were removed from this insect through the quarantine screening process. Such natural enemies of *D. elongata* work to suppress its population in Eurasia, as do some generalist predators both in its homeland, and now in its introduced range. Parasitism in Eurasia is thought to limit populations, and thus defoliation in the countries of origin does not reach the levels noted in North America.

However, since tachinid parasites of other similar insects – including the elm leaf beetle and its introduced tachinid parasite from China – exist in areas where salt cedar leaf beetles will eventually spread it may be that, once exposed to these natural enemies, *D. elongata* will be negatively impacted. Such natural enemies would be expected to reduce its population vigour and salt cedar suppression capabilities. Likewise, other more specific natural enemies, such as other parasites and pathogens, might also become adapted to use *D. elongata*, with similar negative consequences.

The effects that meso-level spatial dynamics of this system may have upon the interaction of salt cedar and the leaf beetle are also unknown. In China and other areas of origin, significantly lower densities of both the plant and this insect exist. Heavy defoliation does occur, but insects seem to change geographical locations between years, allowing the salt cedar to recover between defoliation events. In this way, there is always some non-affected salt cedar available to sustain the relationship. In areas where we have had dramatic defoliation results no such plant escape has yet been observed to occur, as *D. elongata* populations reach very high levels due to little or no top-down population regulation.

Plant availability as a food source (bottom-up regulation), therefore, is the dominant force controlling insect numbers. Such an undamped system may actually cause highly cyclic population dynamics and unstable behaviour in the longer term. Clearly, massive die-offs and localized extinction of beetles at some of our research sites suggests that this may be one potential outcome. A solution to this problem could be to introduce a second control insect that may dampen the population fluctuations. Alternatively, very low beetle numbers could be reintroduced (naturally or with our help) to re-establish a dominant force in controlling salt cedar growth and development. In reality, only time will prove the actual sustainability of this programme (with or without human intervention), and we watch with great interest.

Summary and Future Direction

Expectations from control and future directions

In the short term, we expect AWPM using biological control to reduce salt cedar over a period of 3–5 years and permanently to reduce the abundance – over many years – to below the economic or environmental thresholds of damage. We expect that both the salt cedar and beetles will remain at low fluctuating population levels but nearly always be present in the local environment, where a new ecological balance will be maintained. This is the situation that we expect under the variable and diverse habitats that we see all across the western USA; however, many different local outcomes are to be expected. It is actually very likely that we will need to develop and use additional natural enemies of salt cedar in areas where *D. elongata* does not adequately control this pest plant due to a combination of environmental or human factors. Several such insects have already been investigated overseas in case they are needed later in this programme (Jashenko and Mityaev, 2007).

Under the conditions that we now see, we expect the native plant communities to re-establish naturally in most areas where depth to water table and soil salinity are not too great. This should improve wildlife habitat and allow the recovery of many species of birds and fish, and of some mammals and reptiles, including several threatened and endangered species. Control of salt cedar is expected to increase the amount and quality of water available for irrigated agriculture, and municipal and environmental use, and to help fulfil the water rights agreements between states and between the USA and Mexico. Control also is expected to increase recreational usage of parks and wildland areas, to reduce wildfires and to allow the gradual reduction of salinity levels of surface soils in presently infested areas.

Large-scale revegetation projects are currently under development by the USDI Bureau of Reclamation for areas where natural revegetation may be insufficient. We expect that this will be needed in some areas of AZ, CA and NM. Thus, we feel that results to date indicate that the programme of AWPM of salt cedar using biological control has a high probability of providing good control of salt cedar over much of the infested area of the USA. Salt cedar has also invaded large areas in northern Mexico; the US programme easily could be extended into Mexico at very low cost, through the cooperation of Mexican scientists. Approval of the Mexican government, natural area managers and authorities was granted in the spring of 2007, and plans are now being made to begin testing of this programme at sites along the Rio Grande, in both Texas and Mexico.

We will continue to work to assess and improve the AWPM of salt cedar using biological control as the keystone method of control. In doing so, we will both monitor and assess the situation and work with implementation agencies such as USDA-APHIS and several state Departments of Agriculture, university scientists, private groups, Native Americans and other groups to make and improve action programmes using *D. elongata*. Also, several other potential biological control agents of salt cedar are being investigated and developed by our overseas cooperators. These natural enemies may be necessary for use in fringe climatic areas when *D. elongata* beetles may not provide sufficient control or where predators may limit control, or where their release may not be allowed. Additionally, we will seek natural enemies for other weeds of importance to our customers and cooperators that work for the protection of American agriculture and the environment.

References

BASF (2004) *Arsenal Herbicide, Specimen Label, NVA 2004- 04- 104- 0185*. BASF Corporation, Agricultural Products, Triangle Park, North Carolina.

Baum, B.R. (1967) Introduced and naturalized tamarisks in the United States and Canada (Tamaricaceae). *Baileya* 15, 19–25.

Baum, B.R. (1978) *The Genus* Tamarix. Israel Academy of Sciences and Humanities, Jerusalem.

Bean, D.W., Dudley, T.L. and Keller, J.C. (2007a) Seasonal timing of diapause induction limits the effective range of *Diorhabda elongata deserticola* (Coleoptera: Chrysomelidae) as a biological control agent for tamarisk (*Tamarix* spp.). *Environmental Entomology* 36, 15–25.

Bean, D.W., Wang, T., Bartelt, R.J. and Zilkowski, B.W. (2007b) Diapause in the leaf beetle *Diorhabda elongata* (Coleoptera: Chrysomelidae), a biological control agent for tamarisk (*Tamarix* spp.). *Environmental Entomology* 36, 531–540.

Brown, F.B., Ruffner, G. Johnson, R., Horton, J. and Franson, J. (1989) *Economic Analysis of Harmful and Beneficial Aspects of Saltcedar*. Final report to USDI Bureau of Reclamation, Lower Colorado Region, Boulder City, Nevada. Prepared by Great Western Research, Inc., Mesa, Arizona, 261 pp.

Carruthers, R.I. (2003) Invasive species research in the USDA Agricultural Research Service. *Pest Management Science* 59, 827–834.

Carruthers, R.I. (2004) Biological control of invasive species, a personal perspective. *Conservation Biology* 18, 54–57.

Carruthers, R., Spencer, D., DeLoach, J., D'Antonio, C., Dudley, T., Kazmer, D. and Knutson, A. (2000) *Biologically-based Control for the Area-wide Management of Exotic and Invasive Weeds*. USDA-CSREES-IFAFS Grant, Assistance Transaction Number 1408536, Washington, DC.

Carruthers, R.I., Anderson, G., DeLoach, C.J., Knight, J.B., Ge, S. and Pong, P. (2005) Monitoring saltcedar biological control impact. *Proceedings Monitoring Science and Technology Symposium*, US Forest Service, RMRS P-37CD, Fort Collins, Colorado.

Carruthers, R.I., Herr, J.C., Knight, J. and DeLoach, C.J. (2006) A brief overview of the biological control of salt cedar. In: Hoddle, M. and Johnson, M. (eds) *Proceedings Fifth California Conference on Biological Control*, University of California, Riverside, California, pp. 71–77.

Carruthers, R.I., Herr, J.C. and DeLoach, C.J. (2007) An overview of the biological control of saltcedar. *Proceedings of the California Weed Science Society.* 59 <In press>.

Cossé, A.A., Bartelt, R.J., Zilkowski, B.W., Bean, D.W. and Petroski, R.J. (2005) The aggregation pheromone of *Diorhabda elongata*, a biological control agent of saltcedar (*Tamarix* spp.): identification of two behaviourally active components. *Journal of Chemical Ecology* 31, 657–670.

Crins, W.L. (1989) The Tamaricaceae in the southeastern United States. *Journal of Arnold Arboretum* 70, 403–425.

DeLoach, C.J. (1991) Saltcedar, an Exotic Weed of Western North American Riparian Areas: a Review of its Taxonomy, Biology, Harmful and Beneficial Values, and its Potential for Biological Control. Final report, USDI Bureau of Reclamation, Lower Colorado Region, Boulder City, Nevada, 433 pp.

DeLoach, C.J. and Gould, J.E. (1998) *Biological Control of Exotic, Invading Saltcedar (*Tamarix *spp.) by the Introduction of* Tamarix *– Specific Control Insects from Eurasia.* Research proposal to US Fish and Wildlife Service, 28 August 1998. 45 pp.

DeLoach, C.J. and Tracy, J.L. (1997) *Effects of Biological Control of Saltcedar* (Tamarix ramosissima) *on Endangered Species: Draft Biological Assessment*, 17 October 1997. USDA/ARS, Temple, Texas, 524 pp. + appendices.

DeLoach, C.J., Gerling, D., Fornasari, L., Sobhian, R., Myartseva, S., Mityaev, I.D., Lu, Q.G., Tracy, J.L., Wang, R., Wang, J.F., Kirk, A., Pemberton, R.W., Chikatunov, V., Jashenko, R.V., Johnson, J.E., Zeng, H., Jiang, S.L., Liu, M.T., Liu, A.P. and Cisneroz, J. (1996) Biological control programme against saltcedar (*Tamarix* spp.) in the United States of America: progress and problems. In: Moran, V.C. and Hoffmann, J.H. (eds) *Proceedings of the IX International Symposium on the Biological Control of Weeds*, University of Cape Town, South Africa, pp. 253–260.

DeLoach, C.J., Carruthers, R.I., Lovich, J.E., Dudley, T.L. and Smith, S.D. (2000) Ecological interactions in the biological control of saltcedar (*Tamarix* spp.) in the United States: toward a new understanding. In: Spencer, N.R. (ed.) *Proceedings of the X International Symposium on Biological Control of Weeds*, Montana State University, Bozeman, Montana, pp. 819–873.

DeLoach, C.J., Lewis, P.A., Herr, J.C., Carruthers, R.I., Tracy, J.L. and Johnson, J. (2003) Host specificity of the leaf beetle, *Diorhabda elongata deserticola* (Coleoptera: Chrysomelidae) from Asia, a biological control agent for saltcedars (*Tamarix*: Tamaricaceae) in the western United States. *Biological Control* 27, 117–147.

DeLoach, C.J., Carruthers, R.I., Dudley, T.L., Eberts, D., Kazmer, D.J., Knutson, A.E., Bean, D.W., Knight, J., Lewis, P.A., Milbrath, L.R., Tracy, J.L., Tomic-Carruthers, N., Herr, J.C., Abbott, G., Prestwich, S., Harruff, G., Everitt, J.H., Thompson, D.C., Mityaev, I., Jashenko, R., Li, B., Sobhian, R., Kirk, A., Robbins, T.O. and Delfosse, E.S. (2004) First results for control of salt cedar (*Tamarix* spp.) in the open field in the western United States. In: Cullen, J.M., Briese, D.T., Kriticos, D.J., Lonsdale, W.M., Morin, L. and Scott, J.K. (eds) *Proceedings of the XI International Symposium on Biological Control of Weeds*. CSIRO Entomology, Canberra, Australia, pp. 505–513.

DeLoach, C.J., Knutson, A.E., Thompson, D.C. and Nibling F. (2005) *Saltcedar Biological Control Consortium: Texas, New Mexico, Mexico Section, First (Organizational) Meeting: Minutes, Reviews of Research and Resource Guide*, 29–30 March 2005, El Paso, Texas, 103 pp. ⟨http://bc4weeds.tamu.edu/weeds/rangeland/salt cedar-bcc-2005.pdf⟩.

DeLoach, C.J., Milbrath, L.R., Carruthers, R.I., Knutson, A.E., Nibling, F., Eberts, D., Thompson, D.C., Kazmer, D.J., Dudley, T.E., Bean, D.W. and Knight, J.B. (2005) Overview of saltcedar biological control. *Proceedings Monitoring Science and Technology Symposium*, US Forest Service, RMRS P-37CD, Fort Collins, Colorado.

DeLoach, C. Jack, Knutson, A.E., Moran, P., Thompson, D.C., Everitt, J.H., Eberts, D., Michels, J., Lewis, P.A., Milbrath, L.R., Herr, J.C., Carruthers, R.I., Sanabria, J.,

Muegge, M., Richman, D., Tracy, J.L., Robbins, T.O., Hudgeons, J., Carney, V., Gardner, K., Fain, T., Donet, M., Jashenko, R., Li, B., Sobhian, R., Kirk, A., Mityaev, I., McMurry, M., Nibling, F., Diaz Soltero, H. and Delfosse, E.S. (2007a) Research and initial success on biological control of saltcedar in the United States, with emphasis on Texas and New Mexico. In: *Proceedings of the Sixth Symposium on the Natural Resources of the Chihuahuan Desert*, 15–16 October 2007, Sul Ross State University, Alpine, Texas. (In press).

DeLoach, C.J., Moran, P.J., Knutson, A.E., Thompson, D.C., Carruthers, R.I., Michels, J., Muegge, M., Eberts, D., Randall, C., Everitt, J E., O'Meara, S. and Sanabria, J. (2007b) Beginning success of biological control of saltcedars (*Tamarix* spp.) in the Southwestern United States. In: *Proceedings of the Seventh International Symposium on Biological Control of Weeds*, 22–27 April 2007, La Grande Motte, France. (In press).

Dudley, T.L. and Kazmer, D.J. (2005) Field assessment of the risk posed by *Diorhabda elongata*, a biocontrol agent for control of saltcedar (*Tamarix* spp.), to a nontarget plant, *Frankenia salina*. *Biological Control* 35, 265–275.

Everitt, B.L. (1980) Ecology of saltcedar – a plea for research. *Environmental Geology* 3, 77–84.

Everitt, H. and DeLoach, C.J. (1990) Remote sensing of Chinese tamarisk (*Tamarix chinensis*) and associated vegetation. *Weed Science* 38, 273–278.

Gaskin, J.F. and Schaal, B.A. (2003) Molecular phylogenetic investigation of US invasive *Tamarix*. *Systematic Botany* 28, 86–95.

Ge, S., Carruthers, R.I., Herrera, A.M. and Gong, P. (2006) Texture analysis for invasive *Tamarix parviflora* mapping using aerial photographs along Cache Creek, California. *Environmental Monitoring and Assessment* 114, 65–83.

Gerling, D. and Kugler, J. (1973) *Evaluation of Enemies of Noxious Plants in Israel as Potential Agents for the Biological Control of Weeds, 1 Sept. 1970–31 Aug. 1973*. Final Technical Report to USDA Agricultural Research Service, P.L. 480 (Project #A10-ENT-36), Tel Aviv University, Department of Zoology, Tel Aviv, Israel. 197 pp.

Gould, J.R., Dudley, T.E. and White, L. (2000) *Plan for Monitoring the Effects of Releasing the Saltcedar Leaf Beetle,* Diorhabda elongata *for Biological Control of Saltcedar*. Prepared by the insect, vegetation and wildlife monitoring subcommittee of the Saltcedar Consortium, 29 March 2000, Temple, Texas 18 pp.

Habib, R. and Hassan, S.A. (1982) *Insect Enemies Attacking Tamarisk,* Tamarix *spp., in Pakistan*. Final Report, June 1975–June 1980, Commonwealth Institute of Biological Control, Pakistan Station, Rawalpindi, Pakistan, 138 pp.

Hart, C.R. (2004) *The Pecos River Ecosystem Project Progress Report*. Texas Cooperative Extension/ Texas A&M University System and Texas Water Resources Institute, Annual Report for 2004, Section D, 10 pp. (http://pecosbasin.tamu.edu).

Hart, C.R. (2006) *Pecos River Basin Assessment Programme*. Annual Report to Texas State Soil and Water Conservation Board, under US Environmental Protection Agency, Clean Water Act Section 319, TSSWCB Project # 04–11, December 2006. 17 pp.

Hart, C.R., Clayton, L. and Lee, B. (2000). Salt cedar control in Texas. In: *Proceedings Rangeland Weed and Brush Management: the Next Millenium; Symposium and Workshop*. Agricultural Experiment Station, College Station, Texas, p. 185–190.

Hart, C.R., White, L.D., McDonald, A. and Sheng, Z. (2005) Salt cedar control and water salvage on the Pecos River, Texas, 1999–2003. *Journal of Environmental Management* 75, 399–409.

Herr, J.C., Carruthers, R.I., Bean, D.W., DeLoach, C.J., Kashefi, J. and Sanabria, J. (2007) Host preference between saltcedar (*Tamarix* spp.) and native non-target *Frankenia* spp. within the *Diorhabda elongata* species complex (Coleoptera: Chrysomelidae). *Environmental Entomology* (in press).

Herrera, A.M., Dahlsten, D., Tomic-Carruthers, N. and Carruthers, R.I. (2005) Estimating temperature-dependent developmental rates of *Diorhabda elongata*, a biological control agent of saltcedar. *Environmental Entomology* 34, 775–784.

Jashenko, R.V. and Mityaev, I.D. (1999–2005) *The Biocontrol of* Tamarix *Annual Reports*. Tethys Scientific Society/Laboratory of Entomology, Institute of Zoology, Almaty, Kazakhstan (on file at USDA-ARS, Temple, Texas).

Kovalev, O.V. (1995) Co-evolution of the tamarisks (Tamaricaceae) and pest arthropods (Insecta; Arachnida: Acarina) with special reference to biological control prospects. *Proceedings of the Zoological Institute, Russian Academy of Sciences at St Petersburg*, Vol. 29. Pensoft Publishers, Moscow, 110 pp.

Lair, K.D. and Wynn, S.L. (2002) *Research Proposal: Revegetation Strategies and Technology Development for Restoration of Xeric* Tamarix *Infestation Sites*. Technical Memorandum No. 8220- 02- 04, USDI Bur. Reclamation, Technical Service Center, Denver, Colorado, 48 pp.

Lewis, P.A., DeLoach, C.J., Herr, J.C., Dudley, T.L. and Carruthers, R.I. (2003a) Assessment of risk to native *Frankenia* shrubs from an Asian leaf beetle, *Diorhabda elongata deserticola* (Coleoptera: Chrysomelidae), introduced for biological control of saltcedars (*Tamarix* spp.) in the western United States. *Biological Control* 27, 148–166.

Lewis, P.A., DeLoach, C.J., Knutson, A.E., Tracy, J.L. and Robbins, T.O. (2003b). Biology of *Diorhabda elongata deserticola* (Coleoptera: Chrysomelidae), an Asian leaf beetle for biological control of saltcedars (*Tamarix* spp.) in the United States. *Biological Control* 27, 101–116.

Li, B. and Ming, L. (2001–2002) *Annual Reports on Biological Control of Saltcedar, 2001–2002, Xinjiang Agricultural University, Urumqi, China;* 2003–2005, Nanjiang Agricultural University, Nanjing, Jiangsu, China (on file at USDA-ARS, Temple, Texas).

Longland, W.S. and Dudley, T.E. (2007) Effects of a biological control agent on the use of saltcedar habitat by passerine birds. *Restoration Ecology* (in press).

Lovich, J.E. and de Gouvenain, R.C. (1998) Saltcedar invasion in desert wetlands of the southwestern United States: ecological and political implications. In: Majumdar, S.K., Miller, E.W. and Brenner, F.J. (eds) *Ecology of Wetlands and Associated Systems*, Pennsylvania Academy of Science, Easton, Pennsylvania, pp. 447–467.

Milbrath, L.R. and DeLoach, C.J. (2006a) Host specificity of different populations of the leaf beetle *Diorhabda elongata* (Coleoptera: Chrysomelidae), a biological control agent of salt cedar (*Tamarix* spp.). *Biological Control* 36, 32–48.

Milbrath, L.R. and DeLoach, C.J. (2006b) Acceptability and suitability of athel, *Tamarix aphylla*, to the leaf beetle, *Diorhabda elongata* (Coleoptera: Chrysomelidae), a biological control agent of salt cedar (*Tamarix* spp.). *Environmental Entomology* 35, 1379–1389.

Mityaev, I.D. (1958) A review of insect pests of *Tamarix* in the Balkhash-Alakul depression. Trudy Inst. Zool. Akad. Nauk. Kazakh. SSR 8, 74–97 [in Russian].

Mityaev, I.D. and Jashenko, R.V. (2007) *Insects Damaging Tamarisk in Southeastern Kazakhstan*. Tethys Scientific Society, Almaty, Kazakhstan, 184 pp. [in Russian].

Pemberton, R.W. and Hoover, E.M. (1980) Insects Associated with Wild Plants in Europe and the Middle East: Biological Control of Weeds Surveys. Miscellaneous Publication No. 1382, USDA, Washington, DC.

Sala, A., Smith, S.D. and Devitt, D.A. (1996) Water used by *Tamarix ramosissima* and associated phreatophytes in a Mojave Desert floodplain. *Ecological Applications* 6, 888–898.

Shaforth, P.B., Cleverly, J.R., Dudley, T.L., Taylor, J.P., Van Riper III, C., Weeks, E.P. and Stuart, J.N. (2005) Control of *Tamarix* in the western United States: implications for water salvage, wildlife use, and riparian restoration. *Environmental Management* 35, 231–246.

Sisneros, D. (1991) *Herbicide Analysis: Lower Colorado River Saltcedar Vegetation Management Study*. Report R-91- 96, Applied Sciences Research, USDI-Bureau of Reclamation, Denver, Colorado.

Smith, S.D., Devitt, D.A., Sala, A., Cleverly, J.R. and Busch, D.E. (1998) Water relations of riparian plants from warm desert regions. *Wetlands* 18, 687–696.

Stenquist, S.M. (2000) Saltcedar integrated weed management and the Endangered Species Act. In: Spencer, N.R. (ed.) *Proceedings of the X International Symposia on Biological Control of Weeds*, Montana State University, Bozeman, Montana, pp. 487–504.

Thompson, D.A., Petersen, B.A., Bean, D.W. and Keller, J.C. (2007) Hybridization potential of salt cedar leaf beetle, *Diorhabda elongata*, ecotypes. In: *Proceedings of the XII International Symposium on Biological Control of Weeds*, 22–27 April 2007, La Grande Motte (Montpellier), France. (In press).

Zavaleta, E. (2000) The economic value of controlling an invasive shrub. *Ambio* 29, 462–467.

16 The Hawaii Fruit Fly Areawide Pest Management Programme

ROGER I. VARGAS,[1] RONALD F.L. MAU,[2] ERIC B. JANG,[1] ROBERT M. FAUST[3] AND LYLE WONG[4]

[1]*US Department of Agriculture, Agricultural Research Service, Pacific Basin Agricultural Research Center, Hilo, Hawaii, USA*
[2]*University of Hawaii at Manoa, College of Tropical Agriculture and Human Resources, Department of Plant and Environmental Protection Sciences, Gilmore, Honolulu, Hawaii, USA*
[3]*US Department of Agriculture, Agricultural Research Service, Beltsville, Maryland, USA*
[4]*Hawaii Department of Agriculture, Division of Plant Industry, Honolulu, Hawaii, USA*

Introduction

Description of the problem and need for an AWPM approach

Significance of the pest management problem

Fruit flies (Diptera: Tephritidae) are among the most economically important pests attacking soft fruits worldwide (White and Elson-Harris, 1992). Four invasive species – Mediterranean fruit fly or medfly (*Ceratitis capitata*), melon fly (*Bactrocera cucurbitae*), oriental fruit fly (*Bactrocera dorsalis*) and the so-called Malaysian fruit fly or solanaceous fruit fly (*Bactrocera latifrons*) – have been devastating to Hawaiian agriculture for over 100 years by infesting more than 400 different host plants. These fruit flies:

- Jeopardize development of a diversified tropical fruit and vegetable industry.
- Require that commercial fruits undergo quarantine treatment prior to export.
- Provide a breeding reservoir for their introduction into other parts of the world due to unprecedented travel and trade between countries.

Hawaii is not the only state in the USA troubled by fruit flies. Every year exotic fruit flies are accidentally introduced from various parts of the world into California and Florida. One species, the olive fruit fly (*Bactrocera oleae*), introduced into California in 1998, has become permanently established and has caused serious economic losses to olive growers (Yokoyama and Miller, 2004). Due to continuous introductions, current annual costs incurred in excluding medfly from California and Florida total over US$15 million (http://www.cdfa.ca.gov). If the medfly became permanently established in California, projected losses would exceed US$1 billion per year due to lost revenues, export treatment costs, trade and crop damage (Faust, 2004).

Bactrocera is a genus of 440 described species, widely distributed throughout tropical Asia, the south Pacific and Australia. Relatively few species exist in Africa, and only the olive fly, *B. oleae*, occurs in southern Europe (White and Elson-Harris, 1992). Recently, two species in the *B. dorsalis* complex became established on two new continents: *B. carambolae*, the carambola fruit fly, in South America (Suriname) and *B. invadens* in Africa (Kenya) (Drew *et al.*, 2005; Rousse *et al.*, 2005). The oriental fruit fly is found throughout Asia, including Bhutan, southern China, India and Thailand, and has been recorded from over 173 host plant species (White and Elson-Harris, 1992).

The oriental fruit fly was introduced into Hawaii in 1945 and is now the most abundant and widely distributed fruit fly in the islands. Studies suggest that 95% of the population develops in common guava, *Psidium guajava* and strawberry guava, *P. cattleianum*, and that population cycles are determined primarily by wild guava fruiting (Newell and Haramoto, 1968; Vargas *et al.*, 1983). Commercial and backyard fruits are severely damaged by *B. dorsalis* population increases in nearby guava patches. Because of the abundance of common and strawberry guava throughout Hawaii, *B. dorsalis* has played a direct role in inhibiting the development of a profitable and diversified tropical fruit industry (Vargas *et al.*, 2000).

The melon fly, the second most abundant and widely distributed fruit fly species in Hawaii, is a serious agricultural pest of cucurbits. It has been recorded from over 125 plant species (Weems, 1964) and is found in India, Myanmar, Malaysia, Thailand, the Philippines, southern China, Taiwan, East Africa, Guam, the Commonwealth of the Northern Mariana Islands, Papua New Guinea, Solomon Islands and the Hawaiian Islands (Nishida, 1953; White and Elson-Harris, 1992). In 1895 it was discovered in Hawaii (Back and Pemberton, 1917), where it causes serious economic damage to cultivated species of Cucurbitaceae (e.g. cucumber, *Cucumis sativus*; watermelon, *Citrullus lanatus*; cantaloupe, *Cucumis melo*; pumpkin, *Cucurbita maxima*; cultivated bitter melon (balsam pear), *Momordica charantia*; and courgette, *Cucurbita pepo*) (White and Elson-Harris, 1992). When populations are high and cucurbits scarce, *B. cucurbitae* also attack, with less frequency, other species of vegetables and fruits, such as papaya (*Carica papaya*).

Bactrocera latifrons is a less common dacine species, introduced about 1983 from South-east Asia. It is associated primarily with patches of wild and cultivated solanaceous fruits (Vargas and Nishida, 1985). Economic damage can be extensive in community gardens and farms where crops such as tomato (*Solanum lycopersicon*), aubergine (*Solanum melogena*) and pepper (*Capsicum annuum*) are cultivated (Vargas and Nishida, 1985).

Ceratitis is a genus of 65 species that originated in tropical and southern Africa (White and Elson-Harris, 1992). The medfly, *C. capitata*, was accidentally introduced into Hawaii from Australia in 1907, and it became a serious pest of tree fruits. When oriental fruit fly was introduced into Hawaii in 1945, it displaced medfly throughout most of its range, except in small patches with commercial and wild coffee (*Coffea arabica*), strawberry guava and a variety of upper-elevation fruits (i.e. peaches (*Prunus persica*), loquats (*Eriobotrya japonica*) and persimmons (*Diospyros kaki*)) (Vargas *et al.*, 2001).

In summary, fruit flies are both local and global pests, and areawide procedures developed in Hawaii have both local and worldwide applications.

Description of current management systems and approaches

In Hawaii, a transition from plantation agriculture to a more diversified agricultural economy has changed the diversity of crops grown and the size of farms. Instead of large monocultures such as sugarcane and pineapple, smaller plots with a variety of fruits and/or vegetables (even mixed crops on small acreage) are commonly planted. With few exceptions, independent farmers currently practise agriculture in Hawaii on small farms. Farming is a difficult and risky occupation due to the high cost of land, labour and equipment and the lack of an established marketing infrastructure for distribution of products.

For these reasons, farmers are reluctant to accept new pest control technologies unless they have been demonstrated to be successful and convenient. Growers are often unaware of crops grown on adjacent neighbours' land, and non-cultivated areas, where wild fruit fly host plants support breeding fruit fly populations. These areas are ideal breeding locations which, combined with the mild climatic conditions over much of the Hawaiian Islands, result in large population build-ups nearby, but outside cropping areas. Because of the natural tendency of fruit flies to disperse, the programme, as it expanded, included the whole range of producers, from backyard growers to community growers and on up to large commercial growers across the counties and islands, to make the programme a true areawide approach, as opposed to a farm-to-farm approach. None the less, with increases in the population of Hawaii and expansion of the tourist industry, the demand for fresh fruits and vegetables is greater than ever.

Overwhelmingly, pesticides have been the most popular control practices used against fruit flies. Calendar sprays are routinely used directly on crops to control fruit fly infestation. However, the heavy use of pesticides has been implicated in the reduction of natural enemies and, in some cases, secondary pest outbreaks. In addition, because of the non-traditional types and relatively small value of many crops grown in Hawaii, many pesticides are not registered for use on these crops. Use of non-registered pesticides and overuse of registered pesticides have renewed concerns regarding food safety and groundwater quality in many parts of the world. Because of the complexity of agroecosystems in Hawaii and the pest complexes that can occur on a given crop, areawide pest management (AWPM) approaches to fruit fly suppression were proposed as an alternative to the current practices.

Limitations of current management approaches

Fruit fly eradication programmes have been proposed for Hawaii on many occasions. However, demonstration eradication programmes against medfly conducted in Hawaii in the early 1990s identified several problems associated with the eradication technology available at that time. These included the high economic cost of large-area programmes, planting of large areas with coffee (the preferred host of medfly), lack of sufficient information on the effects upon non-target fauna, environmental concerns, quarantine issues and the lack of a large-scale sterile fly-rearing facility.

Therefore, many scientists suggested that because environmental and economic costs of fruit fly eradication programmes were so high, emphasis should shift toward AWPM programmes and away from eradication. Although scientists in Hawaii have developed most of the technologies over the years to combat accidental fruit fly outbreaks on the US mainland (e.g. California and Florida), the technologies were never packaged and transferred to Hawaiian farmers. The Hawaii Areawide Pest Management programme was designed to transfer these technologies to Hawaiian farmers and residents.

Anticipated benefits of AWPM

The Hawaii AWPM programme was not aimed at eradication of fruit flies, but predicated on a pest management strategy that would reduce the entire population in and around cropping areas where economic damage occurred; or, at least, form part of a comprehensive business plan where potential pest problems (including fruit flies) were identified and factored into an economic cost–benefit analysis to facilitate production of fruits and vegetables for local consumption and export. It was envisioned that integration of new and old technologies into a pest management package would facilitate development of a well-defined agricultural production and marketing plan that would result in a better understanding of the potential of Hawaii agriculture in local, national and international markets. Furthermore, in the absence of eradication programmes in Hawaii, systems approaches using IPM methodologies may be one of the best strategies for reducing the environmental costs of continued high pesticide usage. These methods would also help in achieving quarantine security (Jang and Moffitt, 1994) while at the same time aid in producing higher-quality, safe fruits and vegetables for local consumption and possible niche export markets.

Description of the AWPM Programme and Approaches

AWPM management technologies and approaches

In 1999, the US Department of Agriculture (USDA), Agricultural Research Service (ARS) initiated the Hawaii Fruit Fly AWPM programme to suppress fruit flies below economic thresholds while reducing the use of organophosphate insecticides (Vargas *et al.*, 2003b). The programme included developing and integrating biologically based pest technology into a comprehensive IPM package that was economically viable, environmentally friendly and sustainable. It included operational, research, education and assessment components. The technologies included (see Fig. 16.1):

- Field sanitation (Klungness *et al.*, 2005).
- Application of protein bait sprays (Peck and McQuate, 2000; Vargas *et al.*, 2001, 2002; Prokopy *et al.*, 2004).
- Male and female annihilation with male lures and other attractants (Steiner *et al.*, 1965; Koyama *et al.*, 1984; Vargas *et al.*, 2000, 2003a).
- Sterile insect releases (Steiner *et al.*, 1970; McInnis *et al.*, 1994; Vargas *et al.*, 1994, 1995, 2004; Koyama, 1996).

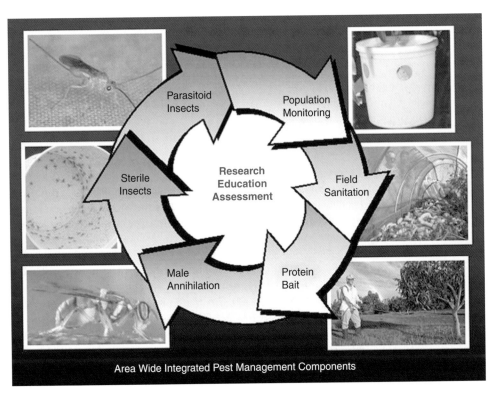

Fig. 16.1. The six components of the Hawaii AWPM programme.

- Conservation or release of beneficial parasitoids (Wong *et al.*, 1991, 1992; Purcell *et al.*, 1994a, b; Knipling, 1995; Vargas *et al.*, 2004, 2007a).

Field sanitation

Field sanitation is a technique that either prevents fruit fly larvae from developing or sequesters young emerging adult flies so that they cannot return to the crop to breed. In the past it was assumed that smashing fruits or rotor tilling the soil would kill most of the fruit fly larvae. However, preliminary tests demonstrated that only a small proportion of the flies were killed in this manner. Likewise, herbicide treatment of uncultivated host plants can stop plant growth, thereby reducing subsequent infestation of young fruit, but had little effect on larvae already developing in the fruit.

Consequently, the Hawaii AWPM programme promoted various methods of either killing the larvae in the fruit or preventing the larvae from developing into adult flies. Among the methods recommended by the programme were bagging or deep-burying infested fruit, drowning larvae in the fruit or sequestering emerging adult flies in tents or under plastic screens. Data suggest that larvae can go through window screens. However, if the screen is under the fruit, the larvae will crawl through it and pupate in the soil, but the emerging adults cannot escape back into the crop environment. Mechanization was recommended for some large farms.

Grinding the fruit into fine pulp, thus macerating the larvae, is the surest method of destroying infested fruit, but may not be the most cost effective for small farms.

One novel sanitation device used in the programme was the augmentorium (Klungness *et al.*, 2005). These tents, called augmentoria, were made with a screen material that restricted the dispersal of fruit fly adults emerging from the fruit placed in the tent, but allowed smaller fruit fly parasitoids that emerged from fruit culls to escape. Further details on methods of sanitation used in the Hawaii AWPM programme are to be found in Klungness *et al.* (2005).

GF-120 Fruit Fly Bait spray

Over the 50 years plus that organophosphate pesticides have been used to control fruit flies, they have been ineffective in the control of egg and larval development within the fruit (Keiser, 1968). Nishida and Bess (1950) recognized the inadequacy of spraying pesticide on the crop to control melon fly, because adult flies enter cultivated fields from surrounding areas to oviposit. Ebeling *et al.* (1953) suggested applying pesticides to maize borders surrounding the crop where flies congregate to invade the cultivated area. Nishida *et al.* (1957) then developed an effective technique for combining a food bait with a pesticide and applying it to border vegetation where flies roost.

Since their discovery, fruit flies have been controlled in agricultural areas of Hawaii using protein bait sprays. Most female flies need protein for full ovarian development and egg production, thus they readily feed on a protein source containing a toxicant. The bait spray strategy dramatically reduces the amount of pesticide needed for fruit fly control and has been used successfully in eradication campaigns (Steiner *et al.*, 1961; Roessler, 1989).

Since the late 1950s, the most common toxicant used in fruit fly bait spray formulations has been the organophosphate insecticide, malathion (Roessler, 1989). Nu-Lure has been the most popular protein bait mixed with malathion for fruit fly control (Prokopy *et al.*, 1992). However, organophosphate insecticides have been implicated in negative effects on natural enemies and human health. Prior to the AWPM programme, new bait spray formulations containing reduced-risk insecticides, such as spinosad or phloxine B, were developed and tested for use in Central America and the USA (McQuate *et al.*, 1999; Peck and McQuate, 2000).

Spinosad, a toxin derived from the soil-dwelling actinomycete bacterium, *Saccharopolyspora spinosa* Mertz and Yao, has low mammalian toxicity and reduced environmental impact on natural enemies (Stark *et al.*, 2004). A hydrolysed protein bait with spinosad that attracted, induced feeding and killed fruit flies was initially developed by Moreno and Mangan (1995). The first bait contained liquid Mazoferm E802™ (an enzymatically hydrolysed protein from maize processing; Corn Products, Argo, Illinois, USA) as the protein component. This bait was shown to have little impact against a series of beneficial hymenopteran parasitoids (Dowell, 1997) and reduced impact against honeybees (Dominguez *et al.*, 2003). The Mazoferm bait was field tested against medfly in Hawaii (Peck and McQuate, 2000; Vargas *et al.*, 2001).

A second bait, known as Solbait and composed of the protein attractant Solulys, a dried and more purified product processed from Mazoferm, was developed and successfully tested with a series of toxicants. Solbait has since been produced as GF-120 Fruit Fly Bait (Dow AgroSciences, Indianapolis, Indiana, USA) (DowElanco, 1994).

Spinosad has extremely low vertebrate and environmental toxicity, with reduced risk to humans and wildlife when compared with traditional insecticides and is effective at much lower doses. It is effective against tephritids in doses as low as 1 ppm in the laboratory. The low toxicity of spinosad towards beneficial insects allows it to be incorporated into many integrated pest management programmes that rely heavily on predators and parasitoids (Vargas *et al.*, 2000, 2002).

The AWPM programme provided farmers with the new commercial formulation GF-120 Fruit Fly Bait as a substitute for Nu-Lure and malathion for control of fruit flies. This novel product, combined with sanitation in an IPM approach, became the major technology transfer to farmers participating in the Hawaii programme and provided the foundation from which the Hawaii AWPM programme grew. The initial successes of spinosad bait sprays were demonstrated with medfly (Peck and McQuate, 2000) and, subsequently, with melon fly (Prokopy *et al.*, 2003). GF-120 was in the top group of proteins screened and generally rated higher than Nu-Lure, particularly when tested with protein-deprived flies (Vargas *et al.*, 2002, 2007a; Prokopy *et al.*, 2003; Vargas and Prokopy, 2007).

Aged baits, when compared with fresh baits, were unattractive to *B. dorsalis* and *B. cucurbitae*. Since attractiveness of bait droplets is short-lived, baits should be applied at short intervals or other ingredients added to baits to extend the period of attractiveness. Weekly applications of baits would probably be the shortest spray interval that is economically feasible for farmers.

Male annihilation

Worldwide, the Dacinae are astonishingly responsive to kairomone lures (Metcalf and Metcalf, 1992). At least 90% of the Dacinae species (comprising the two major genera *Bactrocera* and *Dacus*) are strongly attracted to either methyl eugenol (ME) or cue-lure (C-L/raspberry ketone) (Hardy, 1979). For instance, at least 176 species of the male Dacinae are attracted to C-L/raspberry ketone, and 58 species to ME (Metcalf, 1990). Of the 73 *Bactrocera* and *Dacus* species that are agricultural pests, 41 respond to C-L/raspberry ketone, 22 to ME, and ten to neither (White and Elson-Harris, 1992).

Many tests have indicated that male annihilation (Steiner *et al.*, 1970; Cunningham *et al.*, 1975; Koyama *et al.*, 1984; Cunningham and Suda, 1985, 1986; Vargas *et al.*, 2000) is environmentally sound (Kido *et al.*, 1996), cost effective and has excellent potential for areawide suppression of both melon fly and oriental fruit fly males. Vargas *et al.* (2000, 2003a) found that enclosing a wick inside bucket traps not only provided protection from the weather but also made the device visible, retrievable and reusable with limited environmental contamination.

However, in spite of being used in California for the last 25 years, and for many eradication programmes throughout the Pacific, male annihilation is still not legally available in Hawaii for control purposes, except on an experimental basis. Likewise, C-L has never been registered for control use in Hawaii. It was envisioned that development of simple, practical technologies for areawide use of ME and C-L would have important applications to suppression of fruit flies, not only in the Hawaii programme but also throughout the southern and western Pacific, Australia and tropical Asia, where *Bactrocera* are serious economic pests. Therefore, a special local needs permit was obtained for use of male annihilation in the AWPM programme.

For the first time, the AWPM programme provided farmers and homeowners with male annihilation dispensers for control of oriental fruit fly and melon fly in Hawaii.

Sterile insect technique

Staff from ARS in Hawaii carried out the original pilot tests of the sterile insect technique (SIT) to eradicate oriental fruit fly in the western Pacific (Steiner *et al.*, 1970) and to suppress or eradicate medfly in Hawaii (Harris *et al.*, 1986) and California (Cunningham *et al.*, 1980). In Japan, melon fly was eradicated by SIT (Koyama, 1996) using the Hawaiian approach. Subsequent SIT demonstration tests in Hawaii indicated significant reductions in fruit fly populations infesting large monocultures (Vargas *et al.*, 1994, 1995; Vargas, 1996). However, the use of bisexual strains (males and females) precluded the application of SIT to fruit and vegetable farms with the AWPM programme where crops were susceptible to sting damage.

The potential advantages of an SIT programme through the release of solely males not only included avoidance of 'sting-damage' by sterile females but also avoidance of matings between sterile males and sterile females. The effect of eliminating sterile females translated into increased efficiency of SIT by maximizing matings between sterile males and wild females. In the absence of sterile females, sterile males could find more wild female mates and improve the overall efficiency of an SIT programme (McInnis *et al.*, 1994; Rendon *et al.*, 2004). Development of males-only lines of melon fly and oriental fruit fly sexing strains allowed for the application of SIT to small-farm situations in the Hawaiian programme.

As part of the AWPM programme, a new strain of melon fly was developed and tested, which allowed for colour separation of males and females at the pupal stage through use of high-speed sorting machines. Known as the T-1 strain (McInnis *et al.*, 2006), the resulting males were released in selected areas and significantly reduced the local melon fly population to near extinction. A similar strain was developed for the oriental fruit fly and evaluated in small-scale AWPM demonstration tests (McInnis *et al.*, 2007). Although this approach proved very successful, the need for a large mass-rearing facility in Hawaii and more cost-effective 'sexing strains' limited its implementation.

Fruit fly parasitoids

The role of parasitoids in the Hawaiian AWPM fruit fly programmes was examined at three levels of application: (i) conservation; (ii) classical releases; and (iii) augmentative releases. An overall goal of the AWPM programme was to conserve biological control in economic crops through the use of reduced-risk insecticides such as GF-120 Fruit Fly Bait and male annihilation bucket traps, while using an AWPM approach (Vargas *et al.*, 2001, 2003b; Prokopy *et al.*, 2003; Stark *et al.*, 2004). The programme succeeded in both reducing the use of organophosphates and conserving biological controls, such as *Fopius arisanus* and related braconid species, while suppressing fruit flies below economic injury levels.

Perhaps no fruit fly parasitoid has been as successful in suppressing host populations as *F. arisanus* (Rousse *et al.*, 2005). Because of its habit of attacking host eggs, which are more exposed to parasitism than larvae, it can achieve high levels of parasitism, often surpassing 50% in the field (Vargas *et al.*, 1993, 2007a; Purcell *et al.*, 1996).

The success of classical biological control against fruit flies in Hawaii, in particular with *F. arisanus*, has been thoroughly reviewed by Rousse *et al.* (2005).

In Hawaii, the impact of *F. arisanus* introductions resulted in a 95% reduction in the oriental fruit fly population, from the 1947–1949 peak abundance of oriental fruit fly (DeBach and Rosen, 1991). Furthermore, *F. arisanus* became the major parasitoid of medfly in Hawaii (DeBach and Rosen, 1991; Vargas *et al.*, 2001). Haramoto and Bess (1970) reported that the mean number of fruit fly pupae (oriental fruit fly and medfly) collected from coffee fruits in Kona, Hawaii decreased from 23.6 pupae per 100 fruits (8.7% parasitism) in 1949 to 5.2 (66.6% parasitism) in 1969. With this level of impact on infestation level, establishment of *F. arisanus* has reduced the threat of movement of fruit flies to the mainland from Hawaii.

Since *F. arisanus* was already established in Hawaii, it was not possible to test classical releases of parasitoids in Hawaii. However, just before the AWPM programme was initiated in Hawaii, oriental fruit fly became established in French Polynesia, the most likely source being Hawaii. As part of an AWPM and a Foreign Agricultural Service (FAS) initiative to extend the AWPM programme outside of Hawaii, *F. arisanus* was introduced into French Polynesia. During the project, fruit samples before and after releases of *F. arisanus* on Tahiti Island were compared. From 2002 (before parasitoid releases) to 2006 (after parasitoid releases), there was a decline in numbers of fruit flies emerging (per kg of fruit) for oriental fruit fly, Queensland fruit fly (*B. tryoni*) and *B. kirki* of 75.6, 79.3 and 97.9%, respectively. It is recognized that much of the decline in numbers of Queensland fruit fly and *B. kirki* may have been due to competitive interactions with oriental fruit fly. However, *F. arisanus* probably also played a role in the decline.

French Polynesia consists of over 118 islands and atolls scattered over approximately 2,500,000 km^2 of ocean. Currently, oriental fruit fly is confined to the Society Islands. Initially it was envisioned that *F. arisanus* could be mass reared at an estimated cost of US$2,000 per 1,000,000 parasitoids (Harris *et al.*, 2000) and transferred to other islands as oriental fruit fly spread throughout French Polynesia. However, when *F. arisanus* became numerous in fruits infested with oriental fruit fly on Tahiti Island, it became more cost-effective to recover wasps from fruits held inside screened cages and ship them to the outer islands than to mass rear them in the laboratory on artificial diets. This approach is now being used for shipments to islands where oriental fruit fly has spread in French Polynesia.

None the less, for approximately US$100,000, the shipment and establishment of *F. arisanus* in French Polynesia has provided a sustainable programme to reduce the impact of oriental fruit fly, which was not obtained with much more expensive eradication programmes. Consequently, establishment of *F. arisanus* has reduced the threat of movement of fruit flies to new areas from French Polynesia. Finally, the present programme in French Polynesia has reduced damage by oriental fruit fly and developed a biological base for further development of IPM programmes in conjunction with sanitation, reduced-risk protein bait sprays and male annihilation treatments.

In Hawaii, augmentative release of parasitoids was selected as one of the major technologies to be transferred to farmers in the original project proposal. Numerous studies had demonstrated the feasibility of parasite augmentation to control fruit flies. In Hawaii, release of *Diachasmimorpha tryoni* (at 20,000/km^2 per week over a 14 km^2 area) more than tripled medfly parasitism rates (Wong *et al.*, 1991). In studies with melon fly, augmentatively released *Psytallia fletcheri* significantly enhanced parasitism rates in vegetables (i.e. courgette and cucumber) compared with background populations in commercial fields (Purcell and Messing, 1996). Therefore, during the AWPM programme, *P. fletcheri* and *F. arisanus* were reared and released in wild cucurbit and guava patches, respectively, near agroecosystems (Vargas *et al.*, 1993), with the objective of demonstrating a cost-effective, sustainable technology that could be integrated with bait sprays and male annihilation.

In releases of *P. fletcheri* against melon fly inside field cages, the numbers of melon flies emerging from fruits placed inside treatment cages were reduced up to 21-fold and numbers of parasitoids were increased by 11-fold (Vargas *et al.*, 2004). In open-field releases of *P. fletcheri* into ivy gourd patches throughout the Kailua-Kona area, parasitism rates were increased 4.7 times in release plots compared with those in control plots. However, there was no significant ($P > 0.05$) reduction in emergence of flies from fruits. Similarly, in releases of *P. fletcheri* in courgette plots in Waimea, there was an increase in parasitoid recovery rates; however, there was no reduction in melon fly damage (R.I. Vargas, Hilo, Hawaii, unpublished data). *F. arisanus* was also tested as an augmentative tool in small plots of guava in Waimea where the existing population of *F. arisanus* was low. Levels of parasitism were increased, but infestation was not reduced (R.I. Vargas, Hilo, Hawaii, unpublished data).

Although augmentative releases of parasitoids were shown to increase parasitism in the field, limited rearing capacity and high cost limited their level of implementation into a sustainable AWPM programme. On the other hand, classical biological control was demonstated to be very cost-effective and sustainable in the French Polynesian programme. Establishment of *F. arisanus* in French Polynesia against oriental fruit fly is now the most successful example of classical biological control of fruit flies in the Pacific area outside of the Hawaiian Islands, and serves as a model for introduction of the parasitoid into South America and Africa, where the carambola flies, *B. carambolae* and *B. invadens* (Drew *et al.*, 2005), have recently become established. In addition, *F. arisanus* is being studied as a possible candidate for classical biological control of the peach fruit fly, *B. zonata* (Saunders), in Africa and in the Indian Ocean region (e.g. FAO/IAEA, 2005).

Compatibility of the AWPM programme with crop management of co-occurring pests

The use of environmentally friendly approaches for control of fruit flies created few problems for management of co-occurring pests and was generally compatible with other practices. Implementation of sanitation for fruit fly management also improved control of other pests. GF-120 Fruit Fly Bait received an all-crops label and GF-120 NF Naturalyte Fruit Fly Bait was approved for use in the production of certified organic fruits and vegetables. However, one major issue with the use of lures for male annihilation treatments was the perception that these treatments may be a threat to non-target organisms. Previous studies suggested that methyl eugenol was attractive to numerous non-target insects. However, more recently, in non-target studies of male annihilation funded by the AWPM programme, attraction to most non-targets was not to the male lures but, instead, to rotting insects in traps (Uchida *et al.*, 2004, 2007; L. Leblanc, personal communication, Honolulu, Hawaii, February 2007).

Development and implementation of the AWPM programme

In order to promote and implement the Hawaii AWPM programme, partnerships were created with representation from the federal, state and industrial sectors. These partners included: (i) the USDA (ARS and Agricultural Plant Health Inspection Service (APHIS)); (ii) the University of Hawaii (UH); (iii) the Hawaii Department of Agriculture (HDOA); and (iv) industry (Dow AgroSciences, Farmatech International, Scentry Biologicals, Better World Manufacturing and United Agricultural Products).

Industry provided the technologies (bait sprays, solid lures and traps), and ARS the initial research and development of these technologies. The UH Extension Service provided lists of stakeholders, potential cooperators, grower training, cooperative extension and community-based education on fruit fly issues. HDOA issued permits necessary to implement the new technologies. In addition to local partnerships, a management team and secondary technical advisory group was established to help guide the programme through its initial stages. Each year an annual review meeting was held to evaluate progress of the programme and recommend adjustments when necessary.

The four fruit fly pest species affected specific crops grown by different groups of small farmers, so it was necessary to implement the programme sequentially by pest species. Each species required a customized AWPM programme. The melon fly, the first species targeted, caused highest losses throughout the year to cucurbit, melon and solanaceous crops. These crops were commonly grown in small clusters of farms; medfly suppression was undertaken at the same time because of requests by fruit growers near the melon farmers. Medfly is a serious pest of persimmons grown at upper elevations, but the pest develops in uncultivated fruits that are found throughout the year. We were able to undertake medfly suppression at the same time as that for melon fly because of the enthusiastic assistance of persimmon growers on Maui and the support of the UH Extension Service.

Suppression demonstration programmes were implemented on three islands – Hawaii, Maui and Oahu. At four sites different cropping systems were used to evaluate the various technologies proposed. The four major sites chosen to demonstrate fruit fly suppression technologies included Waimea (Hawaii Island), Kula (Maui Island), Kunia/Ewa (Oahu Island) and Puna (Hawaii Island). Melon fly and oriental fruit fly were the predominant species at all of the sites. Medfly occurred at low and moderate densities at Waimea and Kula, respectively; *B. latifrons* occurred in low numbers at each site.

The first demonstration project was initiated on Hawaii Island in the Waimea region. The 3800 ha demonstration zone (cucurbits and melons) was surrounded by pastures and characterized by homes and a small town that separated two farming areas. Melon fly was the principal species suppressed. The second implementation zone (4400 ha) (cucurbits, melons, tomatoes and persimmons) was at Kula on Maui Island. This zone was characterized by clusters of small farms (*c.* 7–10 ha) surrounded by wild fruit fly hosts. Melon fly, oriental fruit fly and medfly were the principal species controlled. Central Oahu was the third demonstration site; this area encompassed more than 1600 ha of farmland adjacent to large residential and industrial areas. Crops included watermelon, honeydew melon, cantaloupe, courgette, squash and pumpkin. Melon fly was the principal species suppressed. The fourth implementation zone was at Puna, where the programme was applied to approximately 400 ha of papaya orchards. The cultivated area was surrounded by dense stands of unculti-vated strawberry and common guava and fruit trees that sustained a very large orien-tal fruit fly population.

Suppression of oriental fruit fly in papaya orchards proved very challenging. The sequence of sites selected as the programme progressed turned out to be fortu-itous and added to the credibility of the eventual success of the programme. The suc-cess of the melon fly control programme, and subsequently the medfly programme, allowed for development of an oriental fruit fly programme in areas of high infesta-tion such as Puna. Development of the programme to suppress oriental fruit fly, application of the programme to Puna and registration of the necessary chemicals required an extension of 3 additional years. Expansion of the programme beyond the original demonstration sites is discussed under the prospects for sustainability section.

Development and implementation of education and technology transfer programmes

Previous IPM pilot tests in Hawaii had shown potential for local applications, but had never been partnered with a good extension programme. The critical ingredient to the success of the Hawaii AWPM programme was an organized, coordinated and comprehensive outreach educational programme. The Hawaii AWPM programme used the 'logic model' approach to organize, plan, execute and evaluate farmer and community educational programmes state-wide (Mau *et al.*, 2007). The logic model approach was an outcome-driven rather than activity-based method that used a linear sequence that developed relationships between programme inputs, outputs and outcomes.

A 5-year outreach education plan was devised (Mau *et al.*, 2003a). One of the most important outcomes was development of empowered participants who could make informed decisions based on retained knowledge and skills. This effective trans-fer of knowledge and skills helped to assure sustainability of the AWPM programme. Four important types of outputs were established early in the educational programme. The AWPM video provided an overview of the suppression programme in lay terms for commercial and community cooperators (Mau *et al.*, 2003b). This video is shown frequently on public access television.

A series of brochures that described the suppression programme, identification and life cycle of the four targeted species of fruit flies and suppression elements were developed soon thereafter. The brochures included photographs and described in lay terms the importance of species monitoring, crop sanitation, male lures, male annihilation, protein baits and biological control. An Internet web site was created to provide ready access to information and updates (http://www.fruitfly.hawaii.edu).

A newsletter was established and published monthly for cooperators and part-ners who did not have Internet access. Other teaching materials were created and distributed when they were needed. The extension service marketed fruit fly suppres-sion to farmers in the format of: 'As easy as 1 (population monitoring), 2 (sanitation), 3 (protein baits), 4 (male annihilation)'. This fruit fly programme became known as the '1-2-3-4 programme'. More details on the extension and education programme are to be found in Mau *et al.* (2007).

Evaluation of the AWPM Programme

Effectiveness of the AWPM programme at controlling target pests

The effectiveness of the AWPM programme was determined primarily on the basis of lower fruit fly trap captures, reduction in fruit infestation and reduction in organophosphate pesticide use. Depicted in Figs 16.2, 16.3 and 16.4 are the impact on trap captures through implementation of the different programme components against melon fly, medfly and oriental fruit fly at three different demonstration sites located at Waimea, Kula and Puna, respectively.

At Waimea on Hawaii Island, implementation of sanitation reduced captures of melon flies at managed farms to approximately one melon fly/trap/day. Subsequent implementation of GF-120 Fruit Fly Bait sprays, male annihilation (cue-lure traps), sterile flies and parasitoids reduced the melon fly population to nearly zero at treated farms. At Kula on Maui Island, implementation of Biolure traps and GF-120 Fruit Fly Bait sprays reduced captures of medflies to fewer than 0.1 flies/trap/day. At Puna on Hawaii Island, implementation of a combination of sanitation, male annihilation traps and GF-120 sprays reduced captures of oriental fruit fly by tenfold in treated papaya orchard traps when compared with untreated control area traps.

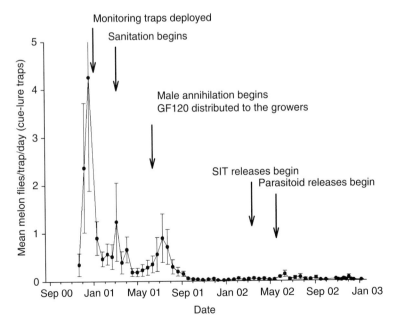

Fig. 16.2. Captures of melon flies on AWPM farrns at Waimea, Hawaii Island, Hawaii.

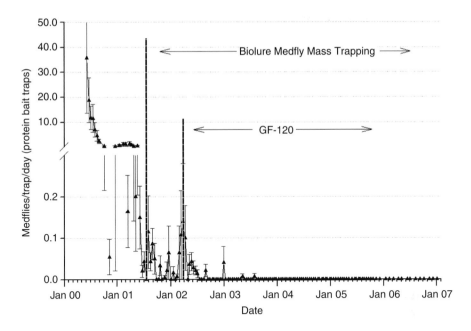

Fig. 16.3. Captures of medflies on AWPM farms at Kula, Maui Island, Hawaii.

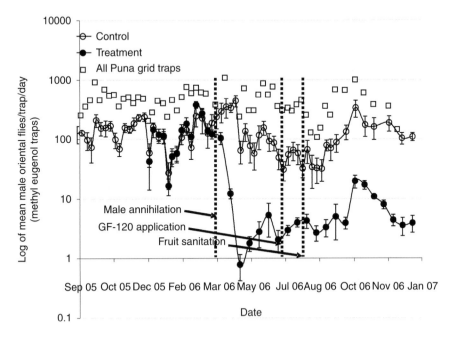

Fig. 16.4. Captures (log scale) of oriental fruit flies at Puna, Hawaii Island, Hawaii, in AWPM orchard traps, control traps and survey traps, with the sequential application of male annihilation, GF-120 fruit fly bait and sanitation.

Table 16.1. Papaya fruit infestation sampled from treated and non-treated orchards by stage of ripeness for oriental fruit fly.

Ripeness index	n	Infested fruit (n)	Infested fruit (%)	Mean flies/g	SEM
AWPM treatment site					
Colour break	84	0	0.00	0.0000	0.0000
¼ ripe	84	0	0.00	0.0000	0.0000
½ ripe	82	2	2.44	0.0022	0.0018
Fully ripe	82	7	8.54	0.0014	0.0067
Control site					
Colour break	90	0	0.00	0.0000	0.0000
¼ ripe	90	0	0.00	0.0000	0.0000
½ ripe	86	3	3.49	0.0005	0.0003
Fully ripe	83	20	24.39	0.0245	0.0069

Comparison of papaya fruit infestation sampled from treated orchards with those from non-treated orchards suggests that riper papaya fruits could be marketed from treated fields, providing higher-quality fruit for local consumption (see Table 16.1). The potential impact of these preharvest suppression measures on quarantine regulations for export of papaya fruit because of reduced infestation is presently being examined.

Unintended negative and positive consequences of the AWPM programme

The major positive feature of the AWPM programme was the close and effective collaboration between the various AWPM lead agencies in Hawaii. The programme's close collaboration is being considered as a template for future agricultural research and technology transfer programmes in Hawaii (Jang, 2003). Furthermore, California and Florida have also shown a keen interest in the programme. California alone would suffer a US$1.5 billion annual loss in export sanctions, treatment costs, lost markets and reduced crop yields if the medfly became established there. Development and application of environmentally friendly areawide fruit fly controls, as performed in the Hawaiian AWPM programme, are of critical importance in keeping the US mainland free of the fruit flies already established in Hawaii.

Finally, unique to the Hawaiian AWPM programme has been development of international collaborations. There have been close interactions with officials and researchers from many other countries, including Taiwan, the People's Republic of China (PRC), Australia, French Polynesia, Fiji, Guam and the Commonwealth of the Northern Mariana Islands. Taiwan has been at the forefront of adopting the technologies that were implemented in Hawaii. The Taiwan Agricultural Research Institute has initiated a programme that includes 5% of Taiwan's land, 172 cooperating towns and villages and 149,713 ha involving 449 districts (McGregor, 2007). The Taiwan AWPM programme is now larger in scope than the Hawaii programme. Similarly, through a partnership between Hawaii and French Polynesia, introduction of *F. arisanus* into French Polynesia has resulted in 50% parasitism of fruit flies infesting a variety of tropical fruits, and reduced numbers of oriental fruit fly emerging from fruits by as much as 75% (Vargas *et al.*, 2007a).

Establishment of *F. arisanus* is the most successful example of classical biological control of fruit flies in the Pacific area outside of the Hawaiian Islands, and serves as a model for introduction of the wasp into South America, Africa and China (PRC), where species of the *B. dorsalis* complex are established, in many cases without effective natural enemies. In summary, success of the fruit fly AWPM programme has not only helped other countries control their fruit fly problems but also helped protect US agriculture from fruit fly spread through a regional containment approach.

Economic evaluation of costs and benefits of the AWPM programme

An agricultural economist evaluated the costs and benefits of the Hawaii AWPM programme through interviews with stakeholders, farmer surveys and visits to demonstration sites and farms (McGregor, 2007). The consolidated estimated industry benefits of the AWPM programme are presented in Table 16.2 for production of cucurbits, tomato, citrus, persimmon, mango, dragon fruit, papaya and a possible new fruit. These benefits were extrapolated to the year 2014. Forecast benefits are projected to increase from US$2.6 million in 2006 to US$3.5 million in 2007. A cost–benefit analysis of the programme is summarized in Table 16.3. Further details on an economic analysis of the Hawaii AWPM programme can be found in McGregor (2007). The substantial non-industry benefits are not included in the formal benefit–cost analysis, but are discussed under sociological benefits.

Sociological evaluation of the AWPM programme

The strengthening of Hawaii's agricultural industry, weakened by the downsizing of the pineapple and sugarcane industries, has had a positive effect on the state economy. New jobs have been created in diversified agriculture and additional income generated as growers have expanded their acreage, sometimes reclaiming acres previously abandoned by growers unable to deal with fruit fly damage. The production of more high-value food crops has helped consumers in an island state that imports fruits and vegetables at considerable cost; many of these fruits and vegetables could be produced locally. If the AWPM programme helps increase local fruit production, consumers benefit from increased availability of quality fruit, lower fruit prices and low chemical residues in fruit. Ultimately, better fruit fly control could lead to new possibilities for export of high-value commodities.

Adoption of the AWPM programme has also benefited the unique, fragile Hawaiian environment by reducing the amount of organophosphate and carbamate

Table 16.2. Consolidated quantifiable industry benefits from the Hawaii fruit fly AWPM programme (US$, 000).

	2000	2001	2002	2003	2004	2005	2006	2007	2008	2009	2010	2011	2012	2013	2014	2015
Benefits based on actual and forecast outputs																
Cucubits	–	200	400	600	1000	1000	1000	1100	1100	1200	1200	1300	1300	1300	1300	1300
Vine-ripened tomatoes	–	–	–	–	200	500	700	800	900	1000	1000	1000	1000	1000	1000	1000
Citrus	–	–	–	300	100	200	300	300	400	400	400	500	500	500	500	500
Persimmons	–	–	–	200	200	300	300	300	400	400	400	500	500	500	500	500
Mango	–	–	–	50	50	100	100	150	150	200	200	300	300	300	300	300
Dragon fruit	–	–	–	30	40	50	60	70	80	90	100	110	110	110	110	110
Papaya	–	–	–	30	40	50	70	50	1000	1500	2000	2000	2000	2000	2000	2000
New 'highly susceptible' fruit	–	–	–	–	20	30	40	60	80	100	200	300	300	300	300	300
Subtotal	–	200	400	1210	1650	2230	2570	3280	4110	4890	5500	6010	6010	6010	6010	6010
Benefits from 'likely' outputs over the next 5 years																
Increased returns to papaya growers from harvesting riper fruit								200	300	400	500	600	600	600	600	600
Reduced quarantine costs for Puna papaya growers										500	500	500	500	500	500	500
Reduced quarantine cost for outshipment of 'low risk' products										140	161	185	213	245	282	324
Subtotal								200	300	1040	1161	1285	1313	1345	1382	1424
Benefits from 'possible' outputs over the next 10 years																
Papaya from Puna control area without quarantine treatment												500	500	500	500	500
Outshipments of breadfruit										300	315	331	347	365	383	402
Exports of high-value melons to Japan										300	330	363	399	439	483	531
Sub total										600	645	1194	1247	1304	1366	1433
Total consolidated benefits	–	200	400	1210	1650	2230	2570	3480	4410	6530	7306	8489	8570	8659	8758	8867

Table 16.3. A comparison of the consolidated programme benefits with costs[a] (US$,000).

	2000	2001	2002	2003	2004	2005	2006	2007	2008	2009	2010	2011	2012	2013	2014	2015
Total consolidated benefit (B)	–	200	400	1210	1650	2230	2570	3280	4110	4890	5500	6010	6010	6010	6010	6010
Programme costs (C)	860	1300	1600	1960	1980	1970	1900	2000	800	750	750	750	250	250	250	250
B–C	(860)	(1100)	(1200)	(750)	(330)	260	670	1280	3310	4140	4750	5260	5760	5760	5760	5760
Internal rate of return (IRR) (%)	28															
Programme net present value (NPV) (@ 10% rate of interest)	34000															

[a]USDA-ARS Internal report by McGregor (2007).

pesticides while still promoting an increase in agricultural production. Because of this programme, there has been tremendous support by growers and the public in utilizing technologies offered by the programme. Grower yields have increased, while organophosphate insecticide use has decreased. Tools are now legally available to control fruit fly and provide high-quality safe fruits and vegetables in Hawaii. Finally, the fortunes of expansion of diversified agriculture are closely linked to those of tourism; some 4.5 million people visit Hawaii annually. That creates a major market and the aircraft in which they arrive provide the freight capacity to outside markets at competitive rates. However, this relationship is not just one way. Diversified agriculture contributes significantly to the value of the tourism product: flowers, pineapples, tropical fruits, the open space created by farms that grow produce and an appealing environment are all part of the visitor experience (McGregor, 2007).

Prospects for the long-term sustainability of the AWPM programme

More than 2648 cooperating growers over five islands, representing more than 8449 ha (see Table 16.4), have joined the '1-2-3-4 programme'. They have been able to cut organophosphate pesticide use by 75–90%. While using the AWPM programme that reduced environmental risks, growers have still cut fruit fly infestation by 30–40% to < 5% (Vargas *et al.*, 2007b). Farmers have enthusiastically embraced the '1-2-3-4 programme'.

Surveys conducted to test grower perceived knowledge of fruit fly control on the Big Island (Hawaii Island) indicated that 85% of growers had a good or fair understanding of the technology and fewer than 10% had poor or no understanding (McGregor, 2007). Technologies have been demonstrated that work, are user friendly and increase financial returns. To introduce the technology to farmers and home growers, monitoring traps with lures, male annihilation traps and GF-120 Fruit Fly Bait spray have been highly subsidized. Interviews with farmers indicate that they will have to meet these costs and are willing to do so after the ARS funding ends.

An 'all crops label' was obtained for GF-120 Fruit Fly Bait and an organically certified formulation, GF-120 NF Naturalyte Fruit Fly Bait, was marketed. Manufacturer's use permits (MUPs) were obtained in 2005 and 2006 for cue-lure and methyl eugenol, respectively. Major research and development efforts are presently under

Table 16.4. Total number of cooperators, number of farms and area impacted by the Hawaii Areawide Pest Management programme.

Island	Total cooperators (*n*)	Farms (*n*)	Area (ha)
Oahu	436	108	2283
Maui	1270	62	2775
Molokai	31	26	141
Kauai	144	63	348
Hawaii	767	394	2902
Statewide	2648	653	8449

way by ARS, UH and industry (FarmaTech, Sentry, BASF and ISCA Technologies) to provide methyl eugenol and cue-lure 'end products' for use in male annihilation treatments when USDA-ARS funding of the programme ends.

Registration of male annihilation end products is critical for programme sustainability. For smaller farmers to continue with male annihilation strategies, these lures must be available in their local farm supply store. It is hoped that the registration of methyl eugenol and cue-lure end products for fruit fly control will be approved by EPA in 2007. It would be ideal if these products could be demonstrated on local farms while the AWPM programme is still being funded.

The ongoing research and extension and public education programme will also need to be continued after programme funding ends, to consolidate and expand the benefits that have been achieved thus far. The cost of the AWPM programme extension and education components have been relatively modest compared with the benefits that have been achieved. The University of Hawaii has conducted an effective extension effort on Oahu, Maui, Kauai and Molokai, channelled through the Cooperative Extension Service. The extension programme on the Big Island has also been effective, where ARS has taken the lead. To ensure sustainability of the AWPM programme, particularly among smaller farmers and new cooperators, there needs to be a future commitment to continued research on these pests, as well as extension support for training and distribution of information on control technologies and products.

The small-scale SIT releases for melon fly, medfly and oriental fruit fly have been effective but, without federal support and a rearing facility, this technology will not be sustainable. Similarly, classical releases of parasitoids were shown to be cost-effective where natural enemies were non-existent, but augmentative releases, although promising, are not a proven technology and cannot be sustained without a rearing facility.

Adherence to the '1-2-3-4 programme' at the four demonstration sites for the three species of fruit fly was shown to be effective in its own right in suppressing fruit flies below economic thresholds. However, the level of suppression will not be as great as with sterile flies. From all accounts the cost of a '1-2-3-4 programme' is far lower without the sterile flies and parasitoids. Furthermore, the responsibility for meeting these costs lies with the farmer. With the farmers controlling their own destiny, greater sustainability can be expected than with programmes relying on continuous public expenditure.

Summary and Future Directions

In summary, ensuring adoption of the programme by Hawaiian farmers required far more than just research and development of the technology. Partnerships were created with the Hawaii Department of Agriculture (HDOA) and the University of Hawaii (UH). ARS researchers and UH personnel and extension agents worked tirelessly with growers to help them take control of the technology package. HDOA provided the impetus to register control products. Other partners were then enlisted to enhance cooperation and give the programme the best chance of success, including the USDA, APHIS, the IR-4 pesticide programme, the US Environmental Protection Agency, private industry (including Dow AgroSciences Inc., BASF, FarmaTech International, Sentry and ISCA Technologies) and local community action groups.

At the heart of the programme, however, were the Hawaiian farmers and gardeners who have participated as cooperators in demonstrating the benefits of the programme and then spread the word to others. The AWPM team had to overcome growers' reluctance to put themselves at economic risk by trying technologies they perceived as experimental. The team also had to overcome growers' disappointment with previous, unsuccessful eradication attempts during the past 25 years.

Extension agents, ARS researchers, UH researchers and HDOA officials met with grower groups to explain the idea and procedures. Extensive educational and 'how-to-do' materials have been created, including videos, a web site, public service announcements, pamphlets, handouts and posters to help growers and gardeners adopt the programme. But personal communication with growers was the real basis for the successful adoption of the programme. Growers were empowered to make informed decisions about adopting and continuing the programme.

Future plans include: (i) expansion of the oriental fruit fly programme to include other crops besides papaya; (ii) training of avocado and papaya growers in Puna and Kona in the '1-2-3-4 programme' approach for fruit fly suppression; (iii) demonstration of the effectiveness of the '1-2-3-4 programme' for control of *B. latifrons*; (iv) continued research to address problems which inhibit implementation of the AWPM programme, such as non-target and economic issues; (v) expansion into other agricultural areas of the state not part of the present AWPM programme; and (vi) promotion of sustainability through registration of methyl eugenol and cue-lure end products with the EPA.

Acknowledgements

The Hawaii Fruit Fly AWPM programme is a USDA-Agricultural Research Service-funded partnership with the University of Hawaii Cooperative Extension Service and Hawaii State Department of Agriculture. Principal investigators (USDA, ARS, USPBARC) on the project included Roger Vargas, Eric Jang, Don McInnis, Grant McQuate, Ernest Harris and Stella Chang. The original project proposal was written and submitted by Roger Vargas and Mary Purcell in 1997, revised by the principal investigators listed above and funded in 1999. The authors especially acknowledge the help of Kim Kaplan (ARS, USDA, information staff), Mike Klungness, Jill Grotkin, Ferol White and Nancy Chaney (ARS, USDA, PBARC, Hilo) with this manuscript. Finally, we would like to recognize the past and present employees of ARS, UH, HDOA and APHIS who worked tirelessly to make this programme a success.

References

Back, E.A. and Pemberton, C.E. (1917) The melon fly in Hawaii. *USDA Bulletin* 491.

Cunningham, R.T. and Suda, D.Y. (1985) Male annihilation of the oriental fruit fly (Diptera: Tephritidae): a new thickener and extender for methyl eugenol formulations. *Journal of Economic Entomology* 78, 503–504.

Cunningham, R.T. and Suda, D.Y. (1986) Male annihilation through mass-trapping of male flies with methyl eugenol to reduce infestation of oriental fruit fly (Diptera: Tephritidae) larvae in papaya. *Journal of Economic Entomology* 79, 1580–1582.

Cunningham, R.T., Chambers, D.L. and Forbes, A.G. (1975) Oriental fruit fly: thickened formulations of methyl eugenol in spot applications for male annihilation. *Journal of Economic Entomology* 68, 861–863.

Cunningham, R.T., Routhier, W., Harris, E.J, Cunningham, G., Tanaka, N., Johnston, L., Edwards, W., Rosander, R. and. Vettel, J. (1980) A case study: eradication of medfly by sterile-male release. *Citrograph* 65, 63–69.

DeBach, P. and Rosen, D. (1991) *Biological Control by Natural Enemies.* Cambridge University Press, Cambridge, UK.

Dominguez, V.M., Leyva, J.L., Moreno, D.S., Trujillo, F.J., Alatorre, R. and Enrique Becerril, A. (2003) Toxicidad sobre *Apis mellifera* de cebos empleados en el combate de moscas de la fruta. *Manejo Integrado de Plagas y Agroecologia* 69, 66–72.

DowElanco (1994) *Spinosad Technical Guide.* DowElanco, Indianapolis, Indiana.

Dowell, R.V. (1997) Laboratory toxicity of a photo activated dye mixture to six species of beneficial insects. *Journal of Applied Entomology* 121, 271–274.

Drew, R.A.I., Tsuruta, K. and White, I.M. (2005) A new species of pest fruit fly (Diptera: Tephritidae: Dacinae) from Sri Lanka and Africa. *African Entomology* 13, 149–154.

Ebeling, W., Nishida, T. and Bess, H.A. (1953) Field experiments on the control of melon fly, *Dacus cucurbitae. Hilgardia* 31, 563–591.

FAO/IAEA (2005) Alert notice USDA-APHIS: elevated risk of peach fruit fly (*Bactrocera zonata*) from Egypt. *Insect Pest Control Newsletter* 65, 18.

Faust, R.M. (2004) Local research, but everyone's watching. (Forum, Hawaii Areawide Fruit Fly Control Program, Pacific Basin Agricultural Research Center). *Agricultural Research* 1 February.

Haramoto, F.H. and Bess, H.A. (1970) Recent studies on the abundance of the oriental and Mediterranean fruit flies and the status of their parasites. *Proceedings of the Hawaiian Entomological Society* 20, 551–566.

Hardy, D.E. (1979) Economic fruit flies of the South Pacific Region. Book Review. *Pacific Insects* 20, 429–432.

Harris, E.J., Cunningham, R.T., Tanaka, N., Ohinata, K. and. Shroeder, W.J. (1986) Development of the sterile-insect technique on the island of Lanai, Hawaii, for suppression of the Mediterranean fruit fly. *Proceedings of the Hawaiian Entomological Society* 26, 77–88.

Harris, E.J., Bautista, R.C. and Spencer, J.P. (2000) Utilization of the egg-larval parasitoid, *Fopius (Biosteres) arisanus*, for augmentative biological control of fruit flies. In: Tan, K.H. (ed.) *Areawide Control of Fruit Flies and Other Insect Pests.* Penerbit Univesiti Sains Malaysia, Pulau Pinang, Malaysia, pp. 725–732.

Jang, E.B. (2003) A prescription against fruit flies. *Agriculture Hawaii* 4, 6–7.

Jang, E.B. and Moffitt, H.F. (1994) Systems approaches to achieving quarantine security. In: Sharp, J.L. and Hallman, G.J. (eds) *Quarantine for Pests of Food Plants.* Westview Press, Boulder, Colorado, pp. 225–239.

Keiser, I. (1968) Residual effectiveness of foliar sprays against the Oriental fruit fly, melon fly, and the Mediterranean fruit fly. *Journal of Economic Entomology* 61, 438–443.

Kido, M.H., Asquith, A. and Vargas, R.I. (1996) Nontarget insect attraction to methyl eugenol used in male annihilation of the oriental fruit fly (Diptera: Tephritidae) in riparian Hawaiian stream habitat. *Environmental Entomology* 25, 1279–1289.

Klungness, L.M., Jang, E.B., Mau, R.F.L, Vargas, R.I., Sugano J.S. and Fujitani, E. (2005) New sanitation techniques for controlling tephritid fruit flies (Diptera: Tephritidae) in Hawaii. *Journal of Applied Sciences and Environmental Management* 9, 4–14.

Knipling, E.F. (1995) *Principles of Insect Parasitism Analyzed from New Perspectives. Practical Implications for Regulating Insect Populations by Biological Means.* Handbook No. 693, United States Department of Agriculture, Pittsburgh, Pennsylvania.

Koyama, J. (1996) Eradication of the melon fly, *Bactrocera cucurbitae* by the sterile insect technique in Japan. *Proceedings of IAEA Training Course on the Use of Sterile Insect and Related Techniques for the Areawide Management of Insect Pests*, Gainesville, Florida, 8 May–19 June 1996.

Koyama, J., Teruya, T. and Tanaka, K. (1984) Eradication of the oriental fruit fly (Diptera: Tephritidae) from the Okinawa Islands by a male annihilation method. *Journal of Economic Entomology* 77, 468–472.

Mau, R.F.L., Sugano, J.S and Jang, E. (2003a) Farmer education and organization in the Hawaii areawide fruit fly pest management program. In: Inamine, K. (ed.) *Recent Trends on Sterile Insect Technique and Areawide Integrated Pest Management: Economic Feasibility, Control Projects, Farmer Organization and Dorsalis Complex Control Study*. Research Institute for Subtropics, Okinawa, Japan, pp. 47–57.

Mau, R.F.L., Sugano, J.S. and Hamasaki, D. (2003b) *Prescription for Fruit Fly Suppression* (videotape). Video Series No. 164, College of Tropical Agriculture and Human Resources, Honolulu, Hawaii.

Mau, R.F.L., Jang, E.B. and Vargas, R.I. (2007) The Hawaii fruit fly area-wide fruit fly pest management programme: influence of partnership and a good education programme. In: Vreysen, M.J.B., Robinson, A.S. and Hendrichs, J. (eds) *Area-wide Control of Insect Pests: from Research to Field Implementation*. Springer, Dordrecht, Netherlands. (In press).

McGregor, A.M. (2007) *An Economic Evaluation of the Hawaii Fruit Fly Area-wide Pest Management Program*. Draft final report, a study funded by the College of Tropical Agriculture and Human Resources, University of Hawaii. Honolulu, Hawaii.

McInnis, D.O., Tam, S., Grace, C. and Miyashita, D. (1994) Population suppression and sterility rates induced by variable sex ratio, sterile insect releases of *Ceratitis capitata* (Diptera: Tephritidae) in Hawaii. *Annals of the Entomological Society of America* 87, 231–240.

McInnis, D.O., Shelly, T.E. and Mau, R.F.L. (2006) All male strains and chemical stimulants: two ways to boost sterile males in SIT programs. In: Ku, T.Y. (ed.) *Proceedings of International Symposium on Area-wide Management of Insect Pests*, Naha, Okinawa, Japan, 8–11 October 2006, pp. 53–62.

McInnis, D., Leblanc, L. and Mau, R.F.L. (2007) Development and field release of q genetic sexing strain of the melon fly, *Bactrocera cucurbitae*, in Hawaii. *Proceedings of the Hawaiian Entomological Society* (in press).

McQuate, G.T., Cunningham, R.T., Peck, S.L. and Moore, P.H. (1999) Suppressing oriental fruit fly populations with phloxine B-protein bait sprays. *Pesticide Science* 55, 566–614.

Metcalf, R.L. (1990) Chemical ecology of Dacinae fruit flies (Diptera: Tephritidae). *Annals of the Entomological Society of America* 83, 1017–1030.

Metcalf, R.L. and Metcalf, E.R. (1992) Fruit flies of the family Tephritidae. In: Metcalf, R.L. and Metcalf, E.R. (eds) *Plant Kairomones in Insect Ecology and Control*. Routledge, Chapman & Hall Inc., New York, pp. 109–152.

Moreno, D.S. and Managan, R.L. (1995) Responses of the Mexican fruit fly (Diptera: Tephritidae) to two hydrolyzed proteins and incorporation of phloxine B to kill adults. In: Heitz, J.R. and Downum, K. (eds) *Light Activated Pest Control*. ACS Symposium Series 616, American Chemical Society, Washington, DC, pp. 257–259.

Newell, I.M. and Haramoto, F.H. (1968) Biotic factors influencing populations of *Dacus dorsalis* in Hawaii. *Proceedings of the Hawaiian Entomological Society* 20, 81–139.

Nishida, T. (1953) Ecological study of the melon fly, *Dacus cucurbitae* Coquillett, in the Hawaiian Islands. PhD dissertation, University of California, Berkeley, California.

Nishida, T. and Bess, H.A. (1950) Applied ecology in melon fly control. *Journal of Economic Entomology* 43, 877–883.

Nishida, T., Bess, H.A. and Ota, A. (1957) Comparative effectiveness of Malathion and Malathion-yeast Hydrolysate bait sprays for control of melon fly. *Journal of Economic Entomology* 50, 680–684.

Peck, S.L. and McQuate, G.T. (2000) Field tests of environmentally friendly malathion replacements to suppress wild Mediterranean fruit fly (Diptera: Tephritidae) populations. *Journal of Economic Entomology* 93, 280–289.

Prokopy, R.J., Papaj, D.R., Hendrichs, J. and Wong, T.T.Y. (1992) Behavioral responses of *Ceratitis capitata* flies to bait spray droplets and natural food. *Entomologia Experimentalis et Applicata* 64, 247–257.

Prokopy, R.J., Miller, N.W., Pinero, J.C., Barry, J.D., Tran, L.C., Oride, L.K. and Vargas, R.I. (2003) Effectiveness of GF-120 fruit fly bait spray applied to border area plants for control of melon flies (Diptera: Tephritidae). *Journal of Economic Entomology* 96, 1485–1493.

Prokopy, R.J., Miller, N.M., Pinero, J.C., Oride, L., Chaney, N., Revis, H.C. and Vargas, R.I. (2004) How effective is GF-120 fruit fly bait spray applied to border area sorghum plants for control of melon flies (Diptera: Tephritidae)? *Florida Entomologist* 87, 354–360.

Purcell, M.F. and Messing, R.H. (1996) Effect of ripening fruit in three vegetable crops on abundance of augmentatively released *Psyttalia fletcheri*: improved sampling and release methods. *Entomophaga* 41, 105–115.

Purcell, M.F., Jackson, C.G., Long, J.P. and Batchelor, M.A. (1994a) Influence of guava ripening on parasitism of the oriental fruit fly, *Bactrocera dorsalis* (Hendel) (Diptera: Tephritidae) by *Diachasmimorpha longicaudata* and *Psyttalia fletcheri* (Hym: Braconidae) and other parasitoids. *Biological Control Theory and Application* 4, 396–403.

Purcell, M.F., Daniels, D.M., Messing, L.M., Whitehand, L.C. and Whitehand, R.H. (1994b) Improvement of quality control methods for augmentative releases of the fruit fly parasitoids *Diachasmimorpha longicaudata* and *Psyttalia fletcheri* (Hym.: Braconidae). *Biocontrol Science and Technology* 4, 155–166.

Rendon, P., McInnis, D., Lance, D. and Stewart, J. (2004) Medfly (Diptera: Tephritidae) genetic sexing: large-scale field comparison of males-only and bisexual sterile fly releases in Guatemala. *Journal of Economic Entomology* 97, 1547–1553.

Roessler, Y. (1989) Insecticidal bait and cover sprays. In: Robinson, A.S. and Hooper, G. (eds) *Fruit Flies. Their Biology, Natural Enemies and Control*. Elsevier, Amsterdam, vol. 3A, pp. 329–335.

Rousse, P., Harris, E.J. and Quilici, S. (2005) *Fopius arisanus*, an egg–pupal parasitoid of Tephritidae. Overview. *Biocontrol News and Information* 26, 59 N–69 N.

Stark, J.D., Vargas, R.I. and Miller, N.W. (2004) Toxicity of spinosad in protein bait to three economically important tephritid fruit fly species (Diptera: Tephritidae) and their parasitoids (Hymenoptera: Braconidae). *Journal of Economic Entomology* 97, 911–915.

Steiner, L.F., Rohwer, G.G., Ayers, E.L. and Christenson, L.D. (1961) The role of attractants in the recent Mediterranean fruit fly eradication programme in Florida. *Journal of Economic Entomology* 54, 30–35.

Steiner, L.F., Mitchell, W.C., Harris, E.J., Kozuma, T.T. and Fujimoto, M.S. (1965) Oriental fruit fly eradication by male annihilation. *Journal of Economic Entomology* 58, 961–964.

Steiner, L.F., Hart, W.G., Harris, E.J., Cunningham, R.T., Ohinata, K. and Kamakahi, D.C. (1970) Eradication of the oriental fruit fly from the Mariana Islands by the methods of male annihilation and sterile insect release. *Journal of Economic Entomology* 63, 131–135.

Uchida, G.K., McInnis, D.O., Vargas, R.I., Kumashiro, B.R. and Jang, E.B. (2004) Non-target arthropods captured in cue-lure baited bucket traps at area-wide pest management implementation sites in Kamuela and Kula, Hawaiian Islands. *Proceedings of the Hawaiian Entomological Society* 36, 135–143.

Uchida, G.K., Mackey, B.E., Vargas, R.I., Beardsley, J.W., Hardy, D.E., Goff, M.L. and Stark, J.D. (2007) Response of nontarget insects to methyl eugenol, cue-lure, trimedlure, and protein bait traps in the Hawaiian Islands. *Proceedings of the Hawaiian Entomological Society* 38, 61–72.

Vargas, R.I. (1996) Suppression of Mediterranean fruit fly populations with SIT in two habitats: a coffee agroecosystem with a braconid parasitoid and a forest with scattered patches of coffee in the understory. *Proceedings of the Fourth International Symposium on Fruit Flies of Economic Importance*, Sand Key, Florida.

Vargas, R.I. and Nishida, T. (1985) Survey for *Dacus latifrons* (Diptera: Tephritidae). *Journal of Economic Entomology* 78, 1311–1344.

Vargas, R.I., and Prokopy, R.J. (2007) Attraction and feeding responses of melon flies and oriental fruit flies (Diptera: Tephritidae) to various protein baits with and without toxicants. *Proceedings of the Hawaiian Entomological Society* 38, 49–60.

Vargas, R.I., Nishida, T. and Beardsley, J.W. (1983) Distribution and abundance of *Dacus dorsalis* (Diptera: Tephritidae) in native and exotic forest areas on Kauai. *Journal of Environmental Entomology* 12, 1185–1189.

Vargas, R.I., Stark, J.D., Uchida, G.K. and Purcell, M. (1993) Opiine parasitoids (Hymenoptera: Braconidae) of Oriental fruit fly (Diptera: Tephritidae) on Kauai Island, Hawaii: islandwide relative abundance and parasitism rates in wild and orchard guava habitats. *Environmental Entomology* 22, 246–253.

Vargas, R.I., Walsh, W.A., Hsu, C.L., Spencer, J., Mackey, B. and Whitehand, L. (1994) Effects of sterile Mediterranean fruit fly (Diptera: Tephritidae) releases on the target species, a nontarget tephritid, and a braconid parasitoid (Hymenoptera: Braconidae) in commercial coffee fields. *Journal of Economic Entomology* 87, 653–660.

Vargas, R.I., Walsh, W.A., Hsu, C.L., Whitehand, L. and Spencer, J.P. (1995) Aerial releases of sterile Mediterranean fruit fly (Diptera: Tephritidae) by helicopter: dispersal, recovery, and suppression. *Journal of Economic Entomology* 88, 1279–1287.

Vargas, R.I., Stark, J.D., Kido, M.H., Ketter, H. and Whitehand, L.C. (2000) Methyl eugenol and cue-lure traps for suppression of male oriental fruit flies and melon flies (Diptera: Tephritidae) in Hawaii: effects of lure mixtures and weathering. *Journal of Economic Entomology* 93, 81–87.

Vargas, R.I., Peck, S.L., McQuate, G.T., Jackson, C.G., Stark, J.D. and Armstrong, J.W. (2001) Potential for areawide integrated management of Mediterranean fruit fly (Diptera: Tephritidae) with a braconid parasitoid and a novel bait spray. *Journal of Economic Entomology* 94, 817–825.

Vargas, R.I., Miller, N.W. and Prokopy, R.J. (2002) Attraction and feeding responses of Mediterranean fruit fly and a natural enemy to protein baits laced with two novel toxins, phloxine B and spinosad. *Entomologia Experimentalis et Applicata* 102, 273–282.

Vargas, R.I., Miller, N.W. and Stark, J.D. (2003a) Field trials of spinosad as a replacement for Naled, DDVP, and malathion in methyl eugenol and cue-lure bucket traps to attract and kill male oriental fruit flies and melon flies (Diptera: Tephritidae) in Hawaii. *Journal of Economic Entomology* 96, 1780–1785.

Vargas, R.I., Jang, E.B. and Klungness, L.M. (2003b) Areawide pest management of fruit flies in Hawaiian fruits and vegetables. In: Inamine, K. (ed.) *Recent Trends on Sterile Insect Technique and Areawide Integrated Pest Management*. Research Institute for Subtropics, Okinawa, Japan, pp. 37–46.

Vargas, R.I., Long, J., Miller, N.W., Delate, K., Jackson, C.G., Uchida, G.K., Bautista, R.C. and Harris, E.J. (2004) Releases of *Psyttalia fletcheri* (Hymenoptera: Braconidae) and sterile flies to suppress melon fly (Diptera: Tephritidae) in Hawaii. *Journal of Economic Entomology* 97, 1531–1539.

Vargas, R.I., Leblanc, L., Putoa, R. and Eitam, A. (2007a) Impact of introduction of *Bactrocera dorsalis* (Diptera: Tephritidae) and classical biological control releases of *Fopius arisanus* (Hymenoptera: Braconidae) on economically important fruit flies in French Polynesia. *Journal of Economic Entomology* 100, 670–679.

Vargas, R.I., Mau, R.F.L. and Jang, E.B. (2007b) The Hawaii fruit fly area-wide pest management programme: accomplishments and future directions. *Proceedings of the Hawaiian Entomological Society* (in press).

Weems, H.V. (1964) Melon fly (*Dacus cucurbitae* Coquillett) (Diptera: Tephritidae). *Entomology Circular, Division of Plant Industry, Florida Department of Agriculture and Consumer Services* 21, 1–2.

White, I.M. and Elson-Harris, M.M. (1992) *Fruit Flies of Economic Significance: their Identification and Bionomics.* CABI, Wallingford, UK.

Wong, T.T.Y., Ramadan, M.M., McInnis, D.O., Mochizuki, N., Nishimoto, J.I. and Herr, J.C. (1991) Augmentative releases of *Diachasmimorpha tryoni* (Hymenoptera: Braconidae) to suppress a Mediterranean fruit fly (Diptera: Tephritidae) population in Kula, Maui, Hawaii. *Biological Control* 1, 2–7.

Wong, T.T.Y., Ramadan M.M., Herr, M.M. and McInnis, D.O. (1992) Suppression of a Mediterranean fruit fly (Diptera: Tephritidae) population with concurrent parasitoid and sterile fly release in Kula, Maui, Hawaii. *Journal of Economic Entomology* 85, 1671–1681.

Yokoyama, V.Y. and Miller, G.T. (2004) Quarantine strategies for olive fruit fly (Diptera: Tephritidae): low-temperature storage, brine, and host. *Journal of Economic Entomology* 97, 1249–1253.

17 Areawide Pest Management for Non-rice Food Crops in South-east Asia

M.D. HAMMIG,[1] B.M. SHEPARD,[2] G.R. CARNER,[2] R. DILTS[3] AND A. RAUF[4]

[1]Department of Applied Economics and Statistics, Clemson University, Clemson, South Carolina, USA
[2]Department of Entomology, Soils, and Plant Science, Clemson University, Clemson, South Carolina, USA
[3]Regional Advisor, Environmental Services Programme, Medan, North Sumatra, Indonesia
[4]Department of Plant Protection, Bogor Agricultural University, Bogor, West Java, Indonesia

Introduction

This chapter focuses on pest management in South-east Asia, specifically Indonesia. The authors of this chapter have been working with colleagues in Indonesia since the mid-1980s, and have been involved with the development of pest management policy by the government of Indonesia as well as the implementation of pest management strategies by Indonesian farmers. Their experience includes activities in the Philippines and other countries in the region, but home base – away from home – has predominantly been Indonesia. Most of the discussion that follows emanates from field activities in Indonesia in which one or more of the authors have been engaged; however, the circumstances with respect to farming systems, crops, agronomic issues and pest complexes are very similar across the region. Therefore, extrapolation of recommended approaches has potential for benefit far beyond Indonesia.

This chapter contains a compendium of pest management practices for food crops other than rice; mainly vegetable crops. While the most important food crop in Indonesia is rice, where over one-third of Indonesians' food budget is spent, nearly one-fourth is spent on fruits and vegetables (USDA/ERS, 2006). Rice is the staple food throughout the country but, since the late 1980s, as average incomes have grown and Indonesians recognize the nutritional importance of vegetables in the diet, vegetable consumption has been steadily growing.

Major vegetables produced include cabbage, potatoes, tomatoes, shallots, chillies, beans, assorted greens, aubergine and onions. Production of all these vegetable crops is found throughout the Indonesian archipelago. Many vegetables are grown in highland areas where vegetable rotations dominate the agricultural landscape, but many are also produced in lowland areas, in rotation with rice during relatively dry periods when rice is not produced.

Price volatility and losses due to pest infestations are major concerns for vegetable growers. On average, vegetables are high-value crops, but the likelihood of losses – either in the marketplace or in the field – causes vegetable production to be relatively risky compared with rice. These uncertainties are important factors affecting farmer choices of crop mix. Pest infestations occasionally wipe out production of certain vegetables, forcing farmers to opt for different crops. For example, leafminers, *Liriomyza huidobrensis*, exotic to Indonesia, substantially reduced plantings of potatoes in many areas in the early 2000s until natural enemy populations became established to suppress the pest. Farmers, fearing losses due to the pest, altered their crop mix to replace potatoes, even when market prices were high, with other vegetables.

Cropping choices are also driven by anticipated returns from vegetable markets. Price volatility is mitigated by diversity of cropping patterns in most areas. Spreading the market risk across crops gives the farmer a cushion to protect against the danger of frequently low prices. Thus the vegetable production landscape, particularly in highland regions, is characterized by a mosaic of small (about 1000 m^2) plantings of a variety of vegetable crops in changing rotations throughout the year.

There are notable exceptions to this cropping pattern in certain areas; farmers in a few areas of Indonesia concentrate on one particular crop. Shallots, for example, are produced along the north-central coast of Java in rotations with rice, with little variation in cropping systems year after year. In some locations, large landholdings are dedicated to a particular crop such as potatoes or cabbages to serve export markets, to other islands within Indonesia or to neighbouring countries.

Other important non-rice food crops also are found in these production systems, including soybeans and maize. These are grown for animal as well as human food, and are most often found in rotation with rice during the dry season.

Management of pests is a persistent problem for all these crops, regardless of where they are found in the country. Because vegetables are potentially very profitable as compared with rice, farmers are typically willing to invest significant resources on chemical pesticides, often to the detriment of a sustainable IPM system. Excessive use of these chemicals is widespread and pervasive. Spraying as often as every 2–3 days is common practice for most vegetable crops. Pesticides are applied as 'cocktails' of assorted insecticides, fungicides and others, with little regard to label instructions and complete disregard for proper application procedures.

The attendant environmental, health and food safety problems become serious issues for government agencies, university researchers, extensionists, NGOs, international institutions and others who are concerned with the broad context of agricultural sustainability. Thus, efforts are driven by government policy and implemented by a large number of local and international experts, working in concert with farmers to find better, more ecologically and economically sustainable ways to manage pests.

Areawide Pest Management in Indonesia: the National IPM Training Programme

Areawide pest management (AWPM) in Indonesia is centred on its National IPM Training Programme, initiated in 1989. The story of the development and implementation of the National Programme is well documented (Hammig, 1998; Thornburn, 2007). The need for a different approach to crop pest management became an important national political issue as a result of widespread rice crop failures in the 1970s and 1980s. Indonesian scientists, working with international experts, met with the then-President Suharto and explained that excessive use of chemical insecticides was the root cause of outbreaks of the brown planthopper, *Nilaparvata lugens*, because of destruction of indigenous natural enemies that normally kept the planthopper in check. Shortly after that meeting, Suharto promulgated a Presidential Instruction in March 1986 that banned 57 insecticides from use on rice, eliminated large government subsidies for pesticides, and declared that IPM would be the national pest management policy. The impacts of this policy are dramatically demonstrated in Figs 17.1 and 17.2, which illustrate the coincidence of reduced pesticide use, reduced numbers of planthoppers and steadily increasing rice production.

To implement this policy, a major effort was undertaken to develop a training programme that would reach as many farmers as possible in the most important rice-growing regions of the country. With assistance from the US Agency for International Development, the United Nations Food and Agriculture Organization and the World Bank, a programme was developed that marshalled the resources of key agencies of Indonesia's Ministry of Agriculture and several universities to implement a programme to train farmers in the principles of IPM, with the goal of reducing farmers' dependence on synthetic chemical pesticides and improving pest management in farmers' fields.

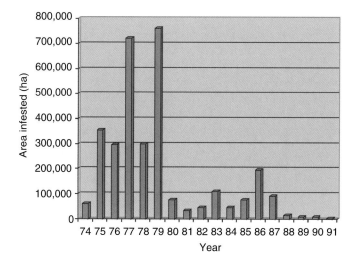

Fig. 17.1. Brown planthopper infestation levels in rice in Indonesia, 1974–1991 (from Kenmore, 2006).

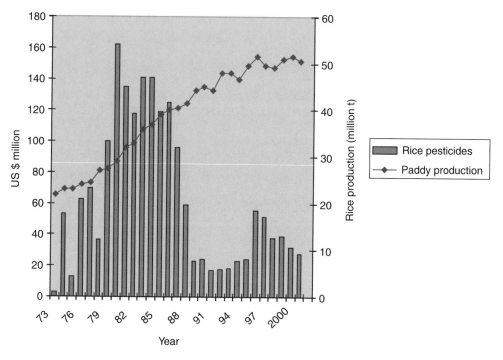

Fig. 17.2. Rice pesticide expenditures and paddy production in Indonesia, 1973–2001 (from Kenmore, 2006).

The National IPM Training Programme used curricula developed specifically for local conditions. Trainers who had, themselves, received training from national and international experts and who recognized the value of farmers' indigenous knowledge, worked with farmer groups, meeting weekly and taking careful note of the ecological conditions of the field: pests, natural enemies, plant health, etc.

The Farmer Field School (FFS) approach was based on four basic principles: (i) grow a healthy crop; (ii) conserve natural enemies; (iii) visit the field regularly (Gallagher, 1990); and (iv) farmers become IPM experts in their own fields (Dilts, 1990). The main paradigm shift was that farmers carried out research/demonstrations in their own fields in a participatory manner rather than receiving recommendations from extension workers, as is the model in most Western societies, and was the norm in Indonesia prior to IPM training.

By the end of the World Bank-supported training effort in 1998, over 1,100,000 farmers were trained in rice IPM, along with over 2300 pest observers and 4000 agricultural extension workers. Furthermore, and a key factor for the subsequent expansion of IPM training, over 21,000 farmers were trained as IPM trainers and, by 1998, were conducting a majority of the FFSs (Hammig, 1998).

Following the National Programme's focused training efforts of the 1990s, IPM training programmes continue to thrive throughout Indonesia, though support for these efforts is no longer from central government authorities as much as from local agencies, some universities and farmer groups with assistance from NGOs and experienced FFS veterans (Pontius *et al.*, 2002). The scope of FFSs also broadened to include plantation crops, soil ecology, participatory plant breeding and local policy advocacy.

Areawide Pest Management in Indonesia: Non-rice Food Crops

Integrated pest management training in rice was the first step toward an areawide approach to pest management in Indonesia that significantly differed from the traditional command and control approach through which packages of inputs, including pesticides, were forced upon farmers by poorly trained extension workers with a common list of instructions that did not take into account variation of local conditions. These command and control programmes succeeded in introducing improved crop varieties and rice production expanded, but they can also be blamed for the excessive pesticide use that led to periodic pest-induced crop failures.

Though the IPM training programme was initially targeted at rice, by the mid-1990s training had been expanded, largely due to farmer demand, to address pest management issues associated with secondary crops grown in rice rotations, and highland vegetables (Oka, 2003). Because pest problems are typically more complex than with rice, in general, farmers' methods for controlling crop pests include exceedingly high frequencies of chemical spray applications, so impacts of training programmes can be even more dramatic than for rice. The National Programme included nearly 3000 FFSs for over 30,000 vegetable farmers (FAO, 2000).

As the FFS model became well known across Indonesia, expansion of training to non-rice areas has emerged through local community efforts. Training for non-rice food crops follows the same basic principles as for rice. However, vegetables – which are for the most part exotic plant species – present different challenges for crop scientists and trainers. IPM approaches differ among crops and pests as trainers and technical support experts seek effective alternatives to chemical controls. Participatory farmer training continues throughout Indonesia, and farmer enthusiasm for training continues to expand. Teams of scientists are working in collaboration with farmer groups to develop pest control approaches appropriate to local conditions that emphasize the importance of natural control agents (Shepard *et al.*, 1999).

The post-Suharto era has significantly changed the political landscape in Indonesia. For 2007, the national Ministry of Agriculture has allocated budget support for over 2000 IPM FFSs, to be implemented by local governments and local farmer organizations. Local governments have much more authority than they had in the past, and programmes for issues like IPM training have devolved to local officials. In some areas, such as West Sumatra, the provincial government has aggressively pursued IPM programmes with a target of reaching all farmers in the province. The Provincial Director of Agriculture has voiced his desire for 'a completely organic Province'. 'Farmer Organic Institutes' have been established in several locations in West Sumatra to support the organic initiative, and over 20 farmer field laboratories producing biological agents are functioning in support of IPM programmes. FFSs for both organic rice and vegetable production are being implemented with government support.

In other areas, such as West Java, subdistrict governments specify budget allocations for vegetable IPM training with more limited ambitions (Hammig *et al.*, 2006). And, in some areas, university scientists, in collaboration with international experts and local farmers, are pursuing research and training activities in important vegetable-producing regions.

The challenges facing farmers, trainers and scientists include development of effective, economical pest management approaches across the host of crops that have been subjected to pest control programmes almost exclusively dependent on synthetic chemicals. These challenges are being addressed in a variety of ways, always with farmer fields as the focus of attention.

Areawide Pest Management Approaches

Secondary food crops in Indonesia include those non-rice food crops (*palawija*) that are grown in rotation with rice and vegetable crops in upland areas where rice is not present. Development of sound, integrated pest management (IPM) practices for these *palawija* and vegetable crops in Indonesia is much more challenging than for rice. Unlike rice, which has been cultivated for thousands of years and has co-evolved into a relatively stable system, *palawija* and vegetable crops arrived in Indonesia much later.

On a per unit area basis, these latter crops receive more pesticides than any other crop. Applications of pesticides every 2–3 days are common, and this has created all of the classic symptoms of pesticide overuse that have occurred in many crops around the world. Vegetable farmers are led to believe through local habit and aggressive marketing that pesticides reduce risk, and are likely to try any new pesticide in hopes of higher yields or less damage. Frequent calendar-based 'cocktail' applications in Indonesia stem from a general lack of understanding of diseases, insect pests, weeds, natural enemies, crop compensation and agronomic factors. Many plant-feeding species targeted by farmers do not cause yield losses but instead serve as food for a large complex of natural enemies (Shepard *et al.*, 1999, 2001). Also, the impact of pesticides on the health of Indonesian farmers can be significant (Murphy *et al.*, 1999; Kishi, 2002). Thus, expanded farmer training in IPM is essential for sustainable vegetable and soybean production.

Soybean

Description of the problem and need for an AWPM approach

Soybean in Indonesia is considered a major *palawija* crop. It is the most important *palawija* crop, with over 1,407,000 ha grown in Indonesia. A large proportion of soybean is used for human consumption. Even with this large planting area, Indonesia still imports between 600,000 and 800,000 t of soybean annually. This crop is an important protein source and includes food items such as tofu, tempe and others.

The most important insect pests of soybean in Indonesia are the pod-boring insects, *Etiella zinckenella* and *Helicoverpa armigera*, the corn earworm (CEW). Of these,

E. zinckenella is by far the most important, based on field surveys carried out by personnel and collaborators with the Clemson University Palawija IPM Project. The stemfly, *Melanagromyza sojae* and the seedling fly, *Ophiomyia phaseoli* can also be locally important, as can pod-sucking bugs such as *Nezara viridula* and *Riptortus linearis*. Foliage feeders such as *Spodoptera litura*, *Omiodes indicata* and loopers – mainly *Chrysodeixis chalcites* – are often targeted by farmers for insecticide sprays because they are large and conspicuous, but these insects rarely cause yield losses. Application of broad-spectrum insecticides that target this foliage-feeding complex may have a profound effect on indigenous biological control agents that may be keeping pod-borers and pod-sucking bugs under control in the absence of these chemicals.

There is an abundance of natural enemies in the soybean systems, which effectively regulate populations of most of the plant-feeding species. The parasitoid complex is particularly rich on some plant-feeding species (Shepard and Barrion, 1998). However, it is apparent from our pest and natural enemy surveys that some key pests are lacking an effective complement of parasites, predators and pathogens. For example, the pod-boring pyralid, *E. zinckenella*, has relatively low levels of parasitism. Only three parasitoid species (*Phanerotoma philippinensis*, *Baeognatha javana* and *Temelucha etiellae*) were frequently encountered (Shepard *et al.*, 1999). No pathogens were found in larval populations of *E. zinckenella*.

In other parts of the world, it has been shown that chemical insecticides can cause resurgence of several species of lepidopteran pests of soybean (Shepard *et al.*, 1977). In most cases, the rapid increase in pest populations was due to the destruction of natural enemies. Therefore, it is important to understand that application of chemical insecticides, whether targeting pod- and stemborers or foliage feeders, may cause non-pests to be elevated to primary pest status.

Farmer participatory research to test IPM strategies

Field studies were carried out with farmers to identify strategies for inclusion in IPM training programmes (Shepard *et al.*, 2001). Soybean is usually grown after rice (during the dry season). The following results were obtained from field studies.

- *Etiella* and *Helicoverpa* populations were higher in late-planted soybeans.
- In the late-planted soybean, applications of insecticides caused an increase in pest insect populations and a concurrent increase in damage by *Helicoverpa* and *Etiella*.
- Yields were lower in late-planted soybeans.
- Insecticides reduced populations of *Etiella* but also decreased numbers of several important insect predators, such as *Pardosa pseudoannulata*, *Paederus fuscipes* and ants.

Conclusions from these studies underline the importance of planting as early as possible to escape build-up of pod-boring pests such as *Etiella* and *Helicoverpa*.

Little information is known about stem flies and their importance in soybean production. Results from our studies with this potential pest revealed the following:

- Yield reductions by the stem fly, *M. sojae*, could not be detected except when the stem was attacked below the hypocotyl, but there was little justification for insecticide treatments.

- Although chemical insecticide treatments were aimed at stem flies, these chemicals caused populations of *S. litura* to resurge, which adversely affected populations of major predators.
- No yield reductions were caused by *S. litura*. Therefore, it may be a beneficial insect providing food for natural enemies that attack more serious insect pests.

Any IPM strategy that is used in soybean, as with other crops, must involve farmers. Field exercises developed with and carried out by farmers will serve to illustrate the principles and practices of IPM. Of all the secondary food crops, soybean stands out as the one in which good agronomic practices are sadly lacking. Often, entomologists focus their research on methods for controlling pests without first understanding that 'growing a healthy crop' is often the most important constraint to production.

Secondly, devising field studies with farmers illustrated that foliage-feeding pests, such as loopers, *Omiodes* sp., geometrids, lymantriids and *S. litura*, rarely cause yield losses. Understanding the role of natural enemies in soybean also is key to the success of a sustainable IPM programme on soybean. Only then can strategies be devised to conserve those biological control agents that normally keep pests under control and which cost the farmer nothing.

Many data sets are available showing that farmers who plant their crops later than most farmers around them are most likely to be impacted by pod-feeders such as *E. zinckenella* and *Helicoverpa* (van den Berg *et al.*, 2000). Thus earlier, more synchronized planting among farmers could result in significant reduction in losses due to pod-boring insects.

Evaluation of the AWPM programme

About 1.5 million ha of soybean are harvested annually in Indonesia. Government programmes that encourage planting and development of improved production systems will stimulate increased plantings in the future. Though we cannot estimate the economic impact of a specific IPM strategy for soybean, we can make some rough estimates of what impacts will be as strategies are developed.

Currently, over 10% of production costs are for pesticides and their application. According to Indonesia's Central Bureau of Statistics data, about Rp 46.5 billion (~ US$5 million as of September 2007) are spent annually on insect pest control on soybean. Most of these pest control expenditures occur in the major growing areas in Java and Sumatra where IPM programmes are concentrated.

Data from field surveys in East Java conducted in 1996 provide a frequency distribution of sprays, showing that farmers applied from zero to eight sprays through the season, with an average of 3.4 for the 100 farmers surveyed (van den Berg *et al.*, 1998). The majority (70%) of insecticidal sprays were applied during the first 45 days after planting, before pod-set, and aimed mainly at defoliators. Because these pests cause little if any yield reduction, it is clear that use of insecticidal sprays can be reduced without causing economic losses. Thus, IPM strategies that reduce the number of pesticide applications have immediate and direct pay-offs to farmers by reducing their costs of production and consequently increasing their profits. If IPM strategies enhance yields as well – which is likely as a result of the programme focus to 'grow a healthy crop', including attention to basic agronomy – then the benefit is increased.

Cabbage

Cabbage is planted to over 67,000 ha in Indonesia, second only to soybean among *palawija* crops, with a total of 1,417,000 t produced annually. This crop is produced mainly in the upland areas of Java, Sumatra and Sulawesi.

The major pest of cabbage in Indonesia is the cabbage head caterpillar (CHC), *Crocidolomia pavonana* (Sastrosiswojo and Setiawati, 1992). The diamondback moth (DBM), *Plutella xylostella*, is generally kept under good control by the parasitoid, *Diadegma semiclausum*, when chemical insecticides are avoided. *Hellula undalis* and the looper complex can be important locally but are not widespread problems.

Diadegma semiclausum was first introduced into Malaysia (Ooi, 1986), and later into Indonesia for DBM control. This parasitoid was later distributed to most of the major cabbage-growing areas (Sastrosiswojo, 1996). Even in areas where the parasitoid is firmly established, farmers do not recognize its importance in biological control of DBM, and routine applications of mixtures of chemical insecticides are still made. This action invariably causes resurgence of populations of DBM by reducing parasitoids and other important natural enemies that normally keep it under control.

Interestingly, one can ascertain the spray history of a cabbage field depending on the presence or absence of dense populations of DBM. High populations are usually indicative of heavy chemical sprays; the presence of high levels of CHC usually indicates few or no chemical sprays. This underscores the importance of considering CHC along with DBM in developing an effective IPM programme for cabbage (Sastrosiswojo and Setiawati, 1992). When chemical sprays for DBM are decreased or terminated altogether, CHC often cause heavy damage. Therefore, as Sastrosiswojo and Setiawati pointed out, the key to successful IPM in cabbage must include a strategy for dealing with CHC: indigenous natural enemies are not able to keep CHC in check.

Another major challenge for cabbage IPM is development of strategies that can help suppress the corn earworm (CEW), *Helicoverpa armigera* and *Hellula*. The latter mostly occurs in lowland cabbage. Although these pests are sporadic and localized, our extensive surveys throughout major vegetable-growing areas revealed heavy populations of CEW in central Java, south Sulawesi (Malino) and east Java (Batu).

Farmer participatory research to test IPM strategies

Field tests were conducted to determine whether hand-picking egg masses and larval clusters of CHC, along with spot applications of *Bacillus thuringiensis* (*Bt*), was a practical approach for CHC control (Shepard and Schellhorn, 1997). Applications of *Bt* and concurrent elimination of chemical sprays would allow *D. semiclausum* and other natural enemies to operate fully against DBM and CHC.

Tests were carried out in Alahan Panjang, West Sumatra (April–July 1995) and in Jaringan Tani, Tanah Karo, near Berastagi, north Sumatra (2 May–28 July 1995). Treatments included: (i) collection of CHC egg masses and larval clusters up to

30 days after transplanting seedlings, then hand-picking plus spot spraying with *Bt* (after about 30 days, the egg masses are difficult to find); (ii) hand-picking throughout the season and spraying the entire plot with *Bt*; (iii) standard farmer practice; and (iv) untreated control.

Use of the first of these options resulted in over 90% of the cabbage heads being rated as marketable. The farmers' usual practice provided the highest yields (about the same as hand-picking and spot spraying), but eight *Bt* and chemical sprays were applied, as compared with only seven spot *Bt* treatments when egg masses and larval clusters were hand-picked. We concluded that results might have been better by applying *Bt* using a backpack sprayer rather than using the small, hand-held sprayer that we used for these studies. In the untreated control plots, nearly 40% of the heads were severely damaged by CHC and were considered unmarketable.

Results were more impressive in north Sumatra, where a backpack sprayer was used to apply spot sprays of *Bt*. This study, planned and executed with personnel from World Education, revealed that yields and marketability of cabbage were significantly lower in the untreated plots. However, the usual practice of farmers in the area called for weekly applications (12) of chemicals. Only seven spot sprays with *Bt* were required in the hand-picking/*Bt* spot spray treatment. Thus, the profitability of hand-picking eggs and larval clusters plus spot spraying with *Bt* may be a viable approach in areas where cabbage fields are small and not much time is required to search the field for egg masses and larval clusters. Considerable build-up of natural enemy communities of both DBM and CHC should result from this approach.

Shallot/Onion

Of all the vegetable crops in Indonesia, shallots are most heavily sprayed with chemical pesticides. In the large shallot-producing areas of Brebes, in central Java, it is common for farmers to apply chemical insecticides every other day. This has resulted in high levels of resistance in the target pest (*Spodoptera exigua*), and the only viable control tactic is hand-picking larvae from the plants.

The major pest of shallots during the dry season is *S. exigua*. During the rainy season, fungi are most important, notably *Alternaria*, *Colletotrichum* and *Peronospora destructor* (Meity Sinaga, personal communication). Weeding is normally carried out on an 'as needed' basis, most often simultaneously with hand-picking of *S. exigua* larvae. Aphids (*Neotoxoptera formosana*) can be locally important in the highlands, but we have not observed them in high numbers in most major production areas.

A heavy infestation of an agromyzid was found on shallots in Alahan Panjang, west Sumatra. This leaf-mining fly was first recognized as a pest in Indonesia in the mid-1990s (Shepard *et al.*, 1997), and is now widespread on the islands of Java, Sumatra, Bali and South Sulawesi. The fly was identified as *Liriomyza huidobrensis* (Blanchard). The extent to which these infestations affected yields has not been determined but, judging from the severity of infestations, yield losses were substantial. More recently, heavy infestations of another exotic leaf miner species, *L. chinensis*, were observed in Brebes, central Java in 2000.

In the Philippines, the most important soil-borne diseases are *Sclerotium cepivorum*, *Fusarium oxysporum* and *Phoma terrestris*, and the use of fungicides for control of these pathogens has been unsuccessful (Gapasin *et al.*, 2003). Anthracnose is one of the most important diseases of onions in the Philippines and, at present, the only effective means of control is through intensive fungicide applications. Purple nutsedge and horse purslane are the most important weed pests of onion (Miller *et al.*, 2005).

Farmer participatory research to test IPM strategies

Major outbreaks of *S. exigua* occurred in the Brebes (central Java) area, where chemical insecticides were often applied every other day. Heavy damage by *S. exigua* was also prevalent in west Sumatra (Alahan Panjang), Batu (east Java), Probolinggo (east Java), Cisantana and Pangalengan (west Java). A microbial control agent, a nucleo-polyhedrovirus (SeNPV), was discovered in populations of *S. exigua* in Cimacan, in the Puncak, west Java, through routine field surveys of shallots (Hammig and Rauf, 1998).

Results from preliminary tests in the Puncak revealed that damage to leaf onion (*bawang daun*) was significantly lower when the SeNPV was applied in farmers' fields. We then tested the virus in the Brebes area in collaboration with Pak Karsum, a shallot farmer in Ciledug, central Java. Results were so impressive that Karsum asked for the SeNPV from our laboratories at Bogor. We worked with him closely to develop production techniques, and soon after he was able to mass-produce the material and carried out tests in his own fields. A unique feature of the biological control system is that the microbe is easily mass-produced because of a ready supply of *S. exigua* larvae that are collected daily by women, an activity that is carried out as part of an effort, along with chemical insecticides, to control the pest.

Field applications of the SeNPV against *S. exigua* by Karsum have been highly successful. Because of these tests, he changed his pest control strategy to SeNPV, instead of chemical insecticides, for control of *S. exigua*. He has now shared this technology with farmers from six other villages. The FAO Action Research Facility in Brebes also worked closely with other farmers in the area to help them understand how the microbial agent works and how to best use it in a programme that helps restore other natural enemies for long-term stability of the system.

We carried out field tests using randomized, replicated plots and farmer-level production and application techniques in Ciledug using a crude preparation of the SeNPV. Six experiments (three pairs) were carried out from July to September 1996, to assess the SeNPV's potential at different *S. exigua* population levels.

Yields were compared from the six treatments: (i) SeNPV plus hand-picking of larvae; (ii) chemical insecticide plus hand-picking; (iii) SeNPV alone; (iv) hand-picking of larvae alone; (v) insecticide alone; and (vi) untreated control. Yields from these plots are shown in Fig. 17.3. Clearly, SeNPV and SeNPV plus hand-picking of larvae provided the best control methods for *S. exigua*.

During the late-season planting, with heavy pressure from *S. exigua*, yields from the untreated control plots were nearly zero. Plots using hand-picking alone significantly improved yields, but highest yields were obtained when SeNPV was applied along with hand-picking. The SeNPV treatment alone was as good as insecticides

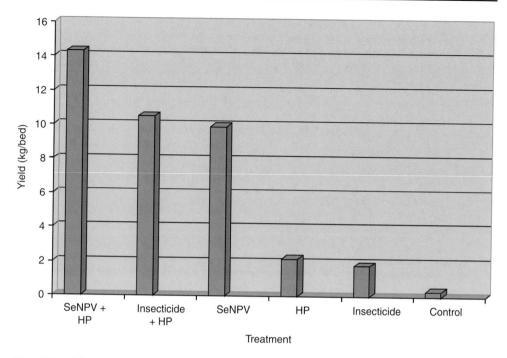

Fig. 17.3. Yields of shallots treated with combinations of SeNPV, insecticides and hand-picking, Ciledug, West Java, September 1996. HP, hand-picking.

plus hand-picking, which was common farmer practice before the IPM system was introduced.

A programme for farmer production and use of SeNPV has been developed and is being carried out in Alahan Panjang, west Sumatra (Zamzami and Djoni, personal communication). Our project supplied the inoculum and training for the West Sumatra Plant Protection Agency staff from Padang and Bukittinggi. They, in turn, have trained 150 shallot farmers. These farmers are currently testing the SeNPV in their own fields. A farmer field seminar for biological control was conducted in Alahan Panjang to bring together farmers, trainers and researchers from all of the major shallot-growing areas of Indonesia, to share experiences and design plans for expanding the understanding and use of SeNPV. Over 10,000 farmers currently use the SeNPV as part of their control programme on shallots in West Sumatra (Zamzami, personal communication).

In summary, the use of SeNPV has excellent potential for providing long-term control of *S. exigua*, while stabilizing the shallot ecosystem by allowing natural enemies to recolonize the areas. Farmer training in IPM is the key to the success of the programme.

The fundamental comparison of common farmer practice with the IPM alternative, based on the use of SeNPV virus, is shown in Table 17.1. The data were obtained from field studies conducted in the Ciledug subdistrict of Cirebon District, central Java. This area is typical of the major shallot-growing areas of Indonesia that

Table 17.1. Comparison of common farmer insect control practice with IPM system for shallots in Java, per hectare, autumn 1996.

Activity	Common practice	IPM
Insect control sprays (*n*)	21	14
Hand-picking (*n*)	49	12
Cost of insecticide (US$)	647	0
Cost of virus (US$)	0	4.30
Spray application labour (US$)	452	302
Hand-picking labour (US$)	528	130
Yield (t)	9.38	15.71
Price/t[a] (US$)	172	280
Net benefit of IPM[b] (US$)	0	3976

[a] The price/t used here is the price quoted for the farm level, based on quality, as of November 1996.
[b] The net benefit of IPM is calculated by summing the cost differences between IPM and common practice and the difference in total return based on yields and price premiums.

include Tegal and Brebes districts, as well as Cirebon. The irrigated production system used is also common in Probolinggo, east Java; another important shallot area. In combination, these areas account for about one-third of all shallot production in Indonesia.

The use of fungicides is largely unsuccessful for the control of soil-borne pathogens of onions. However, *Trichoderma* spp. are know for their antagonistic effects against these fungal pathogens. In the Philippines, *Trichoderma* isolates were as effective as chemical fungicides in reducing the incidence of these soil-borne diseases. VAM (vesicular arbuscular mycorrhizae) has been found to be an economically and environmentally friendly supplement that can help reduce fertilizer input and assists onion plants in tolerating infection from soil-borne pathogens and nematodes (Gergon *et al.*, 2003). For control of anthracnose, the combination of cultural and chemical control reduced the number of fungicide applications (Alberto *et al.*, 2003).

For weed control, IPM CRSP in farm studies showed that one application of the correct herbicide followed by timely hand-weeding controlled weeds as well as the farmer practice of two herbicide treatments followed by three hand weedings. Weed control costs were reduced by 15–70% without reducing weed control efficacy. Another weed management technique, rice straw mulching, was shown to be effective in on-farm studies. Weed growth was reduced by 60%, yields were increased by 70% and weed control costs reduced by 50% (Miller *et al.*, 2005).

Evaluation of the AWPM programme

The IPM system, based on the use of the SeNPV virus together with hand-picking, provides a dramatic opportunity for economic benefits to farmers. Insecticide costs are eliminated and hand-picking requirements are reduced. These factors alone imply that production costs can be reduced by US$1100/ha. In addition to these cost savings, evidence from field studies implies that crops produced under the IPM system have higher yields and improved quality over the common farmer practice. The combination of the yield boost and the price premium paid for high-quality product results in an additional US$2800/ha gain from IPM. Thus, the net benefit is about US$4000/ha.

These data showing the economic benefits from the IPM system in shallots were obtained during the dry season, when insect pests are the major problem for shallot growers. In the regions where the irrigated production system is found, shallots are considered primarily a *palawija* crop and rice is planted during the wet season. Therefore, these very dramatic economic benefits of IPM will be realized on the majority of the shallot crops produced in these key production areas. In addition, health benefits from development of IPM should be substantial (Kishi *et al.*, 1995).

A study was conducted, in 1998, comparing farmers using SeNPV with conventional growers. The results of this study show that many farmers have adopted SeNPV as a viable option for insect control in shallots, and that those farmers realize a significantly higher profit margin compared with farmers who still rely on chemical pesticides. These results show actual farmer practice, rather than experimental results. The difference using the SeNPV was > US$700/ha. Details of the 1998 study are given in Table 17.2.

Chilli

This crop is by far the most important of all the vegetable crops in Indonesia, with production of nearly 900,000 t in 2005 (FAO, 2007). However, development of a sound IPM programme for chillies is the most challenging. Numerous pests, including insects, mites and plant pathogens (Vos and Frinking, 1998) attack the crop. In addition, inappropriate agronomic practices are often major constraints to achieving maximum production. For example, farmers often plant in low, wet areas, thus hindering healthy growth due to poor soil drainage.

Major arthropod pests include mites (*Polyphagotarsonemus latus*) and *Helicoverpa armigera* (CEW). Thrips and aphids may be important locally as vectors of plant viruses. Occasionally, *S. litura* causes farmers to spray insecticides, but this insect feeds mostly on leaves and probably causes little damage in most cases. CEW, on the other hand, selectively feeds on the pods. The gall fly, *Asphondylia* sp., can cause significant pod loss but the impact of this pest, as with CEW, is highly variable between seasons and locations. Recent information from West Sumatra suggests that parasite levels build up during the season and only early-season fruits are affected. The fruit fly, *Bactrocera* (= *Dacus*) *dorsalis* seems to be ubiquitous, but the incidence of pod attack is usually not high in the major chilli-growing areas of Indonesia. More details of the agronomic factors and pests of chillies on Java were reported by Vos (1994). *Colletotrichum, Phytophthora, Alternaria, Cercospora* and

Table 17.2. Comparison of shallot growers who use the microbial agent (SeNPV) to control insect pests with growers who follow conventional chemical-based practice, Cirebon, west Java and Brebes, central Java, September 1998.

Item	SeNPV users (*n* = 17)	Conventional growers (*n* = 52)
Area and yield		
Area harvested (m²)	1847.1**	1473.1
Yield (kg/1000 m²)	678.4	597.0
Pest control		
Pesticide applications/season	11.9***	17.4
SeNPV applications/week	2.4	0
Production costs (US$/1000 m²)		
Land rent	6.10	6.75
Irrigation fee	0.42	0.33
Total fertilizer cost	10.27	12.26
Insecticide	2.05***	9.88
Fungicide	3.43***	7.27
Herbicide	0.50***	0.97
Seed	115.00**	88.82
Labour costs (US$/1000 m²)		
Land preparation	11.58	10.68
Planting	0.66	0.70
Cultivation	1.82	1.61
Hand-picking	6.90	5.87
Pesticide application	4.23***	7.68
Fertilizer application	1.35***	2.03
Watering	10.88	10.94
Weeding	2.61	3.33
Irrigation maintenance	1.54	1.53
Other costs	1.06	0.83
Harvesting		
Transportation	1.02	0.75
Security	3.47	2.05
Tying labour and materials	0.37	0.40
Returns		
Price received (US$/kg)	0.62	0.57
Gross return (US$/1000 m²)	393.48**	309.26
Profit (US$/1000 m²)	205.80*	134.58

*significantly different at 90% confidence, **significantly different at 95% confidence, ***significantly different at 99% confidence.

Pseudomonas – and viruses – are usually among the most important groups of pathogens.

Farmer participatory research to test IPM strategies

Field tests conducted in western Sumatra demonstrated that seedbed height and control of soil pH with lime were effective in reducing the incidence of bacterial wilt. Insecticide sprays were not effective in increasing yields. This study was carried out in an area where CEW was not an important pest. In other areas, the use of HaNPV (*Helicoverpa armigera* nucleopolyhedrovirus) might be a viable tactic for replacing chemical pesticides. Virus diseases that are prevalent in many parts of South-east Asia may be managed using resistant varieties currently under development at the Asian Vegetable Research and Development Center in Taiwan.

Yardlong Bean

Also known as the snake, asparagus or Chinese long bean, and second only to chilli in terms of area planted among vegetables, yardlong beans are an important part of the Indonesian diet. Major insect pests are the pod borer, *Maruca vitrata* (= *testulalis*) and aphids (usually *Aphis craccivora*). The extent to which *M. vitrata* causes economic losses is not understood, and varies widely according to location and market supply and demand. *M. vitrata* damaged an average of only about 3 cm along the length of maturing pods, but caused much more severe damage in younger ones. Economic losses from *M. vitrata* damage in yardlong beans in west Sumatra were estimated at about 25% (Zamzami, personal communication). Aphids are important both as direct feeders on blooms and pods, and also as virus vectors. Sucking bugs are usually present in the crop, but their importance may be overemphasized. *Ophiomyia phaseoli* also can cause yield reductions locally.

Farmer participatory research to test IPM strategies

A field study, carried out by the Clemson Palawija IPM Project, FAO and the Provincial Plant Protection laboratory in Padang, west Sumatra, compared: (i) farmers' usual practice; (ii) no treatment but with good cultural practices; and (iii) designated 'action windows' that we 'generated' for aphids (over 100 per hill), pod borer (over 10% of pods damaged), anthracnose (over 10% infected leaves) and leaf spot (over 10% infected leaves).

Yields from the farmers' practice treatment and the IPM 'action windows' treatment were about the same, although the farmers' treatment called for eight sprays as compared with two in the IPM treatment. The major difference was in the 'untreated' control, where *O. phaseoli* and aphids seriously reduced yields.

Another field study carried out with personnel from the University of Lampung revealed that late-planted yardlong beans were more severely attacked by CEW than those planted early. This difference was not as obvious for *M. vitrata*. In tests at the Muara field station in Bogor, mosaic virus reduced the plant population by 50%. IPM strategies must include tactics for dealing with aphid-borne viruses. Untreated longbean plots near Ciloto resulted in over 50% losses due to direct feeding by the aphid, *A. craccivora*. Recent results indicate that 'spot' treatments versus treatment of the entire plot with aphicides may conserve natural enemies, but this approach requires that the crop be monitored at least twice weekly.

Aubergine

The major insect pest of aubergine is the fruit and shoot borer, *Leucinodes orbonalis* (EFSB). In some areas, this insect is the limiting factor to aubergine production. Farmers may apply insecticides 50 times or more in a single growing season (Miller *et al.*, 2005). In spite of frequent pesticide applications, yields are reduced by more than one-third due to this pest. Also, leafhoppers may be important locally. Of the plant diseases, bacterial wilt is most common. Losses to bacterial wilt in central Luzon consistently reached 30–80%.

Farmer participatory research to test IPM strategies

Data from on-farm research in the Philippines showed that simply removing damaged fruits and shoots reduced infestations by EFSB and, if carried out at harvest time, labour costs were reduced. This resulted in a net incremental benefit of US$2500/ha when conducted weekly, and of US$1000/ha for biweekly removal (Miller *et al.*, 2005). The second approach is the identification of aubergine resistant varieties.

Bacterial wilt-susceptible aubergine grafted on to resistant rootstock (EG 203) increased resistance to the disease by 30% and yields were higher. The stale seedbed technique, which includes sequential harrowing or harrowing followed by a non-selective herbicide at biweekly intervals carried out during the fallow period between the rice and onion crops, was effective in reducing purple nutsedge tuber populations by 80–90% (Miller *et al.*, 2005).

Tomato

Diseases such as early and late blight, powdery mildew, bacterial wilt, *Alternaria* and viruses are the major constraints to tomato production. Insects that vector viruses include thrips, aphids and whiteflies. CEW and, sometimes *S. litura*, often feed directly on the fruit.

Farmer participatory research to test IPM strategies

In some parts of Indonesia staking of tomatoes is not a common cultural practice, and incidence of fungal diseases is high due to contact of plants with soil. Field tests in farmers' fields have demonstrated that staking decreases disease incidence and increases yield. Many farmers in some areas have readily accepted this cultural practice.

In the Philippines, tomato plants do not survive well under the constant high-moisture conditions during the rainy season. Farmer participatory field tests have shown that grafting tomato on to resistant aubergine rootstock greatly increases survival and crop yields.

Citrus

Surveys were carried out in a large citrus-growing area in the Karo District of north Sumatra. Heavy infestations of fruit flies (20% of the fruit was infested) were observed. The fruit fly was identified as the papaya fruit fly, *Bactrocera papayae*. All growers in the area were reporting high levels of fruit loss from this pest. In addition to fruit flies, we observed lepidopteran larvae, *Citripestis sagittiferella*, in about 3% of the fruit.

Farmer participatory research to test IPM strategies

A fruit fly management plan must include participation by all the citrus growers in the area. Due to the intensity of citrus growing in the region, the only effective strategy would be an areawide approach. Without participation by all farmers, reinfestations of fruit flies would continue to occur in IPM-managed areas. Tactics to be included in this management plan should include sanitation activities, spot spraying of protein bait, traps to monitor adult fruit fly populations and early harvesting of fruit.

Summary

Field research and demonstration projects for most of the crops listed above have shown that substantial reductions of pesticide applications are possible without jeopardizing yields. Table 17.3 summarizes results from field tests conducted in the mid-1990s in west Java, central Java and Sumatra, applying IPM principles with specific recommendations for each crop, and with broad applicability throughout the country. Given the potential reductions in pesticide applications, shown in Fig. 17.4, associated environmental and human health benefits justify a policy commitment to expand IPM training to areas where these crops are concentrated.

Some of the IPM tactics that could have a major impact if adopted on a wide area are listed in Table 17.3. One constraint is that information about tactics found useful in one area is often not transferred to another. A mechanism to transfer information from one area to another would greatly expedite adoption on a wider scale. The IT component of the IPM CRSP will be helpful in this regard. We have found

Table 17.3. IPM tactics[a] that could have major impacts if adopted areawide in South-east Asia.

Crop	Tactics
Tomato	Varieties resistant to viruses and fungi
	Staking (with appropriate variety)
	Pruning
	Grafting
Chilli	Resistance to viruses
Aubergine	Straw mulch
	Grafting
	Host plant resistance to nematodes
	Bt-transgenic plants
	Monitoring EFSB with pheromones for timing treatments
Shallot	Microbial control of *Spodoptera exigua* with SeNPV
Onion	VAM and *Trichoderma*
	Straw mulch
	Pheromone traps for timing interventions (*Spodoptera litura*)
	SINPV (insect virus of *S. litura*)
Cabbage	Field scouting for *Crocidolomia*
	Microbial control/spot spraying
	Hand-picking egg masses/larval clusters
	(also conserves natural enemies for DBM)
Soybean	Early planting in the dry season
	(to avoid pod borers/*Etiella* and *Helicoverpa*)
	Avoiding needless sprays for defoliators
Yardlong bean	Spot treatments for aphids
Citrus	Sanitation (removal/destruction of infected fruit)
	Protein bait sprays (for papaya fruit fly control)
	Traps to monitor population
	Early harvest
All vegetable crops	Weed control using stale seedbed technique

[a] The tactics listed here are not all-inclusive.

that workshops that allow participants from different countries, and regions within countries, to come together and exchange information may be one way of exchanging information among researchers. At another level, participatory field studies with farmers will be the most appropriate way to determine whether the various tactics are applicable for specific locations and socio-economic settings.

Socio-economic and Environmental Impacts of IPM

The impact on crop yields of IPM systems that reduce use of chemical inputs is positive in most cases, translating into higher gross economic returns. Evidence from field

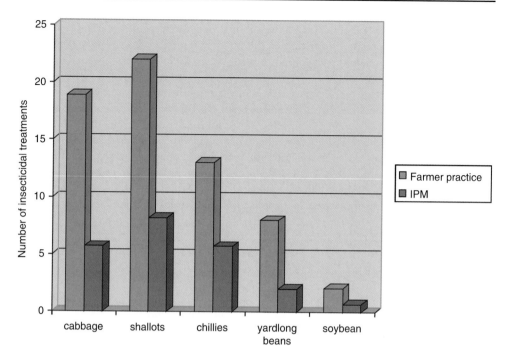

Fig. 17.4. Numbers of pesticide applications per crop per season made by vegetable farmers using their normal practice as compared with IPM practices.

sites where farmer groups use the IPM approach shows that costs of inputs decrease because of dramatically reduced outlays for pesticides. Thus, IPM farmers may enjoy higher profits than their traditional counterparts (Hammig *et al.*, 1997). Much of the economic evidence is anecdotal, but results from Indonesia consistently show improved returns by IPM farmers. A report by van den Berg (2004) synthesized 25 IPM impact evaluations. Although most of the examples featured rice, they also included vegetables. The conclusion was that farmers who had participated in FFSs reported substantial and consistent reductions in pesticides attributable to the effects of the IPM training. Further, more pesticide reductions and higher farm-level revenues were realized in vegetables than with rice.

Clearly, the anticipated economic bottom line is a key determinant of farmers' adoption of alternative production practices. Therefore it is important that analysts address long-term adoption patterns and the persistence of the benefits of IPM training. A survey of west Java vegetable farmers, some with IPM training and others without, showed that the former employed more sustainable farming practices as compared with the latter, even years after the training had occurred (Norvell and Hammig, 1999).

The IPM training routine includes comparisons of fields employing IPM and traditional farmer practice. These comparisons are not just of what is happening to pests and crop yields, but also of the impact on market returns. Farmer groups keep careful accounts of their expenses, and the comparison of IPM and non-IPM results is the focus of group discussions during the training process. There is no doubt that areawide adoption of IPM is contingent on positive results in the marketplace. If adequate returns are not assured, then the traditional practice will dominate, even after training. Fortunately, sound application of IPM principles invariably results in better bottom lines for farmers.

There are greater benefits from IPM than simply those offered by higher market returns. Farmers relying on chemical pesticides pose a significant danger to themselves, their families and their neighbours, not to mention the environment of rural areas. Sustainable IPM systems reduce human health and ecological risks by reducing the volumes of many of the most toxic chemicals applied to crops. In a study from Vietnam, Murphy (2002) showed the correlation between frequency of pesticide applications and farmer illness. Kishi *et al.* (1995) found that IPM-trained farmers in Indonesia apply fewer pesticide applications to their crops and, when they do apply pesticides, they use less toxic chemicals than comparable farmers who have not participated in IPM training.

Impacts beyond the farm gate are meaningful components of comprehensive impact assessments. The highland vegetable areas of Indonesia, in almost all cases, are situated upstream from major population centres. USAID/Jakarta has recognized the critical importance of upland water catchment areas and agricultural practices on urban water systems and has funded an Environmental Services Project (ESP) that, in cooperation with the government of Indonesia, is mounting a comprehensive effort to improve water quality in selected urban centres through improved land and agricultural management.

IPM programmes are integral components of the ESP effort. Examples of this linkage include IPM training in west Java focused in the watershed feeding Jakarta and surrounding communities. Jakarta's fresh vegetable markets are served from the mountainous region immediately to the south of the city. Local governments, Bogor Agricultural University and international collaborators have been working with farmers in that region to reduce the runoff of harmful chemicals through IPM training for selected vegetable crops. Evidence from Shepard *et al.* (2001) shows the potential. Unfortunately, meaningful changes are constrained by the relatively slow process of farmer education. Local government budgets for IPM training are limited, so reaching large numbers is a slow process (Hammig *et al.*, 2006). Similar programmes are in development in watershed areas in central Java (Progo River) and east Java (the Malang vegetable area and the Brantas River).

In North Sumatra, the headwaters of the Deli River that provide water to Medan, the provincial capital, are another area of concentrated vegetable production. The *Lembah Gulen* (vegetable valley) at the foot of Sibayak volcano is a relatively small area composed of two villages, where the population is almost entirely dependent on vegetable production for its livelihood. Traditional production systems are similar to those observed elsewhere. Tomatoes, cabbage, shallots, chillies and other vegetables are grown in continuous rotations.

Prior to the ESP project, production systems were chemical-intensive, and farmers were frustrated by poor response to their control efforts. IPM training was first introduced in 2006, and farmers are now eagerly adopting different approaches (Hammig *et al.*, 2006). At the time of this writing, budget cuts to the ESP have reduced resources available for IPM in *Lembah Gulen*; however, farmers themselves are carrying on the programme with some continuing support from the NGO, Farmer Initiatives for Ecological Literacy and Democracy (FIELD) (Weinarto, personal communication).

In North Sulawesi, the Lake Tondano watershed provides another example of the link between upland vegetable production systems and urban centres. The lake is located in a mountain valley, and the Tondano River flows from the lake to Manado, another provincial capital. It drains into the Molucca Sea at Bunakan, an Indonesian National Marine Park. The mountain slopes surrounding Lake Tondano are covered by vegetable fields, with the usual mix of crops growing all year round.

An earlier USAID/Jakarta watershed management project, focused on environmental stewardship by local communities, spearheaded an effort to motivate local groups to seek better ways of improving the conditions of their environment. IPM training formed a part of this effort and, with assistance from scientists at Sam Ratulangi University, training programmes were initiated for onion, cabbage and tomato growers in the area in 1997 (Sembel, 1998). Vegetable IPM continues to be a high-priority activity for farmers and university scientists working in the area.

Once farmers experience training for one of the selected crops, they recognize the need for training on the other crops they plant as well. This presents a challenge to IPM farmers and IPM trainers. Each crop has different pest management problems, and proposed alternatives for one crop may not be applicable to another. Therefore, the key to obtaining significant widespread impact is to establish a continuous process of field monitoring, research and experimentation, with the farmer as the central figure. Farmers can be introduced to IPM principles through training, and they can access technical support in critical times of need, but the greatest impact occurs when farmers themselves perform their own experiments, and learn with experience how an ecological balance can be maintained in their fields while they continue to obtain positive economic returns.

Gender roles are important in determining socio-economic impacts of IPM. In South-east Asia, gender roles in agriculture vary from region to region. In some areas, field work activities are differentiated by gender. For example, planting, weeding, harvesting and/or pest management tasks such as spraying or hand-picking of pests may be jobs for which gender is the first order of selection. In the shallot fields of central Java, the traditional pest control practice is for women to hand-pick egg masses from the plants at the same time that men apply chemical sprays. Within the household, women are responsible for child rearing and common household tasks. Men do most of the heavy lifting, but by no means all.

Data from the Philippines indicate that, in Nueva Ecija province, women manage the household and farm budgets in the overwhelming majority of cases (Hamilton *et al.*, 2005). Therefore, IPM training, if it is to be effective, must be sensitive to gender roles in the production and management of the crops. Decision making is the essential foundation of IPM, so it is essential that the key decision makers be informed of the ramifications of their choices. In the South-east Asia context, trainers include both men and women, and they are sensitive to the need to ensure that there is no gender bias in selection of training groups. However, recognizing the need does not mean that overcoming obstacles to attaining the ideal gender mix in IPM training is easily done.

Tanzo (2006) highlights some of the key gender-related issues she has observed for IPM training in the Philippines. Domestic tasks overlap with field activities. In the field, women are frequently found weeding, handling pesticides and hand-picking pests. Household tasks include clothes washing, food preparation and family health care. Women are more involved in vegetable pest management than with rice because of their frequent roles in field monitoring and hand-picking. They are exposed to pesticides both in the field and while washing pesticide-soaked clothing. Daily schedules often conflict with IPM training programmes, so few women, relative to their importance in the decision-making process, take advantage of IPM training opportunities.

Impacts of AWPM efforts in Indonesia, implemented over the nearly 20-year history of IPM training, have yielded important benefits to farmers, the environment and consumers of farm products. Evidence of these benefits is apparent from many studies addressing a range of issues. However, in Indonesia and other countries in South-east Asia with similar demographics, where over 40% of the population is involved in production agriculture, the process of spreading the IPM message is slow. The best sign suggesting that there is significant momentum within the IPM paradigm comes from farmers who have embraced IPM and who are the primary motivators for engaging their peers.

Conclusions

Areawide pest management (AWPM) in South-east Asia is addressed through massive training of farmers in the region in the principles of integrated pest management. South-east Asia is characterized by large numbers of farmers operating on small plots. Production systems involve substantial amounts of labour input, which often puts farm labourers at risk from exposure to harmful chemicals. Mechanical devices that replace labour in western agriculture are rare in the South-east Asia region. Technological advances have made an impact, mainly through improved plant varieties and cultural practices to enhance yields. IPM programmes in the region seek to reduce the harm caused by some of these practices that have become entrenched because of misguided national programmes that disregarded local peculiarities. IPM training has taken hold throughout the region as a means of establishing the farmer as the primary decision maker and to equip him or her with an understanding of the critical relationship between agricultural output and field ecology. Training programmes in all South-east Asian countries are aggressively spreading the message to 'grow a healthy crop'.

References

Alberto, R.T., Duca, M.V. and Miller, S.A. (2003) Management of Anthracnose (*Colletotrichum gloeosporiodes*) a disease of increasing importance in onion. In: Management Entity of the IPM CRSP (eds) *IPM CRSP Integrated Pest Management Collaborative Research Support Programme, Tenth Annual Report, 2002–2003*, Virginia Tech University, Blacksburg, Virginia, pp. 48–51.

Dilts, R. (1990) IPM and farmer empowerment: the marriage of social approaches and technological content. In: *Third International Conference on Plant Protection in the Tropics*, Malaysian
Plant Protection Society, Genting Highlands, Pahang, Malaysia, p. 24.

FAO (2000) *Vegetable IPM Evaluation Mission Report*. FAO, Rome.

FAO (2007) *Major Food and Agricultural Commodities and Producers*. FAO, Rome (http://www.
fao.org/es/ess/top/country.html).

Gallagher, K.D. (1990) Implementation of integrated pest (plant?) management. In: *Third International Conference on Plant Protection in the Tropics*, Malaysian Plant Protection Society,
Genting Highlands, Pahang, Malaysia, p. 100.

Gapasin, R.M., Donayre, D.K.M. and Miller, S.A. (2003) Biological control of soil-borne
pathogens using *Trichoderma* spp. in rice–vegetable systems. In: Management Entity of the
IPM CRSP (eds) *IPM CRSP Integrated Pest Management Collaborative Research Support
Programme, Tenth Annual Report, 2002–2003*, Virginia Tech University, Blacksburg, Virginia,
pp. 62–66.

Gergon, E.B., Gapasin, R.M., Padua, L.E., Donayre, D.K.M., Antolin, M.M. and Miller, S.A.
(2003) Management of soil-borne diseases of onion with selected biological control agents
and soil amendments. In: Management Entity of the IPM CRSP (eds) *IPM CRSP Integrated
Pest Management Collaborative Research Support Programme, Tenth Annual Report, 2002–2003*,
Virginia Tech University, Blacksburg, Virginia, pp. 67–77.

Hamilton, S., Moore, K., Harris, C., Erbaugh, M., Tanzo, I., Sachs, C. and Asturias de
Barrios, L. (2005) Gender and IPM. In: Norton, G.W., Heinrichs, E.A., Luther, G.C.
and Irwin, M.E. (eds) *Globalizing Integrated Pest Management: a Participatory Research Process*.
Blackwell, Ames, Iowa, pp. 263–293.

Hammig, M.D. (1998) *USAID and Integrated Pest Management in Indonesia: the Investments and the Payoffs*.
Project report (http://pdf.usaid.gov/pdf_docs/PNACD254.pdf).

Hammig, M.D. and Rauf, A. (1998) West Java soybean and vegetable production efficiency.
Journal of Plant Protection in the Tropics 11, 15–21.

Hammig, M.D., Rauf, A. and Carner, G.R. (1997) Integrated pest management in non-rice food
crops in Indonesia: opportunities to reduce chemical pesticide use. In: Lockeretz, W. (ed.)
Agricultural Production and Nutrition. Tufts University, Boston, Massachusetts, pp. 131–139.

Hammig, M.D., Shepard, B.M. and Carner, G.R. (2006) *Ecologically-based Participatory IPM
for Southeast Asia*. IPM CRSP Trip Report, Clemson University, Clemson, South
Carolina.

Kenmore, P.E. (2006) Delivering on a promise outside the USA, but inside the same globalized
world. Presentation at the *Fifth National IPM Symposium*, 4–6 April, St Louis, Missouri.

Kishi, M. (2002) Farmers' perceptions of pesticides, and resultant health problems from expo-
sures. *International Journal of Occupational and Environmental Health* 8, 175–181.

Kishi, M., Hirschorn, N., Djajadistra, M., Satterlee, I.N., Strowman, S. and Dilts, R. (1995)
Relationship of pesticide poisoning to signs and symptoms in Indonesian farmers. *Scandi-
navian Journal of Work and Environmental Health* 21, 124–133.

Miller, S.A., Razzaul Karim, A.M.N., Baltazar, A.M., Rajotte, E.G. and Norton, G.W. (2005)
Developing IPM packages in Asia. In: Norton, G.W., Heinrich, G.C., Luther, E.A. and
Irwin, M.E. (eds) *Globalizing Integrated Pest Management: a Participatory Research Process*.
Blackwell, Ames, Iowa, pp. 27–50.

Murphy, H.H. (2002) *A Farmer Self-surveillance System of Pesticide Poisoning*. FAO Community Integrated Pest Management in Asia, Rome [Mimeo].

Murphy, H.H., Sanusi, A., Dilts, R., Djajadisastra, M., Hirshhorn, N. and Yuliatingsih, S.
(1999) Health effects of pesticide use among Indonesian women farmers. Part 1: Exposure
and acute health effects. *Journal of Agromedicine* 6, 61–85.

Norvell, S.D. and Hammig, M.D. (1999) Integrated pest management training and sustainable
farming practices of vegetable growers in Indonesia. *Journal of Sustainable Agriculture* 13, 85–101.

Oka, I.P.G.N.J. (2003) Integrated Pest Management in Indonesia: IPM by Farmers. In: Maredia, K.M., Dakouo, D. and Mota-Sanchez, D. (eds) *Integrated Pest Management in the Global Arena*. CABI, Wallingford, UK, pp. 223–237.

Ooi, P.A.C. (1986) Diamondback moth in Malaysia. In: Talekar, N.S. (ed.) *Diamondback Moth Management*. Asian Vegetable Reserarch and Development Center, Taiwan, pp. 25–34.

Pontius, J., Dilts, R. and Bartlett, A. (2002) *From Farmer Field Schools to Community IPM: Ten Years of IPM Training in Asia*. Food and Agriculture Organization of the United Nations Regional Office for Asia and the Pacific, Bankok, Thailand.

Sastrosiswojo, S. (1996) Achievements in vegetable IPM training. *Bogor: Palawija and Vegetable Newsletter* 1, 3–4.

Sastrosiswojo, S. and Setiawati, W. (1992) Biology and control of *Crocidolomia binotalis* in Indonesia. In: Talekar, N.S. (ed.) *Diamondback Moth and Other Crucifer Pests*. Asian Vegetable Research and Development Center, Taiwan, pp. 81– 87.

Sembel, D.T. (1998) IPM for vegetable production in ecologically sensitive areas. Presentation to the *Annual Meeting of the University Development Linkage Programme*, Washington, DC.

Shepard, B.M. and Barrion, A.T. (1998) Parasitoids of insects associated with soybean and vegetable crops in Indonesia. *Journal of Agricultural Entomology* 15, 239–272.

Shepard, B.M. and Schellhorn, N.A. (1997) A Plutella/Crocidolomia management programme for cabbage in Indonesia. *Proceedings of the Third International Symposium on Management of Diamondback Moth and Other Crucifer Pests*, Kuala Lumpur, Malaysia, 29 October–1 November 1996, pp. 262–266.

Shepard, B.M., Carner, G.R. and Turnipseed, S.G. (1977) Colonization and resurgence of insect pests of soybean in response to insecticide and field isolation. *Environmental Entomology* 6, 501–506.

Shepard, B.M., Carner, G.R., Barrion, A.T., Ooi, P.A.C. and van den Berg, H. (1999) *Insects and Their Natural Enemies Associated with Vegetables and Soybean in Southeast Asia*. Quality Printing Company, Orangeburg, South Carolina.

Shepard, B.M., Shepard, E.F., Carner, G.R., Hammig, M.D., Rauf, A. and Turnipseed, S.G. (2001) Integrated pest management reduces pesticides and production costs of vegetables and soybean in Indonesia: field studies with local farmers. *Journal of Agromedicine* 7, 31–66.

Tanzo, I. (2006) IPM and social science. Presentation at *Ecologically Sensitive Participatory IPM for SE Asia Workshop*, Ciloto, West Java, Indonesia.

Thornburn, C. (2007) The arthropod revolution: community IPM in Indonesia. In: Batterbury, S. and Horowitz, L. (eds) *Applied Political Ecology*. Blackwell, London. (In press).

USDA/Economic Research Service (2006) *International Food Consumption Patterns. Food Budget Shares for 114 Countries*. USDA, Washington, DC (http://www.ers.usda.gov/Data/InternationalFoodDemand/).

Van den Berg, H. (2004) *IPM Farmer Field Schools: a Synthesis of 25 Impact Evaluations*. Wageningen University for the Global IPM Facility, Wageningen, Netherlands.

Van den Berg, H., Shepard, B.M. and Nasikin (1998) Damage incidence by *Etiella zinckenella* and *Helicoverpa armigera* in soybean in East Java, Indonesia. *International Journal of Pest Management* 44, 153–159.

Van den Berg, H., Aziz, A. and Machrus, M. (2000) On-farm evaluation of measures to monitor and control soybean pod borer *Etiella zinckinella* in East Java, Indonesia. *International Journal of Pest Management* 46, 219–224.

Vos, J.G.M. (1994) Integrated crop management of hot pepper (*Capsicum* spp.) in the tropical lowlands. PhD thesis, Wageningen Agricultural University, Wageningen, Netherlands.

Vos, J.G.M. and Frinking, H.D. (1998) Pests and diseases of hot pepper (*Capsicum* spp.) in tropical lowlands of Java, Indonesia. *Journal of Plant Protection in the Tropics* 11, 53–71.

18

Areawide Rice Insect Pest Management: a Perspective of Experiences in Asia

JAMES A. LITSINGER

Dixon, California, USA

Introduction

Rice constitutes the primary staple food for more than two billion people in Asia, where it supplies half of their diet and 80% of the caloric intake (IRRI, 1985). Of the total 145 million ha planted to rice worldwide, 92% is in Asia, where it is also consumed to such an extent that very little enters into international trade. Asian societies are so wedded to rice that when it becomes unavailable, either from losses in production due to weather events or pest outbreaks, this leads to surges in prices in urban centres resulting in social upheaval in the form of famine and riots. Insect pests have been one of the sources of high chronic or epidemic losses in production, usually in combination with other stresses (Litsinger *et al.*, 2005). Rice, due to the large cultivated area worldwide, is beset by a wide array of pests, which literally infest all rice fields from the time of sowing until harvest, potentially causing significant losses. Set against this constraint is the need to feed a fast-growing human population, and thus crop protection plays a vital role.

Integrated pest management (IPM) has been adopted by most Asian countries as the reigning strategy for insect pest control since the 1980s, when it was recognized that reliance on insecticides and genetic resistance measures would not suffice (Gallagher *et al.*, 1994). IPM strategies for insect pests are based on using the best mix of genetic, biological, cultural and chemical control tactics that together provide a durable and sustainable rice production system (Kenmore *et al.*, 1985). The first principle is for the farmers to achieve agronomic optimization of the cultivar they are sowing. Rice insect control tactics have been summarized in Reissig *et al.* (1986). These include numerous cultural practices that can have a dual role in strengthening tolerance to losses in general, as well as for pest suppression. Litsinger (1994) divided the various cultural control methods into those that are effective if carried out on a single field versus those that are known to be more effective when carried out on a community-wide basis.

© CAB International 2008. *Areawide Pest Management: Theory and Implementation* (eds O. Koul, G. Cuperus and N. Elliott)

Strictly speaking, it may be said that most insect pest management methods provide greater suppressive activity if carried out areawide. Assessment of alternative control measures and determination of the cost–benefit ratio are the deciding factors on whether such tactics are to be recommended. The mechanism in terms of areawide effect for all practices is the same: to prevent rapid reinvasion from adjacent fields where farmers did not conduct the same control practice at about the same time (Joyce *et al.*, 1970). Rice, *Oryza sativa*, being an annual crop, is constantly invaded by species that have dispersive powers of varying degrees, and even more so by those that are monophagous or oligophagous. In temperate climates or in the rainfed tropics, stronger dispersive powers are needed than in the irrigated rice in the tropics, where rice is multi-cropped and more apparent in time and space.

This review examines a wide array of control practices that, at one time or another, have been conceptualized or implemented community-wide over a spatial scale, such as minimally an irrigation turnout or in a village. Complicating adoption of such practices in most of Asia is the small field size, which in many countries is becoming smaller as population pressure rubs against a non-expanding crop area. Thus, many practices such as hand removal of insect pests from the field that seem out of date and absurd in developed countries (except in the context of a small urban home garden) are being carried out as field sizes in the most densely populated areas of Asia are now equal to those urban home gardens, and have to provide food for six to eight people. Such labour-intensive practices make economic sense in areas where landless labourer populations and unemployment are high. Insecticides, which have been the mainstay of rice insect pest control since the 1950s, are no longer appropriate for all situations, and small-scale farmers cannot afford to use them or their use creates more problems than are solved.

Green Revolution-led Crop Intensification

As a background for a number of the areawide technologies that will be covered, it is useful to review the series of changes that have occurred because of the Green Revolution in rice in Asia. In the decades after the mid-1960s, when high-yielding, modern varieties (MVs), developed by the International Rice Research Institute (IRRI) in Laguna province, Philippines, were introduced, the immediate benefit to adopters of the new MVs, particularly in favourable areas, was quadrupled production. In order to entice farmers to adopt the new technologies, governments (through extension officers) offered low-interest credit or subsidized prices to small-scale farmers to purchase inputs via national programmes such as Masagana 99 in the Philippines and BIMAS in Indonesia. There was an urgency for governments in less-developed Asian countries to increase basic food production at the time due to predictions of imminent famine (Erlich, 1968); thus, as soon as the new varieties were tested in small-field trials for several seasons, they were released. The first MV (IR8) was approved in 1967, with rapid adoption of MVs by Asian farmers, reaching over 99% in many of the large, irrigated regions by 1980 (Pingali *et al.*, 1997). As a result, most Asian countries reached self-sufficiency in food production. Production has slipped somewhat in recent years, with some countries such as Indonesia having to import rice but, as a country develops, farmers find more lucrative crops; thus it is often cheaper to import from other countries such as Vietnam.

Research trials at IRRI showed dramatically increased yield in IR8 over traditional varieties in response to inorganic fertilizers, herbicides and insecticides. However, traditional varieties, particularly in the wet season readily lodged (toppled over) in response to agro-inputs. Even without inorganic fertilizer, yields of MVs were superior to traditional rices. An experiment in Central Luzon, Philippines conducted by us showed MVs significantly out-yielded traditional rices by a factor of almost two, even when both were grown without fertilizer, in the wet season (see Table 18.1).

Modern varieties responded to increased crop management, and thus it became economically attractive for farmers to improve cultivation practices. In addition, one of the qualities of MVs is photoperiod insensitivity, in contrast to traditional rices that are photoperiod-bound – i.e. flower only once a year which, in monsoon Asia where rice evolved, occurs in October–November with the advent of short daylengths (Yoshida, 1981). Photoperiod insensitivity meant that the growth duration was reduced from 9 to 4 months, and thus rice could be sown and harvested more than once a year. Increased yields prompted governments, wanting to promote rice production, to construct and expand irrigation systems not only to supplement water delivery during the fickle rainy season but also to grow a crop in the normally fallow lands in the dry season. A period of intensification of rice culture occurred, which meant that the number of rice crops per year usually doubled, but in limited areas farmers grew five rice crops in 2 years or even triple-cropped rice. A Filipino farmer even developed a method called the rice garden, where four rice crops per year could be grown by weekly plantings in 13 plots (Morooka et al., 1979). This is rarely done, as few farmers have year-round availability of irrigation water.

Transplantation has been the traditional crop establishment method and was continued with the MVs and thus field time was reduced, as a seedbed could be started while the main field was being prepared. Only in the late 1980s, when agricultural labour costs increased, did direct-seeding technology gain favour. However, omitting the seedbed meant greater need for irrigation water, and weed control even by herbicides was not as effective when rice seedlings had to compete directly with weed seedlings. Transplanting older and thus more competitive seedlings, therefore, is regaining favour. There was great motivation to adopt the new technologies, which

Table 18.1. Comparison of yield potential of irrigated traditional and modern rice, Zaragoza and Jaen, Nueva Ecija, central Luzon, Philippines, 1984 wet season[a].

Cultivar	n	Yield (t/ha ± SD)
Modern rice (IR36, IR42)	14	4.10 ± 1.07
Traditional rice (Milagrosa, Wagwag, Pulang bigas)	12	2.20 ± 1.01

[a] ANOVA analysis, F = 21.95, df = 25, P = < 0.0001.
None of the crops received either organic or inorganic fertilizer. Plots 100 m^2 in size were established in the fields of modern rice and grown without added fertilizer. Farmers' yields are those reported in local units of *cavans* per hectare and converted to t/ha. Some of the farmers grew both traditional and modern varieties. 1 *cavan* = 50 kg of unmilled rice.

had increased from one to two crops per year, each averaging double the previous yield.

When the 88 ha experimental farm, provided to IRRI on the campus of the University of the Philippines at Los Baños, was opened in 1961, the main insect pests were notably striped (*Chilo suppressalis*), yellow (*Scirpophaga incertulas*) and pink (*Sesamia inferens*) stemborers. Insecticide trials against mostly stemborers documented in the 1963 IRRI Annual Report averaged 37% yield loss (1.6 t/ha) with granular and spray formulations at high dosages and spray volumes (400–500 gal/ha) on Milfor 6 variety. No other insect pests were significant, although in the Asian literature seed bugs, gall midge and assorted polyphagous defoliators (armyworms, butterflies, leaf beetles, grasshoppers) were most mentioned. Milfor 6, a tall thick-stemmed dryland variety, was a popular choice for insecticide trials as it is highly susceptible to stemborers. Such losses would not occur with cultivars preferred by farmers. Thus from the outset insect pests were wrongly pegged as major constraints to rice production requiring insecticide protection.

Changes from Intensification

Pests affected

The dissemination of MVs has been accompanied by significant ecological changes related to arthropod pests and some natural enemies. One change was the appearance of new insect pests that previously were of only minor importance. Some well-known pests of rainfed rice culture became less abundant while others increased in importance. A number of pests were encouraged by the more permanent ponding afforded by the large-scale irrigation systems. Rice whorl maggot, which was identified only in 1968 (Ferino, 1968) as *Hydrellia philippina*, soon damaged up to over 90% of leaves. As with other hydrophilic species, whorl maggot is more prevalent in the vegetative stage. Two other vegetative-stage Lepidoptera – the green semi-looper *Naranga aenescens* and another new species, hairy caterpillar, *Rivula atimeta* (Malabuyoc, 1977) – added to the defoliation damage. The former is widespread throughout many parts of temperate and tropical Asia, but in the Philippines has rarely attained pest status. Prolonged ponding also encouraged population build-up of the brown planthopper, *Nilaparvata lugens*, and related leafhoppers (Dyck *et al.*, 1979). One reason is the increased fecundity when reared under such conditions, but as these species disperse at dusk and particularly in the full moon, they may use the reflective surfaces of ponded fields from moonlight as a host-seeking mechanism (Perfect and Cook, 1982). Improved irrigation would encourage greater colonization.

Wetland culture discourages soil insect pests in general, some of which have endured even under intermittent irrigation such as the mole cricket, *Gryllotalpa orientalis*, but with more prolonged flooding their prevalence has decreased, probably as a result of smaller-sized bunds that reduced habitat. The rice caseworm, *Nymphula depunctalis*, which requires standing water for its larvae that respire by gill structures, did not appreciably increase in importance as would have been predicted, probably due to its high sensitivity to insecticides (Litsinger and Bandong, 1992). Kiritani (1992) reported that a number of insect pests in Japan disappeared after farmers had adopted insecticides in the 1950s. Univoltine and monophagous species were most affected, such as black bug (*Scotinophara*), grasshoppers (*Oxya*) and leaf beetles (*Oulema* and *Dicladispa*). Heavy insecticide usage in the Philippines, spurred by subsidies in MV production programmes (Kenmore *et al.*, 1987), reduced the presence of other pests such as the larger polyphagous rice butterflies (green horned caterpillar, *Melanitis leda ismene* and rice skippers, *Pelopidas* spp. and *Parnara* spp.). Dry-season rice culture was instrumental in increasing the prevalence of the gall midge, *Orseolia oryzae*, which adapted to the new season and increased its geographic range, which hitherto had needed wild rice to sustain it year-round (Loevinsohn, 1994).

On the other hand, the improved water delivery also favoured beneficial aquatic arthropod predators such as *Microvelia douglasi atrolineata*, *Mesovelia vittigera*, lycosid wolf spiders and the dwarf spider, *Atypena formosana*, all highly effective generalist predators (Ooi and Shepard, 1994). Also by now more abundant were katydids, *Conocephalus longipennis*, sword-tailed crickets, *Metioche vittaticollis* and *Anaxipha longipennis* (Gryllidae) and *Cyrtorhinus lividipennis*, the mirid bug egg predator which responded to the high Homoptera populations that had increased in abundance. The levels of many of these predators rose in response to greater insect pest numbers. One reason the whorl maggot increased may have been that it was held in check in single crop systems by a predacious fly, *Ochthera sauteri*, which captures whorl maggot flies in mid-air, killing up to 20 adults per day (Barrion and Litsinger, 1987). The predatory fly is most effective in wetland fields that are only saturated and not ponded. Thus, the better ponding as a result of irrigation has reduced predation.

Multiple rice cropping meant that there was a reduction in the area planted to rotational crops such as maize and grain legumes, as well as in the cover of grassy weeds during the fallow period, which increased the availability of rice as a host (Loevinsohn *et al.*, 1988). This vegetational change has favoured rice pests with monophagous and oligophagous habits at the expense of the polyphagous ones. Therefore, one observes the new dominance of monophagous pests such as brown planthopper (BPH), green leafhopper (*Nephotettix virescens*) and *Scirpophaga* stemborers, which feed almost exclusively on *Oryza* spp. Rainfed rice is often dominated by *N. nigropictus* over *N. virescens*, as the former has a wider host range, presumably needed to span the off-season (Ishii-Eiteman and Power, 1997). Also, the whitebacked planthopper, *Sogatella furcifera*, which has a wider host range, occurs with BPH.

Shifts in the abundance of species are illustrated from light trap data from Titi Serong in the Krian Irrigation System of Malaysia, where yellow stemborer (YSB) became more favoured over the polyphagous dark-headed stemborer (*Chilo polychrysus*) (Way and Heong, 1994). Likewise the oligophagous striped stemborer (SSB) has been replaced by YSB in Laguna province, Philippines (see Table 18.2). From 1971 to 1979 light trap data from the IRRI farm during the period when rice cultivation was shifting from single to double-cropping, YSB increased by an average of 176% per year, whereas the polyphagous SSB increased only by 64% (Loevinsohn, 1994; Table 18.2).

As there does not appear to have been any significant increase in the abundance of the alternative host plants of *Chilo* spp. (mainly maize and sugarcane), such a pattern is to be expected if it is assumed that monophagous pests are more efficient exploiters of a host plant than their polyphagous guild-mates. SSB is virtually absent

Table 18.2. Rice stemborer abundance measured from daily light trap collections, IRRI farm, Los Baños, Laguna, Philippines, 1965–1979[a].

	Mean annual catch/trap		Ratio
Years	Yellow stem-borer (YSB)	Striped stem-borer (SSB)	(YSB:SSB)
1965–1970	4,370	670	6.49
1971–1979	12,042	1,107	10.9
Rate of increase (x)	1.8	0.7	

[a] Annual catches are the means of four electrically operated light traps counted daily on the IRRI farm (data adapted from Loevinsohn, 1994).

from central Luzon perhaps because of the greater dominance of rice as well as the inability of larvae to withstand flooding (Jahn *et al.*, 2007). SSB, being a larger species, attacks only the thin-stemmed MVs during the later plant growth stages as tiller lumens enlarge (Chaudhary *et al.*, 1984). Kiritani (1992) also reported the reduction of SSB in Japan after farmers had adopted insecticides. In China, YSB greatly increased after irrigation and double-cropping (NAS, 1977).

There are also differences in the species make-up in different rice cultures. Dryland (unbunded rainfed culture) has the most diverse stemborer species assemblage, due to the dominant plant types with wider-diameter tillers and to the lesser dominance of rice in the ecosystem, which is often overshadowed by the maize, sugarcane and wild grasses preferred by *Chilo* and *Sesamia* spp. Stem dissection data reinforced the conclusion of the light trap results, in that dryland rice had the greatest species richness, from four to six species (Jahn *et al.*, 2007, Table 1). Iloilo, a rainfed wetland site in the Visayas, came the closest with four stemborer species. Two irrigated sites in central Luzon registered only YSB, while most sites registered three species. In terms of numbers of moths collected annually in light traps, both wetland irrigated and rainfed cultures heavily favoured *Scirpophaga* (see Table 18.3).

Leptocorisa spp. rice seed bugs became less common in the transformed treeless landscapes that made way for large-scale irrigation systems, particularly in central Luzon. Although there was no difference in mean densities ($P > 0.05$) across the three main rice cultures, there was a trend of lower densities in irrigated rather than in rainfed areas (see Table 18.4). Some irrigated sites, such as in Laguna (1.4 bugs/m^2), with more mixed floral landscapes had populations equivalent to rainfed wetland and dryland areas. The highest mean incidence occurred in the 'slash and burn' rice culture of Siniloan (4.1 bugs/m^2) due to the very small field sizes and mobility of the bugs.

By 1966, BPH and *Nephotettix* green leafhoppers (GLH) and their vectored tungro disease became prevalent on the IRRI farm, causing higher losses than those from stemborers (IRRI, 1967). These pests had not been mentioned in earlier IRRI reports. Rice was grown continuously in the experimental farm, and tungro virus became endemic due to the continuous availability of rice crops. In 1970–1972 there was an extensive outbreak of GLH and tungro in Laguna province (Sogawa, 1976), which caught plant protectionists off guard as to its cause. This was followed by many

Table 18.3. Relative abundance of *Scirpophaga* spp. stemborers as determined from light trap catches in four rice cultures, Philippines, 1979–1991.

Culture	Sites (*n*)[a]	Trap-years (*n*)	*Scirpophaga* spp. in annual stemborer catch (%)[b]
Dryland	4	15	37.0 ± 5.3[d]
Rainfed wetland	3	13	86.3 ± 4.5[c]
Irrigated wetland (synchronous planting)	2	6	80.1 ± 4.4[c]
Irrigated wetland (asynchronous planting)	2	4	75.8 ± 7.8[c]

[a]Dryland sites: Siniloan, Laguna; Claveria, Misamis Oriental; Tanauan, Batangas; Tupi, South Cotabato; rainfed wetland sites: Oton and Tigbauan, Iloilo; Manaoag, Pangasinan; Solana, Cagayan; irrigated synchronous sites: Zaragoza, Jaen and Guimba, Nueva Ecija; Victoria and Santa Maria, Laguna; Koronadal, South Cotabato; irrigated asynchronous sites: Zaragoza, Nueva Ecija; Koronadal, South Cotabato; synchronous planting areas are those sites where neighbouring farmers plant within one month of one another, creating extensive rice-free fallow periods. In asychronous sites one can see rice fields in various plant ages side by side. For more detailed information see Jahn *et al.*, 2007, Table 2.
[b]In a column, means ± SE followed by a common letter are not significantly different ($P > 0.05$) by LSD test.

Table 18.4. Mean rice seed bug incidence as a mean of different sites by rice culture, Philippines, 1981–1991[a].

Rice culture	Rice bugs/m² (*n*)	Sites (*n*)	Crops (*n*)
Dryland	1.3 ± 0.3	4	15
Rainfed wetland	1.2 ± 0.2	3	13
Irrigated	0.6 ± 0.4	4	45

[a]Each crop was an average of 4–6 fields, total area 25 m² (five locations per field of 5 m² each) by sweepnet at milk stage. Dryland sites: Tanauan, Batangas; Claveria, Misamis Occidental; Tupi, South Cotabato; Siniloan, Quezon; rainfed wetland sites: Oton and Tigbauan, Iloilo; Manaoag, Pangasinan; Solana, Cagayan de Oro; irrigated wetland sites: Santa Maria and Caluan, Laguna; Zaragoza, Nueva Ecija; Guimba, Nueva Ecija; Koronadal, South Cotabato. Rice culture means ± SE are not significantly different ($P > 0.05$) by LSD test.

such outbreaks, some of which caused political alarm in a number of countries concerned with providing a stable food supply (Litsinger, 1989; Gallagher *et al.*, 1994). There was no prior expectation that the Green Revolution technology would lead to the ecological perturbations that followed in the succeeding decades following large-scale adoption of MVs. These damaging epidemics prompted a concerted effort to understand the reasons behind them and to come up with answers, as farmers were knocking at IRRI's door. Several decades of research were required to pinpoint the causes and to come up with sustainable control methods. At the present time the number of the causes and remedies are still being debated.

Probable causes

BPH was the quintessential insect pest with regard to causes of outbreaks after the advent of MVs. A conference held at IRRI on the BPH problem (Dyck *et al.*, 1979) identified several possible reasons (see Table 18.5). Although the conference delegates felt that landscape-level factors were believed to be more important, there had been no attempt to prove this through experimentation. Lacking, no doubt, was a method to do so rather than the will. Landscape size experiments are very difficult to carry out, especially in developing countries with low research budgets and few field staff. But, as pointed out by Loevinsohn (1984), without scientific experimentation and sound ecological underpinnings, the proposed causes of the problem were only best guesses.

The following section is mainly based on the studies of Loevinsohn (1984) and Loevinsohn *et al.* (1988, 1994), who examined the response of major rice insect pests to landscape-scale factors, which increased crop permanence through multiple cropping and asynchrony. The studies were guided by the assumption that processes acted at: (i) a spatial scale greater than the length of one field; or (ii) a temporal scale longer than one crop season.

For the most part, rice entomology has been preoccupied with the dynamics of individual species at the smallest of spatial and temporal scales. As one's attention

Table 18.5. Agricultural changes linked with rice leafhopper and planthopper epidemics in Asia and their probable causes (adapted from Loevinsohn, 1984).

Parameter	Suspected mode of action
Acting at field level	
Increased nitrogenous inorganic fertilizer	Increased pest nutrition, increased fecundity, increased survival (Litsinger, 1994)
Decreased plant spacing	Favourable humidity (Dyck *et al.*, 1979)
Improved water management	Favourable humidity (Dyck *et al.*, 1979)
Injudicious use of insecticide	Resurgence or secondary outbreak via greater insecticide-induced mortality of natural enemies (Heinrichs *et al.*, 1982a, b) or direct physiological reproductive stimulation of the pest (Chelliah *et al.*, 1980)
Susceptible varieties	Less mortality from lack of natural plant resistance (Gallagher *et al.*, 1994)
Increased crop duration (plant maturity and mixing of different planting methods at one site)	More pest generations (Loevinsohn *et al.*, 1988)
Acting at landscape level	
Increase in rice area	More pest generations (Loevinsohn *et al.*, 1988)
Double-cropping (decreased fallow)	More pest generations (Loevinsohn *et al.*, 1988)
Asynchrony (decreased fallow)	More pest generations (Loevinsohn *et al.*, 1988)

turns to forces acting on populations and communities at scales greater than the dimension of a field and longer than a single season, one is struck by the paucity of data and the abundance of speculation. Three factors of rice intensification were investigated:

- The proportion of land devoted to rice cultivation (rice area).
- The number of crops per year.
- The degree of asynchrony with which they are planted.

The first two are spatio-temporal trends, whereas asynchrony is not itself an aspect of intensification, but a product of it. Also recognized as having a bearing on asynchrony were: (i) the effect of planting cultivars representing a mix of different maturity classes; and (ii) farmers using several planting methods (direct seeding and transplanting by wetbed or *dapog*) in one area (Holt and Chancellor, 1997). *Dapog* is a seedbed maintained without soil where the seed is placed on banana leaves and seedlings are transplanted aged 10–14 days.

These hypotheses were tested from data collected in a variety of sites, either by daily light trap collections or by weekly field sampling. The first was a network of sites established by IRRI's Cropping Systems Outreach Programme using paired kerosene light traps (IRRI, 1979) in 14 locations in the Philippines, each representing differing rice areas, rice cultures and cropping intensities. Single-crop sites were four dryland locations (Siniloan, Laguna; Claveria, Misamis Oriental; Tanauan, Batangas; and Tupi, south Cotabato) and four rainfed wetland sites (Solano, Cagayan Valley; Manaoag, Pangasinan; Oton/Tigbauan, Iloilo; and Mapalad, Nueva Ecija); the five irrigated, double-cropped sites were Santa Maria, Laguna; Zaragoza, Jaen and Cabanatuan, Nueva Ecija; Koronadal, South Cotabato, with an additional site in Koronadal that cultivated 2.4 crops in 2 years.

Three other sites in irrigated, double-cropped rice areas complemented this database. The first was the IRRI experimental farm (1965–1981), representing the narrow band of rice growing along the edge of Laguna de Bay Lake (four electric light traps). The second was in Titi Serong, northern Krian state, Malaysia from data collected from 1959 to 1976 courtesy of G.S. Lim of the Malaysian Department of Agriculture (one electric light trap). The third was from the Philippines (1981–1983), in Zones II and III of District III (17,000 ha), tail-end sites of the Upper Pampanga River Integrated Irrigation System (UPRIIS) in Nueva Ecija province in the country's largest rice bowl. There the intensity of cultivation was more or less constant, but there was wide variation in the degree of synchrony, providing a natural laboratory in which to study its impact on pest populations. The first two sites were represented by only one location, whereas the third was a network of 23 locations spanning major discontinuities in planting date (46 kerosene light traps).

The most damaging rice insect pests are highly fecund, attack all rice stages and disperse over relatively long distances – all factors that reduce the likelihood of extinction. Damaging populations in the irrigated tropics generally build up from within a field rather than from immigration (Kenmore *et al.*, 1984; Perfect and Cook, 1987). Generations of rice pests are thought to be less well synchronized than in temperate regions, as there is not a complete overlap, as might be expected. In fact, rice fields as sources of immigrants are only periodically present due to the constraints of irrigation water delivery.

The process of agricultural change in north Krian appears to have included the same key elements as in the Philippines. Traditional, photoperiod-sensitive cultivars were replaced by fertilizer-responsive, short-duration ones. The use of agrochemicals increased markedly. Roughly contemporaneous with these changes, gravity irrigation was introduced, and dry-season cropping over wide areas became possible.

In Nueva Ecija province, the following insect pests were monitored: BPH, green leafhoppers (*Nephotettix virescens*, *N. nigropictus*, *N. malayanus*), YSB, leaf folder, *Cnaphalocrocis medinalis*, semi-looper, *Naranga aenescens* and caseworms (*Nymphula depunctalis* and *Paraponyx* (= *Nymphula*) *fluctuosalis*). The last was later found not to be a pest of rice but rather fed on an aquatic weed in the irrigation canals (Litsinger and Chantaraprapha, 1995). The traps were set along three transects spanning single and multiple rice-cropping areas on the one hand, and synchronously and asynchronously planted areas on the other. In large irrigation systems, the degree of asynchrony generally increases with the distance from the main water source through increased delays in water delivery, thus lengthening planting times. In the other two sites the light traps were set out at experimental stations surrounded by large, irrigated rice systems.

Rice area

It was hypothesized (Loevinsohn, 1984) that rice insects disperse passively like raindrops falling on the landscape, and they therefore land on rice fields in the same proportion that fields occur in the landscape. If dispersal is to any extent a declining function of distance, insects will land in greatest measure on fields near the source and, in particular, in the field in which they originated. As fields are often dependent on factors such as proximity to irrigation sources or settlements, rice fields are more likely to be aggregated. Loevinsohn (1984) predicted that the relationship would be non-linear due to the factor of distance from the dispersal source (see Fig. 18.1).

The effect of rice area was tested from light trap data collected from the 14-site cropping systems network. To calculate the rice area using a compass, a person walked 0.5 km away from the each light trap in five directions 72° apart. The area of rice fields within the 314 ha circumference was determined using a surveyor's tape measure. Together, the sites ranged from 3% rice area in the 'slash and burn' area in Siniloan in the Sierra Madre mountain range to 95% in Zaragoza within the central Luzon rice bowl.

Four rice insect pests commonly monitored over the range of sites and years were included in the analyses. Data were collected seasonally (6-month period) on a per-crop basis and averaged over the number of crops per site over the data collection period, then converted to natural log scale. Reporting the data on a per-crop basis rather than over a calendar year allows more realistic comparisons between single- and multi-cropped sites to measure increases above the doubling that would be expected if yearly totals were used in double-crop sites (Way and Heong, 1994). One additional site in Claveria and two in Iloilo increased the total to 17 sites for the analysis.

Two pests showed a significant linear exponential increase in abundance in annual totals per light trap with increasing rice area (see Fig. 18.2): *Scirpophaga*

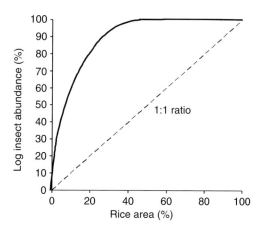

Fig. 18.1. Hypothetical example showing the probability of a pest population successfully locating a field (log insect abundance) during a given generation in relation to the area devoted to rice cultivation. Both variables are expressed as a percentage of maximum attainable values. The dashed line represents a 1:1 linear ratio. (After Loevinsohn, 1984.)

stemborers (F = 9.03, df = 16, P = 0.009) and GLH (F = 9.62, df = 16, P = 0.007). BPH, although showing a similar trend, was just short of the limits set for significance (F = 3.40, df = 16, P = 0.08), which may be due to its greater dispersal behaviour beyond the 1 km limit and strong genetic resistance in MVs. Rice caseworm had fewer data points, as it does not occur in dryland rice culture, but showed a significant linear relationship (F = 17.32, df = 11, P = 0.002). The original analysis of crop area by Loevinsohn (1994) included only one site with over 50% rice area and resulted in a quadratic model. With the addition of three sites with over 20% rice area, the model became a linear function. Increasing rice area, therefore, appears to be a significant factor in augmenting rice pest abundance, as each incremental increase in area results in an exponential increase.

A second series of analyses was undertaken from field trials at the cropping systems programme sites in the Philippines. Averages from weekly samplings were taken for the common pests as described in Litsinger *et al.* (2005). Whorl maggot and lepidopterous defoliators were taken by field from four to eight fields per crop and averaged over crops. Each field was sampled from 20 hills (a hill of rice is five to eight seedlings pushed into the 'pea soup-like' puddled soil). Hills were selected in a stratified manner where damage was assessed in terms of percentage leaves showing signs of feeding 3–5 weeks after transplanting.

Whorl maggot showed a slight but insignficant trend of increasing damage incidence (F = 3.09, df = 12, P = 0.11), whereas *Naranga* and *Rivula* moth larvae defoliators showed a strong level of insignificance. Leaf folder larval damage was similarly assessed during the flag leaf stage (9–12 WAT) depending on the maturity class of the cultivar and showed strong insignificance. Stemborer deadheart (severed tiller) percentage from all stemborers was assessed in the reproductive (6–8 WAT) and ripening (9–12 WAT) stages, resulting in a strong linear relationship during the

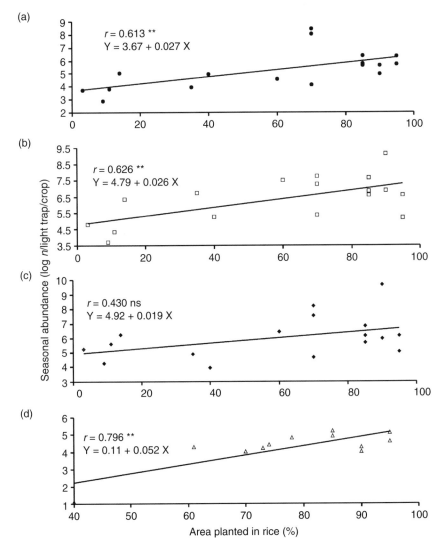

Fig. 18.2. Rice insect pest abundance correlated with area devoted to rice within a 1 km radius of each light trap. Each data point is the annual kerosene trap catch from a pair of traps from each of 14–17 locations in the Philippines, operated from various timespans from 1979 to 1991. (a) *Scirophaga* spp., stemborer; (b) *Nephotettix* spp., green leafhopper; (c) brown planthopper; (d) rice caseworm. (Adapted from Loevinsohn, 1984.)

reproductive stage (F = 9.93, df = 14, P = 0.008) but not in the ripening stage (F = 1.94, df = 16, P = 0.184). Ricebug by contrast showed a weak and insignificant declining linear relationship with increasing rice area (F = 2.30, df = 16, P = 0.15). Yield loss was similarly analysed and showed significant decreasing losses per crop with increasing rice area when expressed in percentage of total loss (F = 4.48, df = 16, P = 0.05), but not when expressed in terms of t/ha.

Number of rice crops per year

With expanded irrigation facilities and photoperiod-insensitive cultivars, more farmers were free to plant throughout the year. The availability of irrigation water and source of irrigation dictate how many times a year that could be. The vast majority of irrigated-rice farmers in Asia plant twice a year. However, in one of the Philippine study sites, farmers have artesian wells that are free-flowing year-round, resulting in farmers trying for five crops in 2 years.

The cropping systems programme outreach light trap data set was used to relate the number of rice crops per year to pest increase. All farmers within the 314 ha circumference of each light trap were surveyed to determine rice crop planting frequency, and the number was averaged over years based on area planted each year. The range between the sites was 1.0–2.4 crops per year.

Three main insect pests responded with positive slopes significantly higher than the expected no change in the null hypothesis (see Fig. 18.3). Data were collected seasonally per crop, as was done for rice area, and converted to natural log scale. The regression for *Scirpophaga* stemborers (F = 28.23, df = 16, $P < 0.0001$), GLH (F = 5.79, df = 16, P = 0.03) and BPH (F = 12.67, df = 16, $P = 0.003$) were exponentially and positively linear. The results for the rice caseworm, however, were insignificant (F = 2.09, df = 11, $P = 0.18$) but showed the same linear trend.

The number of crops grown per year is found to account for the largest amount of variation in the logarithm of aerial density for all four pests and, in each case, in the direction expected on theoretical grounds based on the reduced fallow periods. It is possible that insecticide-induced resurgence and/or stimulation of pest population growth by the adoption of nitrogen fertilizer is in part responsible as well. Further evidence is needed.

As was done with percentage rice area, the major chronic insect pests and yield were regressed with the number of rice crops. Whorl maggot, more than defoliators, showed a tendency for a positive linear increase with the number of rice crops, but was below the significant level (F = 2.99, df = 12, $P = 0.11$). Leaf folders during the flag leaf stage showed no significant relationship, but stemborers did during the reproductive stage (F = 10.34, df = 14, $P = 0.007$) but not the ripening stage. Similarly, rice bug showed a declining but insignificant trend in abundance as the cropping intensity increased. Yield loss relationships showed a trend in lower losses with increased number of crops per year, but were insignificant in terms of percentage ($P = 0.14$) and t/ha ($P = 0.10$).

Using the IRRI light trap data set one can also see a dramatic increase in densities of rice specialists in plotting annual totals during the transition between single and double-cropping in Laguna (see Fig. 18.4). Between 1969 and 1971 the farmers' irrigation system adjacent to the University of the Philippines was expanded to allow dry season cropping, which had risen from 38% in 1969 to almost 100% by 1971. Some structural problems in the irrigation system resulted in diminished command areas during the following 2 years, but this had been repaired by 1974. The increase is most easily seen for YSB since, despite large breeding efforts, IRRI has developed only cultivars with moderate to low levels of resistance, and that mostly based on ovipositional non-preference and tolerance (Chaudhary *et al.*, 1984) rather than on antibiosis, as is the case with Homoptera. Morphological bases of resistance tend to

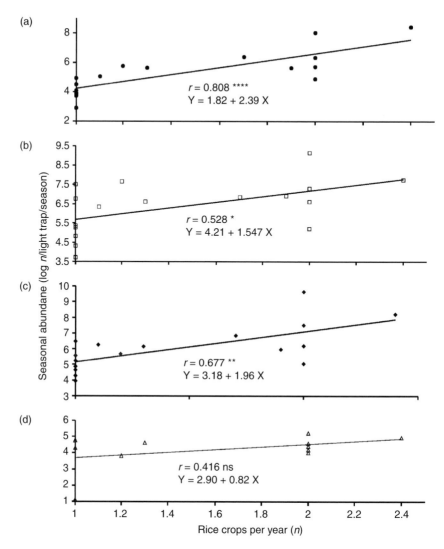

Fig. 18.3. Seasonal light trap collections of key rice pests from 17 sites representing dryland, rainfed wetland and irrigated wetland environments in the Philippines collected over various timespans from 1979 to 1991. Date represent total number collected during the period for one rice crop of 6 months. (a) *Scirpophaga* spp., stemborer; (b) *Nephotettix* spp., green leafhopper; (c) brown planthopper; (d) rice caseworm. (Adapted from Loevinsohn, 1984.)

occur only in specific growth stages rather than throughout the life of the plant (Bandong and Litsinger, 2005).

Yellow stemborer populations rose precipitously between 1970 and 1971 to levels a significant 48% higher (F = 7.97, df = 15, P = 0.01) on a per-crop basis than during the single-crop period. The graphs for GLH and BPH were also dramatic in the macro scale, rising by 920 and 598%, respectively (unfortunately there were not

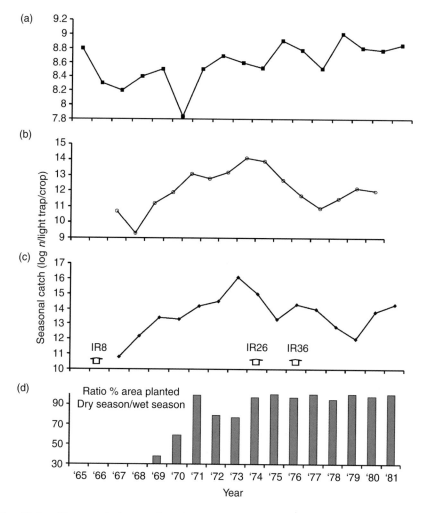

Fig. 18.4. Changes in insect abundance during a period of change from single to double rice cropping in Laguna, Philippines, 1965–1981. Irrigated area data are from the National Irrigation Authority for Mabacan and Santa Cruz River Irrigation System. Data presented on a per-session bassis averaged from four electric light traps on the IRRI farm (after Loevinsohn, 1984). (a) *Scirpophaga* stemborers; (b) *Nephotettix* spp., green leafhopper; (c) brown planthopper; (d) area double cropped

enough years of data before 1970 to allow for statistical testing), even during the presence of resistant varieties against both pests (Khush, 1984). Mean densities of GLH declined sharply when IR36 was released, whereas that for BPH declined when IR26, the first resistant variety, was introduced in 1974, but rose again in 1976 as its field populations overcame the *Bph1* resistance gene in IR26 (Gallagher *et al.*, 1994). However, with the release in 1976 of IR36 having the *Bph2* gene, field populations again declined. Resistance has held steady thereafter throughout the study, which ended in 1991, and has led some to conclude that there may be more than one gene for resistance giving a more sustainable benefit through polygenic gene pyramiding in IR36 and IR64 (Heinrichs, 1994). *Nephotettix virescens* revealed a similar propensity, as it was able to overcome resistant varieties IR26, IR36 and IR54 in three to five generations in the greenhouse (Bottenberg *et al.*, 1990).

The IRRI Laguna light trap data set was further used to determine within-season versus season–season population build-up for the major pests. But before this discussion can begin, some parameters need to be defined. A hypothetical example is used (see Fig. 18.5) where one can discern troughs and peaks in the rise and fall of pest densities over a crop season. Crop senescence, harvesting and land preparation for the following crop give rise to the troughs or lowest densities. Densities decline during the fallow periods between each crop in the double-crop cycles to produce troughs marked 'wet season trough' (WST) and 'dry season trough' (DST) for both the wet and dry seasons. The example covers 2 years, thus WST1 is the wet season trough in the first year and WST2 is the second year trough. Densities tend to rise during the crop period to form peaks, either 'wet season peaks' (WSP) or 'dry season peaks' (DSP), usually late in the crop season. Crops can be of different durations based on the cultivars selected and the degree of synchrony of planting between fields in an area. The crop cycle from trough-to-trough (e.g. WST1 to DST2) in Fig. 18.5 can last from 5 to 7 months.

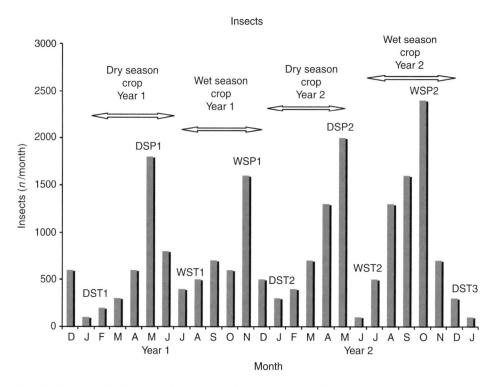

Fig. 18.5. Hypothetical cropping system showing the monthly abundance of a pest over 2 years and four crops. DST1, dry season trough (lower pest density at the beginning of the first year dry season); DSP1, dry season peak (highest density in the first year dry season); WST1, wet season trough; WSP1, wet season peak. (After Loevinsohn, 1984.)

We again use the example of YSB (without much genetic resistance) to illustrate its population characteristics. Different slopes in the relationships caused by increasing rates of different biotic and abiotic variables denote four possible eventualities (b = the slope of the relationship) (Varley *et al.*, 1974): (i) density dependence (b = 1); or (ii) density independence (b = 0) that represent a steady state; as contrasted with (iii) inverse density dependence (b > 1 or unstable); or (iv) undercompensation (b < 1 or stable), which are usually a result of either predation or parasitism. Density-dependent mortality factors are reciprocal, for example because of natural enemies.

Yellow stemborer population density, again expressed on a per-season basis in natural logs, significantly increased in a linear manner (see Fig. 18.6d) over the period from 1965 to 1980 in the Laguna (F = 30.68, df = 15, *P* = < 0.0001). In Fig. 18.6a the log of the WS peak is regressed against the preceding WS trough to measure the within-season increase over the 15-year period. The slope is found to differ significantly from both 0 and 1 (F = 7.35, df = 13, *P* = 0.02), suggesting under-compensating density dependence in its seasonal growth rate.

Populations are affected by what they carry over from one season to the next. Two measures of carry-over are analysed: the first is the carry-over from the peak of the previous dry season (DSP1) to the wet season trough (WST1), just a few months between dry and wet season crops (see Fig. 18.5). The second measures the carry-over from wet season trough in year 1 (WST1) to wet season peak in the second year (WSP2), spanning a year through the intervening wet and dry season peaks.

The within-season carry-over is seen in Fig. 18.6b by the regression of trough numbers prior to the wet season (WST1) on peak number in the dry season (DSP1) immediately preceding it. The slope (b) of significant linear regression (F = 19.47, df = 14, *P* = 0.0007), in contrast to the previous case, is found to differ significantly from 0 but is almost precisely equal to 1, suggesting that carry-over through the fallow between the seasons is essentially density independent.

Regression of trough numbers prior to the wet season (WSP2) on peak numbers in the wet season the previous year (WST1) again results in a linear relationship (F = 11.26, df = 13, *P* = 0.006) (see Fig. 18.6c), the slope obtained not differing significantly from 1, thus suggesting that over this longer period carry-over is still largely independent of density.

Similar analyses were made on light trap data from Titi Serong, Malaysia for YSB for comparison. A more striking similarity in the pattern of response to agricultural change is revealed in the detailed dynamics of YSB (see Fig. 18.7). Here, the analysis is carried out on the same 'within-season' comparison as in Fig. 18.6 in Laguna and a significant regression emerges (see Fig. 18.7a) (F = 18.57, df = 11, *P* = 0.002), with a slope within 12% of the value calculated in Laguna, suggesting again that population increase through the crop is density dependent but under-compensating.

The carry-over coefficient from wet season to wet season is plotted against the proportion of farm area in the vicinity of the trap that is double-cropped (see Fig. 18.7b). A clear positive relationship is apparent (F = 29.17, df = 9, *P* = 0.0006). Interestingly, the mean carry-over rate in the 4 most recent years, when double-cropping was practised on more than 80% of the area, was 10.7% as compared with 12% at IRRI in the period 1971–1979 when double-cropping was dominant.

A somewhat different picture emerges with respect to GLH (see Fig. 18.7c). When one examines the effect of agricultural change on 'between-season' population

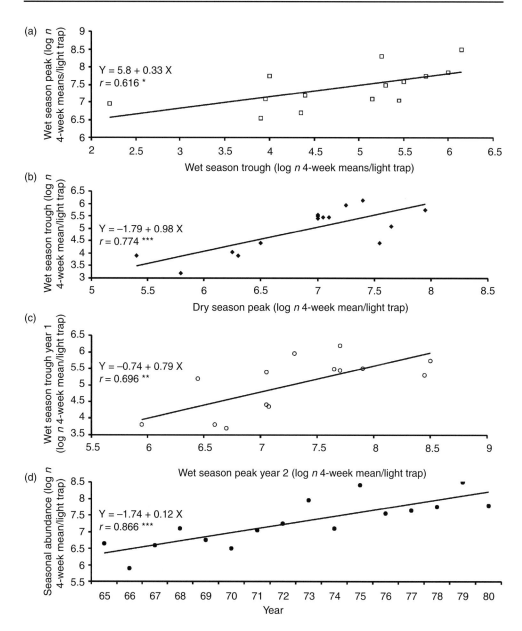

Fig. 18.6. Yellow stemborer population build-up during the period of rice intensification in Laguna, Philippines. Light traps were operated on a daily basis and data were summed on a weekly basis. Four-week periods were sums of the weekly totals for each of four light traps on the IRRI station, 1965–1980. (a) Peak means during the wet seasons as a function of the starting density; (b) wet season trough as a function of the peak in the previous dry season; (c) wet season trough as a function of the peak in the previous wet season; (d) wet season peaks graphed for each year showing steady build-up in density. Data are expressed in densities per crop (6 months). (After Loevinsohn, 1984.)

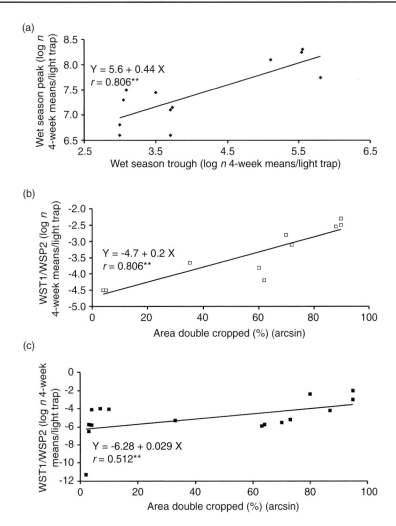

Fig. 18.7. Insect pest population build-up during the period of rice intensification in Titi Serong, northern Krian, Malaysia, 1962–1975. A single electric light trap was operated on a daily basis and the data were summed on a weekly basis. Four-week periods were sums of the weekly total. (a) Peak means during the wet seasons as a function of the starting density for yellow stemborer. Carry-over from wet season to the next as a ratio of densities of the wet season trough of year 1 to the wet season peak of the following for yellow stemborer (b) and green leafhopper (c). (After Loevinsohn, 1984.)

dynamics, *N. virescens* is found to behave much as YSB. When wet season peak numbers are regressed against numbers in the trough preceding it, a slope is obtained that does not differ significantly from 0. In Fig. 18.7c the wet-season-to-wet-season carry-over coefficient was plotted against the proportion of the area double-cropped. The regression is positive and significant (F = 4.96, df = 15, $P = 0.04$), though less variation is accounted for than in the case of YSB. Kuno and Hokyo (1970), who analysed field counts of GLH (*N. cincticeps*) in experimental plots, have reported similar findings – slopes near 0, which they suggested were indicative of strong natural regulatory mechanisms. In the present context, however, under uncontrolled conditions, the impact of human intervention, in the form of insecticide application and the replacement of susceptible by resistant varieties, cannot be ruled out. The latter factor may be particularly important in the case of GLH, as lower levels of genetic resistance occur against YSB.

However, of possibly greater significance than the resistance of the new cultivars to GLH has been their reduced time to maturity. A marked decline was apparent after 1966, when shorter-duration MVs were first grown on an appreciable scale in north Krian, Malaysia. There is no comparable reduction for YSB. GLH is expected to be the more susceptible to this change, as it primarily infests rice during the vegetative stage, which has been truncated by the earlier-maturing rices. YSB usually attacks the crop shortly after transplanting, but continues until ripening and thus would be less affected by the shift in cultivars.

The rate of YSB population increase during 14 wet seasons from 1968 to 1981, measured by the ratio of the annual WS trough to the WS peak in Titi Serong, was highly correlated and illustrates under-compensatory density dependence. The significant carry-over regression from WS to WS over the years suggests that carry-over is largely independent of density (see Fig. 18.7b). These regression analyses show that, over the period of increasing cropping intensification, within-season growth rates of YSB declined while carry-over ratios increased. In Laguna the marked rise in the wet season peak populations of YSB from the early 1970s is related to increased survival between wet seasons, and not within-season increase. This tends to rule out other within-season effects such as fertilizer or resurgence due to insecticide, close spacing or favourable microenvironment of MVs. The greater resistance of MVs to BPH and GLH further supports the observation that pest epidemics in Laguna were mostly related to the change from single to double-cropping made possible by irrigation system improvement from 1969 to 1971. The results are made stronger by the similarity between the two widespread locations, Laguna and Malaysia.

Kenmore *et al.* (1984) also considered critically those factors that had been put forward to account for BPH outbreaks. The MVs, they concluded, were as a group no more susceptible than the traditional ones they replaced, and nitrogen fertilizer had been found significantly to affect BPH densities only at levels considerably above those that farmers had applied. If the resurgence-induced outbreaks were caused by injudicious use of insecticides, BPH damage should have been more severe where the use of resurgence-causing insecticides had been at its greatest. Loevinsohn (1984) concluded that undue attention in the foregoing analysis had been focused on BPH to the exclusion of the other pests that had increased during the period of intensification, few of which have been shown to resurge after pesticide treatment.

Asynchrony

For MVs whose flowering is largely insensitive to photoperiod and, in the lowland tropics, to temperature, variation in the time of planting between neighbours results in equivalent variation in crop maturity. The consequence of high variance of planting dates between farmers and the consequential asynchrony of plant host avail-ability community-wide both extends the period of increase for pests able to disperse between fields and shortens the fallow they must endure. Considering the dispersive (adult) phase of a species, increased populations in one season are expected where the range of planting dates in an irrigation turnout, for example, exceeds a genera-tion length and where this variation is encountered over a distance that adults can readily move. Beyond the generational threshold, the population response to asynchrony should be related to the species' intrinsic powers of increase and to the extent of density-dependent constraints (Loevinsohn, 1984).

Regression analyses of the Nueva Ecija light trap transect data (see Fig. 18.8) showed that the greater the variation between planting dates among neighbouring fields, the greater the pest abundance for YSB males ($F = 6.15$, $df = 13$, $P = 0.02$) and for BPH ($F = 14.74$, $df = 13$, $P = 0.01$), where increasingly higher standard deviations showed an exponential increase in abundance. Data for GLH were not significant.

Using the same data set, a series of regressions was conducted for increasing radii representing different dispersal distances to the trap from the surrounding fields encompassed by those circumferences. When the subsequent regression coefficients

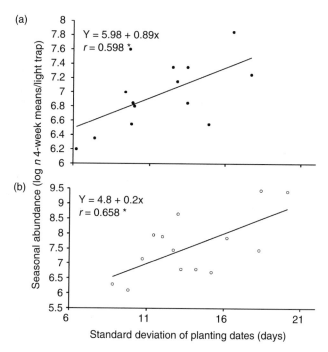

Fig. 18.8. Asynchrony of planting was measured as the standard deviation of planting dates within a radius of 0.5 km from pairs of kerosene light traps. Seasonal abundance is the average of daily catches for pairs of traps converted to natural log. Light traps were placed in 14 locations along a transect along the irrigation system to increasing asynchrony at the lower end of the canal system of the Upper Pampanga River Irrigation System, Zaragoza, Nueva Ecija, Philippines, 1981–1983. (a) Yellow stemborer; (b) brown planthopper. (After Loevinsohn, 1984.)

for each 0.2 km incremental increase in radii were plotted (see Fig. 18.9), the distances where the coefficients showed most significance were related to the known dispersal patterns, with BPH being the most vagile. Best correlations were obtained from YSB at 0.4 km, GLH at 0.6 km and BPH at 2.0 km (the longest distance included in the study equivalent to 1237 ha). Significant distances were obtained for eight of nine comparisons considered (see Table 18.6).

The radius within which the greatest correlation with asynchrony is obtained per pest species will be found to be proportional to the mean distance that adults of the season move within the additional generations that asynchrony makes possible.

Because of an examination of the causes of crop intensification, a number of points are emphasized. First, population growth through the wet season crop has been shown to be density dependent though under-compensating for specialist pests (with the exception of GLH, which increased significantly because of forces of rice intensification). The key dynamic feature is density-independent carry-over between seasons, coupled with under-compensating regulation within seasons. This means that higher pest densities have generally been associated with lower rates of population increase per season, as expected from the numerical and functional response of natural enemies.

As concerns BPH, the results confirm the conclusion of Kenmore *et al.* (1984) with respect to the density-dependence model of population growth in the tropics. Therefore, realized rates of increase are lower at the higher initial population levels that have prevailed. The changes in the agricultural production systems listed in Table 18.5, which were thought to affect pest populations by raising growth rates, are unlikely therefore to have played a central role in increasing pest abundance. Under-compensation implies that changes in the environment that increase pest

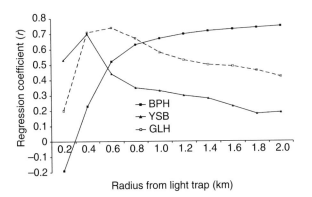

Fig. 18.9. The correlation between seasonal light trap catch and asynchrony of planting as a function of the distance within which the latter is calculated for the brown planthopper (BPH), yellow stemborer males (YSB) and green leafhopper *Nephotettix virescens* (GLH). Asynchrony was measured as the standard deviations of farmers' planting dates from increasing circumferences around each light trap calculated from radii of 0.2–2.0 km. Means of two light traps were taken from each of 14 locations. The distances (radii) associated with the highest correlation coefficients were taken as the effective dispersal distance for each pest. (After Loevinsohn, 1984.)

Table 18.6. Best fitting regression analysis relating dry season total light trap catches to variance in planting dates, measured as standard deviations from increasing radii from each trap in increments of 0.2 km over a range extending to 2 km, Zaragoza, Nueva Ecija, Philippines, 1981[a] (adapted from Loevinsohn, 1984).

Taxon	Distance within which best fit in regression analysis is obtained (km)	F ratio	Significance or P value
YSB	0.4	6.15	≤ 0.05
YSB males	0.4	9.80	≤ 0.01
BPH	2.0	14.74	≤ 0.01
Caseworm	0.4	4.68	≤ 0.01
Naranga aenescens	0.4	2.62	ns
GLH all species	0.6	6.11	≤ 0.05
N. virescens	0.6	3.76	≤ 0.05
N. nigropictus	1.4	3.66	≤ 0.05
N. malayanus	2.0	29.14	≤ 0.001

ns, not significant.
[a]Means of pairs of traps set in a transect of 13 sites from the head to the tail end of the Upper Pampanga River Integrated Irrigation System, leading to increased inter-field asynchrony of planting. Level of significance $P \leq 0.05$.

initial levels infesting the crop would improve carry-over and would have resulted in higher peak levels. Moreover, it can readily be shown that for a population regulated in this fashion, a new equilibrium will be attained within a few years of an increase in carry-over. In Fig. 18.6d, an equilibrium for YSB in Laguna had not yet been reached by 1980, the last data point in the study, as the population increase had continued to be linear since intensification had begun. Curve fitting showed a linear relationship was the best fit on a log scale.

Since 1980, a levelling off is suggested to have occurred as pest outbreaks have been tempered in the Philippines in general (Gallagher *et al.*, 1994), and varieties released at the time, such as IR64, are still being grown, showing durable pest resistance. The most serious pest problems are from new species such as the black bug, *Scotinophara coarctata*, and white stemborer, *Scirpophaga innotata*, which occurred through expansion of distribution in the case of the former (Barrion *et al.*, 1982) and climate change, combined with adoption of early-maturing cultivars, in the case of the latter (Litsinger *et al.*, 2006a).

Areawide Chemical Control

Rice culture in Asia is dominated by small-scale agriculture, with farm sizes averaging 0.5–2.0 ha. This results in a very large number of farmers needing to acquire skills in implementing control methods. Due to sheer numbers and the lack of extension programmes to reach them, management skills, and consequently yields, often vary a great deal between neighbours. Weather events, socio-economic constraints, different pest complexes and uneven management skills are the reasons for the wide range of yields within farm communities. This is borne out by farmer surveys of the sort that were carried out in the four sites in the Philippines by us. From 1981 to 1991, farmers were surveyed in four double-cropped irrigated rice locations in the Philippines. Each season, from 20 to 40 farmers were asked to record their input usage. The number of crops included in the study was: Guimba, 13 crops, 1984–1991; Zaragoza, 15 crops, 1982–1991; Calauan, 17 crops, 1981–1991; and Koronadal, 18 crops, 1983–1991. A description of the study sites is elaborated in Litsinger *et al.* (2005).

In a rice community, yields often range between 1 and 7–9 t/ha, depending on location and season (see Fig. 18.10; Litsinger *et al.*, 2005). It is interesting to note that higher wet season yields occurred in Koronadal, South Cotabato. Mindanao lies

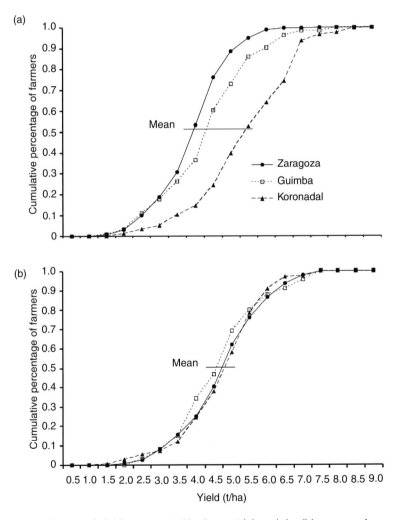

Fig. 18.10. Range of yield as reported in the wet (a) and dry (b) seasons by farmers surveyed in three irrigated rice sites in the Philippines. Some 20 to 40 farmers were interviewed per crop over 14 crops in Zaragoza, 11 in Guimba and 18 in Koronadal, 1981–1991.

outside of the typhoon belt, and yields were almost 2 t/ha greater than in the two Luzon sites – Zaragoza and Guimba – in the wet season within a 10-year period (see Fig. 18.10a). Dry season yields (Fig. 18.10b) in the same sites, on the other hand, showed little difference between sites and a narrower range in yields in each site.

No two farmers manage their rice crop in the same way, varying mostly in input usage regarding material, dosage, timing and frequency. Insecticide usage is a case in point. The results showed that, among the farmers sampled, a wide range of 37–100% used insecticides depending on the location and crop (see Table 18.7). Lowest usage was in Guimba in direct-seeded rice, probably because farmers had to economize more as irrigation water came from a communal electric pump, which was very expensive, particularly in the dry season, and for the greater number of irrigations needed in direct-seeded than in transplanted rice. In Calauan, 99% of farmers used insecticide as compared with 92–98% in Koronadal, 87–91% in Zaragoza and 37–75% in Guimba.

Farmers applied 40 different insecticides comprising 64 distinct brands, to both the seedbed and main crop (see Table 18.8). Most chemicals were organophosphates, followed by 10% organochlorines, 12% carbamates, 9% synthetic pyrethroids and 21% mixtures. The range of formulations was not as varied, as 96% of the applications involved sprayable materials, with the balance being granular. The most popular material by far was monocrotophos (33% of all applications), followed by a mixture of chlorpyrifos + BPMC (15%), endosulfan (9%), methyl-parathion (8%) and cypermethrin (7%). These are all broad-spectrum materials and highly toxic to natural enemies.

Insecticide users applied from one to ten applications per crop, averaging 1.4–3.2 applications (excluding the seedbed), with lowest frequency among Guimba farmers (see Table 18.7). Highest frequency occurred in Koronadal and Zaragoza, sites with the highest pest pressure, as well as in Calauan, a site of low pest incidence (Litsinger *et al.*, 2005). On average, the first application occurred 21–26 days after transplanting (DAT) in the transplanted rice cultures, but occurred earlier in terms of crop stage (26–30 days after sowing) in direct-seeded rice. The second application occurred on average 2 weeks later. Farmers in the three Luzon sites applied insecticides earlier than in the Koronadal site (see Fig. 18.11). This may be the result of lower populations of vegetative stage insect pests such as whorl maggot and lepidopterous defoliators, as well as from the high number of sprays for rice bug. In Luzon half of the applications occurred before 5 WAT, whereas they were later in Mindanao (6–7 WAT).

Seedbed insecticide usage also varied by location and crop culture, but users mostly applied a single application (see Table 18.9). Farmers applying insecticide in wetbed seedbeds ranged from 35% in Koronadal's first crop to 76% in Zaragoza's wet season. *Dapog* seedbeds are located in the field and, surprisingly, 13–21% of Koronadal farmers applied insecticide on these young plants. They may have been instructed to protect the seedbed to prevent virus vectors from infecting the crop. Lowest usage in all rice cultures was in Koronadal (13–43% of farmers), despite farmers' high level of usage in the main crop. Most applications to the wetbed seedbeds occurred during the third week after sowing, but the first application was timed 16–17 days after sowing. For *dapog* seedbeds, most applications took place just before transplanting (12 days after sowing).

Table 18.7. Insecticide usage in the main crops of irrigated, double-cropped rice in four sites, Philippines, 1981–1991[a].

Site	Crops (n)	Users (%)	Applications (n)[b]	Application frequency (%)					Timing (DT/DAS)[c]					Spray volume (l/ha)	Dosage (kg a.i./ha)		
				1x	2x	3x	4x	>4x	1st	2nd	3rd	4th	>4th		Common sprays	Synthetic pyrethroids	Granules
Wetbed transplanted crops																	
Wet season																	
Jaen	8	94	2.4	24	37	25	7	7	22	38	50	56	58		0.20	0.013	0.33
Guimba	5	77	1.4	82	15	1	0	2	26	42	43	36	46		0.25	0.054	0.19
Koronadal	5	93	3.2	11	29	30	18	12	22	37	51	56	63	220	0.23	0.054	0.39
Average		88	2.3	39	27	19	8	7	23	39	48	49	56	220	0.23	0.040	0.30
Dry season																	
Jaen	6	91	2.3	36	29	19	9	8	22	38	49	53	60		0.20	0.041	0.71
Guimba	6	75	1.6	45	46	6	3	0	22	35	44	56			0.19	0.034	
Koronadal	6	92	2.8	21	26	27	17	12	20	36	49	59	67	224	0.23	0.026	0.87
Average		86	2.2	34	34	17	10	7	21	36	47	56	64	224	0.21	0.034	0.79
Dapog transplanted crops																	
Wet season																	
Calauan	4	99	2.8	24	23	30	12	12	23	39	51	61	82	245	0.23	0.013	0.41
Koronadal	4	98	2.8											167	0.25	0.006	
Average		99	2.8	24	23	30	12	12	23	39	51	61	82	206	0.24	0.010	0.41
Dry season																	
Calauan	3	99	2.0	22	26	23	13	17	25	42	49	55	62	245	0.22	0.023	
Koronadal	5	95	3.1											208	0.22	0.013	
Average		97	2.6	22	26	23	13	17	25	42	49	55	62	227	0.22	0.018	

Direct-seeded crops

Wet season																	
Jaen	2	87	2.1	30	43	12	11	4	30	47	52	62			0.21	0.019	0.73
Koronadal	1	100	2.3	40	20	20	10	10	22	36	40	52	78	173	0.19	0.016	0.73
Average		94	2.2	35	32	16	11	7	26	42	46	57	78	173	0.20	0.018	0.73
Dry season																	
Jaen	3	94	2.0	32	38	18	7	4	27	49	57	69	46		0.19	0.021	0.67
Guimba	2	37	1.5	75	9	0	9	4	29	28	36	46	68		0.19	0.045	0.54
Koronadal	3	97	2.5	27	36	11	11	11	31	47	52	53	64	186	0.19	0.010	
Average		67	2.0	51	23	6	10	8	30	38	44	50	66	186	0.19	0.028	0.54

[a]20 to 40 farmers interviewed each season per site.
[b]Insecticide users only.
[c]DT, days after transplanting; DAS, days after sowing for direct seeding.

Table 18.8. Popular insecticides applied to the main crops of irrigated, double-cropped rice in four locations, Philippines 1983–1991.

Common name	Applications per crop/location (%)[a]				
	Zaragoza	Guimba	Koronadal	Calauan	Mean
Crops (*n*)	14	11	11	7	
Years	1981–1991	1984–1990	1983–1991	1986–1990	
Sprayables	97.5	92.3	99.7	92.5	95.5
Organophosphates	44.3	56.7	59.3	25.5	46.4
Monocrotophos	34.5	46.7	25.3	24.5	32.8
Methyl-parathion	5	3.3	24.3	0.5	8.3
Azinphos-ethyl	4	5.0	5.8	0.5	3.8
Chlorpyrifos	0.3	1.7	2.2	0	1.0
Triazophos	0	0	0.8	0	0.2
Fenitrothion	0.8	0	0.3	0	0.3
Methamidophos	0	0.3	0	0	0.1
Malathion	0	0	0.5	0	0.1
Organochlorines	12	4.0	3.3	21.0	10.1
Endosulphan	9.3	4.0	3.3	21.0	9.4
Endrin	1.5	0	0	0	0.4
DDT	1.3	0	0	0	0.3
Carbamates	17.3	2.7	8.5	4.0	8.1
MIPC	15.8	0.3	3.7	4.0	5.9
BPMC	0.5	1.7	3.8	0	1.5
Methomyl	0.8	0.7	0.5	0	0.5
Carbaryl	0	0	0	0	0
Carbofuran	0	0	0.2	0	0
Pyrethroids	11	10.0	14.2	0	8.8
Cypermethrin	10.8	10.0	6.7	0	6.9
Deltamethrin	0	0	5.5	0	1.4
Permethrin	0	0	1.7	0	0.4
Fenvalerate	0	0	0.2	0	0
Cyhalothrin	0	0	0.2	0	0
Other sprayables	0.5	1.3	1.7	0	0.9
Ethofenprox	0	1.3	1.7	0	0.8
Fentin hydroxide	0.3	0	0	0	0.1
Bacillus thuringiensis	0.3	0	0	0	0.1
Spray mixtures	12.3	17.7	12.8	41.5	21.1
Chlorpyrifos + BPMC	9.3	10.7	5.0	35.5	15.1
Monocrotophos + cypermethrin	2.5	4.7	0.7	0	2.0
Fenitrothion + malathion	0	0.3	0.0	0	0.1
Azinphos-ethyl + BPMC	0	2.0	1.8	0	1.0

continued

Table 18.8. *Continued.*

Common name	Applications per crop/location (%)[a]				
	Zaragoza	Guimba	Koronadal	Calauan	Mean
Phenthoate + BPMC	0	0	5.2	0	1.3
MTMC + phenthoate	0.5	0	0	6.0	1.6
Granules	2.5	7.7	0.2	7.5	4.5
Gamma-BHC	0	0.3	0	0	0.1
Diazinon	0.8	0	0	1.0	0.4
Carbofuran	1.8	7.3	0.2	6.5	3.9

[a]Average of wet and dry seasons; each season 15 to 30 farmers were interviewed regarding input usage.

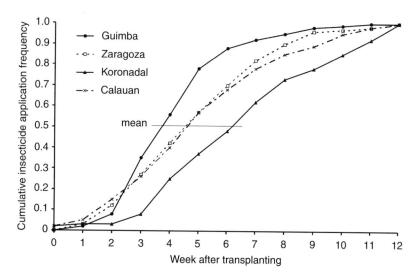

Fig. 18.11. Cumulative frequency of insecticide applications in four irrigated double-crop rice sites as determined from farmer interviews. See text and Fig. 18.7 for description of the sites, 1981–1991.

The volume of water used to apply insecticide per application to the main crop ranged from a low of 173–186 l/ha in Koronadal on direct-seeded rice to 206–227 l/ha on the other rice cultures and locations (see Table 18.7). A cumulative frequency curve revealed that some 10% of applications were sprayed with as little as 50 l/ha, while most were < 400 l/ha (see Fig. 18.12). The recommended spray volume is 500–1000 l/ha (more for taller plants) and farmers reduced this some three- to fivefold, greatly limiting crop coverage. As most of the insecticides used on rice need to be in contact with the insect pest, good coverage is essential for good control. Farmers cut spray volume to save time spraying, with the result of poor control.

Dosages of organophosphate, organochlorine and carbamate sprayables, while recommended at 0.4 kg ai/ha, averaged 0.22 kg ai/ha (see Table 18.9). Insecticide dosage trials conducted at IRRI show that the farmers' spraying level results in over

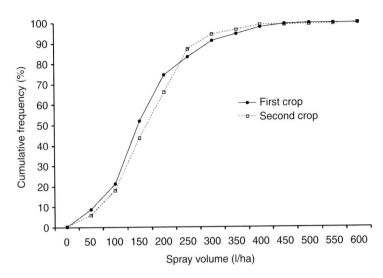

Fig. 18.12. Cumulative frequency of water volumes employed by farmers in applying insecticides, Koronadal, Philippines, 1983–1991. See Table 18.7 for information on the farmers interviewed.

30% mortality. Dosages for synthetic pyrethroids averaged 4 g ai/ha and for granules 0.55 kg ai/ha, again about one-half of the recommended dosage. A cumulative frequency distribution curve for three sites showed that only 10% of applications of organophosphate, organochlorine and carbamate products were at the recommended dosage (see Fig. 18.13). Some one-third of applications were < 0.15 kg ai/ha, producing negligible mortality. In seedbeds, however, which are small in size, dosages were 0.54–0.96 kg ai/ha, or twice the recommended rate for organophosphate, organochlorine and carbamate sprayables (see Table 18.9). Similarly, synthetic pyrethroids were applied at higher levels, 5–11 g/ha, more in line with recommendations. Results for granular formulations were inconsistent, probably because so few farmers applied them.

Interestingly, when we examined individual farmers over a number of seasons, we found as much variation as noted between the farm population as a whole. Mr A. Rombaoa from Koronadal, for example, surveyed over 18 seasons, had sowed eight different rice cultivars. He also applied fertilizer in the seedbed on four of 14 occasions, using three different formulations. He applied insecticide in the seedbed in six of the seasons, while in eight he did not. On the main crop he applied fertilizer in all seasons except one, ranging from one to three times using a variety of formulations and mixtures, including foliar sprays. The timing of the first application varied from 5 to 50 DAT for the granular formulations. Dosages of nitrogen (N) varied from 5 to 68 kg/ha. Herbicide was applied in 12 of 14 seasons using six different products. Dosages varied from 0.13 to 0.69 kg ai/ha. Insecticides were applied to 12 of the 14 seasons, with a range of one to seven sprays in one season. Dosages of organophosphate, organochlorine and carbamate sprayables ranged from 0.06 to 0.50 kg ai/ha. Reasons for spraying varied from prophylactic (calendar-based) applications (53%) to spraying when damage (39%) or insects (8%) were seen. Rice yields per crop ranged widely, from 1.8 to 7.3 t/ha over the seasons.

Table 18.9. Seedbed insecticide usage in irrigated, double-cropped rice in four locations in the Philippines, 1983–1991[a].

Site	Crops (n)	Users (%)	Application No.[b]	Application frequency by application number (%)			Timing (DAS)[c]			Dosage (kg ai/ha)		
				1x	2x	>2x	1st	2nd	>2nd	Common Spray[d]	Synthetic pyrethroids	Granules
Wet/first season wetbed transplanted crops												
Zaragoza	8	76	1.3	75	22	4	19	27	28	0.89	0.079	0.25
Guimba	4	52	1.0	98	2	0	16	27		0.54	0.053	0.34
Koronadal	5	35	1.0	98	2	0	15	22				
Average		54	1.1	90	9	1	17	22		0.71	0.066	
Dry/second season wetbed transplanted crops												
Zaragoza	6	66	1.4	74	14	12	18	24	23	0.96	0.107	0.95
Guimba	5	52	1.0	93	6	0	17	24		0.57		
Koronadal	5	43	1.0	96	4	0	13.8	16				
Average		54	1.2	88	8	4	16	22		0.76		
Wet/first season dapog transplanted crops												
Calauan	3		1.1									
Koronadal	4	21	1.0	95	5	0	12	24		0.12		
Dry/second season dapog transplanted crops												
Calauan	3		1.0									
Koronadal	4	13	1.0	100	0	0	20			0.91		

[a]20 to 40 farmers interviewed each season.
[b]Insecticide users only.
[c]DAS, days after sowing.
[d]Organochlorine, organophosphate, carbamate classes.

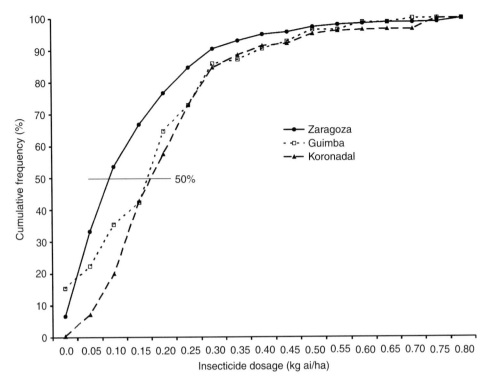

Fig. 18.13. Cumulative frequency of insecticide dosages on irrigated double-crop rice from three sites in the Philippines, 1981–1991. See text and Fig. 18.10 for a description of the sample size of farmers.

Similar results can be seen from a second Koronadal farmer, Mr Nelmeda (16 seasons), who grew 11 rice cultivars. In the seedbed he applied fertilizer four times and did not ten times. Insecticide usage in the seedbed showed him applying six times, while eight times he did not. On the main crop he applied fertilizer in all but one crop, with a range of one to three fertilizer applications per crop, with timings of the first application ranging from 15 to 60 DAT. He applied herbicide to 12 crops and to two he did not, using six different products. Three times he applied herbicide during the first week after transplanting, but mostly applied from 24 to 36 DAT. He sprayed insecticide in 14 of the 16 crops, ranging from one to seven times and using over 12 products, some as mixtures. The reasons he applied were mostly from prophylactic decision making, but also applied when damage was seen. Yield per crop varied from 2.7 to 5.9 t/ha.

This individuality shown among farmers in products used is partly due to having so many choices as, in the Philippines, there is a thriving business in input supplies. Small formulators take advantage of farmers by introducing new brands each year and the farmer does not know that these are not new products, but just new packaging of the same range of insecticides. Farmers, due to more affordable prices, prefer older and cheaper chemicals. Farmers are motivated to experiment with new brand names in the hope of finding better products. This is all done by individual farmers rather than by farm communities as a whole. Goodell (1984a) mentioned the confusion that exists in the range of choices available to farmers and the lack of a more unified process of narrowing down the choices to the better products. Recent studies show that the farmers' preferred method of knapsack sprayer application results in suboptimal levels of control (Litsinger *et al.*, 2005).

Only in Koronadal was there a positive response in yield due to insecticide use in both the first and second crops (see Table 18.10). In the first crop there was a steady increase in yield with each application up to four times in a quadratic function (see Fig. 18.14). In the other sites there was no measured yield response to

Table 18.10. Comparison of yields from farmers who used insecticide with those who did not in three sites, 1981–1991, Philippines[a].

| | Yield (t/ha)[b] | | | |
| | All seasons | | Koronadal | |
Insecticide	Guimba	Zaragoza	First crop	Second crop
Non-user	4.92[c]	4.07[c]	4.49[d]	3.96[d]
User	4.62[c]	4.24[c]	5.19[c]	4.74[c]
df	14	17	8	8
F	0.96		4.98	4.67
P	ns	ns	0.01	0.02

ns, not significant.
[a] Data based on farmer interviews.
[b] In a column, means followed by a common letter are not significantly different (*P* < 0.05) by ANOVA test.

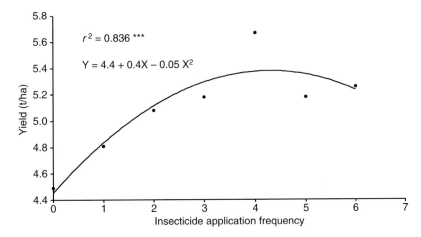

Fig. 18.14. Response surface of yield increase with increasing frequency of insecticide applications, Koronadal, South Cotabato, Philippines, 1983–1991. See Table 18.14 for description of data.

applied insecticides. Most rice cultures in the Philippines would have results similar to those in Calauan and Guimba, not in Zaragoza and Koronadal, as the latter sites were chosen specifically because of the high reported pest incidence.

Stemborers

Rice is known as a political crop, since governments have fallen when supplies gave out, leading to high prices and urban riots. There can be competition between neighbouring countries of different doctrines that use average rice yields as a measure of success (e.g. North and South Korea). Nations concerned with ensuring that shortages will not occur often take more control of the agricultural sector. In South Korea, farmers' fields are scouted by well-paid government extension workers who place code-coloured flags in the field to 'remind' farmers to spray.

Other governments wanting to ensure sufficient rice supplies but mindful of the low management skills of farmers, as exemplified by the description of Filipino farmers, have interceded in various ways. Farmers before the new MVs hardly used fertilizers or pesticides and they had little experience in mastering them. Now they are faced with at least four kinds of fertilizers with abstract names such as 21-0-0. There are usually more than a dozen insecticides in a local agro-input dealer's store, with labels mentioning six or seven pests that are barely visible to the naked eye. Then there are precise instructions governing the timing, with complex formulae determining the rationale for selection, dosage, dilution and spraying. Due to the complexities of applying insecticides, governments at times have initiated areawide spraying programmes. Some of these efforts have occurred to quell epidemics (Fernando, 1975; Ooi et al., 1980; Heong, 1984), while others were initiated to increase average production (Hayshi et al., 1963; Srivastava, 1965).

A campaign for aerial spraying against stemborers in Indonesia in the late 1960s was a case in point. There was not an outbreak, but the goal was to increase national production. Motivation for aerial spraying was that it was felt at the time that MVs, being only recently introduced, were more susceptible to pests, and thus they had a need for chemical control. In addition it was noted that farmers in the BIMAS programme, who had received subsidized fertilizer and insecticide, applied them to more profitable vegetables instead of to rice. The Indonesian government contracted a private company to carry out synchronous control in an area large enough that stemborer recolonization was be minimized. The campaign planners were not sure how large an area that would be, however. The idea was to prevent the normal recolonization of neighbouring fields by moths, which occurs when farmers spray at different times (Joyce, 1985). The planners concluded that destruction of borer larvae in the stem is important only if it decreases the number in the next generation. But if the crop is reinvaded or the next generation comes from a neighbour's field, the control has been wasted. As with migrant pests, insecticide has to be applied as near synchronously as possible to the entire population for best control. Improved efficacy was believed to occur with oil-based formulations, as water droplets evaporate before impacting the crop.

A study in Indonesia (Joyce et al., 1970) showed that spray droplets from Micron sprayers concentrated on leaf tips, a site regularly visited by neonate stemborer larvae.

The optimum droplet size was 80 μ, which gave the highest cover efficiency of 2 ppm. Interestingly, females exposed to insecticide laid eggs haphazardly, leading to further mortality (probably by greater exposure to egg parasitoids). YSB disperses up to 16 km, but as this is only an estimate, the project called for 'as large an area as possible' to be treated. The target area was 300,000 ha in Java. Light traps were set up for synoptic survey, one trap per 1000 ha, and spraying occurred when moth numbers peaked. Insecticide was applied to each field four times during the crop, each representing moth peaks in the wet season. Phosphamidon, being systemic, was thought to kill larvae in the stem and was the material selected (Singh and Sutyoso, 1973).

The result was that significant mean yields of 2.9 versus 1.3 t/ha in the untreated areas occurred over all varieties in Java, with a stemborer reduction of 68% (Singh and Sutyoso, 1973). Aerial spraying eventually failed because farmers planted over wide time intervals in small areas, so that the stemborers varied tremendously within the spray target range of the aircraft. Stemborers persisted after spraying, and repeated aerial applications were uneconomical. The farmers resented this apparent waste of money and became more biased against the overall programme (Hansen, 1978).

Settle *et al.* (1996) stated that the aerial campaigns continued well into the 1970s and, as a result, BPH became an insecticide-induced secondary pest, causing extensive outbreaks. The Indonesian government concluded, however, that the reasons for the BPH outbreaks were from multiple rice cropping and the new varieties, and introduced subsidized insecticides to farmers as a response, which further exacerbated the problem. This large-scale insecticide application also led to environmental concerns from residues in waterways, and farmers objected to being exposed to spray drift (Gorbach *et al.*, 1971). The areawide control was effective, but produced a host of secondary problems.

In Madagascar shortages of rice occurred in the 1970s, and it was decided to solve the immediate problem by treating large areas with insecticides to reduce losses (Schulten, 1989). Contracts were signed with international and local chemical companies for aerial spraying of phosphamidon for the white stemborer, *Maliarpha separatella*. A number of treatments were conducted, but gradually it was realized that very little was known about the rice ecosystem and the effectiveness and economics of the treatments. Field studies showed that the stemborer did not cause as much yield loss as originally thought. Now Madagascar has adopted an IPM programme funded by Switzerland.

Planthoppers

Rice planthoppers are a serious and perennial pest problem in Japan, Korea and northern China, where outbreaks were first recorded in 18 AD (Paik, 1977). Brown planthopper (BPH) and whitebacked planthopper, *Sogatella furcifera* (WBPH) are problem pests, particularly in northern areas where only a single rice crop is grown and where planthoppers cannot survive the winters. The northern limit of overwintering in China is the Tropic of Cancer, 25° N (Zhou *et al.*, 1995).

Their means of recolonizing rice crops each season was a mystery, until 1967, when sailors on a weather ship plying the East China Sea 500 km from Japan noticed swarms of planthoppers, including BPH, WBPH and the small planthopper, *Laodelphax striatellus* (Hirao, 1979). Subsequent studies determined that migration is facilitated by the SW monsoon air masses that prevail in the spring and summer months, particularly during tropical depressions. Planthoppers ride in low-level jet streams at 1.5 km elevation, travelling at < 1 m/s for up to 30 h (Watanabe, 1995). Their point of origin now centres on northern Vietnam and neighbouring provinces in China, especially Hainan Island, where rice is multi-cropped. From these overwintering refuges, the planthoppers spread north and north-eastwards every year, in a series of movements by successive generations. Immigration can cover 2000–3000 km from west to east during the favourable winds. The seasonal migration routes cross southern China into northern China and across the Formosan Strait to Taiwan (Kisimoto and Sogawa, 1995). Some continue across the Yellow Sea and East China Sea reaching Korea and Kyushu, and finally the Honshu Islands of Japan. In China, the northward spring migration into new plantings is more gradual and the impact of long-distance movement is less clear-cut than for Japan, because there is no crop-free zone analogous to the South China Sea serving as a barrier to colonization (Perfect and Cook, 1994).

Radar was used in monitoring planthopper migrations in China (Zhou *et al.*, 1995). Collaboration between Natural Resource Institute, UK and Nanjing Agriculture University, China used high-frequency radar suspended from a kytoon (a hybrid between a kite and a balloon). Radar confirmed mass take-off in the late afternoon, with hoppers staying in the air at temperatures of 16°C. Migration in the warm, moist, night-time air improved survival. The source of the planthoppers was determined by backtracking to Jiangsu province, some 200 km away. Therefore, the planthoppers flew 12 h or some 500 km. Radar density determination agreed with the net catches at landing sites.

Concentrated landing was observed when an air mass carrying migrants encountered a cold front. Landing sites are related to local topographical features and have been identified. Hoppers tended to land en masse in fields located on the east side of hills sheltered from westerly winds along the coast (Noda and Kiritani, 1989). Other landing sites were at the ends of valleys facing windward. Taylor (1974) described the aerodynamics whereby wind-dispersed insects are dumped from the sky from turbulent winds on the leeward side of obstacles. Planthopper migration is intentional, with take-off occurring during two 30 min periods, at dawn and dusk, by the insects actively taking flight. BPH in China can ascend to 700 m or more within 1 h of takeoff in the absence of updraughts (Kisimoto and Rosenberg, 1994). WBPH has greater powers of migration, and is more numerous in collections taken en route on ships and aircraft. Maximum duration of tethered flights is 32 h for WBPH and only 11–23 h for BPH.

Migratory planthoppers have a short generation time (approximately 1 month) and many frequent transient habitats, both of which favour the development of traits that maximize the ability to disperse, arrive successfully in new habitats and exploit them effectively through rapid reproduction before the habitat deteriorates (Kisimoto and Rosenberg, 1994). Both species have two morphological stages in which to maximize efficiency of dispersal (winged macropters) and reproduction (wingless, highly fecund bracrypters). Migrants are sexually mature, so when they arrive they must first feed and develop.

Immigrants guide themselves down from the dispersing winds and alight in a field which, if need be, is followed by a short, active period of inter-field redistribution to locate a more favourable, younger crop. Even though initial immigrant densities are low, they rapidly build up in the absence of significant numbers of natural enemies. It is interesting to note that the mirid predators, *Cyrtorhinus lividipennis* and *Tytthus chinensis*, as well as the leaf folder *C. medinalis*, have been found to migrate along with the planthoppers. No species of *Nephotettix*, on the other hand, is known to migrate.

There is a narrow window through which crop invasion occurs, leading to a high degree of synchrony in the resultant populations. Rise in density is exponential over three distinct generations, whose population peaks are readily discerned. BPH has higher reproductive rates than WBPH as a female can lay over 500 eggs, making it more dominant. Thus, if no control is exerted, hopperburn can occur after heading in the third generation. In tropical double-crop environments, BPH populations peak at the second generation of the first crop and thereafter decline over the second half of the first crop and the entire second crop from pressure of natural enemies. In multi-crop environments, overlapping generations occur as colonization is continuous, depending on local cropping patterns rather than on long-distance migration (Perfect and Cook, 1994). In the tropics, hopperburn is most often linked to negative effects of insecticide on natural enemies, a phenomenon not documented in the temperate, single-crop systems. Hopperburn can occur in the tropics after flooding (Litsinger, *et al.*, 1986) has eliminated natural enemies, but these situations are rare.

In Japan BPH peak densities may be more than 500 times the initial immigrant density, representing an eightfold increase per generation, and this high growth rate can be attributed to high fecundity and lack of significant natural enemy regulation (Perfect and Cook, 1994). In 1995 severe BPH damage occurred after only the first generation. There was no periodicity in the outbreaks, but the number of seasonal depressions correlated with migration rate. Depressions can be held up, however, by other typhoons at sea either stopping movement or redirecting their path.

The largest outbreak occurred in 1966, severely damaging over 780,000 ha. Outbreaks were numerous in the south-western tip in Kyushu, and became less prevalent in a north-easterly direction along the path of tropical depressions over Shikoku and Honshu toward northern Japan. Thus, the south-western corner in Kyushu suffered the most. In China in 1991 BPH infested some 13.3 million ha, causing a loss of 0.5 million t of rice, worth US$400 million. Damage normally occurred 2.5 months after immigration, or two to three generations later.

A national network of traps has been established in Japan to forecast the need for planthopper control, using regression models that can predict population densities from densities earlier in the season. In Japan, net traps are the most effective ground monitoring tools (Watanabe, 1995). Each trap is 1 m in diameter, 1.5 m deep and set on 10 m-high poles. The Plant Protection Division of the Ministry of Agriculture developed the forecasting system. Nets are deployed in several preferred landing sites (observation fields) per prefecture, with catches monitored daily. National bulletins are released every 2 weeks to farmers. Once the planthoppers exceed the 0.01 per hill threshold, farmers are advised to apply insecticides when the population is at a peak of older nymphs before they become adults and oviposit. Timing is critical, as if insecticides are directed to kill adults, there will be no effect against eggs laid in the stems, and chemical is not only wasted but would preferentially kill the few natural enemies that are present.

In China since 1977, weather data have been used to provide long-term forecasts, which are supplemented by field counts by scouts after arrival (Zhou *et al.*, 1995). A network of light traps set next to some 200 monitoring fields has been set up. Fields are sampled (100 hills per 1500 m^2) every 4 days during the migratory season. The government issues chemical control advisories, even specifying the products to be applied. Farmers are advised to synchronize their applications within each village for areawide control. They target the immigrant population in order to retard build-up.

Results show that planthopper numbers have been increasing since 1970 in Japan, probably reflecting the higher numbers of planthoppers at their origins (Kiritani, 1992). Thus, greater damage is to be expected. IRRI has had great success in developing BPH resistant varieties for indica rices, which are now sown throughout tropical Asia (Khush, 1984), and it would seem to be a better policy for Japan and China to augment their breeding programmes to include genetic resistance in order to temper the use of expensive and environmentally destabilizing insecticides.

Locusts

The oriental migratory locust, *Locusta migratoria manilense*, was the most important pest of agriculture in the Philippines even before Spanish colonization (Roffey, 1972). Swarm development typically followed periods of drought and occurred mostly from grassland breeding sites in Mindanao, specifically in South Cotabato near General Santos City. The locust in the non-migratory stage feeds mainly on grasses (*Imperata cylindrica*, *Cynodon dactylon* and an *Aristida*), but pasture grasses are also planted by cattle ranchers. The breeding area falls within the Intertropical Convergence Zone climate known for its El Niño droughts. During droughts the grassland areas contract, causing the once-scattered hoppers (immatures) to aggregate, a behaviour that causes morphological change into the long-winged, dispersal stage (Uvarov, 1936). During the severest droughts larger swarms developed, which migrated north into the Visayas and Luzon Islands and even reached Taiwan and eastern China.

Outbreaks occurred in 1919–1929, 1932–1939, 1941–1949 and 1958–1960. Plagues before the 1960s lasted 7–11 years each but, since the 1960s, have been shorter in duration. Not all outbreaks came from South Cotabato; one swarm in 1912 that alighted in Agusan del Norte province came from the Moluccas and Sulawesi in Indonesia, but the mechanism was the same.

Locust depredations during the Spanish era at times were often followed by famine. Soon after the Americans arrived the Locust Act of 1915 was promulgated, whereby every able-bodied male was to give 2 days per week to locust monitoring and control during declared outbreak periods. Work was divided between scouting and control (including converting grasslands into farmland). Since the 1950s settlement in Mindanao has increased, and particularly so in the 1960s when President

Ferdinand Marcos encouraged Christians from the northern islands to occupy farm-land as a way of dealing with the 'Muslim problem'. Outbreaks became less frequent and more local in nature. The last large swarm to reach China was in 1922.

Local outbreaks have nevertheless still continued, which were combated via aerial insecticide application. In the 1952 outbreak, aldrin and dieldrin were applied to over 500,000 ha in south Mindanao and supplemented by ground crews dispers-ing wet bait. During the El Niño years of 1983–1984, and despite government efforts to undertake aerial and ground insecticide campaigns, the local people pressured the government to desist so they could earn money harvesting them at night to sell them in local markets as a popular snack food. In this case insecticide usage was replaced by mechanical control due to higher population densities of settlers.

Areawide Cultural Control

Synchronous planting

Rationale and expected benefits

The results in the first section suggest an underlying similarity in the annual dynam-ics of rice specialist pests and point to limitation of cropping intensity as a fundamen-tal element in their management. We suggest that efforts to further intensify rice cultivation to more than two crops per year where irrigation makes this possible should be resisted (Loevinsohn, 1984). Extrapolation of available information sug-gests that both pest densities and losses to the crop would increase, and observa-tions in the limited areas where three crops are grown bear this out (Li, 1982). Triple-cropping is regularly carried out in parts of the Mekong Delta of Vietnam and in parts of the Philippines, and such sites are those 'hot spots', where tungro epidem-ics are endemic (Azzam and Chancellor, 2002). In 2005 there was a large BPH and vectored virus outbreak in the Mekong Delta region that led to a loss of 400,000 t (K.L. Heong, personal communication).

As Loevinsohn (1984) predicted, triple-cropping would lead to greater pest levels than with double-cropping, through a reduced non-rice fallow. Figure 18.3 shows that densities of four pests continued to increase in an exponential linear manner as cropping intensity rose from 2.0 to 2.4 crops per year in the communal irrigation sys-tems of Koronadal. Monitoring insect pest damage and yield loss in four Philippine sites showed that the two most asynchronously planted sites resulted in higher chronic infestation levels (Litsinger *et al.*, 2005). We conclude that asynchrony in effect lengthens crop host availability areawide and reduces the rice-free period, to create improved conditions for pest population multiplication on an exponential scale.

The fact that most rice pests responded positively to the duration of rice avail-ability due to increased crop area, multiple rice cropping and to the asynchrony with which crops are established, suggests that reduction of any of these would result in diminished pest populations by reducing the carrying capacity of the environ-ment. Achieving this end would not be politically possible for the first two measures, unless most farmers decided to replace rice with alternative crops such as legumes or vegetables. Best results would accrue if this were done in the dry season, to prevent volunteer rice growth from fallen seeds during harvest or ratoon sprouting. Fields of irrigated rice cannot border fields of non-rice crops because of high water tables. As irrigation increases the level of water tables leading to waterlogging of non-rice crops, all the farmers in an irrigation turnout must agree to the change to non-rice crops. Likewise, retrenchment to one rice crop a year in areas of double-cropping or reduced rice area would have major impacts on the income of farmers and landless labourers, as well as on the availability of food to urban consumers if not replaced wholesale. Clearly, a community-wide decision would need to be achieved, a difficult factor without strong farmer organizations and a good marketing study to be sure that the new crop could be sold at attractive prices.

There are indications that pesticide-induced resurgence was involved in the outbreaks of BPH, GLH and rice leaf folders in the early and mid-1970s (Heinrichs *et al.*, 1982a, b; Litsinger, 1989). Way and Heong (1994), in their Fig. 1, show a close relation in Thailand between insecticide use (in kg/ha) and BPH-damaged areas from 1975 to 1990. It is suggested that such devastating populations would be more prone to build up in agroecosystems in which the duration of the rice host had been extended by multi-rice cropping and asynchrony. From research station and field studies, Kenmore *et al.* (1984) showed that lycosid spiders responded numerically to BPH population increase that strongly suggested that these generalist predators were important in the natural regulation of the pest. Drawing on this and a growing body of other work, they showed that insecticides disrupted this natural control (Kenmore *et al.*, 1987), particularly those chemicals more toxic to spiders than to BPH.

Insecticides selectively caused higher mortality to natural enemies than pests, because they are more mobile due to their searching behaviour that causes them to be exposed to insecticides more frequently. BPH is a sedentary plant feeder whose habitat is the lower portion of tillers, least reached by insecticides sprayed by low volume and low-pressure knapsack sprayers, particularly when the crop has a closed canopy. Some insecticides, such as methyl parathion and diazinon, are known physiologically to stimulate BPH reproduction at sub-lethal doses (Chelliah *et al.*, 1980), but the effect is not enough to explain the scale of the observed pest increase. Our survey data indicated that Filipino rice farmers' use of resurgence-inducing chemicals had increased substantially in the years immediately prior to the first outbreaks of BPH, and that the low dosages typically applied (see Fig. 18.13) were within the range that stimulated reproduction.

Staggered, asynchronous planting of MVs, particularly in multi-cropped areas, has been noted by many researchers as being associated with the occurrence of pest outbreaks, including planthoppers, leafhoppers (and virus diseases vectored by them, see Loevinsohn and Alviola, 1991), stemborers, seed bugs and butterflies (Litsinger, 1994). Thus, to combat these, recommendations for synchronous planting have been made by researchers in most Asian countries (Dammerman, 1929; van Dinther, 1971; MacQuillan, 1974; Kiritani, 1979; Goodell, 1984b; Cabunagan and Hibino, 1989; Sama *et al.*, 1991; Loevinsohn *et al.*, 1993; Koganezawa, 1998). Selecting between the three parameters of cropping intensity, synchronization of planting appears as an inherently less problematic strategy than limitation of cropping intensity or rice area (Loevinsohn, 1984). Synchrony of cropping is a prerequisite for creating a rice-free fallow period, which is the pest-controlling factor. It is believed that efforts should be directed as far as practicable to recovering the strict fallow, the 'tropical winter' characteristic of traditional rice farming, a fallow that for maximum effect should be coordinated across neighbouring farms.

Farmers, when interviewed, state that in their experience fewer pest outbreaks occurred with the long-maturing, single-crop varieties, where rice growing was highly seasonal with sowing during the wet season rains and harvest during the months of short daylengths (Loevinsohn *et al.*, 1993). Traditional varieties are not particularly resistant to insect pests, as had been believed. In plant resistance trials, most of them are in fact highly susceptible to the major insect pests. Stemborers caused significant damage on the low-yielding but dependable rainfed traditional varieties (van der Goot, 1925). With shorter-maturing MVs and irrigation on demand, planting cycles became more random, even field to field. Research has shown tungro incidence and vector populations are high in asynchronous cropping areas and, because viruliferous GLH lose infectivity within a week, a 1-month fallow period, implemented over a wide area, would reduce potential disease sources (Koganezawa, 1998).

Loevinsohn *et al.* (1993) pointed out certain benefits from areawide planting synchrony leading to rice-free periods, such as:

- Reduced pesticide use.
- Prolonged life of pest-resistant varieties.
- More efficient irrigation use by reducing conveyance loss, permitting water to be delivered to areas where it is scarce.
- Providing the incentive to upgrade the irrigation system. Dredging clogged canals, for example, would improve water supply to fields downstream but may not attract sufficient attention until the issue is linked with asynchrony and increased pest levels, which affect farmers over a wide area.
- Fewer drainage problems due to more controlled irrigation delivery.
- More equitable water delivery by introducing 'tail-first' irrigation.
- In many parts of monsoon Asia, and particularly in central Luzon, the frequency of typhoons is highly seasonal and planting now takes place at a time that puts the crop in great risk during the most sensitive stage, from just before flowering until harvest.

Synchronous planting schemes present an opportunity to resolve this wet season problem by advancing the date of planting by 4 weeks to reduce the risk of a typhoon, from almost one in 2 years to one in 3 years. It was estimated that synchronous planting over a 2000 ha block (within 2 weeks) could save as much as US$150,000/2000 ha per season, which includes savings in irrigation water and pesticides as well as higher mean yield (Goodell, 1984b).

Synchronous planting also has benefits against other pest groups, such as rodents, fungal diseases, weeds and nematodes, that are encouraged by continuous rice plantings and extended periods of ponding (Litsinger, 1993). In addition, there are benefits from aerating the soil for several months a year to encourage nutrient uptake and alleviate zinc deficiency (Cassman *et al.*, 1996). Aeration can occur even if an irrigated non-rice crop were grown, as harmful anaerobic conditions occur during ponding when oxygen cannot penetrate into the root zone.

Areawide programmes

There are many examples of synchronous planting, from FAO that called for it as a general recommendation (Brader, 1979), to a law promulgated in Orissa, India for farmers to plant within a restricted period (Banerjee and Srivastava, 1978), to farmers in China (NAS, 1977) and in Burma and Malaysia (Goodell, 1984b). In the latter case in the Muda Scheme, the government began threatening to withhold irrigation water from farmers' fields after expiry of scheduled planting dates formulated to impose synchrony where losses had often exceeded US\$37 million per season.

Examples of rice pest management programmes that have included areawide synchronous planting are presented from six countries. Large-scale commercial rice production ventures have taken place in a number of countries where the rice area was divided into blocks sequentially sown year-round to ensure continuous harvest and even labour distribution. But without a rice-free fallow and with insecticides directed at other pests, secondary pest outbreaks of BPH soon occurred (MacQuillan, 1974; Loevinsohn, 1984). A rice–soybean rotation in the Solomons in which BPH was not a pest was soon changed to a two-rice crop with staggered plantings, which soon emerged into a Malthusian nightmare where BPH could not be controlled with any combination of resistant varieties and insecticides. Overlapping of harvest of one crop with planting of the next resulted in cross-crop infestation. Rice crop ratoon and self-sown rice between crops allowed crop-to-crop carry-over. Continual immigration obscured the extent of reduction of BPH by insecticides. Instituting rice-free fallows by synchronous planting was undertaken as a control measure with success in the Solomons (MacQuillan, 1974) and Indonesia (Oka, 1979). On leased land, with salaried labourers and with control over irrigation, such companies can readily instigate synchronous planting schedules.

In Indonesia, large irrigation systems are controlled by government in a top-down system supported by the BIMAS rice input supply and credit programme. Most of these large systems had experienced pest outbreaks that could not be controlled by pest-resistant cultivars or insecticides. Based on recommendations from government research centres, plans to stop irrigation in the dry season were implemented as a way of creating rice-free fallows (Oka, 1979). Starting in 1976/1977, the government gradually implemented a system of synchronous planting and crop rotation on a large scale to control BPH and virus diseases.

In central and western Java as well as in Bali, for farmers within tertiary turnouts of 300–500 ha, large cadres of extension workers introduced plans for synchronous planting to farmers who had limited input as to the timing of irrigation delivery. This new irrigation scheduling reduced the gap between the wet and dry season rice crops, creating a 1–2-month rice-free period between. The break scheduled during the dry season ensured that sources of rice, such as that coming from ratoons and volunteer seedlings, would not develop. In some areas, farmers planted a non-rice third crop, while in others the fields remained fallow. After initial success, turnout areas were combined into larger geographic units to produce a greater controlling effect. In the top-down system, this was possible and the results showed a significant reduction in BPH, GLH and virus diseases to such a degree that the national variety Cisadane, with moderate resistance to BPH biotype 2, held up for more than eight cropping seasons (Oka, 1988).

The scheduling of rice-free fallows in the dry season has continued to the present day in a number of irrigation systems, including the Jatisari reservoir system in western

Java (Widiarta *et al.*, 1990; Sawada *et al.*, 1991). In central Java, where synchronized cropping was implemented again by control of water delivery, tungro incidence significantly dropped (Koganezawa, 1998). The success in Java soon prompted other regions in Indonesia to embark on similar programmes. One occurred in the western and central Ronboku districts of western Tenggara on the islands east of Bali (Koganezawa, 1998). A second was in southern Sulawesi, which had experienced a continual tungro problem from 1972 to 1975, when damage occurred to 100,000 ha. This outbreak was soon followed by others. Plant resistance, which had been the tactic of choice to control the pests like tungro, proved non-durable (Manwan *et al.*, 1985). Insecticides likewise proved futile on the now susceptible varieties. Like many regions where large pest outbreaks occur with great frequency, southern Sulawesi is an area blessed with ample irrigation water, stemming from sources emanating from two distinct climates to the west and east of the long southern arm of the island. Therefore, irrigation water feeds the central rice bowl year-round.

Research had shown periods during the year when the GLH vector population was lowest, and a large-scale management scheme was initiated in 1982, which lasted until 1988 (Sama *et al.*, 1991). This was a government-run programme during the time of President Soeharto, where top-down instructions were issued to farmers through the effective BIMAS programme. Extension workers were trained in the scheme and the plan was implemented on a subdistrict level (each 3000–20,000 ha), involving local government across the province.

The areawide scheme involved four components: (i) planting each rice crop during a time when GLH populations were historically low; (ii) synchronous planting to create a fallow period; (iii) all farmers selecting cultivars each season from within a recommended group having the same genetic resistance on a rotational schedule among three groupings; and (iv) applying insecticides to fields with tungro to kill GLH adults before they emigrated to infect new fields. The three varietal groups were: group 1 (IR26, IR30, IR46, Seryu), group 2 (IR29, IR34, IR54, IR60, IR64, Kelara) and group 3 (IR36, IR42, Cimanuk). Groups 1 and 2 were rotated in wet seasons, while group 3 was for the dry seasons. Synchronous harvest rather than synchronous planting was emphasized in the dry season so that late-maturing varieties were planted sooner than early-maturing varieties, in order to give a longer rice-free period. The programme was able to achieve 90% of compliance for the new planting schedule and 78–85% compliance in adoption of the varietal group.

The planting schedule created a fallow of over 1 month between dry and wet seasons in each area. In some seasons, farmers were asked to plant 1 month earlier than normal. This method was highly successful, as those planting within the specified months experienced only negligible tungro incidence while those out of the scheme and planting late suffered up to 60% infection, based on area. Even farmers that had grown susceptible varieties within the planting period were spared from the disease. In the final analysis the organizers felt that cultivar rotation was less important than creating the rice-free fallow, which had the effect of reducing tungro acquisition by the pest vectors. Success required effective organization and planning, and a well-functioning seed production system.

Malaysia has also adopted areawide synchronous planting schemes, principally against GLH/tungro and BPH. The first was in the Tuaran district of Sabah, East Malaysia. During 1979 WS there was a severe tungro epidemic and district authorities discontinued the supply of irrigation water, forcing farmers to refrain from planting rice in the 1980 dry season. The incidence of tungro declined in the subsequent wet season. The second was in the largest irrigation system in Malaysia, the Muda Scheme (Koganezawa, 1998) which, in 1981, suffered a large tungro epidemic affecting *c.* 6,000 ha. The government then embarked on a large-scale project that involved: (i) planting resistant or moderately resistant cultivars; (ii) creating a 1-month fallow between the dry season and following wet season so that all inoculum sources dried up (volunteer rice seedlings, rice ratoon and grasses); (iii) destruction of inoculum sources by herbicides, burning and dry rotovation after the wet season harvest; and (iv) judicious use of insecticides to control the vectors only when abundant. There was great resistance among farmers to applying herbicides to rice ratoons and volunteer seedlings on spiritual/religious grounds. Rice in many Asian countries has god-like status as the word rice is often synonymous with food, and thus one should preserve this resource, not kill it. Otherwise farmers complied, and tungro and its vectors significantly declined. From 1986 onwards there has not been any tungro over the entire Muda scheme.

In 1984 a new cropping system of synchronous planting was designed for the Muda II scheme, both for better water use efficiency and for pest control (Nozaki *et al.*, 1984). The first crop must be planted synchronously to create a 1-month fallow between the dry season crop. The dry season crop has staggered planting of blocks in 'waves', also suggested for the Upper Pampanga River System in the Philippines, to be described later. The planting in waves (systematic staggering of planting within a specified time interval) can not only save water but also increase the efficiency of farm machinery that must be moved around. If the staggering is done in a systematic manner, travelling distances to the next field are shorter, saving time. They noted that the length of the break must allow time for rice hosts to die out. Viruliferous hopper vectors can transmit tungro for only 1–5 days after acquisition, and hoppers die within a few days when their host cannot be found (Sogawa, 1976). The greatest need is for the plant hosts to die off, which takes at least 1 month.

A government programme in China achieved enhanced control of rice dwarf by 59–80% with synchronous planting and a rice-free fallow (Cheng *et al.*, 1980), as the mixture of early and late plantings led to greater incidence of the virus. Kiritani (1992) stated that synchronous planting, along with raising seedlings under protected covers to advance planting dates, contributed greatly to the reduced incidence of various virus diseases in Japan.

Finally, two areawide synchronous planting schemes from the author's experience in the Philippines are described in greater detail. The first was carried out in 1981 by Region X of the Philippine Department of Agriculture, which undertook a programme of synchronous planting in Koronadal, southern Mindanao, in response to a number of pests. Simultaneous occurrence of BPH and the grassy stunt II virus that it vectored occurred as well as GLH and tungro disease, resulting from the breakdown of IR36, once resistant to the two insect vectors. There have never been any varieties resistant to rice virus diseases directly, and resistance to the vectors affects control (Azzam and Chancellor, 2002). In addition, the prolonged anaerobic soil status from continuous flooding during rice culture caused high incidence of zinc deficiency. The area involved was of over 2500 ha in the Marbel River Valley. Insecticides or pest-resistant cultivars could not control GLH and BPH. The array of field problems led to many crop failures as, upon injury, rice became stunted, quickly withered and died in all crop growth stages. Individual hills of rice exhibited multiple symptoms of zinc deficiency, tungro, grassy stunt II and BPH burn (hopperburn is the symptom of dried-up plants resulting from the removal of phloem (Kenmore *et al.*, 1984)). Species of *Nephotettix*, although they also feed on phloem, do not cause hopperburn as they are density dependent.

The root of the problem stemmed from an abundance of irrigation water allowing farmers to plant year-round. Asynchrony resulted not only in the Marbel River Diversion system being operated by the National Irrigation Administration (NIA), but also in the eight contiguous small, farmer-run, communal irrigation systems. Irrigation water in the communal systems comes from artesian, free-flowing springs that emerge from the base of mountains lining the 3 km-wide river valley. The NIA system irrigated fields adjoining the river and is a much larger system, whereas each communal system has a command area of only *c.* 50–100 ha. Farmers attempted to grow five crops in 2 years in the communal systems, whereas the NIA system released water for double-cropping only. The Department of Agriculture instituted a synchronized planting scheme by controlling the release in the river diversion system, but had no control over the communal systems. A rice-free period was instituted in the dry season when irrigation supply was lowest.

Research has shown that rice virus diseases are virtually impossible to control by insecticides on susceptible cultivars (Macatula *et al.*, 1987). Out of desperation, however, farmers increased their insecticide application frequency, some spraying twice a week, further exacerbating the problem through insecticide resurgence and secondary pest outbreaks (Heinrichs *et al.*, 1982a, b). A survey of 20 farmers in each of eight villages, equally divided between the two irrigation systems, was conducted in 1983. The results showed that, during the 1979–1983 outbreak, some 64% of rice farmers had suffered an average of 1.5 crop failures (no harvest). There was no difference ($P > 0.05$), however, between the number of failures in the communal (1.5) and river diversion (1.6) systems (F = 0.044, df = 69). Farmers took matters into their own hands and tried planting different cultivars, many obtained without authorization from nearby Department of Agriculture on-farm trials that had tested promising breeding lines for local adaptation. Most lines proved susceptible to either local genotypes of BPH or GLH.

Due to the year-round availability of irrigation in the communal systems, crops in all stages of growth could be seen within easy view. The exploding insect pest populations invariably led to new genetic recombinations that spawned new genotypes which overcame genetic plant resistance (Gallagher *et al.*, 1994). Pests readily spread from the communal systems to the adjoining river diversion areas in the Marbel system. The distance separating the systems often was only a road.

A research team was placed in the site in early 1983 and monitored the pest situation in both irrigation systems until 1991. Each season, 20 to 40 farmers in each of four villages per irrigation system were surveyed to record their cultural practices, including use of inputs. Farmers were given journals for annotations, which were checked monthly by project staff. Pairs of kerosene light traps were established in each village, and each season yield loss trials were set up that included monitoring pests and natural enemies (Litsinger *et al.*, 2005). Weekly sampling of one field per village by DVac® suction machine took place from 3–12 WAT. Five samples were taken per week in each field within a 1 m-diameter plastic cylinder pushed into the soil around rice hills. Arthropods were bagged with a killing solution and identified to species with the use of a dissecting microscope in a rented house in Koronadal. Each season's trials were conducted on the most popular cultivar used by farmers at the time.

The El Niño drought of 1983–1984 severely reduced the flow from the artesian springs, creating rice-free fallows that broke insect pest and disease cycles and dried out the soil, alleviating zinc deficiency and eliminating most plantings. Thus, the area planted to dry season crops in 1983 and 1984 was severely restricted. Farmers had also discovered a BPH-resistant breeding line, '299', later named by the Philippine Seed Board as IR60, to replace IR36. The three factors of: (i) new resistant variety; (ii) the drought; and (iii) synchronized planting in the NIA system, quelled the crop failures. As these factors occurred at the same time, the contribution of each could not be measured.

Over the 8-year study, farmers in the communal system averaged a variance in planting date of 32.4 ± 4.2 days, while for those in the river diversion system the mean was 11.5 ± 1.5 days, significantly different from one another ($F = 22.08$, df = 31, $P < 0.0001$). By 1985 the communal systems had resumed their old intensive cropping schedules and insecticide application frequencies but resistant rices held up, although farmers changed them frequently. Synchronous planting continued to be enforced in the NIA system throughout the study period by regulating water delivery to create a 1–2 month gap in the dry season. NIA was motivated to do this for water conservation goals, as irrigation water was limited and was not enough for all farmers in the dry season.

A limited flare-up of tungro disease reoccurred in 1985–1986 and was documented by the research team. During the first rice crop of 1985, tungro was noted in several fields planted to IR60 in the communal system. Infection levels increased in the following crop in ten fields in the river diversion area and 19 fields in the communal area (see Table 18.11). Incidence was noted in three villages of the communal systems but only in one river diversion system village. By the following 1986 first crop, infection had occurred only in one village of the river diversion system.

Suction sampling showed an increase in *N. virescens*, the main vector of tungro, in the 1985 first crop sown to IR64, attaining 30 to 40 hoppers/m^2, but had declined to 10 to 25 hoppers/m^2 in the second crop (see Fig. 18.15). The less effective vector, *N. nigropictus*, was abundant only in the river diversion system, where it reached 28 hoppers/m^2. *N. virescens* densities rapidly declined by the 1986 first crop with IR60 as the cultivar. Also evident was a rise in BPH and WBPH in the same 1985 first crop but, in the following few crops sown to IR60, planthopper numbers declined. In Fig. 18.16 we see that a third tungro vector, the zigzag leafhopper *Recilia dorsalis*, was consistently abundant in all crops. The second crop of 1984 was during a drought, which explains the low levels of all arthropods.

Two guilds of natural enemies, also suction sampled, were the mirid predator *Cyrtorhinus livipennis* (preys on hopper nymphs and eggs) and spiders. Spiders appeared to respond to the increase in hoppers while *Cyrtorhinus* numbers were numerically high during the 1985 first crop, but not distinctly as compared with all other crops. Spider levels, however, rose distinctly, most likely in response to the higher hopper populations. Both natural enemy guilds responded more to the high

Table 18.11. Tungro incidence during a flare-up in eight villages in synchronously (river diversion) and asynchronously (artesian spring) irrigated areas, Koronadal, Philippines, 1985–1986[a].

Crop/village	Fields with visible damage (n)	Severe loss Fields (n)	Area of fields affected (%)
1985 2nd[b]			
Synchronous			
Bo. 1	8	1	50
Caloocan	2	0	
Conception	0	0	
Santa Cruz	0	0	
Total	10	1	
Asynchronous			
Avancenia	5	1	40
Morales	8	2	60
Namnama	2	0	
Magsaysay	4	1	100
sum	19	4	
1986 1st			
Synchronous			
Bo. 1	0	0	
Caloocan	4	0	
Conception	0	0	
Santa Cruz	0	0	
Total	4	0	
Asynchronous			
Avancenia	0	0	
Morales	0	0	
Namnama	0	0	
Magsaysay	0	0	
Total	0	0	

[a]Noted first in Avancenia in first crop, 1985.
[b]Total area affected was 36.1 ha, predominantly on IR60 (19.6), IR62 (5.4), IR64 (4.9), '–12' (3.5), C-13 (0.8), 206 (0.8), 1609 (0.7) and MRC (0.5 ha).

hopper numbers in the first crop of 1990 in the communal system. The 1990 crop was planted to a breeding line ('–90') popular at the time. IR72 was the most popular released variety, but by that time over 80% of the area had been planted to non-released lines.

Tungro incidence, however, was negligible. The line '–90' was evidently susceptible to all hoppers, as all responded in the same way. Low tungro infection rates, such as evidently occurred in the 1990 crop, can accrue from vectors immigrating

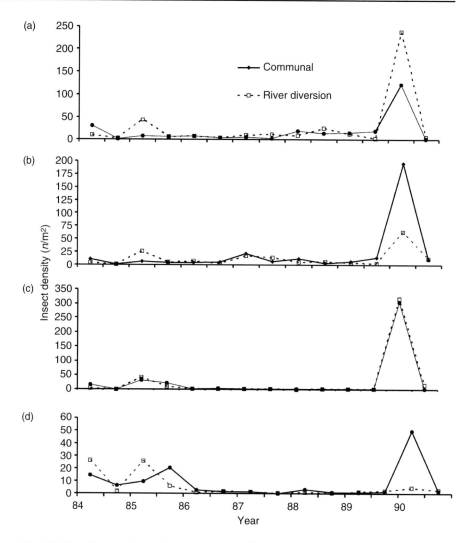

Fig. 18.15. Comparison of population densities of rice hoppers expressed as seasonal means from weekly suction sampling in two irrigation systems over a 7-year span. Data are means of samplings from 3–12 weeks after transplanting in four fields in each of communal (asynchronous planting, 2.4 crops per year) and river diversion (synchronous planting and 2 crops per year) irrigation systems. Koronadal, South Cotabato, Philippines, 1984–1990. (a) Whitebacked planthopper; (b) brown planthopper; (c) *Nephotettix virescens*; (d) *N. nigropictus*.

from older neighbouring rice crops, immigration into an older crop, limited within-field plant to plant spread by infected adult vectors or the inability of the vectors spreading the virus from plant to plant to contract both spherical and bacilliform particles (Chancellor *et al.*, 1996). Tungro is a composite disease and vectors must contract the spherical virus first to be able to contract and transmit the bacilliform

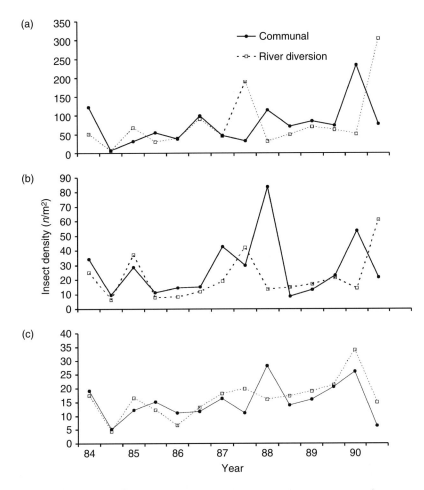

Fig. 18.16. Comparison of population densities of zigzag leafhopper and key predators expressed as seasonal means from weekly suction sampling in two irrigation systems over a 7-year span. Data are means of samplings from 3–12 weeks after transplanting in four fields in each of communal (asynchronous planting, 2.4 crops per year) and river diversion (synchronous planting and 2 crops per year) irrigation systems. Koronadal, South Cotabato, Philippines 1984–90. (a) *Cyrtorhinus*; (b) spider; (c) zigzag leafhopper.

virus (Azzam and Chancellor, 2002). The most severe damage comes if rice plants have both viruses, whereas only negligible loss occurs from plants with only the spherical virus.

In 1983, the last year of the outbreak, IR60 was the dominant cultivar, reaching 31% of the area in the first crop and 48% in the second. IR60 dominated until IR62 reached 60% of the area planted in the first crop of 1986, with IR60 at 23%. Both varieties became infected with tungro, so a loss of varietal resistance to *N. virescens* may have been a contributing factor. IR62 appeared to have higher levels of resistance than IR60.

A characteristic of Koronadal was the propensity of farmers, especially in the communal system, to source unregistered breeding lines, which became most apparent from the second crop of 1988 to the end of the study (see Fig. 18.17). New ones continuously emerged with each crop, and the survey recorded more than ten in any one crop from a sample of *c.* 100 farmers. From 1988 first crop to 1990 second crop, significantly more farmers in the communal system sowed unregistered lines (75.0%) than in the river diversion system (48.2%) (F = 6.50, df = 11, *P* = 0.03). Percentages were first transformed to arcsine in the t-test analysis.

Comparisons among pest populations collected in light traps were made annually from 1983 to 1990 in both irrigation systems (see Table 18.12). Significant differences between some of the insect pest and natural enemy densities were documented between systems. Seasonal light trap catches converted in natural logs were greater in the communal irrigation than in river diversion systems for caseworm, BPH,

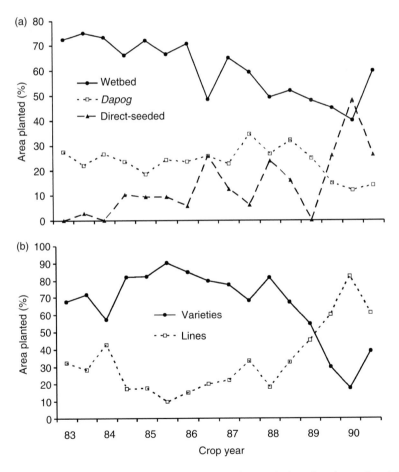

Fig. 18.17. Trend in farmers' planting methods (a) and choice of registered varieties or unauthorized lines (b) as determined from surveys of 20 to 40 farmers per crop, double-cropped rice, Koronadal, South Cotabato, Philippines, 1983–1990.

Table 18.12. Comparison of arthropods collected in light traps in two irrigation sytems, the communal asynchronously planted system and the river diversion synchronously planted system, Koronadal, Philippines, 1983–1991[a].

| Pest | Mean seasonal total (log *n*/light trap)[b] | | Difference | | |
	Communal	River diversion	F	df	P
Caseworm	3.76	1.88	4.66	29	0.04
BPH	9.79	8.50	11.45	29	0.005
Nephotettix virescens	8.93	7.48	5.84	29	0.02
WBPH	8.12	6.58	6.50	29	0.02
Cyrtorhinus lividipennis	8.74	6.40	17.89	29	0.0007
Scirpophaga spp.	8.09	8.10	0.004	29	> 0.05 (ns)

ns, not significant.
[a]Pairs of kerosene light traps placed in four villages of each irrigation system and counted daily.
[b]Seasonal totals converted to natural log for analysis.

N. virescens, WBPH and *Cyrtorhinus lividipennis*, but *Scirpophaga* stemborers showed no difference between the systems.

Crop monitoring likewise recorded greater mean densities per crop in the communal systems for whorl maggot damage and leaf folder damage at the flag leaf stage. Yield loss was higher in terms of both weight and percentage (see Table 18.13). There was no difference for damage levels from the *Naranga* and *Rivula* defoliator complex, nor for mid-crop leaf folder damage and stemborer deadhearts or white-heads, although the latter achieved probabilities close to the limits of significance.

In reviewing pest and natural enemy light trap data over the years, it became apparent that a number had steadily declined over the 9-year period and that there were often differences between the irrigation systems (see Figs 18.18 and 18.19). BPH (F = 7.36, df = 16, P = 0.03), WBPH (F = 17.66, df = 16, P = 0.006), *N. virescens* (F = 4.63, df = 16, P = 0.03) and *Cyrtorhinus* (F = 11.67, df = 16, P = 0.01) significantly declined only in the river diversion system. Caseworm mean densities significantly declined in both irrigation systems following the pest outbreak. In the river diversion area caseworm numbers declined following a quadratic function (F = 85.79, df = 16, P = 0.0001), and a linear function in the communal system (F = 73.53, df = 16, P = 0.0001). *Scirpophaga* stemborer densities did not decline over time when measured by percentage of whiteheads.

The survey results were examined to determine whether any cultural practices might have been associated with the flare-up. One characteristic of Koronadal farmers in both irrigation systems was that they sowed rice using three different methods, a factor which has been reported as increasing cropping asynchrony (Perfect and Cook, 1994). In 1983, over 72% of the crop area was transplanted by wetbed, with 28% using *dapog* wetbed while none direct-seeded (see Fig. 18.17). As the years passed, increasing numbers of farmers direct-seeded at the expense of the two wetbed methods, following a national trend. In the 1990 first crop, 51% of the area was

Table 18.13. Comparison of arthropods monitored in yield loss trials in two irrigation systems, the communal asynchronously planted system and the river diversion synchronously planted system, Koronadal, Philippines, 1983–1991[a].

| | | Per crop | | Difference | | |
		Communal	River diversion	F	df	P
Variable	Unit of measure					
Whorl maggot	Damaged leaves 3–5 WAT (%)	18.2	14.3	15.27	29	0.0005
Defoliators	Damaged leaves 3–5 WAT (%)	5.6	5.9	0.03	29	> 0.05 (ns)
Leaf folder	Damaged leaves 6–8 WAT (%)	3.3	4.3	0.86	29	> 0.05 (ns)
	Damaged flag leaves 9–11 WAT (%)	4.6	1.7	4.12	29	0.04
Stemborer	Deadhearts 6–9 WAT (%)	1.2	1.9	3.57	29	0.07 (ns)
	Whiteheads (%)	2.8	1.5	3.80	29	0.06 (ns)
Yield loss	t/ha	0.93	0.65	4.23	29	0.02
	%	18.0	12.1	5.29	29	0.03

ns, not significant.
[a] Four fields in each irrigation system were monitored each season. Pest data were monitored from samples of 20 hills per field on a weekly basis (WAT, weeks after transplanting) and averaged over the indicated periods. Yield loss was determined by the insecticide check method (Litsinger *et al.*, 2005) as the difference between plots with frequent applications compared with an untreated control.

direct-seeded. The point was that farmers were using three methods within each village and there was no statistical difference between villages among the methods. During the period of the flare-up, direct seeding reached only 10% of the area, and thus it is unlikely that a change in planting method was a contributing factor.

The jury is still out regarding whether direct seeding favours tungro or not. Ishii-Eiteman and Power (1997) suggested that transplanted fields, because of greater GLH inter-field movement and relatively denser populations, would offer greater risk to tungro than would direct-seeded fields, but this does not seem to be a factor. On the other hand, Shepard and Arida (1986) found that dense stands of direct-seeded rice inhibited egg parasitoid host searching, which could lead to greater GLH densities, and Kiritani (1992) suggested that direct-seeded rice would be more prone to virus infection due to its longer period in the field than transplanted rice.

Insecticide application frequency declined over the study period while usage did not (see Fig. 18.18). Highest usage was in 1983, where farmers applied a mean of 3.2 applications per crop and much more frequently during the outbreak years. Farmers in two other sites in Nueva Ecija, central Luzon, applied fewer than three applications per crop. Interestingly, the mean number of applications per crop significantly ($P \leq 0.05$) declined into the 1990s, not only in Koronadal but also in Zaragoza. Guimba continuously had the lowest levels. Prices of insecticides increased by 100%

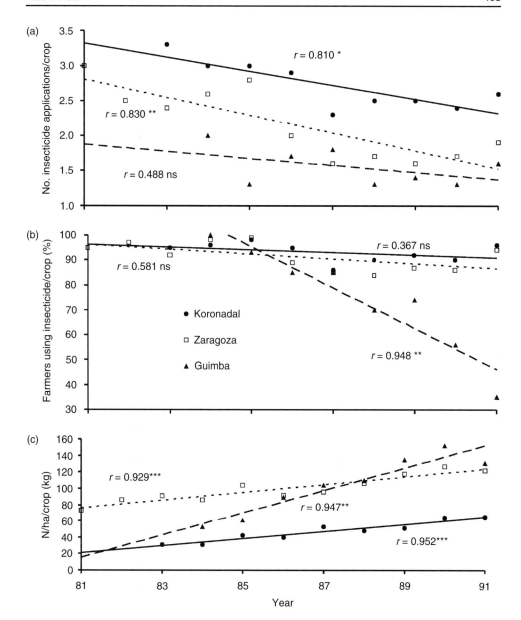

Fig. 18.18. Trends in farmers' agrochemical input usage in three irrigated double-crop rice sites, 1981–1991, Philippines. (a) Mean number of insecticide applications per crop; (b) percentage of farmers using insecticides; (c) mean nitrogen dosage per crop. Data are taken from interviews with 20 to 40 farmers per crop, Koronadal, South Cotabato, Philippines: Zaragoza and Nueva Ecija. The wet and dry season data were averaged for each year for insecticide frequency and percentage of farmers using insecticide.

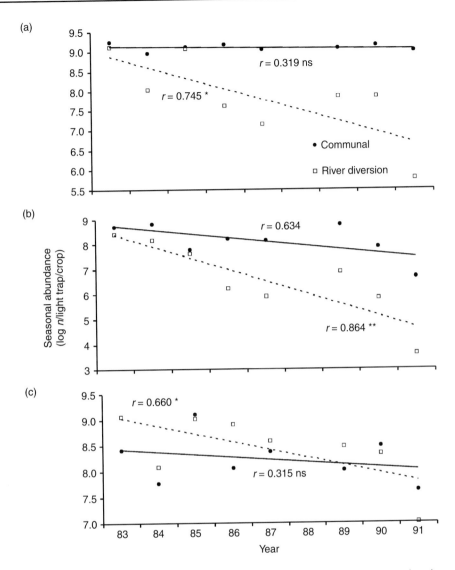

Fig. 18.19. Trend in rice hopper abundance over a 9-year period after a pest outbreak, 1979–1983. Data are seasonal totals of daily light trap collections converted to natural log scale as the means from four villages in each of two irrigation systems: communal artesian wells planted asynchronously averaging 2.4 crops per year and river diversion planted synchronously averaging 2.0 crops per year. Koronadal, South Cotabato, Philippines, 1983–1991. (a) Brown planthopper, *Nilaparvata lugens*; (b) whitebacked planthopper, *Sogatella furcifera*; (c) green leafhopper, *Nephotettix virescens*.

in 1984–1985 when subsidies were removed, which ended up having a significant effect on usage. The lower insecticide pressure in the first crop of 1990 perhaps allowed more natural enemies to act on GLH populations. The percentage of farmers using insecticides did not decline in Koronadal and Zaragoza but did in Guimba, which in 1991 was fewer than 40% of farmers. Despite the generally higher pest inci-dence in the communal irrigation areas, the mean number of insecticide applications per crop was insignificantly different (F = 0.28, df = 37, P = 0.60) over the 9-year period (2.7 in the communal and 2.8 in the river diversion systems). All of this occurred through the will of the farmers themselves, as there were no extension programmes at the time to influence them. Farmers, on the other hand, increased dosages of N fertilizer over the 10-year span, which was the same trend in all three sites, with the greatest rate of increase in Guimba.

Another pest, which should have reached epidemic proportions in Koronadal, was the white stemborer (WSB), *Scirpophaga innotata*, which reappeared in Mindanao in the mid-1980s and is further discussed in the section on delayed planting. Koronadal was spared the high population densities that have occurred in other areas of Mindanao and Indonesia (Litsinger *et al.*, 2006a), and the evidence points to the more favourable habitat for egg parasitoids where four species were found, but the key egg parasitoid/predator, *Tetrastichus schoenobii*, became unusually abundant as compared with other locations (Litsinger *et al.*, 2006b). WSB supplanted YSB, most likely because of the El Niño drought of 1983–1984. Droughts are prevalent in the Intertropical Convergence Zone climate and WSB larvae can aestivate for a year to survive them, whereas YSB lacks this ability. The asynchronous communal areas, although similar to synchronous areas in parasitoid and predator densities, proved to be a refuge for the parasitoids in the dry season and allowed continuous development that suppressed WSB densities by bridging the non-rice season.

In summary, the results showed that synchronous planting reduced most pests, with the notable exception of stemborers, and achieved significantly higher yields than the asynchronous communal irrigation systems. Planthoppers and leafhoppers as a group declined in a linear fashion from the end of the large epidemic of 1979–1983. The epidemic was finally quelled by an El Niño drought and, over the next 9 years, farmers kept ahead of new outbreaks by their continual changing of new cultivars, many of which in the later years were unauthorized breeding lines. The steady decline of pest densities after the outbreak may be the result of an increasingly stabilizing influence of natural enemies after effective insect-resistant cultivars were found and insecticide pressure decreased. Destabilization of the insect fauna leading to tungro build-up first in asynchronously planted areas was suggested by Aryawan *et al.* (1993) as being the result of constant GLH vector emigration in new plantings from adults dispersing from senescing older fields. Young plants are more vulnerable to tungro infection, incubation period is shorter and virulence is higher (see Fig. 18.20).

From 1983 to 1991 two flare-ups were documented. The first was in 1985–1986 on IR60 and IR62 with low incidence of tungro and the second on high-pest popula-tions on a breeding line '–90', which evidently was highly susceptible to insect vec-tors. Even though the pest incidence was much greater in 1990 no tungro appeared. It is interesting that in both flare-up periods all hopper pests increased in incidence at the site and there was no consistent difference between irrigation systems. Widiarta *et al.* (1990) also found that GLH and BPH have similar patterns of population growth in similar cultivation areas. Farmers in Koronadal apply insecticide more fre-quently than in other sites in the Philippines, and in the first flare-up this may have contributed to the high insect pest populations. In the second, farmers frustrated with

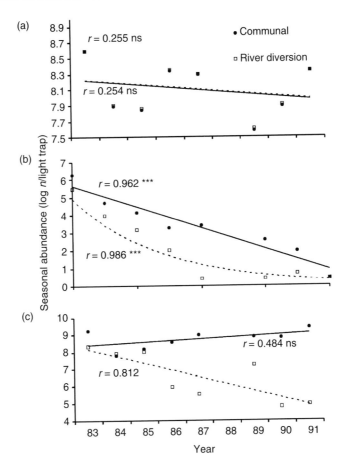

Fig. 18.20. Trend in (a) rice stemborer, (b) caseworm and (c) mirid predator abundance over a 9-year period after a pest outbreak, 1979–1983. Data are seasonal totals of daily light trap collections converted to natural log scale as the means from four villages in each of two irrigation systems: communal artesian wells planted asynchronously averaging 2.4 crops per year and river diversion planted synchronously averaging 2.0 crops per year, Koronadal, South Cotabato, Philippines, 1983–1991.

IR varieties that continued to break down started their own selection process by finding lines in local breeding trials. A new one became popular each season from 1988 onwards, and in 1990 it was '–90'.

The site is unique in a number of ways. First is the existence of some eight small communal systems where irrigation water continuously streams from artesian wells and allows farmers to average 2.4 rice crops per year where the mean standard deviation of planting dates was three times greater than the river diversion synchronous area. The communal systems are juxtaposed around the edges of the larger river diversion system, which synchronously plants two rice crops per year. Suzuki *et al.* (1992) noted that asynchronous areas often serve as sources of tungro for nearby synchronously planted areas. Farmers in both systems sow rice using a mix of three methods, further increasing asynchronous planting, especially since IR36 broke down during the epidemic. Farmers have been constantly replacing varieties and even breeding lines illegally obtained from local breeding trials. Many of the breeding lines are susceptible to the major epidemic pests and they are running a high risk.

Tungro epidemics occur only where susceptible cultivars are grown or when vector populations have become adapted to feed on previously resistant cultivars (Thresh, 1989). Thresh also documented the adaptation that occurred in southern Cotabato over the period of the study. In the 1970s, IR36 and IR42 with resistant gene *Glh6* held up until 1979 in southern Cotabato and elsewhere in the Philippines until 1983. Then IR64 with Gam Pai background came along, but then again in 1986–1987 succumbed in southern Cotabato. It became accepted that local populations could adapt to resistant varieties in a matter of a few years.

Farmers in both systems have been steadily reducing their insecticide usage, which was higher than in average sites in the Philippines after the outbreak. Fertilizer levels were low due to the inherently fertile soils, but farmers gradually increased N levels in response to nutrient depletion, which is particularly grave in double-cropped, high-yielding rice culture. They have avoided major epidemics probably through a combination of reduced insecticide usage and by rotating their cultivars almost every season during the last 4 years of the 9-year study. The asynchronous communal systems, however, by providing year-round refuge for egg parasitoids, have prevented WSB from achieving the epidemic pest status it has in other Mindanao locations and in Java.

The second areawide research project, in Nueva Ecija province, was undertaken by the Department of Agriculture, NIA and IRRI. The purpose was to evaluate synchronous planting on a large scale in a double rice crop area to complement pest-resistant cultivars as a means of reducing insecticide dependence. It was believed that synchronous planting also would have the added benefit of enhancing the field life of pest-resistant cultivars by reducing the pest population increases that magnify the probability of new genotypes developing (Gallagher *et al.*, 1994). Tungro was endemic at the tail end of the irrigation system but there was not the same urgency as was evident in Koronadal, as few farmers were affected. The 200,000 ha Upper Pampanga River Integrated Irrigation System (UPRIIS) supplies water to the largest rice bowl in the Philippines. The system also generates hydroelectric power for the national grid that supplies Manila.

As in many irrigation systems, there is not enough flow to irrigate the whole system at once, thus upstream farmers take first choice. As a result, asynchrony becomes prevalent downstream. The delay in arrival of water downstream is exacerbated by a breakdown in the irrigation infrastructure from poorly sited or unfinished canals and tertiary farm ditches, faulty control structures and reduced flows due to siltation, as well as from excessive use by farmers upstream. Drainage difficulties also affect tail-enders, making cultivation impossible in some low-lying areas in the wet season, while in less severely affected sections farmers are forced to delay planting significantly.

The research project proposed a system of dividing the command area into planting zones that would be sown sequentially and of dimensions that would allow irrigation water to reach broad blocks of land of a predetermined size (Loevinsohn, 1993). One only needs to minimize 'ecologically significant' asynchrony, i.e. variation in the time of planting within a species' dispersal range, great enough for it to be able to complete an additional generation. Holt and Chancellor (1997), for example, found by modelling tungro epidemiology that only implementing slight improvements to achieve greater synchrony of planting had large benefits. That the relationship between planting date and density is exponential suggests that efforts should be directed in the first instance to areas where synchrony is the greatest. The benefits from increased coordination outweigh the costs, as coordination can be achieved at relatively low cost. Irrigation managers regularly stagger releases to minimize water requirements where the supply is limiting in order not to strain the system's delivery and drainage capacity.

The method illustrated in Fig. 18.21 shows the proposed irrigation schedules for the UPRIIS Zones II and III of District III, representing approximately 20,000 ha. The parameters for the width of each 'wave' are a 3-week span within a radius of 5 km, which would permit no more pest generations to develop within the distance traversed by 80% of pests. The arcs are separated by approximately 2.5 km and 10 days of variation in planting date within each consecutive 'wave'. Thus it is not necessary to plant in 'as short a time as possible over as wide an area as possible', as was recommended in the absence of scientific foundation.

Less vagile and longer-lived pests such as YSB would be all the better controlled by such a schedule. This is a formula, therefore, for extreme synchrony yet, within the entire area, planting would take approximately the same amount of time (10 weeks) that it normally does in the wet season. The principal difference is that staggering is arranged in a more ordered fashion, with extreme differences in planting date separated by as large a distance as possible. The scheme is based on the principle of 'tail-first' irrigation (Wickham and Valera, 1978), a measure designed to improve the equitability of water distribution, but the pest suppression effect would be similar were planting to commence in the upstream areas, as at present.

The staggered schedule would also alleviate most of the constraints enumerated in the following section, as planting dates follow an ordered progression over a limited area for each 'wave'. The implementation experience was discussed in detail by Loevinsohn *et al.* (1993). Meetings were held with farmers who were accepting of the synchronization schedule by 'waves'. At the beginning of the project, meetings were held between project staff, farmers and NIA engineers separately. Joint meetings were held once a plan had been drafted. These first meetings were at the local level with field NIA engineers, who brought out long-standing issues in a 'no one wins Catch 22' scenario. Farmers were supposed to pay for the water each season but few did, as they felt that the delivery services were inadequate, and NIA used the meetings to reinforce the need for farmers to pay their irrigation fees. These fees now were to cover not only operational and maintenance costs but also reimbursement of the World Bank loan with which the system had been constructed. Farmers were unaware of this new policy.

Farmers wanted NIA to engage in more drainage works, as the canals that made up the irrigation system dammed the natural drainage ways, which meant that some 20% of the land area in the wet season could not be planted due to flooding. The local engineers were swayed by the argument that, because of the drainage problems and the poor condition of the canal structure, which had not been maintained, farmers could not earn enough profit to pay the fees. If synchronized planting were carried out, yields should increase and farmers would be in a better situation to pay.

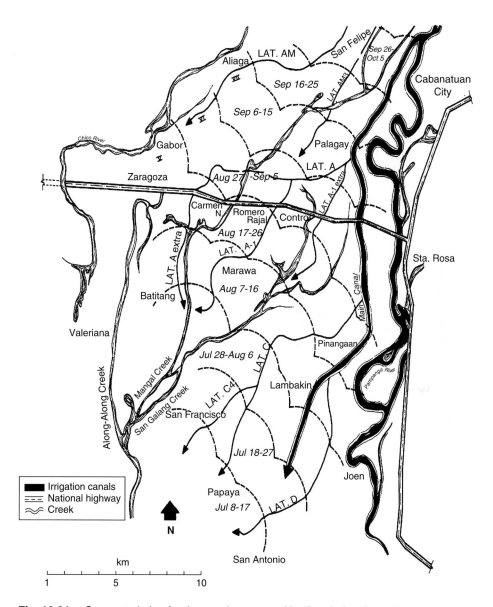

Fig. 18.21. Suggested plan for the synchronous cultivation during the wet season relying on tail-first irrigation for the 17,000 ha of Zones II and III of District III of the Upper Pampanga River Irrigation System, Nueva Ecija, Philippines. The plan is based on the biological characteristics of brown planthopper; areas planted more than 3 weeks apart should be separated by at least 5 km to reduce the number of generations that a given amount of rice will support. The tail-end of the irrigation system would be planted first over a 10-day interval, 8–17 July. Each successive area would be planted 10 days later than the preceding one. The width of each planting area is 2.5 km.

Later meetings were held with NIA supervisors at the system level, and they used the occasion to make demands that farmers pay their fees immediately, and that those in arrears would face police action. Furthermore, agricultural loans released by the banks would deduct the money owed. Because of the hard line pursued by NIA, the project was no longer able to make any progress. In addition, it was noted that the engineers were powerless to schedule water release because this was dictated by the electrical needs of the Manila grid, which could not be known ahead of time. In a conflict between farmers' irrigation needs and the electrical needs of Manila, political realities stated that the latter would be served. Farmers were unorganized and lacked political clout.

Constraints

The experience in UPRIIS brought out many impediments to the implementation of areawide programmes. During the above-mentioned project, IRRI hired an anthropologist, Grace Goodell, who spent a year living in a village within the project area to document the implications of the planned synchronous planting scheme from the farmers and local institutional perspectives and to draw conclusions regarding its viability. Much of this section draws on her experience (Goodell *et al.*, 1982; Loevinsohn, 1984). Goodell (1983) pointed out that any change in the distribution and abundance of rice or the manner in which it was grown would be likely to have manifold impacts in terms of socio-economic consequences beyond altering pest populations. Experience with planning and implementing synchronous planting schemes has led to a number of observations regarding socio-economic and biological constraints during implementation, which are now elaborated. Many of the organizational requirements are summarized in Table 18.14.

One major constraint is that most national research programmes in developing countries have difficulty in field-testing areawide control methods such as synchronous planting, as experimentation on the scale required is not usually possible due to budget constraints, and survey results are often confounded by lack of replication and comparable controls (Holt and Chancellor, 1997). There are few scientific reports of field experimentation involving areawide control measures (Loevinsohn, 1984).

Research disciplines also need to coordinate. The central aim of pest managers is to control pests without excessive use of pesticides, while the central aim of irrigation managers is equitable water distribution and continual reduction of waste, reflecting the value of the water relative to the managerial costs required to reduce the waste further. In research institutes, each branch of science evolves its technology by itself and transfers recommendations up the corresponding bureaucratic channel for extension to farmers. Thus it may happen that years later, when the farmers attempt to integrate in their fields the ingredients that academics and bureaucrats have tried to keep apart, they discover the incompatible organizational forms that are inherent in the components they are told to combine.

Often, characteristically, almost all such programmes are based on one aspect of the technology – usually pest management or irrigation – rather than on a synthesis of their combined implications. Indeed, when one discipline's organizational demands for synchronous planting conflict with those of others, professionals in each field want to dismiss the interests of others as third-generational problems, i.e. problems that are less important than their own.

Table 18.14. Main components of the new rice technology and organizational requirements for large-scale irrigation systems (after Goodell, 1984).

Technology component	Reason
Requiring (or amenable to) large-scale synchronous planting (blocks > 1000 ha)	
Pest control (planting in waves feasible?)	Technical: prevent build-up of pest populations with minimal pesticide use
Irrigation	Managerial: easier for new irrigation
By lateral service areas (500–4000 ha) combining many turnout areas	authority to achieve relative efficiency in early years by delivering in large blocks
Within turnout area (30–50 ha)	Technical: pest and water control
Preferring medium-sized areas synchronously planted (blocks of 500–800 ha) for supervised credit (government or agricultural/banking firms	Managerial/economic: economies of scale in programmes' delivery of services, given limited resources
Preferring staggered planting (by 0.5–50 ha units)	
Rural labour	Economic: to achieve even spread of employment throughout the year
Chemical inputs: government or private suppliers	Managerial/economic: limited storage facilities, etc.
Buffalo distribution and farm machinery: users, owners, operators	Economic: for even spread of demand seasonally
Rice buyers and transport: government or private	Economic: for even spread of demand for limited resources
Small village and provincial middleman services (moneylending, etc.)	Economic: for even spread of demand for limited resources
Requiring staggered planting (by units of 30–50 ha or more) for irrigation	Technical: water conservation Social: equity of resource

One conflict between irrigation management and pest control is the water requirements for land preparation. The crucial stage of irrigated rice cultivation in Asia consumes 40% of the crop's total water needs, as the soil is thoroughly saturated then ploughed, puddled and harrowed until it is finely pulverized. Under optimal conditions a minimum of 6 days/ha is required to complete the process to be ready to transplant. The water delivery requirement over the entire target area of 1000 ha and within the 20-day limit is roughly four times the nominal design capacity of the main canals in modern irrigation systems. Even if the canals are overloaded, these demands push the design to its utmost limit. Providing adequate service to every plot within a 3-week period is not at all difficult during the growing season, but the enormous water requirements of land preparation make this virtually impossible to achieve with any consistent accuracy during the saturation stage.

The desired scale of synchrony to be achieved is also important. The ideal size of an area to be under synchrony is from 600 ha for GLH up to 1200 ha for BPH

(Loevinsohn, 1984). Given the average farm size in the Philippines this would require some 500 farmers, but in Java this would amount to several thousand farmers. Without farmer organizations this becomes an unmanageable task. New technology involving daylength-insensitive cultivars, and double rice cropping has liberated farmers from imposed synchrony throughout all fields across the landscape. If crop synchrony over a large area is technically and economically desirable – everyone's crop reaching the maturation stage at the same time and everyone harvesting together – then the farmers themselves now have to create coordination where previously the sun determined it for them. Furthermore, synchrony across 50-ha tertiary-level turnout blocks is quite different to synchrony across 500-ha lateral canal blocks, or across several thousand-hectare main canal-level blocks. Usually all three conflict. When synchrony is desired at any scale, all farmers planting at the same time and planting varieties of the same maturity class achieve that.

Synchrony places large demands on the water delivery skills of irrigation engineers and the systems themselves. Irrigation systems must fine-tune water delivery on a specific date to very large areas as blocks – namely to areas served by the wide lateral canals – and then cut off water on a specific date. Farmers must start taking what they need on schedule or else they may not receive enough. In South-east Asia, on average it takes 7–13 weeks to plant an area of several hundred to 1000 ha in the best-managed systems. A newly centralized irrigation administration usually needs a decade or more to master this managerial challenge, coordinating its own operations across the entire system and then coordinating with the farmers, delivering punctually and accurately, while firmly enforcing the pre-announced water cut-off dates. In this respect the irrigation and pest management components of the new technology complement each other within large irrigation systems.

A large new system aims to provide accurate scheduling and delivery lateral by lateral in 1000–4000-ha blocks; once the engineers have acquired this expertise they must begin to fine-tune management at lower levels in the system. For optimal efficiency the technology of large-scale gravity irrigation requires not that the water be turned on and then off in 1000-ha blocks, but rather that it be meted out precisely, at least down to the turnout level, to the 30–50 ha sizes, which must be given a week in each turnout before going to the next. In short, staggered planting – the nightmare of pest management – is by far the most efficient way to use a nation's irrigation resources. Irrigation by waves would deliver water to the small rotational areas sequentially, but at the same time completely service any given 1200-ha area (the minimal unit) within 3 weeks. If that could be done, the last field planted within that area would be ready within 21 days after the first field, but all planting would conform to a rotational schedule.

The Taiwanese have been unable to achieve the much less stringent target they set for themselves of achieving rotational land preparation and planting at the 50-ha level without worrying about tight synchronization at the larger level. Physical control of the system exists, as does the bureaucratic organization, but this has been insufficient to control the timing of land preparation and transplanting in the rotational areas. All the farmers want the water as soon as possible, and all transplant as soon as the land has been prepared. But the heavy demand during land preparation, combined with the limited channel capacities, results in extended and variable land preparation dates. The well-disciplined Taiwanese farmers, who have accepted relatively precise scheduling control of the water during the main irrigation season, still refuse to accept such control during land preparation and planting. Undoubtedly many socio-economic and political factors at the village level explain the failure of such carefully made plans, factors which technical perfection cannot overcome. In response, many farmers have installed their own wells within the irrigation system to give greater flexibility.

Holt *et al.* (1996) also noted that insufficient irrigation water was a constraint to the implementation of synchronous planting in many areas. This comment would apply to large irrigation systems in need of repair and unable to supply water in a timely fashion to all areas of the system. What is often lacking is not water but the articulated canal system. Synchronous planting actually makes water use more efficient in irrigation systems in good repair.

The Muda irrigation system in Malaysia began in 1970, and actual cropping periods were markedly delayed year after year due to insufficient irrigation water, delayed arrival of water to fields and farm labour shortage (Nozaki *et al.*, 1984). This led to an inefficient use of irrigation water as well as unstable yields. Low canal density, insufficient irrigation water and poor water management cause delay of irrigation water to fields. To irrigate a flat area it is necessary to keep a certain constant water level at the head of the intake in river diversion, gravity-fed systems. Due to water shortage and poor water management the water level at the head was extremely unstable.

Muda II was probably the most advanced irrigation system in South-east Asia; this government-designed, system-wide rehabilitation programme had a target planting regime of 1000 ha within 3 weeks in mind, specifically for improved pest control. As in Taiwan, the administration could count on universally mechanized ploughing, government credit arranged in advance and carefully supervised farmers' groups. Synchrony has been achieved along the lateral canals but now within the required areas. Notwithstanding the severe infestations of BPH, the government has found it politically impossible to impose water cut-off dates on farmers, which apparently will be necessary in assuring their cooperation.

Most of the examples given in the following section are those imposed by governments, with farmers taking a passive role. We saw no examples of farmer' implementing synchronous planting on their own initiative. This is because governments and not farmers are normally in charge of the irrigation systems. Strong farmers' organizations would be needed to keep their members from planting outside of the scheduled periods. In areas where farmers' organizations are weak implementation is more difficult, as it is the farmers themselves who should make plans for synchronized planting, with agreed-upon rules and enforcement in place.

Synchronous planting, because it involves so many farmers whose needs come at the same time, strains local resources during crop establishment. Farmers often delay in planting because of a lack of credit, as the paperwork necessary to receive a loan takes time to process and banks can become overwhelmed. Agro-inputs represent an enormous expense to a farmer who requires agricultural loans. Most governments consider supervised credit indispensable to the new technology. The reality of administering credit to tens of thousands of farmers requires certain forms of organization at the field level (Goodell, 1984b). First, it is far easier to deal with a group of farmers than individuals, if nothing else but to simplify administration. Group crop management permits an extensionist or bank representative to review the need, examine and advise on the products with greater care to see whether they are appropriate. Many countries and banks lack the numbers of personnel to be able to carry this out; most have low managerial skills that would prohibit the supervision of 1000-ha blocks. At present, the paperwork needed to request a loan that will ensure timely delivery of inputs is a constraint where economies of scale seem to decline seriously after several hundred hectares, even in the most professionally managed programmes.

Because provincial agricultural supply dealers and moneylenders (including those in the village) have much less capital than government programmes, synchronous planting and synchronous harvesting strongly favour government monopolies in credit for small farmers. Yet when sufficient competition exists, village and provincial middlemen are far better managers than their government counterparts and, in many regions, poor repayment rates to government programmes signal the farmers' preference for the traditional sector's quicker and more tailor-made loans. The rapid turnover of small amounts of capital drawn only when needed from local moneylenders, suppliers and rice buyers – which staggered planting allows – is a more efficient use of this scarce resource than the government 'packaged credit' that synchronous planting more or less requires.

Agro-inputs are provided either through the private sector (which cannot afford to keep large inventories) or through the government (which cannot provide the right items where they are needed, on time and on a large scale). Synchronous planting makes large demands on a region's present supply systems, even in centrally located rice bowls, often causing crop losses that may outweigh the gains. Demand also strains input suppliers who need to pre-order sufficient fertilizers and other inputs for delivery within a short timespan.

Another potential bottleneck is the availability of animal power and the repair services for farm machinery. Once farmers double-crop it is usually uneconomical to maintain draught animals. There is difficulty in finding fodder, especially in large rice bowls where only small, weedy lands are available, and thus they must travel great distances to find better lands. Small hand-tractors and even mechanical threshers are owned or rented. Renters, however, are at a disadvantage with synchronous planting as the window for business becomes smaller. Concentrating demand over a short period may drive up the price of renting tractors and labour to establish the crop (Oka, 1988). Many would probably go out of business, and the smaller landholders would be forced to purchase hand-tractors, which may be uneconomical. Larger businesses would move in at the expense of family operators, who would lose business.

There is also a series of bottlenecks that can appear at the village level. One of the most critical constraints is the supply of landless labourers to do the extensive transplanting and later harvesting within the required period (less than 2 weeks). Even if they were present and could provide all the labour needed, this would leave them jobless for many months. In contrast, staggered planting enables them to move about within close proximity to their home, finding employment in planting for several months, after which they can begin the harvesting cycle. Farmers can forego hiring daily labour by direct seeding, thus making them jobless. In Malaysia the reaction to the labour shortages for transplanting was for farmers to seed directly

(Nozaki *et al.*, 1984). The landless make up as much as 50% of the rural population in regions like Java. A tight schedule may force farmers to seed directly, thus depriving labourers of earning potential or causing farmers to look outside the village for labour (e.g. contractual migratory labour gangs) (Loevinsohn *et al.*, 1993). This is particularly important in areas where the landless transplant and thus secure their 'right to harvest', usually 1/6th or 1/7th, of all they harvest (Loevinsohn *et al.*, 1993). Even neighbouring farmers whose fields are too small to provide for their food needs must seek local work to supplement their income.

Synchronized planting leads to synchronized harvesting and drying, thus straining local harvest labour and storage facilities. Drying facilities and storage areas are limited (Oka, 1988). The sudden supply of rice (glut) often decreases the price to farmers (Loevinsohn *et al.*, 1993).

There are also increased demands on transport services to take the harvest to buyers, even when both the public and private sectors combine to meet demand. Private sector trucking business requires 45 truck-days in the best of weather simply to haul the harvest of a 500-ha block a mere 15 km to town, and the government would take 50% longer. When hauling and harvest sales are left to the initiative of the farmers themselves and the provincial entrepreneurs of all sizes, this bottleneck can be resolved over a matter of several years.

Synchronous planting is not always compatible with natural biocontrol (Lim, 1970a; Rombach and Gallagher, 1994; Way and Heong, 1994; Settle *et al.*, 1996). Although the ultimate limitation to the densities of rice pests appears to be set by host plant availability, natural enemies play a major role in their dynamics. As we saw with WSB and its egg parasitoids in Koronadal, there are strong ecological arguments in favour of asynchrony for the benefit of natural enemies. A number of researchers have noted that, although creating a rice-free break in the cropping pattern is very effective in controlling viral diseases, BPH populations often increased to levels that caused hopperburn. Wada and Nik (1992), in the Muda irrigation scheme, documented high BPH populations in the wet season crop, peaking in the second generation planted after a 1–2 month dry fallow. It was concluded that the fallow depressed not only BPH but also its natural enemies. Immigrant BPH arrived in a crop depleted on natural enemies and with its high reproductive potential quickly built up on susceptible varieties. Natural enemies began to overtake BPH by the third generation of the wet season crop and continued on the rapidly established dry season crop. Widiarta *et al.* (1990) corroborated this phenomenon along the northern coast in western Java in studies that compared synchronous and asynchronous sites, also on pest-susceptible cultivars. By contrast, in asynchronously planted irrigation systems in Bali, natural enemies curtailed BPH population densities during the first generation without hopperburn in either crop.

Integration in IPM programmes

Let us sum up the argument to date regarding the role of areawide synchronous planting schemes as a pest management tool. Experience in the Asian countries that implemented areawide schemes that created rice-free periods indicated that both the sowing of pest-resistant varieties and over use of insecticides when combined with staggered planting that did not create rice-free periods lacked durability of pest suppression, particularly for viral diseases and their vectors. IR66, for example, lost resistance to its vector and hence to tungro within 1 year in asynchronous areas. Loevinsohn (1984) discussed the pest problems that occurred in large-scale commercial rice production ventures in the Philippines that had adopted the rice garden approach, which essentially is growing four crops per year.

The three most successful areawide schemes were in Java and southern Sulawesi in Indonesia and in the Muda Scheme in Malaysia, which were directed particularly against tungro and GLH but also negatively affected BPH. All three attributed success in suppressing the virus to synchronous planting and the creation of rice-free periods that removed the viral source. Planting a non-rice crop in the fallow period was attractive to Javanese farmers as added income and incentive. Tillage of the soil for the non-rice crop would remove sources of disease inoculum, and may be more preferable than undertaking soil rotovation and herbicide application to the ratoon with no crop being grown. Insecticide usage was limited in all three programmes to spare natural enemies. Gene rotation of resistant varieties extended their durability.

Probably less important was planting time based on historical data, as well as ensuring synchronous harvest to prolong the rice-free period. However, researchers noted that each year the planting times with least pest abundance shifted. It is well documented that leafhoppers and planthoppers in the tropics colonize from neighbouring fields and are thus not responding to a calendar date (Perfect and Cook, 1994; Chancellor *et al.*, 1996).

The main biological drawback to creating a rice-free period is the reduction of natural enemies, particularly on the first crop planted after the break. Holt *et al.* (1996) felt that synchronous planting might be warranted only in areas with a history of virus diseases. Tungro hot spots or endemic areas are well known in most countries (Cabunagan *et al.*, 2001), and forecasting programmes use them as a basis for making predictions (Suzuki *et al.*, 1992). These are usually areas with ample irrigation water and high planting asynchrony. However, as a caveat it is also known that tungro can suddenly become prevalent throughout whole regions where incidence was previously insignificant or none recorded (Thresh, 1989).

There is ample evidence that synchronously planted areas leading to long, dry fallows suffer from too few natural enemies, particularly early in the season, and become unstable due to high mortalities over the fallow period (Sawada *et al.*, 1992; Settle *et al.*, 1996). Corroborating evidence is that pest outbreaks, such as the tungro epidemic in the Muda Scheme in 1981, followed a drought that lasted from November 1980 to March 1982 (Hirao and Ho, 1987). That outbreak was thought to have been caused by natural enemy populations being decimated and, when the vectors re-established, they exploded as they were able to multiply more rapidly than the natural enemies. Way and Heong (1994) support the contention that natural enemies could suppress viral disease epidemics associated with leafhoppers and planthoppers vectors if free from insecticide perturbation, and that asynchronous planting should be encouraged in order to sustain beneficials. This may be too ambitious.

While data show that natural enemies are negatively affected by long, dry, rice-free fallows, there is no evidence that they can stop a virus disease epidemic in an asynchronous hot spot area (Bottenberg *et al.*, 1990). Transmission studies of viral diseases by vectors also argue against such a success. There is not a demonstrable incubation period of tungro in the vector, and the vector can transmit the virus after an acquisition period as short as 7–30 min. In fact, a single probing by an infected hopper can cause a seedling to become infected even on a resistant variety. Host plant recognition by virus vectors occurs after probing, and viruses may be acquired or inoculated even if the plant is later rejected as an inadequate feeding host (Sogawa, 1976), and thus many weeds can also act as carriers (Bottenberg *et al.*, 1990). *Nephotettix virescens* becomes viruliferous by feeding on virally infected host plants, including many grassy weeds, for only 5–30 min, even if the latter cannot sustain them as hosts.

In addition, Holt *et al.* (1996) concluded that a relatively small population of tungro vectors could give rise to high rates of plant-to-plant spread within the crop, so any attempts to kill vectors would have to be very efficient. Viruliferous vectors may also enter the crop from elsewhere and initiate infection over a long period, so any vector control would probably need to be repeated several times during the early stage of the crop. This is particularly important, as rice varieties have not been found resistant to insect-vectored diseases but only to the vectors. There is ample evidence that there are no durable rice cultivars against viral diseases or their vectors (Azzam and Chancellor, 2002), and it would be a lot to expect that natural enemies could curtail an epidemic on susceptible varieties.

To mitigate the problem of reduced natural enemies, synchronous planting programmes should concentrate on reducing insecticide use to only high populations of chronic insect pests. Even the use of insecticides to kill tungro vectors to prevent their spread should be avoided. Insecticide usage as practised by small-scale farmers has been shown to under-perform if applied by low-volume knapsack sprayers – control efficacies are known to average below 50% mortality of target pests (Litsinger *et al.*, 2005). Registered yield gains from more effective usage are now believed to come from the crop's ability to compensate, not only from insect injury but also from other crop stresses (Litsinger *et al.*, 2005, 2006c). With this new view in mind, farmers should concentrate on removing non-pest crop stresses and on reallocating resources to improving agronomic practices, in place of attempting to kill insect pests directly. We saw Filipino farmers doing exactly this in Fig. 18.20, where insecticide usage was declining and N usage increasing per crop. In addition, improvements in irrigation delivery and drainage that would make synchronization possible are likely to have independent and positive effects on yield (Loevinsohn, 1984).

Natural enemies can be encouraged in other ways. Settle *et al.* (1996) showed the role of non-rice-feeding aquatic insects early in the season, feeding on detritus and plankton. These are important as food to encourage natural enemy colonizers to gain advantage with the pests. Cook and Perfect (1989) believed alternative food sources were critical during the first 20 DAT to determine BPH densities later on in the crop. Settle *et al.* (1996) argue for encouraging natural enemies through landscape design that increases biodiversity by use of organic matter, such as compost or manure, to encourage food items for early-season natural enemy colonizers, as well as screening out herbicides that are detrimental to them. Encouraging vegetation to thrive on the rice bunds is another method. MacQuillan (1974) found that a systematic programme to remove vegetation from rice bunds reduced predators, contributing to BPH population growth. Conserving rice straw mulch in the dry fallow may also benefit generalist predators such as spiders. Growing a non-rice crop in the dry fallow period (Oka, 1988) would sustain generalist natural enemies to span the dry season.

In Koronadal, the results of the study indicate that the communal systems should fall in synchrony with the river irrigation system and reduce the number of rice crops to only two, and should plant a short-duration, non-rice crop such as a legume or green maize as a third crop. If WSB becomes a problem, there are two options. The first is to reduce the period of the rice-free fallow by planting longer-maturing varieties, and the second is to embark on a delayed planting scheme (Litsinger *et al.* 2006a).

The examples given show that synchronous planting is a powerful tool in quelling pest epidemics – particularly vectors of rice viral diseases. Many of the itemized socio-economic constraints outlined above would be minimized in the proposal to plant in 'waves', which is more appropriate in large-scale irrigation systems (Loevinsohn *et al.*, 1993). With farm sizes becoming smaller the need to intensify crop production remains high, and there is ample scope to encourage crop rotation during the dry season with non-rice crops under supplemental irrigation that should overcome the constraint on natural enemies. The need to conserve water during rice culture is growing, especially if a third non-rice crop is contemplated, and methods such as the system for rice intensification (SRI) show great promise in conservation of irrigation water (Uphoff, 2002). Further research should focus on whether this is the case or not.

An equitable distribution of water is of paramount importance, because if some cultivated areas receive more water than they need, a shortage results in other areas and valuable agricultural resources are wasted. Water may not be totally wasted if farmers also tap into shallow groundwater, which in large-scale irrigation systems brings up the groundwater level to shallow depths, particularly in the rainy season. Thus the role of the central irrigation system is to recharge the groundwater supply, as well as to distribute river water to the area. In order to supply water in a timely fashion to the number of fields required for synchronous planting – even in the irrigation by waves method – this may be the only practical solution.

Nozaki *et al.* (1984), keeping in mind the above issues, pointed out a number of measures to increase water use efficiency and greater synchrony of planting:

- Use of shorter-maturing cultivars.
- Creation of a systematic staggered cropping in the whole area. In each unit area, cultivars with common growth durations must be simultaneously sown.
- Separation of irrigation to seedbeds from the main fields.
- Shorter pre-saturation period, preferably 15 days for seedbeds and 20 days for fields.
- Exclusive seedbeds for the dry season.
- Recycling of irrigation water.
- Suspension of tertiary drainage construction, except in areas of very poor drainage.

Further comments on ways to minimize socio-economic constraints are warranted. For the lack of areawide research, various donors need to support such work by emulating the Japanese government, who supported many of the studies quoted herein. More disciplinary coordination will come from greater involvement by national governments in such schemes. Problems of scale are primarily directed at large-scale systems where demands on water management skills, irrigation system design and capacity need to be addressed. Irrigation systems, often due to political pressure, are required to expand their command areas to other areas not served. Often systems are running at maximal capacity, where the limiting factor is water availability. The only way to achieve such demands is to foster water use efficiency technologies, of which there are many.

Synchronous planting has been undertaken in large-scale systems, as we have seen. One of the miracles in China has been the development of sophisticated irrigation systems and associated water management, which should be used as models for the rest of Asia. They successfully irrigate three crops per year, including a non-rice crop during the fallow (NAS, 1977). Perhaps a lesson should be derived from their experiences.

Goodell (1984) neglected to include the role of rainfall in synchronizing wet season plantings. As she mentioned, land preparation presents many difficulties if farmers have to rely solely on irrigation water, as in the dry season they truly must. Rainfall, especially early rainfall, overcomes most of those problems as all farmers can begin land preparation using only rainfall. Irrigation managers need to develop methods to use rainfall in generating irrigation water savings to be able to respond to all possible rainfall patterns where the irrigation system provides supplementary water needs in the rainy season and saving water for the dry season. Farmers should be encouraged to grow a third non-rice crop in order to generate income and perhaps to afford their own pumps, as was the case in Taiwan. A pump can be shared among a group of farmers to make its use more affordable. Having a pump to tap seasonal groundwater as supplemental irrigation when the main system is running behind will do much to improve achieving planting targets. The key to this is for farmers to plant high-value crops in the dry season fallow period.

Improving farmer organizations is a key to overcoming many of the constraints as well. Participatory irrigation management was in fact developed to fill this need, and currently is becoming more popular in Asia (Mosse, 2003). Banks and local money-lenders prefer to work with cohesive farmers groups, such as exist in microcredit schemes that rely on the same principle and are a proven success. Input dealers will emerge in the private sector and create local jobs. Many will be small scale but located in the villages themselves. To make these more effective, the government must regulate them and assure quality control to prevent product adulteration. Goodell (1984) was correct to state that existing dealers cannot cope, but what one sees in many countries is the flourishing of small dealers to fill the need. Farmers often obtain credit from buyers and do not need to fill out much paperwork to obtain the loans, which they pay off in kind when they sell their crop, thus expediting the system.

Animal power to meet the needs for land preparation in double-rice crop systems is uneconomical, as proven by farmers' actions in purchasing mechanized equipment. Finding fodder is the limiting factor, as well as the labour requirements involved in animal husbandry. Small hand-tractors are more convenient and, like pumps, can be shared by groups of farmers. Experience with the landless is that they are taking over the job of weeding in the rural communities, so this gives them more work, as weeds are highly competitive with rice, causing high loss; thus, it is economical for weeds to be controlled by hand labour. If weeding is done at the early stages, weeds can be pushed into the mud and can act as a source of green manure. As most farmer families consume their own rice, the need for trucking is not as great as portrayed by Goodell (1984). In addition, much of the harvest is used to pay off labourers in kind. Most haulage is from the field to under the bed at home, and can be done with family labour.

Synchronous flowering

Rice seed bugs (*Leptocorisa* spp.) and rice stink bugs (*Nezara* spp. and *Cletus* spp.) are among the most recognized of rice insect pests by farmers, due to their relatively large size and characteristic odour. They are unique in two ways. First, they feed directly on the rice grain in the field (few rice pests do), and secondly they can actively locate fields that are in the milk or soft dough stage (most rice pests have poor host-seeking abilities). These bugs either inject their sucking mouthparts between natural openings in the grain (seed bugs) or bore through the outer coat (stink bugs) to pre-digest and imbibe carbohydrate. Farmers often rank them as among the most important rice insect pests, more so because of their large size and the fact that farmers observe them feeding on the grain, thus concluding that they are causing high losses. All rice cultivars have a similar length of the ripening growth stage (Yoshida, 1981), but double-cropping, one would think, would have increased the importance of this guild by doubling the annual food supply. Much of the damage they do, however, can be compensated by modern rice varieties (Litsinger *et al.*, 1998). The crop is only vulnerable during the milk stage, which lasts less than 2 weeks and, as a 3 t/ha rice crop has some 10,000 grains/m^2, the percentage that can be damaged is small and plants can compensate for damage. Farmers have more experience with traditional varieties that have lower grain densities per m^2, and thus seed bugs are relatively more important on these types.

There are several dynamics at work that affect their pest status. During the era of rainfed traditional rice, fallow areas were high in proportion to rice areas, and thus the alternate weedy grasses that sustained seed bugs (as they mature at different times) were plentiful in between the discontinuous periods of milk-stage rice. Traditional rices, being low-yielding, had lower densities of grains, which resulted in higher bug densities per grain and leading to relatively greater damage (Litsinger *et al.*, 1998). This was partially offset by the fact that traditional cultivars were photoperiod sensitive, and so all fields flowered during the same time, thus diluting the impact. Seed bugs also can aestivate or overwinter and are therefore more prevalent near forested areas where this takes place.

As fallows are less prevalent in double-cropped irrigated areas, the densities of alternative weed hosts have diminished, reducing overall seed bug population build-up. Their impact has been reduced since the introduction of MVs. MVs, having greater grain densities per area, tolerate greater seed bug densities. In many areas the pecky rice that results from feeding injury is not discounted at sale. In developed countries pecky rice is a much greater problem, as rice is a cash crop. Most Asian farmers consume their own rice and therefore tolerate the off-colour grains. Water delivery schedules typically irrigate large areas at the same time, tending to synchronize crop maturation. In addition, farmers tend to grow the same maturity class of cultivars in a given location, and so synchronicity of flowering between fields is often the norm.

Rice seed bugs aestivate during rice-free periods in wooded areas (Sands, 1977) that are normally removed during irrigation system construction, and so much of their natural habitat has been removed to reduce local populations. Irrigated areas, which have the greatest seed bug problems, are those where leguminous crops such as soybeans are grown in nearby fields that are alternate hosts to *Nezara* and *Cletus* spp. (Lim, 1970b; Ito, 1978). In the case of *Leptocorisa*, which has only grassy weed alternative hosts, asynchrony allows the bugs to build up (Pathak, 1968). Late-flowering fields therefore can suffer high densities as a result, due to concentration.

Therefore, it has been recommended that farmers in a rice-growing region undertake areawide management through sowing their rice so that adjacent fields flower and ripen at the same time to dilute the damage from the grain-sucking guild (Uichanco, 1921; Rothschild, 1970). There are, however, no reports of the success in implementing such a recommendation.

Delayed planting

White stemborer dominates over YSB in areas where the last-instar larvae can aestivate in the rice stubble undisturbed during long dry seasons. Its distribution overlaps the areas within the Intertropical Convergent Zone climate subject to El Niño droughts, including Indonesia and the southern Philippines (Litsinger, 2006a). WSB was first recorded in 1903 as a serious pest in Java, destroying at times tens of thousands of hectares over the following decades and was intensively studied by van der Goot (1925). With the advent of MVs its distribution became highly restricted to single-crop systems and was supplanted by YSB when double-cropping arrived as aestivating larvae that had accumulated in the wet season crop suffered high mortality during land preparation for the dry season crop. The popularity of early-maturing varieties in the mid-1980s opened the door for WSB to make a comeback in its former habitats, especially when following an El Niño drought (Litsinger *et al.*, 2006a).

WSB larvae have incredible powers of aestivation and can survive for up to 12 months in dry conditions (van der Goot, 1925). Aestivating larvae accumulate at the base of rice stubble in response to short days (conditioning response) and when they feed on rice older than the panicle initiation stage (triggering response) (Triwidodo, 1993). If rain occurs before the 4 months of the aestival diapause period is over, then mortality rates can become high due to drowning and disease. But once the larvae are predisposed, dormancy can be terminated from a minimum of 10 mm of rain. Rainfall synchronizes adult emergence in a 'stubble flight', which is normally the largest flight in the year as recorded by light traps (van der Goot, 1925). In a long dry season, the numbers of mature larvae accumulate and emerge as adults en masse, a situation that can overwhelm natural enemies in the early wet season.

Such outbreaks were characterized by large flights at the beginning of the wet season. Deadheart and whitehead damage became severe, especially in Java (Rubia *et al.*, 1996). The rice plant is particularly susceptible during panicle exsertion, leading to whiteheads (Bandong and Litsinger, 2005). Whiteheads are empty panicles that come about when the larvae sever the base during tunnelling, and grain filling is prevented.

Moth emergence, as detected in light traps, begins 17 days after the first rains and peaks 4–5 weeks afterwards. However, if the dry season was particularly long (e.g. 6 months) the stubble flight peaked earlier, at 3 weeks rather than at 4–5 weeks (van der Goot, 1925). Based on van der Goot's findings, the colonial Dutch government embarked on an areawide, legislated control programme involving delayed sowing (van der Goot, 1948). This programme, started during the mid-1920s in Java, was a successful management strategy that was strictly enforced. During this period only photoperiod-sensitive, long-maturing traditional rices were grown. Successive generations of WSB developed and, if well timed so that older larvae were present during panicle exsertion stage, whitehead incidence became very high, at times resulting in almost total destruction of the crop.

The programme was strictly enforced. Farmers were only allowed to sow their seedbeds and irrigation was withheld until after the 'suicide flight' was over. If the farmer established his seedbed before the specified time, it was destroyed by government agents. Light traps were installed to collect information on the time and abundance of moths, complemented by rain gauges to make the prediction each year. These were combined with close cooperation by the irrigation authorities (who withheld water until advice was received from the entomologists at Bogor). Like YSB, WSB is monophagous to *Oryza* spp., and thus, without their host, emerging larvae from eggs laid on grasses soon perished. This was an effective method of control, lasting from 1929 to 1941 over several thousand hectares of rice in Java each wet season (van der Laan, 1959; Kalshoven, 1981).

The reasons that areawide delayed planting worked in Java were that there was abundant irrigation water, abundant transplanting labour and that farmers were able to tolerate loss of field time caused by the delayed planting (Jepson, 1954). Similar large-scale delayed planting schemes have occurred in the rice-growing areas of northern Australia in more recent years, which are also home to WSB (Li, 1971). Farmers organized themselves in these irrigation schemes and all voted to follow the community planting schedule. The larger the area in the schemes, the greater was the control. This is the only modern-day example of areawide control known to be generated by the farmers themselves.

Areawide Mechanical and Physical Control

Rice is grown and consumed by small-scale cultivators throughout the world, many of whom have little capital for purchased inputs such as insecticides. Faced with many types of pest problems and being inventive, farmers have developed many local control methods that rely on indigenous resources such as labour and materials. Some of the most successful technologies have travelled by word of mouth but many are local in nature and still await discovery by the scientific community. They are also most practised in areas where field sizes are small (< 0.5 ha) and landless labour is plentiful and inexpensive, such as in the Indian subcontinent and Java. These methods are most effective when entire villages can be mobilized, such as in rodent control, to supply the necessary labour, which can occur in some rural societies. Often prizes are given in the Philippines for the most rats caught per farmer, and rats are also valued for their meat.

A number of labour-intensive mechanical and physical control practices lend themselves to areawide management when taken up by rural communities. The upper portions of rice plants can be clipped during the vegetative and reproductive stages and fed to livestock as fodder without harming yield, particularly in nutrient-rich bottom lands that produce more leafy plants (Tirumala Rao, 1948). This method provides control of hispa, *Dicladispa armigera* or leaf folders whose larvae feed near the leaf tips (Otanes, 1947; Alam, 1967). Hand weeding is still the mainstay of weed control in many countries. Less practised, however, is hand removal of insects. Hand removal of pests village-wide was legislated and made mandatory in Japan in the era before commercial insecticides (Miyashita, 1963). Sedentary, large and aggre-gated insect stages are the most economical when using manual methods, which may involve trapping and if carried out in a small area such as a seedbed or during a spe-cific crop age (Litsinger, 1994). There is also incentive for hand removal of pests which have food value such as orthopterans.

Hand collection of rice caseworm can still be relied upon when farmers cannot afford pesticides. Larvae float on the water surface in rolled-up leaves and are often concentrated due to field water currents or wind, when they can be netted with little effort (Nanta, 1935). Netting is an attractive option, as caseworm is damaging only during the early vegetative stages, and thus netting can be carried out with less effort. Rice hispa beetles are often hand-netted (Prakasa Rao *et al.*, 1971). The golden apple snail, *Pomacea canaliculata*, an introduced pest into many Asian countries through escapes from the home aquarium industry, is hand collected and either consumed as a tasty meal or fed to ducks after crushing the shells (Litsinger and Estaño, 1993). Farmers make shallow canals to trap them, as they are attracted to deeper water.

During the recent outbreak of WSB in Java, hand removal of egg masses became the spearhead of its control (Ooi, 1998). Extension workers rallied villages in entire irrigation schemes. Insecticides not only failed to control the pest but may have con-tributed to the cause of the problem (Triwidodo, 1993). Seedbeds may be con-structed with alleyways to allow children entry to achieve complete coverage. To make the method more effective, the collected egg masses are placed in special con-tainers that allow egg parasitoids to issue forth, while preventing neonate larvae from escaping (Otanes, 1947).

Farmers in the Orient (Nanta, 1935; Iso, 1954) remove large larvae such as skip-pers and green-horned caterpillars by hand. Farmers discovered that *Leptocorisa* rice bugs are attracted to decaying protein and make traps using snail and other dead ani-mal flesh (Otanes, 1937; Srivastava and Saxena, 1967).

The physical control method of setting out networks of light traps has been used in China and Japan to control stemborers, leafhoppers and planthoppers, under the guidance of local governments. Khan and Murthy (1995b) reported a 33% yield increase from gall midge control in India, while in China damage was said to be sig-nificantly reduced (Wang, 1931). In China over 1 million traps were set out in over 400,000 ha (NAS, 1977) and, in a second example, the highest yields ever were regis-tered in over 60,000 ha because of 31,000 traps (Chen and Wang, 1978). Decades ago in Japan, a network of 1.2 million electric light traps were set out at 1.0–1.5 m above the crop canopy (Kaburaki, 1938). It was noted, however, that although electrically powered traps attracted high insect numbers, they were expensive to operate. Inexpensive, low-voltage electric light traps can be set out near seedbeds as trap crops or during critical periods such as immediately following transplantation for stemborers (Ballard, 1923; Puttarudriah, 1945) and gall midge (Murthy, 1957) and at booting for stemborers (Ramakrishna Ayyar and Annantanarayanan, 1937). In addition, costs can be reduced by operating the traps only during the first few hours after sunset (Kaburaki and Kamito, 1929). In India, light trapping is still recommended; however, a number of reports state that light traps are neither effective nor economical for stemborers (Kondo, 1917; Shiraki, 1917; Pang, 1932; Stewart, 1934; Tirumala Rao *et al.*, 1956; Li, 1982; Litsinger, 1994) because they catch mostly males; females caught have already laid their eggs, and kill many parasitoids.

In the above examples local communities organized themselves, while for others community action was legislated by government. There are few specific reports nowadays, however, that document farmer-organized mechanical or physical control practices. In China during the era of communes, villages agreed to adopt community-wide practices and enforced their own farmers (NAS, 1977). Many villages in Java undertook hand removal of WSB egg masses from seedbeds as a result of farmer groups being trained by government extensionists (Dilts, 1990).

Areawide Host Plant Resistance

Large areas planted to insect-resistant varieties can cause a significant pest population reduction. This is validated by annual light trap collections that show areawide reductions in pest abundance after the release of insect-resistant cultivars whose basis is antibiosis rather than tolerance (Heinrichs, 1984; Fig. 18.4). Also apparent is the rise of populations that can overcome genetic resistance with time, forcing new solutions to pest problems (Gallagher *et al.*, 1994).

Durable varietal resistance would play a central role in rice IPM programmes. There is a lack of consensus, however, regarding the types of resistance and the means of using cultivars with different genes that would increase durability. It is apparent that monogenic resistance deployed in uniform varieties over wide areas is not a viable long-term strategy for most pests. Polygenic or multigenic inheritance may prove more difficult for pests to overcome and there is also evidence that moderate resistance may help to conserve natural enemies by maintaining low pest numbers.

In the absence of durable resistance, governments have embarked on schemes to rotate varieties with different resistant genes each season, as discussed in an earlier section. Limiting the exposure of new genotypes and complementing control with other measures will increase durability. Selecting cultivars with moderate resistance would also enhance durability of specific genes. The most preferred complementary measures are reduction of insecticide use to encourage natural enemies and adoption of areawide cultural controls, such as synchronous planting. Gene rotation was successfully done in Indonesia for green leafhopper and tungro control in the early

1980s (Manwan *et al.*, 1985), but there are few examples of this method being used due to the problem of having to organize farmers throughout the whole community.

Areawide Extension Programmes

In areawide programmes, particularly those run by the government, someone needs to explain the practices to farmers in order to engage their support. Irrigators' associations sometimes exist, but in the main Asian rice farmers are either unorganized or only loosely organized, very few attending regularly scheduled meetings. Extension workers are the key link between the government and the farmers, and would themselves need to be trained on the specifics of the programme.

The result has seen limited success in reaching farmers with pest control technologies of any kind (Morse and Buhler, 1997). Unfortunately, extension workers, particularly in the less well-off countries, are under-educated, under-trained, underpaid, under-budgeted and understaffed. They usually have some form of transportation, perhaps a bicycle or motorcycle, but normally they lack an adequate budget to carry out programmes on a regular basis.

Since the effort by the World Bank to organize extension workers through the training and visit system (Ganguly *et al.*, 2006), there have been few efforts to improve the system of extension delivery. In some countries, extension workers have been devolved to local governments; elsewhere, attempts are being made to privatize them. In India there has been a freeze on the hiring of government workers over the past decade, so attrition has reduced their numbers to the extent that each worker has to cover some 20 villages, representing > 20,000 farm families. The result is that often only the better-off farmers even know the name of their extension worker. Through inheritance, farms are continually being divided up among the male children, so reducing farm size to less economical units and increasing the workload for extension workers.

Some innovative extension methods have been introduced in the past few decades that offer hope in improving the rural economic sector, particularly in countries where job creation for rural workers is minimal. The farmer field school (FFS) extension method was developed in the mid-1980s in Indonesia to promote IPM (Matteson, 2000). FFS is very effective but costly and so has not been widely adopted in Asia, despite its documented success at educating farmers on using modern agricultural practices. The reason it is expensive is that the number of contact hours required to introduce new concepts is high due to the low educational attainment of most farmers. A farmer who is a high school graduate can be taught in fewer contact hours than one with only a third-grade education. Unfortunately, most farmers in developing countries are under-educated. Governments that are not willing to support rural education adequately are generally also those who do not adequately support farmer-training programmes.

There is a cruel irony in that those countries which are better off, such as Malaysia and Japan, where rural populations have high secondary education graduation rates, have declining farmer populations due to the fact that the younger generations can seek more remunerative opportunities outside of farming. The greatest need is in those nations with the highest proportion of farmers and high numbers of landless living in the villages who seek seasonal work.

Additional constraints to areawide programme implementation are the dispersed rural population, limited resources and the time to undertake training. There is a need for a more efficient delivery system to bridge the knowledge gap between what farmers know and what they need to know to make informed management decisions using modern cultivation practices, cultivars, fertilizers and pesticides. A new participatory learning method has been introduced that focuses on simple messages which could be adopted for areawide IPM technologies. In order for farmers to use scientific information underpinning sound IPM decisions, there is a need for a synthesis and distillation of research results into usable information, such as decision rules or heuristics that farmers can be motivated to test (Heong, 1998).

Over zealous Green Revolution-inspired extension programmes, as well as pesticide company promotions, have influenced farmers to develop unfavourable attitudes toward insect pests and have instilled the need to apply insecticides for higher yields in much the same mode as fertilizers are 'required' for crop growth. Pest epidemics have reinforced farmers' fear of insect pests, and many farmers spray when they observe only a few insects in the field (Heong *et al.*, 1994; Bandong *et al.*, 2002). Most Asian farmers in irrigated rice areas have first-hand experience of an outbreak in either their own field or that of a neighbour (Litsinger *et al.*, 1982). Farmers have continued their use of insecticides even when insect pest-resistant MVs are used, as they believe insecticides are necessary to prevent outbreaks. Due to weak extension services, most farmers are unaware of the utility of resistant cultivars, especially regarding which pests are being controlled. Farmers' distorted perceptions therefore have greatly influenced insecticide usage. Farmers commonly feel that: (i) all insects in their fields are pests that cause loss; (ii) any amount of plant injury translates into a concomitant loss; and (iii) insecticides are a kind of 'medicine' that helps the plant become healthy in the same way that immunization protects humans (Settle *et al.*, 1996).

Heong (1995) found that 80% of sprays from Leyte farmers were being misused, either wrong pest or wrong timing. Some 78% of farmers sprayed in the early crop stages despite low pest infestation or threat. This was not only wasteful but reduced the predator:prey ratio, posing risks of enticing secondary pest outbreaks. Despite their fear of insect pest losses, more Filipino farmers base their decisions to spray on the presence of damage or insect pests than use of prophylactic guidelines. Bandong *et al.* (2002) in central Luzon reported that decisions to spray planthoppers, leafhoppers, stemborers and other moths were based on seeing the pests in the field, while for whorl maggot, *Naranga* and *Rivula* defoliators and leaf folders, applications were based on damage symptoms. Most early-season sprays were timed with fertilizer application, which farmers believe 'softens' the plants, making them more susceptible. This latter observation turns out to be supported by research, but the increased fertility also increases the tolerance to pest damage, and so there is less need to spray (Litsinger, 1993). Other farmers spray when they see their neighbour spray, as they believe that insecticide 'protects' a field and that insects will fly to unsprayed fields. Field evidence against this belief is reported in Litsinger *et al.* (1987).

Farmers' attitudes were found more important in determining insecticide overuse than from those that had received prior training (Lazaro and Heong, 1995).

Farmers' overestimation of yield loss due to pests and their perceptions that insecticides are remedies probably account for much of the unnecessary insecticide use on rice. This thinking challenges the paradigm of herbivore impact, which assumes that the direct effect of insect feeding reduces plant fitness and yield. Waibel (1986) found that farmers overestimated the effect of leaf feeding on yield, and so minor leaf removal would stimulate an insecticide response.

Natural enemies build up from early in the crop to mid-season (Cook and Perfect, 1989; Fowler *et al.*, 1991), but early-season insecticide usage retards the natural build-up. Many studies have shown that broad-spectrum insecticides reduced the chain length of local food webs (Cohen *et al.*, 1994), caused losses of general and specific predators and delayed the build-up of natural enemy populations, slowing their recovery times (Way and Heong, 1994). The ecological costs from indiscriminate use of the most egregious insecticides, such as synthetic pyrethroids or organophosphates, have been estimated as creating an additional 4 million herbivores/ha per crop (Heong, 1998). Litsinger (2005) showed that insecticide usage, even when based on action thresholds, was mostly uneconomical due to the low kill coefficient of insecticides (Waibel, 1986) applied by knapsack sprayers and to the high tolerance of MVs to pest damage. When environmental and health costs are factored in, insecticide usage becomes even less attractive.

Surveys showed that farmers who regarded insecticides favourably were prone to spray early (before 40 DAT) and frequently (more than four times). There is experimental evidence showing that natural enemy populations early in the season are most important in containing vector populations (Holt *et al.*, 1996). This led to a plan to train farmers with the simple messages that rice can tolerate high levels of damage before significant yield loss occurs, and that insecticides are not necessary inputs in the same manner as fertilizers in obtaining high yields. Surveys in the Philippines and in the Mekong Delta, Vietnam, produced similar results (Heong and Lazaro, 1995). Those farmers who sprayed insecticide during the first 40 days were likely to spray in the later stages as well. Therefore, if farmers were taught not to spray early there would be a compound benefit. As seen in Fig. 18.11, from 45 to 80% of applications in four Philippine sites occurred before 40 DAT.

More effective and more time-consuming was the engagement of several farmers in a village to undergo a test trial, where they left a plot without spraying. After several seasons, farmers reduced the frequency of insecticide applications after seeing that there was no yield difference. Field days were organized for farmers to show their neighbours the results. In Leyte, where beforehand the majority of farmers (68 of 101, 67%) had applied insecticide during the first 40 days, this level was reduced to 11% and the mean number of sprays was reduced from 3.2 to 2.0. Attitudes changed, where formerly 77% believed that leaf insects could cause severe damage and 75% that yield loss resulted from the damage; those still holding those beliefs were now reduced to 28% and 9%, respectively. The percentage of those who believed that early-season spraying was needed dropped from 62 to 10. Similar experiments in the Mekong Delta produced comparable results.

In a village-level project conducted jointly by IRRI and the Philippine Department of Agriculture in central Luzon (Heong, 1995), both FFS and participatory learning methods were tried and both gave similar results for the single message of not applying insecticides during the first 40 days. In both extension methods, 80% of farmers stopped using insecticides versus 20% in the control. It was contended that participatory learning is much cheaper than FFS, but the comparison is not equal as FFS teaches much more than just the simple message and gives farmers a greater knowledge base to make their own pest control decisions.

A further effort used mass media alone to deliver the simple messages in Leyte (Escalada *et al.*, 1999). The two key statements were: (i) natural enemies are beneficial; and (ii) insecticides should be applied based on need. Mass media have been shown to be effective in conveying these two messages using a poster, a leaflet and a radio drama. Some 21,000 leaflets were distributed to all households, and 4000 posters were placed in village billboards, coffee shops, pesticide supply shops, government offices and stores. The radio drama was produced on cassette tape and played twice a week on local radio stations over one crop season. After that success, a third simple message was delivered: 'Spraying during the first 40 days is unnecessary'. As a result of these three messages, farmers reduced spraying frequency from 3.4 to 1.6 times per crop, and the percentage of farmers spraying at the early and late tillering and booting stages was reduced from 59, 84 and 85% to 0.2, 19 and 30%, respectively. This shows that further savings can be realized in areawide IPM extension through the delivery of simple messages via mass media to a wide audience of farmers.

Conclusion

Areawide insect pest management schemes have been carried out since ancient times in communities that have cultivated rice for some 10,000 years in Asia. Initially, villagers got together to combat various pests that had emerged to undertake often highly tedious control methods, all using local resources such as family labour for handpicking and setting out traps made of local materials. In more recent times governments, concerned about securing enough food for the nation, organized various schemes involving units either the size of local irrigation systems or the whole country to coordinate various pest control efforts, forecasting systems or extension programmes, often focusing on specific pest outbreaks or perennial occurrence of serious pest problems.

The comparative advantage of coordinating efforts over spatial scales of villages or larger units was to focus resources and educational efforts, as well as to prevent the normal pest recolonization that would take place from field to field when farmers acted individually. The most successful efforts, aside from undocumented local, labour-intensive community campaigns such as rodent or armyworm control, in terms of both involving the cooperation of large groups of farmers and producing the desired result were:

- National forecasting of rice planthopper immigration in Japan, Korea and China.
- Delayed planting to control WSB in Java.
- Locust forecasting and control in the Philippines.
- Aerial insecticide application against stemborers in Indonesia, and in other countries against a wide array of pests during outbreaks.
- Synchronous planting to control tungro vectored by leafhoppers in Indonesia and Malaysia.
- Mass media extension programmes with simple messages in Vietnam.

Synchronous flowering may have been carried out by undocumented efforts in villages to control seed bugs but is found mostly in recommendations and would be effective only if carried out in units at least the size of a village.

The longest documented areawide campaign is planthopper forecasting in eastern Asia, which has been ongoing since the early 1970s. Its purpose is to provide a warning to farmers as to when to apply insecticides against BPH and WBPH. Given the success of incorporating genetic resistance against BPH in tropical rice in Asia, it is believed that such efforts could also be accomplished for japonica rice, which would negate the need for application of so much insecticide. Chemical control lobbies in Japan have, unfortunately, prevented even such research from being carried out.

The first heavy rains of the wet season stimulate mass flights of WSB. The Dutch used this behaviour to mandate farmers in Java by law to delay planting until after the flight. As the irrigation systems were under the control of the government, this programme could be enforced and WSB was successfully controlled for 20 years after World War II. With the development of Green Revolution photoperiod-insensitive rice, WSB was controlled during land preparation of the second rice crop, then disappeared from sight until an El Niño drought in the early 1980s, combined with the adoption of short-maturing varieties, caused it to rise once again as a serious pest. Delayed planting would be a good technology to reintroduce but, like so many other areawide practices, the lack of cohesive farmers' organizations and weak extension systems prevent such measures from being adopted. In Northern Australia, where WSB is prevalent, farmers are more organized and have themselves carried out this areawide strategy successfully.

During the American occupation of the Philippines, locusts were combated by community action legislated by law. The locust problem has diminished due to the conversion of grasslands to plough-based farming, but outbursts occasionally occur, spurred on again by El Niño droughts. The Philippine government has an effective early warning system, but occasionally outbreaks occur. The local residents, however, were opposed to aerial spraying, as they prefer to collect the locusts themselves to sell in local markets as food.

Aerial spraying has been a means of choice among governments to combat epidemics of not only locusts but also rice hispa, stemborers, planthoppers and leafhoppers. The interest of the governments was to ensure sufficient supplies of rice for the nation, but in recent years it has been shown that insecticides are a poor insect control method in rice, as their use has been linked to epidemics of rice pests by the killing off of beneficials. There are also grave human and environmental hazards associated with spraying areas where people live and rivers flow that have discouraged this method in recent times.

Synchronous planting schemes to create rice-free periods in controlling insect pest and associated viral disease build-up have been carried out by government-operated irrigation systems. The rice-free period eliminated the rice host and inoculum source to quell outbreaks, especially in endemic areas. The downside of synchronous planting is that it also kills off natural enemies and pests, such as BPH, that can colonize early rice plantings and quickly multiply to outpace natural enemies; this leads to hopperburn in a number of instances. Synchronous planting should therefore be considered not as a general recommendation but to be used in endemic virus disease locations. To offset the BPH problem, rice varieties with different resistance genes can be rotated to increase durability, and natural enemies encouraged by the use of organic fertilizers, maintaining vegetation on bunds, growing a non-rice crop in the dry season and using insecticides only sparingly. Many of the socio-economic and technical problems of delivering water on a tight schedule can be overcome, but difficulties multiply exponentially as the command area scales up.

Mass media can be effectively used regionwide to transfer simple messages, such as farmers delaying application of insecticides during the first 40 days after transplanting. High rates of adoption were achieved using such cost-effective methods that overcome the problem of reaching so many farmers. Synchronous flowering has been recommended to minimize the damage by rice seed bugs, which are highly mobile and build up over a season, especially in asynchronous plantings, and which cause serious damage to late plantings. This method, along with many others, has not been implemented because of the difficulty of organizing farmers. Few countries in Asia have effective extension services and farmers have difficulty in forming functional groups that can plan and enforce efforts such as areawide management. The problem is becoming more difficult through land inheritance, resulting in smaller farm sizes each successive generation and the increasing numbers of farmers in countries lacking other opportunities besides farming.

Circumstances would indicate that, in the case of small-scale agriculture in Asia, and particularly with reference to areawide management, the input of social scientists and NGOs is needed to form a bridge between stakeholders to anticipate the conflicts in the early stages. They would work with farmers to assist in the formation of cohesive and effective organizations and be able to articulate farmers' needs to bureaucracies and scientists when such contradictions become evident.

Complicating adoption in most cases is that the size of an area where a technology would be more effective now encompasses hundreds of farmers and, in most areas, these are unorganized and would not necessarily follow a leader's advice. This is slowly changing, however, as there are many initiatives now in the rural areas spearheaded by local NGOs to organize farmers for the purposes of obtaining low-interest credit or to make irrigation more efficient. Implementation is more likely if farmers have ownership of the irrigation system, such as recommended in Participatory Irrigation Management, which is becoming more widespread in many countries (Mosse, 2003). It is also important to involve the landless, who often provide vital labour, and thus all stakeholders' voices need to be heard. In many countries landless farmers account for more than half of the population in some villages. Thus we can expect more areawide practices to be taken up as this occurs.

Farmers must collaborate as field neighbours. But almost all groupings by the government (cooperatives, extension meetings, etc.) are based on residential neighbourhoods and groupings of farmers themselves, either from this principle or from kinship. Proposing that lowland farmers must collaborate as 'field' neighbours raises a number of potential problems. From bureaucracy's point of view it is far easier for government agents to meet farmers neighbourhood by neighbourhood than according to field locations. Furthermore, field neighbours can be from different villages. Not being residential neighbours may mean they have previously clashed over boundaries, water or location of canals, etc. In addition, individual farmers may cultivate plots in several different locations. Thus coordination by field location may pose more difficult organizational requirements. The main precedent, however, is the nascent irrigators' associations. Areawide management requirements call for management of consolidated fields.

If small-scale farmers could irrigate their fields, control pests, etc. and receive technological supervision as individuals, then a straightforward marriage of technology and agricultural economics would suffice for evaluating the various components of the new technology in the research stage. But agricultural intensification requires the intensification not only of new rice technology and several of its main components – particularly areawide control, which has inbuilt requirements for farmers' organizational configurations. Unlike the organizational preferences of bureaucracies serving the small-scale Asian farmer, these organizational requirements are integral to the technology itself. Without them the method simply will not work, or its costs will greatly exceed its return.

Acknowledgements

Many locally hired project staff were responsible for the conduct of the research in Zaragoza, Nueva Ecija and Koronadal, south Cotobato, and their invaluable contributions are acknowledged. Those assisting in Zaragoza were Jovito Bandong, Abraham Alviola II, Luzviminda Paladan, Roselle Paragna, Rodolfo Gabriel, Romeo Sernadilla, Danilo Romero and Catalino Andrion, while in Koronadal we thank Crispin dela Cruz, Bernard Canapi, Hector Corpuz, Joseph Siazon, Beatriz Velasco and Anita Labarinto. We would also like to acknowledge the kind cooperation of the ditch-tenders, water management technicians, division chiefs and zone and district engineers of District III of the Upper Pampanga River Irrigation System. NIA engineers P. Gamad, N. Prieto, C. Angeles and C. Malang also provided valuable information and shared their experience. We are appreciative as well of the many farmer cooperators who gave their time freely and shared with us their commitment to the scientific process, and without whose assistance these studies could not have been undertaken.

References

Alam, M.Z. (1967) Insect pests of rice in East Pakistan. In: *The Major Insect Pests of Rice*. Johns Hopkins Press, Baltimore, Maryland, pp. 643–655.

Aryawan, I.G.N., Widiarta, I.N., Suzuki, Y. and Nakasuki, F. (1993) Life table analysis of the green rice leafhopper, *Nephotettix virescens* (Distant) (Hemiptera: Cicadellidae), an efficient vector of rice tungro disease in asynchronous rice fields in Indonesia. *Researches in Population Ecology* 35, 31–43.

Azzam, O. and Chancellor, T.C.B. (2002) The biology, epidemiology, and management of rice tungro disease in Asia. *Plant Disease* 86, 88–100.

Ballard, E. (1923) An account of experiments on the control of *Siga (Schoenobius) incertellus* in the Godavari Delta. *Memoirs of the Department of Agriculture India, Entomology Series* 7, 257–275.

Bandong, J.P. and Litsinger, J.A. (2005) Rice crop stage susceptibility to the rice yellow stem borer *Scirpophaga incertulas* (Walker) (Lepidoptera: Pyralidae). *International Journal of Pest Management* 51, 37–43.

Bandong, J.P., Canapi, B.L., dela Cruz, C.G. and Litsinger, J.A. (2002) Insecticide decision protocols: a case study of untrained Filipino rice farmers. *Crop Protection* 21, 803–816.

Banerjee, S.N. and Srivastava, D.N. (1978) *Integrated Management of Rice Pests and Diseases: Proceedings of FAO Technical Consultants on Integrated Control of Rice Pests in South and Southeast Asia.* FAO, Bangkok.

Barrion, A.T. and Litsinger, J.A. (1987) *Ochthera sauteri* Cresson (Diptera: Ephydridae), predator of rice whorl maggot flies. *International Rice Research Newsletter* 12 (1), 18.

Barrion, A.T., Mochida, O., Litsinger, J.A. and dela Cruz, N. (1982) The Malayan black bug *Scotinophara coarctata* (F.) (Hemiptera: Pentatomidae): a new rice pest in the Philippines. *International Rice Research Newsletter* 7, 6–7.

Bottenberg, H., Litsinger, J.A., Barrion, A.T. and Kenmore, P.E. (1990) Presence of tungro vectors and their natural enemies in different rice habitats in Malaysia. *Agriculture, Ecosystems and Environment* 31, 1–15.

Brader, L. (1979) Integrated pest control in the developing world. *Annual Review of Entomology* 24, 225–254.

Cabunagan, R.C. and Hibino, H. (1989) Rice tungro and its vector leafhopper development in synchronized-planting areas. *International Rice Research Newsletter* 14 (5), 2.

Cabunagan, R.C., Castilla, N., Coloquio, E.L., Tingco, E.R., Truong, X.H., Fernandez, J., Du, M.J., Zaragosa, B., Hozak, R.R., Savary, S. and Assam, O. (2001) Synchrony of planting and proportions of susceptible varieties affect rice tungro disease epidemics in the Philippines. *Crop Protection* 20, 499–510.

Cassman, K.G., Gines, G.C., Dizon, M.A., Samson, M.I. and Alcantara, J.M. (1996) Nitrogen-use efficiency in tropical lowland rice systems: contributions from indigenous and applied nitrogen. *Field Crops Research* 47, 1–12.

Chancellor, T.C.B., Cook, A.G. and Heong, K.L. (1996) The within-field dynamics of rice tungro disease in relation to the abundance of its major leafhopper vectors. *Crop Protection* 15, 439–449.

Chaudhary, R.C., Khush, G.S. and Heinrichs, E.A. (1984) Varietal resistance to rice stem-borers in Asia. *Insect Science and its Application* 5, 447–463.

Chelliah, S., Fabellar, L.T. and Heinrichs, E.A. (1980) Effect of sublethal doses of three insecticides on the reproductive rate of the brown planthopper, *Nilaparvata lugens*, on rice. *Environmental Entomology* 9, 778–780.

Chen, C.C. and Wang, E-sa. (1978) Life history of and varietal resistance in rice plants to rice leaf folder, *Cnaphalocrocis medinalis* Guenée. *Taichung District Agricultural Improvement Station Bulletin* 2, 1–69.

Cheng, S., Ruan, Y., Jin, D., Cheng, G., Lin, R. and Go, D. (1980) On the integrated control of rice dwarf disease in the second crop of rice. *Acta Phytophylacia Sinica* 7, 77–88.

Cohen, J.E., Schoenly, K., Heong, K.L., Justo, H., Arida, G., Barrion, A.T. and Litsinger, J.A. (1994) A food web approach to evaluating the effect of insecticide spraying on insect pest population dynamics in a Philippine rice irrigated ecosystem. *Journal of Applied Ecology* 31, 747–763.

Cook, A.G. and Perfect, T.J. (1989) The population characteristics of the brown planthopper, *Nilaparvata lugens*, in the Philippines. *Ecological Entomology* 14, 1–9.

Dammerman, K.W. (1929) *The Agricultural Zoology of the Malay Archipelago.* J.H. de Bussy, Amsterdam.

Dilts, R. (1990) IPM in action: a 'crash' training programme implemented against the white rice stem borer outbreak in Indonesia. *FAO Plant Protection Bulletin* 38, 89–93.

Dyck, V.A., Misra, B.C., Alam, S., Chen, C.N., Hseih, C.Y. and Rejesus, R.S. (1979) Ecology of the brown planthopper in the tropics. In: *The Brown Planthopper: Threat to Rice Production in Asia.* IRRI, Los Baños, Philippines, pp. 3–17.

Erlich, P.R. (1968) *The Population Bomb.* Sierra Club-Balantine Books, New York.

Escalada, M.M., Heong, K.L., Huan, N.H. and Mai, V. (1999) Communication and behavior change in rice farmers' pest management: the case of using mass media in Vietnam. *Journal of Applied Communications* 83, 7–26.

Fernando, H.E. (1975) The brown planthopper problem in Sri Lanka. *Rice Entomology Newsletter (IRRI)* 2, 34–36.

Ferino, M. (1968) The biology and control of the rice leaf-whorl maggot, *Hydrellia philippina* Ferino (Ephydridae, Diptera). *Philippine Agriculturist* 52, 332–383.

Fowler, S.V., Claridge, M.F., Morgan, J.C., Peries, I.D.R. and Nugaliyadde, I. (1991) Egg mortality of the brown planthopper, *Nilaparvata lugens* (Homoptera: Delphacidae) and green leafhoppers, *Nephotettix* spp. (Homoptera: Cicadellidae), on rice in Sri Lanka. *Bulletin of Entomological Research* 81, 161–169.

Gallagher, K.D., Kenmore, P.E. and Sogawa, K. (1994) Judicial use of insecticides deter planthopper outbreaks and extend the role of resistant varieties in Southeast Asian rice. In: Denno, R.F. and Perfect, T.J. (eds) *Planthoppers, their Ecology and Management.* Chapman & Hall, London, pp. 599–614.

Ganguly, S., Feder, G. and Anderson, J.R. (2006) *The Rise and Fall of Training and Visit Extension: an Asian Mini-drama with an African Epilogue.* World Bank Policy Research Working Paper No. 3928, Washington, DC.

Goodell, G. (1983) Improving administrators' feedback concerning extension, training and research relevance at the local level: new approaches and findings from Southeast Asia. *Agriculture Administration* 13, 39–55.

Goodell, G.E. (1984a) Challenges to international pest management research and extension in the Third World: do we really want IPM to work? *Bulletin of the Entomological Society of America* 30, 18–26.

Goodell, G.E. (1984b) Bugs, bunds, banks, and bottlenecks: organizational contradictions in the new rice technology. *Economic Development and Cultural Change* 33, 23–41.

Goodell, G.E., Kenmore, P.E., Litsinger, J.A., Bandong, J.P., dela Cruz, C.G. and Lumaban, M.D. (1982) Rice insect pest management technology and its transfer to small-scale farmers in the Philippines. In: *The Role of Anthropologists and other Social Scientists in Interdisciplinary Teams Developing Improved Food Production Technology.* IRRI and the Division for Global and Inter-regional Projects, UNDP, IRRI, Los Baños, Philippines, pp. 25–41.

Gorbach, S., Haarring, R., Knauf, W. and Werner, H.J. (1971) Residue analysis in the water system of East Java (River Brantas ponds, sea water) after continued large-scale application of thiodan on rice. *Bulletin of Environmental Contamination Toxicology* 6, 40–47.

Hansen, G.E. (1978) Bureaucratic linkages and policy making in Indonesia: BIMAS revisited. In: Jackson, K.D. and Pye, L.W. (eds) *Political Power and Communication in Indonesia.* University of California Press, Berkeley, California, pp. 322–342.

Hayshi, K., Kubata, M. and Karima, A. (1963) Collective dusting of malathion with helicopter to the final generation of the larvae of brown planthopper. *Proceedings Kanto-Tosan Plant Protection Society* 10, 49–53 [in Japanese].

Heinrichs, E.A. (1994) Host plant resistance. In: Heinrichs, E.A. (ed.) *Biology and Management of Rice Insects.* Wiley Eastern Ltd, New Delhi, India, pp. 517–547.

Heinrichs, E.A., Aquino, G.B., Chelliah, S., Valencia, S.L. and Reissig, W.H. (1982a) Resurgence of *Nilaparvata lugens* (Stål) populations as influenced by method and timing of insecticide applications in lowland rice. *Environmental Entomology* 11, 78–84.

Heinrichs E.A., Aquino, G.B., Chelliah, S., Valencia, S.L. and Reissig, W.H. (1982b) Rates and effect of resurgence-inducing insecticides on populations of *Nilaparvata lugens* (Homoptera: Delphacidae) and its predators. *Environmental Entomology* 11, 1269–1273.

Heong, K.L. (1984) Pest control practices of rice farmers in Tanjong Karang, Malaysia. *Insect Science and its Application* 5, 221–226.

Heong, K.L. (1995) Misuse of pesticides among rice farmers in Leyte, Philippines. In: Pingali, P.L. and Roger, P.A. (eds) *Impact of Pesticides on Farmer Health and the Rice Environment*, IRRI, Los Baños, Philippines, pp. 111–148.

Heong, K.L. (1998) IPM in developing countries: progress and constraints in rice IPM. In: Zalucki, M., Drew, R. and White, G. (eds) *Proceedings Sixth Australasian Applied Entomological Research Conference*, University of Queensland, Brisbane, Australia, pp. 68–77.

Heong, K.L. and Lazaro, A.A. (1995) Relationship between farmers' early- and late-season insecticide sprays. *International Rice Research Newsletter* 20 (2), 19.

Heong, K.L., Escalada, M.M. and Vo, M. (1994) An analysis of insecticide use in rice: case studies in the Philippines and Vietnam. *International Journal of Pest Management* 40, 173–178.

Hirao, J. (1979) Forecasting brown planthopper outbreaks in Japan. In: *The Brown Planthopper: Threat to Rice Production in Asia*. IRRI, Los Baños, Philippines, pp. 102–112.

Hirao, J. and Ho, K. (1987) Status of rice pests and their control measures in the double-cropping area of the Muda irrigation scheme, Malaysia. *Tropical Agriculture Research Series* 20, 107–115.

Holt, J., Chancellor, T.C.B., Reynolds, D.R. and Tiongco, E.R. (1996) Risk assessment for rice planthopper and tungro disease outbreaks. *Crop Protection* 15, 359–368.

Holt, J. and Chancellor, T.C.B. (1997) A model of plant virus disease epidemics in asynchronously-planted cropping systems. *Plant Pathology* 46, 490–501.

International Rice Research Institute (IRRI) (1967) Entomology Department. In: *IRRI Annual Report for 1966*, Los Baños, Laguna, Philippines, pp. 179–.216.

International Rice Research Institute (IRRI) (1979) An inexpensive light trap to monitor rice insects. *International Rice Research Newsletter* 4 (2), 17.

International Rice Research Institute (IRRI) (1985) *International Rice Research: 25 Years of Partnership*. IRRI, Los Baños, Laguna, Philippines.

Ishii-Eiteman, M.J. and Power, A.G. (1997) Response of green rice leafhoppers to rice-planting practices in Northern Thailand. *Ecological Applications* 7, 194–208.

Iso, T. (1954) Rice culturing method. Prevention and extermination of disease and insect harm. In: *Rice and Crops in Rotation in Subtropical Zones*. Japan-FAO Association, Tokyo, pp. 186–190.

Ito, K. (1978) Ecology of the stink bugs causing pecky rice. *Review of Plant Protection Research* 2, 62–78.

Jahn, G.C., Litsinger, J.A., Chen, Y. and Barrion, A.T. (2007) Integrated pest management of rice: ecological concepts. In: Koul, O. and Cuperus, G.W. (eds) *Ecologically Based Integrated Pest Management*. CAB International, Wallingford, UK, pp. 315–366.

Jepson, W.F. (1954) *A Critical Review of the World Literature on the Lepidopterous Stalk Borers of Tropical Graminaceous Crops*. Commonwealth Institute of Entomology, London.

Joyce, R.J.V. (1985) Migrant pests. In: Haskell, P.T. (ed.) *Pesticide Application: Principles and Practice*. Oxford Science Publications, Oxford, UK, pp. 322–353.

Joyce, R.J.V., Marmol, L.C., Luchen, J., Bale, E. and Avantich, R. (1970) Large-scale aerial spraying of paddy in the Java CIBA-BIMAS project. *PANS* 16, 309–326.

Kaburaki, T. (1938) On the physical characteristics of the light sources of light traps for *Chilo simplex* Butl. *Oyo Dobuts Zasshi* 10, 204–207 [in Japanese].

Kaburaki, T. and Kamito, A. (1929) Attraction of the rice borer moth to lights at different periods. *Journal of the College of Agriculture Tokyo* 10, 151–158.

Kalshoven, L.G.E. (1981) *Pests of Crops in Indonesia* [revised by P.A. van der Laan and G.H.C. Rothschild] Ichtiar Baru, Jakarta, 701 pp.

Kenmore, P.E., Cariño, F.O., Perez, C.A., Dyck, V.A. and Guttierrez, A.P. (1984) Population regulation of the rice brown planthopper (*Nilaparvata lugens* Stål) within rice fields in the Philippines. *Journal of Plant Protection in the Tropics* 1, 19–37.

Kenmore, P.E., Heong, K.L. and Putter, C.A.J. (1985) Political, social and perceptual factors in integrated pest management programmes. In: Lee, B.S., Loke, W.H. and Heong, K.L. (eds) *Integrated Pest Management in Malaysia*. Malaysian Plant Protection Society, Kuala Lumpur, pp. 47–67.

Kenmore, P.E., Litsinger, J.A., Bandong, J.P., Santiago, A.C. and Salac, M.M. (1987) Philippine rice farmers and insecticides in thirty years of growing dependency and new options for change. In: Tait, J. and Napompeth, B. (eds) *Management of Pests and Pesticides: Farmers' Perceptions and Practices.* Westview Studies in Insect Biology, Westview Press, London, pp. 98–108.

Khan, M.Q. and Murthy, D.V. (1955a) Some observations on the rice stem borer (*Schoenobius incertellus* Wlk.) in Hyderabad State. *Indian Journal of Entomology* 17, 175–182.

Khan, M.Q. and Murthy, D.V. (1955b) Some observations on the rice gall fly *Pachidiplosis oryzae* (W.M.). *Journal of Bombay Natural History Society* 53, 97–102.

Khush, G.S. (1984) Breeding rice for resistance to insects. *Protection Ecology* 7, 147–165.

Kiritani, K. (1979) Pest management in rice. *Annual Review of Entomology* 24, 279–312.

Kiritani, K. (1992) Prospects for integrated pest management in rice cultivation. *JARQ* 26, 81–87.

Kisimoto, R. and Rosenberg, L.J. (1994) Long-distance migration in delphacid planthoppers. In: Denno, R.F. and Perfect, T.J. (eds) *Planthoppers: their Ecology and Management*, Chapman & Hall, London, pp. 302–322.

Kisimoto, R. and Sogawa, K. (1995) Migration of the brown planthopper *Nilaparvata lugens* and the white backed planthopper *Sogatella furcifera* in East Asia: the role of weather and climate. In: Drake, V.A. and Gatehouse, A.G. (eds) *Insect Migration: Tracking Resources through Space and Time.* Cambridge University Press, Cambridge, UK, pp. 67–91.

Koganezawa, H. (1998) Present status of controlling rice tungro virus. In: Hadidi, A., Khetarpal, R.K. and Koganezawa, H. (eds) *Plant Virus Disease Control.* American Phytopathological Society, St Paul, Minnesota, pp. 459–469.

Kondo, T. (1917) The two- and three-brooded rice borers. Nagasaki Agricultural Experiment Station [extra report No. 18, 103 pp., in Japanese]. *Review of Applied Entomology A-1916* 8, 234–236.

Kuno, E. and Hokyo, N. (1970) Comparative analysis of the population dynamics of the rice leafhoppers *Nephotettix cincticeps* Uhler and *Nilaparvata lugens* (Stål) with special reference to natural regulation of their numbers. *Researches on Population Ecology* 12, 154–184.

Lazaro, A.A. and Heong, K.L. (1995) Analysis of factors influencing farmers' insecticide sprays. *International Rice Research Newsletter* 20 (2), 20.

Li, Ching-Sing (1971) Integrated control of the white rice borer, *Tryporyza innotata* (Walker) (Lepidoptera: Pyralidae), in Northern Australia. *Mushi* 45, 51–59.

Li, Li-Ying (1982) Integrated rice insect pest control in Guangdong Province in China. *Entomophaga* 27, 81–88.

Lim, G.S. (1970a) Some aspects of the conservation of natural enemies of rice stem borers and the feasibility of harmonizing chemical and biological control of these pests in Malaysia. *Mushi* 43, 127–135.

Lim, G.S. (1970b) The importance and control of *Nezara viridula* Linn. on the rice crop in West Malaysia. *Malaysian Agricultural Journal* 47, 465–482.

Litsinger, J.A. (1989) Second generation insect pest problems on high yielding rices. *Tropical Pest Management* 35, 235–242.

Litsinger, J.A. (1991) Crop loss assessment in rice. In: Heinrichs, E.A. and Miller, T.A. (eds) *Rice Insects: Management Strategies.* Springer-Verlag, New York, pp. 1–65.

Litsinger, J.A. (1993) A farming systems approach to insect pest management for upland and lowland rice farmers in tropical Asia. In: Altieri, M.A. (ed.) *Crop Protection Strategies for Subsistence Farmers.* Westview Studies in Insect Biology, Westview Press, Boulder, Colorado, pp. 45–101.

Litsinger, J.A. (1994) Cultural, mechanical, and physical control of rice insects. In: Heinrichs, E.A. (ed.) *Biology and Management of Rice Insects.* Wiley Eastern Ltd, New Delhi, India, pp. 549–584.

Litsinger, J.A. and Bandong, J.P. (1992) Response of the rice caseworm *Nymphula depunctalis* (Guenée) to insecticides. *Journal of Plant Protection in the Tropics* 9, 169–177.

Litsinger, J.A. and Chantaraprapha, N. (1995) Developmental biology and host range of *Parapoynx fluctuosalis* and *P. diminutalis* ricefield caseworms. *Insect Science and its Application* 16, 1–11.

Litsinger, J.A. and Estaño, D.B. (1993) Management of the golden apple snail *Pomacea canaliculata* (Lamarck) in rice. *Crop Protection* 12, 363–370.

Litsinger, J.A., Canapi, B.L. and Alviola, A.L. (1982) Farmer perception and control of rice pests in Solana, Cagayan Valley, a pre-green revolution area of the Philippines. *Philippine Entomologist* 5, 373–383.

Litsinger, J.A., Alviola III, A.L. and Canapi, B.L. (1986) Effects of flooding on insect pests and spiders in a rainfed rice environment. *International Rice Research Newsletter* 11 (5), 25–26.

Litsinger, J.A., Canapi, B.L., Bandong, J.P., dela Cruz, C.G., Apostol, R.F., Pantua, P.C., Lumaban, M.D., Alviola III, A.L., Raymundo, F., Libetario, E.M., Loevinsohn, M.E. and Joshi, R.C. (1987) Rice crop loss from insect pests in wetland and dryland environments of Asia with emphasis on the Philippines. *Insect Science and its Application* 8, 677–692.

Litsinger, J.A., Gyawali, B.K. and Wilde, G.E. (1998) Feeding behaviour of the rice bug *Leptocorisa oratorius* (F.) (Hemiptera: Alydidae). *Journal of Plant Protection in the Tropics* 11, 23–35.

Litsinger, J.A., Bandong, J.P., Canapi, B.L., dela Cruz, C.G., Pantua, P.C., Alviola III, A.L. and Batay-an, E. (2005) Evaluation of action thresholds against chronic insect pests of rice in the Philippines: I. Less frequently occurring pests and overall assessment. *International Journal of Pest Management* 51, 45–61.

Litsinger, J.A., Bandong, J.P., Canapi, B.L., dela Cruz, C.G., Pantua, P.C., Alviola III, A.L., Batay-an, E. and Barrion, A.T. (2006a) Rice white stem borer *Scirpophaga innotata* (Walker) in southern Mindanao, Philippines. I. Supplantation of yellow stem borer *S. incertulas* (Walker) and pest status. *International Journal of Pest Management* 52, 11–21.

Litsinger, J.A., Bandong, J.P., Canapi, B.L., dela Cruz, C.G., Pantua, P.C., Alviola III, A.L., Batay-an, E. and Barrion, A.T. (2006b) Rice white stem borer *Scirpophaga innotata* (Walker) in southern Mindanao, Philippines. II. Planting synchrony and the role of natural enemies. *International Journal of Pest Management* 52, 23–27.

Litsinger, J.A., Bandong, J.P., Canapi, B.L., dela Cruz, C.G., Pantua, P.C., Alviola III, A.L., Batay-an, E. and Barrion, A.T. (2006c) Evaluation of action thresholds against chronic insect pests of rice in the Philippines: IV. Stem borers. *International Journal of Pest Management* 52, 194–207.

Loevinsohn, M.E. (1984) The ecology and control of rice pests in relation to the intensity and synchrony of cultivation. PhD thesis, University of London, 354 pp.

Loevinsohn, M.E. (1994) Rice pests and agricultural environments. In: Heinrichs, E.A. (ed.) *Biology and Management of Rice Insects*. Wiley Eastern Ltd, New Delhi, India, pp. 487–516.

Loevinsohn, M.E. and Alviola, A.A. (1991) Effect of asynchronized rice planting on vector abundance and tungro infection. *International Rice Research Newsletter* 16 (5), 20–21.

Loevinsohn, M.E., Litsinger, J.A. and Heinrichs, E.A. (1988) Rice insect pests and agricultural change. In: Harris, M.K. and Rogers, C.E. (eds) *The Entomology of Indigenous and Naturalized Systems in Agriculture*. Westview Press, Boulder, Colorado, pp. 161–182.

Loevinsohn, M.E., Bandong, J.P., Alviola, A.L. and Litsinger, J.A. (1993) Asynchrony of cultivation among Philippine rice farming: causes and prospects for change. *Agricultural Systems* 41, 419–439.

Macatula, R.F., Valencia, S.L. and Mochida, O. (1987) Evaluation of 12 insecticides against green leafhopper for preventing rice tungro disease. *IRRI Research Paper Series* No. 128, 9 pp.

MacQuillan, M.J. (1974) Influence of crop husbandry on rice planthoppers (Hemiptera: Delphacidae) in the Solomon Islands. *Agro-Ecosystems* 1, 339–358.

Malabuyoc, L.A. (1977) A new lepidopterous pest of rice. *International Rice Research Newsletter* 2 (2), 6.

Manwan, I., Sama, S. and Rizvi, S.A. (1985) Use of varietal rotation in the management of tungro disease in Indonesia. *Indonesian Agriculture Research and Development Journal* 7, 43–48.

Matteson, P.C. (2000) Insect pest management in tropical Asian irrigated rice. *Annual Review of Entomology* 45, 549–574.

Miyashita, K. (1963) Outbreaks and population fluctuations of insects, with special reference to agricultural insect pests in Japan. *Bulletin of the National Institute of Agricultural Science, Series C* 15, 99–170.

Morooka, Y., Herdt, R.W. and Haws, L.D. (1979) An analysis of the labor intensive continuous rice reproduction system at IRRI. *IRRI Research Paper Series* No. 29.

Morse, S. and Buhler, W. (1997) *Integrated Pest Management: Ideals and Realities in Developing Countries.* Lynne Rienner Publishers, Boulder, Colorado.

Mosse, D. (2003) *The Rule of Water: Statecraft, Ecology, and Collective Action in South India.* Oxford University Press, New Delhi, India.

Murthy, D.V. (1957) Studies on the bionomics of the paddy gall-midge *Pachydiplosis oryzae* (W.M.) Mani. *Mysore Agricultural Journal* 32, 145–153.

Nanta, J. (1935) Pests of rice in Tonkin. In: Dumont, R. (ed.) *Rice Culture in the Tonkin Delta.* Societé d' Édition Geographiques, Maritimes et Colonials, Paris, pp. 395–423.

National Academy of Sciences (NAS) (1977) *Insect Control in the People's Republic of China: a Trip Report of the American Insect Control Delegation. Committee on Scholarly Communication with the People's Republic of China (CSCPRC) Report No. 2.* National Academy of Sciences, Washington, DC.

Noda, T. and Kiritani, K. (1989). Landing places of migratory planthoppers, *Nilaparvata lugens* (Stål) and *Sogatella furcifera* (Horváth) (Homoptera: Delphacidae) in Japan. *Applied Entomology and Zoology* 24, 59–65.

Nozaki, M., Wong, H.S. and Ho, N.K. (1984) A new double-cropping system proposed to overcome instability of rice production in the Muda irrigation area of Malaysia. *JARQ* 18, 60–87.

Oka, I.N. (1979) Cultural control of the brown planthopper. In: *Brown Planthopper: Threat to Rice Production in Asia.* IRRI, Los Baños, Philippines, pp. 357–369.

Oka, I.N. (1988) Role of cultural techniques in rice integrated pest management systems. *International Agricultural Research and Development* 10, 37–42.

Ooi, P.A.C. (1998) *Beyond the Farmer Field School: IPM and Empowerment in Indonesia.* International Institute for Environment and Development (IIED), Gatekeeper Series No. 78, London.

Ooi, P.A.C. and Shepard, B.M. (1994) Predators and parasitoids of rice insect pests. In: Heinrichs, E.A. (ed.) *Biology and Management of Rice Insects.* Wiley Eastern Ltd, New Delhi, India, pp. 585–611.

Ooi, P.A.C., Saleh, A.R. and Huat, Y.G. (1980) Outbreak of the white-backed planthopper in the Muda Irrigation Scheme and its control. *Malaysian Agriculture Journal* 52, 315–331.

Otanes, F.Q. (1937) The rice bug and its control. *Philippine Journal of Agriculture* 8, 463–467.

Otanes, F.Q. (1947) Pests of rice. *Philippine Journal of Agriculture* 13, 36–88.

Paik, W.H. (1977) Historical review of the occurrence of the brown planthopper in Korea. In: *The Rice Brown Planthopper.* Asian Pacific Food and Fertilizer Technology Center, Taiwan, pp. 230–247.

Pang, Hwa Tsai (1932) The rice borer problem in China. *Zeitschrift für Angewandte Entomologie* 19, 608–614.

Pathak, M.D. (1968) Ecology of common insect pests of rice. *Annual Review of Entomology* 13, 257–294.

Perfect, T.J. and Cook, A.G. (1982) Diurnal periodicity of flight in some Delphacidae and Cicadellidae associated with rice. *Ecological Entomology* 7, 317–326.

Perfect, T.J. and Cook, A.G. (1987) Dispersal patterns of the rice brown planthopper, *Nilaparvata lugens* (Stål), in a tropical rice-growing system and their limitations for crop protection. *Journal of Plant Protection for the Tropics* 4, 121–127.

Perfect, T.J. and Cook, A.G. (1994) Rice planthopper population dynamics: a comparison between temperate and tropical regions. In: Denno, R.F. and Perfect, T.J. (eds) *Planthoppers, their Ecology and Management.* Chapman & Hall, London, pp. 282–301.

Pingali, P.L., Hossain, M. and Gerpacio, R.V. (1997) *Asian Rice Bowls: the Returning Crisis?* International Rice Research Institute, Los Banos, Laguna, Philippines and CAB International, Wallingford, UK.

Prakasa Rao, P.S., Israel, P. and Rao, Y.S. (1971) Epidemiology and control of the rice hispa *Dicladispa armigera* Oliver. *Oryza* 8, 345–359.

Puttarudriah, M. (1945) Some observations made on the biology, habits and control of the paddy stem-borer (*Schoenobius incertellus* Wlk.). *Mysore Agricultural Journal* 24, 4–9.

Ramakrishna Ayyar, T.V. and Anantanarayanan, K.P. (1937) The stem borer pest of rice (*Schoenobius incertellus* Wlk.) in S. India. *Agriculture and Livestock India* 7, 171–179.

Reissig, W.H., Heinrichs, E.A., Litsinger, J.A., Moody, K., Fiedler, L., Mew, T.W. and Barrion, A.T. (1986) *Illustrated Guide to Integrated Pest Management in Rice in Tropical Asia.* IRRI, Philippines, 411 pp.

Roffey, J. (1972) *Locusta* outbreaks in the Philippines. *Acrida* 1, 177–188.

Rombach, M.C. and Gallagher, K.D. (1994) The brown planthopper: promises, problems and prospects. In: Heinrichs, E.A. (ed.) *Biology and Management of Rice Insects.* Wiley Eastern Ltd, New Delhi, India, pp. 693–709.

Rothschild, G.H.L. (1970) Observations on the ecology of the rice-ear bug *Leptocorisa oratorius* (F.) (Hemiptera: Alydidae) in Sarawak (Malaysian Borneo). *Journal of Applied Ecology* 7, 147–167.

Rubia, E.G, Lazaro, A.A., Heong, K.L., Nurhasyim and Norton, G.A. (1996) Farmers' perceptions of the white stem borer *Scirpophaga innotata* (Walker), in Cilamaya, West Java, Indonesia. *Crop Protection* 15, 327–333.

Sama, S., Hassanudin, A., Manwan, I., Cabunagan, R.C. and Hibino, H. (1991) Integrated rice tungro disease management in Southern Sulawesi, Indonesia. *Crop Protection* 10, 34–40.

Sands, D.P.A. (1977) *The Biology and Ecology of* Leptocorisa *(Hemiptera: Alytidae) in Papua New Guinea.* Research Bulletin No. 18, Department of Primary Industry, Port Moresby, Australia.

Sawada, H., Subroto, S.W.G., Mustaghfirin and Wijaya, E.S. (1991) Immigration, population development and outbreaks of the brown planthopper, *Nilaparvata lugens* (Stål), under different rice cultivation patterns in Central Java, Indonesia. In: *Migration and Dispersal of Agricultural Insects.* TARC, Tsukuba, Japan, pp. 257–267.

Sawada, H., Subroto, S.W.G., Suwardiwijaya, E., Mustaghfrin and Kusmayadi, A. (1992) Population dynamics of the brown planthopper in the coastal lowland of West Java, Indonesia. *JARQ* 26, 88–97.

Schulten, G.G.M. (1989) The role of FAO in IPM in Africa. *Insect Science and its Application* 10, 795–807.

Settle, W.H., Hartiahyo, A., Endah, Tri A., Widyastama, C., Arief, Lukman H., Dadan, H., Alifah, Sri L. and Pajarningsih, S. (1996) Managing tropical rice pests through conservation of generalist natural enemies and alternative prey. *Ecology* 77, 1975–1988.

Shepard, M. and Arida, G.S. (1986) Parasitism and predation of yellow stem borer *Scirpophaga incertulas* (Walker) (Lepidoptera: Pyralidae) eggs in transplanted and direct seeded rice. *Journal of Entomological Science* 21, 26–32.

Shiraki, T. (1917) *The Paddy Borer* Schoenobius incertellus *Walker*. Taihoku Agricultural Experiment Station, Formosa.

Singh, S.R. and Sutyoso, Y. (1973) Effect of phosphamidon ultra-volume aerial application on rice over a large area in Java. *Journal of Economic Entomology* 66, 1107–1109.

Sogawa, K. (1976) Rice tungro virus and its vectors in tropical Asia. *Review Plant Protection Research* 9, 21–46.

Srivastava, A.S. (1965) Control of gundhi bug of paddy in Uttar Pradesh during 1958. *Plant Protection Bulletin* 13, 28–29.

Srivastava, A.S. and Saxena, H.P. (1967) Rice bug *Leptocorisa varicornis* Fabricius and allied species. In: *The Major Insect Pests of Rice*. Johns Hopkins Press, Baltimore, Maryland, pp. 525–548.

Stewart, H.R. (1934) Entomology. *Report of the Department of Agriculture Punjab 1932–1933*, Part 1, 35–39.

Suzuki, Y., Astika, I.G.N., Widrawan, I.K.R., Gede, I.G.N., Raga, I.N. and Soeroto (1992) Rice tungro disease transmitted by the green leafhopper: its epidemiology and forecasting technology. *JARQ* 26, 98–104.

Taylor, L.R. (1974) Insect flight migration periodicity and the boundary layer. *Journal of Animal Ecology* 43, 225–238.

Thresh, J.M. (1989) Insect-borne viruses of rice and the Green Revolution. *Tropical Pest Management* 35, 264–272.

Tirumala Rao, V. (1948) Some observations relating to natural factors influencing the incidence of insect pests. *Madras Agriculture Journal* 35, 104–110.

Tirumala Rao, V., Perraju, A., Ranga Rao, R.V., Narayana, K.L. and Ramakrishna, R.K. (1956) The rice stem borer in Andhra State. *Andhra Agricultural Journal* 3, 209–219.

Triwidodo, H. (1993) The bioecology of white rice stem borer in West Java Indonesia. PhD thesis, Department of Entomology, University of Wisconsin, Madison, Wisconsin, 208 pp.

Uichanco, L.B. (1921) The rice bug *Leptocorisa acuta* Thunberg in the Philippines. *Philippine Agriculture Review* 14, 87–112.

Uphoff, N. (2002) Opportunities for raising yield by changing management practices: the system of rice intensification in Madagascar. In: Uphoff, N. (ed.) *Agroecological Innovation: Increasing Food Production with Participatory Development*. Earthscan Publications Ltd, Sterling, Virginia, pp. 1456–1461.

Uvarov, B.P. (1936) The oriental locust (*Locusta migratoria manilensis* Meyen 1835). *Bulletin of Entomological Research* 27, 191–104.

Van der Goot, P. (1925) Levenswijze en bestrijding van de witte rijstboorder op Java [Life history and control of the white rice-borer in Java]. *Mededelingen Institute Plziekten Buitenzorg* 66, 1–308 [in Dutch with English summary].

Van der Goot, P. (1948) Twaalf jaren rijstboorderbestrijding door zaaitijdsregeling in West-Brebes [Twelve years of rice borer control by regulating the time of sowing in W Java]. *Mededelingen Institute Plziekten Buitenzorg* 104, 1–32 [in Dutch with English summary].

Van der Laan, P. (1959) Correlation between rainfall in the dry season and the occurrence of the white rice borer (*Scirpophaga innotata* Wlk.) in Java. *Entomologia Experimentalis et Applicata* 2, 12–20.

van Dinther, J.B.M. (1971) A method of assessing rice yield losses caused by the stem borer, *Rupela albinella* and *Diatraea saccharalis* in Surinam and the aspect of economic thresholds. *Entomophaga* 16, 185–191.

Varley, G.C., Gradwell, G.R. and Hassell, M.P. (1974) *Insect Population Ecology – an Analytical Approach*. University of California Press, Berkeley, California.

Wada, T. and Nik, M.N. (1992) Population growth of the rice planthoppers, *Nilaparvata lugens* and *Sogatella furcifera*, in the Muda area, West Malaysia. *JARQ* 26, 105–114.

Waibel, H. (1986) *The Economics of Integrated Pest Control in Irrigated Rice: a Case Study from the Philippines.* Crop Protection Monographs, Springer, Berlin.

Wang, C.N. (1931) An estimation of damages caused by *Schoenobius incertellus* Wlk. and *Chilo simplex* Butl. in the Lin-Ping District, Chekiang. Technical Bulletin of the Bureau of Entomology and Phytopathology, Hangchow, China, No. 1 [in Chinese]. *Review of Applied Entomology* A-21, 104.

Watanabe, T. (1995) Forecasting systems for migrant pests. II. The rice planthoppers *Nilaparvata lugens* and *Sogatella furcifera* in Japan. In: Drake, V.A. and Gatehouse, A.G. (eds) *Insect Migration: Tracking Resources through Space and Time.* Cambridge University Press, Cambridge, UK, pp. 365–376.

Way, M.J. and Heong, K.L. (1994) The role of biodiversity in the dynamics and management of insect pests of tropical irrigated rice – a review. *Bulletin of Entomological Research* 84, 567–587.

Wickham, T.H. and Valera, A. (1978) Practices and accountability for better water management. In: *Irrigation Policy and Management in Southeast Asia.* International Rice Research Institute, Los Baños, Philippines, pp. 61–76.

Widiarta, I.N., Suzuki,Y., Sawada, H. and Nakasuji, F. (1990) Population dynamics of the green leafhopper, *Nephotettix virescens* Distant (Hemiptera: Cicadellidae) in synchronized and staggered transplanting areas of paddy fields in Indonesia. *Researches in Population Ecology* 32, 319–328.

Yoshida, S. (1981) *Fundamentals of Rice Crop Science.* IRRI, Los Baños, Philippines.

Zhou, B.H., Want, H.K. and Cheng, X.N. (1995) Forecasting systems for migrant pests. I. The brown planthopper *Nilaparvata lugens* in China. In: Drake V.A. and Gatehouse A.G. (eds) *Insect Migration: Tracking Resources through Space and Time.* Cambridge University Press, Cambridge, UK, pp. 353–364.

19 Areawide Pest Management of Cereal Aphids in Dryland Wheat Systems of the Great Plains, USA

KRISTOPHER GILES,[1] GARY HEIN[2] AND FRANK PEAIRS[3]

[1]Department of Entomology and Plant Pathology, Oklahoma State University, Stillwater, Oklahoma, USA
[2]Department of Entomology, University of Nebraska Panhandle R&E Center, Scottsbluff, Nebraska, USA
[3]Department of Bioagricultural Sciences and Pest Management, Colorado State University, Fort Collins, Colorado, USA

Introduction: Description of the Problem and Need for an Areawide Pest Management Approach

In the Great Plains of the USA from Wyoming to Texas, dryland winter wheat either is regularly grown continuously or is followed by a year of fallow in semi-arid locales (Royer and Krenzer, 2000). It has been well documented that these continuous monocultures can, over time, lead to increased levels of all types of pests (i.e. insects, diseases and weeds) (Andow, 1983, 1991; Vandermeer, 1989; Cook and Veseth, 1990; Elliott *et al.*, 1998a; Way, 1998; Ahern and Brewer, 2002; Boyles *et al.*, 2004; Brewer and Elliott, 2004; Men *et al.*, 2004). Relative to insect pests, the ephemeral nature of insect host resources in these monoculture systems is assumed to curtail the efficiency of natural enemies, leading to increased pest pressure and reduced yields (Booij and Noorlander, 1992; Tscharntke *et al.*, 2005; Clough *et al.*, 2007).

From an ecological standpoint, the absence of habitats that support natural enemies in these monoculture agricultural systems are considered a primary reason why populations of aphids such as the greenbug (GB, *Schizaphis graminum*) and the Russian wheat aphid (RWA, *Diuraphis noxia*) increase above economic injury levels (EILs) (Elliott *et al.*, 1998b, 2002a; French and Elliott, 1990a; Brewer *et al.*, 2001; French *et al.*, 2001a; Giles *et al.*, 2003; Brewer and Elliott, 2004). Economic losses associated with both GB and RWA average US$150 million annually across the Great Plains of the USA (Webster, 1995; Morrison and Pears, 1998).

Management of aphids in winter wheat has been addressed by the use of resistant cultivars (GB- and RWA-resistant wheat); however, the adoption of these cultivars has been limited. In most dryland systems, the primary management tool for suppression of severe aphid populations is the use of costly broad-spectrum insecticides, which can lead to a cycle of pest resurgence, additional applications and increased risk of insecticide resistance (Trumper and Holt, 1998; Wilson *et al.*, 1999; Wilde *et al.*, 2001; Kfir, 2002; Elzen and Hardee, 2003; Peairs, 2006).

Additionally, producers continue to be concerned with increasing weed and disease problems in monoculture wheat production systems and the costs associated with managing these pests (Keenan *et al.*, 2007a, b). All together, these difficult pest management issues have led some producers to move toward more diverse agricultural systems in an effort to reduce pest pressure, minimize inputs and risks and increase net returns (Peterson and Westfall, 1994, 2004; Lyon and Baltensperger, 1995; Dhuyvetter *et al.*, 1996; Brewer and Elliott, 2004; Keenan *et al.*, 2007a, b).

Over the past decade, changes in the US Farm Programme, primarily in the form of reduced crop price supports, have allowed producers to be more flexible in their choice of crops. These reduced price supports demand that producers incorporate efficient pest management tactics. For the typical dryland winter wheat producer in the Great Plains whose profit margin is often very low, it is essential to use innovative IPM approaches that reduce input costs, optimize production and net profits, conserve soil and non-target organisms and reduce risks to humans and livestock (Helms *et al.*, 1987; Sotherton *et al.*, 1989).

Because of the Food Quality Protection Act, inexpensive insecticides traditionally used for aphid control in wheat may not be available in the future; therefore, wheat producers will have to utilize more ecologically based management approaches in this low-profit margin crop. Because of the costs and environmental concerns associated with insecticide use in these wheat systems in the Great Plains, an areawide pest management (AWPM) strategy may be the only justifiable approach in this region.

Knipling (1980) advocated regional, or areawide, population management of pests like GB and RWA that are dispersive and ubiquitous in agricultural landscapes (Elliott *et al.*, 1998a; Vialatte *et al.*, 2006). It is theorized that if a management approach is used over a broad agricultural landscape, pests such as GB and RWA can be effectively managed by 'environmentally benign' approaches (Knipling, 1980). For GB and RWA, which continue to reach economic levels in the traditional wheat-intensive, dryland winter wheat systems, a suitable alternative management strategy should involve the utilization of suppressive forces within cropping systems and across the agricultural landscape.

One major assumption of the Cereal Aphid AWPM project was that both GB and RWA could be maintained below economic levels across a broad area when both available resistant cultivars and diversified cropping systems were utilized within a landscape. Theoretically, the combined effect of reduced aphid numbers over a broad area via resistance and the increased effectiveness of conserved biological control agents would greatly reduce the economic impact of these pests (Holtzer *et al.*, 1996; Peairs *et al.*, 2005).

Fortunately, research on aphid management in wheat systems in the Great Plains supported our assumption that diversified wheat-cropping systems support non-economic populations of aphids and help to conserve aphid predators and parasitoids (Parajulee *et al.*, 1997; Elliott *et al.*, 1998a, 2002a, b; Brewer *et al.*, 2001; French *et al.*, 2001; Brewer and Elliott, 2004). In these studies, inclusion or rotation of crops into wheat systems such as canola, millet, sorghum, clover, lucerne, cotton and sunflowers provided the diverse landscape structure and resources required to conserve aphid predators and parasitoids in wheat (Elliott *et al.*, 1994a, b, c, 1998a, 1999, 2002a; French and Elliott, 1999a, b; French *et al.*, 1999a). Clearly, the strong evidence that diversification of a farming landscape conserves natural enemies justifies the evaluation of an AWPM programme for aphids in winter wheat.

The cereal aphid AWPM programme in wheat was a multifaceted approach that included detailed sociological and economic evaluations, experimental and demonstration pest studies and a comprehensive education/outreach programme that is still ongoing. In this chapter, much of the discussion will focus on the methodology and findings from the regionwide demonstration sites. Demonstration sites, which included monoculture (continuous wheat or wheat–fallow) systems and diversified wheat production systems, were set up at the farm landscape scale and paired throughout the Great Plains region. Ultimately, economic findings from these demonstration farms will support the justification for increasing adoption of diversified farming systems. However, data on aphid, natural enemy and weed densities at paired demonstration sites provide evidence as to the mechanisms involved for AWPM of cereal aphids in wheat.

Significance of the Pest Management Problem

Dryland wheat monocultures (either continuous or wheat–fallow) dominate production landscapes in the Great Plains (see Fig. 19.1), but often lead to increased pest problems. Producers in this region are regularly faced with aphid pressure in their wheat fields, the most common and damaging of which are the greenbug and Russian wheat aphid (Kelsey and Mariger, 2002; Giles *et al.*, 2003; Mornhinweg *et al.*, 2006; Keenan *et al.*, 2007a, b).

The greenbug is considered the key pest of wheat in much of Oklahoma, Texas and Kansas because of its frequent occurrence and potential for severe damage. In the absence of natural enemies, greenbugs are capable of rapidly reproducing in these warmer locations of the Great Plains, damaging or killing wheat plants and significantly reducing yields (Kieckhefer and Kantack, 1988; Webster, 1995; Kindler *et al.*, 2002, 2003; Giles *et al.*, 2003). The GB occurs sporadically throughout Colorado and Nebraska, and will occasionally exceed EILs. In each state of the Great Plains, GB outbreaks occur somewhere every year. Less frequent regional GB outbreaks occur every 5–10 years and result in greatly reduced yields and heavy insecticide use. The combined economic losses associated with insecticide costs and yield reductions caused by the GB alone have not been calculated for the entire region, but estimates for Oklahoma, where yearly losses in wheat range from US$0.5 to 135 million, illustrate the extent of the problem (Webster, 1995). Extrapolating these losses to the entire Great Plains suggests that GB cause annual losses of US$1.5–405 million.

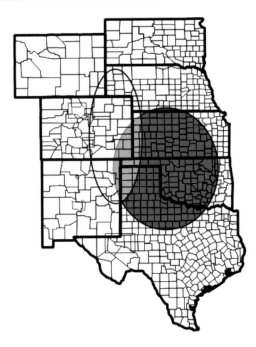

Fig. 19.1. Areas of the Great Plains, USA where RWA and GB are key pests of wheat and other cereals. Dark, GB; white, RWA; grey, the area where both species are severe pests.

Russian wheat aphid continues to be a major problem in the west-central more arid portions of the Great Plains (see Fig. 19.1) and is often the main management focus for wheat producers in this region (Archer *et al.*, 1992, 1998; Peairs, 2006; Keenan *et al.*, 2007a, b). Total economic losses associated with the RWA are estimated to have exceeded US$1.2 billion since its invasion into the USA in 1986. Seventy per cent of these losses have occurred in Texas, Kansas, Oklahoma, Colorado, Nebraska and Wyoming (Elliott *et al.*, 1998a; Morrison and Peairs, 1998).

Limitations and problems associated with current management approaches

Suppression of GB and RWA in the Great Plains has historically relied on curative insecticide use. Resistant wheat cultivars have also been used in some areas where well-adapted varieties have been developed. However, during widespread severe aphid outbreaks, insecticides are applied to prevent crop losses and are often economically justifiable (Crop Profile for wheat in Kansas, 1999; Smith and Anisco, 2000; Crop Profile for wheat in Oklahoma, 2005; NASS, 2005).

During these outbreaks, many fields are treated with compounds that are highly toxic to natural enemies and have been targeted for review by the Food Quality Protection Act (FQPA): chlorpyrifos, dimethoate and methyl parathion (Crop Profile for wheat in Kansas, 1999; Smith and Anisco, 2000; Smolen and Cuperus, 2000; Crop

Profile for wheat in Oklahoma, 2005; NASS, 2005); compounds such as disulfoton and ethyl parathion have recently lost wheat registrations.

During the course of most years, GB populations often remain near or below EILs throughout the region (Giles *et al.*, 2003). However, fields can occasionally be found where GB populations are high enough to kill most plants. These situations are usually localized in fields where natural enemies are absent. High RWA populations are a chronic problem in the more arid wheat-growing areas of the region (see Fig. 19.1), but sporadic throughout most of the Great Plains. Insecticides are the only option to control high RWA populations in fields planted to susceptible cultivars, as infestations can quickly grow and destroy entire fields.

Although severe widespread infestations in this region of the USA are infrequent, these outbreaks have significantly influenced how wheat producers perceive the importance of aphids and approach management. The results from surveys and focus groups conducted to determine producer IPM priorities in wheat (Smolen and Cuperus, 2000; Kelsey and Mariger, 2002; Keenan *et al.*, 2007a, b) indicated that a majority of producers in this region considered aphids a serious to very serious problem. This perception of a potentially serious problem does occasionally lead to an over-reaction to a marginal situation by risk-averse producers.

During non-outbreak years, many acres of wheat have been sprayed to 'protect' fields as aphid populations quickly approach or exceed economic thresholds (ETs) (Giles *et al.*, 2003; NASS, 2005). An important example of this risk-averse aphid management approach was documented in Oklahoma. During the 1995/ 1996 growing season, most greenbug populations in Oklahoma were below EILs; however, over 800,000 acres (320,000 ha, ~US$10 million in costs) were treated with insecticides to 'protect' wheat yields (Crop Profile for wheat in Oklahoma, 2005; NASS, 2005, 2006). The 1995/1996 field season in Oklahoma reinforced findings from several studies which determined that, when greenbug control efforts were geared to protect wheat grain yields independent of economic considerations, losses were closely tied to insecticide costs (Starks and Burton, 1977; Patrick and Boring, 1990; Peairs, 1990; Massey, 1993; Webster, 1995; Giles *et al.*, 2003; Royer *et al.*, 2005).

Because profit margins of dryland wheat production in the Great Plains are very small, the net benefits of regularly suppressing GB and RWA with chemical insecticides are economically questionable. For example, the yield of dryland wheat in Colorado averages 31 bushels per acre (NASS, 1996), and with the price of wheat at US$3.00 (per bushel), the net return is approximately US$25 per acre. If a producer utilizes 1000 acres (400 ha) of a 2000-acre (800 ha) farm in wheat–fallow production, the annual net income would be estimated at US$25,000. If the producer applied just one insecticide treatment at US$11 per acre, annual income would be reduced by 44%. After a single insecticide application, there is little money left to suppress other pest problems if they develop. The common approach of producers to 'protect' wheat fields from aphids with insecticides without adequate knowledge of GB or RWA density seems illogical, but this tactic is often based on the belief that accurate sampling is too expensive and on-farm risks are reduced with the treatment. Clearly, risks are unknown; however, new, highly efficient sampling plans are now available that allow for cost-effective sampling and decision making in wheat production systems (Royer *et al.*, 2007).

Despite cost concerns in dryland wheat systems, insecticides continue to be used regularly throughout the Great Plains to manage GB and RWA. The risk-averse nature of producers in this often harsh region leads to management decisions that are not focused on optimizing economic returns or other potential negative consequences of unjustified insecticide applications (Keenan *et al.*, 2007a, b). Because aphid populations in any given wheat field in the Great Plains are often below the EIL (Giles *et al.*, 2003), this 'protect' approach is likely to result in a significant waste of money. Reliance on insecticides for aphid suppression in many dryland wheat production systems of the Great Plains, without government price supports or high wheat prices, is not economically sustainable. Additionally, this over-reliance on and misuse of insecticides can significantly impact biological control and has led to other problems, including the development of greenbug populations that are resistant to compounds used for control in wheat and concerns about the conservation of migratory birds (Klass, 1982; Grue *et al.*, 1988; Shotkoski *et al.*, 1990; Flickinger *et al.*, 1991; Sloderbeck *et al.*, 1991; Brewer and Kaltenbach, 1995; Wilde *et al.*, 2001).

Despite significant research efforts, winter wheat producers in this region have at their disposal only a few available greenbug-resistant cultivars (Porter *et al.*, 1997). TAM-110 (with the *Gb3* resistance gene) confers resistance to the most abundant greenbug biotypes C, I and E (Porter *et al.*, 1997; Lazar *et al.*, 1998). An Oklahoma-adapted, general-use variety ('OKField') with *Gb3* has been available since the autumn of 2005, but does not perform well in the typical warm soils of Oklahoma or when wheat soilborne and/or spindle streak mosaic viruses are present. TAM-110 is recommended for production in drier climates (e.g. the High Plains) because it is susceptible to leaf rust and therefore is not planted in a widespread fashion across this region (Porter *et al.*, 1997).

The most significant advancement towards management of the RWA was the release of 'Halt', 'Yumar', 'Prairie Red' and 'Prowers 99', which have been followed by several other RWA-resistant cultivars. These cultivars, with the *Dn4* resistance gene, provide protection against RWA biotype 1, but are damaged by the recently described RWA biotype 2 (Peairs, 2006; Wilde and Smith, 2006). To date, there have been no resistant varieties developed with resistance to RWA biotype 2. It is important to note that GB- and RWA-resistant wheat is not immune to infestation, and damage can occur when aphid levels are extremely high; however, resistant cultivars can withstand considerably more feeding than susceptible cultivars (Quick *et al.*, 1996; Lazar *et al.*, 1998; Kindler *et al.*, 2002; Haley *et al.*, 2004). These resistant cultivars are, however, still susceptible to aphids such as *R. padi* (BCOA), which can significantly reduce forage and grain yields (Pike and Schaffner, 1985; Riedell and Kieckhefer, 1995; Riedell *et al.*, 1999; K.L. Giles unpublished data).

Native natural enemies have been shown to play an important role in regulating GB populations in wheat in the Great Plains, often eliminating the need for insecticides (Kring *et al.*, 1985; Giles *et al.*, 2003). Native natural enemies, however, had little impact on the RWA after its introduction, resulting in a multi-year, multi-state classical biological control programme initiated by the USDA to release several exotic parasitoids in the western USA (Meyer and Peairs, 1989; Michels and Whitaker-Deerberg, 1993; Wraight *et al.*, 1993; Prokrym *et al.*, 1994; Elliott *et al.*, 1995; Pike and Stary, 1995; Pike *et al.*, 1996; Brewer *et al.*, 1998a, b).

Subsequent studies demonstrated that these organisms, along with indigenous natural enemies, are usually insufficient to prevent economic damage but are a component of natural suppression of RWA throughout the region (Brewer *et al.*, 1998a, b, 1999, 2001; Michels *et al.*, 2001; Noma *et al.*, 2005; Hein, 2006). Interestingly, wheat cultivars with aphid-resistant genes have been shown to have little to no effect on parasitoids and Coccinellidae predators (Fuentes-Granados *et al.*, 2001; Giles *et al.*, 2005). These tritrophic evaluations indicate that the beneficial effects of resistance and biological control could be synergistic (Boethel and Eikenberry, 1986; Brewer and Elliott, 2004).

Even though effective IPM tools have been developed (presence/absence sampling, resistant cultivars and conservation of biological control) in the Great Plains, many growers in this area are not aware that non-chemical alternatives for aphid control in wheat can be incorporated into their production systems (Keenan *et al.*, 2007a, b). Continuing aphid problems associated with monoculture wheat systems in the Great Plains, and the resulting reliance on insecticides for GB and RWA control, highlight the urgency for development of alternative IPM systems.

Description of the Cereal Aphid Areawide Pest Management Programme in Wheat

According to Keenan *et al.* (2007a, b), a handful of growers in the Great Plains are well aware of the problems associated with traditional management of aphids in continuous or wheat–fallow monocultures. These growers utilize resistant and susceptible wheat cultivars within intensive crop rotations to reduce pest abundance (insect, disease and weeds), conserve natural enemies and conserve moisture in dryland cropping systems. These on-farm examples provide the evidence and justification for the cereal aphid areawide project, which aimed to conserve and stabilize biological control agent populations and reduce yield loss in both resistant and susceptible wheat cultivars within and among farming systems. The maximum impact of a programme based on these technologies will be achieved when it is implemented over broad geographical areas.

The main goal of this programme was to integrate effective non-chemical pest management tactics within a farm-level production setting to prevent economic GB and RWA infestations from occurring. The entire programme included detailed sociological and economic evaluations, experimental and demonstration studies, remote sensing and simulation modelling, and a comprehensive education/outreach programme that is still ongoing. As previously discussed, we will focus on the methodology and findings from the region-wide demonstration sites. These demonstration sites, which included monoculture and diversified wheat systems, were paired throughout the states involved in this study.

Ultimately, the economic findings from these demonstration farms will provide support for adoption of diversified farming systems. The data on aphid, natural enemy and weed densities at paired demonstration sites provide evidence as to the dynamics of pest systems at the farm landscape scale. The individual farm and surrounding agricultural landscape are appropriate spatial scales at which to test the programme. From a logistical and economic standpoint, individual farms were chosen as the most practical spatial unit for evaluation and implementation of IPM tactics.

At the completion of the project we hope to provide an IPM package to wheat producers in the Great Plains that will reduce yield losses caused by aphid pests and that will lower management input costs in wheat and other crops attacked by aphids. Suppression would be accomplished by incorporating host plant resistance when appropriate and the impact of biological control conserved within a diversified system. One of our main assertions was that biological control would be enhanced in diversified cropping systems. Testing this approach on monoculture and diversified farming systems over four consecutive growing seasons was one of the main objectives of the cereal aphid areawide project.

The AWPM programme and co-occurring pests

Because pests often interact at spatial scales larger than individual fields, the effect of diversifying traditional wheat farming systems in the Great Plains on non-target pests must be considered. For example, aphid pests such as bird-cherry oat aphid (BCOA) infrequently reach pest status, but are often at low levels and usually cause little damage to wheat in the region (Riedell and Kieckhefer, 1995; Riedell *et al.*, 1999; K.L. Giles unpublished data).

We expected the impact of diversification to further reduce damage by BCOA and other minor aphid pests. The wheat stem sawfly, *Cephus cinctus*, is restricted as a pest to the northern edge of the region evaluated for this study. Host plant resistance and cultural practices are the main tactics used to control *C. cinctus*. In wheat production systems where tillage is reduced, increased sawfly populations are more likely; however, diversity would be expected to reduce its significance as a pest (Hatchet *et al.*, 1987). Armyworms and cutworms are sporadic pests of small grains in the region, and we anticipated that diversified cropping systems would have little effect on these organisms.

Other arthropod pests such as the wheat curl mite, which transmits wheat streak mosaic virus (WSMV), were considered. WSMV is the most serious cereal disease in the western Great Plains (Brakke, 1987), with widespread epiphytotics occurring every few years. The WSMV situation is complicated by the recent prevalence of High Plains virus (HPV), which is also damaging to wheat and probably interacts with WSMV to impact wheat more severely.

Management of the disease involves managing the mite vector. These mites can survive only on green plant material; therefore, management must focus on reducing mite populations during the period when it must survive between wheat harvest and the subsequent wheat crop (i.e. green bridge period). Volunteer wheat is the most important green bridge host for the mite and virus. Crop diversification with crops that are not hosts to the mites will probably reduce the incidence of the mite and virus unless volunteer wheat is not controlled well in these crops. However, crop rotation with host crops (e.g. maize, foxtail millet) needs to be considered with caution. Delayed planting also reduces the risk of serious WSMV. WSMV/HPV disease was monitored during the programme.

Additionally, producers continue to be concerned with increasing weed problems in monoculture systems and the costs associated with managing these organisms (Boyles *et al.*, 2004; Keenan *et al.*, 2007a, b). Jointed goatgrass, downy brome, volunteer rye and volunteer wheat constitute the most serious weed threats to winter wheat production in the Great Plains. Annual grass weeds reduce wheat yields and cost wheat producers about US$20 million annually in Colorado (Anon., 1990); similar losses occur elsewhere in the Great Plains. Widespread adoption of reduced tillage farming has aided the establishment and spread of annual grass weeds (Anon., 1991).

Winter annual grass control in continuous and wheat–fallow systems is extremely difficult, because the life cycle of grasses is synchronized with that of winter wheat, and few cost-effective available herbicides provide selective grass control in winter wheat. Kochia (*Kochia scoparia*) is the most common spring-germinating annual weed in winter wheat in the Great Plains and has rapidly developed resistance to the primary control strategy (sulphonylurea herbicides). Surveys indicate that over 50% of kochia in dryland sites is herbicide resistant (Westra and Amato, 1995).

Diversification of farming systems by rotation of a second crop will allow for cheaper, less chemically intensive control of grassy weeds and kochia (Lyon and Baltensperger, 1995; Westra and Amato, 1995). Rotation allows for grassy weed germination in a non-grass crop that is highly competitive and allows for use of herbicides that will not damage the non-grass crop. More effective kochia control is possible in the rotational crop by using alternative herbicides (Tonks and Westra, 1997). Because of selective and targeted herbicide use, we anticipate significant reductions of weeds in rotational diversified systems.

Anticipated benefits of Areawide Pest Management

The GB and RWA thrive in the monoculture wheat systems, and other pest problems in general have increased in this system (Way, 1988; Andow, 1991; Lyon and Baltensperger, 1995; Holtzer *et al.*, 1996). Diversification of crops within a production system can have several desirable consequences for farmers. One of the well-documented benefits of diversification is lower insect pest pressure, and evidence is accumulating that diversifying cropping systems increase and support natural enemy populations, and consequently increase the effectiveness of biological control (Parajulee and Slosser, 1999; Guerena and Sullivan, 2003; Brewer and Elliott, 2004). Furthermore, when aphid-resistant wheat cultivars are incorporated into a diversified system, the combined effect of natural enemies and host plant resistance can be interactive, resulting in a reduced probability of aphids reaching EILs (Brewer and Elliott, 2004). Additionally, through crop rotations, these diverse systems can also allow for effective weed management and decreased disease levels (Blackshaw *et al.*, 1994, 2001; Wilson *et al.*, 1999; Boyles *et al.*, 2004). Results from the Kelsey and Mariger (2002) survey and the Keenan *et al.* (2007a, b) focus groups of wheat producers both clearly indicated that suppression of grassy weeds is the most important concern of producers in the Great Plains.

Crop diversification via intensive crop rotation also has agronomic and environmental benefits because, in many systems, rotational crops are increasingly grown no-till, leading to increased water use efficiency and reduced soil erosion (Peterson and Westfall, 1994, 2004; Peterson *et al.*, 1996). Long-term studies confirm that intensive

rotations help to stabilize or increase farm net returns and reduce financial risk compared with monoculture wheat systems. For example, in Colorado annualized grain
production from 1987 to 1993 in dryland wheat–maize–fallow and wheat–maize–
millet–fallow was 72% higher than for wheat–fallow, with a 25–40% increase in net
annual income (Dhuyvetter *et al.*, 1996). However, rotational options in the western
Great Plains are driven by water availability: drier areas will have fewer rotational
options, and this can greatly affect the income potential in these areas (Lyon *et al.*, 2004).

At the beginning of the project we anticipated that if cereal aphid AWPM was fully
implemented, the direct economic benefits of reducing aphid densities in wheat would
average US$75 million per year and that indirect benefits would exceed US$150 million,
for a combined annual total of US$225 million. These figures were based on: (i) expected
reductions in average aphid density across the Great Plains; (ii) documented relationships between aphid numbers and yield loss; (iii) reductions in costs associated with
insecticide use in wheat systems; and (iv) reduced impact of other pests in farming systems. Benefits (~US$102 million) were also expected to result from increased profits
from diversified crop rotations. For example, Boyles *et al.* (2004) suggest that rotations
of winter canola with wheat result in 15% greater wheat grain yields compared with
continuous systems. Difficult to estimate, but clearly important, are the additional
long-term potential benefits of stabilizing farm economies and reduced soil erosion.

Designation of Demonstration Sites and Evaluation Methodology

During initial planning sessions, participants from each state (see Box 19.1) determined that programme evaluation would be conducted at three levels. First, economic data from surveys was collected from a broad pool of producers in each
geographic zone (see Fig. 19.2 and below) using sample survey and focus group
methodology. Secondly, a smaller pool of producers in each zone (three utilizing a
diversified wheat production system and three farms using a monoculture wheat production system) were evaluated using an intensive survey of economic and agronomic variables. Thirdly, and the focus of this chapter, biological data were gathered
from demonstration farms of each type in each zone to gather specific information
on how pest and beneficial organism populations vary between cropping system type
(monoculture versus diversified). The designation of paired demonstration sites
throughout the region was a difficult challenge, but included ecological, environmental
and farming system considerations.

Definition of study areas

The area of interest for the areawide IPM project, i.e. the portion of the Great Plains
where GB and RWA are key pests of wheat (see Fig. 19.1), was divided into three
geographic zones within which agroecological conditions are similar throughout.
The following three zones were delineated (see Fig. 19.2):

- Northern zone (Zone 1): south-east Wyoming, Nebraska Panhandle, north-east
  Colorado; the RWA is the main pest of wheat in this zone. Possibilities for

Box 19.1.   Principal investigators: biologically intensive AWPM of the Russian wheat aphid and greenbug.

USDA, ARS and PSWCRL
John D. Burd (research entomologist)
Norman C. Elliott (research biologist)
Mathew H. Greenstone (research entomologist)
S. Dean Kindler (research entomologist)
David R. Porter (research geneticist)
Kevin A. Shufran (entomologist)

Kansas State University
Department of Entomology
Gerald Wilde (Professor of Entomology, Research)
Southwest Area Extension Office
Phil Sloderbeck (Professor of Entomology, Extension)

Colorado State University
Department of Bioagricultural Sciences and Pest Management
Thomas O. Holtzer (Professor of Entomology, Research)
Frank B. Peairs (Professor of Entomology, Research and Extension)
Crop and Soil Science Department
Gary A. Peterson (Professor of Agronomy, Research and Extension)

Oklahoma State University
Department of Entomology and Plant Pathology
Gerrit W. Cuperus (Professor of Entomology, IPM Coordinator)
Kristopher L. Giles (Associate Professor of Entomology, Research)
Thomas A. Royer (Associate Professor of Entomology, IPM Coordinator and Extension)
Department of Plant and Soil Sciences
Thomas F. Peeper (Professor of Weed Science, Research and Extension)

Texas A&M University, Texas Agricultural Experiment Station
Gerald J. Michels, Jr. (Professor of Entomology, Research)

**University of Wyoming, Department of Renewable Resources –
Entomology**
Michael J. Brewer (Professor of Entomology, Research and Extension);
currently at Michigan State University

University of Nebraska Panhandle R&E Center
Gary L. Hein (Professor of Entomology, Research and Extension)
Drew J. Lyon (Professor of Agronomy, Research and Extension)
Paul Burgener (Agricultural Economist, Research and Extension)

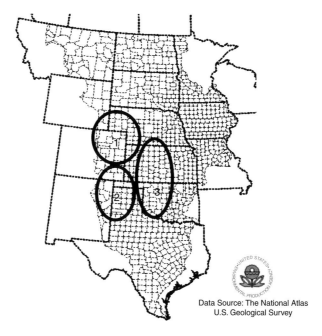

Data Source: The National Atlas
U.S. Geological Survey

Fig 19.2. Geographic zones of the Cereal Aphid AWPM project.

rotational crops with wheat in this region are sunflower, maize, barley and proso millet.

- Southern zone (Zone 2): Texas Panhandle, Oklahoma Panhandle, south-east Colorado; RWA and GB are the main pests of wheat in this zone; sorghum is the only viable rotational crop to use with wheat.
- Eastern zone (Zone 3): central Oklahoma, central Kansas; GB is the main pest of wheat; soybean, sorghum, canola and cotton are the possibilities for rotational crops with wheat.

Selection of sites

With the assistance of county and regional extension professionals, we identified three paired farms for intensive evaluation per zone and, therefore, nine paired farm sites (18 farms) for the programme (see Fig. 19.3). Each pair consisted of one farm and surrounding areas, primarily defined by a monoculture wheat production system and one farm utilizing a diversified rotational wheat production system adapted to the area with resistant wheat cultivars if appropriate. The defined criteria for paired demonstration farms were:

- Each farm had at least 400 contiguous acres farmed, using identical cropping practices throughout.
- Each pair of farms were representative of farms specializing in a particular cropping system for the region, and were similar in terms of factors that determine agronomic and economic potential, such as soil type and topography.

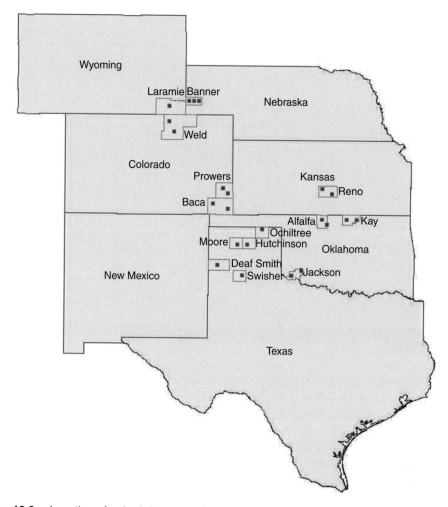

Fig. 19.3. Location of paired demonstration sites for the Cereal Aphid AWPM project.

- Within a particular zone, the rotational crop was standardized among farms. Therefore, farms with diversified farming that use the same alternative crop in rotation with wheat were chosen. One exception was made in Zone 3, where cotton is the most viable rotational crop in the southern portion.
- Farms and fields chosen for inclusion in the project must have had one cycle of the particular rotation completed prior to project initiation.

Insect and Weed Evaluation

Although a typical demonstration farm consisted of many fields, a single wheat field on a farm was deemed sufficient for the evaluation of insects and weeds. The minimum field size for sampling insects for research purposes was determined to be ¼ section (160 acres, 65 ha), with approximately one-half to one-third of the acreage in wheat in any year, depending on whether a monoculture or diversified crop rotation was being monitored. The 160-acre fields were divided into 25 uniformly sized 'quadrants' by using a 5 × 5 grid or other systematic division pattern. On each sampling date, samples were collected randomly near the centre of each quadrant. The pest species being sampled and data collection protocols for demonstration sites fields are outlined in Table 19.1.

Each location was monitored for 4 consecutive years, providing long-term data and information on pest abundance for monoculture (continuous or wheat–fallow) and diversified wheat systems. These data allow us to summarize long-term averages representative of each system and summarize the data by geographic zone.

In this chapter, data on aphid numbers, mummified aphids and visual counts of predators from wheat fields at each location were summarized by identified zone (Fig. 19.2). There are many ways to represent the data (i.e. peak numbers, field averages, seasonal accumulations); however, to account for all of the variability over a 4-year period, our focus will be on a comparison of averages per sample unit. The dynamics within and among growing seasons will be examined in future analyses.

Effectiveness of the Areawide Pest Management Programme at Controlling Target Pests

Over the 4-year period, annual sampling intensity varied among locations ranging from four to ten sampling events for individual fields (see Table 19.2). Low levels of

Table 19.1. Sampling methods for particular classes of pest and beneficial organisms.

Category sampled	Sampling method	Sampling frequency
Cereal aphids	25, 4–tiller counts (cut with scissors at ground level)[a]	Bi-weekly–monthly
	Berlese funnel (25 samples 0.15 m/field; samples included all soil and plant material from 0.1 m-wide shovel; samples left in funnels up to 1 week)[a]	
Cutworms and armyworms	Berlese funnel (25 samples 0.15 m/field)	Monthly
Wheat curl mite	Leaf samples	Seasonally
Natural enemies		
Predators	Sweepnet	Bi-weekly–monthly
	visual counts (25 samples 0.61 m/field)[a]	
Parasitoids	Mummies on stem counts	Bi-weekly–monthly
	Emergence canisters, trap plants[a]	Two times per year (trap plants)
Weeds	Area counts (25 samples 0.5 m²/field)[a]	Once at appropriate time in each crop

[a] Data summarized for this chapter.

Table 19.2. Total samples for each sampling method, 2002–2006.

Sampling method	Zone	System	Sampling events (n)
4-stem counts (n = 25),	1	Diverse	54
visual counts and		Traditional	76
area weed counts	2	Diverse	101
		Traditional	112
	3	Diverse	70
		Traditional	77
Berlese funnel	1	Diverse	46
		Traditional	61
	2	Diverse	83
		Traditional	93
	3	Diverse	65
		Traditional	72

Table 19.3. RWA in each zone, 2002–2006.

Sampling method	Zone	System	RWA (%)
4-stem counts (n = 25)	1	Diverse	85
		Traditional	71
	2	Diverse	61
		Traditional	81
	3	Diverse	0
		Traditional	0
Berlese funnel	1	Diverse	92
(25 samples 0.1 m/field)		Traditional	94
	2	Diverse	52
		Traditional	65
	3	Diverse	0
		Traditional	0

aphids and natural enemies prompted reduced sampling efforts at several locations, whereas in fields with increasing pest levels, participants sampled more frequently to accurately reflect insect activity.

Cereal aphids in wheat

Cereal aphids were the most abundant pests found throughout the study. The relative proportion of RWA varied according to geographic zone; RWA constituted the majority of aphids identified in the more arid regions of the Great Plains (Zones 1 and 2, Fig. 19.1; Table 19.3). GB was the second most common aphid species found, followed by BCOA (*R. padi*), and relatively small numbers of rice root aphids (*Rhopalosiphum rufiabdominalis*), corn leaf aphids (*Rhopalosiphum maidis*) and English grain aphids (*Sitobion avenae*).

The data on aphid abundance (summed for all species) from demonstration plots for both the tiller and Berlese samples provide interesting trends relative to crop diversification and geographic zone. For each approach within a geographic zone, aphid numbers per sample unit were always greater (though not always significant) in wheat fields at 'monoculture' (continuous or wheat fallow) versus 'diverse' demonstration sites (see Figs 19.4 and 19.5).

Reduced aphid levels in the diverse sites in Zones 1 and 2 were also probably influenced by the use of aphid-resistant wheat. Very little difference was observed in Zone 3, where GB-resistant cultivars are not well adapted. The relative discrepancy in aphid numbers between tiller and Berlese sampling may reflect a lack of precision for estimating aphid intensity, especially RWA (Zone 1) with 100 tillers in a field and/or the absolute nature of the Berlese method. Either way, the trends indicate that diversified systems that incorporate aphid-resistant wheat have reduced

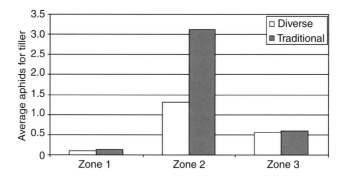

Fig. 19.4.   Four-year (2002–2006) average number of aphids per tiller at AWPM demonstration sites.

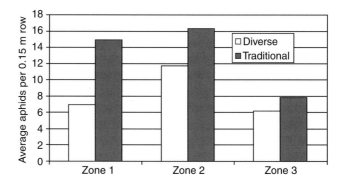

Fig. 19.5.   Four-year (2002–2006) average number of aphids per 0.15 m at AWPM demonstration sites.

infestations of aphids. This appears to be especially true in zones where RWA is most prevalent.

Parasitoids and predators

The data presented on parasitism reflect the current summarization from field tiller sampling and are limited to counts of intact 'gold' and 'black' mummies. Preliminary identification and previous studies in the Great Plains (Gilstrap *et al.*, 1984; Giles *et al.*, 2003; Brewer and Elliott, 2004) suggest that these mummies are represented by *Lysiphlebus testaceipes* (gold mummies) and *Diaeretiella rapae* (black mummies). Data summarized over the 4-year period at demonstration sites indicate that the average number of mummies was quite low, and that no consistent trends were apparent between monoculture and diverse systems (see Fig. 19.6). Potentially, our resolution on measuring parasitism was inadequate, and/or parasitoid populations function at scales different from those evaluated in our study (Brewer and Elliott, 2004) or independent of production system diversity.

Comparing data on aphid abundance with mummy abundance may suggest that parasitoid impact can function independently of aphid densities; the highest average intensity for mummy counts was found in the wheat systems of Zones 2 and 3, where low aphid populations were found (see Figs 19.5 and 19.6). Of course, as suggested by Giles *et al.* (2003), during mild winters local populations of parasitoids in Oklahoma and Texas can function to maintain very low aphid levels; data from Zones 2 and 3 may reflect this cause and effect.

As expected, a common assemblage of predators (adult and immature) were observed in wheat fields throughout the study during visual sampling. These predators included species of Coccinellidae, Nabidae and other Hemipteran predators (species of *Geocoris* and *Orius*, etc.), predatory Carabidae, Staphlyinidae and spiders. Similar to mummies, data on total predators were low and no consistent trends were apparent between monoculture and diverse systems (see Fig. 19.7). The relatively high populations of predators at traditional sites within Zone 1 probably

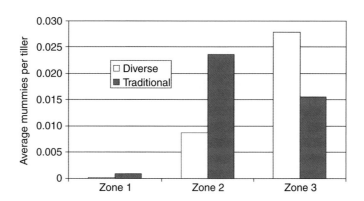

Fig. 19.6. Four-year (2002–2006) average number of mummies per tiller at AWPM demonstration sites.

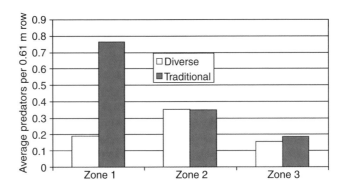

Fig. 19.7. Four-year (2002–2006) average number of all predators per 0.61 m of row at AWPM demonstration sites.

corresponded to the high counts of RWA found in Berlese samples (see Figs 19.5 and 19.7); for more aphidophagous predators such as Coccinellidae, which were often the most abundant group, we might expect this aggregative response to aphids. Hein (2006) demonstrated experimentally (cage exclusion) that this predatory response in Zone 1 was an essential component of RWA natural control; RWA numbers per 25 tillers were up to 40 times greater in cages that excluded natural enemies versus open-field plots.

Our data on predator numbers at demonstration sites do not support findings from studies that have documented increased abundance of predators in diversified systems (Brewer and Elliott, 2004). In fact, it appears as if predators primarily responded to aphid abundance. A careful evaluation of separate predator groups and their dynamics within and among fields is planned for the future.

Weeds

Grass weeds, including *Bromus* species, jointed goatgrass, wild oats and ryegrass, were very common; however, broadleaf weeds such as field bindweed, *Chenopodium*, pigweed and horseweed were prevalent in Zone 2 (see Fig. 19.8). Within a geographic zone, total weed densities were always higher in wheat fields at monoculture (continuous or wheat fallow) versus diverse demonstration sites (see Fig. 19.8). Based on focus group studies with producers in this AWPM programme (Keenan *et al.*, 2007a, b), lower weed densities at diversified sites were expected because producers are very concerned with long-term weed management. Most diversified farmers recommend rotation to a broadleaf crop and selective herbicide use as the only viable long-term strategy in wheat systems. For some time, weed scientists have documented lower weed densities in diversified rotational systems (Blackshaw *et al*, 1994, 2001; Lyon and Baltensperger, 1995; Boyles *et al.*, 2004), and our results provide additional supportive data for producers who are addressing weed problems through crop rotation.

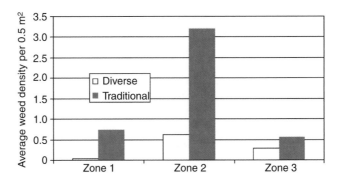

Fig. 19.8. Four-year (2002–2006) average weed densities per 0.5 m² at AWPM demonstration sites.

Unintended Consequences of the Areawide Pest Management Programme

During the project (2001–2006), some major developments occurred that were not necessarily planned. We had planned that all participants in the project would document and quickly communicate any new findings; however, we did not anticipate the rapid build-up of RWA-biotype 2 (Peairs, 2006), which can overcome available RWA-resistant cultivars and is currently the dominant biotype in Colorado and surrounding states. Producers in Zone 1 were made aware of this development, and some participated in documenting the regional prevalence of RWA-2. The focus groups and interviews established an instant and now long-term network of producers who continue to interact directly with AWPM personnel. We believe this group of producers will continue to work with state personnel, providing farming system results and stakeholder recommendations that will drive future research and extension programmes.

This AWPM programme also allowed for delivery of new IPM tools. The '*Glance n' Go*' greenbug + parasitism sampling and management plan was fully developed during this project, and communicated project-wide and throughout the Great Plains as the recommended approach for GB sampling and decision making (Giles *et al.*, 2003; Elliott *et al.*, 2004; Royer *et al.*, 2004a, b, 2007). Participating producers demonstrated the usefulness of this approach on many of their farms.

The project was conducted during a period of severe drought throughout much of the Great Plains. The results of the project during these years demonstrated the impact of drought on monoculture and diversified wheat-cropping systems. In the more arid areas of the project, the benefits of diversity were reduced and the benefits of the monoculture cropping system were enhanced. These differences resulted from the moisture-saving advantages seen in the wheat–fallow (monoculture) systems.

Summary and Future Directions

Relative to AWPM for GB and RWA, lack of information on the dispersal range and extent of migration for aphids and natural enemies hinders the full development of an optimal AWPM strategy (Booij and Noorlander, 1992; Brewer and Elliott, 2004; Vialatte *et al.*, 2006). Based on our methodology for demonstration sites, we could not determine the spatial extent of the suppression area required to minimize colonization of aphids and conserve natural enemies. Further analysis and modelling evaluation of within-season dynamics may help to define regional trends that could be useful in defining appropriate spatial scales for areawide implementation. The sporadic nature of GB and RWA infestations in wheat is also an impediment to the development of areawide programmes focused on aphid management. Producers of this low-value crop are increasingly willing to use low-input strategies such as resistant cultivars (Peairs, 2006) to manage aphids; however, many are reluctant to significantly alter production practices to avoid pests that are not a problem annually (Keenan *et al.*, 2007a, b).

There are four important reasons why we believe that producers will move towards diversification of wheat systems in the Great Plains. First, studies continue to support the idea that diversification of farming systems increases water use efficiency and stabilizes and/or increases farm profits (Peterson and Westfall, 1994, 2004; Dhuyvetter *et al.*, 1996; P. Burgener and S. Keenan, unpublished data). Secondly, fuel and equipment costs related to tillage continue to increase, prompting a shift by producers towards no-till rotational production systems. Thirdly, wheat producers in this region continue to consider weed problems as their most serious pest problem and are becoming increasingly aware of how diversified farming systems allow for more effective long-term selective weed management. Finally, cropping system diversification provides numerous benefits for the management of several other pests. As growers strive to become more cost-efficient, many of these benefits will become more apparent when compared with the alternative of relying on increasingly more costly pesticides.

This anticipated diversification of wheat-farming systems in the Great Plains will probably provide opportunities for evaluation of AWPM of cereal aphids on increasingly larger spatial scales. Findings from this future work may help producers and scientists in designing the most effective areawide approach for each region of the Great Plains.

References

Ahern, R.G. and Brewer, M.J. (2002) Effect of different wheat production systems on the presence of two parasitoids (Hymenoptera: Aphelinidae: Braconidae) of the Russian wheat aphid in the North America Great Plains. *Agricultural Ecosystem and Environment* 92, 201–210.

Andow, D.A. (1983) The extent of monoculture and its effects on insect pest populations with particular reference to wheat and cotton. *Agricultural Ecosystem and Environment* 9, 25–35.

Andow, D.A. (1991) Vegetational diversity and arthropod population response. *Annual Review of Entomology* 36, 561–586.

Anon. (1990) *Jointed Goatgrass Task Force Report.* Colorado Department of Agriculture, February 1990, Denver, Colorado.

Anon. (1991) *Winter Annual Grass Weed Contamintion Survey.* Colorado Department of Agriculture, August 1991, Denver, Colorado.

Archer, T.L. and Bynum, E.D., Jr. (1992) Economic injury level for the Russian wheat aphid (Homoptera: Aphididae) on dryland winter wheat. *Journal of Economic Entomology* 85, 987–992.

Archer, T.L., Johnson, G.D., Peairs, F.B., Pike, K.S. and Kroening, M.K. (1998) Effect of plant phenology and climate on Russian wheat aphid (Homoptera: Aphididae) damage to winter wheat. *Environmental Entomology* 27, 221–231.

Boethel, D.J. and Eikenbary, R.D. (1986) *Interactions of Plant Resistance and Parasitoids and Predators of Insects.* John Wiley & Sons, New York.

Blackshaw, R.E., Larney, F.J., Lindwall, C.W. and Kozub, G.C. (1994) Crop-rotation and tillage effects on weed populations on the semiarid Canadian prairies. *Weed Technology* 8, 231–237.

Blackshaw, R.E., Larney, F.J., Lindwall, C.W., Watson, P.R. and Derksen, D.A. (2001) Tillage intensity and crop rotation affect weed community dynamics in a winter wheat cropping system. *Canadian Journal of Plant Science* 81, 805–813.

Booij, C.J.H. and Noorlander, J. (1992) Farming systems and insect predators. *Agriculture, Ecosystems and Environment* 40, 125–135.

Boyles, M., Peeper, T. and Medlin, C. (2004) *Okanola: Producing Winter Tolerant Canola in Oklahoma.* 2nd edn., OAES and OCES Oklahoma State University, Oklahoma.

Brakke, M.K. (1987) Virus diseases of wheat. In: Heyne, E.G. (ed.) *Wheat and Wheat Improvement.* 2nd edn., American Society of Agronomy, Inc., Crop Science Society of America Inc., Soil Science Society of America Inc., Madison, Wisconsin, pp. 585–624.

Brewer, M.J. and Elliott, N.C. (2004) Biological control of cereal aphids in North America and mediating effects of host plant and habitat manipulations. *Annual Review of Entomology* 49, 219–242.

Brewer, M.J. and Kaltenbach, J.E. (1995) Russian wheat aphid (Homoptera: Aphididae) population variation in response to chlorpyrifos exposure. *Journal of the Kansas Entomological Society* 68, 346–354.

Brewer, M.J., Struttmann, J.M. and Mornhinweg, D.W. (1998a) *Aphelinus albipodus* (Hymenoptera: Aphelinidae) and *Diaeretiella rapae* (Hymenoptera: Braconidae) parasitism on *Diuraphis noxia* (Homoptera: Aphididae) infesting barley plants differing in plant resistance to aphids. *Biological Control* 11, 255–261.

Brewer, M.J., Mornhinweg, D.W., Struttmann, J.M. and Oswald II, C.J. (1998b) Russian wheat aphid (Homoptera: Aphididae) and parasitoids (Hymenoptera: Braconidae and Aphelinidae) found on field-grown barley lines varying in susceptibility to Russian wheat aphid. In: Quisenberry, S.S. and Peairs, F.B. (eds) *Response Model for an Introduced Pest – the Russian Wheat Aphid.* Thomas Say Publications, Entomological Society of America, Lanham, Maryland, pp. 258–269.

Brewer, M.J., Mornhinweg, D.W. and Huzurbazar, S. (1999) Compatibility of insect management strategies: *Diuraphis noxia* abundance on susceptible and resistant barley in the presence of parasitoids. *Biocontrol* 43, 479–491.

Brewer, M.J., Nelson, D.J., Ahern, R.G., Donahue, J.D. and Prokrym, D.R. (2001) Recovery and range expansion of parasitoids (Hymenoptera: Aphelinidae and Braconidae) released for biological control of *Diuraphis noxia* (Homoptera: Aphididae) in Wyoming. *Environmental Entomology* 30, 578–588.

Clough, Y., Kruess, A. and Tscharntke, T. (2007) Local landscape factors in differently managed arable fields affect the insect herbivore community of a non-crop plant species. *Journal of Applied Ecology* 44, 22–28.

Cook, R.J. and Veseth, R.J. (1991) *Wheat Health Management. APS Press, American Phytopathology Society,* St Paul, Minnesota, 152 pp.

Crop Profile for Wheat in Kansas (1998) http://www.ipmcenters.org/cropprofiles/docs/kswheat.html

Crop Profile for Wheat in Oklahoma (2005) http://www.ipmcenters.org/cropprofiles/docs/okwheat.html

Dhuyvetter, K.C., Thompson, C.R., Norwood, C.A., and Halvorson, A.D. (1996) Economics of dryland cropping systems in the Great Plains: a review. *Journal of Production Agriculture* 9, 216–222.

Elliott, N.C., Reed, D.K., French, B.W. and Kindler, S.D. (1994a) Aphid host effects on the biology of *Diaeretiella rapae*. *Southwestern Entomologist* 19, 279–284.

Elliott, N.C., Kindler, S.D., Reed, D.K., and French, B.W. (1994b) Parasitism, adult emergence, sex ratio, and size of *Aphidius colemani* (Hymenoptera: Aphidiidae) on several aphid species. *Great Lakes Entomology* 27, 137–142.

Elliott, N.C., Reed, D.K., French, B.W. and Kindler, S.D. (1994c) Aphid host effects on the biology of *Diaeretiella rapae*. *Southwestern Entomologist* 19, 279–284.

Elliott, N.C., Burd, J.D., Armstrong, J.S., Walker, C.B., Reed, D.K. and Peairs, F.B. (1995) Release and recovery of imported parasitoids of the Russian wheat aphid in eastern Colorado. *Southwestern Entomologist* 20, 125–129.

Elliott, N.C., Hein, G.L., Carter, M.C., Burd, J.D., Holtzer, T.J., Armstrong, J.S. and Waits, D.A. (1998a) Russian wheat aphid (Homoptera: Aphididae) ecology and modeling in Great Plains agricultural landscapes. In: Quisenberry, S.S. and Peairs, F.B. (eds) *Response Model for an Introduced Pest – the Russian Wheat Aphid*. Thomas Say Publications, Entomological Society of America, Lanham, Maryland, pp. 31–64.

Elliott, N.C., Kieckhefer, R.W., Lee, J.H. and French, B.W. (1998b) Influence of within-field and landscape factors on aphid predator populations in wheat. *Landscape Ecology* 14, 239–252.

Elliott, N.C., Lee, J.H. and Kindler, S.D. (1999) Parasitism of several aphid species by *Aphelinus asychis* (Walker) and *Aphelinus albipodus* Hayat and Fatima. *Southwesern Entomologist* 24, 5–12.

Elliott, N.C., Kieckhefer, R.W., Michels, G.J., Jr. and Giles, K.L. (2002a) Predator abundance in alfalfa fields in relation to aphids, within-field vegetation, and landscape matrix. *Environmental Entomology* 31, 253–260.

Elliott, N.C., Kieckhefer, R.W. and Beck, D.A. (2002b) Effect of aphids and the surrounding landscape on abundance of coccinellids in cornfields. *Biological Control* 24, 214–220.

Elzen, G.W. and Hardee, D.D. (2003) United States Department of Agriculture – Agriculture Research Service Research on managing insect resistance to insecticides. *Pest Management Science* 59, 770–776.

Flickinger, E.L., Juenger, G., Roffe, T.J., Smith, M.R. and Irwin, R.J. (1991) Poisoning of Canada geese in Texas by parathion sprayed for control of Russian wheat aphid. *Journal of Wildlife Diseases* 27, 265–268.

French, B.W. and Elliott, N.C. (1999a) Temporal and spatial distribution of ground beetle (Coleoptera: Carabidae) assemblages in grassland and adjacent wheat fields. *Pedobiologia* 43, 73–84.

French, B.W. and Elliott, N.C. (1999b) Spatial and temporal distribution of ground beetle (Coleoptera: Carabidae) assemblages in riparian strips and adjacent wheat fields. *Environmental Entomology* 28, 597–607.

French, B.W., Elliott N.C., Kindler, S.D. and Arnold, D.C. (2001) Seasonal occurrence of aphids and natural enemies in wheat and associated crops. *Southwestern Entomologist* 26, 49–61.

Fuentes-Granados, R., Giles, K.L., Elliott, N.C. and Porter, D.R. (2001) Assessment of greenbug-resistant wheat germplasm on *Lysiphlebus testaceipes* Cresson (Hymenoptera: Aphidiidae) oviposition and development in greenbug over two generations. *Southwestern Entomologist* 26, 187–194.

Giles, K.L., Jones, D.B., Royer, T.A., Elliott, N.C. and Kindler, S.D. (2003) Development of a sampling plan in winter wheat that estimates cereal aphid parasitism levels and predicts population suppression. *Journal of Economic Entomology* 96, 975–982.

Giles, K.L., Dillwith, J.W., Berberet, R.C. and Elliott, N.C. (2005) Survival, development, and growth of *Coccinella septempunctata* fed *Schizaphis graminum* from resistant and susceptible winter wheat. *Southwestern Entomologist* 30, 113–120.

Gilstrap, F.E., Kring, T.J. and Brooks, G.W. (1984) Parasitism of aphids (Homoptera: Aphididae) associated with Texas sorghum. *Journal of Economic Entomology* 13, 1613–1617.

Grue, C.E., Tome, M.W., Swanson, G.A., Borthwick, S.M. and DeWeese, L.R. (1988) Agricultural chemicals and the quality of prairie-pothole wetlands for adult and juvenile waterfowl – what are the concerns? In: Stuber, P.J. (ed.) *Proceedings National Symposium on Protection of Wetlands from Agricultural Impacts.* USDA, Fish and Wildlife Service biology reports 88, pp. 55–66.

Guerena, M. and Sullivan, P. (2003) *Organic Cotton Production* (http://attra.ncat.org/attra-pub/cotton.html).

Haley, S.D., Peairs, F.B., Walker, C.B., Rudolph, J.B. and Randolph, T.L. (2004) Occurrence of a new Russian wheat aphid biotype in Colorado. *Crop Science* 44, 1589–1592.

Hatchett, J.H., Starks, K.J. and Webster, J.A. (1987) *Insect and Mite Pests of Wheat, in Wheat and Wheat Improvement.* Agronomy Monograph 13, 2nd edn., American Society of Agronomy, Madison, Wisconsin.

Hein, G. (2006) Demonstration site summaries: Nebraska and Wyoming. In: Elliott, N.C. (ed.) *Progress Review Report: Areawide Pest Management for Russian Wheat Aphid and Greenbug.* Colby, Kansas, pp. 21–26.

Helms, G.L., Bailey, D., and Glover, T.F. (1987) Government programs and adoption of conservation tillage practices on non-irrigated wheat farms. *American Journal of Agricultural Economics* 69, 786–795.

Holtzer, T.O., Anderson, R.L., McMullen, M.P. and Peairs, F.B. (1996) Integrated pest management of insects, plant pathogens, and weeds in dryland cropping systems of the Great Plains. *Journal of Production Agriculture* 9, 200–208.

Keenan, S.P., Giles, K.L., Elliott, N.C., Royer, T.A., Porter, D.R., Burgener, P.A. and Christian, D.A. (2007a) Grower perspectives on areawide wheat integrated pest management in the southern US great plains. In: Koul, O. and Cuperus, G.W. (eds) *Ecologically Based Integrated Pest Management.* CAB International, Wallingford, UK, pp. 289–314.

Keenan, S.P., Giles, K.L., Burgener, P.A. and Elliott, N.C. (2007b) Collaborating with wheat growers in demonstrating areawide integrated pest management. *Journal of Extension* (http://www.joe.org/joe/2007february/a7.shtml).

Kelsey, K.D. and Mariger, S.C. (2002) *A Survey-based Model for Setting Research, Education, and Extension Priorities at the Land-grant University – a Case Study of Oklahoma Wheat Producers: Final Report.* Agricultural Extension, Communications, and 4-H Youth Development, Oklahoma State University, Stillwater, Oklahoma.

Kfir, R. (2002) Increase in cereal stem borer populations through partial elimination of natural enemies. *Entomologia Experimentalis et Applicata* 104, 299–306.

Kieckhefer, R.W. and Kantack, B.H. (1988) Yield losses in winter grains caused by cereal aphids (Homoptera: Aphididae) in South Dakota. *Journal of Economic Entomology* 81, 317–321.

Kindler, S.D., Elliott, N.C., Giles, K.L., Royer, T.A., Fuentes-Grandaos, R. and Tao, F. (2002) Effect of greenbug (Homoptera: Aphididae) on yield loss of winter wheat. *Journal of Economic Entomology* 95, 89–95.

Kindler, S.D., Elliott, N.C., Giles, K.L. and Royer, T.A. (2003) Economic injury levels for the greenbug, *Schizaphis graminum*, in winter wheat. *Southwestern Entomologist* 28, 163–166.

Klass, E.E. (1982) Effects of pesticides on non-target organisms. In: Dalgren, R.B. (ed.) *Workshop on Midwest Agricultural Interfaces with Fish and Wildlife Resources.* US Fish and Wildlife Services, Agricultural Experimental Station and Cooperative Extension Services, Iowa State University, Ames, Iowa, pp. 7–9.

Knipling, E.F. (1980) Regional management of the fall armyworm – a realistic approach? *Florida Entomologist* 63, 468–480.

Kring, T.J., Gilstrap, F.E. and Michels, G.J., Jr. (1985) Role of indigenous coccinellids in regulating greenbugs (Homoptera: Aphididae) on Texas grain sorghum. *Journal of Economic Entomology* 78, 269–273.

Lazar, M.D., Worrall, W.D., Peterson, G.L., Porter, K.B., Rooney, L.W., Tuleen, N.A., Marshall, D.S., McDaniel, M.E. and Nelson, L.R. (1998) Registration of TAM 110 wheat. *Crop Science* 37, 1978–1979.

Lyon, D.J. and Baltensperger, D.S. (1995) Cropping systems control winter annual grass weeds in winter wheat. *Journal of Production Agriculture* 8, 535–539.

Lyon, D.J., Baltensperger, D.D., Blumenthal, J.M., Burgener, P.A. and Harveson, R.M. (2004) Eliminating summer fallow reduces winter wheat yields, but not necessarily system profitability. *Crop Science* 44, 855–860.

Massey, B. (1993) *Insects on Small Grains and Their Control.* Oklahoma Cooperative Extension Service, No. 7176, Oklahoma.

Men, X.Y., Feng, G.E., Erdal, N.Y. and Parajulee, M.N. (2004) Evaluation of winter wheat as a potential relay crop for enhancing biological control of cotton aphids in seedling cotton. *Biocontrol* 49, 701–714.

Meyer, W. and Peairs, F. (1989) Observations on biological control agents in Colorado. In: Baker, D. (compiler) *Russian Wheat Aphid: Proceedings of the 3rd Russian Wheat Aphid Conference,* New Mexico State University, Albuquerque, New Mexico, pp. 96–98.

Michels, G.J. and Whitaker-Deerberg, R.L. (1993) Recovery of *Aphelinus asychis,* an imported parasitoid of Russian wheat aphid, in the Texas panhandle. *Southwestern Entomologist* 18, 11–17.

Michels, G.J., Elliott, N.C., Romero, R.A., Owings, D.A. and Bible, J.B. (2001) Impact of indigenous coccinellids on Russian wheat aphids and greenbugs (Homoptera: Aphididae) infesting winter wheat in the Texas Panhandle. *Southwestern Entomologist* 26, 97–114.

Mornhinweg, D.W., Brewer, M.J. and Porter, D.R. (2006) Effect of Russian wheat aphid on yield and yield components of field grown susceptible and resistant spring barley. *Crop Science* 46, 36–42.

Morrison, W.P. and Peairs, F.B. (1998) Introduction: response model concept and economic impact. In: Quisenberry, S.S. and Peairs, F.B. (eds) *Response Model for an Introduced Pest – the Russian Wheat Aphid.* Thomas Say Publications, Entomological Society of America, Lanham, Maryland, pp. 1–11.

NASS (1996) *Colorado Agricultural Statistics, 1996.* Colorado Agricultural State Services (http://www.usda.gov/nass/).

National Agricultural Statistics Survey, USDA (2005) *Agricultural Chemical Usage Field Crops Summary* (http://www.usda.gov/nass/).

National Agricultural Statistics Survey, USDA (2006) *Agricultural Chemical Usage Field Crops Summary* (http://www.usda.gov/nass/).

Patrick, C.D. and Boring, E.P. III. (1990) *Managing Insect and Mite Pests of Texas Small Grains.* Texas Agricultural Extension Service Publication B-1251, College Station, Texas.

Parajulee, M.N. and Slosser, J.E. (1999) Evaluation of potential relay strip crops for predator enhancement in Texas cotton. *International Journal of Pest Management* 45, 275–286.

Parajulee, M.N., Montandon, R. and Slosser, J.E. (1997) Relay intercropping to enhance abundance of insect predators of cotton aphid (*Aphis gossypii* Glover) in Texas cotton. *International Journal of Pest Management* 43, 227–232.

Peairs, F.B. (1990) Russian wheat aphid management. In: Peters, D.C., Webster, J.A. and Chlouber, C.S. (eds) *Aphid–Plant Interactions: Populations to Molecules, Proceedings of the International Aphid Symposium,* Oklahoma Agricultural Experimental Station, Oklahoma, pp. 233–241.

Peairs, F.B. (2006) Demonstration site summaries: Colorado. In: Elliott, N.C. (ed.) *Progress Review Report: Areawide Pest Management for Russian Wheat Aphid and Greenbug,* Colby, Kansas, pp. 27–34.

Peairs, F.B., Bean, B. and Gossen. B.D. (2005) Pest management implications of reduced fallow periods in dryland cropping systems in the Great Plains. *Agronomy Journal* 97, 373–377.

Peterson, G.A., and Westfall, D.G. (1994) Economic and environmental impact of intensive cropping systems – semiarid region. In: *Proceedings of the Conference on Nutrient Management of Highly Productive Soils*, Atlanta, Georgia, 16–18 May 1994. PPI/FAR Special Publication 1994–1, pp. 145–158.

Peterson, G.A. and Westfall, D.G. (2004) Managing precipitation use in sustainable dryland agroecosystems. *Annals of Applied Biology* 144, 127–138.

Peterson, G.A., Schlegel, A.J., Tanaka, D.L. and Jones, O.R. (1996) Precipitation use efficiency as affected by cropping and tillage systems. *Journal of Production Agriculture* 9, 180–186.

Pike, K.S. and Schaffner, R.L. (1985) Development of autumn populations of cereal aphids, *Rhopalosiphum padi* (L.) and *Schizaphis graminum* (Rondani) (Homoptera: Aphididae) and their effects on winter wheat in Washington state. *Journal of Economic Entomology* 78, 676–680.

Pike, K.S. and Stary, P. (1995) New species of parasitic wasps attacking cereal aphids in the Pacific northwest (Hymenoptera: Braconidae: Aphidiinae). *Journal of the Kansas Entomological Society* 68, 408–414.

Pike, K.S., Stary, P., Miller, R., Allison, D., Boydston, L., Graf, G. and Miller, T. (1996) New species and host records of aphid parasitoids (Hymenoptera: Braconidae: Aphidiinae) from the Pacific Northwest, USA. *Proceedings of the Entomological Society of Washington* 98, 570–591.

Porter, D.R., Burd, J.D., Shufran, K.A., Webster, J.A. and Teetes, G.L. (1997) Greenbug (Homoptera: Aphididae) biotypes: selected by resistant cultivars or preadapted opportunists? *Journal of Economic Entomology* 90, 1055–1065.

Prokrym, D.R., Gould, J.R., Nelson, D.J., Wood, L.A. and Copeland, C.J. (1994) *Russian Wheat Aphid Biological Control Project, FY 1993 Project Report*. USDA APHIS PPQ report, National Biological Control Laboratory, Niles, Michigan.

Quick, J.S., Ellis, G., Norman, R., Stromberger, J., Shanahan, J., Peairs, F. and Lorenz, K. (1996) Registration of 'Halt' wheat. *Crop Science* 36, 210.

Riedell, W.E. and Kieckhefer, R.W. (1995) Feeding damage effects of three aphid species on wheat root growth. *Journal of Plant Nutrition* 18, 1981–1891.

Riedell, W.E., Kieckhefer, R.W., Haley, S.D., Langham, M.C. and Evenson, P.D. (1999) Winter wheat responses to bird cherry-oat aphids and barley yellow dwarf virus infection. *Crop Science* 30, 158–163.

Royer, T.A., and Krenzer, E.G. (eds) (2000) *Wheat Management in Oklahoma: a Handbook for Oklahoma's Wheat Industry*. Publication E-831, Oklahoma Cooperative Extension Service, Oklahoma.

Royer, T.A., Giles, K.L. and Elliott, N.C. (2004a) *The Cereal Aphid Expert System and Glance 'n Go Sampling, Questions and Answers*. CR-7191, Oklahoma Cooperative Extension Service, Stillwater, Oklahoma.

Royer, T.A., Giles, K.L. and Elliott, N.C. (2004b) *Glance 'n Go Sampling for Greenbugs in Winter Wheat: Fall Edition*. Oklahoma Cooperative Extension Service, Oklahoma State University Extension Facts, L-306, Stillwater, Oklahoma.

Royer, T.A., Giles, K.L., Nyamanzi, T., Hunger, R.M., Krenzer, E.G., Elliott, N.C., Kindler, S.D. and Payton, M. (2005) Economic evaluation of the effects of planting date and application rate of imidacloprid for management of cereal aphids and barley yellow dwarf in winter wheat. *Journal of Economic Entomology* 98, 95–102.

Royer, T.A., Story, S. and Elliott, N.C. (2007) *Cereal Aphid Expert System and Glance n' go Sampling Education Handbook*. Oklahoma Cooperative Extension Service, Oklahoma State University Extension, Stillwater, Oklahoma.

Shotkoski, F.A., Mayo, Z.B. and Peters, L.L. (1990) Induced disulfoton resistance in greenbugs (Homoptera: Aphididae). *Journal of Economic Entomology* 83, 2147–2152.

Sloderbeck, P.E., Chowdhury, M.A., Depew, L.J. and Buschman, L.L. (1991) Greenbug (Homoptera: Aphididae) resistance to parathion and chlorpyrifos-methyl. *Journal of the Kansas Entomological Society* 64, 1–4.

Smith, D. and Anisco, J. (2000) *Wheat in Texas: Crop Brief on Production, Pests, and Pesticides.* TAES, Texas Cooperative Extension (http://extension horticulture.tamu.edu/extension/cropbriefs/wheat.html).

Smolen, M. and Cuperus, G. (2000) *Salt Fork Watershed: Agricultural Producer Survey Results.* Oklahoma Cooperative Extension Service Water Quality Programmes (http://biosystems.okstate.edu/waterquality/publications/project_reports/sfork_survey.pdf).

Sotherton, N.W., Boatman, N.D. and Michael, R.W. (1989) The 'Conservation Headland' experiment in cereal ecosystems. *The Entomologist* 108, 135–143.

Starks, K.J. and Burton, R.L. (1977) *Preventing Greenbug Outbreaks.* USDA Science and Education Administration leaflet No. 309, 11 pp, Washington, DC.

Tonks, D.J. and Westra, P. (1997) Control of sulfonylurea-resistant kochia (*Kochia scoparia*). *Weed Technology* 11, 270–276.

Trumper, E.V. and Holt, J. (1998) Modelling pest population resurgence due to recolonization of fields following an insecticide application. *Journal of Applied Ecology* 35, 273–285.

Tscharntke, T., Rand, T.A. and Bianchi, F.J.J.A. (2005) The landscape context of trophic interactions: insect spillover across the crop–noncrop interface. *Annales Zoologici Fennici* 42, 421–432.

Vandermeer, J. (1989) *The Ecology of Intercropping.* Cambridge University Press, Cambridge, UK.

Vialatte, A., Simon, J.C., Dedryver, C.A., Fabre, F. and Plantegenest, M. (2006) Tracing individual movements of aphids reveals preferential routes of population transfers in agroecosystems. *Ecological Applications* 16, 839–844.

Way, M.J. (1988) Entomology of wheat. In: Harris, M.K. and Rogers, C.E. (eds) *The Entomology of Indigenous and Naturalized Systems in Agriculture.* Westview Press, Boulder, Colorado, pp. 183–206.

Webster, J.A. (1995) *Economic Impact of the Greenbug in the Western United States: 1992–1993.* Volume Publication No. 155, Great Plains Agricultural Council, Stillwater, Oklahoma.

Westra, P. and Amato, T.D. (1995) Issues related to kochia management in western agriculture. *Procedings of the Western Society of Weed Science* 48, 39–40.

Wilde, G. and Smith, M. (2006) Plant resistance to the new RWA biotype. In: Elliott, N.C. (ed.) *Progress Review Report: Areawide Pest Management for Russian Wheat Aphid and Greenbug,* Colby, Kansas, pp. 76–82.

Wilde, G.E., Shufran, R.A., Kindler, S.D., Brooks, H.L. and Sloderbeck, P.E. (2001) Distribution and abundance of insecticide-resistant greenbugs (Homoptera: Aphididae) and validation of a bioassay to assess resistance. *Journal of Economic Entomology* 94, 547–551.

Wilson, J.P., Cunfer, B.M. and Phillips, D.V. (1999) Double-cropping and crop rotation effects on diseases and grain yield of pearl millet. *Journal of Production Agriculture* 12, A198–A202.

Wraight, S.P., Poprawski, T.J., Meyer, W.L. and Peairs, F.B. (1993) Natural enemies of Russian wheat aphid (Homoptera: Aphididae) and associated cereal aphid species in spring-planted wheat and barley in Colorado. *Environmental Entomology* 22, 1383–1391.

20 Boll Weevil Eradication: an Areawide Pest Management Effort

CHARLES T. ALLEN

Programme Director, Texas Boll Weevil Eradication Foundation and Extension Specialist, Texas Cooperative Extension, Abilene, Texas, USA

Introduction

Historians have documented human production and use of cotton from as early as 5000 years ago in India and 4500 years ago in Peru and Arizona (Brown and Ware, 1958; Prentice, 1972; Frisbie *et al.*, 1989). Early cotton production was known from the upper Nile River basin in modern Sudan as early as 500 BC (Brown and Ware, 1958) and from Mayan ruins near Oaxaca, Mexico from 900 AD (Warner and Smith, 1968). In the 16th century, early European explorers reported finding cotton being grown in the lowlands of the Mississippi River and in Texas (Donnell, 1872).

In America, European settlers first began growing cotton about 1600 (Handy, 1896). Cottonseed imported from the West Indies was first planted at Jamestown in 1620 (Anon., 1975). During the American War of Independence the country was supplied with cotton cloth made from cotton grown in Maryland, Delaware and New Jersey (Donnell, 1872). Eli Whitney dramatically changed cotton production with his invention of the cotton gin in 1793 (Donnell, 1872; Linder, 1954). President George Washington signed the patent for Whitney's gin in 1796 (Thomas, 1929). Its use made production of upland cotton commercially feasible (Anon., 1930).

Trelogan (1969) cites the increase in cotton production in the South as a contributing factor to America's rapid population growth between 1840 and 1860. By 1849, American cotton had greater value than any of the country's other agricultural exports. Income from cotton sales paid for two-thirds of all imports coming into the country (Anon., 1850; Phillips, 1850; Haney *et al.*, 1996). By 1850, Watkins (1904) reported that the USA had become the world's largest producer of cotton, providing 85% of the world's production. Additionally, America had become the world's leading exporter, manufacturer and consumer of cotton.

The Civil War devastated the enterprise and infrastructure of the South, including the cotton industry. In 1860, the year before the war began, America produced 2 million bales of cotton. American cotton exported to the UK accounted for 80% of the cotton used in UK textile mills. During the war, production fell to a low of only 2% of the cotton fibre in UK mills. By 1876, the US cotton industry had rebounded, supplying 62% of the UK market (Anon., 1877; Haney, 2001).

Following the Civil War, domestic cotton use increased dramatically. Levis' jeans and other cotton textile products were in demand and cotton fibre became an important component of a number of industrial products, from insulators for telegraph lines and filaments for electric lights to tyre cords and smokeless gun powder. By the end of the century, a threat to the cotton industry was clearly a threat to the US economy. In the South, anything that could cause damage to the cotton industry would have devastating consequences to the economies of the southern states, which were struggling to recover from the destruction left by the Civil War (Haney, 2001). No one would have guessed that the South's recovering cotton industry would soon be driven to its knees again by a small, brown beetle from Mexico.

Unexpected Immigrant: the Weevil Arrives

The earliest specimen of the boll weevil, *Anthonomus grandis*, is from an archaeological site in the Oaxaca Valley of Mexico. In the mid-1960s, a single adult female weevil was found in a cave tangled in the fibre of a cotton boll. It was dated to about 900 AD (Warner and Smith, 1968). The boll weevil was first described and named by the Swedish entomologist, Carl H. Boheman in 1843, from specimens he had received from Vera Cruz, Mexico. However, no host plant was recorded for the species (Worsham, 1914; Parencia, 1978).

In 1880, English botanist Edward Palmer discovered a weevil destroying cotton production near Monclova, Mexico, 190 km south-west of Laredo, Texas. The insect was reported damaging the cotton crop by puncturing immature flower buds and bolls, causing them to fall from the plants. Specimens and samples of damaged bolls were sent to W.G. LeDuc, Commissioner of Agriculture in Washington, DC. The specimens were later identified as *A. grandis* by the French entomologist, August Salle (Helms, 1977; Wagner, 1980; Walker, 1984).

The boll weevil crossed the Rio Grande River into Texas in about 1892 (Newell, 1904) and very quickly revealed its destructive capability. In 1893, C.H. DeRyee wrote to the US Department of Agriculture (USDA) describing the difficulties farmers in the Brownsville area were having with a new cotton pest (Cross, 1976a). In 1894, Townsend (1895) reported that yield losses from the boll weevil had surpassed 90% in cotton fields near Brownsville and San Diego, Texas. Moving at a rate averaging 80–100 km a year, the weevil had infested all US cotton-producing areas east of the Texas High Plains by 1922 (Coad *et al.*, 1922). Long-range movement occurred primarily in the autumn of the year (Hunter, 1911). Clearly, the weevil had a well-developed ability to expand its range.

Weather was an important factor in the year-to-year movement of the weevil. The Galveston hurricane of 8–9 September 1900 scattered boll weevils from the Texas Coastal Plains to the counties south of the Red River (Wagner, 1980). Farmers, ginners and cotton communities in northern areas of the cotton belt debated the ability of the weevil to survive their harsh winters. G.M. Bentley, the Tennessee State

Entomologist, stated in 1917: 'We must face the fact that the boll weevil has shown a wonderful ability to adapt itself to colder climatic conditions. It is plain, therefore, to see how by adaptation the weevil may withstand the Tennessee winter' (Barker, 2001).

The boll weevil was much slower to establish in cotton in western regions. Reproducing weevil populations became established in 1953 at Presidio, Texas (Robertson, 1957; Cross, 1976a). A combination of limited overwintering habitat, cold winters and hot, dry summers prevented the boll weevil from establishing on the High Plains and in the Trans Pecos region for many years. Incipient weevil populations appeared sporadically in these regions for nearly 70 years after the boll weevil had initially entered the state, but populations did not become established (Robertson *et al.*, 1966; Bottrell *et al.*, 1972).

The discovery of reproducing boll weevil populations in the High Plains from 1959 to 1963 was the stimulus for initiation of an areawide diapause control programme in 1964. Several factors were involved in the boll weevil's eventual survival and establishment on the High Plains. Millions of acres of Conservation Reserve Program (CRP) grasses were established in the 1980s and early 1990s, providing overwintering habitat.

Then, in the late 1980s and 1990s a series of mild winters allowed weevil populations to increase. And finally, as weevil populations increased on the High Plains, the suppression programme, which had been initiated to prevent their establishment, was discontinued after the 1996 season (Leser *et al.*, 1997; Carter *et al.*, 2001; Stavinoha and Woodward, 2001). The weevil adapted, as it had done previously in Tennessee. Populations increased and began to cause damage to cotton on the High Plains of Texas. From there, weevils continued to move westward, infesting the High Plains and Pecos Valley regions of New Mexico (Pierce *et al.*, 2001).

In the El Paso region, a localized boll weevil population was detected near the city of Juarez, Mexico in 1987. A cooperative eradication programme successfully eliminated the infestation in 1988. A similar infestation was detected in 1993 near Las Cruces, New Mexico. Cotton growers organized a voluntary, cooperative eradication programme with involvement of federal and state agencies. By the spring of 1997 it was evident that boll weevils had successfully overwintered in and around the city of Las Cruces. A formally organized programme with mandatory participation and access to all cotton fields would be required to achieve eradication. By the late 1990s, the Las Cruces–El Paso–Juarez region was infested with boll weevils (Pierce *et al.*, 2001).

In Arizona the Thurberia weevil, *A. grandis* var. *thurberi*, attracted the attention of cotton growers and entomologists. This biotype of the boll weevil feeds and reproduces preferentially on *Gossypium thurberi*, a wild cotton species that grows in the mountains of Arizona. In 1920, Coad and Moreland (1921) reported an outbreak of Thurberia weevil in cotton near Tucson. State authorities initiated a clean-up campaign and the infestation was eliminated. In 1926, another Thurberia weevil infestation occurred but did not persist. Once again, during the 1960s, an outbreak occurred (Neal and Antilla, 2001).

These outbreaks were associated with the production of 'stub' cotton. In southern Arizona cotton could be grown from the cut stub of the previous year's planting. The production of stub cotton kept host plant material in the field year-round. Stub cotton provided a much more conducive environment for the establishment and adaptation of the Thurberia weevil and boll weevil than did conventionally grown cotton (Bergman *et al.*, 1983; Cross, 1983). During the 1960s, the state of Arizona drafted regulations prohibiting the growing of stub cotton (Neal and Antilla, 2001).

During the 1970s the cotton boll weevil, *A. grandis*, was detected several times in Arizona. Prior to 1978, none of these infestations persisted into the next season. But the ban on stub cotton was temporarily lifted in 1978. This resulted in the cultivation of 40,000 acres (16,000 ha) of stub cotton. Soon afterwards, reproducing boll weevil populations were detected. Within 4 years boll weevil populations had spread across Arizona eastward to south-west New Mexico and westward to southern California (Neal and Antilla, 2001).

The 1 million acres (400,000 ha) of cotton grown in the San Joaquin Valley have been monitored since 1983 by a cooperative state–county–industry trapping programme. The boll weevil has never been detected in the region (Clark, 2001).

The first cotton was planted in Kansas in the early 1990s, and cotton acreage had increased to 50,000–60,000 acres (20,000–24,000 ha) by 2001. Trapping programmes began in 1998. A single weevil was caught in 1998 and six boll weevils were caught in the autumn of 2000. However, the boll weevil never established sustained populations in Kansas (Sim, 2001).

Hell to Pay: Cotton Production, Acreage and Economic Loss

In 1921, A.M. Soule of Georgia wrote: 'The boll weevil has disturbed our economic situation more than any other single factor since the conclusion of the Civil War; it is a pest of as great a magnitude as any which afflicted the Egyptians in the olden days' (Soule, 1921).

Boll weevil damage estimates vary greatly by year and location. In northern production areas, following harsh winters, cotton growers may experience no yield loss at all from boll weevil (Rummel, 1976a; Barker, 2001; Boyd, 2001). In the years before insecticides were available, damage reports ranged from 6.5% statewide in Texas in 1910 (Cook, 1923) to 90% and 100% in the 1890s near Brownsville, Texas (Townsend, 1895). Yield loss estimates in those early years generally varied between 20 and 75% (Worsham, 1914; Lewis, 1920; Isley and Baerg, 1924; Thomas, 1929; Coad, 1930; Wagner, 1999).

The development of insecticide-based control systems stabilized and diminished boll weevil losses, making it possible to produce cotton economically in the South despite the presence of the boll weevil (Parencia, 1978; Lloyd, 2001). By the mid-1920s, calcium arsenate was routinely applied by air to fields across the cotton belt (Hinds, 1926). As farming became mechanized, tractors with power-take-off dusters and sprayers gave farmers other options for efficient application of insecticides (West-brook, 1945; Beckham and Dupree, 1951). Following World War II, the development of synthetic organic insecticides gave farmers even more powerful weapons to use in the war against the boll weevil (Ewing and Parencia, 1950). Cotton yields were protected, but the cost of controlling boll weevils and the secondary pests – released as natural enemies and killed by treatments to control boll weevil – began to erode the profitability of growing cotton.

Trends in planted and harvested cotton acreage provide evidence of the declining profitability of cotton after the boll weevil became firmly established across the cotton belt (see Table 20.1). After the Civil War, the cotton industry in the South began to rebuild. The acreage devoted to cotton production increased steadily each decade for six decades, despite limited improvements in yield and fluctuating prices. The upward trend in cotton acreage peaked in the 1920s, the decade that the boll weevil became established throughout the cotton belt. After the 1920s, planted/harvested cotton acreage moved steadily downward for six decades, from the 1920s to the 1980s, despite yield increases and a strongly upward price trend. Acreage devoted to cotton production in the USA did not increase again until the decade of the 1990s, when boll weevil eradication programmes began eliminating the weevil as a yield-limiting factor over a large part of the cotton belt (Anon., 2002c).

Estimates of annual economic losses from the boll weevil have varied from US$125 million per year up to US$300 million per year since the pest arrived in the USA (Hunter, 1911, 1912a, b; Thomas, 1929; Coad, 1930; Folsum, 1932; Anon., 1950; Barnhart, 1950; Knipling, 1971; Coker, 1973; National Cotton Council, 1974; Perkins, 1980; Deterling, 1992). The consensus for the costs and losses caused by boll weevil during its stay in the USA is US$200 million per year. Numerous authors have referred to the boll weevil as the 'US$10 billion bug', the approximate value of the boll weevil associated losses through about 1950 (Anon., 1950; Coker, 1958;

Table 20.1. Planted and harvested cotton acreage, lint yield and price per pound by decade – from the 1860s to the 1990s[a]

Decade	Acres planted (millions)	Acres harvested (millions)	Yield (lb/acre)	Price (US$/lb)	Notes
1860s[b]	NA[c]	7.56	142.6	NA[c]	
1870s	NA[c]	11.256	174.2	9.17[d]	
1880s	NA[c]	17.598	172.2	9.00	
1890s	NA[c]	22.053	192.2	6.94	BW arrives
1900s	NA[c]	28.886	169.3	8.75	
1910s	34.151	33.301	184.3	17.48	
1920s	39.491	38.250	162.5	19.44	USA infested
1930s	32.952	31.223	205.4	9.37	
1940s	22.380	21.622	266.0	23.26	
1950s	20.079	18.737	362.4	33.70	
1960s	13.538	12.715	477.9	28.12	
1970s	12.660	11.834	475.0	45.36	
1980s	11.351	10.473	576.5	60.99	BWE began
1990s	14.054	12.833	644.7	63.33	

NA, not available,
[a]Source: National Agricultural Statistics Service – Track Records – Crop Production.
[b]Incomplete data for the decade. 1866–1869 data are presented.
[c]Data not available.
[d]Incomplete data for the decade. 1876–1879 data are presented.

Dunn, 1964; Mitlin and Mitlin, 1968). More recent estimates have put the economic loss from boll weevil to 1999 at US$17 billion (Carter *et al.*, 2001; Haney, 2001). Adjusted for inflation to 1999 dollars, Haney (2001) reported that control costs plus losses from boll weevil in the USA from 1892 to 1999 could amount to US$102 billion. Hardee (1972) credited the boll weevil with the distinction of being: 'the most costly insect in the history of American agriculture'.

The Perfect Storm: the Sociological Impact of the Boll Weevil

The economy of the southern and south-eastern USA was largely dependent on cotton in the decades following the Civil War. The arrival of the highly destructive boll weevil and the lack of any means to lessen its damage devastated the economy of the region.

Wherever the boll weevil went, destruction of agricultural communities followed. Soon after the weevil had entered Texas, Seaman A. Knapp arrived in Limestone and Robertson counties and, in 1903, he wrote of the devastation: 'I saw hundreds of farms lying out; I saw wretched people facing starvation; I saw whole towns deserted' (Bailey, 1945).

In 1915 a number of south-eastern farmers and agricultural businessmen toured boll weevil-devastated areas of Louisiana, Mississippi and Alabama. They ended the tour with a meeting in Brookhaven, Mississippi. The proceedings of this meeting, which later became known as the Brookhaven Report, provided a stark picture of the economic and social impacts of the boll weevil in the heavily cotton-dependent communities of the South (Riggs, 1921):

> Wherever the boll weevil has become established the result has been agricultural and economic panic and resulting demoralization. Advances to farmers by banks and merchants on the cotton crop have been greatly curtailed and values have been greatly depressed . . . labor has largely left the country, and the cotton crop, for the first few years at least, has been destroyed. The result has been the loss of lands and homesteads by owners, inability by tenants to pay out and a period of great poverty and distress among all classes of agricultural people.

Off the farm, the Brookhaven Report described how the damage caused by the boll weevil had devastated communities in the South. In 1906, Louisiana had 2076 cotton gins. By 1915, 788 of these gins (38%) had been closed. During the same period, 44% of the state's cotton oil mills had closed. Banks, local businesses and railroads were affected. Farms, homes, even whole towns were abandoned. Demoralized workers and tenants moved east or north (Vietmeyer, 1982).

Haney (2001) described the weevil onslaught as being: 'like an immense hurricane or tornado, but it was a silent storm that did no obvious physical damage as it continued raging for years on end'. Lloyd (2001) also used the storm analogy to describe the appearance of the boll weevil in cotton country:

> The economies of many communities were almost totally based on production and processing of cotton. Control options were limited and inadequate. Destruction of cotton crops by boll weevil infestation collapsed these economies and essentially halted

commerce. The result was economic and social upheaval equivalent to the annual reoccurrence of a widespread natural disaster. Income, possessions and land were lost.

As the weevil spread north and east from southern Texas, it caused cascading waves of social and economic disruption – year after year – for more than 30 years. It was commonly believed that only God could remove the boll weevil. Prayer meetings were held in farming communities across the South, and the Governor of South Carolina proclaimed a day of prayer for deliverance (Haney, 2001).

Weevil Wars: the Battle to Control the Boll Weevil

Cotton-free quarantine zones

As soon as farmers began to experience the boll weevil's destructive capability, they began to search for ways to control the pest or limit the damage it caused. In 1880, Edward Palmer recognized that the weevil depended on the cotton plant for its food source and on the cotton fruit forms in which it reproduced (Wagner, 1980; Walker, 1984). During the period from 1894 to 1895, cotton-free quarantine zones were proposed by three agriculturalists involving land areas that were thought to be wide enough to stop the northward movement of the boll weevil. In 1895, entomologist C.H.T. Townsend (1895) proposed a 50-mile (80-km)-wide no-cotton barrier in Texas. Also in 1895, both entomologist L.O. Howard and Charles Dabney, US Assistant Secretary of Agriculture, proposed similar quarantines. The Texas Legislature did not pass cotton-free zone quarantine legislation, and the boll weevil continued its movement northward through the state (Howard, 1895; Wagner, 1980; Stavinoha and Woodward, 2001). Quarantines to stop the movement of cotton products – which could possibly transport boll weevils – were enacted in Texas, but they did not stop the movement of the weevil (Haney, 2001).

The Louisiana Crop Pest Commission initiated a 30-mile (48-km)-wide, cotton-free quarantine zone from the Arkansas border to the Gulf of Mexico in 1904 to try to prevent further eastward movement of the boll weevil. The quarantine was not successful and this approach was not used again during the movement of the weevil through the South to the Carolinas and Virginia (Barker, 2001; Logan, 2001). In the West, a cotton-free quarantine zone was used in Arizona in 1920 to prevent the establishment of Thurberia weevils (Coad and Moreland, 1921). In 1926, a federal quarantine was instituted in the Marana District of Arizona, but its enforcement was restrained by a lawsuit. Quarantines against movement of boll weevils or materials capable of harbouring boll weevils were enacted by several, as yet uninfested, southern states (Newell, 1904; Hunter, 1912b; Worsham, 1912; Clemson University, 1918, 1919; Barker, 2001).

Cultural control

Very early in the boll weevil battle, entomologists began to develop and implement changes in the way cotton was being grown to give the growers a fighting chance to produce a crop. C.H.T. Townsend, L.O. Howard, E.A. Schwartz and C.L. Marlatt observed and recorded information from which management strategies were developed. F.W. Mally, W.D. Hunter, W.E. Hinds and S.A. Knapp began the work of developing and implementing cultural methods for managing the boll weevil. Mally (1901) recommended that farmers widen the distance between cotton rows to increase the amount of sunlight reaching the soil surface and increase soil surface heating. The higher soil temperatures resulted in greater boll weevil larval mortality in fallen squares. Later researchers found that close spacing between plants could be used to promote earliness, thereby maturing the crop before large, late-season boll weevil populations could develop (Cook, 1924; Hinds, 1928; Ware, 1929, 1930).

Mally (1901) also recognized the value of earliness. He recommended that growers avoid planting cotton after the optimum planting time and suggested that they adopt short-season cotton varieties to escape late-season boll weevils. Mally and others promoted post-harvest stalk destruction as a means of reducing the number of weevils that could survive the winter (Walker and Niles, 1971; Walker, 1984; Klassen and Ridgway, 2001). Hunter (1904b) added to the developing cultural management programme by recommending the use of fertilizers to aid in producing an earlier crop. Emphasis was placed on fertilization practices that would promote fruiting and determinant growth (Hunter and Coad, 1923).

For a short time, it was thought that hand-picking and burning of weevil-damaged squares could lower weevil populations and prevent some of the damage. This practice was soon abandoned as ineffective and too expensive (Wagner, 1980). C.L. Marlot in the late 1890s developed a more practical approach to controlling the larval stage feeding within squares. His strategy was to attach a cross-bar in front of the plough to knock infested squares from the plant during cultivation. Ploughing would then complete the job by burying the larvae within the fallen squares (Haney, 2001). Cultural control was embraced by entomologists and promoted to farmers; it became known as the Government Method. The Government Method was a success in that its use limited boll weevil damage; it provided growers with biologically based, cultural control techniques that remain in use today.

However, stalk destruction and other key components were usually not fully implemented and the boll weevil continued to take its toll. Growers were frustrated with the Government Method because of the difficulty in implementing stalk destruction in the age before mechanization and because, at best, it could accomplish only boll weevil suppression and not control. Proponents of the Government Method were disappointed because stalk destruction was not made mandatory (Helms, 1977; Wagner, 1980; Walker, 1984; Haney, 2001; Stavinoha and Woodward, 2001).

New developments added to the cultural control arsenal over time. Newell and Paulsen (1908) proposed defoliation of cotton plants to arrest the boll weevil's life cycle. Harned (1910) added the concept of using cotton varieties with thick carpal walls to the practices previously recommended by the Government Method. In 1911 the development of the V-shaped stalk cutter made stalk destruction somewhat more achievable (Anon., 1911). The development of power-take-off in 1922 cleared the way for the development and use of effective tractor-powered stalk shredders (Williams, 1987).

Boll weevil remedies

In the early years of the 20th century, hundreds of boll weevil remedies were sold. Many had no positive impact at all. Sprays, dusts, potions and 'weevil machines' were available. Hundreds of machines were tested. Some of them had promise, in theory. They were designed to knock off weevil-punctured squares and remove them from the field. None were successful in providing significant relief from boll weevil damage (Helms, 1977; Haney, 2001).

Predatory arthropods

The search for effective natural enemies of the boll weevil began soon after the weevil crossed into Texas. In 1903, Hunter travelled to Cuernavaca, Mexico to collect and import the predacious mite, *Pediculoides ventricosus*, which had been observed feeding on the boll weevil. The mite was transported to Texas and released, but had negligible effects on boll weevil populations in Texas, reportedly because of differences in the climate (Anon., 1904; Hunter, 1904b). Searches were conducted for effective parasites in Guatemala and Cuba but were unsuccessful (Howard and Galloway, 1904). Newell (1908) reported observing boll weevil larvae being preyed upon by a carabid beetle, and Pierce (1908) published a report of boll weevil larval predation by fire ants. By 1912, Pierce *et al.* (1912) had identified over 50 beneficial species that attacked the boll weevil. More recently, Sterling (1978) documented suppression of boll weevil populations by the red imported fire ant, *Solenopsis invicta*, in eastern Texas. However, predators have never been found capable of consistently maintaining boll weevil populations to below damaging levels (Ables *et al.*, 1983).

Parasitic wasps

Several parasitic wasp species have shown promise in suppressing boll weevil populations. Bottrell (1976) observed that *Bracon mellitor* could suppress weevil populations. Cross *et al.* (1969b) studied introduction and augmentation of *B. kirkpatricki* and Johnson *et al.* (1973) investigated release of *Cattolaccus grandis* to achieve boll weevil suppression. Augmentative releases of *C. grandis* were thoroughly investigated in the early and mid-1990s, and releases of the parasite were successful in reducing boll weevil populations. However, the parasitic wasp could not keep boll weevil populations below damaging levels and could not successfully overwinter in the USA, and laboratory-reared wasps were prohibitively expensive to produce (Morales-Ramos *et al.*, 1992; King *et al.*, 1993; Summy *et al.*, 1993; King, 1995).

Pathogens

The boll weevil can be infected in the field by fungal and protozoan parasites. McLaughlin (1962) demonstrated infectivity by the fungal pathogen, *Beauveria bassiana*.

The protozoans, *Mattesia grandis* and *Glugea gasti*, have been obtained from boll weevil colonies and field tested in bait formulations with high infection levels (McLaughlin, 1965, 1969; Daum *et al.*, 1967). However, the cost of producing the protozoan pathogens in the boll weevil host proved prohibitive (Cross, 1973) and neither the fungal nor the protozoan pathogens acted quickly enough to provide effective control of boll weevil populations in the field (Bell, 1983).

Host plant resistance

Efforts to find plant traits that would provide protection from the boll weevil began soon after the arrival of the weevil. Worsham (1915) proposed development of a plant-breeding programme to develop varieties that could withstand weevil damage. The Georgia State Board of Entomology initiated the plant-breeding programme in 1918 (Lewis and McLendon, 1919). Isley (1928) found that cotton with red leaves and stems provided a degree of protection, apparently because it was less preferred compared with other cotton phenotypes.

The frego bract trait also showed promise. Normal cotton bracts (sepals) are leaf-like structures that cover the flower bud. The space between the bracts and the bud provides the weevil with a snug place to hide, rest, feed and lay eggs. In frego bract cotton the bracts are thin and twist away from the flower bud. As a result, the flower bud is exposed to sunlight and open air. The trait was named after its discoverer, who found it in a field of Stoneville 2-B cotton near Manilla, Arkansas in 1944. Under low to moderate boll weevil pressure, cotton varieties with frego bracts were shown to inhibit boll weevil feeding and reproduction. Research on frego bract cotton varieties demonstrated that non-preference was the mechanism providing the damage reduction (Jones *et al.*, 1964; Lincoln and Waddle, 1965). In addition to damage reduction, frego bract cottons had higher rates of boll weevil larval parasitism by the parasitic wasp, *B. mellitor*, than other cotton types (McGovern and Cross, 1976).

Additionally, insecticide deposition was seven times greater on the flower buds of frego bract cotton than on the flower buds of cotton varieties with conventional bracts (Parrott *et al.*, 1973). Namken and co-workers (1983) observed that varieties with red stems and frego bracts suffered less boll weevil damage and needed fewer insecticide treatments than conventional varieties. However, they also found that the frego bract varieties were especially susceptible to damage by tarnished plant bugs, *Lygus lineolaris*. Although they provided a degree of boll weevil damage reduction, the distinctive red leaf/stem and frego bract traits did not provide enough weevil suppression to be included in modern varieties (McKibben *et al.*, 2001).

In Texas, researchers continued working on the principle of early-maturing cotton varieties and short-season production conceived by Fredrick Mally in 1900 (Walker and Niles, 1971; Parker *et al.*, 1980). Early planting of rapid-fruiting varieties allowed the cotton crop to mature before large, late-season boll weevil populations could develop (Cook, 1906, 1911; Bennett, 1908). In the early 1970s, genotypes with rapid-fruiting characteristics were bred (Niles, 1970; Namken and Heilman, 1973). During this time, Luther Bird and his co-workers developed short-season TAMCOT cotton varieties. Early planting of short-season varieties increased until 1982, when 10% of US production was planted to TAMCOT cultivars. In the late 1970s, 90–100% of the cotton produced in some Texas counties was from TAMCOT lines (Masud *et al.*, 1981; Bowman, 1999).

Inorganic insecticides

After earlier attempts to control the boll weevil using lime and ashes, Paris green, London purple, lead arsenate and many other concoctions, the widespread adoption and use of calcium arsenate for boll weevil control began in about 1920 (Haney *et al.*, 1996; Haney, 2001). The insecticide-dependent pest control system that quickly developed required multiple treatments per season. Regular use of insecticides became the principal method of cotton insect control, and this continued to be the case for decades (Parencia, 1978). In the 1920s, the Georgia State Bureau of Entomology recommended calcium arsenate applications every 4–6 days, nine to ten applications per season (Warren and Williams, 1922; Bass, 1993). Similar programmes with fewer treatments were recommended in areas with lower weevil populations. Repeated calcium arsenate applications sometimes had the undesirable effect of causing increased populations of the cotton aphid, *Aphis gossypii* (Ewing and Ivy, 1943).

In the 1920s, Dwight Isley, a proponent of the Government Method, began integrating cultural controls with judicious insecticide use. He found that use of insecticides on small, strategically located plots of early-planted cotton – trap crops – could reduce weevil populations significantly with minimal inputs. He advocated scouting and the use of economic thresholds to determine when to treat for weevils and other cotton pests. And he showed that spot-treating heavily infested areas of fields was effective against boll weevils in Arkansas. Isley built upon the ideas of the Government Method, developing the early foundations of integrated pest management (IPM) systems (Isley, 1933; Johnson and Martin, 2001; Klassen and Ridgway, 2001).

Synthetic organic insecticides

When the synthetic organic insecticides became available in the late 1940s and 1950s, the number of applications for cotton pests ranged from one or fewer per year in northern and dryland production areas with low pest pressure to 18 or more applications per year in warmer, high-rainfall and irrigated regions (Smith *et al.*, 1964; Haney *et al.*, 1996; Barker, 2001; Boyd, 2001). By the 1950s and 1960s, one-third of the insecticides used in US agriculture was used on cotton (Brazzel *et al.*, 1961; Knipling, 1971; Perkins, 1980).

The discovery of organochlorine insecticide resistance in Louisiana and Texas boll weevil populations in the mid-1950s shook grower confidence in insecticide-based control systems (Roussel and Clower, 1955, 1957; Parencia, 1959; Parencia and Cowan, 1960). The subsequent development of resistance in the tobacco budworm, *Heliothis virescens*, to DDT (Brazzel, 1963) and to methyl parathion (Nemec and

Adkisson, 1969), and in the bollworm, *Helicoverpa zea*, to DDT (Graves *et al.*, 1963) contributed to growers' fears. The extensive use of insecticides and the development of insecticide resistance in pests to chlorinated hydrocarbon, organophosphate and carbamate insecticides were major economic factors in cotton production during the 1960s and 1970s (Herzog *et al.*, 1996). The industry began looking in earnest for alternatives to chemical control of cotton pests.

Birds of the Air, Fish of the Sea: the Environmental Impacts

While repeated use of insecticides was successful in preventing serious losses from insect pests, there were environmental as well as economic costs. Repeated insecticide use led to depletion of the naturally occurring arthropod predators and parasites. Without natural controls, populations of secondary pests such as bollworms (*H. zea*), tobacco budworms (*H. virescens*) and the cotton aphid (*A. gossypii*) increased rapidly. The secondary pest infestations required the use of still more insecticide, a phenomenon called the 'pesticide treadmill' (Stern *et al.*, 1959). The result, over time, was an upward trend in insecticide use, increased environmental damage and higher production costs (Sheppard, 1951; Ridgway *et al.*, 1978).

The reliance on insecticides as the cornerstone for insect control systems in cotton and other crops did not go unnoticed by the developing environmental movement. Rachel Carson (1962) shocked her readers and fuelled the environmental movement with her book, *Silent Spring*, which drew attention to the environmental damage done by pesticides. Her book unified the environmental movement and resulted in extensive study and discussion on the impact of insecticides on the environment. Many long-residual insecticides were banned, and stringent environmental and human health measures were enacted to regulate pesticide use.

Thinking Outside of the Box: the Vision of Boll Weevil Eradication

Through the first half of the 20th century, boll weevil continued to be the dominant concern of cotton growers. It dictated production practices and was the source of both frustration and desperation among growers. Serious losses continued to be experienced by cotton producers. Growers could not or would not consistently destroy stalks; and the research community was not able to bring to bear an effective means of stopping boll weevil damage. In this environment, Hunter (1904b) remarked: 'The work of the Division of Entomology for several years has demonstrated that there is not even a remote possibility that the boll weevil will ever be exterminated.' This view was prevalent among many in the industry and academia for more than 75 years.

It was not until the 1950s that people began to seriously consider changing strategies for management/control of the boll weevil. Cotton production had become less and less profitable due to increasing production costs, and cotton acreage was in decline nationally. The industry had grave concerns about insecticide resistance.

Three critical components were needed to begin bringing about a change in strategy. They were:

- A great science leader and spokesman, E.F. Knipling.
- A strong producer group, the National Cotton Council.
- An attentive and sympathetic US Congress.

Fortunately, all the necessary components were present and fully engaged (Lloyd, 2001).

In 1954, Edward Knipling led a team of scientists who eradicated the screwworm from Curacao using areawide management concepts and the sterile insect technique (SIT) (Klasson and Ridgway, 2001). Knipling's vision and leadership skills were honed while working with Ray Bushland and others to develop the sterile insect approach for screwworm eradication. Orville Freeman, the former US Secretary of Agriculture, recognized the success of Knipling and his group against the screwworm as: 'the greatest entomological achievement of this century' (Anon., 2001). The increasingly desperate need for a change in cotton insect management systems and the success that had been obtained by Knipling and his team emboldened some in the cotton producer and research leadership to embrace the audacious idea that the boll weevil might be eradicated.

Knipling was described by those who knew him as 'a giant in the field of entomology'. He possessed superior intellect and had capabilities for sound and creative thought. In addition, his commanding voice and personality gave him a strong physical presence (McKibben *et al.*, 2001).

Knipling firmly believed that the boll weevil could be eradicated (Anon., 2001). He considered the boll weevil a good target for total population management, his conceptual framework for screwworm eradication. The weevil was an exotic pest, dependent on a single host plant through much of its US range, factors Knipling believed might make eradication feasible. He calculated that the areawide application of uniform controls could achieve eradication (Knipling, 1966, 1967, 1968, 1971). Knipling was particularly sensitive to the limitations of insecticides due to his work with DDT during World War II and the development of DDT resistance in the house fly in 1946 (Brown and Pal, 1971; Cremlyn, 1978; Klassen and Ridgway, 2001). Knipling's presentation at the 1956 Beltwide Cotton Conference (Knipling, 1956) was influenced by the report of boll weevil resistance to organochlorine insecticides (Roussel and Clower, 1955). In it, he outlined the need to broadly expand research in host plant resistance and chemical, biological and cultural control of the weevil. He emphasized the need for imagination to explore and develop untried approaches (Knipling, 1956). B.T. Shaw, Administrator of USDA-ARS, and Knipling agreed on the need for a sharp increase in basic research in support of the needs of agriculture (Shaw, 1956).

Dr Robert Coker was an influential leader of the cotton industry. He worked tirelessly through the National Cotton Council (NCC) to secure resources for the development of technology for boll weevil eradication. Coker envisioned growing cotton without the boll weevil. He shared his vision at the 1958 Beltwide Cotton Conference (Coker, 1958). Coker co-authored with J.F. McLaurin a resolution, which he presented at the NCC's 1958 annual meeting. The resolution called for the development of technology to: 'eliminate the boll weevil as a pest of US cotton at the earliest possible date'. It declared the boll weevil as the number one enemy of efficient cotton production. The resolution was passed unanimously and received a standing ovation (Anon., 2001; Carter *et al.*, 2001). Later that year, his testimony before the House Agriculture Subcommittee was critically important and led to Congressional action instructing the USDA to write a plan to deal with the boll weevil (Perkins, 1982).

Taft Benson, US Secretary of Agriculture, appointed a study group to examine facility needs to develop the necessary technology and report their findings. The report was submitted in 1959. It proposed the establishment of a central boll weevil laboratory and funding for three other USDA boll weevil research locations (USDA, 1958; Davich, 1976).

In 1960, Congress appropriated US$1.1 million for the construction of the USDA Agricultural Research Service (ARS) Boll Weevil Research Laboratory (BWRL) on the campus of Mississippi State University. In addition, it provided US$165,000 for initial staffing. The facility was completed in 1961. Congress also appropriated resources to increase funding for boll weevil research at the ARS laboratories in Florence, South Carolina, College Station, Texas and Baton Rouge, Louisiana. In 1961, President John F. Kennedy pledged his support to the boll weevil eradication effort. The importance of the boll weevil and the eloquence of those presenting the case for fighting it had attracted recognition and support for boll weevil research at the nation's highest level (Carter *et al.*, 2001). The intensified research efforts of the USDA and the states over the next 8 years resulted in a number of extremely promising boll weevil suppression technologies (Davich, 1976).

With leadership from Shaw and Knipling, the research community began shifting its emphasis from the search for new and better insecticides to broadly based biological and technical research. One of the keys to the success of this effort was in the selection of the leader for the new research laboratory. Dr Ted Davich was selected as Director of the BWRL in Mississippi. He had worked with Knipling on the successful screwworm eradication project in Curacao, and at the ARS Cotton Insect Laboratory at College Station, Texas before accepting leadership of the new BWRL (Ridgway and Mussman, 2001). Davich designed the laboratory, specified the equipment and recruited the scientists and support staff (Harris and Smith, 2001). He was a strong leader who valued diversity of thinking and had well-developed abilities to catalyse effective collaboration among research team members (McKibben *et al.*, 2001).

Building Blocks: Developing the Components of Boll Weevil Eradication

Application technology

The development of aerial application, begun in the 1920s by USDA, was of critical importance to later boll weevil eradication programmes (Post, 1924; Hinds, 1926). By 1931, the concept of applying insecticides from aircraft had been widely accepted, and the USDA Bureau of Entomology discontinued aerial application research and sold its aircraft (Parencia, 1978). In 1949, the technology for low-volume application was developed (Parencia, 1959). Ultra-low-volume (ULV) application technology was developed, tested and found to be effective (Hopkins and Taft, 1967).

Aerial application of pesticides was greatly improved with the development of satellite-based, differentially corrected, global positioning system technology (GPS). Adaptation of these systems to agricultural aircraft in the early 1990s greatly improved application precision, tracking and quality control. Joe Hartt with Satloc Inc., John Goodwin of Custom Farm Service in Stanfield, Arizona and Roger Haldenby with Plains Cotton Growers (PCG) in Lubbock, Texas collaborated in the development, field testing and use of this technology in the Texas High Plains Boll Weevil Diapause Suppression Programme. Prototype systems were used in the programme in 1992, and the first commercially available systems were used in 1993 (Haldenby, 2007, personal communication). GPS systems of this type have been used extensively for guidance and tracking of boll weevil eradication flights since that time.

Insect rearing

The primary motivation for developing boll weevil rearing capability was to support the sterile insect component of boll weevil eradication. The presence of a laboratory colony of weevils was, however, critically important in the development of many of the other tools needed for the eradication effort.

The development of systems to efficiently and economically rear boll weevils in the laboratory was a key component to the success of the research effort. Ted Davich collaborated with Erma Vanderzant in the rearing research effort. Together, they developed an artificial diet and oviposition substrate for laboratory-reared weevils (Vanderzant and Davich, 1958). Bob Gast conceived, designed and built the machinery necessary to scale up production of the diet so that large numbers of boll weevils could be reared at reduced cost (Gast, 1961; Gast and Vardell, 1963). Because diseases were a constant threat to the colony, Gast developed the initial processes for disease control (Gast, 1966). Further improved disease prevention within the colony was achieved by Sikorowski (Sikorowski *et al.*, 1977; Sikorowski, 1984).

Sterile insect technique

The concept for sterile insect technique (SIT)-based eradication was to overwhelm the native pest population with sterile males. This would result in non-viable sterile × wild-type crosses (Van der Vloedt and Klassen, 1991). The success of SIT in the screwworm eradication programme led those involved in research and development of boll weevil eradication programme components to believe that SIT would be the primary component in boll weevil eradication.

The components needed for a successful SIT programme against the boll weevil were:

- Efficient mass-rearing techniques to produce vigorous, healthy insects inexpensively.
- An effective sterilization process.

- A process for moving laboratory-reared insects to and from sterilization facilities.
- A process for metering sterilized insects into containers for shipment – a rapid, efficient shipping process.
- A release process that gently meters sterile insects into the environment.
- A system for monitoring programme effectiveness (McKibben et al., 2001).

Research on the components of the SIT programme was conducted concurrently at the BWRL. The sterilization component proved most problematic. Radiation was the first technique to be tried and the last to be abandoned. Doses of radiation high enough to produce sterility caused high mortality (Davich and Lindquist, 1962). Feeding, dipping and fumigation with various sterilizing chemicals were tried with and without irradiation. The results were either insufficient sterility or high mortality (Haynes, 1963; Lindquist et al., 1964; Davich, 1969; Gassner et al., 1974; Earle and Leopold, 1975; Haynes et al., 1975; McHaffy and Borkovec, 1976; Borkovec et al., 1978; McKibben et al., 2001).

Male weevils could be sterilized when treated with a relatively low dose of radiation, while a much higher dose was required to sterilize females. No rapid, reliable means of separating the sexes had been found. In order to use males that had been sterilized with a low dose of radiation, a means of preventing reproduction by the irradiated but still fertile females was needed. Tests with diflubenzuron demonstrated that the eggs of treated female weevils did not hatch (Moore and Taft, 1975; Moore et al., 1978). Methods of applying it with minimal negative effects on males were developed (Earle and Simmons, 1979; Earle et al., 1979; Haynes, 1981; McKibben et al., 2001). Work was conducted to find strains of boll weevils having better post-irradiation viability and competitiveness (Enfield et al., 1981, 1983, 1988; Villavaso et al., 1993). The tests were inconclusive and the project was abandoned (McKibben et al., 2001).

Sterilized and fertile males and virgin females were released into isolated weevil-free 1.5-acre (0.6-ha) cotton plots. Competitiveness was calculated based on egg hatch and sterile:fertile weevil ratios (Fried, 1971). Larger-scale field release studies were carried out in 1983 and 1984. Sterilized weevils were released into large fields of commercial cotton with native weevil populations. Weevil hatch was reduced by 50% against the relatively larger native weevil population in 1983, but only by 15% against the smaller native weevil population present in 1984 (Villavaso et al., 1989). Chemically sterilized boll weevils were used on 20,000 acres (8000 ha) in the Pilot Boll Weevil Eradication Experiment (PBWEE), 1971–1973. The experiment was considered moderately effective (Boyd, 1976a,b; McKibben et al., 2001).

Much time, money and effort were spent working to develop a SIT component for boll weevil eradication. It became clear in the 1980s, however, that SIT was too difficult, too expensive and not sufficiently effective. It became clear that the primary components used in the boll weevil eradication programme would be pheromone traps and malathion (McKibben et al., 2001).

Diapause control

Building on the observations of Coad (1915) and others who had observed that boll weevils underwent winter hibernation, Brazzel and Newsom (1959) described the diapause condition in the boll weevil. Soon afterwards, Brazzel conducted studies that revealed the vulnerability of boll weevil populations to insecticides in the autumn as they prepared to overwinter (Brazzel, 1959, 1962; Brazzel and Hightower, 1960; Brazzel *et al.*, 1961). Brazzel and Newsom's research showed that diapausing weevils stopped laying eggs and fed for an extended period of time in cotton fields to store fat for winter survival. They discovered that only those weevils in diapause could successfully survive the winter. Brazzel's diapause control strategy targeted diapausing adult weevils during their required feeding and fat storage period. His research during 1959 showed that four insecticide applications in the autumn, 14 days apart, reduced boll weevil populations by 90% the following spring (Brazzel *et al.*, 1961).

Knipling expanded on Brazzel's concept, suggesting greater population reduction could be achieved by targeting the last reproductive generation of the boll weevil as well as the diapausing generation (Knipling, 1963). He believed that the more aggressive reproduction-diapause strategy would result in the greater population reduction. And, he calculated that emerging populations in the spring could be reduced to even lower levels if large numbers of pheromone traps were used to remove weevils (Knipling, 1971). Ed Lloyd and co-workers field-tested Knipling's reproduction-diapause concept. Their approach involved seven insecticide applications, applied with 5–7 days between treatments (Lloyd *et al.*, 1964, 1966). The first areawide trials of this concept were conducted on the Texas High and Rolling Plains (Adkisson *et al.*, 1965a, b, 1966; Rummel and Adkisson, 1971).

The pheromone

For many years, entomologists had seen evidence of boll weevil aggregation and suspected the weevil might be making use of some kind of chemical attractant. Dwight Isley (1933) observed the boll weevil aggregation and developed his spot-treatment strategy based on this observation. However, the search for a pheromone, perhaps the most critical component in the development of boll weevil eradication programmes, was not a part of the original research plan when the BWRL was opened in 1961. Pheromones were just beginning to be understood: only the queen substance of honeybees and the sex pheromone of the silkworm moth had hitherto been described (McKibben *et al.*, 2001).

McKibben *et al.* (2001) reported that a worker in the mid-1960s had observed a large aggregation of boll weevils near an exhaust fan in the basement of the BWRL. Though not known at the time, the aggregation had occurred in response to the presence of thousands of male weevils in the rearing laboratory. The boll weevil aggregation was a phenomenon that generated discussion, thought and inquiry. It was soon demonstrated that a substance produced by male boll weevils was attractive to female boll weevils (Keller *et al.*, 1964; Cross *et al.*, 1969a, b) and that male boll weevils were more attractive to other boll weevils than cotton plants (Hardee *et al.*, 1969).

After initial studies attempting to find a chemical attractant in cotton squares, Jim Tumlinson began work in 1966 to isolate a pheromone attractant from the boll weevil. The first unsuccessful attempts to obtain the pheromone were conducted by drawing air over 62,500 male boll weevils to trap the volatile compounds they emitted and by extraction of whole insects. Later, Tumlinson found that steam distillation of an extract of boll weevil faeces produced a substance that was highly attractive to boll weevils. In order to obtain enough of the attractive substance for analysis, 135 pounds (56 kg) of boll weevil faeces were extracted and steam distilled (Tumlinson *et al.*, 1968; McKibben *et al.*, 2001).

An effective bioassay process was developed in Dick Hardee's laboratory (Hardee *et al.*, 1967). The substances Tumlinson isolated were tested using Hardee's bioassay process to identify which substances were attractive to boll weevils. Tumlinson reported having isolated, identified and synthesized the four-component boll weevil sex/aggregation pheromone in 1969 (Tumlinson *et al.*, 1969, 1971). The pheromone, given the name grandlure, was capable of attracting both male and female boll weevils (McKibben *et al.*, 1971; Hardee *et al.*, 1972; Mitchell *et al.*, 1972). In 1974, Hardee and co-workers published the results of studies to find the most active ratio of the four grandlure components (Hardee *et al.*, 1974). To bring grandlure to the development stage took 5 years of research work and cost US$800,000–1 million (Hedin, 1976).

Initially, cigarette filters were used as lure dispensers. They were impregnated with a mixture of polyethylene glycol and grandlure (McKibben *et al.*, 1971). About one million dispensers made from polyester-wrapped cigarette filters were used in the areawide Mississippi and Virginia/North Carolina programmes during the 1970s (McKibben, 1972; McKibben *et al.*, 1974). Later, laminated plastic lure dispensers were developed and shown to be effective (Hardee *et al.*, 1975). Laminated plastic dispensers are currently used in all active boll weevil eradication programmes.

The trap

As work on the boll weevil lure progressed, research was conducted to develop an inexpensive trap that was easy to service and could take full advantage of the powerful boll weevil pheromone. In the late 1960s, initial work was done on trap design (Cross and Hardee, 1968; Cross *et al.*, 1969a, b). At that time, various types of sticky 'wing' traps were being used. They were cumbersome and troublesome to work with because of the extremely sticky coating used on them (McKibben *et al.*, 2001).

In 1970, Joe Leggett came to the BWRL from the USDA Laboratory in Tallulah, Louisiana. He designed a boll weevil trap that combined an inverted floral liner attached to an upward-pointing screen funnel with a capture chamber placed on top of the screen funnel. His trap, called the Leggett trap, was very effective because of the propensity of the weevil to crawl upward after landing near a pheromone source – and it was much easier to use than sticky traps (Leggett and Cross, 1971). Mitchell and Hardee (1974) made improvements to the Leggett trap designing a similar, smaller, but very effective in-field trap. Subsequent minor modifications have been made to the in-field trap (Dickerson *et al.*, 1981; Dickerson, 1986).

Cross (1983) noted that weevils responded best to traps that were of daylight fluorescent colour that looks lemon yellow to our eyes. Technically, he described the colour as being in the blue–green range of the spectrum near 525 nm. He reported that traps of other colours were less attractive and noted that red was not attractive. Cross' observations fit well with those of Isley (1928) that boll weevils did not prefer red cotton.

In addition, Cross (1983) reported that boll weevils do not respond uniformly to traps during all parts of the growing season. They respond very actively in the early season and in the late season, but have a much-reduced response mid-season during peak cotton squaring and blooming. Those running traps in boll weevil-infested areas commonly observe this phenomenon. Some believed that the less effective mid-season detection would keep boll weevil eradication programmes from being successful. Fortunately, this has not been the case.

Those who used the pheromone-baited boll weevil trap quickly recognized its value in monitoring boll weevil populations. The technology was soon being studied as a control component as well. Knipling and McGuire (1966) presented a concept they called the 'competitive theory of attraction', whereby the efficacy of pheromone traps was inversely proportional to the number of competing weevils in the field. Their theory was later confirmed by field tests (Lloyd *et al.*, 1972, 1980). Theory and field evidence suggest that, as the population declines, trap efficacy goes up exponentially. In boll weevil eradication programmes, traps become increasingly effective as boll weevil populations in an area approach zero. At low weevil densities the trap becomes a powerful tool in removing the last weevils (McKibben *et al.*, 2001).

Insecticides

Malathion ULV has been the primary insecticide used in boll weevil eradication because of its excellent activity against the insect, it can be easily and efficiently applied, it is relatively inexpensive and it has low toxicity to most non-target organisms. Harris *et al.* (1999) reported on work with malathion in catfish ponds. Malathion in the water and in the fish was very quickly broken down to undetectable levels.

The early areawide suppression programmes, experiments and trials used 12–16 fluid ounces (355–473 ml) of malathion ULV per acre. A number of researchers investigated the use rate to determine whether efficacy could be maintained at lower use rates. No difference in initial efficacy has been seen for rates from 10 to 16 fluid ounces (296–473 ml) per acre. Several days following application, however, lower use rates tended to provide somewhat lower control and lower malathion residual (Burgess, 1965; Cleveland *et al.*, 1966; Hopkins and Taft, 1967; Mulroony *et al.*, 1995, 1996, 1997; Jones *et al.*, 1996; Villavaso *et al.*, 1996b, 2000). Rainfall quickly reduced efficacy (Hopkins and Taft, 1967; Nemec and Adkisson, 1969). Application of malathion ULV to dew-covered leaves did not reduce effectiveness or malathion residues (Kirk *et al.*, 1997).

Guthion® ULV (azinphosmethyl) was used in place of malathion for the autumn diapause treatments in 1987, the initial programme year in north Florida, south Georgia and south Alabama and under high-lines, etc. in South Carolina. Lloyd *et al.* (1967) compared rates of Guthion® ULV and the 16 fluid ounces (473 ml) per acre rate was selected for use in the programme. Many environmental problems resulted from the use of Guthion® ULV in areawide boll weevil eradication in Florida, Georgia and south Alabama, and the product was not used in US boll weevil eradication programmes after 1987 (Haney *et al.*, 2001a).

Dimilin® (diflubenzuron) was used in the Boll Weevil Eradication Trial (BWET) in Virginia and northern North Carolina (to be discussed later). Dimilin® is an insect growth regulator that acts to inhibit chitin production. Female boll weevils treated with the compound lay eggs that do not hatch. Dimilin® provided suppression of boll weevil populations in field tests (Taft and Hopkins, 1975; Lloyd *et al.*, 1977).

Its use in boll weevil eradication programmes was discontinued after the BWET because it was too expensive and it was not sufficiently effective (Roof, 2001).

Buffer zones

Insecticide-treated buffer zones were set up to protect gains made in the BWET and subsequent eradicated areas in the south-eastern USA. The question of how wide to make the buffer zones was addressed using information from the original infestation of the US cotton belt (Coad *et al.*, 1922) and from subsequent studies on boll weevil movement (Davich *et al.*, 1970; Knipling, 1971).

Bait sticks

Boll weevil bait stick technology combines a high dose of grandlure with a mala-thion-coated, yellow, hollow cardboard tube. The tubes are placed around cotton fields to attract and kill boll weevils in the area. Some research has indicated that use of the bait stick technology provided effective boll weevil control (McGovern *et al.*, 1993, 1996; Villavaso *et al.*, 1998), while other research has found them to be ineffective (Fuchs and Minzenmayer, 1992; Karner and Goodson, 1995; Parker *et al.*, 1995; Spurgeon *et al.*, 1999; Spurgeon, 2001). They have been used in boll weevil eradica-tion, but only on a very limited basis.

Mapping and trapping information systems

Boll weevil eradication programmes have developed computer systems to facilitate programme operations. The systems allow field mapping, field and trap numbering and geo-referencing of fields. They provide means of overlaying aerial application information on field maps for quality control purposes. Data management systems allow trapping data to be efficiently downloaded from hand-held electronic scanners and integrated with maps to facilitate field treatment. In addition to the programme efficiencies provided by hand-held scanners, scanner records provide the trap loca-tion, time of inspection, crop stage, lure change and kill strip change data. This infor-mation can be used to provide quality control of trapping operations (El-Lissy and Moschos, 1999; Harris and Smith, 2001; Goswick *et al.*, 2007).

Together We Fight: Implementation of Areawide Boll Weevil Management Systems

Areawide efforts in the calcium arsenate era

By nature, cotton growers are a strongly independent group. However, the boll weevil's effects were sufficiently severe that grower independence began to be replaced by cooperation. Areawide projects were designed to manage more efficiently the weevil problem. Soon after the discovery that calcium arsenate could provide boll weevil control, large-scale, areawide 'Extension' trials were conducted with cooperating growers in Arkansas (Coad, 1918). In 1939, the majority of Georgia cotton growers participated in an areawide treatment programme in the spring to reduce boll weevil populations (Westbrook, 1939). Boll weevil populations were suppressed and cotton yields increased (Anon., 1940). Treatment of overwintered boll weevils became standard practice, but boll weevils continued to cause damage each year. Profitable production of cotton required in-season treatment of boll weevils in most of the cotton-producing regions of the USA.

Areawide efforts after World War II

The introduction of synthetic organic insecticides in the 1940s was the motivation for a revival of the areawide control concept. In the late 1940s, Ewing and Parencia conducted 3 years of areawide, early-season boll weevil control tests in the Wharton and Waco (Texas) areas. Treated fields made an average of 82% more lint cotton than fields not in the treated area. They proposed the use of the new synthetic organic insecticides in early-season, community-wide boll weevil suppression programmes against overwintered boll weevils (Ewing and Parencia, 1949, 1950). Although sustained areawide boll weevil control programmes did not develop, farmers in heavily boll weevil-infested areas routinely applied one or more insecticide applications to early square-stage cotton for control of overwintered boll weevil and other pests from the 1960s through to the mid-1990s. State cooperative extension services recommended using scouting or traps to determine the need for overwintered weevil applications (Rummel *et al.*, 1980; Benedict *et al.* 1985; Moore *et al.*, 2003; Johnson *et al.*, 2006).

The Texas High Plains boll weevil diapause suppression programme

Brazzel's work on diapause control (Brazzel, 1959; Brazzel *et al.*, 1961) was the catalyst for the development of an areawide diapause control against the boll weevil on the Texas High and Rolling Plains.

While most US cotton farmers had been in a fight to survive the depredations of the boll weevil, farmers on the Texas High Plains had farmed for 70 years without the weevil. The High Plains, comprised of almost 20 million acres (8 million ha), is a relatively level, high plateau separated from the Rolling Plains to the east and south by the caprock escarpment. The land in the Llano Estacado region has been used to grow some 3 million acres (1.2 million ha) of cotton a year, earning it the title of the 'world's largest cotton patch'. Cotton growers in the region had remained weevil free primarily due to the environmental conditions – cold winters and hot, dry summers. Cotton growers in the region, represented by PCG, had no desire to join other Southern cotton growers in having their fields infested by the boll weevil.

Boll weevils first moved on to the High Plains from the Rolling Plains in the late 1950s. The first confirmed boll weevil reproduction on the Texas High Plains was observed in several fields in Crosby and Dickens Counties by R.S. Conner in 1959 (Rummel, 1970; Bottrell *et al.*, 1972; D.R. Rummel, 2007, personal communication). During the period from 1961 to 1963, boll weevils moved westward from heavily infested Rolling Plains fields east of the caprock into fields along the entire eastern edge of the High Plains. By 1963, boll weevil-infested fields were reported as far west as the Lubbock County line (Rummel *et al.*, 1975; Leser *et al.*, 1997). The growers in the region were shocked that the weevil had overcome the perceived caprock barrier (Stavinoha and Woodward, 2001).

In the autumn of 1963, a number of cotton producers in the eastern counties of the Texas High Plains asked the Texas Agricultural Extension Service and PCG to organize a meeting to discuss the problem. The meeting was held in Floydada, Texas in October of that year. The growers had read about Brazzel's diapause control experiments. They wanted to organize an areawide boll weevil diapause suppression programme using Brazzel's methods to suppress weevil populations on the High Plains (D.R. Rummel, 2007, personal communication). APHIS and PCG, with technical guidance from Texas A&M, Texas Tech and Texas Department of Agriculture, developed the programme. The plan was to apply ULV malathion in the autumn to fields in the western Rolling Plains and eastern High Plains region. ULV malathion would be applied at a rate of 16 fluid ounces per acre at 10–14 day intervals. Initially, two to four applications were made before and during harvest; PCG and APHIS would share the cost of the operation; day-to-day programme supervision was initially the responsibility of APHIS (Stavinoha and Woodward, 2001). The programme goals were, first to prevent the westward movement of the boll weevil and secondly to prevent the boll weevil from becoming an economic pest on the High Plains (Adkisson *et al.*, 1965a).

APHIS operated the programme from 1964 to 1967. During that time, 1.1 million cumulative acre applications were made at a cost of just over US$5 million. The programme reduced the overwintering boll weevil population by 99% in 1965 and successfully maintained boll weevil populations at below economically damaging levels on the Texas High Plains for almost 30 years (Haney *et al.*, 1996; Adkisson, 1968; Knipling, 1968; Stavinoha and Woodward, 2001; D.R. Rummel, 2007, personal communication).

In 1965 and 1966, limited spring applications were made to the eastern edges of fields along the caprock to control overwintered weevils (Adkisson *et al.* 1965b; Haldenby, 1992; Leser *et al.*, 1997) and, in 1965, the programme was modified, adding three reproduction-diapause treatments at 5-day intervals. A number of treatment strategies were used from 1966 to 1968 and alternative insecticides were evaluated. But, by 1972, ULV malathion at 12 fluid ounces per acre had become the treatment of choice due to its effectiveness, application efficiency and low cost (Stavinoha and Woodward, 2001).

PCG took over day-to-day programme operations in 1968 and the programme was continued until 1997. The programme kept boll weevils in check from 1968 to 1992. However, boll weevils established a foothold on the High Plains after 1992. Five consecutive mild winters and the establishment in Texas of 4.2 million acres (1.7 million ha) of new boll weevil overwintering habitat in the form of Conservation

Reserve Programme (CRP) grass, most of which was in the High and Rolling Plains regions, contributed to the weevil's success (Carroll *et al.* 1993; Zinn, 1994). By the late summer of 1995, 2.3 million of the 3.2 million cotton acres on the High Plains were boll weevil infested. Funds were limited and not all of the infested acreage could be treated. The programme ended in 1997 (Leser *et al.*, 1997; Stavinoha and Woodward, 2001).

The High Plains Boll Weevil Suppression Programme was the first cooperative attempt in the USA to control the boll weevil on a large-scale, areawide basis (Adkisson *et al.*, 1965b). It operated for 32 years, successfully preventing economically important boll weevil damage and stopping the westward movement of the boll weevil during that time. Through the mid-1980s the economic benefits of the programme were US$273 million, with an overall net cost of US$17 million, a net cost:benefit ratio of 1:16 (Carlson *et al.*, 1989; Frisbie *et al.*, 1989). The High Plains programme and the diapause control programmes that followed set the stage for boll weevil eradication.

Diapause control programmes spread

Other areawide control programmes were initiated in the 1960s and 1970s. In 1969 and 1970, the Foundation for Cotton Research and Education, the forerunner to the Cotton Foundation, provided grants to Alabama, Arkansas, Georgia, Louisiana, Mississippi, North Carolina, Tennessee and Texas to help fund diapause control trials.

Texas cotton growers in areas other than the High Plains approached the Texas Legislature for matching funds for diapause control programmes. Funds were appropriated, beginning in 1965, for cooperative diapause control programmes involving cotton grower associations and the Texas Department of Agriculture. In 1968, the St Lawrence Cotton Growers Association conducted a diapause programme on 46,000 acres (19,000 ha) of cotton south-east of Midland. Infested acres, primarily on the east side of the area, received an average of three treatments in the autumn. This programme reduced in-season treatment costs by 98.4% compared with prior years. The programme was continued in the St Lawrence area until boll weevil eradication began there in 2004 (Rummel, 1976b; Stavinoha and Woodward, 2001). Also in 1968, the Schleicher County Cotton Producers Board and South Texas Cotton and Grain Association applied for matching funds and conducted diapause control programmes in their areas. These programmes were continued until the late 1980s (Stavinoha and Woodward, 2001).

In Mississippi, areawide reproduction-diapause programmes were conducted in 1968 and 1969 in Monroe and Sharkey counties. The Monroe County programme used pheromone-baited sticky traps and autumn insecticide treatments. Very good results were obtained in the Monroe County trial in spite of less than 100% grower participation. The areawide reproduction-diapause control concept was later promoted state-wide (Lloyd *et al.*, 1972; Harris and Smith, 2001; McKibben *et al.*, 2001).

An areawide diapause control programme using ULV azinphosmethyl was conducted from 1969 to 1971 in the Coosa River region of Alabama. Insecticide applications were made in the autumn of each year. By 1970, in-season treatments had been reduced from an average of ten applications per year to one to three applications per year. Ed Lloyd and Floyd Gilliland of the BWRL provided technical assistance and the NCC provided financial assistance (Ledbetter, 1971; Haney *et al.*, 1996; Curtis, 2001).

In 1969, the Georgia Agricultural Commodity Commission for Cotton and the NCC conducted a cooperative boll weevil diapause study in Randolph County, Georgia (Womack, 1970; Haney *et al.*, 2001a).

Rummel (1976b) noted two limitations of areawide diapause boll weevil control programmes. He reported that programmes with limited producer participation invariably failed. And, referring to a diapause control trial in Frio County in South Texas, he noted that successful boll weevil population reduction did not ensure programme success. If large numbers of weevils existed near enough to the suppressed area to move in and cause significant crop damage and/or control cost, a diapause control programme had little chance of being effective.

Other areawide insecticide treatment programmes

In the late 1960s, North Carolina State University Cooperative Extension began organizing 'spray groups'. They trained scouts and helped organize and manage the spray groups. Marshall Grant – who later became one of the premier leaders of the boll weevil eradication effort in the North Carolina programme, the south-east programme and the national programme – was elected chairman of the Gaston Spray Group. All area cotton was scouted by Spray Group scouts and the chairman of the Spray Group made the treatment decisions based on thresholds recommended by North Carolina State University cotton entomologists. Cotton was scouted weekly and treated at 5-day intervals when worm or weevil populations were high enough to justify treatment. Decisions were made on a field-by-field basis when pest populations were low. During the first year of the programme 80% of the area farmers participated, but by the third year 95% of the cotton farmers in the area were participating.

It was apparent that treating cotton on a community-wide basis was advantageous. Pests were controlled quickly and uniformly and treatment costs were lower than when growers were on their own. North Carolina's Spray Groups were valuable as boll weevil eradication plans moved forward. The successful experience of the growers working cooperatively to manage their cotton insect pest problems had preconditioned them for areawide boll weevil eradication. Marshall Grant believed this was a major factor in Jim Brazzel's decision to implement the Boll Weevil Eradication Trial in north-eastern North Carolina (Dickerson *et al.*, 2001).

Community cotton pest management programmes were organized by J.R. (Jake) Phillips in 1972 and operated through the early 1980s in several cotton-growing areas in Arkansas. These programmes, conducted by the Cooperative Extension Service, demonstrated to cotton growers the value of managing cotton insect pests in areawide programmes (Phillips, 1978; Phillips *et al.*, 1981).

Areawide cultural control programmes

Since the early 1900s, the benefits of early, thorough stalk destruction have been advocated by entomologists (Mally, 1901; Newell, 1904). In the early part of the 20th century, the destruction of cotton stalks without mechanized equipment or herbicides was difficult to achieve. As mechanical and chemical tools became available in later years early, thorough stalk destruction became easier, but it was not often achieved. Grower leaders in several states became frustrated with the lack of complete stalk destruction and approached their state legislatures and regulatory agencies to request laws and regulations making stalk destruction mandatory. Many cotton-producing states developed and now have laws requiring stalk destruction by specified dates (Clark, 2001; Neal and Antilla, 2001; Stavinoha and Woodward, 2001).

In the 1970s, southern Texas cotton producers found that short-season cotton production practices gave them the advantage of being able to produce a cotton crop before large, late-season boll weevil populations could develop. Timely planting of early-maturing varieties, use of proper fertilizer rates, proper irrigation where available, control of early-season insects when necessary, crop management for early maturity, early harvest and early, thorough stalk destruction were the practices recommended in the short-season approach. The short-season approach had the added benefit of allowing the crop to be harvested by late August or early September: prior to the mid-September–October hurricane season. Areawide, short-season production practices continue to be recommended and widely used in the Texas Coastal Bend and in the Lower Rio Grande Valley (Walker *et al.*, 1977; Parker *et al.*, 1980; Frisbie *et al.*, 1983).

Another areawide cultural control strategy was developed for the Texas Rolling Plains region in the early 1970s. Cotton production in the region is predominately dryland. The area has a long growing season and much of the crop is not harvested until after the first freeze in November. The late harvest renders implementation of an effective, early stalk destruction programme of little value. Alternatively, Rolling Plains growers could uniformly delay planting so that cotton did not square until after the majority of the weevil population had emerged and died (Slosser, 1978; White and Rummel,1978; Rummel and Carroll, 1985). When planting was uniformly delayed, the larvae of the few surviving weevils were subjected to higher soil temperatures as the squares in which they fed fell to the ground. Since the early 1970s, delayed uniform planting has been widely practised on some 1 million acres (400,000 ha) of cotton grown in the Rolling Plains region (Slosser, 1995; Walker and Smith, 1996; Fuchs *et al.*, 1998).

Currents in the Stream: the Development of Cotton Integrated Pest Management

Insecticide resistance, resurgence of secondary pests and the rising cost and environmental impact of insecticide use drew attention to the need for a change in the way insect pests were being dealt with in cotton and other crops. State and USDA entomologists along with many farmers began to recognize that changes had to be made. Beginning in the mid-1950s, new problems with the strategy of near exclusive reliance on insecticides as the solution to insect pest problems were exposed each year. Scientists began integrating biological, cultural and chemical control methods resulting in production systems which were more economically and environmentally sustainable.

In 1971 USDA-APHIS funded a pilot cotton IPM project in North Carolina and Arizona (Frisbie *et al.* 1989). The project was expanded to 14 states in 1972, and the Cooperative Extension Service (CES) with funding from USDA-Extension assumed, shortly thereafter, administration of the project. The objectives were to reduce the amount of DDT and other insecticides entering the environment, to implement as-needed spraying based on field scouting and to promote use of insecticides against diapausing boll weevil populations in the autumn so that early season spraying could be reduced and natural enemies could be conserved. These IPM projects were highly successful and were continued in 1975 with additional USDA and CES funding. In 1976 an additional US$1.2 million was provided to CES to develop statewide programmes in the 11 southern cotton producing states. These federal funds were the nucleus for expanded IPM programmes. Through the 1970s and 1980s cotton IPM programmes expanded to include other crops and plant protection disciplines (R.E. Frisbie, 2007, personal communication).

Cotton IPM, though typically thought of as being practised on individual farms and fields, demonstrated the value of areawide insect suppression and grower co-operation to manage pest problems. IPM programmes introduced and promoted areawide practices for managing boll weevil to many cotton farmers. Through IPM programmes farmers gained experience with autumn diapause control, spring insecticide programmes for overwintered boll weevil control and areawide cultural control efforts. Some of the areawide cultural control components promoted by IPM programmes included stalk destruction, short season cotton production systems, and optimum or uniform delayed planting strategies. Positive grower experience with IPM programmes led to improved grower confidence in areawide, cooperative approaches to cotton insect problems. As plans for boll weevil eradication developed and trial eradication programmes were conducted, the confidence gained in IPM programmes helped growers to believe that it might be possible to eradicate the boll weevil (R.E. Frisbie, 2007, personal communication).

Testing the Water: the Pilot Boll Weevil Eradication Experiment

Planning

The 1960s brought continuing boll weevil losses, increased insecticide use and control costs, increased risks, declining cotton profitability and acreage, the development of insecticide resistance in several cotton pests, and escalating environmental pressures. In the wake of these troubles, cotton growers and the NCC were ready to take action against the boll weevil. The proven effectiveness of diapause control tactics and progress in the development of the pheromone and the trap along with advancements in rearing and sterilizing boll weevils gave the industry hope that boll weevil eradication could soon become a reality. The severity of the problems facing growers was cause for urgency and impatience. Any delay in starting boll weevil eradication would do further damage to an industry that had already endured years of struggle and sacrifice to produce cotton in fields infested with boll weevils.

In 1969 Jim Mays, president of the NCC, appointed a Special Study Committee to review the current technology and consider the feasibility of taking action to eliminate the boll weevil as a pest of cotton (Smith, 1998). The committee was headed by Robert Coker and included grower leadership, research leaders and representation from NCC, Cotton Incorporated (CI) and the agricultural press. It was assisted by a Technical Advisory Group. The Special Study Committee concluded that adequate technology had been developed to expand boll weevil research to large-scale field testing. It recommended that a Pilot Boll Weevil Eradication Experiment (PBWEE) be conducted in 1970. The objective of the experiment was to assess the technical and operational feasibility of boll weevil eradication. The Special Study Committee appointed a Site Selection Subcommittee to choose a site for the experiment. The subcommittee interviewed people from all cotton-producing states. It then recommended that the PBWEE be conducted in a 50-county area in south Mississippi, Alabama and Louisiana – an area that had over 20,000 acres (8000 ha) of cotton. The Special Study Committee approved the site and recommended that the PBWEE be conducted over a 3-year period (Davich, 1976; Carter *et al.*, 2001; Harris and Smith, 2001; McKibben *et al.*, 2001).

APHIS was given the responsibility of carrying out the experiment, while research support was the responsibility of ARS and state universities. Jim Brazzel with APHIS and Edward Knipling with ARS were co-chairmen of the Technical Guidance Committee (TGC), which was composed of representatives from USDA, universities, state departments of agriculture and the cotton industry. The TGC was responsible for setting policy for the experiment and for interpreting the results (Davich, 1976; Perkins, 1982).

Pilot boll weevil eradication experiment

The PBWEE was conducted in 30 counties in south Mississippi, five parishes in Louisiana and two counties in Alabama. This area was selected because fields were small, 5–12 acres (2–5 ha) average size and surrounded by tall trees. The rationale was that the area selected would provide as great a challenge to boll weevil eradication as would be faced in any US cotton-producing area. If the boll weevil could be eradicated from this area, a strong argument could be made that eradication could be successful beltwide (Boyd, 1976b). Funding for the PBWEE included US$2.5 million from USDA-ARS, US$1.08 million from the USDA Cooperative States Research Service, US$1.08 million from CI and US$520,000 from the state of Mississippi – a total of US$5.2 million (Boyd, 1976b; Carter *et al.*, 2001; Harris and Smith, 2001). The PBWEE was designed to determine whether technology was available to eradicate the boll weevil from the USA (Perkins, 1982).

The programme was initiated in 1971 on about 24,000 acres (9700 ha) of cotton. In both 1972 and 1973 there were about 19,000 acres (7600 ha) in the experiment. The programme design consisted of an outer buffer zone 80 km wide, to reduce the effects of migration, and an inner core, where the evaluation was conducted.

The technology tested was spring, in-season (as needed) and autumn insecticide applications; pheromone traps; trap crops and release of sterile boll weevils. The PBWEE was terminated 1 year earlier than planned, on 10 August 1973, because the funds for the programme had been spent (Cross, 1973; Smith, 1973; Boyd, 1976b; Carter *et al.*, 2001; Harris and Smith, 2001). Encompassing an area of approximately 20,000 square miles (54,800 km²), the PBWEE was the largest entomological experiment ever conducted (Davich, 1976).

In spite of the discovery of an untreated field in the programme area, which infested 1800 nearby acres (728 ha) and in-season applications by only 50% of the growers, good progress was made during the first year of the programme. By the end of the experiment, weevil populations were below detectable levels in 203 of the 236 fields in the eradication zone's inner core. All of the infested fields were located in the northern third of the eradication zone inner core, being < 40 km from substantial populations of boll weevils in fields further north (Boyd, 1976b; Perkins, 1982; Brazzel *et al.*, 1996; Harris and Smith, 2001). Cross (1976b) presented convincing evidence from pheromone trap lines that boll weevils migrated into the core area from infested cotton outside the eradication zone.

Post-pilot boll weevil eradication experiment controversy

E.F. Knipling's draft interpretive statement of the PBWEE results concluded that it was 'technically and operationally feasible to eradicate the boll weevil from the USA by use of techniques that are ecologically acceptable'. Some members of the TGC did not agree with Knipling's interpretation. The focus of the debate was the source of the boll weevils that were found in the core area of the eradication zone. Knipling had concluded that the weevils had migrated into the zone from infested fields in the buffer area. Some of the committee questioned the idea that a reproductive infestation could be ruled out. If the weevils originated from reproduction within the zone, it could be concluded that the available technology and process were not capable of achieving eradication. After much debate, the TGC amended Knipling's original language in the interpretive statement of the report. The term 'eradicate' could not be agreed upon because weevils were found in the core area of the eradication zone. The committee agreed upon use of the words 'eliminate the boll weevil as an economic pest' in place of 'eradicate' in their report. The report recommended continued improvements be made in the technology for boll weevil eradication (Knipling, 1976; Perkins, 1982; Harris and Smith, 2001).

Two other committees reviewed the PBWEE results. A committee from the Entomological Society of America (ESA) chaired by William Eden of the University of Florida determined that eradication had not occurred, but the ESA committee could not agree on whether there was a difference between accomplishing eradication and demonstrating the feasibility of eradication. The committee provided neither guidance nor a recommendation on the question of whether beltwide eradication should be attempted. It had reservations about conducting such a massive undertaking as boll weevil eradication without improvements in the suppressive techniques (Eden, 1976; Perkins, 1982).

A National Academy of Sciences (NAS), National Research Council Cotton Study Team chaired by Stanley Beck, an insect physiologist from Wisconsin, also reviewed the PBWEE. The Cotton Study Team expressed strong doubts about the technical feasibility of eradicating the boll weevil. It did not thoroughly review the PBWEE reports, but instead relied heavily on the interpretation of Perry Adkisson, the only cotton entomologist on the team. The revised final report of the NAS Cotton Study Team expressed strong reservations about the feasibility of eradicating the boll weevil, but it approved the conducting of a trial eradication programme in North Carolina (NAS, 1975; Perkins, 1982).

The Secretary of Agriculture was authorized by the Congress in the NCC-supported Agriculture and Consumer Protection Act of 1973 to carry out an eradication programme against the boll weevil if it was feasible to do so (Carter *et al.*, 2001; McKibben *et al.*, 2001). The NCC's Special Study Committee prepared a plan for a national programme to eliminate the boll weevil and submitted it to the Secretary of Agriculture on 12 December 1973 (Carter *et al.*, 2001).

A Choice to Make: Manage the Weevil or Try to Eradicate It

Planning, organization and funding for BWET and OPMT

The NCC's Beltwide Action Committee on Boll Weevil Elimination was appointed in late 1973 and met in St Louis, Missouri in January 1974. The committee, chaired by Robert Coker, advocated action to obtain an early, favourable decision from the Secretary of Agriculture to implement a national boll weevil eradication programme at the earliest possible time. They advocated starting the programme no later than 1975 and stressed the importance of continuing diapause and other suppression programmes in advance of initiation of programmes in the various regions of the Cotton Belt (Coker, 1976).

In February 1974, a meeting of industry leaders, government officials and state and federal scientists was held in Memphis, Tennessee. A group of entomologists at the conference raised concerns about the plans for an immediate start-up of a national boll weevil eradication programme. They contended that elimination of the boll weevil from a defined area had not been proved and could not be achieved. Their concerns brought a temporary halt to implementation of the national boll weevil eradication programme called for in the 1973 Farm Bill.

In April 1974 industry leaders met with USDA to develop an alternative plan. USDA proposed a trial programme and asked NCC to set up meetings in Texas, Oklahoma and North Carolina to get a better understanding of local interest. From these meetings, a strong consensus developed that the trial should be conducted in the south-east. In October 1974 a meeting of industry leaders, university officials and state departments of agriculture officials was held in Memphis to present the plan for a 3-year boll weevil eradication trial to be held in north-eastern North Carolina and south-eastern Virginia. The plan called for a 25:25:50 percentage cost-sharing ratio for USDA, the states and the growers, respectively (Carter *et al.*, 2001).

The North Carolina General Assembly passed the Uniform Boll Weevil Eradication Act in 1975, the bill enabling boll weevil eradication in the state. North Carolina began a pattern that was repeated in practically every zone considering starting a boll weevil eradication programme, by requiring a two-thirds majority voting for eradication before a programme could begin. North Carolina growers passed their referendum by a majority of 76% in December 1976 (Carter *et al.*, 2001; Dickerson *et al.*, 2001). In December 1977 the Board of the Virginia Department of Agriculture and Commerce adopted the Virginia Cotton Boll Weevil Quarantine, which set out the requirements for cotton production in Virginia and authorized the eradication programme (Tate, 2001).

Concurrently with the Boll Weevil Eradication Trial (BWET), plans called for conducting an Optimum Pest Management Trial (OPMT) on about 44,000 acres (17,800 ha) of cotton located in Panola County in north Mississippi. The OPMT would involve cooperating producers on a voluntary basis and was conducted by the Mississippi Cooperative Extension Service (Harris and Smith, 2001).

Growers and NCC staff met with Secretary of Agriculture, Earl Butz, in November 1975 to urge inclusion of USDA cost-sharing funds in the 1977 budget. In January 1976, NCC staff and growers met with Office of Management and Budget Director, James Lynn, to secure the funding. President Ford's 1977 budget contained the funds for both the BWET and the OPMT (Carter *et al.*, 2001). Congress approved the budget but the House Agriculture Committee, led by Congressman Jamie Whittten of Mississippi, added a stipulation requiring the development of additional technology for boll weevil eradication before the funds for the programme would be released (Ridgway and Mussman, 2001).

The objective of the BWET was to evaluate the technical and operational feasibility of eradicating an established population of boll weevils from a geographically defined area. The objective of the OPMT was to test the technical and operational feasibility of conducting an areawide cotton insect management programme in which boll weevil was managed, but not eradicated (Dickerson *et al.*, 2001).

USDA planned for comprehensive evaluations of both the BWET and OPMT trials. The evaluations were to be conducted under the leadership of the Economic Research Service (ERS). A Biological Assessment Team, an Environmental Assessment Team and an Economic Assessment Team were organized. They were to evaluate, interpret and report on the BWET, OPMT and other boll weevil/cotton insect management programmes. USDA organized the evaluation teams composed of both USDA and state scientists. In addition, USDA organized an Overall Assessment Team, which had the responsibility of analysing, interpreting and reporting on the collective data of the Biological, Environmental and Economic teams (Ridgway and Mussman, 2001).

The results of the biological, economic and environmental assessments were to be submitted to USDA administrators and an independent Board of Agriculture and Renewable Resources – the National Research Council (NRC) evaluation committee. At the conclusion of the trials, the NRC committee was responsible for reviewing the results and identifying the most viable strategy for a solution to the boll weevil problem and providing an assessment of the most efficient means of managing other cotton pests in the Cotton Belt (Ridgway and Mussman, 2001).

The controversy continued between scientists who supported an IPM approach to management of the boll weevil and those who favoured eradication as the solution to the boll weevil problem. In response to the need for policy oversight and improved communications, the USDA appointed Harry Mussman, the Administrator of APHIS; Anson Bertrand, the Director of Science and Education Administration; and John Lee, the Administrator of ERS to serve on a USDA Boll Weevil Policy Group. Mussman was designated as chairman of the group. Kenneth Keller, former Director of the North Carolina Agricultural Research Service, was named as Executive Coordinator. The responsibility of this group was to guide the implementation of the OPM and BWET trials and evaluations, and to provide follow-up policy guidance (Ridgway and Mussman, 2001).

The Congressional requirement that new technology be developed before federal funding could be released for the start of the BWET was satisfied by the discovery that diflubenzuron (Dimilin®) treatment curtailed reproduction in the boll weevil. Congressman Jamie Whitten was briefed on diflubenzuron, its activity and its label status. He was assured that the compound would be registered in time for use in the BWET. The federal funding for the programme was released on 27 September 1977. Through a collaborative effort of USDA, the Environmental Protection Agency (EPA) and land grant university scientists, Dimilin® received its federal label from EPA in time for use in the 1979 season (Brashear and Brumley, 2001; Dickerson *et al.*, 2001; Ridgway and Mussman, 2001).

BWET and OPMT operations

A Boll Weevil Eradication Trial Executive Committee was established to provide oversight and support for the programme: North Carolina cotton grower, Marshall Grant; Al Elder from the North Carolina Department of Agriculture (NCDA); Jim Brazzel of APHIS; and Jack Bacheler from the North Carolina Agricultural Extension Service were the members. Ed Lloyd was chosen to chair a Technical Advisory Committee (TAC), which reported to the Executive Committee. TAC meetings were held monthly, March–October. The meetings were open to cotton producers and the public. All aspects of the programme were discussed at these meetings and grower questions were answered (Brashear and Brumley, 2001; Dickerson *et al.*, 2001).

Jim Brazzel was the Programme Director of the BWET and Milton Ganyard was the Programme Manager. The programme had four APHIS Work Unit Supervisors who were responsible for implementing programme activities on the cotton acres in their work units. Field labourers, cotton scouts, trappers and operators of ground equipment were temporary state employees. ARS scientists Ed Lloyd, Bill Dickerson and Gerald McKibben provided research, data collection and technical support (Brashear and Brumley, 2001; Dickerson *et al.*, 2001).

Operationally, the BWET was a mandatory programme for all cotton in the 50-county area included in the programme. The programme made areawide diapause control treatments, monitored and suppressed weevils with pheromone traps, selectively used the insect growth regulator Dimilin®, released sterile boll weevils, conducted field scouting, provided input into quarantine procedures and provided grower education and technical assistance (Ridgway and Lloyd, 1983). The programme had 12,040 cotton acres (5000 ha) in the eradication zone and 3532 cotton acres (1450 ha) in the buffer zone in 1978. By 1980 there were 24,565 acres (10,000 ha) in the eradication zone and 6090 acres (2500 ha) in the buffer zone (Dickerson *et al.*, 2001).

The OPMT was a voluntary programme on about 44,000 cotton acres (17,800 ha) in north-east Mississippi. It provided full reimbursement to growers for diapause control and pinhead square treatments. Overwintered boll weevil applications were applied as needed. Weevils were monitored with pheromone traps. Intensified field scouting provided for improved timing of insecticide treatments and in addition, educational and technical assistance for growers was enhanced (Andrews, 1981; Hamer *et al.*, 1983).

BWET and OPMT results

The BWET was completed in the autumn of 1980. In April 1981 the Biological Evaluation Team reported that there was a 0.9983 probability that the boll weevil had been eradicated from the BWET area (Dickerson *et al.*, 2001). Only two boll weevils were caught in the eradication area by the end of 1979, and only one weevil was caught in April–June 1980; it had no head and was from a re-used trap (Carter *et al.*, 2001). Subsequent Congressional Hearings and USDA decisions opened the possibility of programme expansion. Growers from South Carolina, Georgia, Florida and Alabama visited the BWET area to talk with growers who had been involved with the programme and to learn all they could about boll weevil eradication (Dickerson, 2001).

Insecticide applications were reduced during the OMPT from up to 18 per acre in years prior to the programme to 3.3, 3.4 and 3.0 in 1978, 1979 and 1980, respectively. Yield increased by an average of 84 pounds per acre (94 kg/ha) during the trial, or 34 pounds per acre (38 kg/ha) above the 10-year average. Cotton acreage increased from 32,075 to 39,000 acres (13,000–16,000 ha) over the course of the trial (Hamer *et al.*, 1983).

Post-BWET controversy

The USDA Biological, Environmental and Economic reports of the BWET effort were positive. The programme had eliminated the boll weevil from the eradication area with a very high degree of certainty. The USDA reports determined that the trial had been technically and biologically successful and that it had had a highly favourable environmental impact (Carter *et al.*, 2001).

The NRC report criticized the planning and implementation of the trials, the evaluation of the trials and the attempts to extrapolate results from across the Cotton Belt. However, they complimented the advances that had been made in insect control technology and management during the trials. They recognized the contribution made to cotton entomology by the technical monograph, which USDA had developed as a part of the effort. They commended USDA on its imaginative approach and creative efforts to project future impacts of cotton management programmes. And they recognized that the OPMT and BWET data clearly showed areawide programmes to be more successful in maintaining low boll weevil populations than the existing insect control programmes. They suggested it would be logical to consider ways to encourage growers to accept a beltwide programme (Ridgway and Mussman, 2001).

The NRC committee recommended that IPM practices be the thrust of boll weevil and other cotton pest control programmes for the next few years. And, they recommended an indefinite postponement of both optimum pest management and boll weevil eradication programmes and encouraged the private sector, the academic community and government agencies to assist in the development and adoption of private IPM programmes to more fully realize their potential (Ridgway and Mussman, 2001).

Perkins (1983) noted that the NRC committee simply stated that eradication was not demonstrated, making no reference to the USDA Biological Assessment Team's statement that, with 0.9983 probability, boll weevil had been eradicated in the eradication zone. He noted that members that had opposed eradication heavily influenced the NRC committee; thus the composition of the committee may have been such that critical evaluation of the boll weevil eradication trial and the feasibility of boll weevil eradication were not possible. Ridgway and Mussman (2001) wrote that the focus of the discussion was on whether eradication had been achieved. A range of circumstances complicated the question of whether technology was available to achieve eradication.

The Boll Weevil Policy Group reviewed all the OPMT and BWET reports and reported to the Secretary of Agriculture in May 1982. Their report recommended postponing beltwide implementation of boll weevil eradication because of budget constraints and lack of personnel. They recommended assisting the maintenance effort in North Carolina to protect programme gains and they recommended evaluation of containment technology. Further, they recommended that USDA facilitate testing and expansion of areawide cotton insect management trials and programmes, including future expansion of boll weevil eradication in the south-east. Finally, they recommended that USDA provide leadership in the decision-making process and coordination of programme activities. John Block, Secretary of Agriculture, concurred with the Boll Weevil Policy Group's report and recommendations, which became the basis for cooperative USDA–state–industry collaboration in the development of regionally specific programmes for eradication/management of cotton pests (Ridgway and Mussman, 2001).

The approval of the Boll Weevil Policy Group report by the Secretary of Agriculture put the future of cotton insect management into the hands of the producers. USDA was prepared to contribute its resources to the achievement of producer initiatives on boll weevil suppression and/or eradication (Mussman, 1982; Ridgway and Mussman, 2001).

The controversy over the results of the PBWEE and the OPMT/BWET was not a simple disagreement over the interpretation of the data. It was a larger intellectual battle about which of two paradigms provided the more favourable framework for the management of insect pests (Rabb, 1972; Perkins, 1982). The IPM paradigm held that pest populations should be intelligently managed through an ecologically based system that integrated cultural, mechanical, biological control, host plant resistance and, lastly, chemical means. Its supporters believed that the extensive use of insecticides over wide areas in boll weevil eradication would be prohibitively expensive, increase the probability of insects becoming resistant to insecticides, cause serious secondary pest resurgence and cause needless environmental contamination and would not be successful.

The Total Population Management (TPM) paradigm involved the use of integrated systems to achieve high-level suppression or pest elimination from a region. Its proponents believed that consistent, areawide application of the available technology could bring about progressive population reduction, culminating in eradication. The eradication concept and operations did not fit within the IPM paradigm of the time (Lloyd, 1972; Perkins, 1980, 1982; Ridgway and Mussman, 2001). The critics could not accept the evidence from the PBWEE and the BWET that the programme might work. They felt that the risks and costs were unacceptable.

In later years, Ridgway and Lloyd (1983) proposed a merger of paradigms. Since areawide programmes had been recognized as effective, they felt that the IPM and TPM paradigms could be merged into an areawide population management paradigm. Hardee and Harris (2003) contended that IPM and TPM were compatible and synergistic. Rabb (1972) suggested that the decision to eradicate was a grave responsibility and that eradication should only be attempted after careful study involving diverse perspectives.

Against the background of successful boll weevil eradication programmes, the ideas promoted by Rabb (1972) and Ridgway and Lloyd (1983) need further development. Eradication programmes for screwworm and boll weevil have been successful. They occur during a relatively short window of time, while IPM is a conceptual framework for managing pests over a long-range time perspective. The IPM concept is adaptable to and can accommodate changes in technology such as transgenic crop cultivars – which incorporate insecticidal characteristics, changes in pest status and eradication of primary pests. A carefully considered TPM (eradication) programme should be thought of as a component part of IPM. Eradication of the boll weevil over a large geographic area has provided an environment in which entomologists can develop and implement IPM programmes for cotton that are sustainable well into the future (Hardee and Harris, 2003).

Winds of Change: Increasing Grower Involvement

The early programme expansion in the south-east was led by APHIS. By the 1990s, programme expansion had greatly increased the acreage involved in the programme but federal funding had remained static. The federal percentage of cost-sharing had, consequently, dropped from the 30% level committed to earlier programmes to less than 5% by 1999 (Brashear and Brumley, 2001). In fiscal year 2001, Congress increased federal cost-sharing funding for boll weevil eradication to 20–30% of the programme cost (Klaussen and Ridgway, 2001; Smith, 2001). Federal support at about this level continued for several years (Anon., 2002b, 2003, 2004).

In 1973, NCC President, Mike Moros established the NCC's Beltwide Action Committee on Boll Weevil. The committee worked with USDA, the various states and Congress to complete research that would demonstrate the feasibility of boll weevil eradication. In the mid-1980s, the Boll Weevil Action Committee (BWAC) was active in the effort to complete boll weevil eradication in the Carolinas and expand the programme into Georgia, Florida and Alabama. The BWAC was reorganized in 1994 and Missouri cotton producer, Charles Parker, was selected as chairman, replacing the previous chairman, Marshall Grant – who remained active on the committee. The BWAC was given the difficult and contentious job of allocating the federal funds appropriated for boll weevil eradication among the active programmes. The division of the funds was based on budgeted expenses and acres. Extra allocations were occasionally made for programmes in serious financial condition (Carter *et al.*, 2001).

With reduced cost-sharing funding, there was a need for other sources of funding. Initially, some states secured loans from Production Credit Associations (Stavinoha and Woodward, 2001). The BAWC asked NCC to investigate the possibility of acquiring a federal loan or revolving fund. John Maguire, Vice President of NCC's Washington Operations, began exploring this possibility. A programme in which federal loans could be used to re-establish grasslands was available, administered by USDA's Farm Service Agency (FSA). Texas' US Congressman, Charles Stenholm, led the effort to adapt this programme to provide badly needed, low-interest loans to boll weevil eradication programmes (Carter *et al.*, 2001).

As the programme progressed, growers' leaders became willing to use their own programme directors and began asking APHIS for technical coordination. APHIS' role began to change from day-to-day management of programmes. Increasing liability concerns led the agency to revise its policies for boll weevil eradication programme staffing. When the California–Arizona programme began in 1983, day-to-day programme management was the responsibility of personnel hired by the grower-run foundation. By 1996, federal employee staffing had been reduced considerably in the south-east programmes and, by 1997, the South-east Boll Weevil Eradication Foundation (SEBWEF) managers had assumed responsibility for daily programme operations. APHIS personnel provided technical assistance and programme oversight. In addition, they helped monitor areas in post-eradication (McKibben *et al.*, 2001).

The Tools of the Trade: the Boll Weevil Eradication Process

Screwworm eradication successfully demonstrated the eradication concept and introduced effective methods of areawide population suppression that could be used to eradicate other pests (USDA-ARS, 1999). Since the screwworm programme, at least 11 species of fruit flies have been eradicated in the USA and many tropical areas of the world. Eradication efforts in the USA are ongoing against several introduced pests of trees, including the citrus longhorned beetle, Asian longhorned beetle, Asian gypsy moth and emerald ash borer. Local eradication of populations of sweetpotato weevil, pecan weevil, khapra beetle, painted apple moth, Mexican bean beetle and Japanese beetle have been reported. In Africa, a successful tsetse fly eradication programme has been conducted on the island of Zanzibar, and tsetse fly eradication in other areas is under way. And, pink bollworm eradication is ongoing on cotton in the south-western USA.

To date, the majority of the eradication efforts have been initiated to eliminate a local, recently introduced pest that has not become well established in its new environment. Screwworm, tsetse fly, pink bollworm and boll weevil are exceptions. When eradication programmes for these pests were initiated, the target pests had been well established for many years over large areas. Screwworm, tsetse fly and pink bollworm eradication programmes had area-specific SIT technology as a primary component of the programmes. In contrast, boll weevil eradication has relied primarily on field-specific trapping, insecticide treatment and stalk destruction to achieve eradication. A high degree of precision is required in locating, mapping, trapping, crop stage determination and treatment of fields. Thoroughness and precision at all stages of boll weevil eradication are critical to the success of the programme and, to some degree, set boll weevil eradication apart from the other large-scale eradication programmes which have been undertaken to date. Boll weevil eradication is the largest insect elimination programme ever undertaken in the USA (Frisbie, 2001), and it has been considered one of the most important agricultural programmes in US history (Smith, 2006).

Through time and across the cotton-growing areas of the USA, programme directors and technical advisory committees have instituted protocols they believed were best suited to boll weevil eradication in the region. The specific protocols have not been uniform across all programmes. However, the primary programme components and the way in which they were used have been essentially unchanged.

Securing supplies and services

As the programme expanded into new areas, office locations had to be chosen. The location of cotton acreage was one of the primary considerations in the location of offices, but the availability of a sufficient workforce and resources to handle automotive repairs and other needs had to be considered as well. Offices were located and rented. Utilities, office equipment, computers and vehicles were obtained; full-time and seasonal staff were hired; aerial application services were secured. As the technology developed, aircraft were required to have on-board GPS tracking systems. In some programmes, contract trapping services were secured. ULV malathion, traps, lure, kill strips and stakes for supporting traps were purchased.

Since the 1990s, hand-held scanners have been purchased to allow for electronic recording of data in the field and downloading of the field information into computers in the offices. The most commonly used scanners have been the TimeWand®II (Videx, Inc., Corvalis, Oregon) and, more recently, the Symbol MC 1000® (Motorola, Inc., Holtsville, New York), though others have been used. GPS equipment and software were acquired to allow eradication programmes to produce electronic maps. The most commonly used GPS equipment was the Geo-Explorer® (Trimble, Sunnyvale, California) hand-held units with post-processing differential correction using PATHFINDER® software (Trimble, Sunnyvale, California).

Programmes developed proprietary software, which allowed the integration of the maps with the data from trapping operations. These Boll Weevil Systems provided supervisors with the ability to manage more easily large amounts of data. They facilitated the use of the trapping data to trigger fields for treatment, provided means for controlling the quality of trapping and provided long-term data storage and retrieval for use in programme review and research (El-Lissy and Moschos, 1999; Harris and Smith, 2001; McGarigle, 2002; Goswick *et al.*, 2007).

Mapping

Since the programme began, managers have recognized the need for good field maps. Local FSA offices provided access to maps and assistance in contacting producers. Farmers, gins, consultants and agricultural supply businesses provided programme employees with information about where cotton fields would be located as growers made their planting plans, seed was purchased and planting operations began. Before the development of electronic mapping systems, the production of maps required a great deal of handwork using coloured pencils to delineate cotton fields and map sensitive locations. As GPS mapping equipment and systems were developed, the work moved from pencil and paper to GPS units and computers (El-Lissy *et al.*, 1996; Harris and Smith, 2001; Kiser and Catanach, 2006). Employees drove to the field, collected way-points describing the field perimeter, downloaded the information to the computer and used mapping software to construct the maps. Each field was given a unique number for identification purposes. The mapping software was capable of adding layers for streets, roads, railroads, rivers, lakes and county/state lines. Prior to and during the deployment of traps, trap maps were constructed showing the location of all traps on all cotton fields (El-Lissy *et al.*, 1996; Goswick *et al.*, 2007).

Trapping

In the active stages of boll weevil eradication, traps were deployed around all sides of all cotton fields as they were planted and emerged. Traps were placed on 4-foot (120-cm) bamboo reeds, wooden stakes, fibreglass rods or 1.5 inch (38 mm)-diameter PVC plastic pipe. Lure and kill strip were placed inside the capture cylinder of the trap. Originally, lures were 3 mg of grandlure impregnated into cigarette filters (Hardee *et al.*, 1972). Soon, 10 mg of grandlure was being impregnated into one inch (25 mm)-square laminated PVC pieces. PVC lures have been the standard boll weevil lure since the mid-1970s (Hardee *et al.*, 1974, 1975). Kill strips were 1.0×0.5 inch (25×12 mm) PVC strips impregnated with 0.6 g of DDVP insecticide. Long-duration lures with grandlure and eugenol were introduced, these allowed trap inspection intervals to be extended to 3 weeks post-eradication (Armstrong *et al.*, 2006). They were used to kill boll weevils caught in the traps. The lure pieces depleted most of their pheromone and had to be changed every 2 weeks. The kill strip lasted for 4 weeks. The first full season trap densities used in eradication programmes have varied from one trap per acre (2.5 traps/ha) to one trap per 8 acres (one trap to 3.2 ha); higher densities were typically used in the south-east and mid-south, and lower densities were typically used west of the Mississippi River (El-Lissy *et al.*, 1996; El-Lissy and Grefenstette, 2006; Goswick *et al.*, 2007).

Traps were placed around fields, preferably near obstructions such as power poles, fence posts, trees, etc. to provide them with protection from farm equipment and road maintenance equipment. Traps were consecutively numbered around fields. Taken together with the field number, each trap had a unique number, which was not repeated within the programme. A unique bar code was attached to each trap. As traps were deployed, the bar codes were scanned and the work unit number, the field number and trap number were entered into the scanner using the keypad. The scanner recorded the time and date of the trap deployment and each subsequent trap inspection. Downloading the scanners entered the deployed traps into the Boll Weevil System. Deployment of the traps in the field and in the Boll Weevil System and creation of the Trap Map established a permanent record of the trap location and deployment history, and set up the format for recording data as traps were inspected each week during the season (El-Lissy *et al.*, 1996; Goswick *et al.*, 2007).

Trap inspection on a field began at least 1 week before cotton began squaring and continued each week until all cotton plants in the field were no longer hostable. When fields were wet, employees were instructed to run all the traps that they could reach without damaging the grower's fields or foundation vehicles (Goswick *et al.*, 2007).

Employees inspecting traps first scanned the bar code. They were then prompted by the scanner to identify the task (i.e. remove, inspect, non-functional, missing/replace, missing/wet or deploy), then to determine and enter the number of boll weevils caught in the trap, then to inspect the field and enter the crop stage and finally to enter whether lure or kill strip was changed. Some programmes used crop stage codes for scanner entry to numerically record the critical crop stage information. Accurate collection of crop stage information was essential because treatment was based on two criteria, weevil capture and hostable crop stage (Goswick *et al.*, 2007).

Since scanners recorded the time and date of each trap inspection, scanner information could be used to check the efficiency of trappers. The time between trap inspections was used to evaluate the performance of the personnel assigned to trap inspection duties. Other trapping quality control systems were developed. Some programmes required employees to write on traps the dates of inspection, numbers of weevils captured and the dates lure and kill strip were changed. In addition, programmes tracked lure and kill strip change schedules in and/or on the trap (either by varying the colour of the lure or kill strip or by writing the date of change on the lure, kill strip and trap). This was very helpful to farmers, supervisors and others to ensure trapping was done properly. Visual inspection of the traps to determine whether they were undamaged, in good locations, were properly spaced and had been cleaned out and properly serviced was another key component to a well-run programme.

In addition, programmes planted dead boll weevils that had been treated with fluorescent dye in traps to determine whether employees were detecting and appropriately reporting boll weevils. Trapping quality control was treated seriously, and employees that could not or would not follow the proper procedures in deploying and servicing traps, detecting boll weevils and properly reporting boll weevils were released. Environmental Monitoring Specialists (EMSs) conducted many of the field quality control inspections in addition to monitoring insecticide applications to sensitive sites (Goswick *et al.*, 2007).

At the end of the work day those employees with responsibilities for inspecting traps returned to the office and downloaded their scanners into the system. They were instructed to bring any boll weevils or insects they suspected might be boll weevils back to the office for verification. The Field Unit Supervisors then used the information and the appropriate level to trigger fields (or parts of fields) for treatment the next day. They then constructed their field treatment maps and prepared the proper documents for treatment of the field (Goswick *et al.*, 2007).

Treatment

Field treatment maps were taken to the aerial contractor early in the morning to provide the pilots with information on which fields were to be treated that day. At the airport, the previous treatment records, GPS information and forms documenting the work done were picked up from the aerial contractor. The electronic treatment map supplied by the contractor was overlaid on the GPS field maps to verify that all fields that were triggered had been treated. Seasonal employees served as airport recorders to record flight time and the amount of insecticide used. Other seasonal employees conducted ground observation of flights. These employees communicated field hazards, wind and weather status and the presence of people in or near fields to pilots. They helped improve application safety and provided documentation of field conditions at the time of treatment (Goswick *et al.*, 2007).

EMSs were hired to monitor and control insecticide applications to fields near sensitive sites. These included hospitals, schools, nursing homes and other areas where people with chemical sensitivities were known to be. Areas defined as habitat for threatened and endangered species were handled as sensitive sites. Special treatment and monitoring procedures were set up through consultations with APHIS and the US Fish and Wildlife Service (FWS) for treatment and monitoring of these areas. Honeybee colonies were also considered as sensitive sites. EMSs monitored wind speed and direction before and during applications near sensitive sites. They placed dye cards on stakes between fields and sensitive sites before applications and monitored the cards during these applications to assure that drift was not occurring into the sensitive areas. When necessary or required they collected plant tissue, soil, water and swab samples from structures, equipment or vehicles. Dye cards, water samples, soil samples, swab samples and plant tissue samples were sent to diagnostic laboratories to test for the presence of malathion.

Thorough records were kept of flights near sensitive sites. In addition, EMSs also took samples of malation ULV and lure for laboratory testing to assure the quality of these critical programme components (USDA-APHIS, 2007a). Programmes developed databases to record and store environmental monitoring information. Programme employees were required to undergo testing for possible exposure to insecticides. Regularly scheduled blood cholinesterase monitoring was performed on those employees with potential exposure to insecticides.

Training

Considerable time and effort were invested in properly training employees. They were provided with training on how to do their jobs properly, how to work safely and how to interact professionally with co-workers. Staff safety meetings and trainings were ongoing through the season.

Oversight, information and guidance

Eradication Foundation Boards of Directors, APHIS, state departments of agriculture, FWS and EPA all had oversight on the manner in which boll weevil eradication programmes were conducted. Audits of financial operations, programme operations, procurement systems and compliance with state and federal regulations were conducted. Grower steering committee meetings were held on a regular basis in eradication zones to share programme and financial information and so that programme personnel could receive input from grower representatives on operational and programme funding matters. Steering committees made policy recommendations to the programme management and to the foundation Boards of Directors.

Programme status reports were made at many Cooperative Extension Service grower information meetings and field days. Newsletters, newspaper articles, grower meetings and individual contacts were used to keep growers and communities informed about programme activities and progress. Pamphlets, fact sheets and newspaper articles/advertisements were used to inform growers and the public about boll weevil eradication programmes prior to referenda.

For technical assistance, almost every programme had a TAC made up of professional entomologists, agronomists and other university personnel, USDA, state departments of agriculture, etc. to provide guidance and direction on matters of a technical nature. TACs worked with programme management on setting zone boundaries, establishing trap densities, recommending trap triggers, insecticide-related issues, organic production issues, sensitive site issues, issues relating to the needs of unique areas and quarantine and post-eradication issues.

Getting the Show on the Road: National Boll Weevil Eradication Begins

After the BWET ended in 1980 and the Boll Weevil Policy Group report was completed and approved by Secretary of Agriculture, John Block, many of the obstacles to initiating cooperative boll weevil eradication programmes were removed. Growers, USDA and others quickly began organizing for programme expansion. Jim Brazzel prepared a plan for the eradication programme expansion in North and South Carolina (Brazzel, 1983, unpublished).

North Carolina and South Carolina

Programme expansion could not occur unless cotton growers in both North and South Carolina passed referenda to enter into and fund eradication. North Carolina passed a referendum to enter the programme in February 1982, but the referendum in South Carolina failed by 13 votes (Brashear and Brumley, 2001; Dickerson *et al.*, 2001). A second referendum was required in both states. On the second attempt in January 1983, both North and South Carolina passed their referenda (Brashear and Brumley, 2001; Dickerson *et al.*, 2001).

The Grower Foundations from North and South Carolina met on 31 May 1983 and formed the South-east Boll Weevil Eradication Foundation (SEBWEF). The SEBWEF was responsible for purchasing insecticide and supplies, issuing aerial contracts, paying employees and conducting the other business of the programme. Federal funds for the North Carolina/South Carolina programme expansion did not become available until late July 1983. With short preparation time, the expansion programme began in the Carolinas with the first applications made by mid-August 1983 (Dickerson *et al.*, 2001). Jim Brazzel was Programme Director and Fred Planer was Programme Manager. The North Carolina programme had 15,000 cotton acres (6000 ha) and the South Carolina programme had 69,000 acres (28,000 ha) of cotton (Roof, 2001; El-Lissy and Grefenstette, 2006).

Both North Carolina and South Carolina organized TACs. The committees served as panels of experts to provide technical recommendations and a forum for information sharing to the programme management. The programme in the Carolinas benefited from unusually cold winters in 1983 and 1985. Much of the expansion area was free of boll weevils by the spring of 1985 (Brashear and Brumley, 2001). The programme was completed in North Carolina in 1987 and in South Carolina by 1990 (El-Lissy and Grefenstette, 2006). Growers from Georgia, Florida and Alabama began visiting farmers in North and South Carolina to learn about the programme and begin considering how they could begin eradication programmes in their states. The Carolinas went into a post-eradication 'holding pattern' to protect their weevil-free status until the Georgia programme could suppress boll weevil populations and end weevil movement into South Carolina. Initially, large migrations of boll weevil had to be controlled in the buffer zone along the western side of South Carolina (Brashear and Brumley, 2001).

California, Arizona and north-western Mexico

In 1983, the California Department of Food and Agriculture (CDFA) initiated a state–county–industry boll weevil eradication programme on the 60,000 acres (24,000 ha) of cotton in Imperial, Riverside and San Bernardino counties (Clark, 2001; El-Lissy and Grefenstette, 2006). APHIS personnel from the Methods Development Laboratory at Mission, Texas provided technical oversight and advice. The components of the programme were a host-free period, trapping, visual surveys and insecticide treatments.

The programme followed the plan set out by Jim Brazzel (1983, unpublished). The programme goals were to eradicate the weevil from the southern desert counties and to protect the San Joaquin Valley's billion-dollar cotton industry from becoming infested by the boll weevil. In 1983, 12,888 boll weevils were trapped and 76,616 cumulative acre (31,000 ha) treatments were applied. In 1984, 27,920 weevils were caught and 90,459 acres (37,000 ha) were treated. Most of these weevils were migrating from Arizona or Mexico. By 1987 no in-field infestations of weevils were found, but 4068 weevils were caught. Most were caught within 1 mile (1.6 km) of the Arizona state line. The last boll weevil trapped in California was caught in 1990 (Clark, 2001).

In response to increasing numbers of boll weevils in Arizona and the start of eradication in California, the Arizona Legislature created the Arizona Cotton Research and Protection Council (ACRPC) in September 1984. Its purpose was to provide an organizational structure and a funding mechanism for boll weevil eradication in Arizona. In 1985, the south-western Boll Weevil Eradication Programme was established so that a cooperative boll weevil eradication programme could be run in California, Arizona and north-west Mexico. The ACRPC ran the day-to-day operation of boll weevil eradication in Arizona, the CDFA ran boll weevil eradication in California and Sanidad Vegetal conducted the programme in Mexico. Frank Meyers was Programme Director and Larry Antilla was Programme Specialist for the Arizona programme. A TAC was created to provide technical guidance for the Arizona programme (Neal and Antilla, 2001).

The programme began on the 70,000 acres (28,000 ha) of cotton in western Arizona in 1985 and expanded to the 420,000 acre (170,000 ha) central Arizona zone in 1988 (Neal and Antilla, 2001; El-Lissy and Grefenstette, 2006). The programme in Arizona did not include diapause phase treatments in the initial year: instead, it began with full-season treatments (Brazzel, 1989). In November 1988, a statewide referendum was held in Arizona to determine whether the programme would continue: it passed with a 75% favourable vote. The programme continued and, by the autumn of 1989, weevil captures were down by 83%. In 1990, a plough-down incentive programme, the Plower Programme, was introduced and it was very successful. By 1991, the eradication programme in Arizona was complete. Frank Meyers retired in 1991 and Larry Antilla was named Programme Director. The post-eradication phase of the programme began in 1992 (Neal and Antilla, 2001).

In the cotton-growing areas of Mexico adjacent to California and Arizona, the 160,000 acre (65,000 ha) Mexicali area and the 5000 acre (2000 ha) Sonoita area began programmes in 1988. These areas successfully completed eradication in 1991 (El-Lissy and Grefenstette, 2006).

Georgia, Florida and south-eastern Alabama

Organizational meetings for expansion into the 287,500 acre (116,000 ha) Georgia zone, the 107,000 acre (44,000 ha) Florida zone and the 61,000 acre (25,000 ha) south-east Alabama zone began in the autumn of 1984. The NCC requested that APHIS write a plan for eradication in the expansion area. Jim Brazzel prepared and submitted the plan for boll weevil eradication in Georgia (excluding five counties against the north-west boundary with Alabama), Florida and 18 counties in south-east Alabama in September 1985 (Brashear and Brumley, 2001; Carter *et al.*, 2001). Enabling legislation, modelled after the legislation passed in North Carolina and South Carolina, was passed in Alabama in 1984, Georgia in 1985 and in Florida in 1987 (Carter *et al.*, 2001; Curtis, 2001; Haney *et al.*, 2001b). Boll weevil eradication foundations were established in each state. The state foundations contracted with the SEBWEF to operate their programmes.

Boll weevil populations and losses were high in the region from 1985 to 1987 (Haney *et al.*, 2001b). This was undoubtedly an important factor in the grower referenda. Alabama passed its referendum in December 1985 (Carter *et al.*, 2001; Haney *et al.*, 2001b). The destruction of the Georgia cotton crop by Hurricane Kate in 1985 was thought to have been a factor in Georgia's failing to pass their December 1985 referendum. Georgia passed its second referendum in November 1986 (Planer, 1988; Lambert, 1991; Brashear and Brumley, 2001; Haney *et al.*, 2001a). Florida growers passed their referendum in June 1987 (Carter *et al.*, 2001). With the time for the programme expansion start-up approaching, USDA released the necessary federal cost-sharing funds at the end of July 1987 (Haney *et al.*, 2001b).

Bids were let for the purchase of the insecticide that would be used in the programme. Mobay Corp. won the bid and Guthion® ULV insecticide was purchased for use in the programme. Use of Guthion® ULV in the 1987 programme was projected to save about US$1 million, so the SEBWEF elected to use it. The decision to use Guthion® caused many environmental problems and concerns about human health (Curtis, 2001). After the 1987 season, malathion ULV became the standard insecticide for boll weevil eradication. After years of carefully controlled and monitored use on millions of cotton acres in boll weevil eradication, malathion ULV's record of human safety and minimal disruption of non-target organisms has been very good.

Fred Planer was the APHIS SEBWEP manager as the programme moved into the autumn diapause phase of the programme in the expansion area. Spraying began later than planned because the federal cost-share funding was delayed. The first treatments were applied on 2 September 1987. There were staffing problems, and soon after treatments began so did problems associated with the decision to use Guthion® ULV (Brashear and Brumley, 2001).

The state foundations vigorously encouraged post-harvest stalk destruction to prevent cotton regrowth. Florida and Alabama offered credits to their boll weevil eradication assessments for stalk destruction (Curtis, 2001; Haney *et al.*, 2001b), while Georgia established penalties for failure to destroy cotton stalks by the deadline (Haney *et al.*, 2001a).

Federal loans were available to the programmes through the Commodity Credit Corporation. These loans were critical to the success of the programme in Georgia, Florida and south-eastern Alabama. The states also provided economic support for the effort. Up to 2000, the Alabama legislature had provided US$21.7 million (Curtis, 2001), to 1991 the Florida legislature had provided US$46.5 million (Haney *et al.*, 2001b) and the Georgia legislature had provided US$3 million toward the eradication effort by 1990 (Haney *et al.*, 2001a).

By 1988, the programme in the Georgia–Florida–Alabama expansion area had developed a number of problems: (i) a large number of weevils emerged in the spring; (ii) an outbreak of the beet armyworm, *Spodoptera exigua* had occurred; (iii) a lawsuit was filed in Alabama questioning operation of the programme without an environmental impact statement; (iv) a large debt had been incurred; and (v) the programme had reduced treatments in order to lower costs (Brashear and Brumley, 2001; Curtis, 2001). In 1989, neither the boll weevil population nor the budget had improved, but more aggressive treatment criteria were used than in 1988. Temperatures in December 1989 went into single figures in several areas. By 1990/1991, only pockets of boll weevils remained (Brashear and Brumley, 2001). In 1991 the environmental lawsuit was settled. The judge ruled that the programme could continue, but an environmental impact statement was required. The judge also required a set of mitigating guidelines to prevent misapplication of insecticides; and, the judge ruled that the public had to be informed that the programme would be spraying (Curtis, 2001).

To deal with the financial situation, the SEBWEF made the bold decision in 1990 to increase the assessment up to US$20 per acre for a 5-year period. This required the passage of new referenda in each of the three states. Alabama failed its first referendum in April, and then passed a second referendum in June 1990. Georgia passed its referendum in May and Florida passed its referendum in July 1990. In addition, North Carolina, South Carolina and Virginia increased their assessments by US$2.30 per acre to help the Georgia–Florida–Alabama expansion programme pay its debt. By December 1993, the debt had been paid (Brashear and Brumley, 2001).

Because of some of the problems that had developed in 1987/1988, a south-eastern TAC was formed in 1990 to provide technical guidance in helping to avoid problems. Jim Brazzel chaired it and remained chairman of the south-eastern TAC until 1993, when Bill Dickerson was named chairman. Dickerson was chairman until he resigned in 1996. At that time, Gerald McKibben was selected to serve as chairman of the committee (Brashear and Brumley, 2001).

In 1993, APHIS determined that a Boll Weevil Eradication National Coordinator was needed. The role of the National Coordinator was to coordinate and facilitate the distribution of federal cost-sharing funds, equipment, etc. and to provide technical assistance to the programmes. Gary Cunningham was named as the National Coordinator (Brashear and Brumley, 2001). Cunnigham served until he retired in 1998. Bill Grefenstette was named National Coordinator in July 1999 (W.J. Grefenstette, 2007, personal communication).

In early 1989, Johnny Paul DeLoach was selected as the Executive Director of the SEBWEF, and Bob Alred became the Chief Financial Officer for the SEBWEF later that year (Brashear and Brumley, 2001). Boll weevil eradication was completed in Georgia (excluding north-west Georgia) in 1992, and in Florida and south-east Alabama in 1993 (El-Lissy and Grefenstette, 2006).

Central Alabama, north-east Alabama, Alabama's Tennessee River Valley, north-west Georgia and middle Tennessee

In February 1992, referenda were held in the north-west Georgia zone, the north-east Alabama zone and the central Alabama zone. None passed. Growers cited cost over-runs, lawsuits and high assessments compared with what it cost the growers to control boll weevils on their own (Brashear and Brumley, 2001).

The 28,000 acre (11,000 ha) north-east Alabama zone passed its second referendum in April 1992 and the programme was initiated (El-Lissy and Grefenstette, 2006). They did not begin with a diapause phase programme but, instead, with spring treatments in 1992. Some of the local sources of information judged the weevil populations to be low, but Extension Service trapping indicated boll weevil numbers were as high in north-east Alabama as they were in the central part of the state. The start-up of the programme in the spring, the higher than anticipated weevil numbers and the proximity of cotton not in a programme in nearby north-west Georgia led to extensive spraying and cost over-runs for the first two full seasons. The programme faced a serious problem when FWS imposed wide buffers that could not be sprayed near the habitats of several threatened and endangered species.

Consultations with FWS convinced them that the long-range benefits of the programme outweighed the short-term risk of treating the buffer areas, and the regulations were amended to allow the treatment of the buffer areas. Strict monitoring and controls on spraying near the endangered species habitats were imposed. Studies of the effects of programme treatments in buffer areas on the threatened and endangered species concluded that the effects were negligible (Brashear and Brumley, 2001).

Central Alabama experienced a severe boll weevil infestation in 1992. The referendum held in the zone in February 1993 passed, clearing the way for start-up of the programme on the 90,000 cotton acres (37,000 ha) in the zone in late summer 1993 (El-Lissy and Grefenstette, 2006). The diapause phase was well organized and, in 1993, it was conducted without budget over-runs. However, in the spring of 1994 substantial populations of weevils remained in the zone. Excessive rains occurred in June and July, negatively affecting trapping and treatment efforts. The result was more boll weevils than expected, added treatments and cost over-runs (Brashear and Brumley, 2001).

The north-west Georgia zone, with 10,000 cotton acres (4000 ha) in five counties (El-Lissy and Grefenstette, 2006), voted its second referendum in early 1993. The referendum passed easily because the assessment was much lower than that in nearby Alabama. The rate in north-west Georgia was the same as for the rest of Georgia, which by 1993 was in post-eradication. North-west Georgia began its programme with the autumn diapause phase in 1993 (Brashear and Brumley, 2001).

The 170,000 acre (49,000 ha) Tennessee River Valley, Alabama zone began its programme with autumn diapause in 1994 (El-Lissy and Grefenstette, 2006). By 1995, boll weevil populations there were reported to be moderate. But, the zone experienced organizational and personnel difficulties; and beet armyworm, *S. exigua* and tobacco budworm, *H. virescens* outbreaks occurred during the 1995 season (Brashear and Brumley, 2001).

The state of Tennessee passed the enabling legislation allowing growers to initiate a boll weevil eradication programme in 1989. The Tennessee Boll Weevil Eradication Foundation was sanctioned by the state as the certified grower organization for boll weevil eradication in July 1993. At the first Tennessee Boll Weevil Eradication Foundation meeting in August 1993, by-laws were adopted and Allen King was elected chairman of the Tennessee Foundation. A Tennessee TAC was appointed and Paulus Shelby, the State Cotton Agronomist with Tennessee Cooperative Extension was selected as chair. In November 1993, a referendum was held in Middle Tennessee. It did not pass, but a second referendum in January 1994 did pass. In March 1994, the Tennessee Foundation signed a cooperative agreement with the SEBWEF to operate their programme (Barker, 2001).

Middle Tennessee was divided into two zones for assessment purposes. The southern middle Tennessee zone consisted of five counties and the central middle Tennessee zone had six counties. In 1995, cotton acreage in both middle Tennessee zones totaled 20,940 acres (8600 ha). Southern middle Tennessee began its programme with the diapause phase in 1994 on 10,669 acres (4700 ha). Central middle Tennessee began eradication on approximately 10,000 acres (4000 ha) with spring trapping and treatment in 1995 (Barker, 2001).

The programme in middle Tennessee benefited from severe winters in 1995 and 1996. However, in 1996 and 1997 cotton acreage decreased by 63 and 44%, respectively. This resulted in financial difficulty because fewer assessment dollars were collected. The state helped with US$300,000 in cost-sharing funds. At the time, the federal cost-sharing was 30% and the availability of state and federal funds helped pay programme expenses (Barker, 2001).

Marshal Grant served as Chairman of the SEBWEF Board of Directors in 1983, 1985 and 1987. Robert Lee Scarborough was Chairman in 1984 and 1986. Johnny Paul DeLoach was Chairman in 1988 and W.L. (Sonny) Corcoran was Chairman in 1989. Bobby Webster was selected SEBWEF Board Chairman in 1990. Copland Griswold became Chairman in 1991, followed by Billy Sanders in 1992 and 1993. Claude Buchanan served as SEBWEF Chairman from 1994 to 1997. Allen King was named Chairman of the SEBWEF Board of Directors in 1998 and has since continued to serve as Board Chairman.

In 1993, Fred Planer left his position as APHIS SEBWEF Director to assume an APHIS Staff Officer position. Pat McFadden was named APHIS SEBWEF Director, replacing Planer. In mid-1995 SEBWEF Executive Director, Johnny Paul Deloach, resigned and Jim Brumley filled the Executive Director position (Brashear and Brumley, 2001).

By 2000, boll weevil eradication in the northern Alabama, north-west Georgia and middle Tennessee areas was complete. By 2005, remarkable progress had been made in the south-east region. No weevils were caught in Virginia (110,000 acres, 45,000 ha), South Carolina (330,000 acres, 13,300 ha), Georgia (1,500,000 acres, 607,000 ha), Florida (130,000 acres, 53,000 ha) and Alabama (610,000 acres, 25,000 ha). A single weevil was caught in North Carolina (970,000 acres, 400,000 ha) (El-Lissy and Greffenstette, 2006). For the region, one weevil was caught on 3.65 million cotton acres (1.47 million ha).

Plan for expansion to the Mid-South and South-west

In October 1993, an ad hoc working group on boll weevil eradication met in Memphis, Tennessee to plan for further expansion of the eradication programme. The group endorsed the goal of eradication of the boll weevil from the USA as rapidly as possible. They agreed on a 'dual front' approach to complete the process sooner; Mid-South and Texas would start simultaneously. The group agreed to seek funding for 30% federal cost-sharing. NCC Executive Vice President, Phil Burnett, wrote a letter to leaders and interested organizations asking them for letters of support for the programme. Support letters from over 50 organizations were written (Carter *et al.*, 2001).

In the summer of 1993, NCC held a series of meetings in the Mid-South to discuss coordination of boll weevil eradication in the region. The consensus from these meetings was that the region was a large, contiguous area of cotton production with no natural breaks and should be regarded as one region for the purposes of planning and implementation. A plan was agreed upon to start in Mississippi's hill and eastern regions, then Louisiana, then Arkansas, south of Interstate 40. The Mississippi Delta would start next, followed by west Tennessee, north-east Arkansas and Missouri (Carter *et al.*, 2001).

Mississippi

News of the success of boll weevil eradication in the south-east stimulated a variety of responses among Mississippi cotton growers. While many anticipated the benefits that were being enjoyed in eradicated areas, some were anxious about secondary pests, insecticide resistance and/or the cost of the programme. In this atmosphere of mixed opinions, the grower leadership began making preparations. The Mississippi Farm Bureau and the Delta Council were key organizations in the preparations. The overriding concern was that the programme in Mississippi should be designed to fit the conditions in the unique cotton-growing regions of the state (Harris and Smith, 2001).

Mississippi took the first step toward boll weevil eradication when Mississippi State University administrators appointed the State Technical Advisory Committee (STAC) in 1989. Bob Head was appointed to chair the committee. The winter of 1989/1990 was especially harsh. The STAC immediately began writing a boll weevil population maintenance plan to maintain the population reduction, which had resulted from the hard winter, until an eradication programme could be implemented. The plan probably resulted in some boll weevil population suppression, but an organized, sustained management programme did not develop (Harris and Smith, 2001). STAC wrote proposals for the enabling legislation that would be needed for programme start-up. The committee also evaluated options for organization, regulation and operation of the eradication programme (Harris and Smith, 2001; Harris and Clark, 2006).

The first step toward grower organization for boll weevil eradication in Mississippi was the appointment of the Mississippi Boll Weevil Management Committee by an organization of Mississippi cotton growers. The committee functioned to manage boll weevil eradication affairs prior to the incorporation of the entity with authority to administer the programme, in the state. Secretary of State, Dick Molpus, incorporated the Mississippi Boll Weevil Management Corporation (MBWMC) in February 1992. George Mullendore was employed in March 1992 as Project Coordinator for the corporation. The MBWMC was recognized by Commissioner Jim Buck Ross, Mississippi Department of Agriculture and Commerce, as the certified cotton grower's organization to conduct boll weevil eradication. Board members were appointed by region to fully represent all the cotton growing areas of the state. Bobby Miller was elected President of the MBWMC Board of Directors in July 1993.

The Board was challenged to unify growers from the different regions, develop an organization that was representative of all cotton growers, assist in conducting petitions and holding referenda and effectively initiate boll weevil eradication in the state. The committee developed by-laws and proposed the enabling legislation necessary for implementation of the programme. A new MBWMC Board of Directors was elected in January 1995. The new Board of Directors elected Bobby Miller Chairman and Kenneth Hood President. The MBWMC contracted with the SEBWEF to conduct boll weevil eradication activities in Mississippi in 1998 (Harris and Smith, 2001; Harris and Clark, 2006).

The Mississippi Boll Weevil Technical Advisory Committee (MBWTAC) was appointed by the MBWMC in early 1993. Their charge was: 'to provide guidance in developing and conducting effective boll weevil management programmes'. Specifically, they were to provide technical programme guidance, determine costs and budgets and define specific operational regions for holding referenda and conducting programmes. Chairmen of the MBWTAC have been C.D. Ranney (1993–1994) and J.W. Smith (1994–1995). Aubrey Harris was named Chairman of the MBWTAC in 1995 (Harris and Smith, 2001; Harris and Clark, 2006).

The enabling legislation was passed in 1993. The Mississippi expansion plan called for start-up in the eastern region of the state as the Alabama programme neared completion. The eastern zone, Region 4, was a strip of land one to three counties wide running from the Tennessee state line to the Gulf of Mexico. The selection of counties for inclusion in Region 4 was based on the ability to manage the programme with the available personnel and equipment and the geographic location of the county (Harris and Smith, 2001).

A referendum for boll weevil eradication was conducted in Region 4 in December 1993. Those in favour of initiation of a programme accounted for a 76% majority, but fewer than 50% of the eligible voters voted. In accordance with the referendum requirements, the referendum was declared invalid. A second referendum was held in January 1994; more than 50% of the eligible voters voted and the referendum passed, with 88% voting to start an eradication programme (Harris and Smith, 2001).

The diapause phase of the boll weevil eradication programme began on the 70,000 acres (28,000 ha) in Region 4 on 1 August 1994 (El-Lissy and Grefenstette, 2006). In the spring of 1995, trapping began. Malathion ULV was applied during the pinhead square stage of crop development, followed by in-season applications based on weevil captures in the traps and using the trap triggers that had been established. Unusually heavy populations of tobacco budworms, *H. virescens*, developed in the hill region of Mississippi, including Region 4. The infestation resulted in control failures, high control costs and severe crop damage in some areas. Growers in the eradication zone blamed the eradication programme for their crop losses and high control costs. Mississippi State University scientists investigated the outbreak. They concluded that the outbreak was not caused by any one factor, but instead that many variables were involved (Williams and Layton, 1996; Luttrell *et al.*, 1997; Harris and Smith, 2001).

A petition drive was conducted to initiate a recall referendum. The petition was signed by more than 20% of the growers in the zone, and a recall referendum was scheduled for March 1996. More than 50% of the eligible voters cast ballots. A 66.6% majority vote was required to continue the programme. Only 57.5% of the growers voted to continue and the programme was ended. Kenneth Hood and others worked out a voluntary 'stop-gap' programme. Growers could either enter into a memorandum of agreement with the SEBWEF for buffer zone treatments to be applied on their farms or purchase malathion ULV at a reduced price for pinhead square applications. The voluntary programme provided for continuation of the areawide suppression programme and kept in place a framework from which boll weevil eradication in Region 4 could be restarted at a later time (Harris and Smith, 2001).

Mississippi's Region 3 consisted of 400,000 acres (162,000 ha) of cotton and included all of the cotton in the Mississippi Hill Country that was not in Region 4. It stretched from the Tennessee border to the Gulf of Mexico and the Louisiana border. And, from Mississippi's south-western border with Louisiana it extended up the Mississippi River as far north as Vicksburg, then wrapped around the east side of the Mississippi Delta. Growers in Region 3 faced severe boll weevil problems, and the multiple insecticide treatments required to control the boll weevil often triggered outbreaks of other pests. Region 3 growers generally favoured boll weevil eradication.

They were concerned, however, about whether Region 4 would re-enter the programme and whether Regions 1 and 2, the Delta Regions, would vote to participate in the programme. The Mississippi Delta, a leaf-shaped plain 400 km long by 100 km wide on the north-west side of the state, had a history of low boll weevil populations and low control costs. Growers in the Delta Region typically planted 825,000 acres (334,000 ha) of cotton. They had not communicated much interest in the programme. Region 3 growers were fearful that boll weevil eradication might stop at the Delta. If eradication stalled they would be left in a buffer zone. In a buffer zone their cotton would receive multiple insecticide applications each year, potentially, for many years. They feared their risk of secondary pest outbreaks would increase, driving production costs up and contributing to economic instability of cotton production in the region (Harris and Smith, 2001; El-Lissy and Grefenstette, 2006).

Referenda were held in December 1996 for regions 2, 3 and 4. Regions 3 and 4 were voting to start eradication in August 1997, while Region 2 was voting on starting in August 1998. Regions 2 and 3 passed their referenda with 71 and 81% favourable votes, respectively, but in Region 4 the proposition failed with 63% of the vote. Another referendum was held in Region 4 in February 1997; this time it passed with a 71% positive vote, allowing August 1997 start-up in Region 3 and re-start in Region 4 (Harris and Smith, 2001).

The SEBWEF leadership team in the Mississippi programme was Jim Brumley, Executive Director of the SEBWEF; Farrell Boyd, Mississippi Programme Manager and A.L. Brashear, APHIS Programme Director. In 2000, Jeannine Smith was promoted to Executive Director of MBWMC, assuming the duties of the retiring Executive Director, George Mullendore (Harris and Smith, 2001).

The southernmost counties in the Mississippi Delta made up Region 2. The region had 225,000 acres (91,000 ha) of cotton and a history of more severe boll weevil infestations than the more northerly Delta counties. Region 2 growers anticipated substantial benefits from boll weevil eradication. The diapause phase of the programme began in early August 1998 (Harris and Smith, 2001; El-Lissy and Grefenstette, 2006).

Cotton growers in the 600,000 acre (243,000 ha) North Delta, Region 1, did not hold a referendum until after the passage of referenda in the other regions. The first referendum was held in April 1997. It did not pass, but 62% of the voters were in favour of starting a programme. A second referendum was held in February 1998: the referendum failed a second time. This time only 59% voted in favour of conducting boll weevil eradication in the region (Harris and Smith, 2001; El-Lissy and Grefenstette, 2006).

The executive committee of the MBWMC asked the MBWTAC to study the possibility of dividing Region 1 into two areas, which would hold separate referenda. After much study and discussion, the MBWMC Board of Directors divided Region 1 into Region 1A on the eastern side and Region 1B on the western side of the Mississippi River (Harris and Smith, 2001). A referendum on boll weevil eradication was held in Region 1A in February 1999: it passed with a 70% favourable vote. A referendum was held in Region 1B in March 1999: it passed with a 79% favourable vote. Boll weevil eradication began in Regions 1A and 1B on 1 August 1999 (Harris and Smith, 2001).

The 11 September 2001 terrorist attacks caused aerial applications to be temporarily halted across the country. Aerial applications near the large commercial airport in Memphis, Tennessee were curtailed through much of the autumn of 2001. This resulted in impaired programme effectiveness in north-east Mississippi, south-west Tennessee and east-central Arkansas.

By the end of 2005, El-Lissy and Grefenstette (2006) reported that boll weevil eradication had been completed in Regions 2, 3 and 4. Region 4 caught no weevils in 2005. In Regions 2 and 3 the number of weevils caught per trap inspection had been reduced by 99.19 and 97.84%, respectively. In Region 1 weevil captures had been reduced by 94.86%. By the end of 2006, only 44 weevils had been caught in Region 1A; three weevils were caught in Region 1B and three weevils were caught in Region 2. In Region 3, 1198 weevils were caught and in Region 4, 35 weevils were caught (Brumley *et al.*, 2007).

Louisiana

In the late 1980s, the Louisiana Farm Bureau and the Louisiana Cotton Producers Association passed resolutions in support of boll weevil eradication. In the early 1990s these groups, working with Commissioner Bob Odom of the Louisiana Department of Agriculture and Forestry (LDAF), developed an Enabling Legislation Committee to draft the legislation that would be needed to begin boll weevil eradication. This committee had the responsibility of recommending in the draft legislation what entity would carry out boll weevil eradication in Louisiana. They could: (i) create a non-profit foundation for Louisiana; (ii) develop a Mid-South Foundation on the model of the SEBWEF; or (iii) create a commission within LDAF. The LDAF proposal offered cost savings to growers through reduced overhead and shared programme costs under the Louisiana Agricultural Finance Authority of LDAF. The Enabling Legislation Committee decided to organize Louisiana's boll weevil eradication programme with a grower-run Boll Weevil Eradication Commission and eradication operations conducted by the LDAF (Logan, 2001).

The committee's work culminated in the enactment of the Louisiana Boll Weevil Eradication Law in 1992. This law prepared the way for Louisiana to participate in the cooperative federal/state/industry effort to eradicate the boll weevil. In addition, it also authorized the creation of an eight-member Boll Weevil Eradication Commission (BWEC), composed primarily of cotton growers to provide programme oversight. Dan Logan, Jr. was elected Chairman of the BWEC (Logan, 2001).

A timetable was developed to organize and prioritize the objectives that needed to be met before a programme could begin. A budget committee was named. Fred Planer, with APHIS, led the budget committee, which took on the task of creating a realistic budget. One of the most difficult issues was balancing fairness to growers who would not or could not abide by the regulations with the need to run a firm and successful programme. After studying the successes and failures in other states, regulations were developed. Since the largest programme expenditures occurred in the first two programme years, the budget committee developed a plan for 'up-front' programme financing (Logan, 2001).

A statewide referendum was held in March 1995. It proposed a US$30 per acre assessment for 5 years – US$150 total programme cost. Some north-east Louisiana growers were not enthusiastic about the assessment – their boll weevil control costs were < US$30 per year. In some areas growers were opposed to any 'government programme'. In the Red River region, where the weevil problem was more severe, the US$30 per acre assessment was received more positively. FSA assisted in conducting the referendum. The referendum failed, primarily because some of the large cotton-growing parishes in north-east Louisiana did not support it. A second statewide referendum was planned for the autumn of 1996, but it was cancelled because of a lawsuit brought by a few north-east Louisiana farmers (Logan, 2001).

Because support for eradication was strong in the Red River Valley, the BWEC made the decision to divide the state. The 66,000 acre (27,000 ha) Red River eradication zone was created in 1996. It consisted of 19 parishes in north-western and central Louisiana. Budgets and timetables were developed for boll weevil eradication in the zone (Logan, 2001; El-Lissy and Grefenstette, 2006).

A referendum on boll weevil eradication was conducted for Red River Valley farmers in late November and December 1996. It passed with an 83% favourable vote. Growers in adjacent south-west Arkansas' Red River Valley area passed a referendum to commence boll weevil eradication, with operation by the Louisiana programme, in October 1995 (Logan, 2001; Kiser and Catanach, 2005).

John Andries was hired to serve as Boll Weevil Eradication Programme Director; Ken Pierce was named as the APHIS Programme Director. The diapause phase of the programme began on 18 August 1997 (Logan, 2001).

Agriculture Commissioner Bob Odom felt that the eradication programme would have difficulty passing in the north-eastern part of the state without supplemental funding from the state. Commissioner Odom, the BWEC and former NCC president, Jack Hamilton met many times with state legislators to secure cost-sharing assistance. Their efforts were successful and the state committed US$50 million to the boll weevil eradication effort in 1998. Because of Louisiana's success, other states increased their efforts to obtain state funding to support their programmes (Logan, 2001).

With nearly half of the programme's cost supplied by the state, the BWEC began planning for another referendum in north-east Louisiana. The referendum was held in September 1998: it passed with a 79% favourable vote. John Andries and his managers went to work immediately acquiring equipment and obtaining aerial contractors to begin boll weevil eradication in the 545,000 acre (220,000 ha) north-east zone in August 1999 (Logan, 2001; El-Lissy and Grefenstette, 2006).

Louisiana growers received US$12 million in additional funds from the state in 2003 to finance a 50:50 state:grower cost-sharing to fund the maintenance phase of the programme (Logan, 2001). John Andries retired in May 2005 and Marc Bordelon was named Director of the programme (M. Bordelon, 2007, personal communication).

By the end of 2005, boll weevil trap captures had been reduced by 99.92% from 2000 levels in the Red River zone; and trap captures had been reduced by 99.99% from levels present in 2000 in the north-east zone (El-Lissy and Grefenstette, 2006).

Arkansas

As boll weevil eradication made progress in the South-east and South-west, discussions about the programme began in Arkansas. Gerald Musick, Dean of the University of Arkansas, College of Agriculture, appointed an Arkansas TAC in July of 1989. Don Johnson and Jake Phillips were co-chairmen. Initial planning meetings were held to develop plans and guidelines for the enabling legislation that would be needed to start a programme. The committee initiated research in hopes of developing a lower-cost approach for use in north-east Arkansas (Johnson *et al.*, 1999). Phil Tugwell replaced Jake Phillips as co-chairman in 1995. Later in 1995, the Boll Weevil Eradication TAC was reorganized with William Yearian as Chairman (Johnson and Martin, 2001). Yearian served as chairman of the Technical Advisory Committee until January, 2002; Don Johnson was named to chair the Technical Advisory Committee following Yearian; Johnson resigned in March 2004. Gus Lorenz was named Chairman of the TAC in May 2004 (D. Kiser, 2007, personal communication).

Work to secure the necessary enabling legislation began in the early 1990s and the bill was introduced in the Arkansas General Assembly in 1991; the Act passed. It established procedures for naming a nine-member Boll Weevil Eradication Board of Directors. Governor Bill Clinton appointed the Board and asked Jack Carey to serve as Chairman. The Board noted that there was no lien provision in the enabling legislation to ensure collection of assessments. After several attempts to secure lien authority, the Board found another way of securing payment of the assessments. They arranged for the issuance of ginning certificates to growers who had paid their assessments. Without a ginning certificate cotton could not be ginned. An amendment to the enabling legislation authorized the ginning certificate process in 1993 (Johnson and Martin, 2001).

Also in 1993, Fred Planer with APHIS met with research and Extension personnel to begin development of the budget for boll weevil eradication. Informational meetings were held during this time as well. Marshall Grant, a cotton grower and influential advocate of boll weevil eradication, attended the grower meetings and shared his knowledge and experience (Johnson and Martin, 2001).

In December 1994, the Arkansas State Plant Board recognized the Arkansas Cotton Growers Organization, giving them the official status to conduct referenda and promulgate the rules necessary to conduct a boll weevil eradication programme. The Arkansas Farm Bureau and Arkansas Cooperative Extension Service sponsored a grower tour to Georgia in 1995, providing Arkansas cotton growers with the opportunity to learn about the programme (Johnson and Martin, 2001).

As Louisiana prepared to conduct a referendum in the Red River Valley zone, the Arkansas Boll Weevil Eradication Foundation (ABWEF) Board of Directors prepared to hold a referendum to determine whether growers in south-west Arkansas wished to participate with Louisiana producers in boll weevil eradication. The producers voted in favour of the referendum in October 1995, and plans were made for the 6000 acre (2400 ha) south-west Arkansas zone to begin eradication in conjunction with the Red River Valley zone in Louisiana. The eradication programme began in the south-west zone in 1997 (Johnson and Martin, 2001). In 2000, the ABWEF assumed operation of the south-west zone (Kiser and Catanach, 2006).

A referendum on boll weevil eradication for the 972,000 acres (393,000 ha) of cotton in eastern Arkansas was held in March 1996. With only 52% of votes in favour, it did not pass. Many growers in north-east Arkansas did not support the plan (Johnson and Martin, 2001).

Jack Carey resigned as ABWEF Board Chairman in 1997 and Joe Burns replaced him. The Board made plans for regional referenda to allow growers in south-east and central Arkansas to move forward. The referenda for south-east Arkansas and central Arkansas were held in September 1997. The referenda passed and the south-east and central zones began to plan for the diapause phase of the programme in 1999 and 2000, respectively (Johnson and Martin, 2001; Kiser and Catanach, 2006). Doug Ladner was hired as Executive Director in November 1998. In 1999, the Arkansas legislature appropriated US$5 million in cost-sharing funding (Johnson and Martin, 2001).

In August 1999, the 300,000 acre (121,000 ha) south-east zone began the diapause control phase of the programme. In the autumn of 2000, the 212,000 acre (86,000 ha) central zone began the diapause phase. In February 2000 a referendum was held in the 135,000 acre (55,000 ha) north-east Ridge zone. It passed by 74% and the zone began diapause control in autumn 2001 (Johnson and Martin, 2001; Kiser and Catanach, 2006). In 2000 Danny Kiser, with several years experience in the Texas programme, was hired by ABWEF as Director of Operations. Doug Ladner retired as Executive Director in 2000 and Kiser was promoted to Executive Director in November 2002 (Anon., 2002a).

In the north-east Delta zone, unsuccessful referenda were held in November 2001, February 2002, December 2002 and February 2003. Following the February 2003 referendum, the Arkansas State Plant Board exercised its authority to initiate boll weevil eradication in the north-east Delta zone. The Arkansas Plant Board entered into an agreement with ABWEF to conduct the eradication activities in the zone. The diapause phase of the programme was initiated in the north-east Delta zone in August 2003 (Kiser and Catanach, 2995). Sixty growers from Mississippi and Craighead Counties filed suit claiming that collection of assessments without passage of a referendum constituted an illegal tax. The courts disagreed. The Arkansas Supreme Court decided in favour of the Arkansas Plant Board in September 2005 (Bennett, 2005).

Joe Burns resigned as Board Chairman in April 2002 and Perry Stratton was named Board Chairman. Stratton served until December 2003. Ritter Arnold was named Chairman of the Board, filling the vacancy left after Stratton's resignation (D. Kiser, 2007, personal communication).

By the end of 2006, the south-west zone was eradicated. Weevil captures in the south-east zone had been reduced by 99.89% since the programme began. Weevil captures in the central zone had been reduced by 99.9%. In the north-east Ridge zone weevil captures had been reduced by 99.9%. And, weevil captures in the north-east Delta zone had been reduced by 99.85% (Kiser and Catanach, 2007).

West Tennessee

The enabling legislation needed for starting boll weevil eradication in West Tennessee had been completed in 1989. No further legislation was needed to allow programme start-up in the West Tennessee programme. The Tennessee Boll Weevil Eradication Foundation Board of Directors, organized in 1993, was in place as well (Barker, 2001).

The 19 cotton-growing counties in West Tennessee, with some 345,000 cotton acres (140,000 ha), had been planning and developing strategies for boll weevil eradication since the early 1990s. At the 1993 National Cotton Council Boll Weevil Action Committee Meeting, Jim Brazzel stated that when creating regions for boll weevil eradication, natural boundaries were much preferred to political boundaries. West Tennessee cotton grower, Allen King, concluded that the Hatchie River would serve as a good natural boundary between regions in West Tennessee. His rationale was that growers south of the river were normally fighting weevils 2 weeks before he did north of the river. Based on this rationale, West Tennessee was divided into two regions: Region 1, the South-west Region, lying south of the Hatchie River and adjacent to the state of Mississippi; and Region 2, the North-west Region, north of the Hatchie. The West Tennessee plan was for Region 1 to begin their eradication programme one year after the Hill Country of Mississippi (Barker, 2001).

A referendum on boll weevil eradication for Region 1 was conducted in March 1997. It passed with 68% of the vote, just above the two-thirds required for passage. Programme start-up was planned for late summer 1998. As they had done in the middle Tennessee programme, the Tennessee Foundation signed an agreement with the SEBWEF to conduct programme operations in West Tennessee. Ron Seward was hired as SEBWEF Programme Manager for West Tennessee (Barker, 2001).

The diapause phase of the programme was started on the 117,610 acres (48,000 ha) of cotton in Region 1 during the first week of August 1998. Farmers noted the exceptional top crop made in 1998, but the killing freeze was later than normal, adding to programme spray costs (Barker, 2001).

In December 1998, educational meetings and other preparations were made for a referendum in north-west Tennessee scheduled for January 1999. The decision had been made to split north-west Tennessee into two regions: Region 3, the northern tier of counties bordering Kentucky; and Region 2 in the central portion of West Tennessee. Operationally, they would be conducted as one region but the assessment would be higher in Region 2 than in Region 3 (Barker, 2001).

The referendum passed in January 1999 with nearly 78% of the vote. The programme in West Tennessee Regions 2 and 3 could start as early as August 1999, but only if the February 1999 referenda in Mississippi Delta Regions 1A and 1B did not pass. Passage in the Delta would preclude a start in north-west Tennessee because the agreed-upon plan for boll weevil eradication in the Mid-South Region had scheduled start-up in the Delta before north-west Tennessee; and the availability of federal cost-sharing funds was limited. Both zones in the Mississippi Delta passed their referenda, delaying programme start-up in north-west Tennessee until 2000 (Barker, 2001).

North-west Tennessee growers were disappointed about their delayed start, and south-west Tennessee growers (West Tennessee Region 1) were left in a buffer zone with high weevil populations along their long northern boundary. In response, the programme managers temporarily changed their goal. They adjusted trap triggers to limit the influx of migrating weevils, but did not aggressively pursue eradication. They tried to control the weevils and, at the same time, avoid decimation of the beneficial insects, which might have led to a secondary pest outbreak. In spite of the trigger adjustments, on average, almost twice as many treatments were applied than were budgeted (Barker, 2001).

West Tennessee Regions 2 and 3 began the diapause phase of their programme in 2000. Once again, migrating weevils limited programme success. Boll weevils entered the westernmost cotton fields in West Tennessee Regions 1, 2 and 3 for a number of years. This resulted in slower than anticipated weevil population reduction, particularly along the Mississippi River (Brumley *et al.*, 2007).

As boll weevil populations declined in the adjacent north-east Delta zone in Arkansas and in Missouri, they also declined in West Tennessee (Brumley *et al.*, 2006, 2007). By the end of 2005, boll weevil captures in Region 1 had been reduced by 94.86% from 2000, captures in Region 2 had been reduced by 99.19% from 2001 and captures in Region 3 had been reduced by 97.84% from 2001 (El-Lissy and Grefenstette, 2006).

Missouri

Because of the cold winters, boll weevil populations in the 405,000 cotton acres (164,600 ha) in Missouri have historically been sporadic. However, during the period from 1980 to 2000, winter temperatures in Missouri were generally mild. This resulted in greater survival of boll weevil during the winter, higher populations during the growing season and increasing yield losses from weevils (Boyd, 2001; El-Lissy and Grefenstette, 2006).

As the eradication programme approached Missouri from Mississippi, Arkansas and West Tennessee, the enabling legislation needed to commence a programme was passed in 1995. The legislation gave the Missouri Department of Agriculture (MDA) authority to regulate and provide oversight for the programme. A nine-member Missouri Cotton Growers Organization (MCGO) was formed to administer the programme (Boyd, 2001). The Missouri Legislature provided US$400,000 to establish a boll weevil trapping programme in the Bootheel region to generate information about boll weevil population densities in the region (Grundler and Sorenson, 1997).

The first meeting of the MCGO Board of Directors was in August 1996; Charles Parker was elected Chairman. The MCGO emphasized that the Missouri programme would be initiated, approved and funded by Missouri cotton producers. In March 1997, several MCGO Board Members met with representatives of the SEBWEF to discuss contracting with them to operate the Missouri programme. In August 1997, the MCGO Board of Directors drafted and approved a letter requesting that SEBWEF operate the Missouri programme (Boyd, 2001).

In September 2000, Jim Brumley, Executive Director of the SEBWEF, outlined to the MCGO Board the proposed eradication procedures and plan for Missouri. The Board expressed the need to operate a low-cost programme in Missouri: it approved a 7-year programme. February 2000 dates were set for the educational meetings preceding the referendum. The referendum on boll weevil eradication in Missouri was held in March 2000. It failed, receiving 56% of the vote. The MCGO Board responded by changing the assessment due date from 5 August to 15 October. A second referendum was scheduled for August 2000. It also failed, but 61% of those voting favoured starting a boll weevil eradication programme. Because federal cost-sharing funds were available for the 2001 season, the MCGO Board was in favour of holding a third referendum in 2000. The referendum was held in November 2000, and this time the referendum passed with 74% of the vote and Missouri began preparing for the diapause phase of the programme to begin in August 2001 (Boyd, 2001).

Dewey Wayne King was hired as SEBWEF Programme Manager for Missouri in 2001. The diapause phase of the programme was initiated in August 2001. Migration of boll weevils from Arkansas' north-east Delta zone, which did not enter the programme until 2003, contributed to the slow initial progress in the zone. Also, the zone had a large number of bee hives located near cotton fields. Treatments were frequently delayed to avoid drift on to bee hives. This further slowed programme progress. Following programme start-up in the north-east Delta zone, weevil trap captures began to decline in Missouri (Brumley, 2003, personal communication).

Dewey Wayne King resigned and Jaye Massey was hired as SEBWEF Programme Manager for Missouri in 2005. In October 2006, MCGO Board Chairman Charles Parker was quoted as saying: 'The Missouri programme is solid as a rock. We are financially sound and we are eradicating the boll weevil' (Laws, 2006).

By the end of 2006, boll weevil populations had been reduced by 89.72% in Missouri (Brumley *et al.*, 2007). The highest weevil captures were in the western Bootheel area. In that area, Crowley's Ridge, Malden's Ridge and the St Francis River provide some of the best overwintering habitat in south-east Missouri (Boyd, 2001).

Oklahoma and Kansas

In December 1992, Oklahoma cotton growers met to begin organizing for boll weevil eradication. They requested the formation of a Legislative Task Force to begin developing the enabling legislation for a cooperative boll weevil eradication programme. In a poll of 125 cotton growers from across the state, only two producers said they would not support statewide boll weevil eradication. Those not in favour of eradication feared that equitable assessment rates for dryland and irrigated production would not be proposed. They favoured a system of pro-rated assessments for areas with low boll weevil populations (King and Pfenning, 2001).

In 1993, the Oklahoma Legislature passed the Boll Weevil Eradication Act. It established a Board comprised of four Directors to oversee the eradication effort. The Board would determine programme costs, develop a plan for assessing growers to pay for the programme and develop a referendum process. The Board was charged with dividing the state into five voting districts and establishing voting procedures and voter eligibility. Governor David Walters appointed the first Board of Directors. The Board divided the state's cotton growing area into five districts so that a representative from each district could be elected to serve on the Board of Directors of the Oklahoma Boll Weevil Eradication Organization (OBWEO) (Karner and Cuperus, 2000). Nominations were made and ballots were mailed in January 1994. In February 1994, the new Board members were sworn in. At the first meeting, Jerry McKinley was elected as Chairman. After much debate, the Board decided that the assessment would be US$75 per harvested acre spread over a 10-year period. A grower referendum was planned for December 1996 (King and Pfenning, 2001).

During the summer of 1996, two groups of Texas cotton producers sued the Texas Foundation. The first group claimed that boll weevil eradication treatments had caused a severe beet armyworm, *S. exigua* outbreak, which had caused extensive crop damage. The second group challenged the constitutional authority of the Texas Foundation to collect assessments. Petition drives to end the programme were initiated in some Texas zones. The problems in Texas had an unsettling effect on Oklahoma cotton growers. If the Texas programme fell apart, there would be no chance for success in Oklahoma (King and Pfenning, 2001).

Despite the turmoil in Texas, the Oklahoma Legislature appropriated US$750,000 to help fund boll weevil eradication in Oklahoma in 1996. A referendum was held in October 1997. The referendum passed with 88% of the vote in favour of conducting a programme. OBWEO hired Jerry Coakley as the Executive Director for the programme. Employees were hired and trained in preparation for programme start-up in 1998. The state of Oklahoma provided economic assistance through a capital bond programme. The state provided US$750,000 per year for 5 years (King and Pfenning, 2001; State of Oklahoma, 2004).

The diapause phase of the eradication programme began on 250,000 acres (101,000 ha) of Oklahoma cotton during the first week of September 1998. The Oklahoma programme, with USDA funding, conducted trapping operations on 90,000 acres (36,000 ha) of Kansas cotton for several years. More recently, the Kansas Department of Agriculture and APHIS have conducted the trapping operation in Kansas. By the end of 2005, no boll weevils were caught in the 252,000 acres (102,000 ha) of cotton in western Oklahoma and none were caught in Kansas (El-Lissy and Grefenstette, 2006). A few weevils were still present in the 1000 cotton acres (405 ha) in south-east Oklahoma. Eradication could not be achieved in south-east Oklahoma until an eradication programme was initiated and control achieved in the Northern Blacklands (NBL) of Texas. Boll weevil eradication was initiated in the Texas NBL zone in 2005, allowing the programme in south-east Oklahoma to proceed (Allen *et al.*, 2006).

Jerry Coakley resigned as Executive Director in the spring of 2004 and Bill Massey became Executive Director; Massey retired in April 2007; Joe Harris was hired as Executive Director in 2007 (B. Massey, 2007, personal communication).

New Mexico

New Mexico began preparing for boll weevil eradication soon after the boll weevil moved into the state from Texas and Arizona in the early 1990s. The Mesilla Valley Pest Management Association was established in 1995. It attempted a voluntary boll weevil eradication programme near Las Cruces in 1995 and 1996 (Ford, 1995). Enabling legislation was passed establishing the legal basis for a cooperative boll weevil eradication programme in 1996 (New Mexico State Assembly, 1996). In response to the boll weevil problem, the New Mexico Cotton Growers' Association was established in 1997 (Pierce *et al.*, 2001). Seven individual Boll Weevil Control Districts were formed as the grower entities, which would carry out boll weevil eradication in New Mexico (J. Friesen, 2007, personal communication).

The Luna County District entered into an agreement with the South Central New Mexico District to operate their programme. The 32,000 acres (13,000 ha) in these districts along the Mexican border initiated the diapause phase of the eradication programme in 1998 (El-Lissy and Grefenstette, 2006).

Much of the cotton grown in the 16,000 acre (6500 ha) Lea County Control District started the programme through individual grower contracts with the Texas programme. Operations began on these farms as the Western High Plains zone (WHP) started its diapause operations in 1999. In 2001, both the Lea County Boll Weevil Control District Board of Directors and the Board of Directors for the 4000-acre (1600-ha) Central Lea County Boll Weevil Control District signed agreements which allowed the Texas Boll Weevil Eradication Foundation (TBWEF) to operate eradication programmes in their districts (Allen *et al.*, 2004; TBWEF, 2007).

In 2001, the 16,000-acre (6500-ha) Curry/Roosevelt County Boll Weevil Control District Board of Directors signed an agreement which authorized the TBWEF to operate their eradication programme in conjunction with operations in the north-west Plains zone (NWP) in Texas. The diapause control phase of the Curry/Roosevelt New Mexico programme began during the first week of September 2001 (Allen *et al.*, 2004).

The 10,000-acre (4000-ha) Pecos Valley Control District started its programme in 2000 (El-Lissy and Grefenstette, 2006). In 2003 the Pecos Valley Control District authorized the TBWEF to operate the programme in the district (Allen *et al.*, 2004). The Pecos Valley programme was operated under the control and supervision of Texas' WHP zone from 2003 to 2006 (Allen *et al.*, 2006).

The Quay County Control District, with 2000 acres (800 ha) of cotton, has run a trapping programme since 2003 without catching a single boll weevil (Miller, 2006). During the 2006 season, no weevils were trapped in the state of New Mexico and all programme debt in New Mexico had been paid (El-Lissy and Grefenstette, 2006; Miller, 2006).

Texas

Interest in boll weevil eradication in Texas began to increase in the early 1990s. The interest was fuelled by the success of the programme in other areas, continuing high control costs, yield losses, resurgence of secondary pests and the necessity to manage every aspect of cotton production systems around the boll weevil. The Texas Cotton Producers (TCP) requested that the Department of Entomology at Texas A&M University draft an eradication plan for Texas. Ray Frisbie and Jim Brazzel (APHIS) chaired a joint committee, which developed 'A Plan for Boll Weevil Eradication in Texas'. The plan was presented to TCP in 1992 (R.E. Frisbie, 2007, personal communication).

Enabling legislation passed in both the Texas Senate and House in 1991, but was vetoed by Governor Ann Richards. Since the Texas Legislature meets only every alternate year, an improved boll weevil eradication bill was put on the priority list for the 1993 legislative session. Growers in the High Plains and the St Lawrence areas were running their own diapause control programmes and they lobbied to have their areas designated as statutory zones. Producers in other areas could petition the Texas Department of Agriculture (TDA) to establish a programme. Each zone would conduct a referendum to decide whether to enter an eradication programme. The enabling legislation gave programme oversight to TDA. The legislation passed and Governor Richards signed it in March 1993. No state funding was provided in the enabling legislation (Stavinoha and Woodward, 2001).

The law required a cotton growers' organization to petition the Commissioner of Agriculture for the right to create the organization responsible for managing eradication activities. In May 1993, TCP petitioned the Texas Agriculture Commissioner, Rick Perry, to authorize the formation of the Texas Boll Weevil Eradication Foundation (TBWEF). The petition delimited nine boll weevil eradication zones and named an individual from each zone to serve on the Board of Directors. Woody Anderson was chosen as the Chairman of the Board. The TBWEF Board of Directors began operations with financial support provided by producer organizations and a US$50,000 grant from TDA in 1993 (Stavinoha and Woodward, 2001).

The enabling legislation required TDA to develop rules for holding referenda and to protect humans, wildlife and honeybee colonies. By the end of 1993, rules for conducting referenda had been completed. The referendum rules required a two-thirds favourable vote of qualified voters or favourable vote by operators who farmed over 50% of the cotton acreage in the zone. The referendum ballot had to specify the annual assessment, the number of years the assessment would be paid and the yearly due date for assessments. The rules required that a board member be elected to represent each zone. Growers were required to be notified that a referendum was going to be held, and educational meetings were required prior to the referendum.

Frank Meyers, the former director of the Arizona programme with 30 years experience in APHIS, was hired to serve as Executive Director of the Texas programme; Osama El-Lissy was hired as Programme Director; he came to the Texas programme with several years of valuable experience in the Arizona programme (Stavinoha and Woodward, 2001). Deborah McPartlan was named the APHIS Co-Programme Director for the Texas Programme in 1995. She served in that position until March 2003.

TBWEF Board Chairman, Woody Anderson, appointed a TAC to provide recommendations and plans to the TBWEF management and Board of Directors. Ray Frisbie was appointed chairman of the TAC; he served as chairman until May 2002.

Jim Leser was appointed chairman in 2002 and served until December 2005. Tom Fuchs was appointed chairman of the TBWEF TAC in 2006. In 2006, Fuchs was also appointed to chair a National TAC to deal with post-eradication issues. The post-eradication TAC reported to the NCC BWAC.

The original plan for Texas was to eradicate from the south to the north, one zone per year. The plan was later modified to operate in a dual-front approach, starting in the Lower Rio Grande Valley zone (LRGV) and the Southern Rolling Plains zone (SRP) and moving north from these zones. The SRP, located in the vicinity of San Angelo in western Texas, held its referendum in March 1994. It passed with an 85% positive vote. In September 1994, the 200,000-acre (81,000-ha) SRP zone began the diapause treatment phase of the programme (Stavinoha and Woodward, 2001).

The 250,000-acre (101,000-ha) LRGV zone held its referendum in April 1994: it passed with a 73% positive vote. However, the assessment did not pass. A new assessment schedule was developed and a second referendum was held in October 1994: it was approved by 70% of the growers.

Responding to the wishes of the growers and against the recommendation of the TAC, the Board agreed to start the programme in the spring, with autumn stalk destruction in lieu of autumn diapause control treatments (R.E. Frisbie, 2007, personal communication). With cotton planting due to start in February 1995, a frantic effort was made to prepare for programme start-up. A lawsuit challenging the procedures for protecting human health, wildlife and endangered species was filed. Fortunately, APHIS had prepared and filed an extensive Environmental Impact Statement, which had determined that the programme would have no significant impact on human health or the environment (USDA-APHIS, 1991; Stavinoha and Woodward, 2001).

Foundation treatments began in April 1995 in the LRGV. The weather was very unfavourable: extremely hot, dry conditions prevailed. Cotton aphid, *A. gossypii* and cotton fleahopper, *Pseudatomoselis seriatus*, populations were high in most fields early in the season and multiple insecticide treatments were made by growers to control these pests. In spite of the programme, growers continued to spray for weevils and other pests. Poor programme organization, poor communications between the programme employees and growers and poor programme responsiveness to growers were cited as problems. The LRGV area experienced a severe outbreak of beet armyworms, *S. exigua*, with loopers, *Trichoplusia ni*, bollworms, *H. zea*, and budworms, *H. virescens* present as well. Although other areas not in eradication programmes also had moderate to severe beet armyworm outbreaks, the programme was blamed for killing the natural enemies and causing the outbreak (Stavinoha and Woodward, 2001).

Texas Commissioner of Agriculture, Rick Perry, appointed a special task force chaired by Ray Frisbie to evaluate the causal factors of the beet armyworm outbreak in the LRGV. This multi-member task force representing scientists, producers and regulatory personnel determined that multiple factors could have caused the outbreak. Some of the factors they identified were severe drought conditions favouring beet armyworm movement from wild hosts, grower-applied, early-season and mid-season insecticide sprays – not associated with the eradication programme – which suppressed beet armyworm natural enemies, as well as malathion sprays applied for boll weevil eradication. The task force concluded that the eradication programme was not the primary causal factor for the beet armyworm outbreak (R.E. Frisbie, 2007, personal communication).

A severe sweet potato whitefly, *Bemisia* sp., outbreak occurred following the beet armyworm outbreak. As with the beet armyworm outbreak, severe sweet potato whitefly outbreaks occurred that year in other US cotton-producing areas (Arizona and southern California), and malathion sprays for boll weevil eradication were not applied in those areas (R.E. Frisbie, 2007, personal communication). The available insecticides were not sufficiently effective and the drought continued without relief. Boll weevil treatment triggers were relaxed to reduce the number of malathion applications. The autumn boll weevil peak occurred and, because of the earlier damage, most of the cotton plants in the zone were highly susceptible to weevil damage, with many immature fruit forms. By autumn, the devastating drought and the severe insect outbreaks had destroyed much of the crop (Stavinoha and Woodward, 2001).

The LRGV assessment was reduced by 50% to lessen some of the financial burden being experienced by growers. Many growers did not pay their assessments by the deadline. TBWEF requested that TDA assess late payment penalties. Many of the growers responded negatively to the penalties (Stavinoha and Woodward, 2001).

A recall petition was circulated in the LRGV in the autumn of 1995. The movement was called 'Sweep Out'. Along with the losses from 1995, the uncertainty of Mexico's participation in boll weevil eradication was given as a reason for signing the petition and voting the programme out. The petition was signed by more than 40% of the growers, surpassing the number needed for a recall referendum. The recall referendum was held in January 1996. The Sweep Out movement was successful. Seventy-four per cent of the eligible voters voted to end the programme. Collection of assessments was problematic due to the economic hardship among the growers and because of legal challenges to the Boll Weevil Eradication law. The LRGV zone's vehicles and equipment were sold and its offices were closed. The Production Credit Association loan went unpaid and interest continued to accumulate (Stavinoha and Woodward, 2001).

In other areas of the state the weather during the summer of 1995 was also hot and dry. Beet armyworm infestations broke out, seriously damaging cotton in the SRP zone and causing damage in other areas as well. Programme operations in the SRP were much smoother than in the LRGV because SRP programme managers had enough time for proper organization and training. In spite of severe beet armyworm damage, SRP growers stuck with the eradication programme (Stavinoha and Woodward, 2001).

In response to the beet armyworm outbreak in 1995, growers in the SRP zone opted to use Vydate® CLV for overwintered weevil control and Phaser® 3 EC for mid-season weevil populations in 1996. The trigger for treating fields was raised to ten weevils per field mid-season. The result was poorer control, partially because of the shorter residual effectiveness of the alternative products (England *et al.*, 1997). By autumn the SRP had not achieved the weevil population reduction desired (El-Lissy *et al.*, 1997). The alternative insecticides were more expensive, and were not used again as a component of US boll weevil eradication programmes.

In spite of the development of the serious problems in the LRGV and other zones, programme expansion continued in Texas. In December 1994 growers in the

500,000 acre (202,000 ha) Rolling Plains Central zone (RPC) passed a referendum with an 85% positive vote to begin eradication. Growers in the RPC, north and west of Abilene, decided to pay an assessment of US$5 per acre one year before programme start-up. The diapause control phase of the programme began in the RPC zone in the autumn of 1996 (Stavinoha and Woodward, 2001).

The South Texas/Winter Garden zone (ST/WG), initially stretching from Houston south to Corpus Christi and west to Uvalde, passed its referendum in February 1995 with a 74% favourable vote. Diapause phase treatments began in the 350,000 acre (142,000 ha) ST/WG zone in the autumn of 1996. Growers could use alternative insecticides if they wished. Rebates from the assessments were provided to growers for early stalk destruction in the ST/WG zone, but wet weather late in the summer prevented early stalk destruction on most farms. Regrowth on harvested fields became hostable and the budget suffered from treatment expenses on cotton stalks after harvest (Stavinoha and Woodward, 2001).

On the High Plains, funding constraints prevented PCG's Diapause Control Programme from being able to treat sufficient acreage to prevent weevils from spreading and becoming established. PCG's leadership proposed a referendum to allow them, through a cooperative agreement with the TBWEF, to conduct an enhanced diapause control programme, 1995–1997. A referendum was held in April 1995 to establish the programme: it passed with more than 77% favourable vote. The programme established three tiers of counties for assessment purposes: the highest assessment rate was in the southernmost counties, and the lowest assessment rate was in the northernmost counties (Stavinoha and Woodward, 2001).

The St Lawrence (STL) zone, a 150,000 acre (61,000 ha) cotton-producing area south-east of Midland, followed PCG's lead. They proposed a cooperative agreement with the Foundation, which would allow one year of enhanced diapause, then entry into the eradication programme. The referendum was held in August 1995. The growers voted to conduct the programme, but the assessment did not pass. A second St Lawrence referendum was held in September 1996. It received 65.7% of the vote, yet it still passed by virtue of positive votes of farmers representing 52.7% of the cotton acreage. The St Lawrence zone was scheduled to start its diapause programme in the autumn of 1997 (El-Lissy et al., 1999; Stavinoha and Woodward, 2001).

A group of Hale County farmers expressed their concerns about the plan for the programme in the High Plains. They felt that the tiered assessment plan was unfair because their farms were not sufficiently infested by boll weevils to warrant the higher assessment they would be required to pay under the plan. In September 1995, ten Hale County farmers filed a lawsuit challenging the constitutionality of the legislation under which the Foundation operated. It claimed that the assessments were an occupational tax, and therefore prohibited by the Texas Constitution. In July 1996, the State District court ruled against the Foundation and the Texas Boll Weevil Eradication Law. In November 1996, the Foundation appealed the case directly to the State Supreme Court (Stavinoha and Woodward, 2001).

While issues of law were being argued, the programme continued in the SRP, the RPC and the ST/WG; and PCG continued its diapause control programme with assessments collected by TBWEF. Due to the legal uncertainties, some farmers did not pay their 1996 assessments. The debt in these zones increased (Anon., 1998; Stavinoha and Woodward, 2001).

In Brazoria County, farmers organized against the programme in the ST/WG zone in 1996. The movement, led by Lorin Batchelor, was called Sweep Out II. A pro-eradication group, called Keep Going, led by Jimmy Dodson, was organized to counter Sweep Out II. The NCC supported Keep Going through providing information on the positive aspects of eradication to both farmers and landowners. Sweep Out II's petition was not signed by the required 40% of the growers and was declared invalid in September 1996 (Carter *et al.*, 2001; Stavinoha and Woodward, 2001).

The Texas Legislature was scheduled to open its 1997 legislative session in mid-January 1997. TBWEF expected the Texas Supreme Court to rule on the Hale County case in time for changes to be made in the Boll Weevil Eradication Law, if needed, before the session closed at the end of May. However, Senator Robert Duncan of Lubbock filed a bill that would allow changes to be made to the law if the court did not rule before the deadline for introducing new bills. On 30 April 1997, 30 days before the end of the session, the court ruled: they found the boll weevil law unconstitutional because it gave the non-profit Foundation authority similar to that reserved for government agencies. The court decision clearly stated what changes would be needed to allow continuation of the programme (Bryant, 1997; Stavinoha and Woodward, 2001).

A new law was written addressing the problems identified by the Texas Supreme Court. Growers testified during the process and Valley growers lobbied not to be held responsible for their boll weevil eradication programme debt. They were partly successful in that the new law did not give the Foundation authority to collect assessments to pay the debt that the LRGV growers owed the Production Credit Associations. Sweep Out II had failed in its bid for a recall referendum, but succeeded in ending the eradication programme in seven counties south-west of Houston. They successfully lobbied to have the 200,000 acre (81,000 ha) 'Seven Counties' area removed from the new eradication programme. Nothing in the new law absolved the LRGV and Seven Counties cotton growers from paying the debt they owed (Stavinoha and Woodward, 2001).

The rewritten boll weevil eradication legislation was approved by the Texas Legislature, and Governor George W. Bush signed it on 30 May 1997, just 1 month after the Supreme Court ruling (Stavinoha and Woodward, 2001). The Legislature ruled that diapause programmes could be resumed in the three remaining active zones: the SRP, ST/WG and RPC, but that confirmation referenda were needed in these zones. The Legislature voided the High Plains April 1995 referendum, split the region into the Northern High Plains (NHP) (which included Hale County) and the Southern High Plains/Caprock (SHP/C) zones and set a referendum for SHP/C growers to vote again in August 1997. In addition, the new law specified that zones must hold retention referenda every 4 years (Stavinoha and Woodward, 2001).

Texas Agriculture Commissioner, Rick Perry, appointed new members to the TBWEF Board of Directors in time for the first meeting of the new board on 3 June 1997. Woody Anderson was elected Chairman; Executive Director, Frank Meyers resigned after the new law passed. Osama El-Lissy was named interim Executive Director and continued to serve as Programme Director. Lindy Patton, who had previously worked as District Representative on Congressman Charlie Stenholm's staff and Executive Director of Rolling Plains Cotton Growers Association, was selected as Executive Director in September 1997 (Smith, 1997; Stavinoha and Woodward, 2001).

The most pressing issue before the newly reorganized Texas Foundation was of finding a way to deal with the debt and to begin paying the notes to the Farm Credit System. Congressman Stenholm took steps to address this need by working to make a USDA-FSA low-interest loan programme available to the programme. In September 1997, the Texas programme obtained a US$25 million loan from FSA. Another US$29.4 million was loaned to the Foundation by three Production Credit Association districts and the Farm Credit Bank of Texas. In December 1997, another US$10 million loan was provided by FSA (Smith, 1997; Stavinoha and Woodward, 2001).

The SRP had managed to keep its programme operating in 1997, but programmes in the ST/WG and RPC had been shut down by the Supreme Court decision. They re-started their programmes with a second diapause control phase in the autumn of 1997. The 1996 diapause programmes in these zones had accomplished little because the progress that had been made was lost when the programme was shut down in the spring and summer of 1997.

In response to the requirement that the SHP/C growers vote again by mid-August, PCG submitted a plan to run an eradication programme in the zone. Commissioner Perry appointed a committee of grower representatives to develop a proposed programme, budget and assessment rate. The grower committee decided to present the growers with a suppression programme with no in-season treatments. TBWEF would assume responsibility for boll weevil eradication activities if the referendum failed. The referendum did not pass (Stavinoha and Woodward, 2001).

The programme confirmation referenda required in the new boll weevil eradication law passed in the ST/WG, SRP and RPC zones by strong margins in late 1997 and early 1998 (El-Lissy *et al.*, 1999; Stavinoha and Woodward, 2001).

The financial difficulties the Foundation faced were made worse by delinquent assessment accounts. To help remedy this situation, TDA adopted rules in 1999 providing for a lien to be placed on cotton when producers were delinquent on their TBWEF accounts. In addition, the Board decided to file individual collection suits to help assure that assessments would be collected (Stavinoha and Woodward, 2001). Collection rates quickly improved. After these changes, collection rates increased to 97–99% in every zone (TBWEF, 2007).

Don Parrish, a Yoakum County farmer, led an organizational effort in the south-western region of the High Plains. This group thought that the original SHP/C encompassing 3 million cotton acres (1.2 million ha) was too large. They wanted a separate zone, the 850,000 acre (344,000 ha) Western High Plains (WHP) zone, which would involve cotton grown in the vicinity of Seminole, Plains and Brownfield. TDA defined the boundaries of the new zone by rule in January 1998 (Stavinoha and Woodward, 2001).

The grower leadership in the counties to the east of the WHP zone initiated the process to establish their own zone and, in January 1998, the 750,000 acre (304,000 ha) Permian Basin (PB) zone was created. The PB zone included cotton acreage in the area of Midland, Stanton, Big Spring and Lamesa (Stavinoha and Woodward, 2001).

In the Northern High Plains region, the gins from the ten-county area appointed representatives to discuss boll weevil eradication. It became clear that only the counties in the western part of the region wanted to proceed. Growers in Bailey, Lamb,

Deaf Smith, Castro Counties and the western parts of Swisher County petitioned the Commissioner of Agriculture to form the 500,000 acre (202,000 ha) North-west Plains (NWP) zone, which would conduct boll weevil eradication in the area from Hereford and Littlefield west to New Mexico (Stavinoha and Woodward, 2001).

In the Texas Blacklands/Brazos River Bottom zone, growers in the south-eastern part of the zone wanted the Commissioner to designate their area a zone by law. The commissioner appointed a committee of growers from each county to advise him on the best way to establish the boundaries. The committee recommended dividing the area into three zones. However, after study of the zone from maps and by air, Foundation personnel felt that because of the lack of natural boundaries the area should be divided into two zones. The 100,000 acre (40,000 ha) northern Blacklands (NBL) zone would take in the area from Waco north through Fort Worth to the Red River and east to Texarkana; and the 100,000 acre southern Blacklands (SBL) zone would include cotton acreage from Luling to Waco, east to Louisiana. In July 1998 the Commissioner established the SBL zone and, in January 1999, the NBL zone was established (El-Lissy *et al.*, 1999; Stavinoha and Woodward, 2001).

Agriculture Commissioner, Rick Perry, appointed an advisory committee to provide direction for the establishment of zone boundaries in the northern region of the Texas Rolling Plains in December 1997. Some Hall and Childress County farmers were not in favour of boll weevil eradication. Since the 100,000 acres of cotton in question was contiguous with the cotton that would be included, their request was not considered feasible. In June 1998, Commissioner Perry designated the northern Rolling Plains (NRP) boll weevil eradication zone. The 300,000 acre (121,000 ha) zone stretched from Paducah north to Pampa and east to Wichita Falls (El-Lissy *et al.*, 1999; Stavinoha and Woodward, 2001).

In the far western region of the state, growers requested their area be designated a zone via the rule-making process. The El Paso/Trans Pecos zone (EP/TP) with 45,000 cotton acres (18,000 ha) was designated in August 1998 (El-Lissy *et al.*, 1999; Stavinoha and Woodward, 2001).

Representatives to the TBWEF Board of Directors were appointed by the Commissioner from each new zone until referenda could be held to elect Board representatives (Stavinoha and Woodward, 2001).

Texas legislators realized the plight of cotton farmers after a severe drought in 1998 and pledged to help with the expenses of boll weevil eradication. In response, committees of farmers in many zones began working on budgets and start-up plans for their boll weevil eradication programmes. TDA developed a schedule of referenda, which allowed zones to vote prior to the normal planting time in their zone. The state legislature approved a US$25 million appropriation for cost-sharing funds. The legislation called for the state to pay up to 50% of the 1999 assessments in the three zones with active programmes (SRP, ST/WG and RPC), and TBWEF received compensation for expenses incurred when acreage was being ploughed up after having been sprayed through the mid-season but on which the growers did not pay an assessment (Stavinoha and Woodward, 2001).

The WHP zone held a referendum in December 1998. The referendum passed with 79% of the vote, allowing the diapause control phase of the programme to begin in the 850,000 acre (344,000 ha) WHP zone in August 1999 (Allen *et al.*, 2001; Stavinoha and Woodward, 2001).

The SHP/C zone held its next referendum in February 1999: the referendum failed. Another referendum was held in November 2000: it passed with over 80% positive vote. The 1.2 million acre (486,000 ha) SHP/C zone, stretching from Crosbyton and Slaton through Lubbock to the New Mexico state line, began eradication with diapause control treatments in the autumn of 2001 (Allen *et al.*, 2001; Stavinoha and Woodward, 2001).

In February 1999, the 100,000 acre (40,000 ha) SBL zone held a referendum. The growers voted in favour of forming a zone and in favour of the programme but did not pass the assessment. A second referendum was held in April 2000 to consider establishing an assessment for the zone. The referendum passed and the diapause phase of the programme began in the autumn of 2001 (Allen *et al.*, 2001; Stavinoha and Woodward, 2001).

In March 1999, growers in the NWP zone held a referendum and approved the programme and the assessment, with 75% in favour of starting a programme. The diapause phase of the programme began in the autumn of 1999 (Allen *et al.*, 2001; Stavinoha and Woodward, 2001).

The EP/TP zone held a referendum in March 1999. In addition to boll weevil eradication, growers in the EP/TP zone were also voting on a pink bollworm suppression programme that would start activities in 2001. Eighty per cent of those voting approved the programme. The diapause phase of the programme began in 1999 and the pink bollworm eradication programme began in 2001 (Allen *et al.*, 2001; Stavinoha and Woodward, 2001).

After the entry of five new zones into the programme, boll weevil eradication was being conducted in eight Texas zones totalling 3.6 million land acres (1.5 million ha) of cotton in 1999. The SRP, ST/WG and RPC zones were operating full-season programmes and the WHP, NWP, EP/TP, PB and NRP zones were operating programmes in the diapause phase (El-Lissy *et al.*, 2000). The state of Texas provided cost-sharing assistance to support boll weevil eradication from 1999 to 2006. State cost-sharing assistance has contributed 23.9% of the operating funds.

Programme Director Osama El-Lissy resigned in 1999 to accept an APHIS programme management position in Beltsville, Maryland. Charles Allen, with 20 years experience as an extension entomologist in Texas and Arkansas, was hired as Programme Director in June 2000. Richard Newman was hired as TBWEF Chief Administrative Officer in the spring of 2000. Newman came to the Texas Foundation with 38 years of experience with FSA, where he had served most recently as Deputy Administrator for Farm Programmes. Danny Kiser resigned as Assistant Programme Director in the Texas programme to accept the Director of Operations position for the Arkansas Boll Weevil Eradication Programme in July 2000. Larry Smith was hired as Assistant Programme Director for the Texas programme later that year. Smith came to the Assistant Programme Director position after having worked for the Texas Foundation for 5 years as zone manager in the RPC zone. Smith had farmed, consulted and worked overseas with USAID prior to working with the Foundation. Aaron Miller was named the APHIS Senior Boll Weevil Coordinator for Texas and New Mexico in August 2004. He replaced Deborah McPartlan, who resigned as APHIS Co-Programme Director in 2003.

The Texas programme made strong progress in 1999 and 2000. The SRP zone was declared functionally eradicated by Agriculture Commissioner Susan Combs in

September 2000. This declaration established quarantine protection for the SRP to help prevent reinfestation (Allen *et al.*, 2001). ULV malathion supplies were critically short during 2000, with intermittent shortages in 2001. Through careful monitoring of use, management of deliveries and air shipment of the insecticide from the production facilities in Denmark, no serious programme interruptions occurred (Allen *et al.*, 2001).

A beet armyworm outbreak occurred across a large area of the High Plains region in 2000; as in 1995, hot, dry weather contributed to the problem. Treatment triggers were adjusted in the WHP and NWP zones. Outbreaks also occurred in the SHP/C and NHP zones, which were not in eradication programmes. Several new insecticides with very good effectiveness against beet armyworms were available. TBWEF trigger adjustments allowed *Cotesia* sp. parasites and arthropod predator populations to stabilize, providing suppression of the beet armyworm populations. The improved insecticides and trigger adjustment process contributed to the successful management of the 2000 beet armyworm outbreak, and a crisis was avoided (Allen *et al.*, 2001).

In October 2000, the 600,000 acre (243,000 ha) NHP zone held a referendum to consider entering the programme. The referendum passed with a 76% positive vote. The diapause control phase of the programme began in the autumn of 2001 (Smith *et al.*, 2002).

In 2000, TBWEF conducted boll weevil eradication operations on 4.3 million land acres (1.7 million ha) of cotton in eight Texas zones (Allen *et al.*, 2001). No referenda to open new zones were conducted in 2001. However, because referenda had been passed in the SBL, SHP/C and NHP zones in 2000, these zones initiated programmes with diapause control phase operations in 2001. The addition of these three zones increased the Texas cotton acreage in boll weevil eradication to 5.8 million acres (2.3 million ha) in 11 zones. Migration of weevils into the NWP zone from neighbouring zones had serious effects on programme operations and costs in the zone in 2000 and 2001. The 2001 initiation of programmes with diapause control phase treatment in the SHP/C and NHP zones provided substantial relief from late-season boll weevil migration into the NWP zone, but diapause treatments in the SHP/C and NHP did not immediately stop the migration (Smith *et al.*, 2002).

A referendum was held in January 2002 in the Upper Coastal Bend (UCB) zone south-west of Houston. Growers in the zone passed the referendum. The region, referred to earlier as the Seven Counties area, had been a part of the ST/WG but was removed by the legislature in 1997. The diapause control phase of the programme began on the 188,000 acres (76,000 ha) in the UCB zone in July of 2002 (Allen, *et al.*, 2003). In February 2002, Agriculture Commissioner Susan Combs declared the RPC zone functionally eradicated (Allen *et al.*, 2003). In December 2002, the NBL zone, located north of Waco, held a referendum to enter boll weevil eradication. The referendum failed, with only 57% of the growers voting for the programme (Allen *et al.*, 2003). A second referendum to conduct boll weevil eradication in the NBL zone was conducted in December 2003. This referendum also failed, with 65% of the growers voting for the programme. A third referendum in the NBL zone in January 2005 passed, with an 84% positive vote. The diapause phase of the programme began in late summer 2005 (Allen *et al.*, 2006).

In western Texas, migration of boll weevils from the St Lawrence area began causing significant problems for the surrounding zones in 2002. Boll weevil migration from the area affected the PB, RPC, WHP, SRP, SHP/C, NHP and NRP zones, some 4.5 million land acres (1.8 million ha) of cotton, from 2002 to 2006. In southern Texas, migration of weevils into the Victoria area of the ST/WG from the Seven Counties area (later the UCB zone) to the north-east, and into the Kingsville and Uvalde districts in the southern and south-western parts of the zone from the LRGV, caused serious problems from 2001 to 2005 (Allen *et al.*, 2003, 2004, 2005, 2006, 2007).

During the 2002 season, TBWEF conducted boll weevil eradication on 5.7 million land acres (2.3 million ha) of cotton in 12 Texas zones (Allen *et al.*, 2003). In August 2003, a second referendum to attach the northern part of Glasscock County to the PB zone was conducted. The referendum failed, as had a previous attempt in August 1998. Boll weevil eradication programmes were operating in 12 Texas zones with 5.7 million land acres of cotton in the programme in 2003 (Allen *et al.*, 2004).

In March 2004, Commissioner Susan Combs declared the EP/TP, NWP, NHP, NRP, SHP/C, WHP and PB zones suppressed. These seven zones joined the functionally eradicated SRP and RPC in having their quarantined status lifted. This action provided protection from reinfestation. Restricted items could not be legally moved into the suppressed or functionally eradicated zones from areas of the state that still had significant weevil populations and were under quarantine. Taken together, 5.2 million acres (2.1 million ha) had been declared either suppressed or functionally eradicated after the Commissioner's 2004 declarations (Allen *et al.*, 2005).

In April 2004, a referendum in the Panhandle (PH) zone passed with a nearly 93% positive vote. Trapping was conducted in the zone from pinhead square to freeze with no weevils caught. Also in April 2004, the STL zone voted. Growers in the zone approved starting a programme, with 83% voting in favour of having a programme. The diapause control phase of the programme began in September 2004 (Allen *et al.*, 2005).

In August and September 2004, large boll weevil migrations occurred from the LRGV zone, which was not yet in an eradication programme, into southern and south-western portions of the ST/WG zone (the Kingsville and Uvalde areas). The result was increased boll weevil populations and control costs in the ST/WG zone. The state of Tamaulipas in north-eastern Mexico began boll weevil eradication with diapause phase treatments in 2004. In the southern tip of Texas, the LRGV zone conducted a referendum in November 2004. The referendum passed with 74% of the vote and preparations were made to begin eradication in the diapause phase as bolls opened in 2005. The presence of an active boll weevil eradication programme in the state of Tamaulipas had a positive influence on the referendum in the LRGV. In 2004 TBWEF was conducting boll weevil eradication operations in 14 Texas zones and 6.3 million land acres (2.5 million ha) of cotton (Allen *et al.*, 2005).

In January 2005 the NBL zone, the area north of Waco, conducted its third referendum on boll weevil eradication. The referendum passed with an 84% favourable vote and preparations were made to begin the diapause phase of the programme in the autumn of 2005. Passage of the referendum in the NBL brought all Texas cotton into the boll weevil eradication programme. The NBL and LRGV entered diapause operations and full-season operations were under way in the other 14 zones. The diapause control phase went smoothly in both the NBL and the LRGV zones.

Nearly 6.8 million land acres (2.8 million ha) of cotton in 16 zones were in the boll weevil eradication programme in Texas in 2005. Weevil numbers had been dramatically reduced in the western part of the state (Allen *et al.*, 2006).

During 2006, nearly 7 million land acres (2.8 million ha) of cotton in Texas were involved in boll weevil eradication (Allen *et al.*, 2007). By the end of the 2006 season, the programme was nearing completion in western Texas. Only 4524 weevils were caught during the year from 5.6 million land acres (2.3 million ha) of cotton in the 11 western Texas zones. No weevils were caught from the 710,000 acres (287,000 ha) in the PH, EP/TP and NWP zones. And, from the 1.9 million acres (769,000 ha) of cotton in the NRP, NHP and WHP zones, only nine weevils were caught. In the entire region, reproducing populations of boll weevils could be found only in the southern half of Howard County and the northern half of Glasscock County, some 50,000 acres (20,000 ha) of cotton. Ninety-seven per cent of the weevils caught in western Texas were caught in this relatively small area. In 2006, weevil trap captures in western Texas had been reduced by over 97% from the previous year. Since the programme began, weevil populations in the region had been reduced by well over 99.99% (Allen *et al.*, 2007).

The 2006 growing season was exceptionally hot and dry. Growers, consultants and others had serious concerns about secondary pest population development in the southern and eastern Texas zones – especially in the LRGV – where many acres were being treated each week. The plan proposed by the TAC, accepted by the TBWEF Board and TDA, and approved by the LRGV growers in their start-up referendum called for no treatment in the month of May to allow the populations of natural enemies to increase. Sweet potato white flies, *Bemesia* sp., became a threat, and treatment triggers were adjusted in the affected work units. A few fields were damaged, but large, areawide populations were avoided. In a few areas, high densities of beet armyworm, *S. exigua*, egg masses were observed, but large larval populations did not develop (Allen *et al.*, 2007; J.W. Norman, 2007, personal communication). Good progress was made in 2006, with weevil populations reduced by 82% from the previous year in the LRGV and by 96% from the previous year in the NBL (Allen *et al.*, 2007).

Weevil populations were considerably higher in the southern and eastern Texas zones than in western Texas at the end of the 2006 season. But, as in western Texas, weevil populations in the five eastern and southern Texas zones had been greatly reduced from the previous year. A total of 4.5 million weevils was caught from the 1.3 million cotton acres (526,000 ha) in the five eastern and southern Texas zones. Over 82% of the weevils caught in these five zones were caught in the LRGV zone. On a per-trap inspected basis, weevil numbers had been reduced by 87% from 2005 to 2006 in the eastern and southern Texas zones. Strong reductions during the year had been seen in the NBL, ST/WG and LRGV. During the same period, weevil populations had been reduced by lesser amounts in the SBL and UCB zones (Allen *et al.*, 2007). Overall, since the inception of the programme, weevil populations in the eastern and southern Texas zones had been reduced by 95.3%.

Comparing captures in the initial programme year with captures in 2006, the percentage boll weevil reductions are: EP/TP, PH and NWP, 100%; NHP, NRP, PB, RPC, SHP/C, SRP and WHP, 99.99%; STL, 99.82%; ST/WG, 99.65%; SBL, 99.28%; UCB, 99.04%; NBL, 96.29%; and LRGV, 81.63%.

Mexico

The cotton-producing areas in north-western Mexico, the Mexicali Valley, the Caborca area and the Sonoita region completed boll weevil eradication in 1991 in conjunction with the California and Arizona programmes. The boll weevil eradication programme in the state of Chihuahua began in 2002 and the eradication programme was begun in the state of Tamaulipas in 2004. Boll weevil populations have been reduced substantially in the Mexican boll weevil eradication programme areas.

The Tally: Economic and Environmental Impacts of Boll Weevil Eradication

A national study on the economic and environmental impact of boll weevil eradication has not yet been conducted. However, regionally specific information on the economic and environmental impacts of boll weevil eradication is available from some of the participating states.

In the Carolinas, following boll weevil eradication, foliar insecticide use was reduced by 60–90% (Carter *et al.*, 2001; Ridgway and Mussman, 2001). Carlson and Suguiyama (1985) reported 53 pounds per acre (59 kg/ha) more yield and US$28/acre less insecticide cost after boll weevil eradication for eight North Carolina counties. The rate of return for boll weevil eradication was 187% over producer costs and 97% over all private and public costs (Carlson *et al.*, 1989). The acreage planted to cotton in the Carolinas started to increase for the first time in 40 years (Haney *et al.*, 1996; Carter *et al.*, 2001; McKibben *et al.*, 2001). Overall economic benefits were calculated at US$112/acre per year (Carlson *et al.*, 1989; Frisbie *et al.*, 1989; Szmedra *et al.*, 1991).

In Georgia, Roberts (1999) reported that insecticide costs on cotton had been reduced by US$65 per acre per year compared with the years before boll weevil eradication. Haney and co-authors (2001) reported that treatments for all pests had been reduced from 14.4 to 3.4 per acre per year and treatment costs had decreased by 76%. The gross value of the crop had increased from US$312 to 467 per acre and the net crop value had increased by 54% after eradication. The net benefit in Georgia to 1999, after deducting the cost of the programme, was about US$3.3 billion. Acreage planted to cotton also increased significantly in Georgia, from an average of 342,000 acres (138,000 ha) during active eradication and clean-up (1987–1992) to 1,194,000 acres (483,000 ha) during post-eradication (1993–1999) (Haney *et al.*, 2001a).

In Alabama, Curtis (2001) reported that insecticide usage had dropped by 50–60%, a saving of US$36 per acre, and yields had increased by 100 pounds (45 kg) or more per acre. In Florida, dramatic reductions in losses have been seen. The financial benefits of boll weevil eradication derived from the increased cotton acreage and yield in just one season have been greater than the entire cost of the Florida programme (Haney *et al.*, 2001b).

In the southern Rolling Plains of Texas, a benefit of US$1.45 was realized for every dollar spent in that low-input production region (Johnson *et al.*, 2001). In a recent Texas study, the increase in net returns above variable costs after boll weevil eradication in 2005 varied by zone from US$22 to $48 per acre. The study, which did not include the LRGV, NBL and EP/TP zones, estimated the increase in net returns at the farm level to be US$206 million in 2005 for the 13 Texas zones studied. Since 1996, the estimated cumulative increase in net returns at the farm level, after accounting for boll weevil eradication costs, has been US$947 million (McCorkle *et al.*, 2007).

Junk Yard Dog: Protecting Boll Weevil Eradication Investment

The eradication programmes, the NCC and APHIS have led the national effort to develop plans and procedures that will provide farmers with protection from becoming re-infested in future years. The individual state departments of agriculture have developed quarantines to help protect eradicated or suppressed areas within the state from reinfestation. The state quarantines define and control the movement of regulated articles into zones in which quarantines have been lifted (because weevils are not present or are present at very low levels). They provide protection, both from the movement of regulated articles within the state and from movement of regulated articles into the state from another state.

A subcommittee of the NCC BWAC has developed national minimum standards for programmes in post-eradication. The plans call for the development of new and less costly mapping systems, and trapping programmes with required minimum trap densities and trap inspection intervals. They require programmes to maintain systems for conducting quality control of trapping programmes.

In the area of boll weevil identification, the minimum standards plan calls for APHIS to develop a rapid reply system to identify insects that are suspected of being boll weevils. The minimum standards protocol describes actions that will be taken when weevils are caught. It specifies how sources of critical supplies and equipment are to be maintained. The protocol requires programmes to be able quickly to acquire aerial application services to treat reinfestations. Finally, the plans require programmes to maintain sufficient financial resources to provide for the initial treatment of reinfestations. The plan includes a contingency fund held by APHIS to assist programmes with the cost of eradicating reinfestations. Only those programmes that have met the minimum standards in their post-eradication operations will be eligible to receive monetary assistance from USDA to eradicate reinfestations (NCC, 2006).

In addition, APHIS has developed Federal Quarantine Rules to help prevent reinfestation. The quarantine defines and regulates the interstate movement of regulated articles, such as cotton fruit, boll weevils, gin trash, etc. which might move boll weevils into an eradicated area. It provides for the issuance of certificates, compliance agreements and permits for the movement of regulated articles. It establishes that states can be declared regulated areas if they become infested with boll weevils or fail to maintain post-eradication programmes at levels that meet the minimum standards described by the NCC BWAC (USDA-APHIS, 2007b).

Summary

Eradication of the boll weevil, the most costly insect pest in the history of American agriculture (Hardee, 1972), has required strong and sustained commitment by growers, state and federal legislatures, USDA, the state universities, state departments of agriculture, grower-run foundations and others for many years. The commitment by cotton growers to eradicate this successful and well-entrenched pest is indicative of the optimistic, can-do attitude that has prevailed among the cotton producer leadership ever since Robert Coker and J.F. McLaurin first proposed the idea of boll weevil eradication in 1958 (Coker, 1958). The US boll weevil eradication programme has been a massive project. Its completion in many of the cotton-growing areas of the USA has resulted in cotton production systems with greatly improved economic and environmental sustainability. The benefits will continue to accrue far into the future.

References

Ables, J.R., Goodenough, J.L., Hardstack, A.A. and Ridgway, R.L. (1983) Entomophagous arthropods. In: Ridgway, R.L., Lloyd, E.P. and Cross, W.H. (eds) *Cotton Insect Management with Special Reference to the Boll Weevil.* USDA–ARS Agricultural Handbook No. 589, Washington, DC, pp. 103–127.

Adkisson, P.L. (1968) Problems and progress in controlling diapausing boll weevils. In: Brown, J.M. (ed.) *Proceedings of the Beltwide Cotton Conference,* National Cotton Council, Memphis, Tennessee, pp. 18–20.

Adkisson, P.L., Davis, J.W., Owen, W.L. and Rummel, D.R. (1965a) Evaluation of the 1964 diapause boll weevil control program on the High Plains of Texas. *Texas Agricultural Experiment Station, Department of Entomology Technical Report* No. 1, 59 pp.

Adkisson, P.L., Rummel, D.R. and Sterling, W.L. (1965b) A two phased control program for reducing diapausing boll weevil populations on the High Plains of Texas, 1965. *Texas Agricultural Experiment Station, Department of Entomology Technical Report* No. 2, 5 pp.

Adkisson, P.L., Rummel, D.R., Sterling, W.L. and Owen, W.L., Jr. (1966) Diapause boll weevil control: a comparison of two methods. *Texas Agricultural Experiment Station Bulletin* 1054, 11 pp.

Allen, C.T., Patton, L.W., Smith, L.E. and Newman, R.O. (2001) Texas boll weevil eradication update. In: Dugger, P. and Richter, D.A. (eds) *Proceedings of the Beltwide Cotton Conference,* National Cotton Council, Memphis, Tennessee, pp. 934–937.

Allen, C.T., Smith, L.E., Patton, L.W. and Newman, R.O. (2003) Status of boll weevil eradication in Texas. In: Richter, D.A. (ed.) *Proceedings of the Beltwide Cotton Conference,* National Cotton Council, Memphis, Tennessee, pp. 1340–1345.

Allen, C.T., Smith, L.E., Patton, L.W. and Newman, R.O. (2004) Update on boll weevil eradication in Texas. In: Richter, D.A. (ed.) *Proceedings of the Beltwide Cotton Conference,* National Cotton Council, Memphis, Tennessee, pp. 1470–1477.

Allen, C.T., Patton, L.W., Smith, L.E. and Newman, R.O. (2005) Texas boll weevil eradication report. In: Richter, D.A. (ed.) *Proceedings of the Beltwide Cotton Conference,* National Cotton Council, Memphis, Tennessee, pp. 1196–1205.

Allen, C.T., Smith, L.E., Patton, L.W. and Newman, R.O. (2006) Boll weevil eradication in Texas and eastern New Mexico: a status report. In: Richter, D.A. and Huffman, M. (eds) *Proceedings of the Beltwide Cotton Conference,* National Cotton Council, Memphis, Tennessee, pp. 1297–1306.

Allen, C.T., Smith, L.E., Patton, L.W. and Newman, R.O. (2007) Boll weevil eradication: the Texas program. In: Boyd, E.S., Huffman, M., Richter, D.A. and Robertson, B. (eds) *Proceedings of the Beltwide Cotton Conference*. National Cotton Council, Memphis, Tennessee, pp. 1141–1147.

Andrews, G.L. (1981) Special insect projects: optimum insect management trial. In: Brown, J.M. (ed.) *Proceedings of the Beltwide Cotton Conference*, National Cotton Council, Memphis, Tennessee, pp. 41–44.

Anon. (1850) Cotton. *Report of the Commissioner of Patents for the Year 1849*. House of Representatives, Washington, DC, Exec. Doc. No. 20, pp. 307–313.

Anon. (1877) Cotton investigation, *Report of the Commissioner of Agriculture of the Operations of the Department for the Year 1876*. Government Printing Office, Washington, DC, pp. 114–117.

Anon. (1904) Work on the cotton boll weevil. *Yearbook of the USDA for 1903*. Government Printing Office, Washington, DC.

Anon. (1911) Fighting the boll weevil by clean farming methods. *USDA Farmers Bulletin* No. 45, *Experiment Station Work LXIV*, 11–14.

Anon. (1930) Georgia and her resources, yearbook of agriculture, Georgia. *Department of Agriculture Quarterly Bulletin* 117, 34–36.

Anon. (1940) 1940 Annual Report. *Annual Report of the Georgia Agricultural Extension Service*, 5–7.

Anon. (1950) Research and related services in the United States Department of Agriculture. *Report to the Committee on Agriculture of the House of Representatives*, Eighty-first Congress, Second Session l, pp. 756–760.

Anon. (1975) 1975 Annual report. *University of Georgia College of Agriculture Experiment Station Report* No.1, 13–14.

Anon. (1998) Weevil eradication takes a hit in the plains. *Land of Cotton Archives*, September 1998 (http://landofcotton.com/archives/arc3.shtml).

Anon. (2001) Dedication: in memory of Dr Robert C. Coker, Dr Edward F. Knipling, and Dr James R. Brazzel. In: Dickerson, W.A., Brashear, A.L., Brumley, J.T., Carter, F.L., Grefenstette, W.J. and Harris, F.A. (eds) *Boll Weevil Eradication in the United States Through 1999*. Reference Book Series No. 6, The Cotton Foundation, Memphis, Tennessee, pp. xxxix–xlii.

Anon. (2002a) Board selects Danny Kiser to be acting executive of programme. *Boll Weevil Bugle* November 2002 (http://www.arkansas weevil.org/newsletters/2002/11–2002% 20 newsletters pdf).

Anon. (2002b) Making appropriations for agriculture, rural development, Food and Drug Administration, and related agencies. *House report 107–275*, Programmes for the Fiscal Year Ending September 30, 2002 (http://congress.gov/cgi-bin/cpquerry/?&sid= cp107gbc6D&refer=&r_n=hr275.107&db_id= 107&item=&sel=TOC_190038&).

Anon. (2002c) *Track Records United States Crop Production – Cotton*. National Agricultural Statistics Service (http://www.usda.gov/nass/pubs/trackrec/track02a.htm#cotton).

Anon. (2003) Alexander Frist secure support for Tennessee agricultural projects. *Press Release*, November 7, 2003 (http://alexander.senate.gov/index.cfm?FuseAction=PressReleases. Details&PressRelease_id=317&month=11&Year=2003).

Anon. (2004) Alexander Frist announce support for Tennessee agriculture projects. *Press Release*, November 22, 2004 (http://www.alexander.senate.gov/index.cfm/FuseAction= PressReleases.Delta&PressRelease_id=663&Month=11&Year=2004).

Armstrong, J.S., Spurgeon, D.W. and Suh, C.P.-C. (2006) Comparisons of standard and extended-life boll weevil (Coleoptera: Curculionidae) pheromone lures. *Journal of Economic Entomology* 99, 323–330.

Bailey, J.C. (1945) *Seaman A. Knapp: Schoolmaster of American Agriculture*. Columbia University Press, New York.

Barker, B. (2001) History of the boll weevil eradication program in Tennessee. In: Dickerson, W.A., Brashear, A.L., Brumley, J.T., Carter, F.L., Grefenstette, W.J. and Harris, F.A. (eds) *Boll Weevil Eradication in the United States Through 1999*. Reference Book Series No. 6, The Cotton Foundation, Memphis, Tennessee, pp. 431–450.

Barnhart, F. (1950) *Cotton*. Floyd Barnhart, Caruthersville, Missouri.

Bass, M.H. (1993) *The UGA Coastal Plain Experiment Station . . . the First 75 Years*. University of Georgia Coastal Plain Experiment Station, Lang Printing Company, Tifton, Georgia.

Beckham, C.M. and Dupree, M. (1951) Progress report: summary of cotton insect control experiments during 1951. *Georgia Experiment Station Mimeo Series* No. 38, 1–10.

Bell, M.R. (1983) Microbial Agents. In: Ridgway, R.L., Lloyd, E.P. and Cross, W.H. (eds) *Cotton Insect Management with Special Reference to the Boll Weevil*. USDA-ARS Agriculture Handbook, Washington, DC, pp. 129–151.

Benedict, J.H., Urban, T.C., George, D.M., Segers, J.C., Anderson, D.J., McWhorter, G.M. and Zummo, G.R. (1985) Pheromone trap thresholds for management of overwintered boll weevil. *Journal of Economic Entomology* 78, 169–171.

Bennett, D. (2005) Arkansas Supreme Court sides with Plant Board in boll weevil ruling. *Delta Farm Press* 26 September 2005 (http://deltafarmpress.com/news/050926-arkansas-weevil/).

Bennett, R.L. (1908) A method of breeding early cotton to escape boll weevil damage. *USDA Farmers Bulletin* 134, 28 pp.

Bergman, D., Henneberry, T.J. and Bariola, L.A. (1983) Overwintering boll weevil populations in Southwestern Arizona cultivated cotton. In: Brown, J.M. (ed.) *Proceedings of the Beltwide Cotton Conference*. National Cotton Council, Memphis, Tennessee, pp. 182–185.

Borkovec, A.B., Woods, C.W. and Terry, P.H. (1978) Boll weevil: chemosterilization by fumigation and dipping. *Journal of Economic Entomology* 71, 862–866.

Bottrell, D.G. (1976) The boll weevil as a key pest. In: Davich, T.B. (ed.) *Boll Weevil Suppression, Management, and Elimination Technology – Proceedings of a Conference*. USDA-ARS-S-71, Memphis, Tennessee, pp. 5–8.

Bottrell, D.G., Rummel, D.R. and Adkisson, P.L. (1972) Spread of the boll weevil into the High Plains of Texas. *Environmental Entomology* 1, 136–140.

Bowman, D.T. (1999) Contemporary issues – public cotton breeders – do we need them? *Journal of Cotton Science* 3, 139–152 (http://www.cotton.org/journal/1999- 03/3/).

Boyd, F.J. (1976a) Boll weevil population levels during the in-season and reproduction-diapause control phases of the Pilot Boll Weevil Eradication Experiment. In: Davich, T.B. (ed.) *Boll Weevil Suppression, Management and Elimination Technology – Proceedings of a Conference*. USDA-ARS-S-71, Memphis, Tennessee, pp. 75–81.

Boyd, F.J. (1976b) Operational plan and execution of the Pilot Boll Weevil Eradication Experiment. In: Davich, T.B. (ed.) *Boll Weevil Suppression, Management and Elimination Technology – Proceedings of a Conference*. USDA-ARS-S-71, Memphis, Tennessee, pp. 62–69.

Boyd, M.L. (2001) The role of the boll weevil in Missouri's cotton production. In: Dickerson, W.A., Brashear, A.L., Brumley, J.T., Carter, F.L., Grefenstette, W.J. and Harris, F.A. (eds) *Boll Weevil Eradication in the United States Through 1999*. Reference Book Series No. 6, The Cotton Foundation, Memphis, Tennessee, pp. 345–360.

Brashear, A.L. and Brumley, J.T. (2001) The evolution of program management, USDA/grower responsibilities and funding in the southeastern boll weevil eradication program, In: Dickerson, W.A., Brashear, A.L., Brumley, J.T., Carter, F.L., Grefenstette, W.J. and Harris, F.A. (eds) *Boll Weevil Eradication in the United States Through 1999*. Reference Book Series No. 6, The Cotton Foundation, Memphis, Tennessee, pp. 77–100.

Brazzel, J.R. (1959) The effect of late-season applications of insecticides on diapausing boll weevils. *Journal of Economic Entomology* 52, 1042–1045.

Brazzel, J.R. (1962) Diapause as related to boll weevil control. In: Brown, J.M. (ed.) *Proceedings of the Beltwide Cotton Conference*, National Cotton Council, Memphis, Tennessee, pp. 19–20.

Brazzel, J.R. (1963) Resistance to DDT in *Heliothis virescens*. *Journal of Economic Entomology* 56, 571–574.

Brazzel, J.R. (1989) Boll weevil eradication – an update. In: Brown, J.M. and Richter, D.A. (eds) *Proceedings of the Beltwide Cotton Conference*, National Cotton Council, Memphis, Tennessee, pp. 218–219.

Brazzel, J.R. and Hightower, B.G. (1960) A seasonal study of diapause, reproductive activity and seasonal tolerance to insecticides in the boll weevil. *Journal of Economic Entomology* 53, 41–46.

Brazzel, J.R. and Newsom, L.D. (1959) Diapause in *Anthonomus grandis* Boheman. *Journal of Economic Entomology* 52, 603–611.

Brazzel, J.R., Davich, T.B. and Harris, L.D. (1961) A new approach to boll weevil control. *Journal of Economic Entomology* 54, 723–730.

Brazzel, J.R., Smith, J.W. and Knipling, E.F. (1996) Boll weevil eradication. In: King, E.G., Phillips, J.R. and Coleman, R.J. (eds) *Cotton Insects and Mites: Characterization and Management*, Reference Book Series 3, The Cotton Foundation, Memphis, Tennessee, pp. 625–652.

Brown, A.W.A. and Pal, R. (1971) *Insecticide Resistance in Arthropods*. World Health Organization Monograph Series No. 38, Geneva.

Brown, H.B. and Ware, J.O. (1958) *Cotton*. 3rd edn., McGraw-Hill, New York.

Brumley, J., Boyd, F.J. and Seward, R. (2006) Update of Southeastern Boll Weevil eradication. In: Richter, D.A. and Huffman, M. (eds) *Proceedings of the Beltwide Cotton Conference*, National Cotton Council, Memphis, Tennessee, pp. 1294–1296.

Brumley, J., Boyd, F.J., Seward, R. and Massey, J. (2007) Update on Southeastern boll weevil eradication programs. In: Boyd, E.S., Huffman, M., Richter, D.A. and Robertson, B. (eds) *Proceedings of the Beltwide Cotton Conference*, National Cotton Council, Memphis, Tennessee, p. 1135.

Bryant, J. (1997) Weevil onslaught threatens 'biggest cotton patch'. *Texas Agricultural Extension Service*, News Release 9 May 1997 (http://agnews.tamu.edu/stories/ENTO/May0997a.htm).

Burgess, E.D. (1965) Control of the boll weevil with technical malathion applied by aircraft. *Journal of Economic Entomology* 58, 414–415.

Carlson, G.A. and Suguiyama (1985) *Economic Evaluation of Areawide Cotton Insect Management: Boll Weevils in the Southeastern United States*. North Carolina Agricultural Research Service Bulletin 473, North Carolina State University, Raleigh, North Carolina.

Carlson, G.A., Sappie, G. and Hammig, M. (1989) *Economic Returns to Boll Weevil Eradication*. USDA-Economic Research Service Report 621, Washington, DC, 31 pp.

Carroll, S.C., Rummel, D.R. and Sagarra, E. (1993) Overwintering by the boll weevil (Coleoptera: Curculionidae) in conservation reserve program grasses on the Texas High Plains. *Journal of Economic Entomology* 86, 382–393.

Carson, R. (1962) *Silent Spring*. Fawcett Books Group, Brooklyn, New York.

Carter, F.L., Nelson, C., Jordan, A.G. and Smith, J.R. (2001) US cotton declares war on the boll weevil. In: Dickerson, W.A., Brashear, A.L., Brumley, J.T., Carter, F.L., Grefenstette, W.J. and Harris, F.A. (eds) *Boll Weevil Eradication in the United States Through 1999*. Reference Book Series No. 6, The Cotton Foundation, Memphis, Tennessee, pp. 25–54.

Clark, D.O. (2001) Boll weevil establishment and eradication in California. In: Dickerson, W.A., Brashear, A.L., Brumley, J.T., Carter, F.L., Grefenstette, W.J. and Harris, F.A. (eds) *Boll Weevil Eradication in the United States Through 1999*. Reference Book Series No. 6, The Cotton Foundation, Memphis, Tennessee, pp. 235–244.

Clemson University (1918) *Thirty-first Annual Report of the South Carolina Experiment Station of Clemson Agricultural College for the Year Ending 30 June 1918*, p. 11.

Clemson University (1919) *Thirty-second Annual Report of the South Carolina Experiment Station of Clemson Agricultural College for Year Ending 30 June 1919*, p. 11.

Cleveland, T.C., Scott, W.P., Davich, T.B. and Parencia, C.R. (1966) Control of the boll weevil on cotton with ultra low volume (undiluted) technical malathion. *Journal of Economic Entomology* 59, 973–976.

Coad, B.R. (1915) Recent studies of the Mexican cotton boll weevil. *USDA Bulletin* No. 231, 34 pp.

Coad, B.R. (1918) Recent experimental work on poisoning cotton boll weevils. *USDA Bulletin* No. 731, 1–15.

Coad, B.R. (1930) The entomologist in relation to cotton insect problems of today. *Journal of Economic Entomology* 23, 667–672.

Coad, B.R. and Moreland, W. (1921) Dispersion of the boll weevil in 1920. *USDA Circular* 163, 1–2.

Coad, B.R., Tucker, E.S., Williams, W.B., Bondy, F.R. and Gaines, R.C. (1922) Dispersion of the boll weevil in 1921. *USDA Circular* 210, 1–3.

Coker, R.R. (1958) The impact of the boll weevil on cotton production costs. In: *Proceedings of the Beltwide Cotton Conference*, National Cotton Council, Memphis, Tennessee, pp. 3–5.

Coker, R.R. (1973) Review of boll weevil eradication program. *House of Representatives, Committee on Agriculture, Subcommittee on Cotton*, Washington, DC, pp. 1–14.

Coker, R.R. (1976) Report of industry Action Committee. In: Davich, T.B. (ed.) *Boll Weevil Suppression, Management and Elimination Technology – Proceedings of a Conference*. USDA-ARS-S-71, Memphis, Tennessee, pp. 167–168.

Cook, O.F. (1906) Weevil-resisting adaptation of the cotton plant. *USDA Bureau of Plant Industries Bulletin* 88, 87 pp.

Cook, O.F. (1911) Relation of drought to weevil resistance in cotton. *USDA Bureau of Plant Industries Bulletin* 220, 30 pp.

Cook, O.F. (1923) Boll-weevil cotton in Texas. *USDA Department Bulletin* 1153, 20 pp.

Cook, O.F. (1924) Cotton improvement under boll weevil conditions. *USDA Farmers Bulletin* 501, 18 pp.

Cremlyn, R. (1978) *Preparation and Mode of Action*. John Wiley and Sons Ltd, New York.

Cross, W.H. (1973) Biology, control and eradication of the boll weevil. *Annual Review of Entomology*, 18, 17–46.

Cross, W.H. (1976a) History of the boll weevil problem. In: Davich, T.B. (ed). *Boll Weevil Suppression, Management and Elimination Technology – Proceedings of a Conference*, USDA-ARS-S-71, Memphis, Tennessee, pp. 1–2.

Cross, W.H. (1976b) Relative populations and suggested long range movements of boll weevils throughout the area of the Pilot Boll Weevil Eradication Experiment as indicated by traps in 1973. In: Davich, T.B. (ed.) *Boll Weevil Suppression, Management and Elimination Technology – Proceedings of a Conference*, USDA-ARS-S-71, Memphis, Tennessee, pp. 103–107.

Cross, W.H. (1983) Ecology of cotton insects with special reference to the boll weevil. In: Ridgway, R.L., Lloyd, E.P. and Cross, W.H. (eds) *Cotton Insect Management with Special Reference to the Boll Weevil*. USDA-ARS Agricultural Handbook 589, Washington, DC, pp. 53–70.

Cross, W.H. and Hardee, D.D. (1968) Traps for survey of overwintered boll weevil populations. *Cooperative Economic Insect Report* 8, 430.

Cross, W.H., Hardee, D.D., Nichols, F., Mitchell, E.B., Huddleston, P.M. and Tumlinson, J.H. (1969a) Attraction of female boll weevils to traps baited with males or extracts of males. *Journal of Economic Entomology* 62, 154–161.

Cross, W.H., McGovern, W.L. and Mitchell, H.C. (1969b) Biology of *Bracon kirkpatricki* and field releases of the parasite for control of the boll weevil. *Journal of Economic Entomology* 62, 448–454.

Curtis, W.C. (2001) Boll weevil eradication in Alabama. In: Dickerson, W.A., Brashear, A.L., Brumley, J.T., Carter, F.L., Grefenstette, W.J. and Harris, F.A. (eds) *Boll Weevil Eradication in the United States Through 1999*. Reference Book Series No. 6, The Cotton Foundation, Memphis, Tennessee, pp. 175–212.

Daum, R.J., McLaughlin, R.E. and Hardee, D.D. (1967) Development of the bait principle for boll weevil control: cottonseed oil, a source of attractants and feeding stimulants for boll weevil. *Journal of Economic Entomology* 60, 321–325.

Davich, T.B. (1969) Sterile male technique for control or eradication of the boll weevil, *Anthonomus grandis* Boheman – Pannel: sterile male technique for eradication and control of harmful insects. In: Brown, J.M. (ed.) *Proceedings of the Beltwide Cotton Conference*, National Cotton Council, Memphis, Tennessee, pp. 65–72.

Davich, T.B. (1976) Foreword. In: Davich, T.B. (ed.) *Boll Weevil Suppression, Management and Elimination Technology – Proceedings of a Conference*. USDA-ARS-S-71, Memphis, Tennessee, pp. i–ii.

Davich, T.B. and Lindquist, D.A. (1962) Exploratory studies on gamma radiation for sterilization of boll weevil. *Journal of Economic Entomology* 55, 164–167.

Davich, T.B., Hardee, D.D. and Alcala, M.J. (1970) Long-range dispersal of boll weevils determined with wing traps baited with males. *Journal of Economic Entomology* 63, 1706–1708.

Deterling, D. (1992) The boll weevil: a century of pestilence. *Progressive Farmer* October, pp. 26–30.

Dickerson, W.A. (1986) Grandlure: use in boll weevil control and eradication program in the United States. *Florida Entomologist* 69, 147–153.

Dickerson, W.A., McKibben, G.H., Lloyd, E.P., Kearney, J.F., Lam, J.J. and Cross, W.H. (1981) Field evaluation of a modified in-field trap. *Journal of Economic Entomology* 78, 280–282.

Dickerson, W.A., Cross, G.B. and Grant, M. (2001) North Carolina boll weevil eradication and post-eradication program. In: Dickerson, W.A., Brashear, A.L., Brumley, J.T., Carter, F.L., Grefenstette, W.J. and Harris, F.A. (eds) *Boll Weevil Eradication in the United States Through 1999*. Reference Book Series No. 6, The Cotton Foundation, Memphis, Tennessee, pp. 375–404.

Donnell, E.J. (1872) *Chronological and Statistical History of Cotton*. James Sutton and Company, New York.

Dunn, H.A. (1964) Cotton boll weevil, *Anthonomus grandis* Boh. abstracts of research publications 1843–1960. *USDA Miscellaneous Publication* No. 986, 194 pp.

Earle, N.W. and Leopold, R. (1975) Sterilization of the boll weevil: vacuum fumigation with hempa combined with feeding bisulfan-treated diet. *Journal of Economic Entomology* 68, 283–286.

Earle, N.W. and Simmons, L.A. (1979) Boll weevil: ability to fly affected by acetone, irradiation and diflubenzuron. *Journal of Economic Entomology* 72, 573–575.

Earle, N.W., Nilake, S.S. and Simmons, L.A. (1979) Mating ability of irradiated male boll weevils treated with diflubenzuron or penfluron. *Journal of Economic Entomology* 72, 334–336.

Eden, W.G. (1976) Report of Entomological Society of America Review Committee on the Pilot Boll Weevil Eradication Experiment. In: Davich, T.B. (ed.) *Boll Weevil Suppression, Management and Elimination Technology – Proceedings of a Conference*. USDA-ARS-S-71, Memphis, Tennessee, p. 126.

El-Lissy, O.A. and Grefenstette, W.J. (2006) Progress of Boll Weevil Eradication in the US, 2005. In: Richter, D.A. and Huffman, M. (eds) *Proceedings of the Beltwide Cotton Conference*, National Cotton Council, Memphis, Tennessee, pp. 1266–1276.

El-Lissy, O. and Moschos, J. (1999) Development of a computerized expert system as a management tool for boll weevil eradication. In: Dugger, P. and Richter, D.A. (eds) *Proceedings of the Beltwide Cotton Conference*, National Cotton Council, Memphis, Tennessee, pp. 834–837.

El-Lissy, O., Myers, F., Frisbie, R., Fuchs, T., Rummel, D., Smathers, R., King, E., Planer, F., Bare, C., Carter, F., Busse, G., Neihues, H. and Hayes, J. (1996) Boll weevil eradication status in Texas. In: Dugger, P. and Richter, D.A. (eds) *Proceedings of the Beltwide Cotton Conference*, National Cotton Council, Memphis, Tennessee, pp. 831–837.

El-Lissy, O., Meyers, F., Frisbie, R., Fuchs, T., Rummel, D., Parker, R., Dipple, D., King, E., Cunningham, G., Carter, F., Boston, J. and Hayes, J. (1997) Boll weevil eradication update – Texas, 1996. In: Dugger, P. and Richter, D.A. (eds) *Proceedings of the Beltwide Cotton Conference*, National Cotton Council, Memphis, Tennessee, pp. 973–980.

El-Lissy, O., Patton, L., Kiser, D., Frisbie, R., Fuchs, T., Rummel, D., Coppedge, J.R., Cunningham, G., Carter, F., Boston, J. and Hayes, J. (1999) Boll weevil eradication update – Texas, 1998. In: Dugger, P. and Richter, D.A. (eds) *Proceedings of the Beltwide Cotton Conference*, National Cotton Council, Memphis, Tennessee, pp. 818–823.

El-Lissy, O., Kiser, D., Patton, L., Frisbie, R., Fuchs, T., Parker, R., Slosser, J., Dippel, D., Coppedge, J.R., Carter, F., Boston, J. and Hayes, J. (2000) Boll weevil eradication update – Texas, 1999. In: Dugger, P. and Richter, D.A. (eds) *Proceedings of the Beltwide Cotton Conference*, National Cotton Council, Memphis, Tennessee, pp. 1076–1082.

Enfield, F.D., North, D.T. and Erickson, R. (1981) Response to selection for resistance to gamma radiation in the cotton boll weevil. *Annals of the Entomological Society of America* 74, 422–424.

Enfield, F.D., North, D.T., Erickson, R. and Rotering, L. (1983) A selection plateau for radiation resistance in the cotton boll weevil. *Theory and Application of Genetics* 65, 277–281.

Enfield, F.D., Sawicki, C. and North, D.T. (1988) Selection for mating propensity in irradiated populations of the cotton boll weevil. *Theory and Application of Genetics* 76, 861–864.

England, M., Minzenmayer, R. and Sansone, C. (1997) Impact of selected insecticide on boll weevil and natural enemies. In: Dugger, P. and Richter, D.A. (eds) *Proceedings of the Beltwide Cotton Conference*. National Cotton Council, Memphis, Tennessee, pp. 989–993.

Ewing, K.P. and Ivy, E.E. (1943) Some factors influencing bollworm populations and damage. *Journal of Economic Entomology*. 36, 602–606.

Ewing, K.P. and Parencia, C.R., Jr. (1949) Experiments in early-season application of insecticides for cotton-insect control. *USDA-ARS* E-792, 9 pp.

Ewing, K.P. and Parencia, C.R., Jr. (1950) Early season application of insecticide on a community-wide basis for cotton insect control in 1950. *USDA Bureau of Entomology and Plant Quarantine* E810, 8 pp.

Folsum, J.W. (1932) Insect enemies of the cotton plant. *USDA Farmers Bulletin* 1688, 27 pp.

Ford, D. (1995) New Mexico cotton growers must quickly deal with destructive boll weevil. *New Mexico State University* (http://cahe.nmsu.edu/news/1995/100495_cotton.html).

Fried, M. (1971) Determination of sterile insect competitiveness. *Journal of Economic Entomology* 64, 869–872.

Frisbie, R.E. (2001) History of boll weevil eradication. In: Falconer, L. and Parker, R. (eds) *Proceedings of the Gulf Coast Cotton Conference*. 7–8 November, Corpus Christi, Texas, pp. 83–86 (http://agfacts.tamu.edu/~lfalcon/Reports/TexasGulfCoastCottonConference).

Frisbie, R.E., Phillips, J.R., Lambert, W.R.A. and Jackson, H.B. (1983) Opportunities for improving cotton insect management programs and some constraints on beltwide implementation. In: Ridgway, R.L., Lloyd, E.P. and Cross, W.H. (eds) *Cotton Insect Management with Special Reference to the Boll Weevil*. USDA-ARS Agricultural Handbook 589, Washington, DC, pp. 521–557.

Frisbie, R.E., El-Zik, K.M. and Wilson, L.T. (1989) *Integrated Pest Management Systems and Cotton Production*. John Wiley and Sons, New York.

Fuchs, T.W. and Minzenmayer, R. (1992) Field evaluation of boll weevil bait sticks in West Texas. In: Herber, D.J. and Richter, D.A. (eds) *Proceedings of the Beltwide Cotton Conference*, National Cotton Council, Memphis, Tennessee, pp. 718–720.

Fuchs, T.W., Rummel, D.R. and Boring, E.P. III (1998) Delayed uniform planting for areawide boll weevil suppression. *Southwestern Entomology* 23, 325–333.

Gassner, G.D., Childress, D., Pomonis, G. and Eaton, J. (1974) Boll weevil chemosterilization by hypobarometric distillation. *Journal of Economic Entomology* 67, 278–280.

Gast, R.T. (1961) Some shortcuts in laboratory mass rearing of boll weevils. *Journal of Economic Entomology* 54, 395–396.

Gast, R.T. (1966) Control of four diseases of laboratory reared boll weevils. *Journal of Economic Entomology* 59, 793–797.

Gast, R.T. and Vardell, H. (1963) Mechanical devices to expedite boll weevil rearing in the laboratory. *USDA-ARS* 33, 89.

Goswick, C., Welch, R. and Broyles, L. (2007) *Field Training Manual.* Texas Boll Weevil Eradication Foundation, Abilene, Texas.

Graves, J.B., Roussel, R.J. and Phillips, J.R. (1963) Resistance to some chlorinated hydrocarbon insecticides in the bollworm, *Heliothis zea. Journal of Economic Entomology* 56, 442–444.

Grundler, J. and Sorenson, C.E. (1997) An intensive boll weevil trapping effort in southern Missouri. In: Dugger, P. and Richter D.A. (eds) *Proceedings of the Beltwide Cotton Conference,* National Cotton Council, Memphis, Tennessee, pp. 971–973.

Haldenby, R.K. (1992) Keeping the High Plains free of boll weevil; a quiet success story – 28 years of containment. In: Herber, D.J. and Richter, D.A. (eds) *Proceedings of the Beltwide Cotton Conference,* National Cotton Council, Memphis, Tennessee, pp. 713–714.

Hamer, J.L., Andrews, G.L., Seward, R.W., Young, D.F., Jr. and Head, R.B. (1983) Optimum Pest Management Trial in Mississippi. In: Ridgway, R.L., Lloyd, E.P. and Cross, W.H. (eds) *Cotton Insect Management with Special Reference to the Boll Weevil.* USDA Agricultural Handbook No. 589, Washington, DC, pp 385–407.

Handy, R.B. (1896) History and general statistics of cotton. In: True, A.C. (ed.) *The Cotton Plant: its History, Botany, Chemistry, Culture, Enemies, and Uses.* USDA Bulletin 33, Washington, DC, pp. 17–66.

Haney, P.B. (2001) The cotton boll weevil in the United States: impact on cotton production and the people of the Cotton Belt. In: Dickerson, W.A., Brashear, A.L., Brumley, J.T., Carter, F.L., Grefenstette, W.J. and Harris, F.A. (eds) *Boll Weevil Eradication in the United States Through 1999.* Reference Book Series No. 6, The Cotton Foundation, Memphis, Tennessee, pp. 7–24.

Haney, P.B., Lewis, W.J. and Lambert, W.R. (1996) Cotton production and the boll weevil in Georgia: history, cost of control and benefits of eradication. *The Georgia Agricultural Experiment Stations Research Bulletin* No. 428, 49 pp. (http://pubs.caes.uga.edu/caespubs/PFD/RB428.pdf).

Haney, P.B., Herzog, G. and Roberts, P.M. (2001a) Boll weevil eradication in Georgia. In: Dickerson, W.A., Brashear, A.L., Brumley, J.T., Carter, F.L., Grefenstette, W.J. and Harris, F.A. (eds) *Boll Weevil Eradication in the United States Through 1999.* Reference Book Series No. 6, The Cotton Foundation, Memphis, Tennessee, pp. 259–290.

Haney, P.B., Sprenkel, R., Clark, R.A. and Griswold, C. (2001b) Boll weevil eradication in Florida. In: Dickerson, W.A., Brashear, A.L., Brumley, J.T., Carter, F.L., Grefenstette, W.J. and Harris, F.A. (eds) *Boll Weevil Eradication in the United States Through 1999.* Reference Book Series No. 6, The Cotton Foundation, Memphis, Tennessee, pp. 245–258.

Hardee, D.D. (1972) A review of literature on the pheromone of the boll weevil, *Anthonomus grandis* Boheman (Coleoptera: Curculionida). *Cooperative Economic Insect Report* 22, 200–207.

Hardee, D.D. and Harris, F.A. (2003) Eradicating the boll weevil (Coleoptera: Curculionidae): a clash between a highly successful insect, good scientific achievement and differing agricultural philosophies. *American Entomologist* 49, 82–97 (http://www.entsoc.org/Pubs/Periodicals/AE/AE-2003/summer/Feature-Hardee.pdf).

Hardee, D.D., Huddleston, M.P. and Mitchell, E.B. (1967) Procedure for bioassaying the sex attractant of the boll weevil. *Journal of Economic Entomology* 60, 169–171.

Hardee, D.D., Cross, W.H. and Mitchell, E.B. (1969) Male boll weevils are more attractive than cotton plants to boll weevil. *Journal of Economic Entomology* 62, 165.

Hardee, D.D., McKibben, G.H., Gueldner, R.C., Mitchell, E.B., Tumlinson, J.H. and Cross, W.H. (1972) Boll weevils in nature respond to grandlure, a synthetic pheromone. *Journal of Economic Entomology.* 65, 97–100.

Hardee, D.D., McKibben, G.H., Rummel, D.R., Huddleston, P.M. and Coppedge, J.R. (1974) Response of boll weevils to component ratios and doses of the pheromone, gossyplure. *Environmental Entomology* 3, 135–138.

Hardee, D.D., McKibben, G.H. and Huddleston, P.M. (1975) Grandlure for boll weevils: controlled release with a laminated plastic dispenser. *Journal of Economic Entomology* 68, 477–479.

Harned, R.W. (1910) Boll weevil in Mississippi. *1909 Mississippi Agricultural Experiment Station Bulletin* 139, 44.

Harris, F.A. and Clark, L.B. (2006) Mississippi people – contributors to boll weevil eradication. *Mississippi Agricultural and Forestry Experiment Station Bulletin* 1154, 23 pp.

Harris, F.A. and Smith, J.W. (2001) Boll weevil eradication in Mississippi. In: Dickerson, W.A., Brashear, A.L., Brumley, J.T., Carter, F.L., Grefenstette, W.J. and Harris, F.A. (eds) *Boll Weevil Eradication in the United States Through 1999.* Reference Book Series No. 6, The Cotton Foundation, Memphis, Tennessee, pp. 305–344.

Harris, F.A., Wise, D.J., Ingram, R.L., Alley, E.G., Mulroony, J.E., Robinson, E.H. and Watson, C.E. (1999) Malathion rate in water and catfish. *Mississippi Agricultural Experiment Station Technical Bulletin* 225, 10 pp.

Haynes, J.W. (1963) Chemical sterility agents as they affect the boll weevil, *Anthonomus grandis* Boheman. MS thesis, Mississippi State University, Mississippi.

Haynes, J.W. (1981) Effects of soil temperatures and chilling on flight and mortality of sterile boll weevils. *Georgia Entomological Society* 16, 254–257.

Haynes, J.W., Mitlin, N., Davich, T.B., Nail, J. and Dawson, J.R. (1975) Mating and sterility of male boll weevils treated with bisulfan plus hempa. *Environmental Entomology* 4, 315–318.

Hedin, P.A. (1976) Grandlure development. In: Davich, T.B. (ed.) *Boll Weevil Suppression, Management and Elimination Technology – Proceedings of a Conference.* USDA-ARS-S-71, Memphis, Tennessee, pp. 31–33.

Helms, D. (1977) Just lookin' for a home: the boll weevil and the South. PhD thesis, University of Florida, Gainesville, Florida.

Herzog, G.A., Graves, J.B., Reed, J.T., Scott, W.P. and Watson, T.F. (1996) Chemical control. In: King, E.G., Phillips, J.R. and Coleman, R.J. (eds) *Cotton Insects and Mites: Characterization and Management.* The Cotton Foundation, Memphis, Tennessee, pp. 447–469.

Hinds, W.E. (1926) Airplane dusting of cotton for boll weevil control. *Journal of Economic Entomology* 19, 607.

Hinds, W.E. (1928) The effect of the spacing of cotton upon the form and height of the plant. *Journal of Economic Entomology* 21, 741–748.

Hopkins, A.R. and Taft, H.M. (1967) Control of cotton pests by aerial application of ultra-low-volume (undiluted) technical insecticides. *Journal of Economic Entomology* 60, 561–565.

Howard, L.O. (1895) The Mexican cotton boll weevil. *USDA Division of Entomology Circular* 6, 2nd Series, 5 pp.

Howard, L.O. and Galloway, B.T. (1904) Annual report of the USDA for fiscal year ending June 30, 1904. *Report of the Secretary of Agriculture, Department Report,* Washington, DC, xxxii–xxxv, lxxvi–lxxx, 74–75, 274–279.

Hunter, W.D. (1904a) Information concerning the Mexican Boll Weevil. *USDA Farmers Bulletin* No. 189, 29 pp.

Hunter, W.D. (1904b) The status of the Mexican cotton boll weevil in the United States in 1903. *USDA Farmers Bulletin* No. 189, 205–214.

Hunter, W.D. (1911) The boll weevil problem with special reference to means of reducing damage. *USDA Farmers Bulletin* 344, 37 pp.

Hunter, W.D. (1912a) The boll weevil problem with special reference to means of reducing damage. *USDA Farmers Bulletin* 512, 46 pp.

Hunter, W.D. (l912b) The movement of the Mexican cotton boll weevil in 1911. *USDA Bureau of Entomology Circular* 167, 4 pp.

Hunter, W.D. and Coad, B.R. (1923) The boll weevil problem. *USDA Farmers Bulletin* 1329, 1–29.

Isley, D. (1928) The relation of leaf color and leaf size to boll weevil infestation. *Journal of Economic Entomology* 21, 553–559.

Isley, D. (1933) Control of the boll weevil in Arkansas. *University of Arkansas Cooperative Extension Service Circular* 162, 9 pp.

Isley, D. and Baerg, W.J. (1924) The Boll Weevil Problem in Arkansas. *University of Arkansas Agricultural Experiment Station Bulletin* No. 190, 22 pp.

Johnson, D.R. and Martin, G. (2001) Boll weevil eradication in Arkansas. In: Dickerson, W.A., Brashear, A.L., Brumley, J.T., Carter, F.L., Grefenstette, W.J. and Harris, F.A. (eds) *Boll Weevil Eradication in the United States Through 1999*. Reference Book Series No. 6, The Cotton Foundation, Memphis, Tennessee, pp. 225–234.

Johnson, D.R., Lorenz, G.M., Studebaker, G.E. and Allen, C.T. (1999). Identification of overwintering habitat and control of boll weevil, *University of Arkansas Cooperative Extension Publication* FSA 7049, 3 pp. (http://ipm.naex.edu/insects/id%20of%20overwintering%20bw.htm).

Johnson, D.R., Studebaker, G.E., Lorenz, G. and Greene, J.K. (2006) Cotton insect management. *Arkansas Cooperative Extension*, Publication FSA 2065 -PD-2- 02 RV (http://www.uaex.edu/other_Areas/publications/html/fsa-2065.asp).

Johnson, J.L., Sansone, C.G., Fuchs, T.W. and Minzenmayer, R.R. (2001) Grower-level benefit cost analysis of boll weevil eradication in the Southern Rolling Plains of Texas. In: Dugger, P. and Richter, D.A. (eds) *Proceedings of the Beltwide Cotton Conference*, National Cotton Council, Memphis, Tennessee, pp. 227–230.

Johnson, W.L., Cross, W.H., McGovern, W.L. and Mitchell, H.C. (1973) Biology of *Heterolaccus grandis* in a laboratory culture and its potential as an introduced parasite of the boll weevil in the United States. *Environmental Entomology* 2, 112–118.

Jones, J.S., Newsom, L.D. and Tipton, K.W. (1964) Differences in boll weevil infestation among several biotypes of upland cotton. In: Brown, J.M. (ed.) *Proceedings of the Beltwide Cotton Conference*, National Cotton Council, Memphis, Tennessee, pp. 48–55.

Jones, R.G., Wolfenbarger, D.A. and El-Lissy, D. (1996) Malathion ULV rate studies under boll weevil eradication field conditions. In: Dugger, P. and Richter, D.A. (eds) *Proceedings of the Beltwide Cotton Conference*, National Cotton Council, Memphis, Tennessee, pp. 717–719.

Karner, M. and Cuperus, G. (2000) *Oklahoma's Boll Weevil Eradication Program*. Oklahoma State University Cooperative Extension Service Bulletin L-255, Oklahoma.

Karner, M. and Goodson, J. (1995) Performance of boll weevil attract and control tubes (BWACT) under Oklahoma conditions in 1994. In: Richter, D.A. and Armour, J. (eds) *Proceedings of the Beltwide Cotton Conference*, Memphis, Tennessee, pp. 1013–1016.

Keller, J.C., Mitchell, E.B., McKibben, G. and Davich, T.B. (1964) A sex attractant for female boll weevils from males. *Journal of Economic Entomology* 57, 609–610.

King, E.G. (1995) Suppression of the boll weevil by mass propagation and innoculative/augmentative releases of the wasp parasite, *Catolaccus grandis*. In: *Minutes and Proceedings of the Southern Plant Board*, Atlanta, Georgia, 10–13 April 1995, pp. 30–34.

King, E.G., Summy, K.R. and Morales-Ramos, J.A. (1993) Integration of boll weevil biological control by innoculative/augmentative releases of the parasite *Catolaccus grandis* in short season cotton. In: Herber, D.J. and Richter, D.A. (eds) *Proceedings of the Beltwide Cotton Conference*, National Cotton Council, Memphis, Tennessee, pp. 910–913.

King, M. and Pfenning, S. (2001) The Oklahoma boll weevil story. In: Dickerson, W.A., Brashear, A.L., Brumley, J.T., Carter, F.L., Grefenstette, W.J. and Harris, F.A. (eds) *Boll Weevil Eradication in the United States Through 1999*. Reference Book Series No. 6, The Cotton Foundation, Memphis, Tennessee, pp. 405–412.

Kirk, I.W., House, V.S. and Mulrooney, J.E. (1997) Influence of dew on leaf surfaces at time of application of ULV malathion – chemical residue and boll weevil mortality. In: Dugger, P. and Richter, D.A. (eds) *Proceedings of the Beltwide Cotton Production Conference*, National Cotton Council, Memphis, Tennessee, pp. 1212–1213.

Kiser, D. and Catanach, M. (2005) Boll weevil eradication update, Arkansas, 2004. In: Richter, D.A. (ed.) *Proceedings of the Beltwide Cotton Conference*, National Cotton Council, Memphis, Tennessee, pp. 1074–1090.

Kiser, D. and Catanach, M. (2006) Boll weevil eradication update – Arkansas, 2005. In: Richter, D.A. and Huffman, M. (eds) *Proceedings of the Beltwide Cotton Conference*, National Cotton Council, Memphis, Tennessee, pp. 1277–1293.

Kiser, D. and Catanach, M. (2007) Boll weevil eradication update, Arkansas, 2006. In: Richter, D.A. and Huffman, M. (eds) *Proceedings of the Beltwide Cotton Conference*, National Cotton Council, Memphis, Tennessee, pp. 1277–1293.

Klassen, W. and Ridgway, R.L. (2001) Foreword. Boll weevil eradication in the United States through 1999. In: Dickerson, W.A., Brashear, A.L., Brumley, J.T., Carter, F.L., Grefenstette, W.J. and Harris, F.A. (eds) *Boll Weevil Eradication in the United States Through 1999*. Reference Book Series No. 6, The Cotton Foundation, Memphis, Tennessee, pp. xxi–xxiv.

Knipling, E.F. (1956) Basic research for control of cotton insects tomorrow. In: *Proceedings of the Beltwide Cotton Conference*, National Cotton Council, Memphis, Tennessee, pp. 20–21.

Knipling, E.F. (1963) *An Appraisal of the Relative Merits of Insecticidal Control Directed against Reproducing versus Diapausing Boll Weevils in Efforts to Develop Eradication Procedures*. A letter to the USDA-ARS, Entomology Research Division, Cotton Insects Research Branch.

Knipling, E.F. (1966) Total suppression of insect populations. In: Brown, J.M. (ed.) *Proceedings of the Beltwide Cotton Conference*, National Cotton Council, Memphis, Tennessee, pp. 15–17.

Knipling, E.F. (1967) Technically feasible approaches to boll weevil eradication. In: Brown, J.M. (ed.) *Proceedings of the Beltwide Cotton Conference*, National Cotton Council, Memphis, Tennessee, pp. 14–18.

Knipling, E.F. (1968) Technically feasible approaches to boll weevil eradication. In: Brown, J.M. (ed.) *Proceedings of the Beltwide Cotton Conference*, National Cotton Council, Memphis, Tennessee, pp.18–20.

Knipling, E.F. (1971) Boll weevil and pink bollworm eradication: progress and plans. In: Brown, J.M. (ed.) *Proceedings of the Beltwide Cotton Conference*, National Cotton Council, Memphis, Tennessee, pp. 23–30.

Knipling, E.F. (1976) Report of the Technical Guidance Committee for the Pilot Boll Weevil Eradication Experiment. In: Davich, T.B. (ed.) *Boll Weevil Suppression, Management and Elimination Technology – Proceedings of a Conference*. USDA-ARS-S-71, Memphis, Tennessee, pp. 122–123.

Knipling, E.F. and McGuire, J.U., Jr. (1966) Population models to the theoretical effects of sex attractants used for insect control. *USDA Agricultural Information Bulletin* 308, 20 pp.

Lambert, W.R. (1991) The Southeastern Boll Weevil Eradication Programme: extension perspective. In: Herber, D.J. and Richter, D.A. (eds) *Proceedings of the Beltwide Cotton Conference*, Memphis, Tennessee, pp. 611–613.

Laws, F. (2006) Weevil buffer zone concerns lead to legal action. *Delta Farm Press* 3 October (http://deltafarmpress.com/news/061003-buffer-zone/).

Ledbetter, R.J. (1971) The Coosa River Valley diapause control program. In: Brown, J.M. (ed.) *Proceedings of the Beltwide Cotton Conference*, National Cotton Council, Memphis, Tennessee, pp. 32–33.

Leggett, J.E. and Cross, W.H. (1971) A new trap for capturing boll weevils. *USDA Cooperative Economic Insect Report* 21, 773–774.

Leser, J.F., Bodden, E.A. and Haldenby, R. (1997) Boll weevil status in the Texas High Plains. In: Dugger, P. and Richter, D.A. (eds) *Proceedings of the Beltwide Cotton Conference*, National Cotton Council, Memphis, Tennessee, pp. 1201–1204.

Lewis, A.C. (1920) Annual Report of the State Entomologist for 1919. *Georgia State Board of Entomology Bulletin* 58, 12 pp.

Lewis, A.C. and McLendon, C.A. (1919) Cotton Variety Tests 1918. *Georgia State Board of Entomology Circular* 52, 5–7.

Lincoln, C. and Waddle, B.A. (1965) Insect resistance of frego bract cotton. *Arkansas Farm Research* 15, 5.

Linder, T. (1954) *Georgia Historical Agricultural Data*. Georgia State Department of Agricultural Special Report to the State Historical Commission, Georgia.

Lindquist, D.A., Gorzycki, L.J., Mayer, J.S., Seales, A.L. and Davich, T.B. (1964) Laboratory studies on sterilization of the boll weevil with apholate. *Journal of Economic Entomology* 57, 745–750.

Lloyd, E.P. (1972) Progress report on the Pilot Boll Weevil Eradication Experiment. In: Brown, J.M. (ed.) *Proceedings of the Beltwide Cotton Conference*, National Cotton Council, Memphis, Tennessee, pp. 46–49.

Lloyd, E.P. (2001) Introduction. In: Dickerson, W.A., Brashear, A.L., Brumley, J.T., Carter, F.L., Grefenstette, W.J. and Harris, F.A. (eds) *Boll Weevil Eradication in the United States Through 1999*. Reference Book Series No. 6, The Cotton Foundation, Memphis, Tennessee, pp. 1–6.

Lloyd, E.P., Laster, M.L. and Merkl, M.E. (1964) A field study of diapause, diapause control and population dynamics of the boll weevil in Mississippi. *Journal of Economic Entomology* 57, 433–436.

Lloyd, E.P., Tingle, F.C., McCoy, J.R. and Davich, T.B. (1966) The reproduction-diapause approach to population control of the boll weevil. *Journal of Economic Entomology* 59, 813–816.

Lloyd, E.P., Tingle, F.C., Merkl, M.E., Burt, E.C., Smith, D.B. and Davich, T.B. (1967) Comparison of three rates of ultra-low-volume azinphosmethyl in a reproduction-diapause control program against the boll weevil. *Journal of Economic Entomology* 60, 1696–1699.

Lloyd, E.P., Merkl, M.E., Tingle, F.C., Scott, W.P., Hardee, D.D. and Davich, T.B. (1972) Evaluation of male baited traps for control of boll weevils following a reproductive-diapause program in Monroe County, Mississippi. *Journal of Economic Entomology* 65, 522–555.

Lloyd, E.P., Wood, R.H. and Mitchell, E.B. (1977) Boll weevil: suppression with TH-6040 applied in season in cotton-seed oil as a foliar spray. *Journal of Economic Entomology* 70, 442–444.

Lloyd, E.P., McKibben, G.H., Knipling, E.F., Witz, J.A., Hardstack, A.W., Leggett, J.E. and Lockwood, D.F. (1980) Mass trapping for detection, suppression, and integration with other suppression measures against the boll weevil. Presented at an *International Colloquium on the Management of Insect Pests with Semio-chemicals*, Gainesville, Florida.

Logan, D.P. (2001) Boll weevil eradication in Louisiana. In: Dickerson, W.A., Brashear, A.L., Brumley, J.T., Carter, F.L., Grefenstette, W.J. and Harris, F.A. (eds) *Boll Weevil Eradication in the United States Through 1999*. Reference Book Series No. 6, The Cotton Foundation, Memphis, Tennessee, pp. 293–304.

Luttrell, R.G., Cooke, F.T., Freeland, T.B., Jr., Gibson, J., Harris, F.A., Laughlin, D.G., Reid, J.T. and Watson, C.E. (1997) Study of cotton insect control in 1995 with special emphasis on boll weevil eradication and tobacco budworm. *Mississippi Agricultural and Forest Experiment Station Technology Bulletin* 217, 29 pp.

Mally, F.W. (1901) The Mexican boll weevil. *USDA Farmers Bulletin* 130, 29 pp.

Masud, S.M., Lacewell, R.D., Taylor, C.R., Benedict, J.H. and Lippke, L.A. (1981) Economic impact of integrated pest management strategies for cotton in the coastal bend region of Texas. *Southern Journal of Agricultural Economics* 15, 47–52.

McCorkle, D.A., Frisbie, R.E., Robinson, J., Allen, C., Fuchs, T., Hanselka, D. and Klose, S. (2007) *Economic Impact Brief: Economic Impact of Boll Weevil Eradication in Texas.* Texas Cooperative Extension Publication MKT-35580, Texas A&M University System, College Station, Texas, 1 p. (http://agecoext.tamu.edu/econimpact/).

McGarigle, B. (2002) Pest patrol. *Government Technology: Solutions for State and Local Government in the Information Age*, 2 pp. (http://www.govtech.net/magazine/story.php?id=8069&story_pg=2).

McGovern, W.L. and Cross, W.H. (1976) Affects of two cotton varieties on levels of boll weevil parasitism (Coleoptera: Curculionidae). *Entomophaga* 21, 123–125.

McGovern, W.L., Smith, J.W., Villavaso, E. and McKibben, G.H. (1993) Boll weevil suppression in Rutherford County, Tennessee with bait sticks. In: Herber, D.J. and Richter, D.A. (eds) *Proceedings of the Beltwide Cotton Conference*, Memphis, Tennessee, pp. 928–929.

McGovern, W.L., Villavaso, E. and McKibben, G.H. (1996) Final evaluation of 1994 boll weevil bait stick test in Noxubee County, Mississippi. In: Dugger, P. and Richter, D.A. (eds) *Proceedings of the Beltwide Cotton Conference*, National Cotton Council, Memphis, Tennessee, pp. 994–997.

McHaffy, D.G. and Borkovec, A.B. (1976) Vacuum dipping: a new method of administering chemosterilants to the boll weevil. *Journal of Economic Entomology* 69, 139–143.

McKibben, G.H. (1972). A device for injecting grandlure into cigarette filters. *Journal of Economic Entomology* 65, 1509–1510.

McKibben, G.H., Hardee, D.D., Davich, T.B., Gueldner, R.C. and Hedin, P.A. (1971) Slow release formulations of grandlure, the synthetic pheromone of the boll weevil. *Journal of Economic Entomology* 64, 317–319.

McKibben, G.E., Davich, T.B., Gueldner, R.C., Hardee, D.D. and Hedin, P.A. (1974) Polymeric compositions for attracting boll weevils. *U.S. Patent* No. 3, 803, 303.

McKibben, G.H., Villavaso, E.J., McGovern, W.L. and Grefenstette, B. (2001) United States Department of Agriculture – research support, methods development and programme implementation. In: Dickerson, W.A., Brashear, A.L., Brumley, J.T., Carter, F.L., Grefenstette, W.J. and Harris, F.A. (eds) *Boll Weevil Eradication in the United States Through 1999.* Reference Book Series No. 6, The Cotton Foundation, Memphis, Tennessee, pp. 101–136.

McLaughlin, R.E. (1962) Infectivity tests with *Beauveria bassiana* (Balsamo) Vuillemin on *Anthonomus grandis* Boheman. *Journal of Insect Pathology* 4, 386–388.

McLaughlin, R.E. (1965) *Mattesia grandis* n. sp., a sporozoan pathogen of the boll weevil, *Anthonomus grandis* Boheman. *Journal of Protozoology* 12, 405–413.

McLaughlin, R.E. (1969) *Glugea gasti* sp. n., a microsporidian pathogen of the boll weevil, *Anthonomus grandis* Boheman. *Journal of Protozoology* 16, 84–92.

Miller, A. (2006) New Mexico boll weevil eradication program weekly status report. *USDA-APHIS Report* (http://www.txbollweevil.org).

Mitchell, E.B. and Hardee, D.D. (1974) In-field traps: a new concept in survey and suppression of low populations of boll weevils. *Journal of Economic Entomology* 67, 506–508.

Mitchell, E.B., Hardee, D.D., Cross, W.H., Huddleston, D.M. and Mitchell, H.C. (1972) Influence of rainfall, sex ratio and physiological condition of boll weevils on their response to pheromone traps. *Environmental Entomology* 1, 438–440.

Mitlin, L.L. and Mitlin, N. (1968) Boll weevil (*Anthonomus grandis* Boheman) abstracts of research publications 1961–1965. *USDA Miscellaneous Publication* No. 1092, 32 pp.

Moore, G.C., Parker, R.D., Fromme, D.D. and Kuntson, A.C. (2003) Managing cotton insects in the southern eastern and Blackland areas of Texas 2003. *Texas Cooperative Extension*, Publication E-5 (http://insects.tamu.edu/extension/bulletins/e-5.html#Overwintered %20Boll%20Weevil).

Moore, R.F. and Taft, H.M. (1975) Boll weevils: chemosterilization of both sexes with bisulfan plus Thompson-Hayward TH-6040. *Journal of Economic Entomology* 68, 96–98.

Moore, R.F., Leopold, R.A. and Taft, H.M. (1978) Boll weevil's mechanism of transfer of diflubenzuron from male to female. *Journal of Economic Entomology* 71, 587–590.

Morales-Ramos, J.A., Summy, K.R., Roberson, J.L., Gast, R.J., Cate, J.R. and King, E.G. (1992) Feasibility of mass rearing *Catolaccus grandis*, a parasite of the boll weevil. In: Herber, D.J. and Richter, D.A. (eds) *Proceedings of the Beltwide Cotton Conference*, National Cotton Council, Memphis, Tennessee, pp. 723–726.

Mulrooney, J.E., Hanks, J.E., Howard, K.D. and Jones, R.G. (1995) Ultra-low-volume application of insecticides for boll weevil control in the eradication program. In: Richter, D.A. and Armour, J. (eds) *Proceedings of the Beltwide Cotton Conference*, National Cotton Council, Memphis, Tennessee, pp. 1046–1048.

Mulrooney, J.E., Howard, K.D., Hanks, J.E. and Jones, R.G. (1996) Efficacy of ULV insecticides against boll weevils. In: Dugger P. and Richter, D.A. (eds) *Proceedings of the Beltwide Cotton Conference*, National Cotton Council, Memphis, Tennessee, pp. 720–721.

Mulrooney, J.E., Howard, K.D., Hanks, J.E. and Jones, R.G. (1997) Application of ultra-low-volume malathion by air-assisted ground sprayer for boll weevil (Coleoptera: Curculionidae) control. *Journal of Economic Entomology* 90, 640–645.

Mussman, H.C. (1982) Cotton Boll Weevil: USDA's response to the National Research Council Evaluation. In: Brown, J.M. (ed.) *Proceedings of the Beltwide Cotton Conference*, National Cotton Council, Memphis, Tennessee, pp. 91–93.

Namken, L.N. and Heilman, M.D. (1973) Determinate cotton cultivars for more efficient cotton production in the Lower Rio Grande Valley. *Agriculture Journal* 65, 953–956.

Namken, L.N., Heilman, M.D., Jenkens, J.N. and Miller, P.A. (1983) Plant resistance and modified cotton culture. In: Ridgway, R.L., Lloyd, E.P. and Cross W.H. (eds) *Cotton Insect Management with Special Reference to the Boll Weevil*. USDA-ARS Agriculture Handbook 589, Washington, DC, pp. 73–101.

National Academy of Sciences, National Research Council, Cotton Study Team (1975) Pest control: an assessment of present and alternative technologies. In: *National Academy of Sciences*, Vol. 3. Cotton Pest Control, Washington, DC, pp. 4–5.

National Cotton Council (1974) Boll weevil losses: value and location of losses caused by the boll weevil; beltwide state cost summaries estimated by the National Cotton Council in cooperation with state extension specialists. *National Cotton Council*, Memphis, OCLC No. 21433429, 52 pp.

National Cotton Council (2006) Minumum standards for boll weevil post-eradication. *National Cotton Council Boll Weevil Action Committee Meeting* 6 October 2006, 6 pp.

Neal, C.R. and Antilla, L. (2001) Boll weevil establishment and eradication in Arizona and northwest Mexico. In: Dickerson, W.A., Brashear, A.L., Brumley, J.T., Carter, F.L., Grefenstette, W.J. and Harris, F.A. (eds) *Boll Weevil Eradication in the United States Through 1999*. Reference Book Series No. 6, The Cotton Foundation, Memphis, Tennessee, pp. 213–224.

Nemec, S. and Adkisson, P.L. (1969) Laboratory tests of insecticides for bollworm, tobacco budworm and boll weevil control. *Texas Agricultural Experiment Station Progress Report* 2674, 4 pp.

New Mexico State Assembly (1996) *Cotton Boll Weevil Control Committees*, Chapter 77, Section 11 (http://legis.state.nm.us/sessions/05%20regular/bills/house/HB0751.pdf).

Newell, W. (1904) The Mexican cotton boll weevil. *Georgia State Board of Entomology Bulletin* 12, 29 pp.

Newell, W. (1908) A new predaceous enemy of the boll weevil. *Journal of Economic Entomology* l, 244.

Newell, W. and Paulsen, T.C. (1908) The possibility of reducing boll weevil damage by autumn spraying of cotton fields to destroy the foliage and squares. *Journal of Economic Entomology* 1, 112–116.

Niles, G.A. (1970) Development of plant types with special adaptation to narrow row culture. In: Brown, J.M. (ed.) *Proceedings of the Beltwide Cotton Conference*, National Cotton Council, Memphis, Tennessee, pp. 63–64.

Parencia, C.R. (1959) Comparative yields of cotton in treated and untreated plots in insect control experiments in central Texas, 1939–1958. *Journal of Economic Entomology* 52, 747–758.

Parencia, C.R., Jr. (1978) One hundred twenty years of research on cotton insects in the United States. *USDA Agricultural Handbook* No. 515, 75 pp.

Parencia, C.R., Jr. and Cowan, C.B., Jr. (1960) Increased tolerance of the boll weevil and cotton fleahopper to some chlorinated hydrocarbon insecticides in central Texas in 1958. *Journal of Economic Entomology* 53, 52–56.

Parker, R.D., Walker, J.K., Niles, G.A. and Mulkey, J.R. (1980) The 'short-season effect' in cotton and escape from boll weevil. *Texas Agricultural Experiment Station Bulletin* B-1315, 45 pp.

Parker, R.D., Wolfenbarger, D.A. and Kniffen, B.A. (1995) Field evaluation of boll weevil Attract-and-Kill devices in the Coastal Bend of Texas. In: Richter, D.A. and Armour, J. (eds) *Proceedings of the Beltwide Cotton Conference*, National Cotton Council, Memphis, Tennessee, pp. 937–940.

Parrott, W.L., Jenkins, J.N. and Smith, D.B. (1973) Frego bract cotton and normal bract cotton: how morphology affects control of boll weevils by insecticides. *Journal of Economic Entomology* 66, 222–225.

Perkins, J.H. (1980) Boll Weevil Eradication. *Science* 207, 1044–1050.

Perkins, J.H. (1982) *Insects, Experts and the Insecticide Crisis.* 1st edn., Plenum Press, New York.

Perkins, J.H. (1983) The boll weevil in North America: scientific conflicts over management of environmental resources. *Agriculture, Ecosystems, and Environment* 10, 217–245.

Phillips, J.R. (1978) Integrated insect pest management systems: an areawide insect management program. In: Phillips, J.R., Yearian, W.C. and Mayse, M.A. (eds) *Proceedings of a Symposium on the Development of Optimum Crop Production Systems for the Mid-South.* University of Arkansas Special Report 67, Arkansas, pp.104–105.

Phillips, J.R., Nicholson, W.F., Teague, T., Bernhardt, J. and Mueller, T.F. (1981) Community insect management programs. In: *Proceedings of the Cotton Biological Control Conference.* Dallas, Texas, 15–16 January, 1981, pp. 44–49.

Phillips, M.W. (1850) *Remarks on the Cultivation of Cotton.* House of Representatives, Report of the Commissioner of Patents for the Year 1850, Washington DC, Exec. Doc. No. 20, pp. 313–316.

Pierce, J.B., Sutherland, C.A., Miller, A.B., Sanderson, S.S. and Calvani, J.A. (2001) Boll weevil establishment and eradication in New Mexico. In: Dickerson, W.A., Brashear, A.L., Brumley, J.T., Carter, F.L., Grefenstette, W.J. and Harris, F.A. (eds) *Boll Weevil Eradication in the United States Through 1999.* Reference Book Series No. 6, The Cotton Foundation, Memphis, Tennessee, pp. 361–374.

Pierce, W.D. (1908) The economic bearing of recent studies of the parasites of the cotton boll weevil. *Journal of Economic Entomology* 1, 117–122.

Pierce, W.D., Cushman, R.A. and Hood, C.E. (1912) The natural enemies of the cotton boll weevil. *USDA Bureau of Entomology Bulletin* 100, 99 pp.

Planer, F.R. (1988) Southeast boll weevil eradication program. In: Brown, J.M. and Richter, D.A. (eds) *Proceedings of the Beltwide Cotton Conference*, National Cotton Council, Memphis, Tennessee, pp. 239–240.

Post, G.B. (1924) Boll weevil control by airplane. *Georgia State College of Agriculture Extension Division Bulletin* 301, 1–22.

Prentice, A.N. (1972) *Cotton, With Special Reference to Africa*. Longman Group Limited, London.

Rabb, R.L. (1972) Principles and concepts of pest management: Implementing practical pest control strategies. In: USDA Cooperative State Extension Services (eds) *Proceedings of a National Insect Pest Management Workshop*, Purdue University, West Lafayette, Indiana, pp. 6–29.

Ridgway, R.L. and Lloyd, E.P. (1983) Evolution of cotton insect management in the United States. In: Ridgway, R.L., Lloyd, E.P. and Cross, W.H. (eds) *Cotton Insect Management with Special Reference to the Boll Weevil*, USDA Agriculture Handbook No. 589, Washington, DC, pp. 3–27 pp.

Ridgway, R.L. and Mussman, H.C. (2001) Integrating science and stakeholder inputs – the pivotal years. In: Dickerson, W.A., Brashear, A.L., Brumley, J.T., Carter, F.L., Grefenstette, W.J. and Harris, F.A. (eds) *Boll Weevil Eradication in the United States Through 1999*. Reference Book Series No. 6, The Cotton Foundation, Memphis, Tennessee, pp. 55–76.

Ridgway, R.L., Tinney, J.C., MacGregor, J.T. and Starler, N.E. (1978) Pesticide use in agriculture. *Environmental Health Perspectives* 27, 103–112.

Riggs, W.M. (1921) Report of the South Carolina Boll Weevil Commission, 67th Congress, In Session, Senate Doc. 76, *Clemson Agricultural College Bulletin* 20, 22 pp.

Roberts, P. (1999) Post eradication insect control: was it worth it? In: Dugger, P. and Richter, D.A. (eds) *Proceedings of the Beltwide Cotton Conference*, National Cotton Council, Memphis, Tennessee, pp. 832–833.

Robertson, O.T. (1957) Occurrence of the boll weevil in the Big Bend of Texas. *Journal of Economic Entomology* 50, 102.

Robertson, O.T., Noble, L.W. and Orr, G.E. (1966) Spread of boll weevil and its control in far west Texas. *Journal of Economic Entomology* 59, 754–756.

Roof, M.E. (2001) Boll weevil eradication and post-eradication in South Carolina. In: Dickerson, W.A., Brashear, A.L., Brumley, J.T., Carter, F.L., Grefenstette, W.J. and Harris, F.A. (eds) *Boll Weevil Eradication in the United States Through 1999*. Reference Book Series No. 6, The Cotton Foundation, Memphis, Tennessee, pp. 413–430.

Roussel, J.S. and Clower, D.F. (1955) Resistance to the chlorinated hydrocarbon insecticides in the boll weevil (*Anthonomus grandis* Boheman). *Louisiana State University and Agricultural and Mechanical College Circular* 41, 9 pp.

Roussel, J.S. and Clower, D.F. (1957) Resistance to the chlorinated hydrocarbon insecticides in the boll weevil. *Journal of Economic Entomology* 50, 463–468.

Rummel, D.R. (1970) Population suppression of the boll weevil, *Anthonomus grandis* Boheman, in the High and Rolling Plains of Texas. PhD dissertation, Texas A&M University, College Station, Texas. 143 pp.

Rummel, D.R. (1976a) An areawide boll weevil suppression program – organization, operation and economic impact. In: *Proceedings of the US–USSR Symposium: the Integrated Control of the Arthropod, Disease and Weed Pests of Cotton, Grain Sorghum and Deciduous Fruit*, Texas A&M Agricultural Experiment Station, College Station, Texas. Miscellaneous Publication 1276, pp. 152–163.

Rummel, D.R. (1976b) Reproduction-diapause boll weevil control. In: Davich, T.B. (ed.) *Boll Weevil Suppression, Management and Elimination Technology – Proceedings of a Conference*. USDA-ARS-S-71, Memphis, Tennessee, pp. 28–30.

Rummel, D.R. and Atkisson, P.L. (1971) A two-phased control program designed for maximum suppression of the boll weevil in the high and rolling plains of Texas. *Journal of Economic Entomology* 64, 919–922.

Rummel, D.R. and Carroll, S.C. (1985) Longevity of overwintered boll weevils (Coleoptera: Curculionidae) following emergence in spring and early summer. *Environmental Entomology* 14, 127–130.

Rummel, D.R., Bottrell, D.G., Adkisson, P.L. and McIntyre, R.C. (1975) An appraisal of a 10-year effort to prevent the westward spread of the boll weevil. *Bulletin of the Entomological Society of America* 21, 6–11.

Rummel, D.R., White, J.R., Carroll, S.C. and Pruitt, G.R. (1980) Pheromone trap index system for predicting the need for overwintered boll weevil control. *Journal of Economic Entomology* 73, 806–810.

Shaw, B.T. (1956) Today's research and tomorrow's agriculture. *Presentation at the Agricultural Research Institute*, Washington, DC, 15 October 1956.

Sheppard, H.A. (1951) *The Chemistry and Action of Insecticides*. 1st edn., McGraw Hill, New York.

Sikorowski, P.P. (1984) Pathogens and microbial contaminants: their occurrence and control. In: Sikorowski, P.P., Griffith, J.G., Roberson, J.L. and Lindig, O.H. (eds) *Boll Weevil Mass Rearing Technology*. University Press of Mississippi, Jackson, Mississippi, pp. 115–169.

Sikorowski, P.P., Wyatt, J.M. and Lindig, O.H. (1977) Method of surface sterilization of boll weevil eggs. *Southwestern Entomologist* 2, 32–36.

Sim, T. (2001) Boll weevil trapping in Kansas. In: Dickerson, W.A., Brashear, A.L., Brumley, J.T., Carter, F.L., Grefenstette, W.J. and Harris, F.A. (eds) *Boll Weevil Eradication in the United States Through 1999*. Reference Book Series No. 6, The Cotton Foundation, Memphis, Tennessee, pp. 291–292.

Slosser, J.E. (1978) The influence of planting date on boll weevil management. *Southwestern Entomologist* 3, 241–246.

Slosser, J.E. (1995) *Cultural Control of the Boll Weevil – a Four-season Approach – Texas Rolling Plains*. Texas Agricultural Extension Service Publication B-1721, 3 pp (http://insects.tamu.edu/extension/bulletins/b-1721.html).

Smith, G.L., Clevland, T.C. and Clark, J.C. (1964) *Cost of Cotton Insect Control with Insecticides at Tallulah, Louisiana*. US Agriculture Service-ARS, No. 33–96, 7 pp.

Smith, J.R. (1973) *Statement on Boll Weevil Eradication*. Cotton Subcommittee of the House Committee on Agriculture, 4 April 1973. National Cotton Council, Memphis, Tennessee, 13 pp.

Smith, J.T. (1997) Lindy Patton named state Executive Director of weevil war. *Abilene Reporter News* 16 September (http://texnews.com/biz97/patt091697.html).

Smith, J.W. (1998) Boll weevil eradication: areawide pest management. *Annals of the Entomological Society of America* 91, 239–247.

Smith, L.E., Allen, C.T., Patton, L.W. and Newman, R.O. (2002) Status of boll weevil eradication in Texas. In: McRae, J. and Richter, D.A. (eds) *Proceedings of the Beltwide Cotton Conference*, National Cotton Council, Memphis, Tennessee, 5 pp.

Smith, R. (2001) Significant headway in 2000. *Southwest Farm Press* 11 January (http://southwestfarmpress.com/mag/farming_significant_headway/).

Smith, R. (2006) Weevil eradication nears the finish line. *Southeast Farm Press* 6 February (http://southeastfarmpress.com/mag/farming_weevil_eradication_nears/).

Soule, A.M. (1921) Some factors affecting the economic production of cotton. *Georgia State College of Agriculture Extension Division Bulletin* 247, 16 pp.

Spurgeon, D.W. (2001) Efficacy of field-aged bait sticks against the boll weevil. *Journal of Cotton Science* 5, 68–73.

Spurgeon, D.W., Coppedge, J.R., Raulston, J.R. and Marshall, H. (1999) Mechanisms of boll weevil (Coleoptera: Curculionidae) bait stick activity relative to pheromone traps. *Journal of Economic Entomology* 92, 960–966.

State of Oklahoma (2004) FY 2005 *Executive Budget – Agriculture* (http://www.osf.state.ok.us/bud05-agriculture.pdf).

Stavinoha, K.D. and Woodward, L.A. (2001) Texas boll weevil history. In: Dickerson, W.A., Brashear, A.L., Brumley, J.T., Carter, F.L., Grefenstette, W.J. and Harris, F.A. (eds) *Boll Weevil Eradication in the United States Through 1999*. Reference Book Series No. 6, The Cotton Foundation, Memphis, Tennessee, pp. 451–502.

Sterling, W.L. (1978) Fortuitous biological suppression of the boll weevil by the red imported fire ant. *Environmental Entomology* 7, 564–568.

Stern, V., Smith, R.F., van den Bosch, R. and Hagen, K.S. (1959) The integrated control concept. *Hilgardia* 29, 81–101.

Summy, K.R., Morales-Ramos, J.A. and King, E.G. (1993) Suppression of boll weevil infestations by augmentative releases of *Catolaccus grandis*. In: Herber, D.J. and Richter, D.A. (eds) *Proceedings of the Beltwide Cotton Conference*, National Cotton Council, Memphis, Tennessee, pp. 908–909.

Szmedra, P.T., McClendon, R.W. and Watzstein, M.E. (1991) Economic risk efficiency of boll weevil eradication. *Southern Journal of Agricultural Economics* 23, 237–245.

Taft, H.M. and Hopkins, A.R. (1975) Boll weevils: field populations controlled by sterilizing emerging overwintered females with TH-6040 sprayable bait. *Journal of Economic Entomology* 68, 551–554.

Tate, J.R. (2001) Cotton production and the boll weevil in Virginia. In: Dickerson, W.A., Brashear, A.L., Brumley, J.T., Carter, F.L., Grefenstette, W.J. and Harris, F.A. (eds) *Boll Weevil Eradication in the United States Through 1999*. Reference Book Series No. 6, The Cotton Foundation, Memphis, Tennessee, pp. 503–508.

TBWEF (2007) Financial reports, 31 December 2007. In: *Texas Boll Weevil Eradication Foundation Board of Directors Meeting Report*, 2 March 2007.

Thomas, F.L. (1929) What does the future hold in store for the South? *Journal of Economical Entomology* 22, 736–743.

Townsend, C.H.T. (1895) Report on the Mexican cotton boll weevil in Texas. *Insect Life* 7, 295–309.

Trelogan, H.C. (1969) The story of U.S. agricultural estimates. *USDA Miscellaneous Publications* No. 1088. 137 pp.

Tumlinson, J.H., Hardee, D.D., Minyard, J.P., Thompson, A.C., Gast, R.T. and Hedin, P.A. (1968) Boll weevil sex attractant: isolation studies. *Journal of Economic Entomology* 61, 470–474.

Tumlinson, J.H., Hardee, D.D., Gueldner, R.C., Thompson, A.C., Hedin, P.A. and Minyard, J.P. (1969) Sex pheromones produced by male boll weevils: isolation, identification, and synthesis. *Science* 166, 1010–1012.

Tumlinson, J.H., Gueldner, R.C., Hardee, D.D., Thompson, A.C., Hedin, P.A. and Minyard, J.P. (1971) Identification and synthesis of the four compounds comprising the boll weevil sex attractant. *Journal of Organic Chemistry* 36, 2616–2621.

USDA (1958) *The Boll Weevil Problem and Research and Faculty Needs to Meet the Problem*. US Department of Agriculture, a report prepared at the request of the US Congress, 30 December 1958, 50 pp.

USDA-APHIS (1991) *Final Environmental Impact Statement – National Boll Weevil Cooperative Control Program*. USDA-APHIS, Hyattsville, Maryland.

USDA-ARS (1999) The history of screwworm eradication. *Special Collections of the National Agricultural Library* (http://www.nal.usda.gov/speccoll/screwworm/history.htm).

USDA-APHIS (2007a) Environmental monitoring plan – boll weevil eradication programme New Mexico, Oklahoma, and Texas. *USDA-APHIS PPQ, Environmental Compliance Team*, 32 pp.

USDA-APHIS (2007b) Proposed rules – boll weevil; quarantine and regulations. *Federal Register* 20 (210), 63707–63717.

Van der Vloedt, A.M. and Klassen, W. (1991) The development and application of the sterile insect technique (SIT) for new world screw worm eradication. *World Animal Review –*

Special Issue (http://www.fao.org/WAICENT/faoINFO/AGRICULT/aga/AGAP/FRG/FEEDback/War/u4220b/u220b00.htm#Contents).

Vanderzant, E.S. and Davich, T.B. (1958) Laboratory rearing of the boll weevil: a satisfactory larval diet and oviposition studies. *Journal of Economic Entomology* 51, 288–291.

Vietmeyer, N.D. (1982) Our 90 year war with the boll weevil isn't over. *Smithsonian Magazine* August, 60–68.

Villavaso, E.J. and McGovern, W.L. (2000) Malathion and boll weevil eradication: application rates, mist blower effectiveness and bait stick evaluation. In: Dugger, P. and Richter, D.A. (eds) *Proceedings of the Beltwide Cotton Conference*, National Cotton Council, Memphis, Tennessee, pp. 1067–1069.

Villavaso, E.J., Roberson, J.L. and Seward, R.W. (1989) Effectiveness of sterile boll weevils (Coleoptera: Curculionidae) against a low density population in commercially grown cotton in north-central Mississippi. *Journal of Economic Entomology* 82, 472–476.

Villavaso, E.J., Bartlett, A.C. and Laster, M.L. (1996a) Genetic control. In: King, E.G. (ed.) *Cotton Insects and Mites: Characterization and Management*. Reference Book Series No. 3, The Cotton Foundation, Memphis, Tennessee, pp. 539–562.

Villavaso, E.J., Mulrooney, J.E., McGovern, W.L. and Howard, K.D. (1996b) Lower dosages of malathion for boll weevil control. In: Dugger, P. and Richter, D.A. (eds) *Proceedings of the Beltwide Cotton Conference*, National Cotton Council, Memphis, Tennessee, pp. 727–729.

Villavaso, E.J., McGovern, W.L. and Wagner, T.L. (1998) Efficacy of bait sticks versus pheromone traps for removing boll weevils (Coleoptera: Curculionidae) from released populations. *Journal of Economic Entomology* 91, 637–640.

Wagner, F. (1980) *The Boll Weevil Comes to Texas*. The Friends of the Corpus Christi Museum, Corpus Christi, Texas.

Wagner, F. (1999) Boll Weevil. *The Handbook of Texas Online* (http://www.tsha.utexas.edu/handbook/online/).

Walker, J.K. (1984) The boll weevil in Texas and the cultural strategy. *Southwestern Entomology* 9, 444–463.

Walker, J.K. and Niles, G.A. (1971) Population dynamics of the boll weevil and modified cotton types. *Texas A&M Agricultural Experiment Station Bulletin* 1109, 14 pp.

Walker, J.K. and Smith, C.W. (1996) Cultural control. In: King, E.G, Phillips, J.R. and Coleman, R.J. (eds) *Cotton Insects and Mites: Characterization and Management*. Reference Book Series No. 3, The Cotton Foundation, Memphis, Tennessee, pp. 471–510.

Walker, J.K., Jr., Gannaway, J.R. and Niles, G.A. (1977) Age distribution of cotton bolls and damage from the boll weevil. *Journal of Economic Entomology* 70, 5–8.

Ware, J.O. (1929) Cotton spacing. I. Studies of the effect on yield and earliness. *Arkansas Agricultural Experiment Station Bulletin* 230, 84 pp.

Ware, J.O. (1930) Cotton spacing. II. Effect of blooming on earliness, fruit set and yield. *Arkansas Agricultural Experiment Station Bulletin* 253, 64 pp.

Warner, R.E. and Smith, Jr., C.E. (1968) Boll weevil found in Pre-Columbian cotton from Mexico. *Science* 162, 911–912.

Warren, D.C. and Williams, I.W. (1922) Results of cotton dusting experiments for 1921, together with summary of the dusting results for the past three years, with recommendations for cotton dusting for coming season. *Georgia State Board of Entomology Bulletin* 62, 1–10.

Watkins, J.L. (1904) Consumption of cotton in the cotton states. *Yearbook of the USDA for 1903*, Government Printing Office, Washington, DC.

Westbrook, E.C. (1939) Cotton culture in Georgia. *University of Georgia College of Agriculture Experiment Station Bulletin* 469, 19–23.

Westbrook, E.C. (1945) Modernizing cotton production. *University of Georgia College of Agriculture Experiment Station Bulletin* 523, 4–7.

White, J.R. and Rummel, D.R. (1978) Emergence profile of overwintering boll weevils and entry into cotton. *Environmental Entomology* 7, 7–14.

Williams, M.R. and Layton, M.B. (1996) Taken by storm: a report of the tobacco budworm problem. In: Dugger P. and Richter, D.A. (eds) *Proceedings of the Beltwide Cotton Conference* National Cotton Council, Memphis, Tennessee, pp. 823–825.

Williams, R.C. (1987) *Fordson, Farmall and Popping Johnny: a History of the Farm Tractor and Its Impact on America.* University of Illinois Press, Illinois.

Womack, H. (1970) Boll weevil diapause study. In: *Proceedings of the Southeastern Cotton Workshop*, pp. 12–13.

Worsham, E.L. (1912) Crop pest law of Georgia and regulations of the State Board of Entomology. *Georgia State Board of Entomology Bulletin* 37, 16 pp.

Worsham, E.L. (1914) The Mexican cotton boll weevil. *Georgia State Board of Entomology Bulletin* 39, 12 pp.

Worsham, E.L. (1915) Annual report of the State Entomologist for 1914. *Georgia State Board of Entomology Bulletin* 42, 6–10.

Zinn, J. (1994) Conservation Reserve Program: policy issues for the 1995 Farm Bill. *Congressional Research Service – Report for Congress* (http://www.cnie.org/nle/crsreports/agriculture/ag-4.cfm).

Appendix: List of Acronyms

ABWEF: Arkansas Boll Weevil Eradication Foundation

ACRPC: Arizona Cotton Research and Protection Council

APHIS: Animal and Plant Health Inspection Service of the US Department of Agriculture

ARS: Agricultural Research Service of the US Department of Agriculture

BWAC: Boll Weevil Action Committee, a committee of the National Cotton Council active in the effort to eradicate the boll weevil

BWEC: Boll Weevil Eradication Commission of Louisiana

BWET: Boll Weevil Eradication Trial, conducted 1978–1980 in North Carolina and Virginia

BWRL: Boll Weevil Research Laboratory, a USDA Agricultural Research Service Laboratory built in Starkville, Mississippi

CDFA: California Department of Food and Agriculture

CES: Cooperative Extension Service

CI: Cotton Incorporated, an organization funded by cotton growers to provide research, development and promotion of cotton products

CRP: Conservation Reserve Program

DDVP: Dichlorvos insecticide

EP/TP: El Paso/Trans Pecos zone, Texas Program. Pecos south to Presidio, west to El Paso and north to New Mexico

EPA: Environmental Protection Agency of the USA

ERS: Economic Research Service of the US Department of Agriculture

ESA: Entomological Society of America, the primary organization of entomologists in the USA

FSA: Farm Service Agency of the US Department of Agriculture

FWS: US Fish and Wildlife Service

GPS: Global positioning system

IPM: Integrated Pest Management, a system for managing pests in which pest populations are intelligently managed through an ecologically based system that integrates cultural, mechanical and biological controls with host plant resistance and, as a last option, uses chemical controls

LDAF: Louisiana Department of Agriculture and Forestry

LRGV: Lower Rio Grande Valley zone, Texas Program: the southern tip of Texas

MBWMC: Mississippi Boll Weevil Management Corporation

MBWTAC: Mississippi Boll Weevil Technical Advisory Committee; appointed by MBWMC in 1993

MCGO: Missouri Cotton Growers Organization

MDA: Missouri Department of Agriculture

NAS: National Academy of Sciences

NBL: Northern Blacklands zone, Texas Program; Waco north to Oklahoma and east to Texarkana

NCC: National Cotton Council

NHP: Northern High Plains zone, Texas Program. Texas Panhandle region, originally Floydada west to New Mexico and north to Oklahoma. Later, the Floydada, Plainview, Tulia area

NRC: National Research Council of the Board of Agriculture and Renewable Resources

NRP: Northern Rolling Plains zone, Texas Program. Paducah north to Pampa and east to Wichita Falls; adjacent to south-west Oklahoma

NWP: Northwest Plains zone, Texas Program. Texas Panhandle region, Dimmitt and Littlefield west to New Mexico

OBWEO: Oklahoma Boll Weevil Eradication Organization

OPMT: Optimum Pest Management Trial, conducted in Panola Co., Mississippi, concurrently with the BWET, 1978–1980

PB: Permian Basin zone, Texas Program. Midland, Big Spring, Lamesa area

PBWEE: Pilot Boll Weevil Eradication Experiment, conducted 1971–1973 in Mississippi, Louisiana and Alabama

PCG: Plains Cotton Growers, the grower organization that represents cotton growers from the Texas High Plains region

PH: Panhandle zone, Texas Program. Texas Panhandle region, Amarillo north to Oklahoma

PVC: polyvinyl chloride

RPC: Rolling Plains Central zone, Texas Program. Located west of Fort Worth, the Abilene, Snyder, Munday area

SBL: Southern Blacklands zone, Texas Program. Located east of Austin, Luling north to Waco and east to Louisiana

SHP/C: Southern High Plains/Caprock, Texas Program. Originally Lubbock south to Andrews, Midland and Big Spring. Later, Crosbyton west through Lubbock to New Mexico

SIT: Sterile insect technique

SRP: Southern Rolling Plains zone, Texas Program. South central portion of West Texas, the San Angelo, Ballinger area

ST/WG: South Texas/Winter Garden zone, Texas Program. Initially, Houston to Corpus Christi to Uvalde; after 1997, Victoria to Corpus Christi to Uvalde

STAC: State Technical Advisory Committee. Appointed 1989 by Mississippi State University to develop plans for boll weevil eradication in Mississippi; served until 1993

STL: St Lawrence zone, Texas Program. Located south-east of Midland in the vicinity of Garden City

TAC: Technical Advisory Committee

TBWEF: Texas Boll Weevil Eradication Foundation

TCP: Texas Cotton Producers

TDA: Texas Department of Agriculture

TGC: Technical Guidance Committee, the committee responsible for setting the policy for the Pilot Boll Weevil Eradication Experiment

TPM: Total Population Management, use of integrated systems over a region to achieve a high level of pest suppression or eradication

UCB: Upper Coastal Bend, Texas Program. The area south-west of Houston to just north of Victoria; known as the Seven Counties area from 1997–2001

ULV: Ultra-low volume insecticide applications applied without diluents

USDA: US Department of Agriculture

WHP: Western High Plains zone, Texas Program. Plains, Brownfield, Seminole area, adjacent to south-eastern New Mexico.

21 Current Approaches to Areawide Pest Management: Obstacles and Future Directions

OPENDER KOUL[1] AND GERRIT W. CUPERUS[2]

[1]Insect Biopesticide Research Centre, Jalandhar, India
[2]Stillwater, Oklahoma, USA

Introduction

The areawide pest management (AWPM) concept was refined by various groups in the 1970s and 1980s (Pruess *et al.*, 1974; Rummel, 1976; Bottrell and Rummel, 1978; Huber *et al.*, 1979; Knipling, 1980; Kunz *et al.*, 1983) to control key pests. However, the AWPM approach pre-dates this work. For example, David McNeal (USDA) found an original manuscript from the 1930s that has a good discussion of what is essentially a multi-tactic AWPM approach for cotton in Arkansas, long before the AWPM and IPM concepts were formally introduced into the literature. The premise behind AWPM is that existing technologies (whether used singly or multiply, in an integrated programme) are most effective when used over a broad geographic area. Crucial to success is to have all or most of the farmers in a large area simultaneously implementing the programme so that pests have no safe haven or alternative food source. Adoption of the technologies by growers and pest control practitioners is a goal of demonstration projects.

The late Edward F. Knipling, an ARS pioneer in insect control, was a strong proponent of the areawide IPM concept. One of his major achievements was development of the sterile-male release technique, which eliminated screwworm and other insect infestations in many parts of the world. In the early 1980s, Knipling developed the concept of using specific insect parasites, predators and other tactics over broad areas to keep pest populations below the point at which they impose a financial burden on farmers and ranchers. When kept at low levels, pests are more responsive to biological rather than chemical controls. Today, the areawide concept has grown to include not only parasites and predators, but also other environmentally friendly tactics, such as mating disruption and insect attracticides – an attractant combined with a pesticide. To be precise, the term areawide is increasingly used in research reports and usually means a cooperative management programme applied to a regional area by using several complementary methodologies of IPM (Kogan, 1995; Calkins, 1998; Brewster *et al.*, 1999).

Areawide programmes using pheromones as a single control tactic have been conducted to disrupt lepidopteran pests or to trap-out bark beetles over large areas (Huber *et al.*, 1979; Schneider, 1989). Areawide programmes using augmentation or release of natural enemies, insect pathogens, sterile males, insecticides, etc. have been implemented for control of pests, including codling moth, cotton bollworm, tobacco budworm, fruit flies, parasitic flies, boll weevil, corn rootworm, fire ants and whiteflies (Carlson and Suguiyama, 1985; Bell and Hayes, 1994; Calkins, 1998; Fuchs *et al.*, 1998; Gray, 2001; Vargas *et al.*, 2001; Hendrichs *et al.*, 2002; Drees and Gold, 2003; Siegfried *et al.*, 2004). A few models have addressed areawide crop- and pest-specific parameters of growth, dispersal, host resistance and predators or parasitoids in attempts at predicting population and damage levels (Schneider, 1989; Bessin *et al.*, 1991; Legaspi *et al.*, 1998).

The objectives of AWPM have been directed to enhance the efficacy of non-pesticidal systems for pest control by reducing non-essential neurotoxins and to systematically reduce a target pest below a low residual level through the use of uniformly applied control measures over large, defined geographical areas. The intent is not to eradicate the pest but to achieve and maintain an overall reduced general equilibrium density of the pest population over a large area. For instance, codling moth mating disruption works more effectively when applied over large areas because less pheromone will be used to control whole populations rather than on a farm-by-farm process (Calkins and Faust, 2003). It has also been emphasized that such programmes will improve chances of biological control, mating disruption and reduce postharvest pesticide residues.

Other major objectives of AWPM have been to establish methods of monitoring insects for making pest management decisions, such as: (i) stored grain AWPM (Flinn *et al.*, 2003); (ii) to measure the efficacy of current pest management practices; (iii) to evaluate sanitation programmes; (iv) to reduce the risk of economic losses; and (v) to develop risk-management strategies. A recent endeavour was to stimulate a mosaic of different crops infested by a polyphagous pest insect to investigate the interplay between pest population growth, dispersal and control treatments termed 'sprays' (Byers and Castle, 2005). In fact, the goal has been to better understand the consequences of asynchronous versus synchronous control in areawide management. Thus, in the recent past, various approaches have been tried to make AWPM a success.

Current Approaches

The Agricultural Research Service, USA launched the first AWPM attacks against the codling moth, a pest in apple and pear orchards, on 7700 acres (3100 ha) in the Pacific north-west. The adoption of areawide mating disruption in Michigan has directly influenced the types of insecticides used and the number of sprays targeting codling moth. Orchards in the second year of areawide disruption reduced use of insecticide sprays targeting codling moth by 24% from the first year to the second.

In 2006, the project will expand to additional farms in the Fruit Ridge region, to Old Mission Peninsula north of Traverse City and possibly a location in south-west Michigan (IPM Report, 2006).

Other programmes include a major assault against the corn rootworm on over 40,000 acres (16,000 ha) in the Corn Belt, fruit flies in the Hawaiian Islands and leafy spurge in the Northern Plains area. For the last few decades, Hawaii has been plagued by four fruit fly species, costing the agricultural industry billions of dollars. A collaboration of federal, state and university scientists has developed a system of field sanitation, biological controls and lures to quell the problem. One of the largest producers on Oahu, Aloun Farms, saw crop losses from melon fly drop from 22 to 1% in one year. The approach was applied on over 200 small farms in 2004 and continues to grow each year under the AWPM strategy (http://www.extento.hawaii.edu/fruitfly/). In 2001, an AWPM project began for fire ants on pastures in Florida, Mississippi, Oklahoma, South Carolina and Texas using natural enemies, microbial pesticides and attracticides.

Pilot studies for areawide management of the western corn rootworm, *Diabrotica virgifera virgifera*, were conducted from 1996 to 2002. The primary management tool in these programmes exploited behavioural adaptations of this insect to feeding stimulants and arrestants to deliver high doses of traditional neurotoxic insecticide to individual insects (Comis, 1997) while minimizing the overall rate of insecticide use. In 2002, ARS scientists in Stillwater, Oklahoma began an AWPM project on Russian wheat aphid and greenbug on wheat in the US Great Plains using customized cultural practices, pest-resistant cultivars, biological control agents and other biologically based pest control technologies. Nearly 150 wheat growers from six states have deliberated on cropping systems, yields, pests and a variety of production practices. In addition to the group meetings, the growers were interviewed individually each year to evaluate the economics of their individual farm enterprise. Data were collected for 4 years so that researchers could look at the variability from year to year. An economic summary of this data should be available in 2007. Also in 2002, an AWPM project began for tarnished plant bug on cotton in the delta of Mississippi and Louisiana using host destruction, host-plant resistance and remote-sensing technology.

The Ecological Areawide Management (TEAM) Leafy Spurge project was a US Department of Agriculture and Agricultural Research Service (USDA–ARS) regional, Integrated Pest Management (IPM) programme focused on the Little Missouri River drainage in the states of North and South Dakota, Montana and Wyoming, USA. The TEAM Leafy Spurge project represented the first large-scale, systematic study and demonstration of weed management alternatives under USDA-ARS's Areawide Pest Management Program. The other three projects previously approved under the programme targeted insect pests. TEAM Leafy Spurge's primary goal had been to demonstrate the use of ecologically based IPM strategies areawide to achieve effective, affordable leafy spurge control. The five components of the TEAM Leafy Spurge research and demonstration project, which are shared by all USDA-ARS supported projects, were: (i) programme management; (ii) operations; (iii) assessment; (iv) supporting research; and (v) technology transfer (Prosser *et al.*, 2002; Anderson *et al.*, 2003).

The areawide concept has also been used for other exotic weeds. The musk thistle, *Carduus nutans*, is an exotic weed from Europe that spread across the USA and was found in Oklahoma in the 1940s (Rodunner *et al.*, 2003). By 1960, it had spread through 29 counties in north-east and central Oklahoma and, by 2001, musk thistle was reported in 61 of 77 counties (Rodunner *et al.*, 2005). By working on an areawide approach, the implementation of biological and cultural controls helped reduce this exotic weed (Rodunner *et al.*, 2003, 2005). This has greatly reduced the use of herbicides for musk thistle control in Oklahoma now – and in the future.

Vreysen *et al.* (2006) have recently emphasized that understanding the principles that govern the mating behaviour of insects that are the target of AWPM programmes by using the sterile insect technique (SIT) is a prerequisite for ensuring optimal efficiency of such programmes. Models were constructed to assess the effect of mating preference of insects that display a female- or male-choice mating system, on the efficiency of SIT programmes that release males only or programmes that release both sexes. The model on preferential mating indicated that in a male-choice mating system (e.g. screwworm, *Cochliomyia hominivorax*), overcoming the discrimination of wild males against mating with sterile females would require a doubling of the number of sterile males compared with male-only releases. The model on female choice was incapable of distinguishing between reduced sterile male competitiveness and female preference for wild males and implied, in addition, that the release of both sexes and male-only releases required the same sterile:wild male overflooding ratio.

Operational SIT projects have, however, shown a significant benefit with male-only releases against insects which have a female-choice mating system (e.g. Mediterranean fruit fly, *Ceratitis capitata*), and models were constructed to assess the potential effect of sterile female presence or absence on some parameters, i.e. reduced sterile sperm quantity with remating, reduced sterile sperm quality with ageing and incomplete redistribution of the sterile males with the wild insects. The model suggests that, in all three cases, male-only releases result in relatively more efficient sterile insects compared with programmes releasing both sexes. The results of the models are discussed in relation to data available from operational screwworm and Mediterranean fruit fly AWPM programmes with an SIT component (Vreysen *et al.*, 2006).

Simulation models are useful in defining problems, understanding the system, identifying the areas to investigate, making predictions, generating hypotheses and acting as standards for comparison (Worner, 1991). The spatially explicit approach used recently by Carrière *et al.* (2006) is based on global positioning system (GPS) and geographic information system (GIS) technologies combined with spatial statistics to assess the maximum distance at which forage and seed lucerne, fallow fields with weeds and cotton affect *Lygus hesperus* population density, which can provide direct information on the effects of agroecosystem heterogeneity on the areawide abundance and distribution of key insect pests. A long-standing question that has been crucial for the management of this highly mobile polyphagous pest has been the scale of movements to cotton from lucerne, weeds and cotton (Stewart and Layton, 2000; Goodell *et al.*, 2002). This strongly suggests that a GIS-based approach can substantially contribute to the development of AWPM (Carrière *et al.*, 2006).

During a workshop held in Rome in May 2002, the concept of AWPM was agreed upon for the African continent. A set of criteria/guidelines for selection of priority areas and joint international action for tsetse and trypanosomiasis (T&T)

intervention was agreed. Additionally, two earlier identified areas were validated against these criteria and sequential steps in project implementation were defined (FAO, 2002). The first project area is located in the Ethiopian Southern Rift Valley system and the other area in the common Burkina Faso–Mali 'cotton belt' zone in West Africa. The AWPM concept, based on data-driven decision making, has profited greatly from the development of geographical information systems (GIS) – tools that allow the spatial analysis of data in a multidisciplinary manner. In addition to the establishment of countrywide multidisciplinary databases, results obtained in West, East and Southern Africa during the 1990s have greatly contributed towards the implementation of an AWPM approach for T&T intervention programmes (http://www.FAO.org).

The IAEA's (International Atomic Energy Agency) work in Africa supporting NEPAD's (New Partnership for Africa's Development) strategic priorities related to agriculture and market access is aimed at poverty alleviation and food security goals. Assistance deals mainly with the application of radiation and isotopes in pest control, with special emphasis on tsetse eradication, and improving crop production and increasing livestock productivity through better disease control, artificial insemination and feed supplementation. In terms of project funding, 20.8% of the TCF (Twin City Federal) resources are allotted under the technical cooperation programme for 2005–2006 to food and agriculture.

Among the many projects identified by African countries, the IAEA is actively supporting the initiative of the African Union (AU) to carry out and coordinate the Pan-African Tsetse and Trypanosomiasis Eradication Campaign (PATTEC), which was launched in Ouagadougou, Burkina Faso in October 2001. The objective is to free sub-Saharan Africa from one of the main persisting constraints to sustainable development. The tsetse infests 37 sub-Saharan African countries, 32 of them among the 42 most heavily indebted poor countries in the world.

The IAEA contributes directly in the field to the implementation of PATTEC's Plan of Action by supporting activities in several countries. Agency support focuses on the transfer of the Sterile Insect Technique (SIT) in the context of AWPM in support of creating tsetse-free zones in selected areas in African Member States. Assistance has been provided to: (i) establish/upgrade tsetse-rearing facilities in Burkina Faso, Ethiopia and the United Republic of Tanzania; (ii) perform test sterile fly release in Mali; (iii) develop a standardized recording, reporting and management system for field operations; (iv) collect entomological and veterinary baseline data in target areas; and (v) carry out genetic studies of tsetse fly populations.

Under the 2005–2006 programme, the Agency will continue providing support through national projects to activities connected to PATTEC in Botswana, Burkina Faso, Ethiopia, Kenya, Mali, Senegal, South Africa, the United Republic of Tanzania and Uganda. Under a regional project, support will be given to the Member States' relevant activities in terms of awareness raising, technical planning, training and institutional capacity building (http://www.IAEA.org).

In Australia, the Wide Area Mating Disruption (WAMD) strategy has been applied to improve the protection of orchards against migration of oriental fruit moth (OFM). During 1997/1998 season the experimental area over 800 ha of orchard in the Murray-Goulburn Valley, Victoria was saturated with OFM sex pheromone to test the concept of WAMD. Mating disruption lures called 'Isomate

OFM Plus' (Biocontrol Ltd) were applied on all fruit trees, including peaches and nectarines, as well as on pears, apples, apricots and plums in which mating disruption (MD) was not normally used. Such a large area was used to ensure that any edge effects and migration of mated OFM females would be overcome. The initial level of OFM population in the 1996/1997 season, before the application of the WAMD experiment, was measured. During the 1997/1998 season WAMD was closely monitored for OFM. More than 230 food traps for OFM monitoring were placed in the area and shoot tip and fruit damage assessments were made.

Results of detailed monitoring showed a reduction of OFM population in the WAMD. Also, shoot tip and fruit damage in the peach blocks were reduced when surrounding pears were treated with MD compared with chemical control. The WAMD experiment showed that the OFM population was greatly reduced in the hot spots, and edge effects that could be explained by migration of OFM were controlled (Il'ichev *et al.*, 1999). Similarly, integrating pest management as a group is paying off in improved profit margins for Australian cotton growers. More growers are combining IPM programmes on an areawide basis for mutual benefit.

Integrated pest management encourages growers to make use of natural predators by delaying or eliminating applications of broad-spectrum sprays such as pyrethroids and organophosphates. The results of this survey, and of others like it, encourage those who believe that sustainable cotton production depends on the extension of this management approach. 'What these surveys are showing is that there doesn't seem to be much relationship between yield and dollars spent on pest management. When you compare IPM fields with conventional management, IPM is coming out in front by up to a few hundred dollars per hectare' (full article at http://www.cottonworld.com.au/cworld/).

Silverleaf whitefly (SLW) is difficult to manage because of its wide host range and because its population can increase extremely rapidly. Once SLW populations explode in a cropping region, the pest cannot satisfactorily be managed with pesticides alone, even where effective products are registered. This is because of sheer weight of numbers (billions/ha) and because SLW can rapidly become resistant to new pesticides (within a single season). Hard pesticides used against other pests are likely to flare SLW as they kill SLW parasites and predators. AWPM strategies in Australia that include planting gaps between susceptible hosts, the conservation of beneficial insects and the use of pesticides only as prescribed in resistance management strategies hold the most promise for SLW. There are no silver bullets and 'pesticide-alone' strategies are doomed to failure! For soybeans, the problem is compounded by the attractiveness of the crop and the fact that soybeans mature later in the season than other susceptible crops such as cotton. The strategies have been well documented for Queensland and for the Australian cotton-growing regions (http://www.cotton.crc.or.au).

Recently, a concerted effort to bring together the Asian perspective of AWPM was demonstrated in a symposium (http://www.pref.okinawa.jp/arc/fftc/dl/highlights.doc; October 2006). Japan shared and exchanged significant results of its long experience in the use of areawide approaches to suppress/control pests like *Chilo suppressalis* and migratory rice planthoppers in rice, solanaceous fruit fly, sweet-potato weevil and sugarcane wireworm. *C. suppressalis*, once considered as one of Japan's major rice insect pests causing significant yield loss, reached a low population status due to density-dependent factors such as: (i) long-duration IPM scheme (10 years); (ii) wide area to cover population displacement of *Chilo* stemborers; and (iii) integrated pest management of key pests in complementation with modern farming technologies.

Simulation models of population dynamics and biotype evolution for rice planthoppers were developed in Japan in order to evaluate the efficacy of conservation biological control combined with endophyte-infected rice plants exhibiting resistance. The results indicated that, if the mortality of rice planthoppers depends on the ratio of natural enemy density to planthopper density, then the combination can successfully maintain the density of planthoppers below acceptable levels and that the development of a biotype resistant to endophyte-infected rice plants can be deterred. A new three-dimensional backward trajectory analysis method of migration prediction for rice planthoppers was also developed in Japan, to find migration sources of rice planthoppers that had immigrated into the country. The results of monitoring for population characteristics such as insecticide resistance, virulence to resistant rice varieties and wing-form ratio of Japanese BPH immigrants suggest that the recent planthoppers differ in population characteristics from those captured in previous years, indicating a different geographic source.

For the control of the sugarcane wireworm, communication disruption using sex pheromone was attempted in Japan, the results of which showed that virtually no males were caught in the monitoring traps, except in a few traps in the treated area. This result shows the effectiveness of communication disruption as a tool in the management of sugarcane wireworm.

Areawide management of oriental fruit fly and sweet potato weevil in Taiwan is well known. The oriental fruit fly seriously infests major fruit crops in all parts of Taiwan. An areawide control programme for oriental fruit fly using methyl eugenol has been conducted for many years, but its population remains at a high level in the field. Culture and protein bait containing spinosad has also been used, while non-chemical efforts are concentrated on: (i) organizing farmers; (ii) establishing geological information; (iii) monitoring fruit fly density; (iv) providing fruit production information; (v) holding training programmes; and (vi) maintaining orchard sanitation. Some studies conducted in recent years include the ecology of the fruit fly in small-scale and diversified farming systems, the development of female-targeted lures and the potential of using PE plastic mulch for fruit fly control in abandoned orchards.

The sweet potato weevil is an important pest of sweet potato in Taiwan. In general, the damage caused by the weevil is now under satisfactory control, with an integrated pest management strategy based on mass trapping of pests in the fields with synthetic sex pheromones and the minimum use of granular insecticides.

In Korea, several environmentally friendly methods have been carried out for the control of pests and diseases of specific crops. In fruit growing, minimum use of chemicals and environmentally friendly control methods are being carried out in full scale. Insect attractant-like sex pheromones are actively used for timely forecasting of major pests and for reducing the timing of pesticide sprays. Moreover, many experiments are in progress concerning mating disruptions and mass trappings of pests in the fields with synthetic sex pheromones.

The Malayan rice black bug is a serious invasive pest of rice affecting some of the islands of the Philippine archipelago. It is being managed effectively with the use of biological control agents such as *Metarhizium anisopliae*, the green muscardine fungus and *Telenomus triptus*, the egg parasitoid. The use of these biocontrol agents is also coupled with suggested cultural management and full cooperation of concerned stakeholders for effective management.

In Thailand, the AWPM approach is being adopted in the control of several insect pests and pest complexes. In addition to rice, which is the major economic crop of the country, there are several insect pests and pest complexes associated with field crops, vegetable crops, plantation crops and fruit trees, including some invasive alien species, insects, weeds and snails. The feasibility of using an areawide approach must be determined based on the suitability of the target pests, the implementing agencies responsible, availability of specific AWPM technologies and the interest and socio-economic condition of the farmers involved.

In Vietnam, the coconut hispine beetle, *Brontispa longissima*, is a very serious insect pest of coconut palms. It invaded the country as recently as 1999 and has spread since then. The parasitoid *Asecodes hispinarum* was imported from Western Samoa into Vietnam in 2003 to control the pest; this parasitoid became established after only 1 year and it suppressed the pest sustainably in a large area of the Mekong Delta. The recovery of coconut palm trees is now complete, with no significant reinfestation.

Obstacles

Wearing (1988) lists the obstacles to IPM implementation under five interrelated headings: technical, financial, educational, marketing/social and organizational. These apply to crucifer IPM in the Changjiang River Valley. Liu and Yan (1998) discussed these obstacles in some detail. For example, with regard to the organizational obstacles, they pointed out that the coordination among organizations, disciplines and personnel will remain a serious problem. IPM can only be implemented effectively on an areawide scale (Morse and Buhler, 1997). This calls for close cooperation of many farmers in an area, which is difficult to achieve. Lack of trained extension workers will continue to be a major obstacle. Many of the state-employed extension workers in developing countries have been directly involved in marketing chemical pesticides since the late 1980s. Consequently, their advice to farmers is no longer IPM-oriented but biased towards increasing pesticide inputs. This unfortunate situation has been seen as being a major reason for the rapid increase of pesticide application in recent years in the Asian continent. It seems unlikely that this organizational problem will be overcome quickly.

Implicit in the argument for areawide treatment is that sources of dispersing insects are suppressed to preclude reinoculation of previously treated areas. Although AWPM programmes generally have used pheromones or biological agents in a synchronized way, treatment thresholds were not explicitly considered and results were not directly compared with single-field IPM. The advantages of the programme are difficult to establish with scientific rigour because treatment and control are confounded by differences in temporal and spatial dimensions (Byers and Castle, 2005). Differences in time or place, as well as the broad scales required, make it difficult to compare an AWPM programme based on traditional IPM thresholds applied simultaneously to a mosaic of fields with a programme based on treatment of fields independently using the same threshold.

Cooperation among private landowners, universities, special-interest groups and local, state and federal governments is critical for any areawide implementation. For instance, in the case of leafy spurge, the invasive weed is so widespread and simple management solutions do not exist. The dynamic expansion of leafy spurge provides a valuable example of why noxious and invasive species require aggressive and comprehensive strategies for control. Education is key to solving many pest management problems. For example, adoption of chemical, biological, cultural and integrated controls requires providing information on where, how, when and what to do. Understanding economic trade-offs among treatment alternatives is another obstacle where provision of education to the public is required in order to develop a broad base of support. Weeds are an environmental problem, shared by everyone, not just individual land owners; target sports groups, environmental groups, farmer organizations and youth groups within the educational process have to come to a common platform simultaneously to solve the problems.

Population genetics does hinder the use of AWPM systems: implementation of AWPM programmes against riverine tsetse in West Africa is one good example of this. Various observations are of prime importance in the context of creating sustainable tsetse-free zones in West Africa using an AWPM approach, i.e. the management and elimination of the entire tsetse populations within a circumscribed area. Whereas available tsetse distribution maps, risk prediction models and population genetics data indicate fragmentation of the tsetse belt in East Africa (Krafsur, 2003), the distribution of riverine tsetse in the humid savannah area of West Africa seems more complex. Riverine tsetse species such as *Glossina palpalis gambiensis* seem to be restricted for most of the year to the riparian forests bordering the various river systems, and this close relationship between their spatial occupation of the habitat and hydrology/drainage systems could be exploited in AWPM intervention.

It has been postulated that populations of riverine tsetse species might be completely confined to the rivers and tributaries of a specific basin because areas between adjacent basins prevent dispersion. According to this hypothesis, the 'primary river basin' could be considered and used as the 'geographical unit area of operation' in AWPM intervention campaigns, which would allow the creation of tsetse-free zones in West Africa. Some data refute this hypothesis, although it needs to be emphasized that the sampling of flies was carried out in only a very small geographical area in Mali. More systematic tsetse sampling on a regional scale in West Africa is needed to allow a better assessment of the degree of isolation of the riverine tsetse residing in the various river basins. Data on the efficiency of the watersheds as barriers for the various river basins will be necessary to decide whether tsetse intervention campaigns can be launched in West Africa according to the AWPM concept (Marquez *et al.*, 2004).

Future Directions

Consumer aversions to pesticide residues and increasing demands for food safety have been major forces driving implementation of IPM in vegetables in many

Asian countries. For instance, in China and India, serious poisoning of humans by insecticide residue on crucifer vegetables and various crops has frequently been reported since the mid-1980s. These poisoning events initiated the demand by consumers for reduced chemical pesticide use on vegetables. Cosmetic standards of vegetables have become less stringent (Liu *et al.*, 1996). Monitoring of pesticide residue has increased in both domestic vegetable supplies and international trade. There is some evidence that consumers are prepared to pay a slightly higher price for 'green and clean' vegetables. As the lifestyle of consumers improves, their demand for eliminating pesticide residue on crops will become stronger and provide increasing opportunities for biological methods of pest control through the areawide approach to IPM. Accordingly, major opportunities for promotion of IPM would be consumers' demands for food safety, development of better-organized farming, increased support for research and extension and policy and legislation support (Liu and Yan, 1998). This clearly emphasizes the need for specific goals for the future if AWPM is to become a successful strategy:

1. One goal should be to bring more and more of a nation's farmland under biointensive integrated pest management. While it will be a difficult task to accomplish, by implementing areawide projects that target key pests the goal should become even more within our grasp.
2. The use of multiple control tactics in a manner consistent with widely accepted resistance management theory and practice should serve to reduce selective pressures and sustain the AWPM concept.
3. When feasible, efforts should be directed at demonstrating the positive impacts and advantages of AWPM through enhanced profits for the grower, safety of the worker, less stressed environment and/or proven superiority of the AWPM strategy over existing control approaches.
4. To meet the goals through the adoption of AWPM, mature, areawide management is the requirement that would emphasize easily understandable access to operational programmes.
5. Development, validation and experimentation with spatially explicit population dynamics simulation models for important pest species could help better to understand the synchronous benefits and provide the confidence necessary to initiate new AWPM programmes.
6. If an areawide approach becomes widely practised, alternative technologies should be incorporated into the programme. The use of multiple control tactics in a manner consistent with widely accepted resistance management theory and practice should serve to reduce selective pressures and sustain the areawide management concept (Siegfried *et al.*, 2004).
7. The success of areawide IPM is very much dependent on the willingness and vigorous participation of farmers jointly in the ongoing areawide control programme. Hence, government and private entities alike must be able to provide proper education/training, hands-on learning and demonstration of successful cases, as well as support services to equip them with the technological and financial resources necessary for them to be actively involved in AWPM programmes. Transfer of technologies to end-users – particularly to small-scale farmers – must be done as a collaborative, public and private partnership mission, with shared responsibilities among the concerned agencies/authorities.

8. In most insect pest management programmes, the conventional strategies and tactics available are suitable mainly for individual, small to large farmers, whether they will be used as a single component or as an integrated pest management strategy. Any or all such strategies could be integrated into an AWPM system, but with a different approach and operational procedure. Each target insect pest or pest complex has a characteristic in itself, and AWPM programmes for each must be specifically designed taking into consideration the availability of resources required, human capacity, technology and implementing authority.

9. Mechanisms for participation, cooperation and implementation need to be developed at both the national and regional level for the maintenance and sustainability of AWPM programmes for the small-scale farming systems that exist most notably in the Asian and Pacific regions and in Africa. Government policies, interest of the farmers and long-term input, effort and commitment of all stakeholders, as well as the socio-economic condition of the farmers involved, are some of the keys to the success of this collaborative mechanism for AWPM.

References

Anderson, G.L., Prosser, C.W., Wendel, L.E., Delfosse, E.S. and Faust, R.M. (2003) The ecological areawide management (TEAM) of leafy spurge program of the United States Department of Agriculture-Agricultural Research Service. *Pest Management Science* 59, 609–613.

Bell, M.R. and Hayes, J.L. (1994) Areawide management of cotton bollworm and tobacco budworm (Lepidoptera: Noctuidae) through application of a nuclear polyhedrosis virus on early-season alternate host. *Journal of Economic Entomology* 87, 53–57.

Bessin, R.T., Stinner, R.E. and Reagan, T.E. (1991) Modeling the areawide impact of sugarcane varieties and predation on sugarcane borer (Lepidoptera: Pyralidae) populations in southern Louisiana. *Environmental Entomology* 20, 252–257.

Bottrell, D. and Rummel, D.R. (1978) Response of *Heliothis* populations to insecticides applied in an areawide reproduction diapause boll weevil suppression program. *Journal of Economic Entomology* 71, 87–92.

Brewster, C.C., Allen, J.C. and Kopp, D.D. (1999) IPM from space: using satellite imagery to construct regional crop maps for studying crop–insect interaction. *American Entomologist* 45, 105–117.

Byers, J.A. and Castle, S.J. (2005) Areawide models comparing synchronous versus asynchronous treatments for control of dispersing insect pests. *Journal of Economic Entomology* 98, 1763–1773.

Calkins, C.O. (1998) Review of the codling moth areawide suppression program in the western United States. *Journal of Agricultural Entomology* 15, 327–333.

Calkins, C.O. and Faust, R.J. (2003) Overview of areawide programs and the program for suppression of codling moth in the western USA directed by the United States department of Agriculture-Agricultural Research Service. *Pest Management Science* 59, 601–604.

Carlson, G.A. and Suguiyama, L. (1985) Economic evaluation of areawide cotton insect management: bollweevils in the southeastern United States. *NC Agricultural Research Service Bulletin* 473, 1–24.

Carrière, Y., Ellsworth, Peter C., Dutilleul, P., Ellers-Kirk, C., Barkley, V. and Antilla, L. (2006) A GIS-based approach for areawide pest management: the scales of *Lygus hesperus* movements to cotton from alfalfa, weeds, and cotton. *Entomologia Experimentalis et Applicata* 118, 203–210.

Comis, D. (1997) Corn belt growers give areawide IPM a try. *USDA, ARS Agricultural Research* 45, 4–7.

Drees, B.M. and Gold, R.E. (2003) Development of integrated pest management programs for the red imported fire ant (Hymenoptera: Formicidae). *Journal of Entomological Science* 38, 170–180.

FAO (2002) *Workshop on PAAT-PATTEC Harmonization*, 2–3 May 2002, Rome.

Flinn, P.A., Hagstrum, D.W., Reed, C. and Phillips, T.W. (2003) United States Department of Agriculture-Agricultural Research Service stored-grain areawide integrated pest management program. *Pest Management Science* 59, 614–618.

Fuchs, T.W., Rummel, D.R. and Boring, E.P. III (1998) Delayed uniform planting for areawide boll weevil suppression. *Southwest Entomologist* 23, 325–333.

Goodell, P.B., Lynn, K. and McFeeters, S.K. (2002) Using GIS approaches to study western tarnished plant bug in the San Joaquin valley. *KAC Plant Protection Quarterly* 12, 3–8.

Gray, M.E. (2001) The role of extension in promoting IPM programs. *American Entomologist* 47, 134–137.

Hendrichs, J., Robinson, A.S., Cayol, J.P. and Enkerlin, W. (2002) Medfly areawide sterile insect technique programs for prevention, suppression or eradication: the importance of mating behavior studies. *Florida Entomologist* 85, 1–13.

Huber, R.T., Moore, L. and Hoffmann, M.P. (1979) Feasibility study of area-wide pheromone trapping of male pink bollworm *Pectinophora gossypiella* moths in a cotton insect pest management program. *Journal of Economic Entomology* 72, 222–227.

Il'ichev, A.L., Hossain, M.S. and Jerie, P.H. (1999) Application of wide area mating disruption for control of oriental fruit moth *Grapholita molesta* Busck (Lepidoptera: Tortricidae) migration in Victoria, Australia. *IOBC WPRS Bulletin* 22, Australia.

IPM (2006) Apple growers succeed with areawide approach to managing codling moth. *IPM Reports* 12, 7.

Knipling, E.F. (1980) Regional management of the fall armyworm, a realistic approach? *Florida Entomologist* 63, 468–480.

Kogan, M. (1995) Areawide management of major pests: is the concept applicable to the Bemisia complex? In: Gerling, D. and Mayer, R.T. (eds) *Bemisia 1995: Taxonomy, Biology, Damage Control and Management*. Intercept Ltd, Andover, UK, pp. 643–657.

Krafsur, E.S. (2003) Tsetse fly population genetics: an indirect approach to dispersal. *Trends in Parasitology* 19, 162–166.

Kunz, S.E., Kinzer, H.G. and Miller, J.A. (1983) Areawide cattle treatments on populations of horn flies (Diptera: Muscidae). *Journal of Economic Entomology* 76, 525–528.

Legaspi, B.C., Jr., Allen, J.C., Brewster, C.C., Morales Ramos, J.A. and King, E.G. (1998) Area-wide management of the cotton boll weevil: use of a spatio-temporal model in augmentative biological control. *Ecological Models* 110, 151–164.

Liu, S.S. and Yan, S. (1998) Brassica IPM in Asia: successes, challenges, and opportunities. In: Zaulcki, M.P., Drew, R.A.I. and White, G.G. (eds) *Pest Management – Future Challenges. Proceedings of the Sixth Australasian Applied Entomological Research Conference*, Brisbane, Australia, Vol. 1, pp. 85–97.

Liu, S.S., Brough, E.J. and Norton, G.A. (1996) *Integrated Pest Management in Brassica Vegetable Crops*. Cooperative Research Centre for Tropical Pest Management, Brisbane, Queensland, Australia.

Marquez, J.G., Vreysen, M.J.B., Robinson, A.S., Bado, S. and Krafsur, E.S. (2004) Mitochondrial diversity analysis of *Glossina palpalis gambiensis* from Mali and Senegal. *Medical and Veterinary Entomology* 18, 288–295.

Morse, S. and Buhler, W. (1997) IPM in developing countries: the danger of an ideal. *Integrated Pest Management Reviews* 2, 175–185.

Prosser, C.W., Anderson, G.L., Wendel, L.E., Richard, R.D. and Redlin, B.R. (2002) TEAM leafy spurge: an areawide pest management program. *Integrated Pest Management Reviews* 7, 47–62.

Pruess, K.P., Witkowski, J.F. and Raun, E.S. (1974) Population suppression of western corn rootworm by adult control with ULV malathion. *Journal of Economic Entomology* 67, 651–655.

Rodunner, M., Cuperus, G.W., Mulder, P., Stritzke, J. and Payton, M. (2003) An historical success story in biological control managing musk thistle, *Carduus nutans*, in Oklahoma using the musk thistle weevil, *Rhynocyllus conicus*, and the rosette weevil, *Trichosirocalus horridus*. *American Entomologist* 49, 112–120.

Rodunner, M., Mulder, P., Cuperus, G.W., Stritzke, J. and Payton, M. (2005) Plant growth parameters of musk thistle, *Carduus nutans* and egg distribution patterns of *Rhinocyllus conicus* on their blooms. *Southwest Entomologist* 30, 93–103.

Rummel, D.R. (1976) An area-wide boll weevil (*Anthonomus grandis*) suppression program. Organization, operation, and economic impact. *Miscellaneous Publication of Texas Agricultural Experimental Station* 1276, 152–159.

Schneider, J.C. (1989) Role of movement in evaluation of area-wide insect pest management tactics. *Environmental Entomology* 18, 868–874.

Siegfried, B.D., Meinke, L.J., Parimi, S., Scharf, M.E., Nowatzki, T.J., Zhou, X. and Chandler, L.D. (2004) Monitoring western corn rootworm (Coleoptera: Chrysomelidae) susceptibility to carbaryl and cucurbitacin baits in the areawide management plot program. *Journal of Economic Entomology* 97, 1726–1733.

Stewart, S.D. and Layton, M.B. (2000) Cultural controls for the management of *Lygus* populations in cotton. *Southwestern Entomologist* 23, 83–95.

Vargas, R.I., Peck, S.L., McQuate, G.T., Jackson, C.G., Stark, J.D. and Armstrong, J.W. (2001) Potential of areawide integrated management of Mediterranean fruit fly (Diptera:Tephritidae) with a braconid parasitoid and a novel bait spray. *Journal of Economic Entomology* 94, 817–825.

Vreysen, M.J.B., Barclay, H.J., and Hendrichs, J. (2006) Modeling of preferential mating in areawide control programs that integrate the release of strains of sterile males only or both sexes. *Annals of the Entomological Society of America* 99, 607–616.

Wearing C.H. (1988) Evaluating the IPM implementation process. *Annual Review of Entomology* 33, 17–38.

Worner, S.P. (1991) Use of models in applied entomology: the need for perspective. *Environmental Entomology* 20, 768–773.

Index